Dictionary of 21st Century Energy Technologies, Financing & Sustainability

Dictionary of
21st Century Energy Technologies,
Financing & Sustainability

Gene Beck, CEM, CLP

THE FAIRMONT PRESS, INC.

CRC Press
Taylor & Francis Group

Library of Congress Cataloging-in-Publication Data

Beck, Gene
     Dictionary of 21st century energy technologies, financing & sustainabiliy / Gene Beck, CEM, CLP.
          pages cm
     Includes bibliographical references and index.
     ISBN 0-88173-736-4 (alk. paper) -- ISBN 0-88173-737-2 (electronic : alk. paper) -- ISBN 978-1-4822-5304-7 (Taylor & Francis distribution : alk. paper) 1. Power resources--Dictionaries. 2. Renewable energy sources--Dictionaries. 3. Energy industries--Dictionaries. 4. Environmental protection--Dictionaries. I. Title.

     TJ163.16.B43 2014
     621.04203--dc23

                                                                        2014003701

Published by The Fairmont Press, Inc.
700 Indian Trail
Lilburn, GA 30047
tel: 770-925-9388; fax: 770-381-9865
http://www.fairmontpress.com

Distributed by Taylor & Francis Ltd.
6000 Broken Sound Parkway NW, Suite 300
Boca Raton, FL 33487, USA
E-mail: orders@crcpress.com

Distributed by Taylor & Francis Ltd.
23-25 Blades Court
Deodar Road
London SW15 2NU, UK
E-mail: uk.tandf@thomsonpublishingservices.co.uk

Printed in the United States of America
10 9 8 7 6 5 4 3 2 1

ISBN 0-88173-736-4 (The Fairmont Press, Inc.)
ISBN 978-1-4822-5304-7 (Taylor & Francis Ltd.)

While every effort is made to provide dependable information, the publisher, authors, and editors cannot be held responsible for any errors or omissions.

The views expressed herein do not necessarily reflect those of the publisher.

This glossary does not necessarily represent the views of any agency, entity or company. We assume no legal liability for the information in this glossary nor does any party represent that the uses of this information will not infringe upon privately owned rights. No regulatory or certification agency has passed upon the accuracy or adequacy of the information in this glossary.

Preface

This book was written for you, the layperson or professional, that is directly or tangentially in the field of energy, sustainability, financial and environmental issues and how they are integrated. Although even veterans of the industries occasionally forget certain terms and acronyms and professionals frequently don't understand the terminology that others in related energy related professions use, we can all use a reference manual to refer to.

This was certainly the case for me. Although a seasoned practitioner in the field of financing equipment and machinery for the commercial and industrial sector of our economy, I had little knowledge of the field of CleanTech Renewable Energy that I found myself getting immersed into at the turn of the century. The opportunities for creative financing was particularly applicable to this emerging financial sector. I needed to be brought up to speed quickly to understand the credit requests and investment opportunities that were being presented to me. Additionally, my clients expected me to understand the language they were using and assumed that I would be conversant in the "jargon" used in the real world of the new energy paradigm.

It was from that perspective that I started writing this book about eight years ago. I researched every source I could think of and quickly found out that there was no single source of information for what I was looking for. Hence, what came as the end product was an amalgamation of a wide variety of bits and pieces from a large number of sources.

I also now know that this book can never be complete. After nearly a decade of making notes and trying to organize my sticky notes on hundreds of papers, I have decided that for better-or-worse, I need to share what I have learned with others and hopefully make your learning period much shorter. Hence, what was originally a manuscript that I thought others might also like to keep in their files when the subjects of "energy, sustainability, financing and the environment" begged for a more in-depth compilation, turned into what you now have in your hands that the reader can enjoy and use as a quick desk reference. The new emerging renewable energy, energy efficiency and the relationships those key sectors have to our global climate, sustainability and environmental stewardship have become critically important for all people of the world.

Enjoy!

Gene Beck, CEM, CLP

Dictionary of 21st Century Energy Technologies, Financing & Sustainability

Symbols

¢/kWh—cents per kilowatt-hour.

A

A&C—Abatement and Control

A&I—Alternative and Innovative (Wastewater Treatment System)

A&R—Air and Radiation

A/R—Afforestation and reforestation. Term given to the class of projects devoted to the planting of trees on unforested land for carbon emissions reduction and other environmental benefits.
- Accounts Receivable (financial)

A/WPR—Air/Water Pollution Report

AA—Accountable Area
- Adverse Action
- Advices of Allowance
- Attainment Area
- Atomic Absorption

AAA—American Arbitration Association www.adr.org

AAAS—American Association for the Advancement of Science www.aaas.org

AAC—Acceptable Ambient Concentration

AAEE—American Academy of Environmental Engineers www.aaee.net

AAL—Acceptable Ambient Limit

AANWR—Alaskan Arctic National Wildlife Refuge

AAP—Affirmative Action Plan
- Asbestos Action Program

AAPCO—American Association of Pesticide Control Officials www.aapco.ceris.purdue.edu

AAQS—Ambient Air Quality Standards

AASHE—Association for the Advancement of Sustainability in Higher Education. AASHE is an association of colleges and universities that are working to create a sustainable future. www.aashe.org

AAU—Assigned Amount Unit. Allowances for carbon emissions allocated to developed countries up to their target level under the Kyoto Protocol. These allowances are tradable under Kyoto's international emission trading mechanisms in place from 2008 to 2012. Each AAU equates to one ton of CO_2e (CO_2 equivalent).

AB—Afterburner
- Assembly Bill

AB 1407—Assembly Bill 1407, codified as California Civil Code section 714, was signed by Governor Davis on September 3, 2003. Among other things, this legislation voids and makes unenforceable any existing covenant, restriction, or condition contained in any deed, contract, security instrument, or other instrument affecting real property, as specified, that prohibits or restricts the installation or use of a solar energy system, excepting provisions that impose reasonable restrictions on solar energy systems. This statute also mandates that whenever approval is required for the installation or use of a solar energy system, that such approval be processed in the same manner as approval of an architectural modification, and not be willfully avoided or delayed. Any Public Entity (see definition) may not receive funds from a state sponsored grant or loan program, including the CSI, for solar energy if it fails to comply with these requirements. A Public Entity must certify that it is meeting these requirements when applying for these grants or loans. See California Civil Code section 714 for full statutory requirements and further detail.

AB 2514—(a.k.a. Public Utilities Code Sections 2835-2839) Legislation enacted in 2010 directing the California Public Utilities Commission to open a proceeding to determine, if appropriate, procurement targets for energy storage by load serving entities.

ABA—American Bar Association www.abanet.org

Abandonment—The retirement from further use of a fixed asset. Also, the relinquishment of salvage to insurers with intention of claiming the full amount of insurance value, or, the giving up of title or the right of property by voluntary surrender or neglect.
- Regulatory authorization for a utility to cease provision of a particular service and/or to shut down a particular facility.

Abatement—Is simply the reduction of emissions—so "abatement costs" are the costs of reducing emissions. Can be a reduction in the quantity or intensity as in greenhouse gas emissions. The reduction of the degree or intensity of pollutants or emissions.

ABEL—EPA's computer model for analyzing a violator's ability to pay a civil penalty.

ABES—Alliance for Balanced Environmental Studies (Solutions)

Abiotic—Having an absence of life or living organisms.

Ablation—The rapid reduction of soil particles by means of a focused laser.

Above Building Standard—Upgraded finishes and specialized designs necessary to accommodate a tenant's requirements.

Above-Market Cost—The cost of a service in excess of the price of comparable services in the market.

ABS—Acrylonitrile-Butadiene Styrene
  • Asset-Backed Securities

Absolute Humidity—The ratio of the mass of water vapor to the volume occupied by a mixture of water vapor and dry air.

Absolute Open Flow (AOF)—The number of cubic feet of gas per 24 hours that would be produced by a well if the only pressure against the face of the producing sand in the well bore were atmospheric pressure.

Absolute Pressure—Gauge pressure plus barometric pressure. Absolute pressure can be zero only in a perfect vacuum.

Absolute Viscosity—The measure of a fluid's tendency to resist flow, without regard to its density. By definition, the product of a fluid's kinematic viscosity times its density.

Absolute Zero—The zero point on the absolute temperature scale. It is equal to -273 degrees C or 0 degrees K (Kelvin), or -459.69 degrees F, or 0 degrees R (Rankine).

Absorbed Glass Mat (AGM) Batteries—Employ a micro-fibrous silica glass mat envelope to immobilize their electrolyte. This makes them non-spillable and gives them a lower self-discharge rate than conventional flooded lead-acid batteries.

Absorbent—A material that extracts one or more substances from a fluid (gas or liquid) medium on contact, and which changes physically and/or chemically in the process. The less volatile of the two working fluids in an absorption cooling device.

Absorber—The component of a solar thermal collector that absorbs solar radiation and converts it to heat, or, as in a solar photovoltaic device, the material that readily absorbs photons to generate charge carriers (free electrons or holes). For solar tube collectors with reflective panels, the entire circumferential surface area of the inner tube is often used when calculating absorber area, as the reflective panel is supposed to reflect light onto the underside of the evacuated tube.

Absorbers—Dark-colored objects that soak up heat in solar collectors.

Absorptance—The ratio of the radiation absorbed by a surface to the total energy falling on that surface described as a percentage.

Absorption—The passing of a substance or force into the body of another substance.
  • In international trade terms, absorption is investment and consumption purchases by households, businesses, and governments, both domestic and imported. When absorption exceeds production, the excess is the country's current account deficit.

Absorption Chiller—A type of air cooling device that uses absorption cooling to cool interior spaces. A water chilling process in which cooling is accomplished by the evaporation of a fluid (usually water), which is then absorbed by a different solution (usually lithium bromide), then evaporated under heat and pressure. The fluid is then condensed with the heat of condensation rejected through a cooling tower.

Absorption Coefficient—In reference to a solar energy conversion devices, the degree to which a substance will absorb solar energy. In a solar photovoltaic device, the factor by which photons are absorbed as they travel a unit distance through a material.

Absorption Cooling—A process in which cooling of an interior space is accomplished by the evaporation of a volatile fluid, which is then absorbed in a strong solution, then desorbed under pressure by a heat source, and then re-condensed at a temperature high enough that the heat of condensation can be rejected to an exterior space.

Absorption Period—The actual or expected period required from the time a property (real estate) is initially offered for lease, purchase, or use by its eventual users until all portions have been sold or stabilized occupancy has been achieved. Although marketing may begin before the completion of construction, most forecasters consider the absorption period to begin after completion of construction.

Absorption Plant—A device that removes hydrocarbon compounds from natural gas, especially casinghead gas. The gas is run through oil of proper character, which absorbs the liquid constituents, which are then recovered by distillation.

Absorption Rate—The rate at which rentable space is filled. Gross absorption is a measure of the total square feet leased over a specified period with no consideration given to space vacated in the same geographic area during the same time period. Net absorption is equal to the amount occupied at the end of a period minus the amount occupied at the beginning of a period and takes into consideration space vacated during the period.

Absorption Refrigeration—A system in which a secondary fluid absorbs the refrigerant, releasing heat, then releases the refrigerant and reabsorbs the heat. Ammonia or water is used as the vapor in commercial absorption cycle systems, and water or lithium bromide is the absorber.

Absorption Type Air Conditioner, Direct Fired—A self-contained device which provides cooling by direct application of heat.

Absorptivity—In a solar thermal system, the ratio of solar energy striking the absorber that is absorbed by the absorber to that of solar energy striking a black body (perfect absorber) at the same temperature. The absorptivity of a material is numerically equal to its emissivity. A measure of the ability of a material to absorb solar radiation.

Abstract—An abridgment; a brief summary.

ABTRES—Abatement and Residual Forecasting Model

Abutting Owner—Owner whose land touches another parcel, i.e., a road, highway, or other parcel of land.

AC—Actual Commitment
- Advisory Circular
- Alternating Current

ACA—Absolute Calibration Audit

ACBM—Acronym for "asbestos-containing building material."

ACCA—The Air Conditioning Contractors of America www.acca.org

ACCA Manual J—The ACCA document entitled "Manual J—Residential Load Calculation, Eighth Edition" (2003)

Accelerated Amortization—Writing down the cost of an asset in a shorter period of time than is customary for the item. This serves to reduce income taxes during the period and is permitted by the federal tax authorities as an incentive to the purchase of needed items for expanding a sluggish economy. Also, a device where the due date on the principal of a debt is accelerated by previous agreement.

Accelerated Cost Recovery System (ACRS)—The tax depreciation, or cost recovery, method for IRS purposes, which was introduced by the 1981 Economic Recovery tax Act and was effective for all depreciable property placed in service after December 31, 1980 and before January 1, 1987. ACRS was replaced by the Modified Accelerated Cost Recovery System (MACRS) of the 1986 Tax Reform Act and is the method used for depreciating personal property in the US today.

Accelerated Depreciation—Any depreciation method that allows for greater deductions or charges in the earlier years of an asset's depreciable life, with charges becoming progressively smaller in each successive period. Examples would include the double declining balance and sum-of-the-years digits methods.

Accelerated Payments—A remedy the lender can execute in the event of contract default. All future payments are due and payable. No additional interest is due.

Acceleration Clause—The clause in a note, bond or mortgage or any financial instrument which stipulates that in the event of default by the debtor, the entire outstanding balance becomes due and payable immediately.

Accent (light)—Is a direction luminaire designed to highlight or spotlight objects. It can be recessed, surface mounted, or mounted to a pendant, stem or track.

Acceptance for Filing (of a Rate Schedule)—Commission action by which a rate schedule is accepted for filing and becomes a legal rate schedule when made effective by the Commission.

Acceptance Requirements for Code Compliance—A description of test procedures in the Nonresidential ACM Manual (Title 24 of the State of California) that includes equipment and systems to be tested, functions to be tested, conditions under which the test shall be performed, the scope of the tests, results to be obtained and measurable criteria for acceptable performance.

Acceptor—A dopant material, such as boron, which has fewer outer shell electrons than required in an otherwise balanced crystal structure, providing a hole, which can accept a free electron.

Access Charge—A charge for a power supplier, or its customer, for access to a utility's transmission or distribution system. It is a charge for the right to send electricity over another's wires.

A charge paid by all market participants withdrawing energy from the ISO controlled grid. The access charge will recover the portion of a utility's transmission revenue requirement not recovered through the variable usage charge. It is a charge for the right to send electricity over another's wires.

Accessible—Is having access thereto, but which first may require removal or opening of access panels, doors, or similar obstructions.

Accommodation Endorsement—This is signed by a person or a company who endorses a note or instrument solely to induce a bank or other lending institution to lend money to a borrower whose credit is not substantial enough to warrant a loan. The endorser (signer) is liable to repay the amount in full, but ordinarily does not expect to do so since the signer derives no benefit from the transaction, but acts as a guarantor or surety for the borrower. Another form of accommodation endorsement is the practice among banks of endorsing the acceptances of other banks for a fee, in order to make them acceptable for purchase in the acceptance market.

Accommodation Paper—This is the actual instrument signed relative to the accommodation endorsement.

Accord and Satisfaction—In law, an agreement that is reached via a formally executed agreement between aggrieved parties whereby one aggrieved party accepts something different from the original right or claim.

Account—The place where allowance or credits are held and transactions recorded (greenhouse gases, etc.)

Account Classification—The way in which suppliers of electricity, natural gas, or fuel oil classify and bill their customers. Commonly used account classifications are "Residential," "Commercial," "Industrial," and "Other." Suppliers' definitions of these terms vary from supplier to supplier. In addition, the same customer may be classified differently by each of its energy suppliers.

Account of Others (natural gas)—Natural gas deliveries for the account of others are deliveries to customers by transporters that do not own the natural gas but deliver it for others for a fee. Included are quantities covered by long-term contracts and quantities involved in short-term or spot market sales.

Account Party—In a commercial letter of credit, the party instructing the bank to open a letter of credit and on whose behalf the bank agrees to make payment. In most cases, the account party is an importer or buyer, but alternatively, may be a construction contractor or a supplier bidding on a contract.

Accountability—Being answerable to all stakeholders, including any natural or social systems affected by a business such as customer, employees, and communities.

Accounting System—A method of recording accounting data for a utility or company or a method of supplying accounting information for controlling, evaluating, planning and decision-making.

Accounts Payable—A current liability representing the amount owed by a debtor to a creditor for merchandise or services purchased on open account or short- or long-term credit.

Accounts Receivable—Money due from customers carried as "open book" accounts. The amount is carried in the current assets section of the firm's balance sheet.

Accredited Investor—Rule 501 of the SEC regulations defines an individual accredited investor as: "Any natural person whose individual net worth or joint net worth with that person's spouse at the time of his purchase exceeds $1,000,000," OR "Any natural person who had an individual income in excess of $200,000 in each of the two most recent years or joint income with that person's spouse in excess of $300,000 in each of those years and has a reasonable expectation of reaching the same income level in the current year."

Accredited Verifier—Individual or firm approved by a regulator to provide verification services for those subject to reporting.

Accreting Swap—A swap in which the notational principal amount increases at a predetermined way over time.

Accrual Basis of Accounting—A method of keeping accounts that shows expenses incurred and income earned for a given fiscal period even though such expenses and income have not been actually paid or received in cash.

Accrued Expense—Expenses incurred but not yet paid.

Accrued Income—Income earned but not yet collected.

Accrued Interest—Interest earned but not yet collected.

Accumulated Deferred Income Taxes—Account(s) shown on a corporation's balance sheet (typically a net liability) that represents a future (deferred) claim by the government against the corporation's assets. Deferred income taxes arise from the use of accelerated or liberalized depreciation for tax purposes instead of straight-line or other non-liberalized depreciation methods used for book purposes, and from other temporary differences in the recognition of revenue and expense items for income tax purposes and for financial reporting purposes.

Accumulated Depreciation—In accounting, the amount

of depreciation expense that has been claimed to date on an asset. A financial reporting term for a contra-asset account that shows the total depreciation charges for an asset since acquisition.

Accumulated Provision for Depletion—The net accumulated credit resulting from offsetting charges to income for the pro-rata cost of extracted depletable natural resources such as coal, gas, oil, etc.

Accumulated Provision for Depreciation and Amortization—The net accumulated credit balance arising from provisions for depreciation and/or amortization of assets, usually utility plant and non-utility property. The net balance reflects current and prior credits less charges but is not a measure of actual depreciation.

Accumulating Shear—A feller-buncher shearhead that is capable of accumulating and holding 2 or more cut stems.

Accumulator—A component of a heat pump that stores liquid and keeps it from flooding the compressor. The accumulator takes the strain off the compressor and improves the reliability of the system.

ACDP—Federal Air Contaminant Discharge Permit.

ACE—Alliance for Clean Energy
- Any Credible Evidence
- Accounting and Controlling Engine
- Army Corps of Engineers

ACEEE—American Council for an Energy Efficient Economy. www.aceee.org Publishes books and papers on industrial, commercial, and residential energy efficiency.

Acetone—A widely used, highly volatile solvent. It is readily absorbed by breathing, ingestion or contact with the skin. Workers who have inhaled acetone have reported respiratory problems.

ACFM—Actual cubic feet per minute. The actual cubic feet per minute of gas flowing in a process at the temperature and pressure of the process at that point.

Achievable Potential—In Demand Side Management (DSM) of energy, an estimate of energy savings based on the assumption that all energy-efficient options will be adopted to the extent that they are cost-effective and possible through utility DSM programs.

ACHP—Advisory Council on Historic Preservation

ACI—Activated Carbon Injection
- American Concrete Institute
- Association for Conservation Information

Acid Hydrolysis—A chemical process in which acid is used to convert cellulose or starch to sugar.

Acid Mine Drainage—This refers to water pollution that results when sulfur-bearing minerals associated with coal are exposed to air and water and form sulfuric acid and ferrous sulfate. The ferrous sulfate can further react to form ferric hydroxide, or yellowboy, a yellow-orange iron precipitate found in streams and rivers polluted by acid mine drainage.

Acid Rain—Also called acid precipitation or acid deposition, acid rain is precipitation containing harmful amounts of nitric and sulfuric acids formed primarily by sulfur dioxide and nitrogen oxides released into the atmosphere when fossil fuels are burned. It can be wet precipitation (rain, snow, or fog) or dry precipitation (absorbed gaseous and particulate matter, aerosol particles or dust). Acid rain has a pH below 5.6. Normal rain has a pH of about 5.6, which is slightly acidic. The term pH is a measure of acidity or alkalinity and ranges from 0 to 14. A pH measurement of 7 is regarded as neutral. Measurements below 7 indicate increased acidity, while those above indicate increased alkalinity.

This should correctly be called acid precipitation because it includes rain, snow, sleet, fog and any other form of precipitation. It is produced as industrial byproducts in emissions of sulfur and nitrogen oxides from burning coal and petroleum products. Found throughout the world, its heaviest concentrations are in urban areas. Among other things, it harms aquatic wildlife, corrodes monuments and bridges, destroys exterior paint, kills forests, damages some agricultural soils, makes drinking water toxic by leaching lead from pipes, and reduces visibility.

The acid rain program in the United States was established under the Clean air Act Amendments of 1990. It established a cap and trade system for reducing SO_2 emissions from power plants.

Acid Test—The ratio between current (totally liquid) assets and current liabilities on the balance sheet. Also, an indication of the ability of a business to pay all of its current liabilities immediately in cash. It is also known as the quick ratio.

Acids—A class of compounds that can be corrosive when concentrated. Weak acids, such as vinegar and citric acid, are common in foods. Strong acids, such as muriatic (or hydrochloric), sulfuric and nitric acid have many industrial uses, and can be dangerous to those not familiar to handling them.

Acknowledgement—An act of authenticating instruments conveying property or otherwise conferring rights; a declaration by the Party executing an instrument that it is the Party's act and Deed; a form used by Notaries Public and other authorized officials to verify the act of Acknowledgement.

ACL—Alternate Concentration Limit

ACM—See *Alternative Calculation Method*
  • Asbestos-containing Material

ACNT—Australian Council of National Trusts

ACO—Administrative Consent Order

ACOP (Adjusted Coefficient of Performance)—A standard rating term that was used to rate the efficiency of heat pumps in California. ACOP was replaced by Heating Seasonal Performance Factor (HSPF) in 1988.

ACP—Average collection period

ACQR—Air Quality Control Region

Acquisition (foreign crude oil)—All transfers of ownership of foreign crude oil to a firm, irrespective of the terms of that transfer. Acquisitions thus include all purchases and exchange receipts as well as any and all foreign crude acquired under reciprocal buy-sell agreements or acquired as a result of a buy-back or other preferential agreement with a host government.

Acquisition (minerals)—The procurement of the legal right to explore for and produce discovered minerals, if any, within a specific area; that legal right may be obtained by mineral lease, concession, or purchase of land and mineral rights or of mineral rights alone.

Acquisition Costs—Direct costs and indirect costs incurred to acquire legal rights to natural resources or capital assets. Direct costs include costs incurred to obtain options to lease or purchase mineral rights and costs incurred for the actual leasing (e.g., lease bonuses) or purchasing of the rights. Indirect costs include such costs as: brokers' commissions and expenses; abstract and recording fees; filing and patenting fees; and costs of legal examination of title and documents.

ACR—Agency Confirmation Agreement

Acre—A two-dimensional measure of land equaling 4,480 square yards or 43,560 square feet.

Acreage—An area (measured in acres) that is subject to ownership or control by those holding total or fractional shares of working interests. Acreage is considered developed when development has been completed. A distinction may be made between "gross" acreage and "net" acreage:
  • Gross—All acreage covered by any working in-

terest, regardless of the percentage of ownership in the interest.
  • Net—Gross acreage adjusted to reflect the percentage of ownership in the working interest in the acreage.

Acre-foot—The volume of water that will cover an area of 1 acre to a depth of 1 foot.
  • A unit of measurement applied to petroleum and natural gas reservoirs. It is equivalent to an acre of production formation one foot thick.

ACRN—Accounting Classification Reference Number

ACRS—A tax acronym for "Accelerated Cost Recovery System." Now replaced by MACRS or "Modified Accelerated Cost Recovery System."
  • Accelerated Cost Recovery System (depreciation). Implemented by ERTA in 1981, this system superseded the Asset Depreciation Range formerly in effect.

ACS—Acid Scrubber

Act of God—An unanticipated grave natural disaster or other natural phenomenon of an exceptional, inevitable, and irresistible character, the effects of which could not have been prevented or avoided by the exercise of due care or foresight.

Action Level—1. The exposure level at which OSHA regulations to protect employees take effect, e.g., workplace air analysis, employee training, medical monitoring, and record keeping. Exposure at or above the action level is termed "occupational exposure."
2. Regulatory level recommended by EPA for enforcement by FDA and USDA when pesticide residues occur in food or feed commodities for reasons other than the direct application of the pesticide. As opposed to "tolerances" which are established for residues occurring as a direct result of proper usage, actions levels are set for inadvertent residues resulting from previous legal use or accidental contamination.
3. In the Superfund program, the existence of a contaminant concentration in the environment high enough to warrant action or trigger a response under SARA and the National Oil and Hazardous Substances Contingency Plan. The term can be used similarly in other regulatory programs.

Activated Shelf Life—The time it takes for the capacity of a charged battery to fall to an unusable level when stored at a specified temperature.

Activated Sludge Process—A biological wastewater treatment process in which a mixture of waste water and activated sludge is agitated and aerated. The activated sludge is then separated from the treated wastewater by sedimentation and disposed of or returned to the process as needed. A term used to describe sludge that contains microorganisms that break down organic contaminants (e.g., benzene) in liquid waste streams to simpler substances such as water and carbon dioxide. It is also the product formed when raw sewage is mixed with bacteria-laden sludge, then stirred and aerated to destroy organic matter.

Activation Voltage—The voltage at which the controller will operate to protect the batteries.

Active Cooling—The use of mechanical heat pipes or pumps to transport heat by circulating heat transfer fluids.

Active Noise Cancelation (ANC)—Involves reducing a sound field through the interaction of a primary sound source with an actively controlled identical secondary sound that is 180 degrees out of phase.

Active Power—The power (in Watts) used by a device to produce useful work. Also called input power. Measured in kilowatts (kW) or megawatts (MW). Also known as "real power." The terms "active" or "real" are used to modify the base term "power" to differentiate it from Reactive Power.

Active Solar—As an energy source, energy from the sun collected and stored using mechanical pumps or fans to circulate heat-laden fluids or air between solar collectors and a building.

Active Solar Energy—Solar radiation used by special equipment to provide space heating, hot water or electricity.

Active Solar Energy System—A solar application that uses electrical or mechanical equipment to assist in the collection and storage of solar energy. A system designed to convert solar radiation into usable energy for space, water heating, or other uses as in electricity. It frequently requires a mechanical device, usually a pump or fan, to collect the sun's energy.

Active Solar Heater—A solar water or space-heating system that uses pumps or fans to circulate the fluid (water or heat-transfer fluid like diluted antifreeze) from the solar collectors to a storage tank subsystem.

Activities Implemented Jointly (AIJ)—AIJ is a UNFCCC (United Nations Framework Convention on Climate Change) established pilot program to allow private entities in one country to reduce, sequester, or avoid emissions through a project in a different country. The pilot phase ended in 2000. AIJ has evolved into Joint Implementation under the Kyoto Protocol.

ACTS—Asbestos Contractor Tracking System

Actual Peak Reduction—The actual reduction in annual peak load (measured in kilowatts) achieved by customers that participate in a utility demand-side management (DSM) program. It reflects the changes in the demand for electricity resulting from a utility DSM program that is in effect at the same time the utility experiences its annual peak load, as opposed to the installed peak load reduction capability (i.e., potential peak reduction). It should account for the regular cycling of energy efficient units during the period of annual peak load.

Actual Usage Profile—The Actual Usage Profile is a profile of energy usage taken directly from utility invoices or meter readings.

Actuarial Interest—A constant interest charge based upon a declining principal balance.

ACUPCC—American College and University Presidents' Climate Commitment. The ACUPCC provides a framework for America's colleges and universities to implement sustainable practices on campus in pursuit of climate neutrality.

Acute—Occurring over a short period of time; used to describe brief exposures and effects which appear promptly after exposure.

Acute Exposure—A single exposure to a toxic substance which results in severe biological harm or death. Acute exposures are usually characterized as lasting no longer than a day.

Acute Hazard—Hazard associated with short-term exposure to relatively large amounts of toxic substances.

ACWA—American Clean Water Association www.amsa-cleanwater.org

ADA—Anti-Deficiency Act. American Disabilities Act.

Adaptive Management—A continuing process of action-based planning, monitoring, researching, evaluating, and adjusting with the objective of improving implementation and achieving the goals of the selected alternative.

Adaptive Management Area—Landscape units designated for development and testing of technical and social approaches to achieving desired ecological, economic, and other social objectives.

ADARD—Acid Deposition and Atmospheric Research Division

ADAS—Acid Deposition Assessment Staff

ADB—Asian Development Bank

ADCO—Alternate Document Control Officer

ADCR—Automated Document Control Register

ADCRMG—Automated Document Control Register Management Group

ADD—Average Daily Dose

Addbacks—Items represented as expenses on tax returns that are "added back" to the company's pretax net income for purposes of arriving at a fair valuation of a company. Typical addbacks are travel, automobile, professional expenses, entertainment, and other quasi-personal expenses that are paid by the company. After the addbacks are calculated, the business will "recast" its financials.

Addition—Any change to a building that increases conditioned floor area and conditioned volume. Addition is also any change that increases the floor area or volume of an unconditioned building of an occupancy group. Addition is also any change that increases the illuminated area of an outdoor lighting application.

Additionality—According to the Kyoto Protocol Articles on Joint Implementation and the Clean Development Mechanism, Emissions Reduction Units (ERUs) will be awarded to project-based activities provided that the projects achieve reductions that are "additional to those that otherwise would occur." A distinction is made between environmental additionality and economic/financial additionality.

Financial additionality means projects will only earn credit if funds additional to existing Official Development Assistance commitments are specifically committed to achieve the greenhouse gas reductions.

Environmental additionality requires that emission reductions represent a physical reduction or avoidance of emissions over what would have occurred under a business as usual scenario.

In the greenhouse gas program, reductions are considered to be additional if they represent reductions that would not have occurred without the credit-producing project.

ADI—Acceptable Daily Intake

Adiabatic—Without loss or gain of heat to a system. An adiabatic change is a change in volume and pressure of According to the Kyoto Protocol Articles on Joint Implementation and the Clean Development Mechanism, Emissions Reduction Units (ERUs) will be awarded to project-based activities provided that the projects achieve reductions that are "additional to those that otherwise would occur." A distinction is made between environmental additionality and economic/financial additionality.

• A parcel of gas without an exchange of heat between the parcel and its surroundings.

• In reference to a steam turbine, the adiabatic efficiency is the ratio of the work done per pound of steam, to the heat energy released and theoretically capable of transformation into mechanical work during the adiabatic expansion of a unit weight of steam.

Adjudication—Judicial determination of a case or controversy.

Adjustable Set Point—A feature allowing the user to adjust the voltage levels at which a charge controller will become active.

Adjustable Speed Drive—An electronic device that controls the rotational speed of motor-driven equipment such as fans, pumps, and compressors. Speed control is achieved by adjusting the frequency of the voltage applied to the motor. Terms used to describe this category include polyphase motors, motor oversizing, and motor rewinding.

Adjusted Baseline Usage Profile—The Adjusted Baseline Usage Profile is the same as the Baseline Usage Profile except that it takes into account Baseline Adjustments, which are significant physical changes or facility occupancy changes that have occurred since the Baseline Period. Changes like this can obscure the impact of energy management projects, so they have to be reflected in the M&V (Monitoring and Verification) process. If there are no Baseline Adjustments, the Adjusted Baseline is exactly the same as the Baseline. During the M&V period, the Adjusted Baseline Usage Profile is compared against the Actual Usage Profile to determine energy savings after correcting for weather and other Independent Variables. Baseline Adjustments are calculated and added to the baseline values on M&V reports.

Adjusted Basis—The undepreciated amount of an asset's original basis that is used, for tax purposes, to calculate the gain or loss on disposition of an asset.

Adjusted Electricity—A measurement of electricity that includes the approximate amount of energy used to generate electricity. To approximate the adjusted amount of electricity, the site-value of the electricity is multiplied by a factor of 3. This conversion factor of 3 is a rough approximation of the Btu value of raw fuels used to generate electricity in a steam-generation power plant.

Adjustment Bid—A bid auction conducted by the independent system operator or power exchange to redirect supply or demand of electricity when congestion is anticipated.

Adjustments: Routine and Non-Routine—Changes made to the baseline and/or the performance period energy use to account for changes in measuring energy production from a renewable energy system. Routine adjustments are used to account for expected variations in independent variables. Non-Routine adjustments used to compensate for unexpected changes unrelated to the ECMs.

Administrative and General Expenses—Expenses of an electric utility relating to the overall directions of its corporate offices and administrative affairs, as contrasted with expenses incurred for specialized functions. Examples include office salaries, office supplies, advertising, and other general expenses. Also known as G&A or General & Administrative Expenses.

Administrative Law Judge—The officer designated by the FERC to conduct the proceedings in a rate or other tariff filing.

Admiralty—The court having jurisdiction over questions of maritime law; the system of law administered by admiralty courts.

Adobe—A building material made from clay, straw, and water, formed into blocks, and dried; used traditionally in the southwestern U.S.

ADP—Automated Data Processing

ADPCE—Arkansas Department of Pollution Control & Ecology www.adeq.state.ar.us

ADPE—Automated Data Processing Equipment

ADPS—Acid Deposition Planning Staff

ADQ—Audits of Data Quality

ADR—Alternative Dispute Resolution see *AAA (American Arbitration Association)*
  • American Depositary Receipt (ADRs). A security issued by a U.S. bank in place of the foreign shares held in trust by that bank, thereby facilitating the trading of foreign shares in U.S. markets.

ADR System—A tax depreciation system that establishes the minimum, midpoint and maximum number of years, by asset category, over which an asset can be depreciated. The midpoint life has become synonymous with the term "ADR class life."

ADS—Asset Depreciation System or ADS (formerly Asset Depreciation Range or ADR). A system of class-lives for various categories of assets, using a minimum, midpoint, and maximum number of years.

Adsorption—The extraction from a mixture of gases or liquids of one or more components, by surface adhesion to that material with which the gases or liquids come in contact. The adsorption or extraction process does not cause and is not accompanied by either a physical or chemical change in the sorbent material.

ADSS—Air Data Screening System

ADT—Average Daily Traffic

Advalorem—Literally: according to value. Any charge, tax, or duty that is applied as a percentage of value.

Advance Payments—One or more Lease payments required to be paid by the lessee to the lessor at the beginning of the lease term. Lease structures commonly require one or more payments to be made in advance when the lease contract is consummated.

Advance Refunding—A transaction involving the issuance of new debt to replace existing debt with the proceeds from the new debt placed in trust or otherwise restricted to retire the existing debt at a determinable future date or dates.

Advance Rental—In a lease contract, any payment in the form of rent made before the start of the lease term. The term also is used to describe a rental payment arrangement in which the lessee pays each rental, on a per period basis, at the start of each rental payment period. For example, a quarterly, in advance, rental program requires the lessee to pay one fourth of the annual rental at the start of each consecutive three-month period during the lease term.

Advance Royalty—A royalty required to be paid in advance of production from a mineral property that may or may not be recoverable from future production.

Advanced Battery Systems—A new generation of batteries characterized by improved efficiency and power, as well as fast charging.

Advances From Municipality—The amount of loans and advances made by the municipality or its other departments to the utility department when such loans and advances are subject to repayment but not subject to current settlement.

Advances To Municipality—The amount of loans and advances made by the utility department to the municipality or its other departments when such loans or advances are subject to current settlement.

Adverse Health Effects—Effects of chemicals or other materials that impair one's health. They can range from relatively mild temporary conditions such as minor eye or throat irritation, shortness of breath or headaches to permanent and serious conditions such as cancer, birth defects or damage to organs.

Adverse Hydro—Water conditions limiting the production of hydroelectric power. In years having below-normal levels of rain and snow, and in seasons having less-than-usual runoff from mountain snow pack, there is then less water available for hydro energy production.

Adverse Water Conditions—Reduced stream flow, lack of rain in the drainage basin, or low water supply behind a pondage or reservoir dam resulting in a reduced gross head that limits the production of hydroelectric power or forces restrictions to be placed on multipurpose reservoirs or other water uses.

Advised Letter of Credit—A commercial letter of credit whose authenticity has been verified by a bank, generally to the beneficiary's location. The bank then advises the beneficiary of the authenticity of the letter of credit but does not take on any payment obligation.

Advisory Level—The level above which an environmental protection agency suggests it is potentially harmful to be exposed to a contaminant, although no action is mandated.

AEA—Atomic Energy Act (1954)

AEC—Army Environmental Center
- Associate Enforcement Counsels

AED—Air Enforcement Division
- Analysis and Evaluation Division

AEE—Association of Energy Engineers www.aeecenter.org
- Alliance for Environmental Education www.ee-alliance.org

AEERL—Air and Energy Engineering Research Laboratory

AEFA—Average Emission factor Approach

AEGL—Acute Exposure Guideline Levels

AEI—Advanced Energy Initiative

AEL—Airborne Exposure Limit

AEM—Acoustic Emission Monitoring

AEO—Annual Energy Outlook—a DOE/EIA publication

Aeration—Passing air through a solid or liquid, especially a process that promotes breakdown or movement of contaminants in soil or water by exposing them to air.

Aeration Basin—A basin where oxygen is supplied by mechanical agitation or pneumatic means to enhance the breakdown of wastes held in suspension.

Aerator—An Aerator is a device installed in a faucet or showerhead that adds air to the water flow, thereby maintaining an effective water spray while reducing overall water consumption.

AERE—Association of Environmental and Research Economists www.aere.org

Aerobic—Life or biological processes that can occur only in the presence of oxygen.

Aerobic Bacteria—Microorganisms that require free oxygen, or air, to live, and that which contribute to the decomposition of organic material in soil or composting systems.

AEROS—Aerometric and Emissions Reporting System

Aerosol—Solid or liquid particles suspended within the atmosphere.

AESA—Association of Environmental Scientists and Administrators

AFA—American Forestry Association www.american-forests.org

AFC—Application for Certification

AFCA—Air Fuel Consumption Allocation

AFDO—Award Fee Determination Official

Affiant—A person who makes an Affidavit

Affidavit—A written declaration under oath made before an authorized official.

Affiliate—An 'affiliate' of, or a person 'affiliated' with, a specific person is a person that directly, or indirectly through one or more intermediaries, controls, or is controlled by, or is under common control with, the person specified. The term 'affiliate' includes any subsidiary or parent of the person specified.

Affiliated Entities Test—A test to determine if the amount paid for gas to an affiliate exceeds the amount paid in comparable first sales between non-affiliated entities.

Affiliated Marketer (Energy)—A marketer that is owned either by a distribution or transmission company, or by a corporation that also owns a distribution or transmission company.

Affiliated Power Producer—A generating company that is affiliated with a utility.

Afforestation—Planting of new forests on lands that have not been recently forested. Conversion of bare or cultivated land into forest. The process of establishing and growing forests on bare or cultivated land which has not been forested in recent history.

AFGE—American Federation of Government Employees www.afge.org

AFI—American Forest Institute

AFTA—American Forest & Timber Association

After-Cooling—The process of cooling a compressed air

or gas immediately after compression.

After-Market—Broad term that applies to any change after the original purchase, such as adding equipment not a part of the original purchase. As applied to alternative fuel vehicles, it refers to conversion devices or kits for conventional fuel vehicles.

Aftermarket Converted Vehicle—A standard conventionally fueled, factory-produced vehicle to which equipment has been added that enables the vehicle to operate on alternative fuel.

Aftermarket Vehicle Converter—An organization or individual that modifies OEM vehicles after first use or sale to operate on a different fuel (or fuels).

After-Tax Cash Flow (ATCF)—The portion of pre-tax cash flow that remains after all income tax liabilities have been deducted.

After-Tax Real Rate of Return—Money after-tax rate of return minus the inflation rate.

AFUE (Annual Fuel Utilization Efficiency)—A measure of heating efficiency, in consistent units, determined by applying the federal test method for furnaces. This value is intended to represent the ratio of heat transferred to the conditioned space by the fuel energy supplied over one year. [See California Code of Regulations, Title 20, Section 1602(d)(1)]

AFV—Alternate fueled vehicle

AFY—Acre-feet Per Year

AG—Attorney General

AGA—American Gas Association www.aga.org

AGC—Associate General Counsel
 • Association of General Contractors www.agc.org

Agent—A person who acts for another by proper authorization.

Agglomerating Character—Agglomeration describes the caking properties of coal. Agglomerating character is determined by examination and testing of the residue when a small powdered sample is heated to 950 degrees Centigrade under specific conditions. If the sample is "agglomerating," the residue will be coherent, show swelling or cell structure, and be capable of supporting a 500-gram weight without pulverizing.

Aggregate Ratio—The ratio of two population aggregates (totals). For example, the aggregate expenditures per household is the ratio of the total expenditures in each category to the total number of households in the category.

Aggregated Unit—A group of two or more Generating Units, each individually capable of producing Energy and operating in parallel with the electric grid in the ISO Balancing Authority Area, which are aggregated for scheduling and Settlement purposes and operated as if they were a single Generating Unit. An Aggregated Unit must first be approved by the ISO and may represent a Physical Scheduling Plant or collection of other Generating Units, which need to be combined for operational purposes due to common metering, electrical connection arrangement, interrelated operating characteristics, or other physical factors or constraints.

Aggregation—The process of organizing small groups, businesses or residential customers into a larger, more effective bargaining unit that strengthens their purchasing power with utilities.

The functional bundling of dispersed resources (or loads) to operate as a combined unit.

Aggregator—An entity responsible for planning, scheduling, accounting, billing, and settlement for energy deliveries from the aggregator's portfolio of sellers and/or buyers. Aggregators seek to bring together customers or generators so they can buy or sell power in bulk, making a profit on the transaction.

An entity that negotiates the purchase of energy in bulk for a group of consumers, and tries to negotiate lower prices. The group of consumers is called a buying group.

AGM (Adsorbed Glass Mat)—A newer type of battery construction that uses saturated adsorbant glass mats rather than gelled or liquid electrolyte. AGM batteries are typically more expensive than flooded (liquid), but offer enhanced reliability.

Agrarian—Promoting agricultural interests.

Agreement and Undertaking—A document which an independent as producer may be allowed to file, at the discretion of the Commission, in lieu of a bond, agreeing to refund that portion of an increased rate which has been made effective subject to refund and is ultimately found not justified by the Commission.

Agricultural Building—A structure designed and constructed to house farm implements, hay, grain, poultry, livestock or other horticultural products. It is not a structure that is a place of human habitation, a place of employment where agricultural products are processed, treated or packaged, or a place used by the public.

Agricultural Waste—Poultry and livestock manure or residual materials in liquid or solid form generated in the production and marketing of poultry, livestock, fur-bearing animals and their products, rice straw, rice husks and other plant wastes.

Agriculture—An energy-consuming sub sector of the industrial sector that consists of all facilities and equipment engaged in growing crops and raising animals.

Agriculture, Mining, and Construction (consumer category)—Companies engaged in agriculture, mining (other than coal mining), or construction industries.

AGST—Above-Ground Storage Tanks

AGW—Anthropogenic Global Warming

AH—Allowance Holder

Ah—Amp hour, a unit of electrical energy.

AHERA—Asbestos Hazard Emergency Response Act (TSCA, 1986)

AHM—Acutely Hazardous Material
- Allowance Monthly Holder

AHQ—Acute Hazard Quotient

AHU—Air Handling Unit

AHW—Acutely Hazardous Waste

AI—Active Ingredient
- Artificial Intelligence

AIA—American Institute of Architects www.aia.org
- Asbestos Information Association

AIAA—American Institute of Aeronautics and Astronautics www.aiaa.org

AIADA—American International Automobile Dealers Association www.aiada.org

AIC—Acceptable Intake—Chronic
- Active to Inert Conversion

AIC—See "amperage interrupt capability."

AIChE—American Institute of Chemical Engineers www.aiche.org

AICPA—American Institute of Certified Public Accountants www.aicpa.org

AICR—Alternative Internal Control Review

AICUZ—Air Installation Compatible Use Zones

AID—Agency for International Development www.us-aid.gov

AIEC—Acute Inhalation Exposure Criteria

AIG—Assistant Inspector General

AIHA—American Industrial Hygiene Association www.aiha.org

AIHC—American Industrial Health Council

AIP—Auto-ignition Point

Air—The mixture of gases that surrounds the earth and forms its atmosphere, composed of, by volume, 21 percent oxygen, 78 percent nitrogen.

Air Change—The replacement of a quantity of air in a space within a given period of time, typically expressed as air changes per hour. If a building has one air change per hour, this is equivalent to all of the air in the building being replaced in a one-hour period.

- A measure of the rate at which the air in an interior space is replace by outside (or conditioned) air by ventilation and infiltration; usually measured in cubic feet per time interval (hour), divided by the volume of air in the room.

Air Cleaner—A device using filters or electrostatic precipitators to remove indoor-air pollutants such as tobacco smoke, dust, and pollen. Most portable units are 40 watts when operated on low speed and 100 watts on high speed.

Air Collector—A medium-temperature collector used predominantly in space heating, utilizing pumped air as the heat-transfer medium.

- In solar heating systems, a type of solar collector in which air is heated in the collector.

Air Conditioner—A device for conditioning air in an interior space. A Room Air Conditioner is a unit designed for installation in the wall or window of a room to deliver conditioned air without ducts. A Unitary Air Conditioner is composed of one or more assemblies that usually include an evaporator or cooling coil, a compressor and condenser combination, and possibly a heating apparatus. A Central Air Conditioner is designed to provide conditioned air from a central unit to a whole house with fans and ducts.

Air Conditioning—Cooling and dehumidifying the air in an enclosed space by use of a refrigeration unit powered by electricity or natural gas. Note: Fans, blowers, and evaporative cooling systems ("swamp coolers") that are not connected to a refrigeration unit are excluded.

Air Conditioning Intensity—The ratio of air-conditioning consumption or expenditures to square footage of cooled floor space and cooling degree-days (base 65 degrees F). This intensity provides a way of comparing different types of housing units and households by controlling for differences in housing unit size and weather conditions. The square footage of cooled floor space is equal to the product of the total square footage times the ratio of the number of rooms that could be cooled to the total number of rooms. If the entire housing unit is cooled, the cooled floor space is the same as the total floor space. The ratio is calculated on a weighted, aggregate basis according to this formula:

$$\text{Air-Conditioning Intensity} = \frac{\text{Btu for Air Conditioning}}{\text{Cooled Square Feet} * \text{Cooling Degree-Days}}$$

Air Control Layer—An Air Control Layer (or air barrier) is a material that can affect energy usage by inhibiting airflow between a conditioned space and an unconditioned space (or between units, in multi-family and apartment construction). It is often a continuous sheet of polyethylene, polypropylene or extruded polystyrene, wrapped around the outside of a building during construction and sealed at the joints to reduce air infiltration and exfiltration.

Air Control Layer System—The Air Control Layer System is the primary air enclosure boundary that separates indoor (conditioned) air and outdoor (unconditioned) air. In multi-unit/townhouse/apartment construction, the Air Control Layer System also separates the conditioned air from any given unit and adjacent units. Air Control Layer Systems also typically define the location of the Pressure Boundary of the building enclosure. In multi-unit/townhouse/apartment construction, the Air Control Layer System is also the fire barrier and smoke barrier in inter-unit separations. In such assemblies, the Air Control Layer System also must meet the specific fire-resistance rating requirement for the given separation. Air Control Layer Systems typically are assembled from materials (such as gypsum board, sealant, etc.) incorporated in assemblies (such as walls, roofs, etc.) that are interconnected to create enclosures. Each of these three elements has measurable resistance to airflow.

Air Diffuser—An air distribution outlet, typically located in the ceiling, which mixes conditioned air with room air.

Air Economizer—A ducting arrangement and automatic control system that allows a cooling supply fan system to supply outside air to reduce or eliminate the need for mechanical cooling.

Air Film—A layer of still air adjacent to a surface which provides some thermal resistance.

Air Film Coefficient—A measure of the heat transfer through an air film. [See ASHRAE Table 1, ASHRAE Handbook, 1985 Fundamentals]

Air Handling Unit (AHU)—Equipment that includes a fan or blower, heating and/or cooling coils, regulator controls, condensate drain pans, and air filters.

Air Heater—Combustion air (fed to burners) can be heated to approximately 500 degrees F by transferring heat from the flue gases to the air.

Air Infiltration Measurement—A building energy auditing technique used to determine and/or locate air leaks in a building shell or envelope.

Air Mass—The air mass relates to the path length of solar radiation through the atmosphere. An air mass of 1.0 means the sun is directly overhead and the radiation travels through one atmosphere thickness. Approximately equal to the secant of the zenith angle, i.e. the angle from directly overhead to a line to the sun.

• The ratio of the mass of atmosphere in the actual observer-sun path to the mass that would exist if the observer was at sea level, at standard barometric pressure, and the sun was directly overhead (sometimes called air mass ratio). Air mass varies with the zenith angle of the sun and the local barometric pressure, which changes with altitude.

Air Mass 1.5 (AM1.5) Standard Reference Spectrum—The solar spectral irradiance distribution (diffuse and direct) incident at sea level on a sin-facing 37-degree tilted surface. The atmospheric conditions for AM1.5 are: precipitable water vapor, 14.2mm; total ozone, 3.4mm; turbidity (base e, l=0.5 mm), 0.27. [ASTM E 8092, Table 2]

Air Pollution—The presence of contaminants in the air in concentrations that prevent the normal dispersive ability of the air, and that interfere with biological processes and human economics.

Air Pollution Abatement Equipment—Equipment used to reduce or eliminate airborne pollutants, including particulate matter (dust, smoke, fly, ash, dirt, etc.), sulfur oxides, nitrogen oxides (NO_X), carbon monoxide, hydrocarbons, odors, and other pollutants. Examples of air pollution abatement structures and equipment include flue-gas particulate collectors, flue-gas de-sulfurization units and nitrogen oxide control devices.

Air Pollution Control—The use of devices to limit or prevent the release of pollution into the atmosphere.

Air Porosity—A measure of the air-tightness of infiltration barriers in units of cubic feet per hour per square foot per inch of mercury pressure difference.

Air Quality Maintenance Area—Specific populated area where air quality is a problem for one or more pollutants.

Air Quality Standards—The prescribed level of pollutants allowed in outside or indoor air as established by legislation.

Air Register—The component of a combustion device that regulates the amount of air entering the combustion chamber.

Air Retarder/Barrier—A material or structural element that inhibits air flow into and out of a building's envelope or shell. This is a continuous sheet composed of polyethylene, polypropylene, or extruded polystyrene. The sheet is wrapped around the outside of a house during construction to reduce air in-and exfiltration, yet allow water to easily diffuse through it.

Air Rights—The right to undisturbed use and control of designated air space above a specific land within stated elevations. Such rights may be acquired to construct a building above the land or build another or to protect the light and air of an existing or proposed structure on an adjoining lot.

Air Shutter—An adjustable device for varying the primary air inlet(s) regulating primary or secondary air.

Air Space—The area between the layers of glazing (panes) of a window.

Air Stripping Tower—Air stripping removes volatile organic chemicals (VOCs) such as solvents from contaminated water by causing them to evaporate. Polluted water is sprayed downward through a tower filled with packing materials while air is blown upwards through the tower. The contaminants evaporate into the air, leaving significantly-reduced pollutant levels in the water. The air is treated before it is released into the atmosphere.

Airflow Across the Evaporator—The rate of airflow, usually measured in cfm (cubic feet per minute) across a heating or cooling coil. The efficiency of air conditioners and heat pumps is affected by the airflow across the evaporator (or condenser in the case of a heat pump). See *Thermostatic Expansion Valves (TXV)*.

Air-Gas Ratio—The ratio of the air volume to the gas volume. A specified ratio is necessary to achieve a desired character of combustion.

Air-Impermeable Material—An Air-Impermeable Material is an Air Control Layer. An Air-Impermeable Material has an air permeance equal to or less than 0.004 cfm/ft^2 at 0.30 in. w.c. (0.02 L/s∑m^2 at 75 Pa) pressure differential when tested according to ASTM standard test method E2178 or E 283.

Airlock Entry—A building architectural element (vestibule) with two airtight doors that reduces the amount of air infiltration and exfiltration when the exterior most door is opened.

AIRMON—Atmospheric Integrated Research Monitoring Network

Air-Permeable Material—An Air-Permeable Material has an air permeance greater than 0.004 cfm/ft^2 at 0.30 in. w.c. (0.02 L/s∑m2 at 75 Pa) pressure differential when tested according to ASTM standard test method E2178 or E 283.

AIRS—Accident and Illness Reporting System
- Aerometric Information Retrieval System
- Air Quality Subsystem

Air-Source Heat Pump—A type of heat pump that transfers heat from outdoor air to indoor air during the heating season, and works in reverse during the cooling season.

Airtight Drywall Approach (ADA)—A building construction technique used to create a continuous air retarder that uses the drywall, gaskets, and caulking. Gaskets are used rather than caulking to seal the drywall at the top and bottom. Although it is an effective energy-saving technique, it was designed to keep airborne moisture from damaging insulation and building materials within the wall cavity.

Air-to-Air Heat Exchanger—A device which will reduce the heat losses or gains which occur when a building is mechanically ventilated, by transferring heat between the conditioned air being exhausted and the unconditioned air being supplied.

Air-to-Air Heat Pump—See *Air-Source Heat Pump*.

Air-to-Water Heat Pump—A type of heat pump that transfers heat in outdoor air to water for space or water heating.

AIS—Asbestos Information system
- Acceptable Intake for Subchronic Exposures

AISI—American Iron and Steel Institute www.steel.org

AL—Acceptable Level
- Action Level

ALA—American Lung Association www.lungusa.org

ALAPCO—Association of Local Air Pollution Control Officers www.capcoa.org

ALARA—As Low as Reasonably Achievable

Alaskan System Coordination Council (ASCC—One of the ten regional reliability councils that make up the North American Electric Reliability Council (NERC).

Albedo—Refers to the ratio of light from the sun that is reflected by the Earth's surface to the light received by it. Unreflected light is converted to infrared radiation (i.e., heat), which causes atmospheric warming (see "radiative forcing"). Thus, surfaces with a high albedo (e.g., snow and ice) generally contribute to cooling, whereas surfaces with a low albedo (e.g., forests) generally contribute to warming. Changes in land use that significantly alter the characteristics of land surfaces can therefore influence the climate through changes in albedo.

ALC—Application Limiting Constituent

Alcohol—A general class of hydrocarbons that contain a hydroxyl group (OH). The term "alcohol" is often used interchangeably with the term "ethanol," even though there are many types of alcohol. (See *Butanol, Ethanol, Methanol*.)

Alcohol Fuels—A class of liquid chemicals that have certain combinations of hydrogen, carbon and oxygen, and that are capable of being used as fuel.

Algae—Primitive plants, usually aquatic, capable of synthesizing their own food by photosynthesis.

Alienation—The voluntary parting with an interest in the Ownership of Real Property.

Alienation Clause—A clause in a Note and Trust Deed permitting the payee to declare the entire unpaid balance immediately due and payable upon subsequent transfer of the property. Also referred to as a "Due On Sale" clause.

Aliphatic—Hydrocarbon that does not contain an aromatic ring structure.

ALJ—Administrative Law Judge

Alkali—A soluble mineral salt.

Alkaline—1. A type of fuel cell that uses alkaline potassium hydroxide as the electrolyte (used by NASA for space missions).
2. Having the properties of a base, a pH greater than 7. Usually used as an adjective, i.e. "alkaline soil."

Alkylate—The product of an alkylation reaction. It usually refers to the high-octane product from alkylation units. This alkylate is used in blending high octane gasoline.

Alkylation—A refining process for chemically combining isobutane with olefin hydrocarbons (e.g., propylene, butylene) through the control of temperature and pressure in the presence of an acid catalyst, usually sulfuric acid or hydrofluoric acid. The product, alkylate, an isoparaffin, has high octane value and is blended with motor and aviation gasoline to improve the antiknock value of the fuel.

All-Electric Home—A residence in which electricity is used for the main source of energy for space heating, water heating, and cooking. Other fuels may be used for supplementary heating or other purposes.

Allergen—A substance capable of causing an allergic reaction because of an individual's sensitivity to that substance.

Alliance of Small Island States (AOSIS)—AOSIS is a group of countries formed during the Second World Climate Conference in 1990 that includes 35 states from the Atlantic, Caribbean, Indian Ocean, Mediterranean and Pacific. AOSIS countries are small islands and low-lying coastal developing countries that are particularly vulnerable to the effects of climate change such as sea level rise, coral bleaching and the increased frequency and intensity of tropical storms. These countries share a common objective on environmental and sustainable development matters.

All-in Cost—Total costs, explicit and other.

All-in Rate—An interest rate on a loan which includes the cost of compensating balances, commitment fees and any other charges.

All-Inclusive Deed of Trust (AITD)—This is a junior Deed of Trust on the property that includes the amount actually due the Beneficiary therein but also includes the unpaid principal balances of Deeds of Trust that are to remain of record and are senior to this Deed of Trust. The Beneficiary agrees that by accepting payment on the amount of the "All-Inclusive" Deed of Trust that they will make the payments, when due, on the senior Deeds of Trust.

• This type of Deed of Trust is often used by a Seller of the property rather than using a Contract of Sale to avoid complex problems of foreclosing on a Contract of Sale.

• The "All-Inclusive" Deed of Trust is also often called a "Wrap-Around" Deed of Trust, a "Hold Harmless" Deed of Trust, or an "Over-riding" Deed of Trust.

Allocated Costs—Costs, systematically assigned or distributed among parties, products, departments, or other elements.

Allocated Pool—A pool in which the total oil or natural gas production is restricted and allocated to various wells therein in accordance with proration schedules.

Allocation—The number of emission trading credits or allowances that an affected source holds for a specific compliance year. Allocation is the number of credit or allowance permits provided to an emissions source (e.g. company with net emissions) by a jurisdictional regulatory body during a specific compliance period. Allocation of permits occurs primarily through grandfathering or auctioning.

The amount of securities assigned to an investor, broker, or underwriter in an offering. An allocation can be equal to or less than the amount indicated by the investor during the subscription process, depending on market demand for the securities.

The process of determining ownership rights to the gas delivered to a meter.

Allocation Method—A method of allocating volumes to affected parties when an imbalance occurs.

Allotment Trading Units (ATU)—In the Illinois ERMs program, an ATU is the tradable unit issued by the Illinois Environmental Protection Agency. An ATU represents 200 pounds of volatile organic material emissions and is a limited authorization to emit 200 pounds of volatile organic material emissions during the seasonal allotment period under the ERMS program.

Allowables—The permitted rate of production from a well or group of wells that is allowed by a particular state or governing body. The rate is set by rules which vary among the various states or governing bodies.

Allowance—Allowances are the unit of trade under closed systems. Allowances grant the holder the right to emit a specific quantity of pollution once (e.g. one ton). The total quantity of allowances issued by regulators dictates the total quantity of emissions possible under the system. At the end of each compliance period each source must surrender sufficient allowances to cover their emissions during that period.

Allowance for Funds Used During Construction—Construction activities may be financed from internally generated funds (primarily earnings retained in the business), or from funds provided by other external sources (short- and long-term debt). The allowance for funds used during construction is intended to recognize the cost of these funds dedicated to construction activities during the construction period. To arrive at the "allowance," a common procedural method makes use of a formula that is based on the assumption that short-term debt is the first source of construction funds. The cost rate for short-term debt is based on current costs. Since a utility plant is subject to depreciation, the allowance for funds used during construction is recovered in the form of depreciation from ratepayers over the service life of the plant to which it applies

Allowed Rate of Return—The rate of return that a regulatory commission allows on a rate base in establishing just and reasonable rates for a utility. It is usually based on the composite cost of financing rate base from debit, preferred stock, and common equity.

Alluvial Deposit—An area of sand, clay or other similar material that has been gradually deposited by moving water, such as along a river bed or shore of a lake.

Alongside—A phrase referring to the side of a ship. Goods to be delivered "alongside" are to be placed on the dock or barge within reach of the transport ship's tackle so that they can be loaded aboard the ship.

Alpha Particle—The least penetrating type of radiation usually not harmful to life. A positively-charged particle emitted by radioactive atoms. Alpha particles travel less than one inch in the air and a thin sheet of paper will stop them. The main danger from alpha particles lies in ingesting the atoms which emit them. Body cells next to the atom can then be irradiated over an extended period of time, which may be prolonged if the atoms are taken up in bone, for instance.

Alpha Radiation—Alpha rays consist of nuclei of the element helium and carry a positive charge. Penetrates the least, but does great damage in a small area. (see radiation)

Alteration—Any change to a building's water heating system, space conditioning system, lighting system, or building envelope that is not an addition.
• Modifications to leased equipment, generally subject to restoration at the conclusion of the lease.

Alternate Energy Source for Primary Heater—The fuel that would be used in place of the usual main heating fuel if the building had to switch fuels. (See *Fuel-Switching Capability*.)

Alternate Fuels (Natural Gas)—Other fuels that can be substituted for the fuel in use. In the case of natural gas, the most common alternative fuels are distillate fuel oils, residual fuel oils, coal and wood.

Alternating Current (AC)—An electrical current flowing at varying potential (voltage). Flow of electricity that constantly changes direction between positive and negative sides. Almost all power produced by electric utilities in the United States moves in current that shifts direction at a rate of 60 times per second.

Alternative—Under the National Environmental Policy Act, a comprehensive management strategy. When a federal agency is considering an action, the agency must develop and analyze a range of alternatives. The alternatives must show a reasonable range of actions, including a "no action" alternative.

Alternative Calculation Method—In California, it is the Alternative Calculation Method (ACM) Approval Manual APPROVAL MANUAL OR ACM MANUAL—for the 2001 Energy Efficiency Standards for Nonresidential Buildings, (P400-01-011) for non-residential buildings, hotels, and multi-family residential buildings with four or more stories and the Alternative Calculation Method (ACM) Approval Manual for the 2001 Energy Efficiency Standards for Residential Buildings, (P400-01-012) for all single family and low-rise multi-family residential buildings.

Alternative Calculation Methods (ACMS)—In California, the Commission's Public Domain Computer Programs, one of the Commission's Simplified Calculation Methods, or any other calculation method approved by the Commission.

Alternative Component Package—In California, one of the sets of low-rise residential prescriptive requirements contained in § 151(f) of Title 24. Each package is a set of measures that achieve a level of performance, which meets the standards. These are often referred to as the prescriptive packages or packages. "Buildings that comply with the prescriptive standards shall be designed, constructed and equipped to meet all of the requirements of one of the alternative packages of components shown in Tables 151-B and 151-C for the appropriate climate zone..."

Alternative Delivery Method (ADM)—The ADM delivers the same services available to all customers through Savings By Design (www.savingsbydesign.com). The purpose of the flexible model is to provide a short-term, focused offering of SBD services to promote the use of a new energy efficient technology or to cultivate participation from a particular market segment or customer type that may not have participated in the program previously.

Alternative Energy—Energy from a source other than the conventional fossil-fuel sources of oil, natural gas and coal (i.e., wind, running water, the sun, geothermal). Also referred to as "alternative fuel."

Alternative Energy Sources—See *Renewable Energy*.

Alternative Fuel—Fuel that is substantially non-petroleum based. Also known as Clean Burning Fuel. Alternative fuels, for transportation applications, include the following:
- Methanol
- Denatured ethanol, and other alcohols
- Fuel mixtures containing 85 percent or more by volume of methanol, denatured ethanol, and other alcohols with gasoline or other fuels—natural gas
- Liquefied petroleum gas (propane)
- Hydrogen
- Coal-derived liquid fuels
- Fuels (other than alcohol) derived from biological materials (biofuels such as soy diesel fuel)
- Electricity (including electricity from solar energy.)
- Any other fuel a governmental body determines, by rule, is substantially not petroleum and would yield substantial energy security benefits and substantial environmental benefits. The term "alternative fuel" does not include alcohol or other blended portions of primarily petroleum-based fuels used as oxygenates or extenders, i.e. MTBE, ETBE, other ethers, and the 10-percent ethanol portion of gasohol.

Alternative Fuel Vehicle Converter—An organization (including companies, government agencies and utilities), or individual that performs conversions involving alternative fuel vehicles. An AFV converter can convert (1) conventionally fueled vehicles to AFVs, (2) AFVs to conventionally fueled vehicles, or (3) AFVs to use another alternative fuel.

Alternative Fuels (Transportation)—As defined by the National Energy Policy Act (EPAct) the fuels are: methanol, denatured ethanol and other alcohols, separately or in mixtures of 85 percent by volume or more (or other percentage not less than 70 percent as determined by U.S. Department of Energy rule) with gasoline or other fuels; CNG; LNG; LPG; hydrogen; "coal-derived liquid fuels;" fuels "other than alcohols" derived from "biological materials;" electricity, or any other fuel determined to be "substantially not petroleum" and yielding "substantial energy security benefits and substantial environmental benefits."

Alternative Fuels—A popular term for "non-conventional" transportation fuels derived from natural gas (propane, compressed natural gas, methanol, etc.) or biomass materials (ethanol, methanol).

Alternative Fuels Vehicle (AFV)—A vehicle designed

to operate on an alternative fuel (e.g., compressed natural gas, methane blend, or electricity). The vehicle could be either a dedicated vehicle designed to operate exclusively on alternative fuel or a non-dedicated vehicle designed to operate on alternative fuel and/or a traditional fuel.

Alternative Method—Any method of sampling and analyzing for an air pollutant which is not a reference method or an equivalent method but which has been demonstrated to the Administrator's satisfaction to produce, in specific cases, results adequate for their determination of compliance.

Alternative Minimum Tax (AMT)—A separate tax calculation required by the IRS for most taxpayers. A taxpayer must pay the higher of its regular tax or the AMT tax in a given tax period and the AMT is generally thought of as a "penalty" tax. The corporate AMT rate, although lower than the regular tax rate, is applied to a different, typically higher, taxable income than for regular taxes. The Tax Reform Act of 1986 substantially modified the AMT, which must be calculated by all taxpayers.

Alternative Transportation—Modes of travel other than private cars, such as walking, bicycling, rollerblading, carpooling, and public transit.

Alternative-Rate DSM Program Assistance—A DSM (demand-side management) program assistance that offers special rate structures or discounts on the consumer's monthly electric bill in exchange for participation in DSM programs aimed at cutting peak demands or changing load shape. These rates are intended to reduce consumer bills and shift hours of operation of equipment from on-peak to off-peak periods through the application of time-differentiated rates. For example, utilities often pay consumers several dollars a month (refund on their monthly electric bill) for participation in a load control program. Large commercial and industrial customers sometimes obtain interruptible rates, which provide a discount in return for the consumer's agreement to cut electric loads upon request from the utility (usually during critical periods, such as summer afternoons when the system demand approaches the utility's generating capability).

Alternator—A generator producing alternating current by the rotation of its rotor, and which is powered by a primary mover. A device for producing Alternating Current ("AC") electricity. Usually driven by a motor, but can also be driven by other means, including water and wind power.

Altitude Angle—The angle of the sun above the horizon, measured in degrees. In winter, the sun is at a low solar altitude, and in the summer, the sun is at a high solar altitude.

AMA—See *Adaptive Management Area*
- American Medical Assn. www.ama-assn.org

AMBA—American Boiler Manufacturers Association www.abma.org

Ambient—The surrounding atmosphere; encompassing on all sides; the environment surrounding a body but undisturbed or unaffected by it.

Ambient Air—The air external to a building or device. Refers to surrounding air.

Ambient Air Quality—The condition of the air in the surrounding environment.

Ambient Air Temperature—Surrounding temperature, such as the outdoor air temperature around a building.

Ambient Charge—A form of tax on non-uniformly mixed pollutants. It is calculated to be the same in terms of the emission's impact on ambient environmental quality at some receptor site. As a result, an ambient charge to a firm closer to the receptor site will normally be higher *per litre* than that charged to firms further away.

Ambient Lighting—Lighting in an area from any source that produces general illumination, as opposed to task lighting.

Ambient Temperature—The temperature of a medium, such as gas or liquid, which comes into contact with or surrounds an apparatus or building element. The temperature of the surrounding area.

Ambient Temperature Sensor—Used to measure the air temperature. Consists of a K-type thermocouple junction encased in protective shielding to minimize variations in temperature caused by wind currents, reflected heat and rain.

Ambient Vaporizer—A vaporizer which derives energy for vaporizing and heating LNG from storage conditions to send out conditions from naturally occurring sources such as the atmosphere, sea water, or geothermal waters.

American Gas Association (AGA)—Trade group representing natural gas distributors and pipelines. www.aga.org

AMI—Advanced Metering Infrastructure. A term denoting electricity meters that measure and record usage data at a minimum, in hourly intervals, and provide usage data to both consumers and energy companies at least once daily.

AMIS—Air Management Information System

Ammeter—A device used for measuring current flow at any point in an electrical circuit.

Ammonia—A colorless, pungent, gas (NH_3) that is extremely soluble in water, may be used as a refrigerant; a fixed nitrogen form suitable as fertilizer.

Amorphous—Devoid of crystallinity. Most plastics are in the amorphous state at processing temperatures; many retain this state under normal conditions.

Amorphous Semiconductor—A non-crystalline semiconductor material that has no long-range order.

Amorphous Silicon—An alloy of silica and hydrogen, with a disordered, noncrystalline internal atomic arrangement, that can be deposited in thin-film layers (a few micrometers in thickness) by a number of deposition methods to produce thin-film photovoltaic cells on glass, metal, or plastic substrates. A thin-film solar PV cell material which has a glassy rather than crystalline structure. Made by depositing layers of doped silicon on a substrate normally using plasma-enhanced chemical vapor deposition of silane.

Amortization—The depreciation, depletion, or charge-off to expense of intangible and tangible assets over a period of time. In the extractive industries, the term is most frequently applied to mean either (1) the periodic charge-off to expense of the costs associated with non-producing mineral properties incurred prior to the time when they are developed and entered into production or (2) the systematic charge-off to expense of those costs of productive mineral properties (including tangible and intangible costs of prospecting, acquisition, exploration, and development) that had been initially capitalized (or deferred) prior to the time the properties entered into production, and thereafter are charged off as minerals are produced.

AMOS—Air Management Oversight System

Amount to Finance—The total asset cost, minus total down payments, plus total sales tax, plus financed fees, plus non-takeout loans. If the trade-in treatment is set to cost reduction then the amount to finance also includes minus total trade-ins, plus total liens.

Ampacity—Refers to the highest safe amount of electrical current through conductors, overcurrent devices, or other electrical equipment. Ampacity is determined by the cross-sectional area and the material of the conductor, or the manufacturer's equipment rating.

Amperage Interrupt Capability (AIC)—Direct current fuses should be rated with a sufficient AIC to interrupt the highest possible current.

Ampere (Amp)—The unit of measure that tells how much electricity flows through a conductor. It is like using cubic feet per second to measure the flow of water. For example, a 1,200 watt, 120-volt hair dryer pulls 10 amperes of electric current (watts divided by volts).

Ampere Hour Meter—An instrument that monitors current with time. The indication is the product of current (in amperes) and time (in hours).

Ampere-Hour (Ah/AH)—The quantity of electrical energy corresponding to the flow of current of one ampere for one hour. The term is used to quantify the energy stored in a battery. Most batteries are rated in Ah.

Amplitude—Generally refers to the maximum and minimum voltage attained by an alternating or pulsed current in each complete cycle or pulse of that current.

AMPS—Automatic Mapping and Planning system

AMR—Automated Meter Reading. A term denoting electricity meters that collect data for billing purposes only and transmit this data one way, usually from the customer to the distribution utility.

AMS—Administrative Management Staff
- American Meteorological Society
- Army Map Service

AMSA—Association of Metropolitan Sewer Agencies www.amsa-cleanwater.org

AMSD—Administrative and Management Services Division

AMSO—Acquisition Management Staff Officer

AMT—Alternative Minimum Tax. A separate federal income tax imposed on individuals and corporations where their alternative minimum tax exceeds their regular tax rate. Alternative Minimum Tax is computed after adjustments to regular taxable income.

AMTI—Alternative Minimum Taxable Income. The amount of income that is used to compute alternative minimum tax.

AMTIC—Ambient Monitoring Technical Information Center

Anadromous Fish—Fish that hatch in freshwater, migrate to the ocean to mature, then return to freshwater to spawn. An example is salmon.

Anaerobic—Life or biological processes that occur in the absence of oxygen.

Anaerobic Bacteria—Microorganisms that live in oxygen deprived environments.

Anaerobic Digester—A device for optimizing the anaerobic digestion of biomass and/or animal manure, and possibly to recover biogas for energy production. Digester types include batch, complete mix, continuous flow (horizontal or plug-flow, multiple-tank, and vertical tank), and covered lagoon.

Anaerobic Digestion—The complex process by which organic matter is decomposed by anaerobic bacteria. The decomposition process produces a gaseous byproduct often called "biogas" primarily composed of methane, carbon dioxide, and hydrogen sulfide.

Anaerobic Lagoon—A holding pond for livestock manure that is designed to anaerobically stabilize manure, and may be designed to capture biogas, with the use of an impermeable, floating cover.

Ancillary Benefits—Complementary benefits of a climate policy including improvements in local air quality and reduced reliance of imported fossil fuels.

Ancillary Services—The services other than scheduled energy that is required to maintain system reliability and meet WSCC/NERC operating criteria. Such services include spinning, non-spinning, and replacement reserves, voltage control, and black start capability. Services that the Independent System Operator may develop, in cooperation with market participants, to ensure reliability and to support the transmission of energy from generation sites to customer loads. Such services may include: regulation, spinning reserve, non-spinning reserve, replacement reserve, voltage support, and black start.

Non-electrical-energy products that generation resources also provide to maintain grid system reliability. Ancillary services include: spinning and non-spinning reserve, frequency regulation, ramping up or down, voltage control, black start capability and other services defined by a grid operator or utility control operator.

ANEC—American Nuclear Energy Council

Anemometer—An instrument for measuring the force or velocity of wind; a wind gauge. Measures wind speed and direction. Wind speed is delivered from the sensor as pulses; each time the cups turn, a pulse is recorded

Angel—A private investor who often has non-monetary motives for investing as well as the usual financial ones.

Angel Financing—Capital raised by a private company from independently wealthy investors. This capital is generally used as seed financing.

Angel Fund—A formal or informal assemblage of active angel investors who cooperate in some part of the investment process. Key characteristics of an angel group are: control by member angels (who manage the entity or have control over the entity's managers), and collaboration by member angels in the investment process.

Angel Investing—An angel investor is an individual who makes direct investments of personal funds into a venture, typically early-stage businesses. Because the capital is being invested at a risky time in a business venture, the angel must be capable of taking a loss of the entire investment, and, as such, most angel investors are high-net-worth individuals. These individuals are nearly always "accredited investors" (*see also Accredited Investors*) as defined under the Securities Act of 1933.

Angle of Incidence—In reference to solar energy systems, the angle at which direct sunlight strikes a surface; the angle between the direction of the sun and the perpendicular to the surface. Sunlight with an incident angle of 90 degrees tends to be absorbed, while lower angles tend to be reflected. Angle between the normal to a surface and the direction of incident radiation; applies to the aperture plane of a solar panel. Only minor reductions in power output within plus/minus 15 degrees occur.

Angle of Inclination—In reference to solar energy systems, the angle that a solar collector is positioned above horizontal.

Angstrom Unit—A unit of length named for A.J. Angstome, a Swedish spectroscopist, used in measuring electromagnetic radiation equal to 10^{-10} m (one ten-billionth of a meter) or 0.1 nm.

Anhydrous—A compound that does not contain water. For example, fuel ethanol is referred to as "anhydrous ethanol" because most of the water has been removed.

Anhydrous Ethanol—One hundred percent alcohol; neat ethanol.

Animal Waste Conversion—Process of obtaining energy from animal wastes. This is a type of biomass energy.

Animal Waste Methane Recovery—Large farms that require animal containment (hogs, dairy, etc.) incorporate large, uncovered lagoons to store manure until is used as fertilizer. Methane produced from the waste decomposition is released into the atmosphere during lagoon storage and after the fertilizer is spread on the field. Recovery technologies include installing an anaerobic digester (microbial breakdown in a controlled environment capturing the Methane) and utilizing the Methane to produce energy technology that involves the injection of the waste under the soil or used to generate electricity or other energy that is used in another process.

Annealing—A process involving controlled heating and subsequent controlled, generally slow, cooling applied usually to induce ductility in metals. The term also is used to cover treatments intended to remove internal stresses, alter mechanical or physical properties, produce a definite microstructure, and remove gases.

Annex A—A list in the Kyoto Protocol of the six greenhouse gases and the sources of emissions covered under the Kyoto Protocol. See also "Basket of Gases.

Annex B Countries—Annex B countries are the 39 emissions-capped industrialized countries and economies in transition listed in Annex B of the Kyoto Protocol. Legally-binding emission reduction obligations for Annex B countries range from an 8% decrease (e.g., various European nations) to a 10% increase (Iceland) in relation to 1990 levels during the first commitment period from 2008 to 2012.

Annex I Countries—Annex I countries are the 40 industrialized countries and economies in transition listed in Annex I of the United Nations Framework Convention on Climate Change (UNFCCC or the Convention). Their responsibilities under the Convention are various, and include a non-binding commitment to reducing their greenhouse gas emissions to 1990 levels by the year 2000. Note that Belarussia and Turkey are listed in Annex I but not Annex B; and that Croatia, Liechtenstein, Monaco and Slovenia are listed in Annex B but not Annex I. In practice, Annex I of the Convention and Annex B of the Kyoto Protocol are used almost interchangeably. However, strictly speaking, it is the Annex I countries which can invest in Joint Implementation (JI)/Clean Development Mechanism (CDM) projects as well as host JI projects, and non-Annex I countries which can host CDM projects. This is true, despite the fact that it is the Annex B countries which have the emission reduction obligations under the Kyoto Protocol.

Annex II Countries—Annex II of the United Nations Framework Convention on Climate Change (UNFCCC or the Convention) includes all original OECD member countries plus the European Union. Under Article 4.2 (g) these countries have a special obligation to help developing countries with financial and technological resources.

Annual—A period of time covering a calendar year from January 1 through December 31. Normally used when reporting emissions, consumptions, etc. to regulators.

Annual Effects—Effects in energy use and peak load resulting from participation in Demand Side Management programs in effect during a given period of time.

Annual Equivalent—An equal cash flow amount that occurs every year.

Annual Fuel Utilization Efficiency (AFUE)—In California, a measure of the percentage of heat from the combustion of (AFUE) gas or oil which is transferred to the space being heated during a year, as determined using the applicable test method in the Appliance Efficiency Regulations or Section 112 of Title 24.

• AFUE measures average annual seasonal efficiency of a gas furnace or boiler and may be expressed as total heating output divided by total energy (fuel) input. AFUEs for furnaces can range from 55% to 97%.

Annual Load Fraction—That fraction of annual energy demand supplied by a solar system.

Annual Maximum Demand—The greatest of all demands of the electrical load that occurred during a prescribed interval in a calendar year.

Annual Operating Factor—The annual fuel consumption divided by the product of design firing rate and hours of operation per year.

Annual Percentage Rate (APR)—The actual cost of borrowing money. It may be higher than other rates because it represents full disclosure of the interest rate, loan origination fees, loan discount points and other credit costs paid to the lender.

Annual Report—An accounting report produced at the end of a company's fiscal year end that includes a balance sheet, income statement, cash flow statement, footnotes and a cover letter from the person or firm preparing the report.

• A report issued annually, typically on the anniversary date of the energy project acceptance, which documents the execution and results of the M&V activities prescribed in the M&V plan. This documentation verifies the continued operation of the ECMs, provides the associated energy savings estimates, demonstrates proper maintenance, and provides M&V results. The energy savings documented in the report serves as the basis for the energy service providers invoice after the regular interval report has been reviewed and approved by the energy purchaser.

Annual Requirement—The reporting company's best estimate of the annual requirement for natural gas to make direct sales or sales for resale under certificate authorizations and for company use and unaccounted-for gas during the year next following the current report year.

Annual Solar Savings—The annual solar savings of a solar building is the energy savings attributable to a solar feature relative to the energy requirements of a non-solar building.

Annual Volume Method—A method to allocate commodity costs by function to customer classes based on the Test Period volume level for that customer.

Annualization—To adjust to a full-year basis any item not included in the Base Period or included in the Base Period for less than a full year.

Annualize—A method of estimating the annual quantity, often of taxable income, based on values part way through the year. Often used for estimating tax payments.

Annuity—A stream of even cash flows occurring at regular intervals, such as even monthly lease payments. An annuity in advance is one in which the annuity payment is due at the beginning of each period. An annuity in arrears is one in which the annuity payment is due at the end of each period.

Annunciated—A type of visual signaling device that indicates the on, off, or other status of a load.

Anode—The positive pole or electrode of an electrolytic cell, vacuum tube, etc. (See also *Sacrificial Anode*).

Anodic Inhibitor—A chemical substance or combination of substances that prevents or reduces the rate of the anodic or oxidation reaction by a physical, physio-chemical or chemical action.

Anodic Protection—A technique to reduce corrosion of a metal surface under some conditions by passing sufficient anodic current to it to cause the electrode potential of the surface to enter and remain in the passive region.

 • An appreciable reduction in corrosion by making a metal an anode and maintaining this highly polarized condition with very little current flow.

ANOPR—Advance Notice of Proposed Rulemaking

ANPR—Advance Notice of Proposed Rulemaking (also ANPRM)

ANSI—American National Standards Institute is the national organization that coordinates development and maintenance of consensus standards and sets rules for fairness in their development. ANSI also represents the USA in developing international standards. ANSI oversees the creation, promulgation and use of norms and guidelines that directly impact businesses in nearly every sector including construction equipment and energy distribution. The organization is also engaged in accrediting programs that assess conformance to standards including internationally recognized cross-sector programs such as the ISO 9000 (quality), ISO 14000 (environmental) management systems, and ISO 50001 (Cenergy management). www.ansi.org

ANSI Assembly Identifier—The serial numbering scheme adopted by the American National Standards Institute (ANSI) to ensure uniqueness of an assembly serial number.

ANSI Z21.10.3—The American National Standards Institute document entitled "Gas Water Heaters, Volume I, Storage Water Heaters with input ratings above 75,000 Btu per hour," 2001 (ANSI 221.10.3-2001).

ANSI Z21.13—The American National Standards Institute document entitled "Gas-Fired Low Pressure Steam and Hot Water Boilers," 2000 ANSI Z21. 13-2000).

ANSI Z21.40.4—The American National Standards Institute document entitled "Performance Testing and Rating of Gas-Fired, Air Conditioning and Heat Pump Appliances," 1996 (ANSI 221.40.4-1996).

ANSI Z21.47—The American National Standards Institute document entitled "Gas-Fired Central Furnaces," 2001 (ANSI Z21.47-2001).

ANSI Z83.8—The American National Standards Institute document entitled "Gas Unit Heaters and Gas-Fired Duct Furnaces," 2002 (ANSI Z83.8 -2002).

Anthracite—The highest rank of coal; used primarily for residential and commercial space heating. It is a hard, brittle, and black lustrous coal, often referred to as hard coal, containing a high percentage of fixed carbon and a low percentage of volatile matter. The moisture content of fresh-mined anthracite generally is less than 15 percent. The heat content of anthracite ranges from 22 to 28 million Btu per ton on a moist, mineral-matter-free basis. The heat content of anthracite coal consumed in the United States averages 25 million Btu per ton, on the as-received basis (i.e., containing both inherent moisture and mineral matter). Note: Since the 1980's, anthracite refuse or mine waste has been used for steam electric power generation. This fuel typically has a heat content of 15 million Btu per ton or less.

Anthropogenic—Made or generated by a human or caused by human activity. The term is used in the context of global climate change to refer to gaseous emissions that are the result of human activities, as well as other potentially climate-altering activities, such as deforestation.

Anticline—An upfold or arch in rock strata in which the beds or layers dip in opposite directions from the crest, permitting possible entrapment of oil and gas.

Antifreeze Solution—A fluid, such as methanol or ethylene glycol, added to vehicle engine coolant, or used in solar heating system heat transfer fluids, to protect the systems from freezing.

Antireflection Coating—A thin coating of a material applied to a photovoltaic cell surface that reduces the light reflection and increases light transmission.

Antitrust—Consisting of laws to protect trade and commerce from unlawful restraints and monopolies or unfair business practice.

ANWR—Arctic National Wildlife Refuge

AO—Administrative Officer
- Administrative Order
- Area Office
- Awards and Obligations

AOC—Abnormal Operating Conditions
- Administrative Order on Consent
- Area of Concern
- Area of Contamination

AOO—Accounting Operations Office
- American Oceanic Organization

AOS—Audit Operations Staff

AOSIS—Refers to the Alliance of Small Island States. It is an ad hoc coalition of low-lying and island countries that are particularly vulnerable to sea-level rise and that also share common public policy positions on climate change. The 42 members and observers are American Samoa, Antigua and Barbuda, Bahamas, Barbados, Belize, Cape Verde, Comoros, Cook Islands, Cuba, Cyprus, Dominica, Federated States of Micronesia, Fiji, Grenada, Guam, Guinea-Bissau, Guyana, Jamaica, Kiribati, Maldives, Malta, Marshall Islands, Mauritius, Nauru, Netherlands Antilles, Niue, Palau, Papua New Guinea, Samoa, Sao Tome and Principe, Seychelles, Singapore, Solomon Islands, St. Kitts & Nevis, St. Lucia, St. Vincent and the Grenadines, Suriname, Tonga, Trinidad and Tobaga, Tuvula, U.S. Virgin Islands, and Vanuatu.

AP—Accounting Point

APA—Acid Precipitation Act (1980)

- Administrative Procedures Act
- American Planning Association

APC—Air Pollution Control

APCA—Air Pollution Control Association

APCD—Air Pollution Control Device
- Air Pollution Control District

APCO—Air Pollution Control Officer

APCS—Air Pollution Control System

APDS—Automated Procurement Documentation System

APER—Air Pollution Emissions Report

Aperture—An opening; in solar collectors, the area through which solar radiation is admitted and directed to the absorber. For evacuated solar thermal tubes this refers to the cross-sectional surface area of the outer clear glass tube measured using the internal diameter, not the outside diameter.

APF—Appropriated Fund

APHA—American Public Health Association www.apha.org

API—The American Petroleum Institute, a trade association. www.api.org

API Gravity—American Petroleum Institute measure of specific gravity of crude oil or condensate in degrees. An arbitrary scale expressing the gravity or density of liquid petroleum products. The measuring scale is calibrated in terms of degrees API; it is calculated as follows:

Degrees API = (141.5/sp.gr.60 deg.F/60 deg.F) - 131.5

APL—Aqueous Phase Liquid

APPA—American Public Power Association www.appanet.org

Apparent Consumption (Coal)—Coal production plus imports of coal, coke, and briquettes minus exports of coal, coke, and briquettes plus or minus stock changes. Note: The sum of "Production" and "Imports" less "Exports" may not equal "Consumption" due to changes in stocks, losses, unaccounted-for coal, and special arrangements such as the United States shipments of anthracite to United States Armed Forces in Europe.

Apparent Consumption, Natural Gas (International)—The total of an individual nation's dry natural gas production, plus imports, less exports.

Apparent Consumption, Petroleum (International)—Consumption that includes internal consumption, refinery fuel and loss, and bunkering. For countries in the Organization for Economic Cooperation and Development (OECD), apparent consumption is derived from refined product output plus refined product imports minus refined product exports plus refined product stock changes plus other oil consumption (such as direct use of crude oil). For countries outside the OECD, apparent consumption is either a reported figure or is derived from refined product output plus refined product imports minus refined product exports, with stock levels assumed to remain the same. Apparent consumption also includes, where available, liquefied petroleum gases sold directly from natural gas processing plants for fuel or chemical uses.

Apparent Day—A solar day; an interval between successive transits of the sun's center across an observer's meridian; the time thus measured is not equal to clock time.

Apparent Power (kVA)—This is the voltage-ampere requirement of a device designed to convert electric energy to a non-electrical form.

Appliance—A piece of equipment, commonly powered by electricity, used to perform a particular energy-driven function. Examples of common appliances are refrigerators, clothes washers and dishwashers, conventional ranges/ovens and microwave ovens, humidifiers and dehumidifiers, toasters, radios, and televisions. Note: Appliances are ordinarily self-contained with respect to their function. Thus, equipment such as central heating and air conditioning systems and water heaters, which are connected to distribution systems inherent to their purposes, are not considered appliances.

Appliance Efficiency Index—A relative comparison of trends in new-model efficiencies for major appliances and energy-using equipment. The base year for relative comparisons was 1972 (1972=100). Efficiencies for each year were efficiencies of different model types that were weighted by their market shares.

Appliance Efficiency Regulations—In California, the regulations in Title 20, Section 1601 et seq. of the California Code of Regulations.

Appliance Efficiency Standards—The National Appliance Energy Conservation Act of 1987 established minimum efficiency standards for major home appliances, including furnaces, central and room air conditioners, refrigerators, freezers, water heaters, dishwashers, and heat pumps. Most of the standards took effect in 1990. The standards for clothes washers, dishwashers, and ranges took effect in 1988, because they required only minor changes in product design, such as eliminating pilot lights and requiring cold water rinse options. The standards for central air conditioners and furnaces took effect

in 1992, because it took longer to redesign these products. Appliance efficiency standards for refrigerators took effect in 1993.

• California Code of Regulations, Title 20, Chapter 2, Subchapter 4: Energy Conservation, Article 4: Appliance Efficiency Standards. Appliance Efficiency Standards regulate the minimum performance requirements for appliances sold in California and apply to refrigerators, freezers, room air conditioners, central air conditioners, gas space heaters, water heaters, plumbing fittings, fluorescent lamp ballasts and luminaries, and ignition devices for gas cooking appliances and gas pool heaters. New National Appliance Standards are in place for some of these appliances and will become effective for others at a future date.

Appliance Energy Efficiency Ratings—The ratings under which specified appliances convert energy sources into useful energy, as determined by procedures established by the U.S. Department of Energy.

Appliance Saturation—A percentage telling what proportion of all households in a given geographical area has a certain appliance.

Appliance Standards—Standards contained in the Appliance Efficiency Regulations. See *Appliance Efficiency Regulations*

Appliance Standards—Standards established by the U.S. Congress for energy consuming appliances in the National Appliance Energy Conservation Act (NAECA) of 1987, and as amended in the National Appliance Energy Conservation Amendments of 1988, and the Energy Policy Act of 1992 (EPAct). NAECA established minimum standards of energy efficiency for refrigerators, refrigerator-freezers, freezers, room air conditioners, fluorescent lamp ballasts, incandescent reflector lamps, clothes dryers, clothes washers, dishwashers, kitchen ranges and ovens, pool heaters, television sets (withdrawn in 1995), and water heaters. The EPAct added standards for some fluorescent and incandescent reflector lamps, plumbing products, electric motors, and commercial water heaters and Heating, Ventilation, and Air Conditioning (HVAC) systems. It also allowed for the future development of standards for many other products. The U.S. Department of Energy (DOE) is responsible for establishing the standards and the procedures that manufacturers must use to test their models. These procedures are published in the Code of Federal Regulations (10 CFR, Ch. II, Part 430), January 1, 1994 (Federal Register).

Applicability Factor—The percentage of end-use energy and demand used by a technology to which the demand-side management (DSM) measure applies. For example, the high-efficiency fluorescent lighting DSM measure applies to fluorescent lighting but not all lighting. Applicability therefore represents the percent of the lighting end-use attributable to fluorescence for which there could be high-efficiency replacements installed.

Appraisal—An opinion as to the fair market value of property.

Approved—As to a home energy rating provider or home energy rating system, is reviewed and approved by the Commission under Title 20, Section 1675 of the California Code of Regulations.

Approved By The Commission—Means approval under 25402.1 of the Public Resources Code in California.

Approved Calculation Method—In California, a Public Domain Computer Program approved under Section 10-109 (a), or any Alternative Calculation Method approved under Section 10-109 (b) of Title 24. See *Alternative Calculation Method*

Approved Opportunity—An Approved Opportunity is an opportunity that has been approved for implementation, but is not yet complete and is not yet expected to produce energy savings. It is in the design or construction stage.

Appurtenant—Belonging to.

APR—Annual Percentage Rate

APS—Automated Personnel System

APT—Associated Pharmacists and Toxicologists

APTI—Air Pollution Training Institute www.epa.gov/apti

AQ-7—Non-reactive Pollutant Modeling

AQCCT—Air Quality Criteria and Control Techniques

AQCP—Air Quality Control Program

AQCR—Air Quality Control Region

AQDHS—Air Quality Data Handling System

AQDM—Air Quality Display Model

AQMA—Air Quality Maintenance Area

AQMD—Air Quality Management District. Also known as an Air Pollution Control District.

AQMP—Air Quality Management Plan
  • Air Quality Maintenance Plan

AQSM—Air Quality Simulation Model

AQTAD—Air Quality Technical Assistance Demonstration

Aqueous—Water-based.

Aquifer—A geological formation or group of formations in the ground which is usually composed of rock, gravel, sand or other porous material and which yields water to wells or springs. Can be polluted by introduction of pollutants through poorly capped wells, injection waste disposal and other entries below ground.

AR—Administrative Record
  • Accounts Receivable

ARA—Assistant Regional Administrator
  • Associate Regional Administrator

ARAC—Acid Rain Advisory Committee

ARAR—Applicable, Relevant and Appropriate Requirements

ARB—Air Resources Board (as in California)

Arbitrage—Simultaneous purchase in one market and sale in another of a security in order to make a profit on relative price differences.

ARC—Agency Ranking committee

ARCS—Alternative Remedial Contract Strategy

ARD—Air & Radiation Division
  • Aquatic Resource Division

ARE—Acute Respiratory Exposures

Area Load—The total amount of electricity being used at a given point in time by all consumers in a utility's service territory.

Areal Heat Capacity—See *Heat Capacity*

AREL—Acute Reference Exposure Level

ARG—American Resources Group

Argon—A colorless, odorless inert gas sometimes used in the spaces between the panes in energy efficient windows. This gas is used because it will transfer less heat than air. Therefore, it provides additional protection against conduction and convection of heat over conventional double -pane windows.

ARI—Air-Conditioning and Refrigeration Institute www.ari.org.

ARI 210/240—The Air-conditioning and Refrigeration Institute document entitled "Unitary Air-Conditioning and Air-Source Heat Pump Equipment," 2003 (ARI 210/240-94).

ARI 310/380—The Air-conditioning and Refrigeration Institute document entitled "Packaged Terminal Air-Conditioners and Heat Pumps," 1993 (ARI 3101380-93).

ARI 320—The Air-conditioning and Refrigeration Institute document entitled "Water-Source Heat Pumps," 1998 (ARI 320-98).

ARI 325—The Air-conditioning and Refrigeration Institute document entitled "Ground Water-Source Heat Pumps," 1998 (ARI 325- 98).

ARI 340/360—The Air-conditioning and Refrigeration Institute document entitled "Commercial and Industrial Unitary Air-Conditioning and Heat Pump Equipment," 2000 (ARI 340/360-2000).

ARI 365—The Air-conditioning and Refrigeration Institute document entitled, "Commercial and Industrial Unitary Air-Conditioning Condensing Units," 2002 (ARI 365-2002)

ARI 460—The Air-conditioning and Refrigeration Institute document entitled "Remote Mechanical-Draft Air-Cooled Refrigerant Condensers."

ARI 550/590—The Air-conditioning and Refrigeration Institute document entitled "Standard for Water Chilling Packages Using the Vapor Compression Cycle," 1998 (ARI 5501590-98).

ARI 560—The Air-conditioning and Refrigeration Institute document entitled "Absorption Water Chilling and Water Heating Packages," 2000 (ARI 560-2000).

ARIP—Accidental Release Information Program

ARL—Air Resources Laboratory

ARM—Air Resources Management
  • Adjustable Rate Mortgage

Arm's-length Transaction—A transaction between unrelated parties under no duress. Frequently used concept when using an appraiser to establish value on real or personal property.

ARO—Alternative Regulatory Option

Aromatics—Hydrocarbons characterized by unsaturated ring structures of carbon atoms. Commercial petroleum aromatics are benzene, toluene, and xylene (BTX).

ARPO—Acid Rain Policy Office

ARPS—Atmospheric Research Program Staff

ARRA—American Recovery & Reinvestment Act of 2009

Array (Solar)—Any number of solar photovoltaic modules or solar thermal collectors or reflectors connected together to provide electrical or thermal energy. Arrays are often designed to produce significant amount of electricity.

Array Current (Solar)—The electrical current output of a PV array when exposed to sunlight.

Array Operating Voltage—The voltage output of a PV array when exposed to sunlight and feeding a load.

Arrears Rental—A rental that is due at the end of each period. Compare to advance rental.

ARRO—Acid Rain Research Program (also ARRP)

ARRPA—Air Resources Regional Pollution Assessment

Arsenic—A gray, brittle and highly poisonous metal. It is used as an alloy for metals, especially lead and copper, and is used in insecticides and weed killers. In its inorganic form, it is listed as a cancer-causing chemical.

ART—Advanced Renewable Tariff. Feed-in tariffs that are differentiated by technology so that there is one price for electricity produced by wind facilities, one price for electricity produced by solar facilities, etc.

Artesian Well—A well that flows up like a fountain because of the internal pressure of the aquifer.

ARZ—Auto-Restricted Zone

ASAP—As Soon As Possible

ASB—American Standards Board

Asbestos—A group of naturally occurring minerals that separate into long, thin fibers. Asbestos was used for many years to insulate and fireproof buildings. In the 1989 CBECS, information on asbestos in buildings was collected (Section R of the Buildings Questionnaire) for the U.S. Environmental Protection Agency (EPA) Asbestos treatment methods include removal, encapsulation or sealing, and enclosure behind a permanent barrier.

As-Billed—A method by which a pipeline includes in its rates, and charges its customers, the costs of gas or transportation services in the same manner as it is billed by its pipeline suppliers or transporter.

ASC—Area Source Category

ASCE—American Society of Civil Engineers www.asce.org.

ASCII—American Standard Code for Information Exchange

ASCP—American Society of Consulting Planners

ASD—Administrative Services Division
  • Analysis and Support Division

ASDWA—Association of State Drinking Water Administrators www.asdwa.org

ASEAN—Association of Southeast Asian Nations

ASES—American Solar Energy Society. ASES is a nonprofit organization with more than 11,000 members dedicated to increasing the use of solar energy, energy efficiency and other sustainable technologies in the U.S. The organization nationally promotes solar energy education, advocacy and public outreach, and also presents the annual ASES National Solar Convention. www.ases.org

ASG—Area Support Group

Ash—Non-organic, non-flammable substance left over after combustible material has been completely burned.
  • Impurities consisting of silica, iron, alumina,

and other noncombustible matter that are contained in coal. Ash increases the weight of coal, adds to the cost of handling, and can affect its burning characteristics Ash content is measured as a percent by weight of coal on an "as received" or a "dry" (moisture-free, usually part of a laboratory analysis) basis.

ASHAA—Asbestos in Schools Hazard Abatement Act (1984)

ASHRAE—The American Society of Heating, Refrigerating and Air Conditioning Engineers. ASHRAE is a nonprofit technical organization of 50,000 members that creates industry standards and recommended guidelines and procedures for heating, ventilation, air-conditioning and refrigerator technology. The organization's areas of expertise include energy efficiency, high performance buildings, indoor air quality, green building design, building codes and standards, and health concerns such as mold growth. www.ashrae.org

ASHRAE 55—The American Society of Heating, Refrigerating and Air Conditioning Engineers document entitled "Thermal Environmental Conditions for Human Occupancy," 1992 (ASHRAE Standard 55-1992).

ASHRAE Climatic Data for Region X—The American Society of Heating, Refrigerating and Air Conditioning Engineers document entitled "ASHRAE Climatic Data for Region X, Arizona, California, Hawaii and Nevada," Publication SPCDX, 1982 and "Supplement," 1994.

a-Si—Amorphous Silicon

Asia-Pacific Partnership on Clean Development and Climate—The Asia-Pacific climate pact is a rival international climate change agreement to the Kyoto Protocol. Its initiators in 2005 were the United States and Australia, the only two industrialized nations not to have ratified the Kyoto treaty at that time (Australia since has ratified in 2007). The group also includes China, India, Japan, South Korea and now Canada. APP rejects Kyoto-style emission reduction targets in favor of encouraging business to invest in clean fossil-fuel technology and renewable energy.

ASID—American Society of Interior Designers. www.asid.org A nonprofit professional society representing the interest of interior designers and the interior design community.

ASIS—American Society for Industrial Security www.asisonline.org

ASIWPCA—Association of State and Interstate Water Pollution Control Administrators www.asiwpca.org

ASMDHS—Airshed Model Data Handling System

ASME—The American Society of Mechanical Engineers www.asme.org

ASPA—American Society of Public Administration www.aspanet.org

Asphalt—A dark brown-to-black cement-like material obtained by petroleum processing and containing bitumen's as the predominant component; used primarily for road construction. It includes crude asphalt as well as the following finished products: cements, fluxes, the asphalt content of emulsions (exclusive of water), and petroleum distillates blended with asphalt to make cutback asphalts. Note: The conversion factor for asphalt is 5.5 barrels per short ton.

Asphalt Blowing—The process by which air is blown through asphalt flux to change the softening point and penetration rate.

ASRL—Atmospheric Sciences Research Laboratory

ASSE—American Society of Sanitary Engineers. www.asse-plumbing.org

Assembly Identifier (Nuclear)—A unique string of alphanumeric characters that identifies an assembly, bundle, or canister for a specific reactor in which it has been irradiated.

Assembly Type (Nuclear)—Each assembly is characterized by a fabricator, rod-array size, and model type. An eight-digit assembly type code is assigned to each assembly type based on certain distinguishing characteristics, such as the number of rods per assembly, fuel rod diameter, cladding type, materials used in fabrication, and other design features.

Assessed Value—The value of equipment as determined by a taxing authority for the purposes of assessing personal property tax. The annual determination of a property's assessed value/state equalized value is done by the local assessor. The Assessor is required by state law to assess at current market value all assessable property each year. Simply stated, market value is the probable price that a property would sell for in an arm's length transaction between a willing buyer and a willing seller. There is no cap or limit on the assessed value change.

Assessment Work—The annual or biennial work performed on a mining claim (or claims), after claim location and before patent, to benefit or develop the claim and to protect it from relocation by third parties.

Asset—The total resources of a person or business, as

cash, notes and accounts receivable, securities, goodwill, equipment or real estate (as opposed to liability).

Asset Based Financing—Financing an enterprise by using its hard assets for collateral to acquire a loan of sufficient size with which to finance operations. Widely used in leveraged buyouts (LBOs).

Asset Class Life—The updated ADR midpoint life as modified by the 1986 Tax Reform Act. An asset class life represents the IRS designated economic life of an asset, and is used as the recovery period for alternative tax depreciation computations.

Asset Controlling Supplier—Any entity that controls the power output from or sales from electricity generating facilities.

Asset Owning Supplier—An entity owning electricity generating facilities that delivers electricity to a transmission or distribution line.

Asset Securitization & Placement—In asset securitization, cash flows from specific assets are used directly for interest and redemption payments of a financing arrangement. Assets are bundled and sold to a non-consolidated special purpose vehicle, which refinances the purchase price of the asset-backed securities. The more precisely future cash flows can be estimated, the more favorable the asset-backed financing based on these cash flows.

Asset Turnover Ratio—A board measure of asset efficiency, defined as net sales divided by total assets.

Asset-Backed Securities (ABS)—A collection of similar receivables. The sale of ABS serves to create liquidity. Receivables with maturities of between 30-90 days are the rule but a portfolio of receivables (such as lease receivables or other contract receivables with longer terms) is not uncommon. ABS is a modern form of corporate financing and can be viewed as a substitute for the traditional loan. This form of financing not only increases a company's liquidity, generally at more favorable financing costs, its off-balance sheet status as well as the diversification of funds are also decisive aspects. To collateralize the securitized loans, financing instruments, which are described as securities backed by assets, are constructed.

Asset-Backed Security—Securities collateralized by a pool of assets. The process of creating securities backed by assets is referred to as "asset securitization."

Assign—To transfer or exchange future rights. Often used in various types of contracts. Most frequently used when the contract is assigned for certain cash values that represent the present value of a future stream of payments to the transferor.

Assigned Amount (AA)—The assigned amount is the total amount of greenhouse gas that each country is allowed to emit during the first commitment period of the Kyoto Protocol. This total amount is then broken down into measurable units.

Assignee—The Party to whom the Assignment is given.

Assignment—A provision within a financial contract agreement that allows either, neither or both parties of a transaction to delivery their obligation to a third party in return for immediate compensation. For example, a lessor may assign all current and future lessee's lease payments to a third party in return for cash today.

Assignor—The Party who makes an Assignment.

Assistance for Heating in Winter—Assistance from the Low-Income Home Energy Assistance Program (LIHEAP). The purpose of LIHEAP is to assist eligible households to meet the costs of home energy, i.e., a source of heating or cooling residential buildings.

Assistance for Weatherization of Residence—The household received services free, or at a reduced cost, from the Federal, State, or local Government. Any of the following services could have been received:

- Insulation in the attic, outside wall, or basement/crawlspace below the floor of the house
- Insulation around the hot water heater
- Repair of broken windows or doors to keep out the cold or hot weather
- Weather stripping or caulking around any windows or doors to the outside
- Storm doors or windows added
- Repair of broken furnace
- Furnace tune-up and/or modifications
- Other home energy-saving devices.

Associated Gas—Natural gas that can be developed for commercial use, and which is found in contact with oil in naturally occurring underground formations.

Associated Liquids—Condensates (liquid hydrocarbons without free water) produced in conjunction with the production of gas to be transported or liquefiable hydrocarbons contained in such gas, but not including oil.

Associated Natural Gas—See *Associated-Dissolved Natural Gas* below and *Natural Gas*.

Associated-Dissolved Natural Gas—Natural gas that occurs in crude oil reservoirs either as free gas (associated) or as gas in solution with crude oil (dissolved gas). Also see *Natural Gas*.

Assumption—An agreement to undertake a debt or obligation contracted by another.

AST—Aboveground Storage Tank
Advanced Secondary (Wastewater) Treatment

ASTHO—Association of State and Territorial Health Officers. www.astho.org

ASTM—The American Society for Testing & Materials. www.astm.org An organization with a membership drawn from engineering, scientific, manufacturing, consumer and governmental groups interested in establishing voluntary test standards for materials, products, systems, and services. ASTM International, originally known as the American Society for Testing and Materials [ASTM]), has been around for more than a century. This international voluntary standards organization develops and publishes consensus technical standards for a diverse range of materials, products, systems, and services.

More than 1,300 ASTM standards cover building and construction industry areas including cement, concrete, roofing and waterproofing. ASTM also sets energy standards in areas such as petroleum products and natural gas as well as alternative renewable energy sources such as such as solar heating and cooling systems and materials; photovoltaic electronic power conversion; and geothermal utilization and materials. Other areas covered by ASTM standards include air and water quality, pollution control and water management.

ASTM standards are the work of over 30,000 ASTM members, who are technical experts representing producers, users, consumers, government and academia from over 120 countries. In the United States, ASTM standards have been adopted (by reference or incorporation) in numerous federal, state, and municipal government regulations, and other governments internationally have referenced ASTM standards.

ASTM C1167—The American Society for Testing and Materials document entitled "Standard Test Method for Aging Effects of Artificial Weathering on Latex Sealants," 2001 (ASTM C732-01).

ASTM C1371—The American Society for Testing and Materials document entitled "Standard Test Method for Determination of Emittance of Materials Near Room Temperature Using Portable Emissometers," 1998 (ASTM C1371-98).

ASTM C272—The American Society for Testing and Materials document entitled "Standard Test Method for Steady-State Heat Flux Measurements and Thermal Transmission Properties by Means of the Guarded-Hot-Plate Apparatus," 1997 (ASTM C177-97).

ASTM C335—The American Society for Testing and Materials document entitled "Standard Test Method for Steady-State Heat Transfer Properties of Horizontal Pipe Insulation," 1995 (ASTM C335-95).

ASTM C518—The American Society for Testing and Materials document entitled "Standard Test Method for Steady-State Thermal Transmission Properties by Means of the Heat Flow Meter Apparatus," 2002 (ASTM C518-02).

ASTM C55—The American Society for Testing and Materials document entitled "Standard Specification for Concrete Brick," 2001 (AST M C55-01).

ASTM C731—The American Society for Testing and Materials document entitled "Standard Test Method for Extrudability, After Package Aging of Latex Sealants," 2000 (ASTM C731-00).

ASTM D1003—The American Society for Testing and Materials document entitled "Standard Test Method for Haze and Luminous Transmittance of Transparent Plastics," 2000 (ANSI/ASTM D1003-00).

ASTM D2824—The American Society of Testing and Materials document entitled "Standard Specification for Aluminum-Pigmented Asphalt Roof Coatings, Non-fibered, Asbestos Fibered, and Fibered without Asbestos," 2002 AS(—TM D2824-02).

ASTM D3805—The American Society of Testing and Materials document entitled "Standard Guide for Application of Aluminum Pigmented Asphalt Roof Coatings," 1997 [ASTM D3805-97 (re-approved 2003)].

ASTM D6848—The American Society of Testing and Materials document entitled, "Standard Specification for Aluminum-Pigmented Emulsified Asphalt Used as a Protective Coating for Roofing Asphalt Roof Coatings," 2002 (ASTM D6848-02).

ASTM D822—The American Society of Testing and Materials document entitled, "Standard Practice for Filtered Open-Flame Carbon-Arc Exposures of Paint and Related Coatings," 2001 (ASTM D822-01).

ASTM E283—The American Society for Testing and Materials document entitled "Standard Test Method for Determining the Rate of Air Leakage Through Exterior Windows, Curtain Walls, and Doors Under Specified Pressure Differences Across the Specimen," 1991 [ASTM F283-91(1999)]

ASTM E408—The American Society for Testing and Materials document entitled, "Standard Test Methods for Total Normal Emittance of Surfaces Using Inspection-Meter Techniques," 1971 [ASTM E408-71(2002)].

ASTM E96—The American Society for Testing and Materials document entitled "Standard Test Methods for Water Vapor Transmission of Materials," 2000 (ASTM E96-00).

ASTS—Asbestos in Schools Tracking System

ASTSWMO—Association of State and Territorial Solid Waste Management Officials www.astswmo.org

ASUS—Administrative Support/Utilization

Asynchronous Generator—A type of electric generator that produces alternating current that matches an existing power source.

AT—Advanced Treatment
• Averaging time
• Ash Trap

at wt—The abbreviation for atomic weight.

ATA—American Trucking Associations www.truckline.com

ATC—Area Training Center

ATCF—After-tax Cash Flows

ATCS—Audit Tracking and Control system

ATD—Air and Toxics Division

ATERIS—Air Toxics Exposure and Risk Information System

Atgas—Synthetic gas produced by dissolving coal in a bath of molten iron. The process was developed by Applied Technology, Inc. Synthetic gas may be used as a substitute for natural gas in industrial and home uses.

ATM—Asynchronous Transfer Mode

atm—Atmosphere

Atmospheric Crude Oil Distillation—The refining process of separating crude oil components at atmospheric pressure by heating to temperatures of about 600 degrees to 750 degrees Fahrenheit (depending on the nature of the crude oil and desired products) and subsequent condensing of the fractions by cooling.

Atmospheric Pressure—The pressure of the air at sea level; one standard atmosphere at zero degrees centigrade is equal to 14.695 pounds per square inch (1.033 kilograms per square centimeter).

ATMS—Advanced Traffic Management Systems

Atom—The smallest unit of an element consisting of a dense positively charged nucleus (of protons and neutrons) orbited by negatively charged electrons.

Atomic Energy Commission—The independent civilian agency of the federal government with statutory responsibility to supervise and promote use of nuclear energy. Functions were taken over in 1974 by the Energy Research and Development Administration (now part of the U.S. Department of Energy) and the Nuclear Regulatory Commission.

Atomic Nucleus—The positively charged core of an atom.

Atomize—To reduce a liquid to a fine spray or mist.

ATP—Anti-tampering Program

ATR—Agency Technical Representative

At-Risk Rules—Federal tax laws that prohibit individuals (and some corporations) from deducting tax losses from equipment leases in excess of the amount they have at risk.

Atrium—A large-volume space created by openings connecting two or more stories and is used for purposes other than an enclosed stairway, an elevator hoist-way, an escalator opening, or as a utility shaft for plumbing, electrical, air-conditioning or other equipment, and is not a mall.

ATRMRD—Air Toxics and Radiation Monitoring Research Division

ATS—Action Tracking System
• Administrator's Tracking System
• Allowance Tracking system
• Assignment Tracking system

ATSDR—Agency for Toxic Substances and Disease Registry

Attachment—A seizure of property by Judicial Process during the Pendency of an Action.

Attainment Area—A geographic region where the concentration of a specific air pollutant does not exceed federal standards.

ATTF—Air Toxics Task Force

Attic—An enclosed unconditioned space directly below the roof and above the ceiling.

Attic Fan—A fan mounted on an attic wall used to exhaust warm attic air to the outside.

Attic Vent—A passive or mechanical device used to ventilate an attic space, primarily to reduce heat buildup and moisture condensation.

Attorney-In-Fact—An Agent authorized to act for another under a "Power-of-Attorney."

Attribute (Green-e)—Descriptive or performance characteristics of a particular generation resource. The characteristics of renewables and other generating types (both positive and negative) not reflected in the price of power are referred to as externalities and include environmental, economic, and social characteristics. As Detailed Below:

- Physical Attributes: Physical characteristics such as size, location, fuel type, time of generation, etc. The value of these characteristics tends to be captured in the price of power.
- Environmental Attributes: Environmental attributes include the environmental benefits and costs associated with the construction and operation of specific types of power generation facilities. For renewable facilities, their environmental attributes might include the benefits of such things as emissions offsets or avoidance, as say from wind-generated electricity. Several air pollutants (e.g. CO_2, NO_X, and SO_X) have separate markets today where the value of a pound of pollution is determined through sales and trade. Trading markets for other power plant pollutants, such as mercury and particulates do not exist today but may come into being later.
- Economic Attributes: Economic attributes might include such things as the development of local jobs and businesses, as well as reductions in the costs of having a secure domestic supply of electricity.
- Social Attributes: Examples of social attributes include health and quality of life factors, the introduction of innovative technologies and technology applications, as well as social equity considerations related to the location and siting of power plants.

Attrition—Erosion of earnings on invested capital resulting from the regulatory practice of setting utility rates based on past costs during an inflationary period.

ATTS—Agency Technology Transfer Staff

Auctioning—Auctioning of emissions permits is a method by which permits for greenhouse gas emissions may be allocated among emitters and firms in a domestic emissions trading regime based upon willingness to pay for these permits. Supporters of this method of emissions trading assert that the advantage of auctioning is that it would provide governments with revenue and provide price signals to the new and developing market for permits. Critics contend that auctioning's disadvantage is that it may be less politically acceptable to those entities that would stand to gain from grandfathering of permits. This method of allocation may be combined with Grandfathering. (*See Grandfather.*)

Audit (Energy)—The process of determining energy consumption, by various techniques, of a building or facility.

Auger Mine—A surface mine in which the coal bed is removed by means of a large diameter drill. Usually operated only when the overburden becomes too thick for economical strip mining.

AUM—Assets Under Management. Total value of owned and third-party assets managed.

AUP Report—Agreed-Upon Procedures Report. An attestation report prepared by an independent accountant in accordance with standards established by the American Institute of Certified Public Accounts (AICPA). The AUP report is required for taxpayers applying for a Treasury Grant payment pursuant to American Recovery and Reinvestment Act of 2009 for which the project costs basis is between $400,000 and $3,333,333.

AUSA—Assistant U.S. Attorney

AUSM—Advanced Utility Simulation Model

AusWEA—Australian Wind Energy Association

Authority or Agency—A state or local unit of government created to perform a single activity or a limited group of functions and authorized by the state legislature to issue bonded debt.

Authorized Cash Distribution To Municipality—The authorized cash distributions to the municipality from the earned surplus of the utility department.

Automatic—Capable of operating without human intervention.

Automatic (or Remote) Meter Reading System (AMR)—A system that records the consumption of electricity, gas, water, etc., and sends the data to a central data accumulation device.

- "Real time" monitoring of energy quantities and characteristics as it passes through a specific location.

Automatic Damper—A device that cuts off the flow of hot or cold air to or from a room as controlled by a thermostat.

Automatic Multi-Level Daylighting Control—A multi-level lighting control that automatically reduces lighting in multiple steps or continuous dimming in response to available daylight. This control uses one or more photocontrols to detect changes in daylight illumination and then change the electric lighting level in response to the daylight changes.

Automatic Set-back or Clock Thermostat—A thermostat that can be set to turn the heating/cooling system off and on at certain predetermined times.

Automatic Time Switch Control Devices—Devices capable of automatically turning loads off and on based on time schedules.

Automobile and Truck Classifications—Vehicle classifications for automobiles and light duty trucks were obtained from the EPA (Environmental Protection Agency) mileage guide book. Almost every year there are small changes in the classifications, therefore the categories will change accordingly. The EPA mileage guide can be found at any new car dealership.

Autonomous System (Solar)—A PV System that operates without any other energy generating source. It may or may not include storage batteries.

Auxiliary Energy or System—Energy required to operate mechanical components of an energy system, or a source of energy or energy supply system to back-up another.

Auxiliary Energy Subsystem—Equipment using conventional fuel to supplement the energy output of a solar system. This might be, for example, an oil-fueled generator that adds to the electrical output of a substitute for the solar system during long overcast periods when there is not enough sunlight.

Auxiliary Equipment—Extra machinery needed to support the operation of a power plant or other large facility.

Auxiliary Power—Power from an secondary source that augments the performance criteria established for the primary power source

Av (Ave)—Average

Availability—Describes the reliability of power plants. It refers to the number of hours that a power plant is available to produce power divided by the total hours in a set time period, usually a year. Solar: The quality or condition of a PV system that is available to provide power to a load. Usually measured in hours per year. One minus availability equals downtime.

Availability Factor—A percentage representing the number of hours a generating unit is available to produce power (regardless of the amount of power) in a given period, compared to the number of hours in the period.

Available But Not Needed Capability—Net capability of main generating units that are operable but not considered necessary to carry load and cannot be connected to load within 30 minutes.

Available Heat—The amount of heat energy that may be converted into useful energy from a fuel.

AVCS—Automatic Vehicle Control Systems

Average Collection Period—The number of days required, on average, to collect accounts receivable.

Average Cost—The revenue requirement of a utility divided by the utility's sales. Average cost typically includes the costs of existing power plants, transmission, and distribution lines, and other facilities used by a utility to serve its customers. It also included operating and maintenance, tax, and fuel expenses.

Average Daily Production—The ratio of the total production at a mining operation to the total number of production days worked at the operation.

Average Delivered Price—The weighted average of all contract price commitments and market price settlements in a delivery year.

Average Demand—The demand on, or the power output of, an electrical system or any of its parts over an interval of time, as determined by the total number of kilowatt-hours divided by the units of time in the interval.

Average Household Energy Expenditures—A ratio estimate defined as the total household energy expenditures divided by the total number of households.

Average Hydro—Rain, snow and runoff conditions that provide water for hydroelectric generation equal to the most commonly occurring levels. Average hydro usually is a mean indicating the levels experienced most often in a 104-year period.

Average Invested Capital—The sum of the capitalization, long-term debt due within one year, and short-term debt outstanding at the end of each month, for a period of time (usually 12 months), divided by the number of such months. The computation may also be made for individual classes of capital, i.e., long-term debt, short-term debt, preferred stock, and common stock equity.

Average Life—The average length of time an issue of serial bonds and/or term bonds with mandatory sinking funds and/or estimated prepayments is expected to be outstanding. It also can be the average maturity of a bond portfolio.
   • The period of time that an average unit of principal is outstanding in an amortization. Used often in reference to debt.

Average Megawatt—(MWa or aMW) One megawatt of capacity produced continuously over a period of one year. 1 aMW = 1 MW x 8760 hours/year =

8,760 MWh = 8,760,000 kWh.

Average Mine Price—The ratio of the total value of the coal produced at the mine to the total production tonnage.

Average Payment Period—The number of days, on average, within which a firm pays off its accounts payables.

Average Production Per Miner Per Day—The product of the average production per miner per hour at a mining operation and the average length of a production shift at the operation.

Average Production Per Miner Per Hour—The ratio of the total production at a mining operation to the total direct labor hours worked at the operation.

Average Rate of Return (ARR)—The ratio of average net earnings to average investment.

Average Revenue per Kilowatt-hour—The average revenue per kilowatt-hour of electricity sold by sector (residential, commercial, industrial, or other) and geographic area (State, Census division, and national) is calculated by dividing the total monthly revenue by the corresponding total monthly sales for each sector and geographic area.

Average Stream Flow—The rate, usually expressed in cubic feet per second, at which water passes a given point in a stream over a set period of time.

Average Vehicle Fuel Consumption—A ratio estimate defined as total gallons of fuel consumed by all vehicles divided by: (1) the total number of vehicles (for average fuel consumption per vehicle) or (2) the total number of households (for average fuel consumption per household).

Average Vehicle Miles Traveled—A ratio estimate defined as total miles traveled by all vehicles, divided by: (1) the total number of vehicles (for average miles traveled per vehicle) or (2) the total number of households (for average miles traveled per household).

Average Water Conditions—The amount and distribution of precipitation within a drainage basin and the run off conditions present as determined by reviewing the area water supply records over a long period of time.

Average Wind Speed (or Velocity)—The mean wind speed over a specified period of time.

Aviation Gasoline (Finished)—A complex mixture of relatively volatile hydrocarbons with or without small quantities of additives, blended to form a fuel suitable for use in aviation reciprocating engines. Fuel specifications are provided in ASTM Specification D 910 and Military Specification MIL-G-5572. Note: Data on blending components are not counted in data on finished aviation gasoline.

Aviation Gasoline Blending Components—Naphthas that will be used for blending or compounding into finished aviation gasoline (e.g., straight run gasoline, alkylate, reformate, benzene, toluene, and xylene). Excludes oxygenates (alcohols, ethers), butane, and pentanes plus. Oxygenates are reported as other hydrocarbons, hydrogen, and oxygenates.

AVL—Automatic Vehicle Location

Avoidance—The avoidance of Greenhouse Gas emissions resulting from the substitution of a high emitting source with a lower or non-emitting source or through improvements in energy efficiency.

Avoided Cost—(Regulatory) The amount of money that an electric utility would need to spend for the next increment of electric generation to produce or purchase elsewhere the power that it instead buys from a co-generator or small-power producer. Federal law establishes broad guidelines for determining how much a qualifying facility (QF) gets paid for power sold to the utility.

Avoided Emissions—Avoided emissions would have been emitted under a business as usual scenario but were avoided due to the implementation of an emission reduction project.

Avoided Energy Use—The reductions in energy use that occurred during the performance period relative to what would have been used during the baseline period of a M&V plan, using actual operating conditions experienced during that period. This may require baseline energy use to be adjusted to actual conditions. This approach in an energy purchase contract is different than calculating normalized savings.

AW&MA—Air and Waste Management Association (also AWMA) www.awma.org

AWC—Area-wide Contract

AWEA—American Wind Energy Association www.awea.org

AWFCO—Automatic Waste Feed Cutoff

AWG—The abbreviation for American Wire Gauge; the standard for gauging the size of wires (electrical conductors). The higher the number, the smaller the wire. Most house wiring is #12 or 14.

AWISE—Association of Women in Science and Engineering www.awis.org

AWMA—Air & Waste Management Association www.awma.com

AWMD—Air and Waste Management Division

Awning—An architectural element for shading windows and wall surfaces placed on the exterior of a building; can be fixed or movable. It serves as a covering to screen persons or parts of buildings from the sun, rain or other elements.

AWQC—Ambient Water-Quality Criteria

AWRA—American Water Resources Association www.awra.org

AWT—Advanced Wastewater Treatment

AWWA—American Water Works Association www.awwa.org

AWWUC—American Water Works Utility Council

AX—Administrator's Office

Axial Fans—Fans in which the direction of the flow of the air from inlet to outlet remains unchanged; includes propeller, tubaxial, and vaneaxial type fans.

Axial Flow Compressor—A type of air compressor in which air is compressed in a series of stages as it flows axially through a decreasing tubular area.

Axial Flow Turbine—A turbine in which the flow of a steam or gas is essentially parallel to the rotor axis.

Azimuth (Azimuth Angle)—The angular distance between true south and the point on the horizon directly below the sun. Typically used as an input for opaque surfaces and windows in computer programs for calculating the energy performance of buildings. As applied to the PV array, 180 degree azimuth means the array faces due south.

B

BAC—Bioremediation Action Committee
  • Biotechnology Advisory Committee

BACER—Biological and Climatological Effects Research

Back—The back side of the building as one faces the front facade from the outside (See *Front*). This designation is used on the Certificate of Compliance (CF-1R form) to indicate the orientation of fenestration (e.g., Back-West)

Back Leveraging—The period of time that an average unit of principal is outstanding in an amortization. Used often in reference to debt.

Back Pressure—Pressure against which a fluid is flowing. May be composed of friction in pipes, restrictions in pipes, valves, pressure in vessels to which fluid is flowing, hydrostatic head, or other resistance to fluid flow.

Backdrafting—The flow of air down a flue/chimney and into a house caused by low indoor air pressure that can occur when using several fans or fireplaces and/or if the house is very tight.

Backfill—1. To refill an excavated area with uncontaminated soils
2. The material used to refill an excavated area

Background Concentration—Represents the average amount of toxic chemicals in the air, water or soil to which people are routinely exposed. More than half of the background concentration of toxic air in metropolitan areas comes from automobiles, trucks and other vehicles. The rest comes from industry and business, agriculture, and from the use of paints, solvents and chemicals in the home.

Background Level—The average amount of a substance present in the environment. Originally referring to naturally occurring phenomena. Used in toxic substance monitoring.

Background Radiation—Radiation that occurs naturally in the environment from cosmic rays and radon or from atomic tests carried out by man.

Back-to-back Letter of Credit—A letter of credit issued on the strength of another letter of credit (backing credit). It is, in effect, an extension of the terms and conditions of the backing credit. To qualify as a back-to-back credit, the terms must be identical with those of the backing credit except for any or all of the following features: the beneficiary's name; the account party; the amount, which cannot be more than that of the backing credit; the validity date; and the insurance amount.

Backup Electricity, Backup Services—Power or services needed occasionally; for example, when on-site generation equipment fails.

Backup Energy System—A reserve appliance; for example, a stand-by generator for a home or commercial building.

Backup Fuel—In a central heat pump system, the fuel used in the furnace that takes over the space heating when the outdoor temperature drops below that which is feasible to operate a heat pump.

Backup Power—Electric energy supplied by a utility to replace power and energy lost during an unscheduled equipment outage.

Backup Rate—A utility charge for providing occasional electricity service to replace on-site generation.

BACM—Best Available Control Measures

BACT—See *Best Available Control Technology*.

Bacteria—Single-celled organisms, free-living or parasitic, that break down the wastes and bodies of dead organisms, making their components available for reuse by other organisms.

Bad Debt—A receivable that is written off because of the obligor's unwillingness or inability to pay.

Bad Debt Reserve—An account offsetting gross receivables on the balance sheet, representing estimated write-offs.

BADCT—Best Available Demonstrated Control Technology

Badge, Meter—A permanent plate, affixed in a conspicuous place on a meter, containing basic meter information.

BADT—Best Available Demonstrated Technology

BAF—Bioaccumulation Factor

Baffle—A device, such as a steel plate, used to check, retard, or divert a flow of a material or to shield a surface from direct or unwanted light.

Baffle Chamber—In incinerator design, a chamber designed to settle fly ash and coarse particulate matter by changing the direction and reducing the velocity of the combustion gases.

Baffles—Plates, louvers, or screens placed in the path of fluid flow to cause change in the direction of flow; these are used to promote mixing of gases or to eliminate undesirable solid or liquid particles in the fluid stream. Sometimes baffles are inserted in a flue to lengthen the travel of flue gases and increase efficiency of operation.

BAFO—Best and final offer

Bagasse—The fibrous material remaining after the extraction of juice from sugarcane; often burned by sugar mills as a source of energy.

Baghouse—A chamber containing fabric filter bags that remove particles from stack gases. Used to eliminate particles greater than 20 microns in diameter.

• An air pollution control device used to filter particulates from processes; a chamber containing a bag filter.

Baghouse Filter—Large fabric bag, usually made of glass fibers, used to eliminate intermediate and large (greater than 20 microns in diameter) particles. This device operates in a way similar to the bag of an electric vacuum cleaner, passing the air and smaller particulate matter, while entrapping the larger particles.

Balance of Plant—Supporting components based on site-specific requirements and integrated into a comprehensive power system.

Balance of System (BOS)—Represents all components and costs other than the photovoltaic modules/array. It includes design costs, land, site preparation, system installation, support structures, power conditioning, operation and maintenance costs, indirect storage, and related costs

Balance Point—An outdoor temperature, usually 20 to 45 degrees Fahrenheit, at which a heat pump's output equals the heating demand. Below the balance point, supplementary heat is needed.

Balance Sheet—An accounting statement showing the financial condition of a company at a point in time: includes assets, liabilities and net worth.

Balanced Schedule—A Scheduling Coordinator's schedule is balanced when generation, adjusted for transmission losses, equals demand.

Balance-of-System—In a renewable energy system, refers to all components other than the mechanism used to harvest the resource (such as photovoltaic panels or a wind turbine). Balance-of-system costs can include design, land, site preparation, system installation, support structures, power conditioning, operation and maintenance, and storage.

Balancing—Making receipts and deliveries of energy into or withdrawals from a company equal. Balancing may be accomplished daily, monthly or seasonally, with penalties generally assessed for excessive imbalance.

Balancing Agreement—A contractual agreement between two or more legal entities to account for differences between chart measured quantities and the total confirmed nominated quantities at a point. They have been used to keep track of over/under production relative to entitlements between producers; over/under deliveries relative to confirmed nominations between operators of wells, pipelines and LDCs.

Balancing Item—Represents differences between the sum of the components of natural gas supply and the sum of the components of natural gas disposition. These differences may be due to quantities lost or to the effects of data reporting problems. Reporting problems include differences due to the net result of conversions of flow data metered at varying temperature and pressure bases and converted to a standard temperature and pressure base; the effect of variations in company accounting and billing practices; differences between billing cycle and calendar period time frames; and imbalances resulting from the merger of data reporting systems that vary in scope, format, definitions, and type of respondents.

Baling—A means of reducing the volume of a material by compaction into a bale.

Ballast—A device that provides starting voltage and limits the current during normal operation in electrical discharge lamps (such as fluorescent lamps).

Ballast Efficacy Factor—The measure of the efficiency of fluorescent lamp ballasts. It is the relative light output divided by the power input.

Ballast Factor—The ratio of light output of a fluorescent lamp operated on a ballast to the light output of a lamp operated on a standard or reference ballast.

Ballast-Electronic—An Electronic Ballast is a lighting Ballast constructed entirely of electronic circuitry. Because Electronic Ballasts generate less heat than traditional Magnetic Ballasts, they are more efficient at converting electrical power to light. Most modern ballasts, especially for fluorescent lamps, are electronic.

Ballast-Magnetic—A Magnetic Ballast is a traditionally designed lighting Ballast constructed with a magnetic core and a winding of insulated wire. Magnetic Ballasts are generally less energy-efficient than Electronic Ballasts.

Balloon Payment—A large payment at the end of the financial obligation allowing smaller payments to be made during the original term.
- A lump payment of principal at the end of a debt amortization, often paid at least in part by the residual value of the asset.

BANANA—Build Absolutely Nothing Anywhere Near Anybody

Band Gap—In a semiconductor, the energy difference between the highest valence band and the lowest conduction band.

Band Gap Energy—The amount of energy (in electron volts) required to free an outer shell electron from its orbit about the nucleus to a free state, and thus promote it from the valence to the conduction level.

Banking—banking entails saving emissions permits or Certified Emissions Reductions for future use in anticipation that these will accrue value over time. Within the Kyoto Protocol, emission permits not used in one commitment period can be saved or 'banked' for future use in a subsequent compliance period.
- The ability under an emissions trading scheme to save emission permits issued in one year for use in later years (banking), or bring forward some of a future year's permit allocation for use in the current year (borrowing).

Bankruptcy—An action taken by a party to legally protect its remaining assets by declaring that it cannot pay its bills. Typically, liabilities exceed assets.

BAP—Benefits Analysis Program

Bar Screen—A screen made of parallel bars set 3/4" to 2" apart used to filter out large objects.

BARF—Best Available Retrofit Facility

Bargain Purchase Option—A lease provision allowing the lessee, at its option, to purchase the equipment for a price predetermined at lease inception that is substantially lower than the expected fair market value at the date the option can be exercised.

Bargain Renewal Option—A lease provision allowing the lessee, at its option, to extend the lease for an additional term in exchange for periodic rental payments that are sufficiently less than fair market value rentals for the property, such that exercise of the option appears, at the inception of the lease, to be reasonably assured.

Barometer—Instrument used for measuring atmospheric pressure.

Barrel—In the petroleum industry, a barrel is 42 U.S. gallons. One barrel of oil has an energy content of 6 million British thermal units. It takes one barrel of oil to make enough gasoline to drive an average car from Los Angeles to San Francisco and back (at 18 miles per gallon over the 700-mile round trip).

Barrel of Oil Equivalent—A unit of energy equal to the amount of energy contained in a barrel of crude oil. Approximately 5.78 million Btu or 1,700 kWh. A barrel is a liquid measure equal to 42 gallons.

Barrels per Calendar Day—The amount of input that a distillation facility can process under usual operating conditions. The amount is expressed in terms of capacity during a 24-hour period and reduces the maximum processing capability of all units at the facility under continuous operation (See ***Barrels per Stream Day*** below) to account for the following limitations that may delay, interrupt, or slow down production.
1. The capability of downstream processing units to absorb the output of crude oil processing facilities of a given refinery. No reduction is necessary for intermediate streams that are distributed to other than downstream facilities as part of a refinery's normal operation;
2. The types and grades of inputs to be processed;
3. The types and grades of products expected to be manufactured;
4. The environmental constraints associated with refinery operations;
5. The reduction of capacity for scheduled downtime due to such conditions as routine inspection, maintenance, repairs, and turnaround; and
6. The reduction of capacity for unscheduled downtime due to such conditions as mechan-

ical problems, repairs, and slowdowns.

Barrels per Day (Operable Refinery Capacity)—The maximum number of barrels of input that can be processed during a 24-hour period after making allowances for the following limitations: the capability of downstream facilities to absorb the output of crude oil processing facilities of a given refinery (no reduction is made when a planned distribution of intermediate streams through other than downstream facilities is part of a refinery's normal operation); the types and grades of inputs to be processed; the types and grades of products to manufactured; the environmental constraints associated with refinery operations; the reduction of capacity for scheduled downtime, such as routine inspection, mechanical problems, maintenance, repairs, and turnaround; and the reduction of capacity for unscheduled downtime, such as mechanical problems, repairs, and slowdowns.

Barrels per Day Equivalent (BPD-Equivalent)—A unit of measure that tells how much oil would have to be burned to produce the same amount of energy. For example, California's hydroelectric generation in 1983 was 58,000 barrels per day equivalent.

Barrels per Stream Day—The maximum number of barrels of input that a distillation facility can process within a 24-hour period when running at full capacity under optimal crude and product slate conditions with no allowance for downtime.

Barrier Energy—The energy given up by an electron in penetrating the cell barrier; a measure of the electrostatic potential of the barrier.

BART—Best Available Retrofit Technology

Basal Metabolism—The amount of heat given off by a person at rest in a comfortable environment; approximately 50 Btu per hour (Btu/h).

Base (cushion) Gas—The volume of gas needed as a permanent inventory to maintain adequate reservoir pressures and deliverability rates throughout the withdrawal season. All native gas is included in the base gas volume.

Base Bill—A charge calculated by taking the rate from the appropriate electric rate schedule and applying it to the level of consumption.

Base Contract Price—The stated per unit price for energy in a contract between a producer and a purchaser.

Base Gas—The gas required in a storage reservoir to provide the pressure to cycle the normal working storage volume.

Base Load—The minimum amount of electric power delivered or required over a given period of time at a steady rate.

Base Load Capacity—The generating equipment normally operated to serve loads on an around-the-clock basis.

Base Load Plant—A plant, usually housing high-efficiency steam-electric units, which is normally operated to take all or part of the minimum load of a system, and which consequently produces electricity at an essentially constant rate and runs continuously. These units are operated to maximize system mechanical and thermal efficiency and minimize system operating costs.

Base Load Unit—A power generating facility that is intended to run constantly at near capacity levels, as much of the time as possible.

Base Period—The period of time for which data used as the base of an index number, or other ratio, have been collected. This period is frequently one of a year but it may be as short as one day or as long as the average of a group of years. The length of the base period is governed by the nature of the material under review, the purpose for which the index number (or ratio) is being compiled, and the desire to use a period as free as possible from abnormal influences in order to avoid bias. A preceding twelve-month period, or longer, selected as the standard for measurement of energy consumption and energy savings due to implementation of energy conservation measures or services.

Base Power—Power generated by a power generator that operates at a very high capacity factor.

Base Pressure (Gas)—The pressure used as a standard in determining gas volume. Volumes are measured at operating pressures and then corrected to base pressure volume. Base pressure is normally defined in any gas measurement contract. The standard value for natural gas in the United States is 14.73 psia, established by the American National Standards Institute as standard Z-132.1 in 1969.

Base Pressure Index—A device which continuously and automatically compensates to correct gas volume at operating pressure to volume at base pressure, without regard for any correction for temperature.

Base Rate—Amount of the total electric or gas rate covering the general expense of business without the fuel expense.

Base Term—The initial, non-cancellable term of the lease (or other similar contract) used by the lessor in computing the payment. The base term is the minimum time period during which the lessee (or purchaser of the energy) has the use and custody of the equipment.

Base Term Rate—The rate that discounts the base rents or payments to the equipment cost at the contract commencement date.

Base Volume Index—A device which continuously and automatically compensates to correct gas volumes measured at operating temperature and pressure to volume at a specified base temperature and pressure.

Base Year—The first year of the period of analysis. The base year does not have to be the current year.

• The year on which escalation clauses in a lease are based.

• According to the IPCC, the Base Year is that year for which a national inventory is to be taken, which is currently 1990 for Annex I countries. In some case (such as estimating Methane from rice production), the base year is simply the middle of a three-year period over which an average must be taken. A base year may also be used as a reference for establishing an emissions baseline. Under the Kyoto Protocol, the base year for hydrofluorocarbons, perfluorocarbons and sulfur hexafluoride is 1995.

Baseboard Heater—As a type of heating equipment, a system in which either electric resistance coils or finned tubes carrying steam or hot water are mounted behind shallow panels along baseboards. Baseboards rely on passive convection to distribute heated air in the space. Electric baseboards are an example of an "Individual Space Heater."

Baseboard Radiator—A type of radiant heating system where the radiator is located along an exterior wall where the wall meets the floor.

Baseline Adjustment—A Baseline Adjustment is a change to a Facility that will have an impact on energy usage, but that is outside of the energy asset upgrade program. Typical Baseline Adjustments include: adding a wing to a building, installing significant new processes or equipment, changing how a large part of the Facility is used, increasing occupancy density (possibly through consolidation of operations) or extending operating hours. Baseline Adjustments are assigned to meter components, along with the start date and monthly impacts (either as physical units or as a percentage of the physical units).

Baseline and Baseline Scenario—The baseline represents the forecast emissions of a company, business unit or project, using a business as usual scenario, often referred to as the 'baseline scenario', i.e. expected emissions if the firm did not implement emission reduction activities. This forecast incorporates the economic, financial, technological, regulatory and political circumstances within which a firm operates.

Baseline and Credit—A type of emissions trading scheme where firms are encouraged to reduce their greenhouse gas emissions below a projected "business as usual" path of increasing emissions. Any reductions below that future path earns credits for the difference which can be sold to other emitters struggling to contain increases to baseline levels.

Baseline Conditions—Physical conditions that existed prior to the implementation of the energy project (such as equipment inventory and conditions, occupancy, nameplate data, energy consumption rate, and control strategies), which are determined through surveys and audits, inspections, spot measurements, and short-term metering activities. Baseline conditions are established for the purposes of estimating savings and are also used to account for any changes that may occur during the performance period of the energy project, which may require adjustments to the baseline energy use.

Baseline Energy or Demand—The calculated or measured energy usage or demand by a piece of equipment or a site prior to implementation of the energy project.

Baseline Forecast—A prediction of future energy needs which does not take into account the likely effects of new conservation programs that have not yet been started.

Baseline Model—A Baseline Model is a mathematical formula used to predict meter energy usage that should appear on an invoice, based on the values of known Independent Variables. The most commonly used Independent Variables are Billing Period (number of days), HDDs (a measure of heating load) and CDDs (a measure of cooling load). When it comes time to measure the performance of energy conservation efforts by determining actual energy savings, the Baseline Model can be thought of as a hypothetical, mathematically defined "what would have happened" utility meter. The energy savings are the difference between this meter and the actual meter.

Baseline Performance Value—Initial values of I_{sc}, V_{oc}, P_{mp}, I_{mp} measured by the accredited laboratory and corrected to Standard Test Conditions, used to validate the manufacturer's performance measurements provided with the qualification modules per IEEE 1262.

Baseline Period—In relation to the quantification of ERCs, the Baseline Period is a time period prior to the reduction of emissions.

Baseline Usage Profile—The Baseline Usage Profile for a meter component describes what energy consumption would have been if no Energy Conservation Measures had been implemented and no Baseline Adjustments made. The baseline modeling process is used to create a mathematical description of past meter performance, and it is this Baseline Model that is used to create the Baseline Usage Profile.

Baseload Capacity—The power output that generating equipment can continuously produce.

Baseload Demand—The minimum demand experienced by an electric utility, usually 30-40% of the utility's peak demand.

Baseload Power—Using the output of a power source at its full continuous rating.

Basement—The conditioned or unconditioned space below the main living area or primary floor of a building.

Bases—A class of compounds that are "opposite" to acids, in that they neutralize acids. Weak bases are used in cooking (baking soda) and cleaners. Strong bases can be corrosive, or "caustic." Examples strong bases that are common around the house are drain cleaners, oven cleaners and other heavy duty cleaning products. Strong bases can be very dangerous to tissue, especially the eyes and mouth.

Basic Service—The four charges for generation, transmission, distribution and transition that all customers must pay in order to retail the electric services of a utility.

Basis—The original cost of an asset plus other capitalized acquisition costs such as installation charges, permit fees, engineering, auditing costs, spare parts and sales tax. Basis reflects the amount upon which depreciation charges are computed.

Basis Point—One one-hundredth of a percent (.01%). Used when quoting interest.

Basket of Gases—This refers to the group six of greenhouse gases regulated under the Kyoto Protocol. They are listed in Annex A of the Kyoto Protocol and include: carbon dioxide (CO_2), methane (CH4), nitrous oxide (N2O), hydrofluorocarbons (HFCs), perfluorocarbons (PFCs), and sulphur hexafluoride (SF6).

BAT—Best Available Technology
  • Best Available Treatment

Batch Heater—This simple passive solar hot water system consists of one or more storage tanks placed in an insulated box that has a glazed side facing the sun. A batch heater is mounted on the ground or on the roof (make sure your roof structure is strong enough to support it). Some batch heaters use "selective" surfaces on the tank(s). These surfaces absorb sun well but inhibit radiative loss. Also known as bread box systems or integral collector storage systems.

Batch Process—A process for carrying out a reaction in which the reactants are fed in discrete and successive charges.

Batch Solar Hot Water Heater—The simplest of solar hot water systems. A tank of water within a glass-covered insulated enclosure aimed at the sun. Water is heated in the tank and then flows to the load or an auxiliary water heater.

BATEA—Acronym for "Best Available Technology Economically Achievable."

Bathroom—A room containing a shower, tub, toilet or a sink that is used for personal hygiene.

Batt/Blanket—A flexible roll or strip of insulating material in widths suited to standard spacings of building structural members (studs and joists). They are made from glass or rock wool fibers. Blankets are continuous rolls. Batts are pre-cut to four or eight foot lengths.

Batten—Small rectangular piece of building material used to provide fixings for tiles or slates. Cover-slip concealing the joint between two boards, or a strip of building material fixed across two parallel boards to join them together.

Batten Seam—Joint in a roof formed over a strip or roll of building material.

Battery—A device that stores energy and produces electric current by chemical action. A system in which stored chemical energy is converted directly into electrical energy. Can be either rechargeable or non-rechargeable. Different to a fuel cell in that it contains a fixed quantity of stored chemical energy rather than a continuous supply of fuel producing a continuous energy flow.

Battery Available Capacity—The total maximum charge, expressed in ampere-hours, that can be withdrawn from a cell or battery under a specific set of operating conditions including discharge rate, temperature, initial state of charge, age, and cut-off voltage.

Battery Capacity—The total number of ampere-hours (Ah) that a fully charged battery can output.

Battery Cell—An individual unit of a battery that can store electrical energy and is capable of furnishing a current to an external load. For lead-acid batteries the voltage of a cell (fully charged) is about 2.2 volts dc. A battery may consist of a number of cells.

Battery Charger—A device used to charge a battery by converting (usually) mains voltage AC to a DC voltage suitable for the battery. Chargers often incorporate some form of regulator to prevent overcharging and damage to the battery.

Battery Cycle Life—The number of times a battery can undergo a cycle of discharge and recharge before failing. Cycle Life is normally specified as a function of discharge rate and temperature.

Battery Energy Capacity—The total energy available, expressed in watt-hours (kilowatt-hours), which can be withdrawn from a fully charged cell or battery. The energy capacity of a given cell varies with temperature, rate, age, and cut-off voltage. This term is more common to system designers than it is to the battery industry where capacity usually refers to ampere-hours.

Battery Energy Storage—Energy storage using electrochemical batteries. The three main applications for battery energy storage systems include spinning reserve at generating stations, load leveling at substations, and peak shaving on the customer side of the meter. Battery storage has also been suggested for holding down air emissions at the power plant by shifting the time of day of the emission or shifting the location of emissions.

Battery Life—The period during which a cell or battery is capable of operating above a specified capacity or efficiency performance level. Life may be measured in cycles and/or years, depending on the type of service for which the cell or battery is intended.

Battery Self Discharge—Energy loss by a battery that is not under load.

Battery State of Charge (SOC)—Extent of battery charge status as a percentage of full charge. Also 100 per cent minus the Depth of Discharge.

bbl—The abbreviation for barrel(s).

bbl/d—The abbreviation for barrel(s) per day.

bbl/sd—The abbreviation for barrel(s) per stream day

BBRS—Board for Building Regulations and Standards

BCC—Bioaccumulative Chemical of Concern

BCF—One billion cubic feet of gas
- Bioconcentration Factor

BCFD—Billion Cubic Feet Per Day

BCPCT—Best Conventional Pollutant Control Technology

BCPT—Best Conventional Pollutant Technology

BCT—Best Control Technology

BDAT—Acronym for "Best Demonstrated Achievable (or Available) Technology"

BDT—Acronym for "Best Demonstrated Technology."
- Bone Dry Tons

BDU—See *Bone Dry Unit*.

BEA—Bureau of Economic Advisors

Beadwall™—A form of movable insulation that uses tiny polystyrene beads blown into the space between two window panes.

Beam Radiation—Solar radiation that is not scattered by dust or water droplets.

Bear Hug—An offer made directly to the Board of Directors of a target company. Usually made to increase the pressure on the target with the threat that a tender offer may follow.

Bearer Security—A security whose owner is not registered on the books of the issuer. A bearer security is payable to the holder.

Bearing Wall—A wall that carries ceiling rafters or roof trusses.

BEI—Biological Exposure Index

BEJ—Best Engineering Judgment
- Best Expert Judgment

Bell Hole—A hole dug to allow room for workmen to make a repair or connection in buried pipe, such as caulking bell-and-spigot pipe or welding steel pipe. In the broad sense, any hole other than a continuous trench opened for working on a buried facility.

Below Grade Wall—The portion of a wall, enclosing conditioned space, that is below the grade line.

Benchmark—Also comparative index or comparative standard. A benchmark is an important reference value that is used to compare investments or to measure the performance of investment funds or even commodities such as energy.

Benefit Cost Analysis—The method of evaluating a proposal by comparing the costs and benefits *in common units* associated with the proposal. The common units are normally dollars.

Benefit Cost Ratio—The Benefit Cost Ratio is an indicator of the attractiveness of an investment which takes equipment life expectancy into account. When comparing investment opportunities, a higher value is better. Benefit Cost Ratio = (Annual Savings x Present Worth Factor)/Price Estimate.

Benefit Transfer Method—A method of putting dollar values on something (e.g. air quality or risks) by assembling other studies of similar situations and somehow `transferring' the results of the other studies to the situation at hand.

Benefits Charge—The addition of a per-unit-tax on sales of electricity, with the revenue generated used for or to encourage investments in energy efficiency measures and/or renewable energy projects.

Benzene—A toxic, six-carbon aromatic component of gasoline. A known carcinogen. A few uses are: synthesis of rubber, nylon, polystyrene, and pesticides. Benzene is a highly volatile chemical readily absorbed by breathing, ingestion or contact with the skin. Benzene is listed as a cancer-causing chemical.

Benzene (C_6H_6)—An aromatic hydrocarbon present in small proportion in some crude oils and made commercially from petroleum by the catalytic reforming of naphthenes in petroleum naphtha. Also made from coal in the manufacture of coke. Used as a solvent in the manufacture of detergents, synthetic fibers, petrochemicals, and as a component of high-octane gasoline.

BEP—Basic Earning Power

Berkeley Energy and Resource (BEAR) Model—A dynamic general equilibrium forecasting model that simulates the way that changes in energy investment, price and use affect how citizens live their lives.

Berlin Mandate—Decision of the Parties reached at the first session of the Conference of the Parties to the UNFCCC (COP-1) in 1995 in Berlin that the commitments made by Annex I countries were inadequate and thus needed to be strengthened. The Berlin Mandate launched the talks that led to the adoption of the Kyoto Protocol in 1997.

Berm—A curb, ledge, wall or mound used to prevent the spread of contaminants. It can be made of various materials, even earth in certain circumstances.

Best Available Control Measures (BACM)—The most effective measures for controlling small or dispersed particulates from sources such as soot and ash from woodstoves and open burning of brush, timber, grasslands, or trash.

Best Available Control Technology (BACT)—That combination of production processes, methods, systems, and techniques that will result in the lowest achievable level of emissions of air pollutants from a given facility. BACT is an emission limitation that the permitting authority determines on a case-by-case basis, taking into account energy, environmental, economic and other costs of control. BACT may include fuel cleaning or treatment or innovative fuel combustion techniques. Applies in EPA designated air attainment areas.

Best Efforts—An offering in which the investment banker agrees to distribute as much of the offering as possible and return any unsold shares to the issuer.

Best Management Practices—A practice or combination of practices that a designated agency determines to be the most effective, practical means of reducing the amount of pollution generated by non-point sources to a level compatible with water quality goals.

Beta Particle—An elementary particle emitted by radioactive decay that may cause skin burns. It is halted by a thin sheet of metal. Very high-energy particle identical to an electron and emitted by some radioactive elements.

BETEC—Building Environment and Thermal Envelope Council

Betterments—A substantial enlargement or improvement of existing structures, facilities, or equipment by the replacement or improvement of parts, which has the effect of extending the useful life of the property, increasing its capacity, lowering its operating cost, or otherwise adding to the worth through the benefit it can yield.

Betz Limit—The theoretical maximum energy that a wind generator can extract from the wind—59.6 percent.

BEV—Battery electric vehicle

BG—Billion Gallons
  • Biomass Gassification

BGS—Below Ground Surface

BI—Background Information

BIA—Bureau of Indian Affairs www.bia.gov

Bid—A bid is the price a prospective buyer is willing to pay. Bidding on forwards indicates the buyers willingness to obligate himself to the purchase of the emission reductions at an agreed upon time in the future. Bidding on a call means that the buyer is willing to purchase a call option. Bidding on a put means that the buyer is willing to purchase a put option.

BID—Background Information Document
  • Buoyancy Induced Dispersion

Biennial Report—The report issued by the California Energy Commission to the Governor and the Legislature every odd-numbered year assessing California's energy industry. The Biennial Report is supported by four policy documents that are issued every even-numbered year: the Electricity Report, the Fuels Report, the Conservation (or Efficiency) Report and the Energy Development Report.

BIF—Boiler and Industrial Furnace

Bi-Fuel Vehicle—A vehicle with two separate fuel systems designed to run on either fuel, using only one fuel at a time. These systems are advantageous for drivers who do not always have access to an alternative fuel refueling station. Bi-fuel systems are usually used in light-duty vehicles. One of the two fuels is typically an alternative fuel.

Big Box Store—A general merchandise store that sells products at a discount.
  • Value retailer

Bi-Gas—A process being developed as a means of making synthetic gas from coal. The synthetic gas would be intended to substitute for natural gas in meeting industrial and home energy needs.

Bilateral Contract—A two-party agreement for the purchase and the sale of energy products and services. A function of retail wheeling in which a generator and a user establish a contract for the supply of electricity bypassing the existing utility structure and producing stranded costs for the utility. In a restructured electrical world, a bilateral contract could be used to establish a price for electrical supply over a certain duration and would remove the risk associated with the spot market.

Bilateral Electricity Contract—A direct contract between an electric power producer and either a user or broker outside of a centralized power pool or power exchange.

Bilateral Transaction—A trade that does not include an intermediary exchange and is made on a direct one-on-one basis.

Bill of Lading—A document issued by a transportation company giving evidence of the movement of the goods from one location to another. The bill of lading is a receipt for the goods and a contract for their delivery and in some forms represents the title to the goods.

Bill of Sale—A written instrument evidencing the transfer of Title to Personal Property.

Billing Credit—In the Pacific Northwest, a payment by the Bonneville Power Administration to a wholesale customer for actions taken by that customer to reduce BPA's obligations to acquire new resources. The payment is usually made by an offset against billings.

Billing Cycle—The regular, periodic interval used by a utility for reading the meters of customers for billing purposes. Usually billing cycles are monthly or bi-monthly.

Billing Determinant—The demand which is used to determine demand charges in accordance with the provisions of a rate schedule or contract. It does not necessarily coincide with the actual measured demand of the billing period.

Billing Period—The time between meter readings. It does not refer to the time when the bill was sent or when the payment was to have been received. In some cases, the billing period is the same as the billing cycle that corresponds closely (within several days) to meter-reading dates. For fuel oil and LPG, the billing period is the number of days between fuel deliveries.

Bimetal—Two metals of different coefficients of expansion welded together so that the piece will bend in one direction when heated, and in the other when cooled, and can be used to open or close electrical circuits, as in thermostats.

Bi-Monthly Billing—A customer billing procedure in a distribution company where bills are rendered every month, but meters are read every other month. An estimate is made of the volume of gas used in months when meters are not read.

Bin Method—A method of predicting heating and/or cooling loads using instantaneous load calculation at different outdoor dry-bulb temperatures, and multiplying the result by the number of hours of occurrence of each temperature.

Binary Cycle—Combination of two power plant turbine cycles utilizing two different working fluids for power production. The waste heat from the first turbine cycle provides the heat energy for the operation of the second turbine, thus providing higher overall system efficiencies.

Binary Cycle Geothermal Plants—Binary cycle systems can be used with liquids at temperatures less than 350 F (177 C). In these systems, the hot geothermal liquid vaporizes a secondary working fluid, which then drives a turbine.

Binding Targets—Environmental standards that are to be met in the future. Binding targets are agreed or mandated emission limits on an entity that are to be met at a specific point of time or period.

Bioaccumulants—Substances in contaminated air, water, or food that increase in concentration in living organisms exposed to them because the substances are very slowly metabolized or excreted.

Bioaccumulation—The process by which the concentrations of some toxic chemicals gradually increase in living tissue, such as in plants, fish, or people as they breathe contaminated air, drink contaminated water, or eat contaminated food. See *Bioconcentration*.

Bioassay—A study of a living organism to measure the effect of a substance, factor, or condition.

Biochar—Carbon-rich charcoal created when plant matter is heated in an oxygen-free environment. Carbon that would otherwise combine with oxygen, burn, and be emitted to the air is contained in the charcoal, which can be used to fertilize soils or make biofuels.

Biochemical Conversion Process—The use of living organisms or their products to convert organic material to fuels, chemicals or other products.

Biochemical Oxygen Demand (BOD)—A standard means of estimating the degree of water pollution, especially of water bodies that receive contamination from sewage and industrial waste. BOD is the amount of oxygen needed by bacteria and other microorganisms to decompose organic matter in water. The greater the BOD, the greater the degree of pollution. Biochemical oxygen demand is a process that occurs over a period of time and is commonly measured for a five-day period, referred to as BOD5.

Bioconcentration—(Bioaccumulation) The accumulation of a chemical in tissues of an organism to levels greater than in the environment in which the organism lives.

Bioconversion—Processes that use plants or micro-organisms to change one form of energy into another. For example, an experimental process uses algae to convert solar energy into gas that could be used for fuel.

Biodegradable—Capable of decomposing rapidly under natural conditions.

BIODG—Biodegradation of Organics or Non-metallic Inorganics

Biodiesel—Any liquid bio-fuel suitable as a diesel fuel substitute or diesel fuel additive or extender. Bio-diesel fuels are typically made from oils such as soybeans, grapeseed, or sunflowers, or from animal tallow. Bio-diesel can also be made from hydrocarbons derived from agricultural products such as rice hulls. A type of fuel made by combining animal fat or vegetable oil with alcohol. Bio-diesel can be directly substituted for diesel (known as B100, for 100% biodiesel), or be used as an additive mixed with traditional diesel (known as B20, for 20% bio-diesel.

Biodiversity—The relative abundance and variety of plant and animal species and ecosystems within particular habitats

Bioenergy—Renewable energy produced from organic matter. The conversion of the complex carbohydrates in organic matter to energy. Organic matter may either be used directly as a fuel or processed into liquids or gases.

Bioethanol—Ethanol, also known as ethyl alcohol, alcohol, or grain spirit derived from biological sources. A clear, colorless, flammable oxygenated hydrocarbon with a boiling point of 78.5 degrees Celsius in the anhydrous state. In transportation, ethanol is used as a vehicle fuel by itself, blended with gasoline, or as a gasoline octane enhancer and oxygenate.

Biofuels—Liquid fuels and blending components produced from biomass (plant) feedstock, used primarily for transportation.

Biofuels are renewable fuels made from plants that can be used to supplement or replace the fossil fuels petroleum and diesel used for transport. The two main biofuels are ethanol and biodiesel. Ethanol is produced from the fermentation of sugar or starch in crops such as corn and sugar cane. Biodiesel is made from vegetable oils in crops such as soybean, or from animal fats. Depending on the processes used to make biofuels, greenhouse emissions from cars and fuel-powered machinery can be substantially reduced by their use.

Biogas—A combustible gas created by anaerobic decomposition of organic material, composed primarily of methane, carbon dioxide, and hydrogen sulfide.

Biogas Digester—Converts animal and plant wastes into gas usable for lighting, cooking, heating, and electricity generation.

Biogasification or Biomethanization—The process of decomposing biomass with anaerobic bacteria to produce biogas.

Biological Assessment—A specific process required as part of an environmental assessment. An evaluation of potential effects of a proposed project on proposed, endangered, threatened and sensitive animal and plant species and their habitats.

Biological Magnification—The process by which substances such as pesticides or heavy metals become concentrated as they move up the food chain.

Biological Oxidation—Decomposition of organic materials by microorganisms.

BIOLOGS—Biological Data Management System

Biomass—As defined by the Energy Security Act (PL 96-294) of 1980, "any organic matter which is available on a renewable basis, including agricultural crops and agricultural wastes and residues, wood and wood wastes and residues, animal wastes, municipal wastes, and aquatic plants."

• Eligible biomass resources for Transferable Renewable Credits (TRC) products include: landfill gas, digester gas, plant-based agricultural, vegetative and food processing waste, bioenergy crops, clean urban waste wood, and mill residues. There are some states that have exclusions to some of these biomass resources used in TRCs.

• Living or recently-dead organic material that can be used as an energy source or in industrial production. Excludes organic material that has been transformed by geological processes (such as coal or petroleum).

Biomass Conversion—Process by which biomass materials are burned for direct energy or by which such materials are converted to synthetic fuels.

Biomass Energy—Energy produced by the conversion of biomass directly to heat or to a liquid or gas that can be converted to energy. See *Bioenergy*.

Biomass Fuel—Biomass converted directly to energy or converted to liquid or gaseous fuels such as ethanol, methanol, methane, and hydrogen.

Biomass Gas—A medium Btu gas containing methane and carbon dioxide, resulting from the action of microorganisms on organic materials such as a landfill.

Biomass Gasification—The conversion of biomass into a gas, by biogasification (see above) or thermal gasification, in which hydrogen is produced from high-temperature gasifying and low-temperature pyrolysis of biomass.

Biomass Generation—Biomass generation is a biomass fuel gasification plant that produces electricity.

Biomass Power & Heat—Power and/or heat generation from solid biomass, which includes forest product wastes, agricultural residues and waste, energy crops, and the organic component of municipal solid waste and industrial waste. Also includes power and process heat from biogas.

Biome—The community of living organisms in a given area.

Biomimicry—A science that studies natural processes and models in order to imitate the designs to solve human problems, i.e. studying a leaf to better understand and design solar cells.

Biophotolysis—The action of light on a biological system that results in the dissociation of a substrate, usually water, to produce hydrogen.

Bioplume—Model to predict the maximum extent of existing plumes

Bioremediation—The use of living organisms to clean up pollutants from soil or water. Changes toxic compounds into non-toxic ones.

BIOS—Natural Biological Information System

Biosolid—Residuals generated by the treatment of sewage, petroleum refining waste and industrial chemical manufacturing wastewater with activated sludge. See *Activated Sludge*

Biosphere—The zone at and adjacent to the earth's surface where all life exists; all living organisms of the earth.

BIOSTU—Bioassay Studies

Biota—The animal and plant life of a region or period.

Biotechnology—Technology that use living organisms to produce products such as medicines, to improve plants or animals, or to produce microorganisms for bioremediation. Btiotech medicines use the human body's own natural defense mechanisms to fight disease, including cells, genes, proteins, enzymes and antibodies. By harnessing natural mechanisms, scientists can find more accurate ways to solve medical problems while producing fewer side effects and unintended consequences for the individual patient.

Biotherapeutic—A drug derived from a living source (human, animal or unicellular). Most biotherapeutics are complex mixtures that are not easily identified or characterized, and many are manufactured using biotechnology.

Biotic—Pertaining to life or living organisms.

Biotic Community—A naturally occurring, interdependent community of plants and animals that live in the same environment.

Biotransformation—Transformation of one chemical to others by populations of microorganisms in the soil.

BIPV (Solar)—Building Integrated Photovoltaics—A term for the design and integration of photovoltaic (PV) technology into the building envelope, typically replacing conventional building materials. This integration may be in vertical facades, replacing view glass, spandrel glass, or other facade material; into semitransparent skylight systems; into roofing systems, replacing traditional roofing materials; into shading "eyebrows" over windows; or other building envelope systems.

Bitumen—A naturally occurring viscous mixture, mainly of hydrocarbons heavier than pentane, that may contain sulphur compounds and that, in its natural occurring viscous state, is not recoverable at a commercial rate through a well.

Bituminous Coal—A dense coal, usually black, sometimes dark brown, often with well-defined bands of bright and dull material, used primarily as fuel in steam-electric power generation, with substantial quantities also used for heat and power applications in manufacturing and to make coke. Bituminous coal is the most abundant coal in active U.S. mining regions. Its moisture content usually is less than 20 percent. The heat content of bituminous coal ranges from 21 to 30 million Btu per ton on a moist, mineral-matter-free basis. The heat content of bituminous coal consumed in the United States averages 24 million Btu per ton, on the as-received basis (i.e., containing both inherent moisture and mineral matter).

BL—Baseline

Black Carbon Aerosols—Particles of carbon in the atmosphere produced by inefficient combustion of fossil fuels or biomass. Black carbon aerosols absorb light from the sun, shading and cooling the Earth's surface, but contribute to significant warming of the atmosphere (see "radiative forcing").

Black Liquor—A byproduct of the paper production process, alkaline spent liquor that can be used as a source of energy. Alkaline spent liquor is removed from the digesters in the process of chemically pulping wood. After evaporation, the residual "black" liquor is burned as a fuel in a recovery furnace that permits the recovery of certain basic chemicals.

Black Lung Benefits—In the content of the coal operation statement of income, this term refers to all payments, including taxes, made by the company attributable to Black Lung.

Blackbody—An ideal substance that absorbs all radiation falling on it, and reflecting nothing.

Blackout—A power loss affecting many electricity consumers over a large geographical area for a significant period of time.

Blackwater—Water rich in humic acids and poor in nutrients. Found in tropical areas especially and supporting a distinct fish fauna.
- Water with human, animal and food wastes
- Blackwater is water discharged from toilets, urinals and kitchen sinks

Blade—The energy-capturing, aerodynamically designed part of a wind turbine, which interacts directly with the wind.

Blast Furnace—A furnace in which solid fuel (coke) is burned with an air blast to smelt ore.

Blast-furnace Gas—The waste combustible gas generated in a blast furnace when iron ore is being reduced with coke to metallic iron. It is commonly used as a fuel within steel works.

Bldg—Building

Blending Plant—A facility that has no refining capability but is either capable of producing finished motor gasoline through mechanical blending or blends oxygenates with motor gasoline.

BLM—Bureau of Land Management

Blocking Diode (Solar)—A diode used to prevent current flow in an undesirable direction e.g. from the rest of the PV array to a failed module or from the battery to the PV array when current generation is low. Alternatively, diode connected in series to a PV string; it protects its modules from a reverse power flow and, thus, against the risk of thermal destruction of solar cells.

Block-Rate Structure—An electric rate schedule with a provision for charging a different unit cost for various increasing blocks of demand for energy. A reduced rate may be charged on succeeding blocks.

Blowdown—Blowdown is the discharge of water from a boiler or a cooling tower sump that contains a high proportion of total dissolved solids.
- The process of reducing gas pressures by means of releasing such pressures to atmosphere.

Blower—The device in an air conditioner that distributes the filtered air from the return duct over the cooling coil/heat exchanger. This circulated air is cooled/heated and then sent through the supply duct, past dampers, and through supply diffusers to the living/working space.

Blower Door—A device used by energy auditors to pressurize a building to locate places of air leakage and energy loss.

Blown In Insulation—(See also *Loose Fill*) An insulation product composed of loose fibers or fiber pellets that are blown into building cavities or attics using special pneumatic equipment.

BLP—Buoyant Line and Point Source Model

BLS—Bureau of Labor Statistics within the U.S. Department of Labor.

Blue Sky Laws—A common term that refers to laws passed by various states to protect the public against securities fraud. The term originated when a judge ruled that a stock had as much value as a patch of blue sky.

BMD—Benchmark Dose

BMP—See *Best Management Practices.*

BMR—Baseline Monitoring Report

BNL—Brookhaven National Laboratory

BNRtm—Bureau of National Affairs www.bna.com

BOA—Basic ordering agreement

Board Feet (BF)—Unit of measure for logs and lumber. One board foot is equivalent to a piece of wood 1 inch thick, 12 inches wide, and 12 inches long.

BOD—See *Biochemical Oxygen Demand.*

BOD5—The amount of dissolved oxygen consumed in five days by biological processes breaking down organic matter. See *Biochemical Oxygen Demand.*

BOE—The abbreviation for barrels of oil equivalent (used internationally).

Boil Off—A natural phenomenon which occurs when liquefied natural gas in a storage vessel warms to its boiling point and gases evolve.

Boiler—A device for generating steam for power, processing, or heating purposes; or hot water for heating purposes or hot water supply. Heat from an external combustion source is transmitted to a fluid contained within the tubes found in the boiler shell. This fluid is delivered to an end-use at a desired pressure, temperature, and quality.

Boiler Efficiency—The ratio of the useful heat output to the heat input, multiplied by 100 and expressed in percent.

Boiler Feedwater—The water that is forced into a boiler to take the place of that which is evaporated in the generation of steam.

Boiler Fuel—An energy source to produce heat that is transferred to the boiler vessel in order to generate steam or hot water. Fossil fuel is the primary energy source used to produce heat for boilers.

Boiler Horsepower—A unit of rate of water evaporation equal to the evaporation per hour of 34.5 pounds of water at a temperature of 212 degrees Fahrenheit into steam at 212 degrees F.

Boiler Plate—Legal clauses routinely included in all contracts that, while important, have little to do with the actual substance of the contract.

Boiler Pressure—The pressure of the steam or water in a boiler as measured; usually expressed in pounds per square inch gauge (psig).

Boiler Rating—The heating capacity of a steam boiler; expressed in Btu per hour (Btu/h), or horsepower, or pounds of steam per hour.

Boiler-Packaged—A Packaged Boiler is a Boiler that is shipped complete with heating equipment, mechanical draft equipment and automatic controls. It is usually shipped in one or more sections.

Boiling-Water Reactor (BWR)—A light-water reactor in which water, used as both coolant and moderator, is allowed to boil in the core. The resulting steam can be used directly to drive a turbine.

BOM (United States)—Bureau of Mines. The primary United States government agency conducting scientific research and disseminating information on the extraction, processing, use, and conservation of mineral resources.

BOMA—Building Owners and Managers Association www.boma.org

Bona Fide Purchaser—One who buys property, in good faith, for a fair value, and without notice of adverse claims or rights of third Parties.

Bond—A bond is a negotiable note or certificate which evidences indebtedness. It is a legal contract sold by one party, the issuer, to another, the investor, promising to repay the holder the face value of the bond plus interest at future dates. Bonds are also referred to as notes or debentures.

Bonded Petroleum Imports—Petroleum imported and entered into Customs bonded storage. These imports are not included in the import statistics until they are: (1) withdrawn from storage free of duty for use as fuel for vessels and aircraft engaged in international trade; or (2) withdrawn from storage with duty paid for domestic use.

Bonded Warehouse—A warehouse authorized by Customs authorities for storage of goods on which payment of duties is deferred until the goods are removed.

Bone (Oven) Dry—In reference to solid biomass fuels, such as wood, having zero moisture content.

Bone Coal—Coal with a high ash content; it is dull in appearance, hard, and compact.

Bone Dry—Having zero percent moisture content. Wood heated in an oven at a constant temperature of 212 degrees F or above until its weight stabilizes is considered bone dry or oven dry.

Bone Dry Ton—(or "oven dry ton"). An amount of wood that weighs 2,000 pounds at zero percent moisture content.

Bone Dry Unit (BDU)—A quantity of wood residue which weighs 2,400 pounds at zero percent moisture content.

Bonus Depreciation—Extra or special depreciation allowed in addition to what is generally available to certain types of tangible, personal property including many types of property used to produce renewable energy.

BOO—Build, Own & Operate—type of project financing

Book Cost—The amount at which property is recorded in plant accounts without deduction of related provisions for accrued depreciation, depletion, amortization, or for other purposes.

Book Value—The portion of the carrying value (other than the portion associated with tangible assets) prorated in each accounting period, for financial reporting purposes, to the extracted portion of an economic interest in a wasting natural resource.
• Book value of a stock is determined from a company's balance sheet by adding all current and fixed assets and then deducting all debts, other liabilities, and the liquidation price of any preferred issues. The sum arrived at is divided by the number of common shares outstanding, and the result is book value per common share.
• The value at which an item is reported in the financial statements.

Book Value per Share of Common Stock—Common stock equity divided by the number of common shares outstanding at the date of the computation.

Booked Costs—Costs allocated or assigned to inter-departmental or intra-company transactions, such as on-system or synthetic natural gas (SNG) production and company-owned gas used in gas operations and recorded in company books or records for accounting and/or regulatory purposes.

Booster—A compressor used to raise pressure in a gas or oil pipeline.

Booster Pump—A pump for circulating the heat transfer fluid in a hydronic heating system.

Booster Station—A facility containing equipment which increases pressure on oil or gas in a pipeline.

Boot—In heating and cooling system distribution ductwork, the transformation pieces connecting horizontal round leaders to vertical rectangular stacks.

Borderline Customer—A customer located in the service area of one utility, but supplied by a neighboring utility through an arrangement between the utilities.

Boring—Usually, a vertical hole drilled into the ground from which soil samples can be collected and analyzed to determine the presence of chemicals and the physical characteristics of the soil.

Boron—The chemical element commonly, semi-metallic in nature, used as the dopant in solar photovoltaic device or cell material.

Boston Box—A square box installed flush with the pavement.

Bottled Gas—The liquefied petroleum gases propane and butane, contained under moderate pressure (about 125 pounds per square inch and 30 pounds per square inch respectively), in cylinders.

Bottled Gas, LPG, or Propane—Any fuel gas supplied to a building in liquid form, such as liquefied petroleum gas, propane, or butane. It is usually delivered by tank truck and stored near the building in a tank or cylinder until used.

Bottleneck Facility—A point on the system, such as a transmission line, through which electricity must pass to get to its intended buyer. If there is a limited capacity at this point, a capacity constraint, some priorities must be developed to decide whose power gets through and in what order. It also must be decided if the owner of the bottleneck may, or must, build additional facilities to relieve the constraint. This assumes that in a restructured electrical world the transmission and distribution functions would remain regulated monopolies, thereby removing the ability for competitors to build additional capacity.

Bottom Ash—Residue mainly from the coal burning process that falls to the bottom of the boiler or other combustion chamber.

Bottom Hole Contract—A contract providing for the payment of money or other consideration upon the drilling of a well to a specified depth.

Bottom-Cycle Plants—An energy system which produces heat first for process use and electricity as a by-product.

Bottom-hole Contribution—A payment (either in cash or in acreage) that is required by agreement when a test well is drilled to a specified depth regardless of the outcome of the well and that is made in exchange for well and evaluation data.

Bottoming Cycle—A means to increase the thermal efficiency of a steam electric generating system by converting some waste heat from the condenser into electricity. The heat engine in a bottoming cycle would be a condensing turbine similar in principle to a steam turbine but operating with a different working fluid at a much lower temperature and pressure.
• A cogeneration system in which steam is used first for process heat and then for electric power production.

Boule—A sausage-shaped, synthetic single-crystal mass grown in a special furnace, pulled and turned at a rate necessary to maintain the single-crystal structure during growth.

BOY—Beginning of Year Violator

Boycott—Absolute restriction against the purchase and importation of certain goods from other countries.

BOYSNC—Beginning of Year Significant Non-Compliers

bp—The abbreviation for boiling point.

BP or BPS—Basis point or basis points (100 bps = 1 percentage point in interest)

BPA—(Short for Bonneville Power Administration)— One of five federal power marketing administrations that sell low-cost electric power produced by federal hydro-electric dams to agricultural and municipal users. BPA serves Idaho, Oregon, and Washington as well as parts of Nevada and Wyoming. It also sells power to California companies in "wheeling" trades.
- Blanket Purchase Agreement

BPATT—Best Practicable Available Treatment Technology

BPCT—Best Practicable Control Technology

BPCTCA—Best Practicable Control Technology Currently Available

BPHE—Baseline Public Health Evaluation

BPJ—Best Professional Judgment

BPT—Best Practicable Treatment
- Best Practicable Control Technology

BPWTT—Best Practical Wastewater Treatment Technology

BRA—Baseline Risk Assessment

Bradenhead—A packer (or fitting) installed on a well at the surface that enables the use of one size pipe inside another, for the subsequent control of products being delivered from either one of the two pipes.

Brake—Device for stopping a wind turbine. This can be an electric brake that shorts the output of the turbine (dynamic braking), or a mechanical brake that physically stops the rotation, as with a brake drum and shoe.

Branded Product—A refined petroleum product sold by a refiner with the understanding that the purchaser has the right to resell the product under a trademark, trade name, service mark, or other identifying symbol or names owned by such refiner.

Brayton Cycle—A thermodynamic cycle using constant pressure, heat addition and rejection, representing the idealized behavior of the working fluid in a gas turbine type heat engine.

Bread Box System—This simple passive solar hot water system consists of one or more storage tanks placed in an insulated box that has a glazed side facing the sun. A bread box system is mounted on the ground or on the roof (make sure your roof structure is strong enough to support it). Some systems use "selective" surfaces on the tank(s). These surfaces absorb sun well but inhibit radiative loss. Also known as batch heaters or integral collector storage systems.

Breaker—A manually operable switching device that also automatically opens a circuit in the event of overcurrent.

Break-Even Cutoff Grade—The lowest grade of material that can be mined and processed considering all applicable costs, without incurring a loss or gaining a profit.

Breakeven Point—In real estate investment analysis, the point at which the cumulative income (effective gross income) of an investment property equals its cumulative loss (normal operating expenses plus debt service.

Breast Wall—Retaining wall, or parapet, that is breast-high from the floor.

Breccia—A coarse-grained clastic rock, composed of angular broken rock fragments held together by a mineral cement or in a fine-grained matrix.

Breeching—Breeching is the section of flue gas venting at the point where the gases exit a Boiler. The formal definition, used by heating plant professionals, is all flue gas venting from the Boiler outlet to the main vertical chimney.

Breeder Reactor—A reactor that both produces and consumes fissionable fuel, especially one that creates more fuel than it consumes. The new fissionable material is created by a process known as breeding, in which neutrons from fission are captured in fertile materials.

Breeze—The fine screenings from crushed coke. Usually breeze will pass through a 1/2-inch or 3/4-inch screen opening. It is most often used as a fuel source in the process of agglomerating iron ore.

BRI—Building Related Illness

Bridge Loan—Short-term, temporary financing used until permanent financing can be secured.

Brine—Water saturated or strongly impregnated with salt.
- A strong saline solution such as common salt and water cooled by a refrigerant and used for the transmission of heat without a change in its state, having no flash point or a flash point above 150 degrees Fahrenheit.

British Thermal Unit (Btu)—The standard measure of heat energy. It takes one Btu to raise the temperature of one pound of water by one degree Fahrenheit at sea level. For example, it takes about 2,000 Btu's to make a pot of coffee. One Btu is equivalent to 252 calories, 778 foot-pounds, 1055 joules, and 0.293 watt-hours. Note: In the abbreviation, only the B is capitalized.

Broadcast Burn—Controlled fire over the entire surface of a designated area.

Broker—A retail agent who buys and sells power. The agent may also aggregate customers and arrange for transmission, firming and other ancillary services as needed. A Broker acts as an intermediary between a buyer and a seller, usually charging a commission for the services provided.

Brown Power/Energy—Electricity generated from the combustion of nonrenewable fossil fuels (coal, oil, or natural gas) which generates significant amounts of greenhouse gases.

Brownfield—With certain legal exclusions and additions, the term "brownfield site" means real property, the expansion, redevelopment, or reuse of which may be complicated by the presence or potential presence of a hazardous substance, pollutant, or contaminant.

Brownout—A controlled power reduction in which the utility decreases the voltage on the power lines, so customers receive weaker electric current. Brownouts can be used if total power demand exceeds the maximum available supply. The typical household does not notice the difference.

BRS—Biennial Reporting System

Bryd-Hagel Resolution—In June 1997, anticipating the December 1997 meeting in Kyoto, Senator Robert C. Byrd (D-WV) introduced, with Sen. Chuck Hagel (R-NE) and 44 other cosponsors, a resolution stating that the impending Kyoto Protocol (or any subsequent international climate change agreement) should not—"(A) mandate new commitments to limit or reduce GHG emissions for the Annex I Parties [i.e. industrialized countries], unless the protocol or other agreement also mandates new specific scheduled commitments to limit or reduce GHG emissions for Developing Country Parties within the same compliance period, or (B) would result in serious harm to the economy of the United States..."

Bryophytes—Plants of the phylum Bryophyta, including mosses, liverworts and hornworts, characterized by the lack of true roots, stems and leaves.

BS—Bilateral Staff

BSAC—Biotechnology Science Advisory Committee

BSAF—Sediment Bioaccumulation Factor

BSI—British Standards Institute www.bsi-global.com

BTF—Beyond the Floor

BTL—Biomass-to-liquid

Btu—British Thermal Unit (Btu) is the standard unit of measure of heat energy. It takes one Btu to increase the temperature of one pound of water by one degree Fahrenheit. One Btu is equivalent to 252 calories and 0.293 watt-hours.

Btu Adjustment Clause—A clause in a gas purchase contract that may adjust the contract price if the heat content of the gas delivered does not fall within a specified range.

Btu Conversion Factors—Btu conversion factors for site energy are as follows:

- Electricity—3,412 Btu/kilowatt hour
- Natural Gas—1,031 Btu/cubic foot
- Fuel Oil No.1—15,000 Btu/gallon
- Kerosene—135,000 Btu/gallon
- Fuel Oil No.2—138,690 Btu/gallon
- LPG (Propane)—91,330 Btu/gallon
- Wood—20 million Btu/cord

Btu Method—A method of allocating costs between different operations or between different products based upon the heat content of products produced in the various operations or of the various produced products.

Btu per Cubic Foot—The total heating value, expressed in Btu, produced by the combustion, at constant pressure, of the amount of the gas that would occupy a volume of 1 cubic foot at a temperature of 60 degrees F if saturated with water vapor and under a pressure equivalent to that of 30 inches of mercury at 32 degrees F and under standard gravitational force (980.665 cm. per sec. squared) with air of the same temperature and pressure as the gas, when the products of combustion are cooled to the initial temperature of gas and air when the water formed by combustion is condensed to the liquid state. (Sometimes called gross heating value or total heating value.)

Btu, Dry—Heating value contained in cubic foot of natural gas measured and calculated free of moisture content. Contractually, dry may be defined as less than or equal to seven pounds of water per Mcf.

Btu, Saturated—he number of Btus contained in a cubic foot of natural gas fully saturated with water under actual delivery pressure, temperature and gravity conditions.

Btu/H—The amount of heat in Btu that is removed or added during one hour. Used for measuring heating and cooling equipment output.

BTX—The acronym for the commercial petroleum aromatics—benzene, toluene, and xylene. See individual categories for definitions.

BTZ—Below the Treatment Zone

BU—Bargaining Unit

Bubble—A regulatory concept whereby two or more emission points are treated as if they were under a hypothetical dome, thus creating a single emissions source. A bubble can involve two or more emission points within a facility, or can extend to two or more different facilities within a limited geographic area. This creates flexibility to apply pollution control technologies to whichever source under the bubble has the most cost effective pollution control options, while ensuring the total amount of emissions under the bubble meets the environmental requirements for the entity. Bubbles are closed systems. Article 4 of the Kyoto Protocol allows a bubble to be formed between Annex B countries, for example the European Union nations.

The generic concept of a "bubble" refers to the idea that emissions reductions anywhere within a specific area count towards a common reduction goal—as if a giant bubble were placed over the various sources to contain them in a common area.

Buck—To cut a log into smaller portions.

Buddy Swap—An arrangement whereby, during a period of severe curtailment, one industrial or commercial customer that can use an alternate fuel agrees to do so temporarily and transfers that part of his gas allocation to another customer that cannot use an alternate fuel.

Budget Plan—An agreement between the household and the utility company or fuel supplier that allows the household to pay the same amount for fuel for each month for a number of months.

Building Commissioning (Cx)—Commissioning is a risk reduction or quality assurance process for new construction projects that operates from pre-design to design, construction and operations. The purpose of commissioning is to ensure that all components of a building have been designed, installed and tested, and capable of being operated and maintained in conformity with the design intent.

Building Cooling Load—The hourly amount of heat that must be removed from a building to maintain indoor comfort and is measured in Btus.

Building Department—The city, county or state agency responsible for approving the plans, issuing a building permit and approving occupancy of the dwelling unit.

Building Efficiency—Building Efficiency relates to energy efficiency for heating, cooling, and lighting and the use of energy-saving appliances and equipment.

Building Energy Efficiency Standards—The California Building Energy Efficiency Standards as set—forth in the California Code of Regulations, Title 24, Part 6. Also known as the *California Energy Code.*

Building Energy Ratio—The space-conditioning load of a building.

Building Envelope—The outside of a building that contains the interior space, including the roof: the skin of waterproof covering of the structure.

• The assembly of exterior partitions of a building that enclose conditioned spaces, through which thermal energy may be transferred to or from the exterior, unconditioned spaces, or the ground. [See California Code of Regulations, Title 24, Section 2-5302]

Building Façade—See *Outdoor Lighting*

Building Heat-Loss Factor—A measure of the heating requirements of a building expressed in Btu per degree-day.

Building Location Data—The specific outdoor design temperatures shown in Joint Appendix II (Title 24 of California code) used in calculating heating and cooling loads for the particular location of the building.

For heating, the outdoor design temperature shall be the Winter Median of Extremes value. A higher temperature may be used, but lower values are not permitted.

For low-rise residential buildings for cooling, the outdoor design temperatures shall be the 1.0 percent Cooling Dry Bulb and Mean Coincident Wet Bulb values. Lower temperatures may be used, but higher values are not permitted. Temperatures are interpolated from the 0.5% and 2.0% values in the ASHRAE publication, Climatic Data for Region X, 1982 edition and 1994 supplement (See Joint Appendix II).

For nonresidential buildings, high-rise residential buildings and hotels/motels for cooling, the outdoor design temperatures shall be the 0.5 percent Cooling Dry Bulb and Mean Coincident Wet Bulb. For cooling towers the outdoor design temperatures shall be the 0.5 percent Cooling Design Wet Bulb values. Lower temperatures may be used, but higher values are not permitted.

If a building location is not listed, the local enforcement agency may determine the location for which outdoor design temperature data is available that is closest to the actual building site.

Building Orientation—The relationship of a building to true south, as specified by the direction of its longest axis.

Building Overall Energy Loss Coefficient-Area Product—The factor, when multiplied by the monthly degree-days, that yields the monthly space heating load.

Building Overall Heat Loss Rate—The overall rate of heat loss from a building by means of transmission plus infiltration, expressed in Btu per hour, per degree temperature difference between the inside and outside.

Building Permit—An electrical, plumbing, mechanical, building, or other permit or approval, that is issued by an enforcement agency, and that authorizes any construction that is subject to Part 6 of Title 24 of California code (or similar codes in other states).

Building Pressurization—The air pressure within a building relative to the air pressure outside. Positive building pressurization is usually desirable to avoid infiltration of unconditioned and unfiltered air. Positive pressurization is maintained by providing adequate outdoor makeup air to the HVAC system to compensate for exhaust and leakage.

Building Shell (envelope) DSM Program—A DSM program that promotes reduction of energy consumption through improvements to the building envelope. Includes installations of insulation, weather-stripping, caulking, window film, and window replacement. (Also see *DSM, Demand-Side Management Programs*.)

Building Shell Conservation Feature—A building feature designed to reduce energy loss or gain through the shell or envelope of the building. Data collected by EIA on the following specific building shell energy conservation features: roof, ceiling, or wall insulation; storm windows or double- or triple-paned glass (multiple glazing); tinted or reflective glass or shading films; exterior or interior shadings or awnings; and weather stripping or caulking. (See Roof or Ceiling Insulation, Wall Insulation, Reflective or Shading Glass or Film, Storm Windows or Triple-Paned Glass, Building Shell (Envelope), Exterior or Interior Shadings or Awnings, and Weather Stripping or Caulking.)

Building Types—The classification of buildings defined by the CBC and applicable to the requirements of the *Energy Efficiency Standards* (state of California).

Building-integrated PV (BIPV)—Used to describe a structure where PV replaces conventional materials and is integrated into the building. *See BIPV*

Built-in Electric Units—An individual-resistance electric-heating unit that is permanently installed in the floors, walls, ceilings, or baseboards and is part of the electrical installation of the building. Electric-heating devices that are plugged into an electric socket or outlet are not considered built in. (Also see *Heating Equipment*.)

Bulb—The transparent or opaque sphere in an electric light that the electric light transmits through.

Bulb Turbine—A type of hydro turbine in which the entire generator is mounted inside the water passageway as an integral unit with the turbine. These installations can offer significant reductions in the size of the powerhouse.

Bulk Charge—The initial phase of battery charging, when the largest amount of energy is put into the battery.

Bulk Density—The weight of a material per unit of volume compared to the weight of the same volume of water.

Bulk Energy Storage—Large-scale energy storage that is interconnected to the grid at transmission-level voltage, and is used primarily for electric supply capacity. Can be generator co-located (storage onsite combustion turbines, or stand-alone (compressed air energy storage, pumped hydro), or aggregated (large-scale aggregated battery storage interconnected at transmission level).

Bulk Power Market—Wholesale purchases and sales of electricity.

Bulk Power Supply—Often this term is used interchangeably with wholesale power supply. In broader terms, it refers to the aggregate of electric generating plants, transmission lines, and related-equipment. The term may refer to those facilities within one electric utility, or within a group of utilities in which the transmission lines are interconnected. Also known as wholesale power supply.

Bulk Power Transactions—The wholesale sale, purchase, and interchange of electricity among electric utilities. Bulk power transactions are used by electric utilities for many different aspects of electric utility operations, from maintaining load to reducing costs.

Bulk Sales—Wholesale sales of gasoline in individual transactions which exceed the size of a truckload.

Bulk Station—A facility used primarily for the storage

and/or marketing of petroleum products, which has a total bulk storage capacity of less than 50,000 barrels and receives its petroleum products by tank car or truck.

Bulk Terminal—A facility used primarily for the storage and/or marketing of petroleum products, which has a total bulk storage capacity of 50,000 barrels or more and/or receives petroleum products by tanker, barge, or pipeline.

Bullet Loan—Gap financing for leased-up properties used when the construction loan has expired and acceptable permanent financing has not yet been found—typically an interest-only loan for two to 10 years that cannot be prepaid.

BUN—Blood Urea Nitrogen

Bundled Lease—A lease that includes many additional services such as maintenance, insurance and property taxes that are paid for by the lessor, the cost of which is built into the lease payment.

Bundled Sales Service (Gas)—Natural gas sold on an as-needed basis, without prior scheduling, to the local distribution company at FERC-approved rates. Prior to implementation of various transportation programs, this constituted all gas delivered to an LDC.

Bundled Utility Service (Electric)—A means of operation whereby energy, transmission, and distribution services, as well as ancillary and retail services, are provided by one entity.

Bunker C Fuel Oil—A very heavy substance, left over after other fuels have been distilled from crude oil. Also called NO. 6 FUEL, it is used in power plants, ships and large heating installations. California's Bunker C fuel oil has high sulfur content, which causes air quality concerns when burned as fuel.

Bunker Fuels—Fuel supplied to ships and aircraft, both domestic and foreign, consisting primarily of residual and distillate fuel oil for ships and kerosene-based jet fuel for aircraft. The term "international bunker fuels" is used to denote the consumption of fuel for international transport activities. Note: For the purposes of greenhouse gas emissions inventories, data on emissions from combustion of international bunker fuels are subtracted from national emissions totals. Historically, bunker fuels have meant only ship fuel.

BUREC—Bureau of Reclamation. Second largest supplier of wholesale water and hydroelectric power in the American West. www.usbr.gov

Burn Days—The number of days the station could continue to operate by burning coal already on hand assuming no additional deliveries of coal and an average consumption rate.

Burn Out/Cram Down—Extraordinary dilution, by reason of a round of financing, of a non-participating investor's percentage ownership in the issuer.

Burn Rate—The rate at which a company expends net cash over a certain period, usually a month.

Burner Capacity—The maximum heat output (in Btu per hour) released by a burner with a stable flame and satisfactory combustion.

Burning Point—The temperature at which a material ignites.

Burnup—Amount of thermal energy generated per unit mass of fuel, expressed as Gigawatt-Days Thermal per Metric Ton of Initial Heavy Metal (GWDT/MTIHM), rounded to the nearest gigawatt day.

Burst Strength—The internal pressure required to cause a pipe or fitting to fail. NOTE: This pressure will vary with the rate of buildup of the pressure and the time during which the pressure is held.

Burst Test—Method of hydrostatic testing plastic pipe by a uniformly increasing internal pressure so that the pipe fails in 60 to 70 seconds. See ASTM D 1599. Also called quick burst test.

Bus—An electrical conductor that serves as a common connection for two or more electrical circuits.

Busbar—In electric utility operations, a busbar is a conductor that serves as a common connection for two or more circuits. It may be in the form of metal bars or high-tension cables.

Busbar Cost—The cost of producing electricity up to the point of the power plant busbar.

Business As Usual Scenario (BAU)—Estimate of a company's future and current emissions under normal operating circumstances. Depending on the scope of the business as usual scenario this may incorporate some emission reduction regulatory controls including carbon taxes, etc.

Business Judgment Rule—The legal principle that assumes the board of directors is acting in the best interests of the shareholders unless it can be clearly established that it is not. If the board was found to violate the business judgment rule, it would be in violation of its fiduciary duties to the shareholders.

Business Plan—A document that describes the entrepreneur's idea, the market problem, proposed solution, business and revenue models, marketing strategy, technology, company profile, competitive landscape, as well as financial data for coming years. The business plan opens with a brief executive summary, most probably the most important element of the document due to the time constraints of venture capital funds and angels.

Buss—An electrical connection component that can accept multiple cables or wires. Also bus, bus bar, or busbar (*see busbar*).

BUST—Bureau of Underground Storage Tanks

Butane (C_4H_{10})—A normally gaseous straight-chain or branch-chain hydrocarbon extracted from hydrocarbon extracted from natural gas or refinery gas streams. It includes isobutane and normal butane and is designated in ASTM Specification D1835 and Gas Processors Association Specifications for commercial butane. Butane turns into a liquid when put under pressure. It is sold as bottled gas. It is used to run heaters, stoves and motors, and to help make petrochemicals.

Butanol or Butyl Alcohol—An alcohol with the chemical formula $CH_3(CH_2)3OH$. It is formed during anaerobic fermentation using bacteria to convert the sugars to butanol and carbon dioxide.

Butt Log—The log taken from the base of a tree; often slightly irregular.

Butylene (C_4H_8)—An olefinic (class of unsaturated open-chain hydrocarbons) hydrocarbon recovered from refinery processes.

Buy Down—A payment of discounts/points in exchange for a lower rate of interest. It has the effect of providing the lender with a greater yield today in exchange for a lower yield in the future.

Buy Out—The amount a lessee must pay to the lessor to terminate a lease early. Usually calculated to include tax recaptures, unpaid property taxes and lost revenues.

Buy Through—An agreement between utility and customer to import power when the customer's service would otherwise be interrupted.

Buy-back Oil—Crude oil acquired from a host government whereby a portion of the government's ownership interest in the crude oil produced in that country may or should be purchased by the producing firm.

Buyer—An entity that purchases electrical energy or services from the Power Exchange (PX) or through a bilateral contract on behalf of end-use customers.
- A legally recognized entity (individual, corporation, not-for-profit organization or government, etc.) who acquires credits, reductions or allowances from another legally recognized entity through a purchase, lease, trade, or other means of transfer.

BVPS—Book Value Per Share

BW—Body Weight

BY—Budget Year

Bypass—An alternative path. In a heating duct or pipe, an alternative path for the flow of the heat transfer fluid from one point to another, as determined by the opening or closing of control valves both in the primary line and the bypass line.

Bypass Diode (Solar)—A diode connected across one or more solar cells in a photovoltaic module such that the diode will conduct if the cell(s) become reverse biased. [UL 1703] Alternatively, diode connected anti-parallel across a part of the solar cells of a PV module. It protects these solar cells from thermal destruction in case of total or partial shading of individual solar cells while other cells are exposed to full light.

Bypass Dust—Discarded dust from the bypass system dedusting unit of suspension preheater, precalciner and grate preheater kilns, consisting of fully calcined kiln feed material as is commonly found in cement plants.

Bypassed Footage—Bypassed footage is the footage in that section of hole that is abandoned as the result of remedial sidetrack drilling operations.

Byproduct—A secondary or additional product resulting from the feedstock use of energy or the processing of non-energy materials. For example, the more common byproducts of coke ovens are coal gas, tar, and a mixture of benzene, toluene, and xylenes (BTX).

C

C—Celsius

C Unit (CCF)—One hundred cubic feet of solid wood. Used as a log measure or as a measure of solid wood content. 1 CCF contains typically 1.4 BDT.

C&D—Construction and Demolition

C&I—Cost & Insurance. A pricing term indicating that the cost of the product and insurance is included in the quoted price. The buyer is responsible for freight to the named port of destination.

C_4H—A mixture of light hydrocarbons that have the general formula C_4H_n, where n is the number of hydrogen atoms per molecule. Examples include butane (C_4H_{10}) and butylene (C_4H_8).

CA—Citizen Act
- Competition Advocate
- Corrective Action
- Cooperative Agreement
- Carbon Absorber

CAA—Clean Air Act (1970, amended in 1977 and again in 1990) *see Clean Air Act*

- Compliance Assurance Agreement

CAAA—Clean Air Act Amendments (of 1990)

CAAAC—Clean Air Act Advisory Committee

CAAQS—California Ambient Air Quality Standards

CAASE—Computer Assisted Area Source Emissions

CAB—Civil Aeronautics Board

Cable Yarding—A term used to describe a means of removing logs from the stump area to a landing or yarding area through use of an overhead system of winch-driven cables to which logs are attached with cables.

Cabotage—Refers to the required use of domestic carriers for shipments in U.S. coastal waters.

CADER—Communities for Advanced Distributed Energy Resources: A nonprofit organization committed to advancing the successful use of highly efficient and environmentally responsible distributed energy resources into competitive energy markets. www.cader.org. www.cader.org

CADMAC—California Demand-Side Management Measurement Advisory Committee: An organization that provides a forum for regional and state energy efficiency programs using Public Goods Charge funds. Members include the PUC, the Office for Ratepayer Advocates, the California Energy Commission, Pacific Gas and Electric, Sempra, Southern California Edison and the National Resources Defense Council.

Cadmium (Cd)—A chemical element used in making certain types of solar cells and batteries. A natural element in the earth's crust, usually found as a mineral combined with other elements such as oxygen. Because all soils and rocks have some cadmium in them, it is extracted during the production of other metals like zinc, lead and copper. Cadmium does not corrode easily and has many uses. Cadmium salts are toxic in higher concentrations.

Cadmium Telluride (CdTe)—A polycrystalline thin-film photovoltaic material.

CAER—Community Awareness and Emergency Response Program
- Chemical Awareness and Emergency Response Program

CAFÉ—Corporate Average Fuel Economy

CAFM (Computer Aided Facility Management)—Using computer technology, clients are helped to improve the efficiency of their building operation and management, provide appropriate building management, and reduce LCC (life-cycle cost management) through facility management (in which a company or other organization implements, from an operational perspective, operational and managerial activities for an entire facility or its environment, for the purpose of comprehensive planning, management, and practical use.)

CAFO—Consent Agreement/Final Order

CAG—Cancer Assessment Group
- Carcinogenic Assessment Group
- Community Advisory Groups

CaGBC—Canada Green Building Council www.cagbc.org

Cage—The component of an electric motor composed of solid bars (of usually copper or aluminum) arranged in a circle and connected to continuous rings at each end. This cage fits inside the stator in an induction motor in channels between laminations, thin flat discs of steel in a ring configuration.

CAGR—Compound Annual Growth Rate. The year-over-year growth rate applied to an investment or other aspect of a firm using a base amount.

CAIDI—Customer Average Interruption Duration Index: The average length of an interruption, weighted by the number of customers affected, for customers affected during a specific time period. It is calculated by adding the customer-minutes off during each interruption and dividing the sum by the number of customers experiencing one or more sustained interruptions (greater than one minute) during the time period. The resulting unit is minutes.

CAIR—Comprehensive Assessment Information Rule

CAIRD—Cohort Analysis of Increased Risks Deaths Model

CAISO—California Independent System Operator. Coordinator of electricity distribution markets in real-time markets and the day-ahead integrated forward market.

Calcination—A process in which a material is heated to a high temperature without fusing, so that hydrates, carbonates, or other compounds are decomposed and the volatile material is expelled. The thermal decomposition of carbonate minerals, such as calcium carbonate (limestone) to form clinker.

Calcine—To heat a substance so that it oxidizes or reduces.

Calcium Sulfate—A white crystalline salt, insoluble in water. Used in Keene's cement, in pigments, as a paper filler, and as a drying agent.

Calcium Sulfite—A white powder, soluble in diluted sulfuric acid. Used in the sulfite process for the manufacture of wood pulp.

CALEP—California Association of Lighting Efficiency

Professionals. An association of companies specializing in energy-efficient lighting that works with the California Board for Energy Efficiency to turn the energy-efficiency marketplace into a market-driven industry.

CalEPA—California Environmental Protection Agency www.calepa.ca.gov

Calibrate—To ascertain, usually by comparison with a standard, the locations at which scale or chart graduations should be placed to correspond to a series of values of the quantity which the instrument is to measure, receive or transmit. Also, to adjust the output of a device, to bring it to a desired value, within a specified tolerance for a particular value of the input. Also, to ascertain the error in the output of a device by checking it against a standard.

Calibration—The set of specifications, including tolerances, unique to a particular design, version or application of a component or components assembly capable of functionally describing its operation over its working range.

Calibration Error—The difference between the pollutant concentration indicated by the measurement system and the known concentration of the test gas mixture.

California Climate Action Registry (CCAR)—A private non-profit organization originally formed by the State of California. The California Registry serves as a voluntary greenhouse gas (GHG) registry to protect and promote early actions to reduce GHG emissions by organizations. www.climateregistry.org

California Endangered Species Act—The state law originally enacted in 1970, expresses the state's concern over California's threatened wildlife, defined rare and endangered wildlife, and gave authority to the Department of Fish and Game to "identify, conserve, protect, restore, and enhance any endangered species or any threatened species and its habitat in California...." The statute is under the state Fish and Game Code as Chapter 1.5.

California Energy Commission—The state agency established by the Warren-Alquist State Energy Resources Conservation and Development Act in 1974 (Public Resources Code, Sections 25000 et seq.) responsible for energy policy (www.energy.ca.gov). The Energy Commission's five major areas of responsibilities are:
• Forecasting future statewide energy needs
• Licensing power plants sufficient to meet those needs

• Promoting energy conservation and efficiency measures
• Developing renewable and alternative energy resources, including providing assistance to develop clean transportation fuels
• Planning for and directing state response to energy emergencies

Funding for the Commission's activities comes from the Energy Resources Program Account, Federal Petroleum Violation Escrow Account and other sources

California Environmental Quality Act (CEQA)—(CEQA—pronounced See' quah) Enacted in 1970 and amended through 1983, established state policy to maintain a high-quality environment in California and set up regulations to inhibit degradation of the environment. The law requires that governmental decision-makers and public agencies study the significant environmental effects of proposed activities, and that significant avoidable damage be avoided or reduced where feasible. CEQA also requires that the public be told why the lead public agency approved the project as it did, and gives the public a way to challenge the decisions of the agency.

California ISO—California Independent System Operator

California Power Exchange—A State-chartered, non-profit corporation which provides day-ahead and hour-ahead markets for energy and ancillary services in accordance with the power exchange tariff. The power exchange is a scheduling coordinator and is independent of both the independent system operator and all other market participants.

California Public Utilities Commission (CPUC)—A state agency created by constitutional amendment in 1911 to regulate the rates and services of more than 1,500 privately owned utilities and 20,000 transportation companies. The CPUC is an administrative agency that exercises both legislative and judicial powers; its decisions and orders may be appealed only to the California Supreme Court. The major duties of the CPUC are to regulate privately owned utilities, securing adequate service to the public at rates that are just and reasonable both to customers and shareholders of the utilities; including rates, electricity transmission lines and natural gas pipelines. The CPUC also provides electricity and natural gas forecasting, and analysis and planning of energy supply and resources. Its main headquarters are in San Francisco. www.

cpuc.ca.gov.

California Utility Research Council (CURC)—Public Utilities Code, Sections 9201-9203 requires the California Energy Commission, the California Public Utilities Commission, and the investor-owned utilities (Pacific Gas and Electric Company, Southern California Edison, and San Diego Gas & Electric) to coordinate and promote consistency of research, development and demonstration (RD&D) programs with state energy policy. The CURC provides coordination for and sharing of information on energy RD&D in California to avoid duplication of efforts.

CALINE—California Line Source Model

Call Option—An option that gives the buyer the right, but not the obligation, to buy a futures contract or physical commodity or asset for a specific price within a specific time period in exchange for a one-time payment period. Should the option be exercised at the designated price, the seller is obligated to sell the futures contract or commodity at that price.

Call-Back—A provision included in some power sale contracts that let the supplier stop delivery when the power is needed to meet certain other obligations.

Calorie—The amount of heat required to raise the temperature of a unit of water, at or near the temperature of maximum density, one degree Celsius (or Centigrade [C]); expressed as a "small calorie" (the amount of heat required to raise the temperature of 1 gram of water one degree C), or as a "large calorie" or "kilogram calorie" (the amount of heat required to raise one kilogram [1,000 grams] of water one degree C); capitalization of the word calorie indicates a kilogram-calorie.

Calorific Value—The heat liberated by the combustion of a unit quantity of a fuel under specific conditions; measured in calories.

Calorimeter—An apparatus for measuring the amount of heat released by the combustion of a compound or mixture.

CALTRANS—California Department of Transportation www.dot.ca.gov

Calvert—An investment firm that highlights socially responsible investing and publishes an annual index of the largest U.S. companies that represent socially responsible investments. www.calvert.com

CAM—Compliance Assurance Monitoring

CAMEO—Computer Aided Management of Emergency Operations

CAMP—Continuous Air Monitoring Program

CAMU—Combined Area Management Units
- Corrective Action Management Unit

CAN—Common Account Number

Canadian Deuterium Uranium Reactor (CANDU)—Uses heavy water or deuterium oxide (D_2O), rather than light water (H_2O), as the coolant and moderator. Deuterium is an isotope of hydrogen that has a different neutron absorption spectrum from that of ordinary hydrogen. In a deuterium-moderated-reactor, fuel made from natural uranium (0.71 U-235) can sustain a chain reaction.

Cancer Risk—A number, generally expressed in exponential form (i.e., 1×10^{-6}, which means one in one million), which describes the increased possibility of an individual developing cancer from exposure to toxic materials. Calculations producing cancer risk numbers are complex and typically include a number of assumptions that tend to cause the final estimated risk number to be conservative.

Candela—The luminous intensity, in a given direction, of a source that emits monochromatic radiation of frequency 540×10^{12} hertz and that has a radiant intensity in that direction of 1/683 watt per steradian.

Candidate Species—See *Threatened, Endangered and Sensitive Species*.

Candle Power—The illuminating power of a standard candle employed as a unit for determining the illuminating quality of an illuminant.

Cannel Coal—A compact, tough variety of coal, originating from organic spore residues, that is non-caking, contains a high percentage of volatile matter, ignites easily, and burns with a luminous smoky flame.

CAO—Acronym for "Corrective Action Order"
- Contract Administration Office
- Control Area Operator: The operator of an electric power system bound by interconnection (tie-line) metering and telemetry. It controls generation to maintain its interchange schedule with other control areas, to maintain instantaneous load/resource balance within its system, and contributes to frequency regulation of the interconnection.

Cap—A contract between a buyer and seller that assures the buyer that he will not pay more than a stated price.
- A layer of clay, or other highly impermeable material installed over the top of a closed landfill to prevent entry of rainwater and minimize production of leachate.

- A limit on emissions.

CAP—Capacity
- Capacity Assurance Plan
- Compliance Audit Program
- Corrective Action Plan
- Cost Allocation Procedure
- Criteria Air Pollutant

Cap and Trade—The Cap and Trade system involves trading of emission allowances, where the total allowance is strictly limited or 'capped'. A regulatory authority established the cap which is usually considerably lower (50% to 85%) than the historic level of emissions. Allowances are created to account for the total allowed emissions (an allowance is a unit of measurement referred to as AAU). Trading occurs when an entity has excess allowances, either through actions taken or improvements made, and sells them to an entity requiring allowances because of growth in emissions or an inability to make cost-effective reductions. Cap and Trade programs are closed systems, but can be modified to allow the creations of new permits by non-capped sources in the manner of credit-based systems.
- A strategy to reduce carbon emissions or other criteria pollutants via financial incentives. "Caps" establish emissions limits and fines for exceeding those limits, while companies operating below their carbon limits can sell or "trade" their offsets to companies that are operating above the limits. Overall, the strategy reduces the emissions for all sources in a region over a period of time.
- An environmental regulatory program that limits (caps) the total emissions of a certain pollutant by issuing tradable allowances and requiring that allowances be surrendered to cover actual emissions. The limit on the number of tradable allowances issued ensures that emissions will not exceed the desired amount.

Capability—The maximum load that a generating unit, power plant, or other electrical apparatus can carry under specified conditions for a given period of time, without exceeding its approved limits of temperature and stress.

Capability Margin—The difference between net electrical system capability and system maximum load requirements (peak load); the margin of capability available to provide for scheduled maintenance, emergency outages, system operating requirements and unforeseen loads.

Capable of Being Fueled—A vehicle is capable of being fueled by a particular fuel(s) if that vehicle has the engine components in place to make operation possible on the fuel(s). The vehicle does not necessarily have to run on the fuel(s) in order for that vehicle to be considered capable of being fueled by the fuel(s). For example, a vehicle that is equipped to operate on either gasoline or natural gas but normally operates on gasoline is considered to be capable of being fueled by gasoline and natural gas.

Capacitance—A measure of the electrical charge of a capacitor consisting of two plates separated by an insulating material.

Capacitor—An electrical device that adjusts the leading current of an applied alternating current to balance the lag of the circuit to provide a high power factor.

Capacity—Electricity-The load carrying capability of generators, transmission or distribution lines, defined in megawatts. The maximum power that a machine or system can produce or carry safely. The maximum instantaneous output of a resource under specified conditions. The capacity of generating equipment is generally expressed in kilowatts or megawatts. Capacity is measured in megawatts and is also referred to as the Nameplate Rating.
- Battery—The measurement of an amount of energy a battery can provide in one discharge.
- Natural Gas—The transportation volume of gas pipelines, expressed in millions of cubic feet per day.

Capacity (Condensing Unit)—The refrigerating effect in Btu/h produced by the difference in total enthalpy between a refrigerant liquid leaving the unit and the total enthalpy of the refrigerant vapor entering it. Generally measured in tons or Btu/h.

Capacity (Effective, of a Motor)—The maximum load that a motor is capable of supplying.

Capacity (Heating, of a Material)—The amount of heat energy needed to raise the temperature of a given mass of a substance by one degree Celsius. The heat required to raise the temperature of 1 kg of water by 1 degree Celsius is 4186 Joules.

Capacity (Purchased)—The amount of energy and capacity available for purchase from outside the system.

Capacity Charge—An element in a two-part pricing method used in capacity transactions (energy charge is the other element). The capacity charge, sometimes called Demand Charge, is assessed on the amount of capacity being purchased.

Capacity Constraint—See Bottleneck Facility.

Capacity Factor—The ratio of the actual sales during any specified period to the maximum amount of sales the system is capable of delivering during that time.

• The ratio of the electrical energy produced by a generating unit for the period of time considered to the electrical energy that could have been produced at continuous full power operation during the same period. The amount of energy that a power plant actually generates compared to its maximum rated output, expressed as a percentage. Capacity factor = [kWh of electricity generated]/[Rated generating capacity (kW) X period (in hours)] For example, typical plant capacity factors range as high as 80 percent for geothermal and 70 percent for cogeneration. In another example, the capacity factor for a wind farm ranges from 20% to 35%. Thirty-five percent is close to the technology potential.

Capacity Rating—The capacity rating is a load that a power generation unit, such as a photovoltaic system, is rated by the manufacturer to be able to meet or supply. For a solar system, this will occur when the system is in direct sunlight with no shade.

Capacity Release—A secondary market for capacity that is contracted by a customer that is not using all of its capacity.

• A mechanism by which holders of firm interstate transportation capacity can relinquish their rights to utilize the firm capacity to other parties that are interested in obtaining the right to use that capacity for a specific price, for a given period of time and under a specifically identified set of conditions. The firm transportation rights may include transmission capacity and/or storage capacity.

Capacity Rights—Refers to the level of firm transportation service to which a customer has a contractual right.

Capacity Transaction—The acquisition of a specified quantity of generating capacity from another utility for a specified period of time. The utility selling the power is obligated to make available to the buyer a specified quantity of power.

Capacity Utilization—Capacity utilization is computed by dividing production by productive capacity and multiplying by 100.

Capacity Value—The avoided cost of new generation capacity that would otherwise be contracted or constructed to meet an incremental resource need.

Capacity, Effective—The maximum load which a machine, apparatus, device, plant, or system is capable of carrying under existing service conditions.

Capacity, Installed—The maximum load for which a machine, apparatus, device, plant, or system is designed or constructed, not limited by existing service conditions.

CAPCA—Carolina Air Pollution Control Association www.capca-carolinas.org

CAPCOA—California Air Pollution Control Officers Association www.capcoa.org

Capex—Capital expenditures

Capital—A term commonly used as a synonym for cash. Goods: material assets, equipment, machinery or tools. Funds: cash assets. The amount funded into a venture.

Capital Asset Recovery—A method to determine the cost of common equity component of return using the rate of risk- free investments plus a risk premium based on the stock market and the company's market volatility.

Capital Budgeting—The process of analyzing projects, such as the acquisition of new equipment, and deciding whether or not the revenue and/or cost savings generated by a specific project are sufficient to justify the costs of the project.

Capital Cost—The total investment needed to complete a project and bring it to a commercially operable status. The cost of construction of a new plant. The expenditures for the purchase or acquisition of existing facilities.

Capital Expenditure—Money spent for the purchase or expansion of plant or equipment. Can also include the cost of an improvement made to extend the useful life of a property or to add to its value, such as adding a room. The cost of repairing a property is not a capital expenditure. Capital expenditures are depreciated over their useful life and repairs are subtracted from income for the current year on the income statement to the company's financial statements.

Capital Gain—The amount by which the net proceeds from the resale of an asset exceed the adjusted cost basis, or book value, of the asset. Used primarily in income tax computations and may be classified as short- or long-term (see IRS Code).

Capital Impairment Rule—A requirement in most states limits the payment of cash dividends. This is to protect the claims of creditors in case of insolvency.

Capital Improvement—Expenditures that stop deterioration of property or add new improvements and appreciably prolong its life.

Capital Lease—From a financial reporting perspective, a lease that has the characteristics of a purchase agreement that is paid over time, and also meets certain criteria established by Financial Accounting Standards Board Statement Number 13 (FASB 13). Such a lease is required to be shown as an asset and a related debt obligation on the balance sheet of the lessee.

Capital Recovery Factor (CRF)—A factor used to convert a lump sum value to an annual equivalent.

Capital Stock (Energy)—Property, plant and equipment used in the production, processing and distribution of energy resources.

Capital Structure—The long-term debt and equity of a company. In ratemaking the capital structure is projected at the end of the test period (or when new rates are expected to go into effect) and used to determine the rate of return on rate base.

Capitalization—The conversion of income into value.

Capitalization Rate—Any rate used to convert income into value. Also known as Cap Rate.

Capitalization Table—Also called a "Cap Table," this is a table showing the total amount of the various securities issued by a firm. This typically includes the amount of investment obtained from each source and the securities distributed—e.g., common and preferred shares, options, warrants, etc.—and respective capitalization ratios.

Capitalize—To record an expenditure that may benefit future periods as an asset rather than as an expense to be charged off in the period of its occurrence or when paid.

• In standard contracts of leasing and PPA companies, the asset is the property of the lessor or PPA energy provider in economic, legal and tax respects. They capitalize it as an asset so it will be subject to depreciation and amortization for tax purposes. The debtor (lessee or host customer) does not capitalize the asset. If, however, the asset is attributed to the debtor, i.e. barring the existence of criteria in the corresponding lease or PPA remission in the contract, the debtor/host customer capitalizes the object. The lessor or PPA provider must then enter the corresponding sales price in their accounts.

Capitalized Cost—The amount of an asset to be shown on the balance sheet, from a financial reporting perspective. The total capitalized cost is also known as the "basis." It is the amount upon which tax benefits, such as MACRS depreciation deductions, are based and may include asset costs plus other defined amounts. See *basis*.

CAPM—Capital Asset Pricing Model

CAPNLOIS—Corrective Action Plans and Loss of Interim Status

Captive Coal—Coal produced to satisfy the needs of the mine owner, or of a parent, subsidiary, or other affiliate of the mine owner (for example, steel companies and electricity generators), rather than for open market sale.

Captive Customer—A customer who does not have realistic alternatives to buying power from the local utility, even if that customer had the legal right to buy from competitors. Captive electricity customers are generally considered to be the residential customers. The commercial and industrial customers, in contrast, are thought to be more mobile. This mobility, or lack thereof, relates to the restructuring debate since the larger customers can threaten to leave the area (causing higher rates as fewer customers share the bill for fixed or sunken costs) or are able to win greater concessions in a negotiated process through their buying power.

Captive Electrolyte Battery—A battery with an immobilized electrolyte (gelled or absorbed in a material).

Captive Lessor—A leasing company that has been set up by a manufacturer or equipment dealer to finance the sale or lease of its own products to end-users or lessees.

Captive Refinery MTBE Plants—MTBE (methyl tertiary butyl ether) production facilities primarily located within refineries. These integrated refinery units produce MTBE from Fluid Cat Cracker isobutylene with production dedicated to internal gasoline blending requirements. MTBE reduces emissions from vehicles, but can greatly increase contamination in water tables.

Captive Refinery Oxygenate Plants—Oxygenate production facilities located within or adjacent to a refinery complex.

Captive-Key Override—A type of lighting control in which the key that activates the override cannot be released when the lights are in the on position.

CAR—Corrective Action Report

CARB—California Air Resources Board www.arb.ca.gov

Carbamates—A group of insecticides related to carbamic acid. They are primarily used on corn, alfalfa, tobacco, cotton, soybeans, fruits and ornamental plants.

Carbohydrate—Chemical compound made up of carbon, hydrogen, and oxygen. Includes sugars, cellulose, and starches.

Carbon Adsorption—An environmental treatment system in which organic contaminants are removed from groundwater and surface water by forcing it through tanks containing activated carbon, a specially-treated material that retains such compounds. Activated carbon is also used to purify contaminated air by adsorbing the contaminants as the air passes through it.

Carbon Black—An amorphous form of carbon, produced commercially by thermal or oxidative decomposition of hydrocarbons and used principally in rubber goods, pigments, and printer's ink.

Carbon Budget—The balance of the exchanges (incomes and losses) of carbon between carbon sinks (e.g., atmosphere and biosphere) in the carbon cycle. Also see *Carbon Cycle* and *Carbon Sink* below.

Carbon Cycle—The natural processes that influence the exchange of carbon (in the form of carbon dioxide, carbonates and organic compounds, etc.) among the atmosphere, ocean and terrestrial systems. Major components include photosynthesis, respiration and decay between atmospheric and terrestrial systems (approximately 100 gigatons/year); thermodynamic invasion and evasion between the ocean and atmosphere, operation of the carbon pump and mixing in the deep ocean (approximately 90 gigatons/year). Deforestation and fossil fuel burning releases approximately 7Gt into the atmosphere annually. The total carbon in the reservoirs is approximately 2,000 Gt in land biota, soil and detritus, 750 Gt in the atmosphere and 38,000 Gt in the oceans. All carbon sinks and exchanges of carbon from one sink to another by various chemical, physical, geological, and biological processes. Also see *Carbon Sink* below.

Carbon Dioxide (CO_2)—A colorless, odorless, non-poisonous gas that is a normal part of Earth's atmosphere. Carbon dioxide is a product of fossil-fuel combustion as well as other processes. It is considered a greenhouse gas as it traps heat (infrared energy) radiated by the Earth into the atmosphere and thereby contributes to the potential for global warming. The global warming potential (GWP) of other greenhouse gases is measured in relation to that of carbon dioxide, which by international scientific convention is assigned a value of one (1). Also see *Global Warming Potential (GWP)* and *Greenhouse Gases*. A product of combustion. It is the most common of the six primary greenhouse gases, consisting of a single carbon atom and two oxygen atoms.

Carbon Dioxide Equivalent (CO_2 **e**)—The amount of carbon dioxide by weight emitted into the atmosphere that would produce the same estimated radiative forcing as a given weight of another radiatively active gas. Carbon dioxide equivalents are computed by multiplying the weight of the gas

being measured (for example, methane) by its estimated global warming potential (which is 21 for methane). "Carbon equivalent units" are defined as carbon dioxide equivalents multiplied by the carbon content of carbon dioxide (i.e., 12/44).

The universal unit of measurement used to indicate the global warming potential (GWP) of each of the 6 greenhouse gases. It is used to evaluate the impacts of releasing (or avoiding the release of) different greenhouse gases. Carbon dioxide equivalents are commonly expressed as "million metric tons of carbon dioxide equivalents (MMTCO$_2$ e)."

Green House Gas	Global Warming Potential (GWP)
CO_2	1
Methane	21
Nitrous Oxide	310
HVC-134a	1,300
HFC-23	11,700
HFC-152a	140
HFC-125	2,800
SF_6	23,900

Carbon Emission Inventory—The process of creating an inventory of the air pollutants released by an entity or community into the atmosphere over a finite period of time.

Carbon Flux—See *Carbon Budget* above.

Carbon Footprint—The total amount of greenhouse gases (GHG) emitted directly or indirectly through any human activity, typically expressed in equivalent tons of either carbon or carbon dioxide.

The global warming impact of human activities in terms of the amount of greenhouse gases they produce. The emissions associated with the use of power, transport, food and other consumption for an individual, family or organization are added up to give one comparable measure in units of carbon dioxide equivalent.

Carbon Intensity—The amount of carbon by weight emitted per unit of energy consumed. A common measure of carbon intensity is weight of carbon per British thermal unit (Btu) of energy. When there is only one fossil fuel under consideration, the carbon intensity and the emissions coefficient are identical. When there are several fuels, carbon intensity is based on their combined emissions coefficients weighted by their energy consumption levels. Also see *Emissions Coefficient and Carbon Output Rate.*

Carbon Monoxide (CO)—A colorless, odorless, highly poisonous gas made up of carbon and oxygen molecules formed by the incomplete combustion of carbon or carbonaceous material, including gasoline. It is a major air pollutant on the basis of weight.

Carbon Neutral—An individual, family or organization that is responsible for no net emissions of greenhouse gases from all its activities is considered "carbon neutral." Emissions must be cut to a minimum and any necessary emissions then offset by emission reducing activities elsewhere. Buying accredited clean electricity helps cut household or office greenhouse emissions, while investing in sustainable energy projects or afforestation schemes are examples of offsets.

Carbon Output Rate—The amount of carbon by weight per kilowatt-hour of electricity produced.

Carbon Positive—An individual, family or organization that is responsible for taking more greenhouse gases out of the atmosphere than it emits is said to be "carbon positive." This requires minimizing one's own emissions and more than offsetting remaining emissions by paying for activities such as forest planting or investing in renewable energy.

Carbon Price—An economic value placed on the emission of greenhouse gases into the atmosphere from human activity. This price is designed to create a disincentive for emissions and incentive to avoid them. A carbon price takes the form of either a carbon tax or an emissions trading scheme.

Carbon Sequestration—The long-term storage of carbon CO_2 in the forests, soils, ocean or underground in depleted oil and gas reservoirs, coal seams and saline aquifers. Examples include the separation and disposal of CO_2 fuel gases or processing fossil fuels to produce H_2O- and CO_2-rich fractions, and the direct removal of CO_2 from the atmosphere through land use change, afforestation, reforestation, ocean fertilization and agricultural practices to enhance soil carbon.

It refers to projects that capture and store carbon in a manner that prevents it from being released into the atmosphere for a specified period of time, the storage area is commonly referred to as a carbon sink.

The absorption and storage of carbon dioxide from the atmosphere. Naturally occurring in plants.

In the greenhouse gas program, a concept that refers to capturing carbon and keeping it from entering the atmosphere for some period of time. Carbon is sequestered in carbon sinks such as forests, soils, or oceans.

Carbon Sequestration projects include:
* Forest Sequestration
* Land Conservation
* Soil Conservation & Land Use
* Waste CO_2 Recovery / Deep Injection

Carbon Sequestration and Storage—Carbon capture and storage (CCS) (or carbon capture and sequestration) is the new emerging technology to process or capture waste carbon dioxide (CO_2) from large point sources, such as fossil fuel power plants, transporting it to a storage site, and depositing it where it will not enter the atmosphere, normally an underground geological formation. The aim is to prevent the release of large quantities of CO_2 into the atmosphere (from fossil fuel use in power generation and other industries). It is a potential means of mitigating the contribution of fossil fuel emissions to global warming and ocean acidification.

Carbon Sink—A reservoir that absorbs or takes up released carbon from another part of the carbon cycle. The four sinks, which are regions of the Earth within which carbon behaves in a systematic manner, are the atmosphere, terrestrial biosphere (usually including freshwater systems), oceans, and sediments (including fossil fuels). Forests are the most common form of sink, as well as soils, peat, permafrost, ocean water and carbonate deposits in the deep ocean.

Carbon Source—A pool (reservoir) that gives up carbon to another reservoir within the Carbon Cycle. For example, if the net exchange between the Biosphere and the Atmosphere is toward the ocean, then the atmosphere is the source. Common human sources include: fossil fuel combustion, solid waste decomposition, land use change, and transport.

Carbon Stocks—Carbon stocks include carbon stored in vegetation (above and below ground), decomposing matter, soils, wood products, and the carbon substituted by burning wood for energy instead of fossil fuels.

Carbon Tariff—Import duty levied by countries with greenhouse gas emission caps in place on carbon-intensive goods from countries without such controls in place. The intention is to protect the competiveness of local industries whose goods have higher prices than their imported rivals because they reflect the cost of carbon.

Carbon Tax—Policy instrument used to discourage the use of fossil fuels and reduce CO_2 emissions by placing a surcharge on carbon content in oil, coal and gas. A surcharge or levy on the carbon content of oil, coal, and/or gas to discourage the use of fossil fuels, with the aim of reducing carbon dioxide emissions.

One form of carbon price on greenhouse gas emissions. Set by governments, a price on emissions is fixed and emitters are allowed to emit whatever they want at that price. Emissions trading prices carbon in the reverse approach; fixing emissions, with price varying.

Carbon Trading—A trading system for countries, companies and individuals designed to offset carbon emissions from one activity with another, whereby those who cannot meet their emissions goals may purchase credits from those who surpass their goals.

Carbon Zinc Cell Battery—A cell produces electric energy by the galvanic oxidation of carbon; commonly used in household appliances.

Carbon/Hydrogen Ratio—The ratio, either on a weight or on a molecular basis, of carbon-to-hydrogen in a hydrocarbon material. Materials with a high carbon/hydrogen ratio (e.g., coal) are solid. The ratio is useful as a preliminary indication of the hydrogen quantity needed to convert the hydrocarbon to a gas and/or liquid.

Carbonaceous Material—A material which contains carbon.

Carburetor—A fuel delivery device for producing a proper mixture of gasoline vapor and air and for delivering it to the intake manifold of an internal combustion engine. Gasoline is gravity-fed from a reservoir bowl into a throttle bore, where it is allowed to evaporate into the stream of air being inducted by the engine. Also see *Diesel Fuel System* and *Fuel Injection*.

Carcinogens—Potential cancer-causing agents in the environment. They include among others: industrial chemical compounds found in food additives, pesticides and fertilizers, drugs, toy, household cleaners, toiletries and paints. Naturally occurring ultraviolet solar radiation is also a carcinogen.

CARES—Conservation and Renewable Energy System

Carnot Cycle—An ideal heat engine (conceived by Sadi Carnot) in which the sequence of operations forming the working cycle consists of isothermal expansion, adiabatic expansion, isothermal compression, and adiabatic compression back to its initial state.

Carry Forward (Tax)—The credit accruing during a period of alternative minimum tax.

Carrying Costs—Costs incurred in order to retain exploration and property rights after acquisition but before production has occurred. Such costs include legal costs for title defense, ad valorem taxes on non-producing mineral properties, shut-in royalties, and delay rentals.

Carryover—A term describing the postponing of tax losses until they can be used. Examples include net operating loss and investment tax credit carryovers.

CARS—Corrective Action Reporting Systems

CAS—Chemical Abstract Service

CASAC—Clean Air Scientific Advisory Committee

Cascade Aeration—Aeration of an effluent stream through the action of falling water.

Cascade Cycle—A liquefaction process in which a series of refrigerants are used to obtain successively lower temperatures.

CASE—Commission for Additional Sources of Energy (India)

CASEREP—Field Office Inspection Data Base

CASETRK—FIFRA and TSCA Cast Tracking System

Cash and Carry—Kerosene, fuel oil, or bottled gas (tank or propane) purchased with cash, by check, or by credit card and taken home by the purchaser. The purchaser provides the container or pays extra for the container.

Cash Budget—A plan or projection of cash receipts and disbursements for a given period of time.

Cash Flow—Describes the amount of dollars in and out of a business or project during a stated period of time. From a credit perspective, cash flow is a critical indicator how well a company can meet its immediate payment obligations. Cash flow can be negative or positive.

Cash Flow Analysis—A study of the anticipated movement of cash into or out of an investment.

Cash Flow Test (Tax)—The IRS test which evaluates the rate at which free cash is received during the term of the lease.

Cash Incentive—An incentive in the form of a rebate or cash payment that is used to induce customers to participate in a DSM program.

Cash Position—The amount of cash available to a company at a given point in time.

Cash Value—The current value of a future cash flow (maturing receivables, etc.). The cash value is determined by discounting the interest accruing until the maturity date, i.e. of a receivable, the redemption amount (cash flow) and the refinancing costs. The cash value amount can be transposed to any point in time. This can occur both by adding on interest (future cash value date) or discounting (past cash value date).

Cash-on-Cash yield—The relationship, expressed as a percentage, between the net cash flow of a property and the average amount of invested capital during an operating year.

Casinghead Gas (or oil well gas)—Natural gas produced along with crude oil from oil wells. It contains either dissolved or associated gas or both.

CASLP—Conference on Alternative State and Local Practices

CASRN—Chemical Abstract Services Registry Number

Cast Silicon—Crystalline silicon obtained by pouring pure molten silicon into a vertical mold and adjusting the temperature gradient along the mold volume during cooling to obtain slow, vertically advancing crystallization of the silicon. The polycrystalline ingot thus formed is composed of large, relatively parallel, interlocking crystals. The cast ingots are sawed into wafers for further fabrication into photovoltaic cells. Cast silicon wafers and ribbon silicon sheets fabricated into cells are usually referred to as polycrystalline photovoltaic cells.

CASTNET—Clean Air Status and Trends Network

CASU—Cooperative Administrative Support Units

Casualty Value—A schedule included in a lease (or other financial contract) that establishes the liability of the lessee to the lessor in the event the leased equipment is lost or rendered unusable during the lease term because of casualty loss. The casualty value is the amount that maintains the lessor's yield in the event of casualty.

Catalyst—A material, such as platinum, which promotes or increases the rate of a chemical reaction without itself undergoing any permanent chemical change. A substance added to a chemical reaction, which facilitates or causes the reaction, and is not consumed by the reaction.

Catalyst Coke—In many catalytic operations (e.g., catalytic cracking), carbon is deposited on the catalyst, thus deactivating the catalyst. The catalyst is reactivated by burning off the carbon, which is used as a fuel in the refining process. This carbon or coke is not recoverable in a concentrated form.

Catalytic Converter—An air pollution control device that removes organic contaminants by oxidizing them into carbon dioxide and water through a chemical reaction using a catalysis, which is a substance that increases (or decreases) the rate of a chemical reaction without being changed itself; required in all automobiles sold in the United States, and used in some types of heating appliances.

Catalytic Cracking—The refining process of breaking down the larger, heavier, and more complex hydrocarbon molecules into simpler and lighter molecules. Catalytic cracking is accomplished by the use of a catalytic agent and is an effective process for increasing the yield of gasoline from crude oil. Catalytic cracking processes fresh feeds and recycled feeds.

Catalytic Hydrocracking—A refining process that uses hydrogen and catalysts with relatively low temperatures and high pressures for converting middle boiling or residual material to high octane gasoline, reformer charge stock, jet fuel, and/or high grade fuel oil. The process uses one or more catalysts, depending on product output, and can handle high sulfur feedstock without prior de-sulfurization.

Catalytic Hydrotreating—A refining process for treating petroleum fractions from atmospheric or vacuum distillation units (e.g., naphthas, middle distillates, reformer feeds, residual fuel oil, and heavy gas oil) and other petroleum (e.g., cat cracked naphtha, coker naphtha, gas oil, etc.) in the presence of catalysts and substantial quantities of hydrogen. Hydrotreating includes de-sulfurization, removal of substances (e.g., nitrogen compounds) that deactivate catalysts, conversion of olefins to paraffins to reduce gum formation in gasoline, and other processes to upgrade the quality of the fractions.

Catalytic Reforming—A refining process using controlled heat and pressure with catalysts to rearrange certain hydrocarbon molecules, thereby converting paraffinic and naphthenic type hydrocarbons (e.g., low octane gasoline boiling range fractions) into petrochemical feedstock and higher octane stocks suitable for blending into finished gasoline. Catalytic reforming is reported in two categories. They are:

• Low Pressure. A processing unit operating at less than 225 pounds per square inch gauge (PSIG) measured at the outlet separator.

• High pressure. A processing unit operating at either equal to or greater than 225 pounds per square inch gauge (PSIG) measured at the outlet separator.

CATC—Clean Air Technology Center www.epa.gov/ttncatc1

Catenary Support—A steel cable or cables strung between two supports and sagged to the point of minimum tension for a given evenly distributed load. Used to support a heavy electrical cable, gas main or other load which is not designed to be self-supporting and is too heavy to be carried by a normal messenger wire strung with minimum sag.

Cathedral Ceiling/Roof—A type of ceiling and roof assembly that has no attic.

Cathode—The negative pole or electrode of an electrolytic cell, vacuum tube, etc., where electrons enter (current leaves) the system; the opposite of an anode.

Cathode Disconnect Ballast—An electromagnetic ballast that disconnects a lamp's electrode heating circuit once is has started; often called "low frequency electronic" ballasts.

Cathodic Protection—A method of preventing oxidation (rusting) of exposed metal structures, such as bridges and pipelines, by imposing between the structure and the ground a small electrical voltage that opposes the flow of electrons and that is greater than the voltage present during oxidation.

CATS—Corrective Action Tracking System

CAU—Carbon Adsorption Unit
- Command Arithmetic Unit

Caulking—Material used to make an air-tight seal by filling in cracks, such as those around windows and doors.

Cause-related Marketing—A business strategy whereby a company aligns its mission and goals to create a specific and tailored partnership with a nonprofit organization or cause.

Caustic—The common name for sodium hydroxide, a strong base. Also used as an adjective to describe highly corrosive bases.

CAV—Constant Air Volume

CBA—Cost Benefit Analysis

CBC—CBC is the 2001 California Building Code
- Complete Blood Count

CBD—Central Business District
- Commerce Business Daily

CBEC—Concentration-Based Exemption Criteria

CBECS—Commercial Building Energy Consumption Survey

C-BED—Community-Based Energy Development. Initiatives set forth by various jurisdictions to optimize the benefits derived from renewable energy development and to encourage and facilitate further development and investment in such renewable energy technologies.

CBEP—Community Based Environmental Protection

CBER—Center for Biologics Evaluation and Research

CBI—Compliance Biomonitoring Inspection
- Confidential Business Information

CBO—Congressional Budget Office
- Collateralized Bond Obligation
- Community-based Organizations: Organizations dedicated to bettering the community through job training, public program information

and assistance, and other help.

CBOD—Carbonacious Biochemical Oxygen Demand

CBPD—Carnegie Mellon University—Center for Building Performance and Diagnostics www.arc.cmu.edu/cbpd

CBT—Computer Based Training

CC List—Contaminant Categories for Land Disposal Restrictions

CC/RTS—Chemical Collection System

CCA—Competition in Contracting Act
- Community Choice Aggregators: A community organization of electricity customers who group together to solicit bids, broker, and contract for electricity and energy services to facilitate the sale and purchase of electricity and other related services on behalf of local citizens, businesses, and itself.

CCAA—Canadian Clean Air Act

CCAI—Coalition for Clean Air Implementation

CCAP—Center for Clean Air Policy
- Climate Change Action Plan

CCB—Continuous Calibration Blank

CCC—Cash Conversion Cycle

CCD—Chemical Control Division

CCEA—Conventional Combustion Environmental Assessment

CCF—One hundred cubic feet. A unit of water; one cf equals 748 gallons.

CCH—Commerce Clearing House

CCHW—Citizens' Clearinghouse for Hazardous Wastes

CCID—Confidential Chemicals Identification Systems

CCL—Contaminant Candidate List
- Construction Completion List

CCM—Certified Construction Manager

CCP—Composite Correction Plan

CCPS—Center for Chemical Process Safety

CCR—California Code of Regulations.

CCS—Carbon capture and storage. A two-step measure to prevent carbon dioxide being emitted to the atmosphere, particularly from power generation and industrial processes. Instead of venting CO_2, it is contained and pumped underground under pressure and sealed off, where it cannot contribute to global warming. This technology is still in its infancy with results largely unproven. Also known as carbon sequestration.

CCTP—Clean Coal Technology Program

CCTV—Closed Circuit Television

CCU—Correspondence Control Unit

CCV—Continuous Calibration Verification

CD—Calibration Drift
- Certification Division

- Climatological Data
- Compliance Division
- Consent Decree

CDB—Waste Management Data Base System

CDBA—Central Data Base Administrator

CDBG—Community Development Block Grant

CDC—Centers for Disease Control & Prevention

CDE—Control Device Evaluation

CDETS—Consent Decree Tracking System

CDF—California Department of Forestry

CDFG—California Department of Fish & Game www.dfg.ca.gov

CDHS—Comprehensive Data Handling system

CDI—Case Development Inspection
- Chronic Daily Intake

CDM—Clean Development Mechanism
- A Kyoto Protocol initiative under which projects set up in developing countries to reduce greenhouse gas emissions generate tradable credits called CERs, the first step towards a global carbon market. These credits can be used by industrialized nations to offset carbon emissions at home and meet their Kyoto reduction targets. The projects include renewable energy generation, reforestation and clean fuels switching.
- Climatological Dispersion Model
- Comprehensive Data Management

CDMQC—Climatological Dispersion Model with Calibration and Source Contribution

CDMS—Cost Development Management System

CDNS—Climatological Data National Summary

CDP—CDP works to accelerate solutions to climate change and water management by putting relevant information at the heart of business, policy and investment decisions. www.cdproject.net
- Census Designated Places
- Common Depth Point

CDRL—Contract Data Requirements List

CDS—Compliance Data System

CdTe—Cadmium Telluride

CE—Calibration Error
- Categorical Exclusion
- Conditionally Exempt Generator
- Cost Effectiveness
- Conditioning Equipment: Equipment modifications or adjustments necessary to match transmission levels and impedances and which equalize transmission and delay to bring circuit losses, levels, and distortion within established standards.

CEA—Canadian Electricity Association www.electricity.ca

- Cooperative Enforcement Agreement
- Cost and Economic Assessment
- Council of Economic Advisors

CEAM—Center for Exposure Assessment Modeling

CEAS—Coastal Environmental Assessment Studies

CEAT—Contractor Evidence Audit Team

CEC—California Energy Commission www.energy.ca.gov
- Commission for Environmental Cooperation www.cec.org
- Commission of European Communities

CEC-AC—California Energy Commission Alternating Current—refers to inverter efficiency rating.

CEC-AC Rating—The calculation that provides a total estimated energy output of a solar system, factoring in the efficiency of the inverter.

CED—Criminal Enforcement Division
- California Energy Demand

CEEM—Center for Energy and Environmental Management

CEERT—Center for Energy Efficiency and Renewable Technologies. Nonprofit organization of concerned scientists, environmentalists, public interest advocates and innovative technology companies involved in developing innovative energy technologies. www.ceert.org

CEG—Community Exposure Guidelines
- Conditionally Exempt Generator

CEI—Compliance Evaluation Inspection
- Comprehensive Emissions Inventory

Ceiling—The interior upper surface of a space separating it from an attic, plenum, indirectly or directly conditioned space or the roof assembly, which has a slope less than 60 degrees from horizontal.

Ceiling Fan—A mechanical device used for air circulation and to provide cooling.

Cell—A component of a electrochemical battery. A 'primary' cell consists of two dissimilar elements, known as 'electrodes,' immersed in a liquid or paste known as the 'electrolyte.' A direct current of 1-1.5 volts will be produced by this cell. A 'secondary' cell or accumulator is a similar design but is made useful by passing a direct current of correct strength through it in a certain direction. Each of these cells will produce 2 volts; a 12 volt car battery contains six cells.
- The basic unit of a photovoltaic system.

Cell Barrier—A very thin region of static electric charge along the interface of the positive and negative layers in a photovoltaic cell. The barrier inhibits the movement of electrons from one layer to the other, so that higher-energy electrons from one side diffuse preferentially through it in one direction, creating a current and thus a voltage across the cell. Also called depletion zone or space charge.

Cell Efficiency—The ratio of the electrical energy produced by a photovoltaic cell (under full sun conditions or 1 kW/m2) to the energy from sunlight falling upon the photovoltaic cell.

Cell Junction (Solar)—The area of immediate contact between two layers (positive and negative) of a photovoltaic cell. The junction lies at the center of the cell barrier or depletion zone.

Cell Temperature Sensor—Used to measure surface temperature, such as for a solar module. This can be used on any other surface, such as a water pipe. Consists of a K-type thermocouple junction encased in a flat pad that can be attached to any flat, clean surface where a surface temperature measurement is desired.

Cells—Refers to the un-encapsulated semi-conductor components of the module that convert the solar energy to electricity.

Cells to OEM (Non-PV)—Cells shipped to non-photovoltaic original equipment manufacturers such as boat manufacturers, car manufacturers, etc.

Cellulase—An enzyme complex, produced by fungi and bacteria, capable of decomposing cellulose into small fragments, primarily glucose.

Cellulose—The main carbohydrate in living plants. Cellulose forms the skeletal structure of the plant cell wall.

Cellulose Insulation—A type of insulation composed of waste newspaper, cardboard, or other forms of waste paper. It is an insulation alternative to glass fiber insulation. Cellulose insulation is most often a mixture of waste paper and fire retardant, and has thermal properties often superior to glass fiber. Glass fiber batt insulation often contains formaldehyde, which can adversely affect indoor air quality and human health, and the glass fibers themselves are hazardous if inhaled and irritating to the skin and eyes. You can specify cellulose insulation with high recycled content for maximum environmental benefit.

Celsius—A temperature scale based on the freezing (0 degrees) and boiling (100 degrees) points of water. Abbreviated as C in second and subsequent references in text. Formerly known as Centigrade. To convert Celsius to Fahrenheit, multiply the number by 9, divide by 5, and add 32. For example:

10 degrees Celsius x 9 = 90; 90/5 = 12;

18 + 32 = 50 degrees Fahrenheit

Celsius Scale—Favored name for centigrade scale, with freezing points and boiling points of water at 0 degrees and 100 degrees, respectively.

CEM—Continuous Emission Monitoring
- Cooperative Environmental Management
- Certified Energy Manager—an international accreditation by the Association of Energy Engineers, www.aeecenter.org

Cement—A building material that is a powder made of a mixture of calcined limestone and clay. Used with water and sand or gravel to make concrete and mortar. Cement is produced by heating mixtures of limestone and other minerals or additives at high temperatures in a rotary kiln, followed by cooling, grinding, and finish mixing.

Cement Kiln Dust (CKD)—The fine-grained, solid, highly alkaline waste removed from cement kiln exhaust gas by air pollution control devices, consisting of a partly calcined kiln feed material. CKD includes all dust from cement kilns and bypass systems including bottom ash and bypass dust.

Cement Plant—An industrial structure, installation, plant, or building primarily engaged in manufacturing Portland, natural, masonry, pozzolanic, and other hydraulic cements. Cement plants mine, quarry, and calcine raw materials to manufacture cement.

Cementious Product—Cement, cement kiln dust, clinker, and clinker dust.

CENR—Committee on Environmental and Natural Resources

Census Division—Any of nine geographic areas of the United States as defined by the U.S. Department of Commerce, Bureau of the Census. The divisions, each consisting of several States, are defined as follows:

- New England: Connecticut, Maine, Massachusetts, New Hampshire, Rhode Island, and Vermont;
- Middle Atlantic: New Jersey, New York, and Pennsylvania;
- East North Central: Illinois, Indiana, Michigan, Ohio, and Wisconsin;
- West North Central: Iowa, Kansas, Minnesota, Missouri, Nebraska, North Dakota, and South Dakota;
- South Atlantic: Delaware, District of Columbia, Florida, Georgia, Maryland, North Carolina, South Carolina, Virginia, and West Virginia;

- East South Central: Alabama, Kentucky, Mississippi, and Tennessee;

- West South Central: Arkansas, Louisiana, Oklahoma, and Texas;

- Mountain: Arizona, Colorado, Idaho, Montana, Nevada, New Mexico, Utah, and Wyoming;

- Pacific: Alaska, California, Hawaii, Oregon, and Washington.

Note: Each division is a sub-area within a broader Census Region. For the relationship between Regions and divisions, See *Census Region/Division* below. In some cases, the Pacific division is subdivided into the Pacific Contiguous area (California, Oregon, and Washington) and the Pacific Noncontiguous area (Alaska and Hawaii).

Census Region—Any of four geographic areas of the United States as defined by the U.S. Department of Commerce, Bureau of the Census. The Regions, each consisting of various States selected according to population size and physical location, are defined as follows:

- Northeast: Connecticut, Maine, Massachusetts, New Hampshire, New Jersey, New York, Pennsylvania, Rhode Island, and Vermont.

- South: Alabama, Arkansas, Delaware, District of Columbia, Florida, Georgia, Kentucky, Louisiana, Maryland, Mississippi, North Carolina, Oklahoma, South Carolina, Tennessee, Texas, Virginia, and West Virginia.

- Midwest: Illinois, Indiana, Iowa, Kansas, Michigan, Minnesota, Missouri, Nebraska, North Dakota, Ohio, South Dakota, and Wisconsin.

- West: Alaska, Arizona, California, Colorado, Hawaii, Idaho, Montana, Nevada, New Mexico, Oregon, Utah, Washington, and Wyoming.

Note: Each region comprises two or three sub-areas called Census divisions.

Center of Glass U-Factor—The U-factor for the glass portion only of vertical or horizontal fenestration and is measured at least two and one half inches from the frame. Center of glass U-factor does not consider the U-factor of the frame. Center of glass U-factor is not used.

Central Chiller—Any centrally located air conditioning system that produces chilled water in order to cool air. The chilled water or cold air is then distributed throughout the building, using pipes or air ducts or both. These systems are also commonly known as "chillers," "centrifugal chillers," "reciprocating chillers," or "absorption chillers." Chillers are generally located in or just outside the building they serve. Buildings receiving district chilled water are served by chillers located at central physical plants.

Central Cooling—Cooling of an entire building with a refrigeration unit to condition the air. Typically central chillers and ductwork are present in the centrally cooled building.

Central Heating System—A system where heat is supplied to areas of a building from a single appliance through a network of ducts or pipes.

Central Physical Plant—A plant owned by, and on the grounds of, a multi-building facility that provides district heating, district cooling, or electricity to other buildings on the same facility. To qualify as a central plant it must provide district heat, district chilled water, or electricity to at least one other building. The central physical plant may be by itself in a separate building or may be located in a building where other activities occur.

Central Power—The generation of electricity in large power plants with distribution through a network of transmission lines (grid) for sale to a number of users. Opposite of distributed power.

Central Receiver Solar Power Plants—Also known as "power towers," these use fields of two-axis tracking mirrors known as heliostats. Each heliostat is individually positioned by a computer control system to reflect the sun's rays to a tower-mounted thermal receiver. The effect of many heliostats reflecting to a common point creates the combined energy of thousands of suns, which produces high-temperature thermal energy. In the receiver, molten nitrate salts absorb the heat energy. The hot salt is then used to boil water to steam, which is sent to a conventional steam turbine-generator to produce electricity.

Central Warm Air Furnace—A type of space heating equipment where a central combustor or resistance unit generally using gas, fuel oil, or electricity provides warm air through ducts leading to the various rooms. Heat pumps are not included in this category. A forced air furnace is one in which a fan is used to force the air through the ducts. In a gravity furnace, air is circulated by gravity, relying on the natural flow of warm air up and cold air down; the warm air rises through ducts and the cold air falls through ducts that return it to the furnace to be reheated and this completes the circulation cycle.

Centralized Sewage Treatment—The collection and treatment of sewage from many sources to remove pollutants and pathogens.

Centralized Water Heating System—Equipment, to heat and store water for other than space heating purposes, which provides hot water from a single location for distribution throughout a building. A residential type tank water heater is a good example of a centralized water heater.

CEO—Chief Executive Officer

CEP—Council on Economic Priorities

CEPP—Chemical Emergency Preparedness Plan

CEPPO—Chemical Emergency Preparedness and Prevention Office

CEPRC—Chemical Emergency Planning and Response Commission

CEQ—See *Council on Environmental Quality*.

CEQA—California Environmental Quality Act: The principal statute mandating environmental assessment of projects in California. The purpose of CEQA is to evaluate whether a proposed project may have an adverse effect on the environment and, if so, if that effect can be reduced or eliminated by pursuing an alternative course of action or through mitigation.

CER—Certified Emission Reduction. A credit generated under Kyoto's Clean Development Mechanism (CDM) for the reduction of emissions of greenhouse gases equal to one ton of CO_2-equivalent. They are designed to be used by industrialized countries to count toward their Kyoto targets but can also be used by EU companies and governments as offsets against their emissions under the EU Emissions Trading Scheme.

Ceramic Radiants—Baked clay devices which become incandescent and radiate heat released to them by a gas flame.

CERCLA—Comprehensive Environmental Response, Compensation and Liability Act (1980—"Superfund")

CERCLIS—Comprehensive Environmental Response, Compensation and Liability Information System

CERES—A national network of investors, environmental organizations and other public interest groups working with companies and investors to address sustainability challenges such as global climate change. Ceres hosts an annual competition to highlight the best examples of sustainability reporting in North America. www.ceres.org

CERI—Center for Environmental Research Information www.cerc.columbia.edu

CERT—Certificate of Eligibility

Certifiable Emission Factor—An emission factor that is based on collected data, and has a well documented and transparent calculation methodology.

Certificate—A type of permit for public convenience and necessity issued by a utility commission, which authorizes a utility or regulated company to engage in business, construct facilities, provide some services, or abandon service.

Certificate Condition—A condition imposed by the FPC or FERC when granting a Certificate of Public Convenience and Necessity.

Certificate of Compliance (CF-1R)—A document with information required by the California Energy Commission that is prepared by the Documentation Author that indicates whether the building includes measures that require field verification and diagnostic testing.

Certificate of Delivery and Acceptance—A document that is signed to acknowledge that the equipment or other asset(s) has been delivered and is acceptable. Used in most lease agreements and usually triggers the commencement of the lease contract.

Certificate of Field Verification and Diagnostic Testing (CF-4R)—A document with information required by the Commission that is prepared by the HERS Rater to certify that measures requiring field verification and diagnostic testing comply with the requirements.

Certificate of Necessity (for Amortization)—A certificate issued by a Federal authority certifying that certain facilities are necessary in the interest of national defense, which permits accelerated amortization of the cost of the facilities or a certain specified percentage thereof, for income tax purposes, over a 60-month period.

Certificate of Occupancy—A document presented by a local government agency or building department certifying that a building and/or the leased area has been satisfactorily inspected and is in a condition suitable for occupancy.

Certificate of Origin—A document, required by certain foreign countries for tariff purposes, certifying the county of origin of specified goods.

Certificate of Participation—A municipal lease or other finance structure fractionalized into shares and assigned or marketed to investors.

Certificate of Public Convenience and Necessity (Gas)—A special permit (which supplements the franchise), commonly issued by a state commission, which authorizes a utility to engage in business, construct facilities, or perform some other service. Also, a permit issued by the Federal Energy Regulatory Commission (FERC) to engage in the transportation or sale for resale of natural gas in interstate commerce or to construct or acquire and operate any facilities necessary therefore, to which certificate the Commission may attach such reasonable terms and conditions as the public convenience and necessity may require.

Certificate Requirement—The maximum annual volume allowed for sales to resale or direct sale customers under certificate authorizations by the Federal Energy Regulatory Commission.

Certification—The process by which an independent accredited body (operational entity) gives written assurance of the emission reductions that have been achieved. In the case of an activity under the Clean Development Mechanism under the Kyoto Protocol, certification also gives assurances that the reductions occurred under the conditions (sustainable development objectives have been met) necessary for recognition by the Parties.

• Under the Kyoto Protocol, the certification process has not been fully defined, it is expected that emission reductions will be certified by independent third parties through a verification process. Certification is likely to endorse the existence, eligibility and title of the emission reduction (in relation to the underlying project). Once certification has occurred the emission reduction then becomes a separate tradable commodity.

• Process by which a motor vehicle, motor vehicle engine, or motor vehicle pollution control device satisfies the criteria adopted by the California Air Resources Board (ARB) for the control of specified air contaminants from vehicular sources (Health & Safety Code, Section 39018). Certification constitutes a guarantee by the manufacturer that the engine will meet certain standards at 50,000 miles; if not, it must be replaced or repaired without change.

• Is certification by the manufacturer to the Energy Commission (in California), as specified the Appliance Efficiency Regulations, that the appliance complies with the applicable standard for that appliance. The Commission's database of certified heating appliances can be accessed by contacting the Commission Energy Hotline or from the Commission's website at www.energy.ca.gov/efficiency/appliances/index.htm

• The term certification is also used in other ways in the standards. Many of the compliance forms are certificates, whereby installers, HERS testers and others certify that equipment was correctly installed and/or tested.

Certified—As to a home energy rater, is having been found by a certified home energy rating provider to have successfully completed the requirements established by that home energy rating provider.

Certified Emission Reductions (CER)—In the greenhouse gas program, CERs are verified and authenticated reductions of greenhouse gas from the abatement or sequestration projects which are certified by the Clean Development Mechanism. Annex I investors in Clean Development Mechanism (CDM) projects can earn Certified emission reduction units (CERs) for the amount of greenhouse emission reductions achieved by their CDM projects, provided they meet certain eligibility criteria. For example, CERs generated under the CDM will only be recognized when:

• the reductions of greenhouse gas emissions are additional to any that would occur in the absence of the certified project (see Additionality)

• requirements of the Host Country are met and

• the CDM Adaptation charge is paid i.e. the Levy to offset climate change adaptation costs in "vulnerable" developing countries. This levy is generally envisioned as an initial percentage of the total financing cost and is paid up front by the project sponsor, in the form of either currency or emission credits, which are then auctioned. Proceeds are held in an adaptation fund for later disbursement.

Certified Tradable Offset (CTO)—A financial instrument that can be used to transfer (sell) Greenhouse Gas offsets in the international marketplace. A Certified Tradable Offset (CTO) represents a specific number of units of greenhouse gas emission expressed in carbon equivalent units reduced or sequestered. The home-country verification process certifies that the offsets are of high enough quality to allow them to count against national and company-level greenhouse gas reduction commitments, if such crediting is eventually permitted under the UNFCCC.

Certified Wood—Wood-based materials used in building construction that are supplied from sources that comply with sustainable forestry practices, protecting trees, wildlife habitat, streams and soil as determined by the Forest Stewardship Council. www.fsc.org

Certifying Organization—An independent organization recognized by the Energy Commission (in

California) to certify manufactured devices for performance values in accordance with procedures adopted by the Energy Commission.

CESA—California Endangered Species Act

CESQG—Conditionally Exempt Small Quantity Generator

Cesspool—An underground reservoir for liquid waste, typically household sewage.

CEST—Community Environmental Service Team

CET—Customer Event Tracking

Cetane—Ignition performance rating of diesel fuel.

Cetane Number—A measure of a fuel's (liquid) ease of self-ignition.

CETRED—Combustion Emissions Technical Resource Document

CEU—Continuous Education Units

CEUS—Commercial End Use Survey

CF—Cash Flow

CFA—Consumer Federation of America www.consumerfed.org

CFB—Circulating Fluidized Bed

CFC—See *Chlorofluorocarbon* below.

CFCA—Core Fixed Cost Account: A balancing account that matches the authorized base revenue requirement and certain other costs for core customers with recorded revenues intended to recover that revenue requirement and the other specified costs. Like balancing accounts, the CFCA provides for any over-collection or under-collection of revenues, plus interest to be refunded to or recovered from ratepayers.

CFCs (Chlorofluorocarbons or Chlorinated Fluorocarbons)—Chlorofluorocarbons or chlorinated fluorocarbons are artificially produced chemicals that are partly responsible for depletion of the ozone layer. CFCs have been used in refrigerants and a variety of other solvents since introduced in mid-1930.

A family of artificially produced chemicals receiving much attention for their role in stratospheric ozone depletion. On a per molecule basis, these chemicals are several thousand times more effective as greenhouse gases than carbon dioxide. Since they were introduced in the mid-1930s, CFCs have been used as refrigerants, solvents and in the production of foam material. The 1987 Montreal protocol on CFCs sought to reduce their production by one-half by the year 1998.

CFEE—California Foundation on the Environment and the Economy: The California Foundation on the Environment and the Economy brings together business, labor, community, and environmental

leadership with legislative and regulatory officials and expert academicians, in forums designed to address complex economic and social issues. www.cfee.net

CFI—Carbon Financial Instrument. The name of the futures contract through which parcels of emission permits are traded on the European Climate Exchange and the Chicago Climate Exchange. Each CFI consists of 100 permits (mandatory EUAs in Europe and voluntary allowances and offsets on the Chicago market) covering the emission of 100 tons of CO_2.

CFM—Certified Facility Manager. The certification is issued through the International Facility Management Association. www.ifma.org

Cfm—Cubic feet per minute is a measure of flow rate.
• Measurement of the volume of air being moved through an air duct.

CFPS—Cash Flow Per Share

CFR—Code of Federal Regulations
• A pricing term indicating that the cost of the goods and freight charges are included in the quoted price. The buyer arranges for and pays insurance.

CFS—Cubic feet per second.

CFSG—Citizen Forum for Self Government

CGA—Compressed Gas

CGBP—Certified Green Building Professional. Certification issued through Build It Green. www.builditgreen.org

CGI—Combustible Gas Indicator

CGL—Comprehensive General Liability

CH4—Methane, a greenhouse gas.

Chain of Title—A chronological list of recorded instruments affecting the Title to land.

Chained Dollars—A measure used to express real prices. Real prices are those that have been adjusted to remove the effect of changes in the purchasing power of the dollar; they usually reflect buying power relative to a reference year. Prior to 1996, real prices were expressed in constant dollars, a measure based on the weights of goods and services in a single year, usually a recent year. In 1996, the U.S. Department of Commerce introduced the chained-dollar measure. The new measure is based on the average weights of goods and services in successive pairs of years. It is "chained" because the second year in each pair, with its weights, becomes the first year of the next pair. The advantage of using the chained-dollar measure is that it is more closely related to any given period covered and is therefore subject to less distortion over time.

CHAMP—Community Health Air Monitoring Program

Chapter 11—The part of the Bankruptcy Code that provides for reorganization of a bankrupt company's assets.

Chapter 7—The part of the Bankruptcy Code that provides for liquidation of a company's assets.

Char—The remains of solid biomass that has been incompletely combusted, such as charcoal if wood is incompletely burned.

Characterization—Sampling, monitoring, and analysis activities to determine the extent and nature of contamination at a facility or site. Characterization provides the necessary technical information to develop, screen, analyze, and select appropriate cleanup techniques.

Charcoal—A material formed from the incomplete combustion or destructive distillation (carbonization) of organic material in a kiln or retort, and having a high energy density, being nearly pure carbon. (If produced from coal, it is coke.) Used for cooking, the manufacture of gunpowder and steel (notably in Brazil), as an absorbent and decolorizing agent, and in sugar refining and solvent recovery.

Charge—The process of inputting electrical energy to a battery.

Charge Capacity—The input (feed) capacity of the refinery processing facilities.

Charge Carrier—A free and mobile conduction electron or hole in a semiconductor.

Charge Controller—A component that controls the flow of current to and from the battery subsystem to protect the batteries from overcharge and over discharge. Essential for ensuring that batteries obtain maximum state of charge and longest life. The charge controller may also monitor system performance and provide system protection. Charge Controllers are also sometimes called Regulators.

Charge Coupler—A connector and vehicle receptacle for hybrid and electric vehicle charging.

Charge Factor—A number corresponding to the time (in hours) for which a battery can be charged at a constant current without damaging it. Usually expressed as a function of battery capacity, e.g. C/10 indicates a charge factor of 10 hours. Related to Charge Rate.

Charge Rate—A measure of the current used to charge a battery as a proportion of its capacity. The current applied to a cell or battery to restore its available capacity, specified in relation to total battery size. A C/20 rate is a charge rate that is 1/20th of the total battery capacity. Also called a "20-hour rate."

Charge/Discharge Cycle—The operational profile of an energy storage device that defines how much of the time it must be used to store electrical energy versus how much time it is available to supply electrical energy or other services.

Charrette—A meeting held early in the design phase of a project, in which the design team, contractors, end users, community stake-holders, and technical experts are brought together to develop goals, strategies, and ideas for maximizing the environmental performance of the project.

Chase—A Chase is an enclosure designed to hold ducts, plumbing, electric, telephone, cable or other linear components. To maintain energy efficiency, a Chase designed for ducts should be in conditioned space and include airflow retarders and thermal barriers between it and unconditioned spaces such as attics.

Chattel—Personal Property.

Chemical Energy—The energy generated when a chemical compound combusts, decomposes, or transforms to produce new compounds.

Chemical Oxygen Demand (COD)—The amount of dissolved oxygen required to combine with chemicals in wastewater. A measure of the oxygen equivalent of that portion of organic matter that is susceptible to oxidation by a strong chemical oxidizing agent.

Chemical Separation (Nuclear)—A process for extracting uranium and plutonium from dissolved spent nuclear fuel and irradiated targets. The fission products that are left behind are high-level waste. Chemical separation is also known as reprocessing.

Chemical Vapor Deposition (CVD)—A method of depositing thin semiconductor films used to make certain types of solar photovoltaic devices. With this method, a substrate is exposed to one or more vaporized compounds, one or more of which contain desirable constituents. A chemical reaction is initiated, at or near the substrate surface, to produce the desired material that will condense on the substrate.

CHEMNET—Chemical Industry Emergency Mutual Aid Network

CHEMTREC—Chemical Transportation Emergency Center

CHESS—Community Health and Environmental Surveillance System

CHIEF—Clearinghouse for Inventories and Emission Factors www.epa.gov/ttn/chief

Chill Factor—The temperature (at zero wind velocity)

which would produce the same chilling effect as a particular combination of temperature and wind velocity.

Chiller—A water-cooling device that generally reduces water temperature to approximately 45 degrees Fahrenheit and is used to cool air.

Chimney—A masonry or metal stack that creates a draft to bring air to a fire and to carry the gaseous by-products of combustion safely away.

Chimney Effect—The tendency of heated air or gas to rise in a duct or other vertical passage, such as in a chimney, small enclosure, or building, due to its lower density compared to the surrounding air or gas.

China Walls—An expression that refers to the complete separation of operations for affiliated companies within a corporation to prevent undue business advantages. Pipeline companies, for example, are expected to have "china walls" separating their transportation departments from marketing affiliates to ensure that all customers moving gas on the pipeline get equal treatment.

CHIP—Chemical Hazard Information Profile

Chip Vans—Special construction bottom dump trucks (grain trucks) or conventional tractor- trailer vans used for hauling pulp chips, mill residues, hog fuel, and other biomass of smaller piece sizes.

Chipper—A machine that produces wood chips by knife action.

Chips—Woody material cut into short, thin wafers. Chips are used as a raw material for pulping and fiberboard or as biomass fuel.

Chlorobenzene—A volatile organic compound that is often used as a solvent and in the production of other chemicals. It is a colorless liquid with an almond-like odor. It is toxic.

Chlorofluorocarbon (CFC)—CFCs are synthetic industrial gases composed of chlorine, fluorine, and carbon. They have been used as refrigerants, aerosol propellants, cleaning solvents and in the manufacture of plastic foam. There are no natural sources of CFCs. CFCs have an atmospheric lifetime of decades to centuries, and they have 100-year "global warming potentials" thousands of times that of CO_2, depending on the gas. In addition to being greenhouse gases, CFCs also contribute to ozone depletion in the stratosphere and are controlled under the Montreal Protocol.

Chloroform—Chloroform was once commonly used as a general anesthetic and as a flavoring agent in toothpastes, mouth wastes and cough syrups. It is now listed as a cancer-causing chemical.

Choker—A cable loop that is attached to a log during the yarding process.

CHP—Combined Heat and Power. The simultaneous production of electricity and heat from a single fuel source such as natural gas, biomass, biogas, coal, waste heat, or oil used in a turbine, internal combustion engine or other type of equipment producing energy.

CHRED—Chemical Reduction

CHRIS—Chemical Hazard Response Information System

Christmas Tree—The valves and fittings installed at the top of a gas or oil well to control and direct the flow of well fluids.

Chromium—A hard, brittle, grayish heavy metal used in tanning, in paint formulation, and in plating metal for corrosion protection. It is toxic at certain levels and, in its hexavalent (versus trivalent) form, chromium is listed as a cancer-causing agent.

Chronic Exposure—Repeated contact with a chemical over a period of time, often involving small amounts of toxic substances.

CHU—See *Critical Habitat Unit*.

CHWMP—County Hazardous Waste Management Plan

CHWR—Chilled Water Return

CHWS—Chilled Water Supply

CI—Compression Ignition
  • Confidence Interval

CIAC—Contributions in Aid of Construction: Non-refundable contributions in cash or properties from individuals, governmental agencies, or others for construction or property additions. The utility may not earn a return on such contributions.

CIAQ—Council on Indoor Air Quality

CIBL—Convective Internal Boundary layer

CIBO—Congress of Industrial Boiler Owners

CIC—Chemical Industry Council

CICA—Competition in Contracting Act

CICS—Customer Information Control System

CIF (Cost, Insurance, Freight)—This term refers to a type of sale in which the buyer of the product agrees to pay a unit price that includes the f.o.b. value of the product at the point of origin plus all costs of insurance and transportation. This type of a transaction differs from a "delivered" purchase, in that the buyer accepts the quantity as determined at the loading port (as certified by the Bill of Lading and Quality Report) rather than pay based on the quantity and quality ascertained at the unloading port. It is similar to the terms of an f.o.b.

sale, except that the seller, as a service for which he is compensated, arranges for transportation and insurance.

CIGS—Copper Indium Gallium (di)Selenide

CIH—Certified Industrial Hygienist

CII—Criminal Investigation Index

CIP—Capital Improvement Program

Circuit—One complete run of a set of electric conductors from a power source to various electrical devices (appliances, lights, etc.) and back to the same power source.

Circuit Breaker—A device used to interrupt or break an electrical circuit when an overload condition exists; usually installed in the positive circuit; used to protect electrical equipment.

Circuit Lag—As time increases from zero at the terminals of an inductor, the voltage comes to a particular value on the sine function curve ahead of the current. The voltage reaches its negative peak exactly 90 degrees before the current reaches its negative peak thus the current lags behind by 90 degrees.

Circuit-Mile—The total length in miles of separate circuits regardless of the number of conductors used per circuit.

Circulating Fluidized Bed—A type of furnace or reactor in which the emission of sulfur compounds is lowered by the addition of crushed limestone in the fluidized bed thus obviating the need for much of the expensive stack gas clean-up equipment. The particles are collected and re-circulated, after passing through a conventional bed, and cooled by boiler internals.

CIS—Chemical Information System
- Contracts Information System
- Contact Information System
- Copper Indium (di)Selenide

CISR—Chemical Inventory System

CISWI—Commercial Industrial Solid Waste Incinerator

City Gate Rate—The rate charged a distribution company by its supplier(s). It refers to the cost of the gas at the point at which the distribution utility takes title to the gas.

Citygate—The delivery point between a major natural gas pipeline and the local distribution company.

CIWMB—California Integrated Waste Management Board www.calepa.ca.gov

CJO—Chief Judicial Officer

CKRC—Cement Kiln Recycling Coalition

Cladding—External face or skin of a building.

Claim Dilution—A reduction in the likelihood that one or more of the firm's claimants will be fully repaid, including time value of money considerations.

Clarifier—A tank used to remove solids by gravity, to remove colloidal solids by coagulation, and to remove floating oil and scum through skimming.

Class "C" Building—Buildings that offer few amenities but are otherwise in physically acceptable condition and provide cost-effective space to tenants who are not particularly image conscious.

Class I Area—Any area designated for the most stringent protection from air quality degradation.

Class II Area—Any area where air is cleaner than required by federal air quality standards and designated for a moderate degree of protection from air quality degradation. Moderate increases in new pollution may be permitted in Class II areas.

Class Rate Schedule—An electric rate schedule applicable to one or more specified classes of service, groups of businesses, or customer uses.

Classes of Service—Customers grouped by similar characteristics in order to be identified for the purpose of setting a common rate for electric service. Usually classified into groups identified as residential, commercial, industrial, and other.

Claw Back—A British term to describe a taxpayer paying back to the government an amount equal to tax benefits previously claimed, such as depreciation benefits or possibly tax credits received.

Clawback—A clawback obligation represents the general partner's promise that, over the life of the fund, the managers will not receive a greater share of the fund's distributions than they bargained for. Generally, this means that the general partner may not keep distributions representing more than a specified percentage (e.g., 20%) of the fund's cumulative profits, if any. When triggered, the clawback will require that the general partner return to the fund's limited partners an amount equal to what is determined to be "excess" distributions.

CLC—Capacity Limiting Constituents

Clean Air Act (CAA)—National law establishing ambient air quality emission standards to be implemented by participating states. Originally enacted in 1963, the CAA has been amended several times, most recently in 1990. The CAA includes vehicle emission standards regulating the emission of criteria pollutants (lead, ozone, carbon monoxide, sulfur dioxide, nitrogen oxides and particulate matter). The 1990 amendments added reformulated gasoline (RFG) requirements and oxygenated gasoline provisions.

Clean Development Mechanism (CDM)—A Kyoto Protocol program that enables industrialized countries to finance emissions-avoiding projects in developing countries and receive credit for reductions achieved against their own emissions limitation targets. CDM was established by Article 12 of the Kyoto Protocol. The CDM is designed to meet two main objectives: to address the sustainable development needs of the host country, and to increase the opportunities available to Parties to meet their reduction commitments. Also see *Kyoto Protocol*.

The CDM allows industrialized countries with a greenhouse gas reduction commitment (called Annex 1 countries) to invest in projects that reduce emissions in developing countries as an alternative to more expensive emission reductions in their own countries. A crucial feature of an approved CDM carbon project is that it has established that the planned reductions would not occur without the additional incentive provided by emission reductions credits, a concept known as "additionality." The CDM allows net global greenhouse gas emissions to be reduced at a much lower global cost by financing emissions reduction projects in developing countries where costs are lower than in industrialized countries.

The text of Article 12 currently describes more of an idea than an operational entity. Highly innovative, it has the potential to meet the needs of both developing and industrialized countries. It could help solve non-Annex I needs for capital for the financing of technology transfer for clean, energy efficient economic development and for addressing environmental issues such as loss of biodiversity, while also providing a lower cost, more flexible alternative for Annex I countries to meet emissions reduction targets.

According to Article 12 of the Kyoto Protocol, the "purpose of the Clean Development Mechanism shall be to assist Parties not included in Annex I in achieving sustainable development, and in contributing to the ultimate objective of the Convention, and to assist Parties included in Annex I in achieving compliance with their quantified emission limitation and reduction commitments under Article 3."

Clean Fuel Vehicle—Is frequently incorrectly used interchangeably with "alternative fuel vehicle." Generally, refers to vehicles that use low-emission, clean-burning fuels. Public Resources Code Section 25326 defines clean fuels, for purposes of the section only, as fuels designated by ARB (California) for use in LEVs, ULEVs or ZEVs and include, but are not limited to, electricity, ethanol, hydrogen, liquefied petroleum gas, methanol, natural gas, and reformulated gasoline.

Clean Letter of Credit—A letter of credit payable upon presentation of a draft, not requiring the presentation of documents.

Clean Power Generator—A company or other organizational unit that produces electricity from sources that are thought to be environmentally cleaner than traditional sources. Clean, or green, power is usually defined as power from renewable energy that comes from wind, solar, biomass energy, etc. There are various definitions of clean resources. Some definitions include power produced from waste-to-energy and wood-fired plants that may still produce significant air emissions. Some states have defined certain local resources as clean that other states would not consider clean. For example, the state of Texas has defined power from efficient natural gas-fired power plants as clean. Some northwest states include power from large hydropower projects as clean although these projects damage fish populations. Various states have disclosure and labeling requirement for generation source and air emissions that assist customers in comparing electricity characteristics other than price. This allows customers to decide for themselves what they consider to be "clean." The federal government is also exploring this issue.

Clean Production—A concept developed under the Kyoto Protocol in which manufacturing processes reduce environmental impact and decrease ecological problems by minimizing energy and raw materials use, and making user emissions and waste are as minimal and as non-toxic to environmental and human health as possible.

Clean Water Act—A federal law passed in 1977 and enforced by U.S. EPA. A key provision is that "any person responsible for the discharge of a pollutant or pollutants into any waters of the United States from any point source must apply for and obtain a permit." This is reflected by the National Pollutant Discharge Elimination system (NPDES), through which the permits are issued by Regional Water Quality Control Boards. Permits are now being required for stormwater runoff from cities and other locations.

Clear Gas—Tar free gas occurring between the carbonization and gasification zones in a coal gasification plant.

Clearcut—The removal, in a single cutting, of the entire stand of trees within a designated area. Stand regeneration is accomplished by planting the site or by natural seeding from adjacent stands.

Cleavage of Lateral Epitaxial Films for Transfer (CLEFT)—A process for making inexpensive Gallium Arsenide (GaAs) photovoltaic cells in which a thin film of GaAs is grown atop a thick, single-crystal GaAs (or other suitable material) substrate and then is cleaved from the substrate and incorporated into a cell, allowing the substrate to be reused to grow more thin-film GaAs.

Clerestory—A window located high in a wall near the eaves that allows daylight into a building interior, and may be used for ventilation and solar heat gain.

Climate—The prevailing or average weather conditions of a geographic region.

Climate Change—A term used to refer to all forms of climatic inconsistency, but especially to significant change from one prevailing climatic condition to another. In some cases, "climate change" has been used synonymously with the term "global warming"; scientists, however, tend to use the term in a wider sense inclusive of natural changes in climate, including climatic cooling. Changes in global climate patters (such as temperature, precipitation, or wind) that last for extended periods of time as a result of either natural processes or human activity. The contemporary concern is that human activity is now transcending natural processes in causing the most prevalent climate changes of our time.

Climate Neutrality—The effort to balance out the total amount of carbon output based on the notion that unavoidable emissions in a specific location can be neutralized by protective measures taken in another specific location.

Climate Reserve Ton (CRT)—The unit of offset credits used by the Climate Action Reserve. Once Climate Reserve Ton is equal to one metric ton of GHG reduced/sequestered.

Climate Sensitivity—The average global air surface temperature change resulting from a doubling of pre-industrial atmospheric CO_2 concentrations. The IPCC estimates climate sensitivity at 1.5-4.5°C (2.7-8.1°F).

Climate Variability—Refers to changes in patterns, such as precipitation patterns, in the weather and climate.

Climate Zone—A geographical area is the state that has particular weather patterns. These zones are used to determine the type of building standards that are required by law.

Climate Zones (California)—The 16 geographic areas of California for which the Energy Commission has established typical weather data, prescriptive packages and energy budgets. Climate zone boundary descriptions are in the document "California Climate Zone Descriptions" (July 1995).

CLIN—Contract Line Item Number

Clinker—Powdered cement, produced by heating a properly proportioned mixture of finely ground raw materials (calcium carbonate, silica, alumina, and iron oxide) in a kiln to a temperature of about 2,700 degrees Fahrenheit.

Clip and Strip Bond—In this type of bond, the principal and coupon portion of the bonds may be split apart and sold separately.

CLMA—Conservation Load Management Adjustment: Tracks an electricity utility's actual expenditures and compares them with the amount allowed for conservation and load management programs authorized in a general rate case.

CLO—Collateralized Loan Obligation

Clone—A genetically identical duplicate of an organism.

Close Coupled—An energy system in which the fuel production equipment is in close proximity, or connected to, the fuel using equipment.

Closed Cycle—A system in which a working fluid is used over and over without introduction of new fluid, as in a hydronic heating system or mechanical refrigeration system.

Closed End Lease—A true lease in which the lessor assumes the risk of depreciation and residual value. The lessee bears little or no obligation at the conclusion of the lease term. Usually a net lease in which the lessee maintains, insures and pays property taxes on the equipment. The term is used to distinguish a lease from an open-end lease, particularly in automobile leasing.

Closed Loop Recycling—A process of utilizing a recycled product in the manufacturing of a similar product or the remanufacturing of the same product. In almost all cases a large of amount of energy is saved in that process.

Closed Loop System—A solar hot water system of which no part is vented to the atmosphere or fed with fresh liquid. The system liquid, usually some form of antifreeze solution, is recirculated. Closed loop solar systems are also known as glycol systems and indirect systems.

Closed-Loop Biomass—Any organic material from a plant that is planted exclusively for use at a qualified facility to produce electricity.

As defined by the Comprehensive National Energy Act of 1992 (or the Energy Policy Act; EPAct): any organic matter from a plant which is planted for the exclusive purpose of being used to produce energy." This does not include wood or agricultural wastes or standing timber.

Closed-Loop Geothermal Heat Pump Systems—Closed-loop (also known as "indirect") systems circulate a solution of water and antifreeze through a series of sealed loops of piping. Once the heat has been transferred into or out of the solution, the solution is recirculated. The loops can be installed in the ground horizontally or vertically, or they can be placed in a body of water, such as a pond.

Closely Held Corporation—A corporation owned by a few individuals, who also own all the outstanding stock. No stock in the corporation is publicly traded. State regulations administer the establishment of corporations.

Closing—An investment event occurring after the required legal documents are implemented between the investor and a company and after the capital is transferred in exchange for company ownership or debt obligation.

Closing Costs—The expenses which borrowers incur to complete a loan transaction. These costs may include title searches, title insurance, closing fees, recording fees, processing fees, documentation fees, environmental fees, permit fees, audit fees, loan fees, etc.

Closing Documentation—The final documents designed to complete a business transaction.

Closing Statement—An accounting of funds from a real estate transaction, also known as a HUD.

Cloud Condensation Nuclei—Aerosol particles that provide a platform for the condensation of water vapor, resulting in clouds with higher droplet concentrations and increased albedo.

Cloud Enhancement (Solar)—The increase in solar intensity due to reflected light from nearby clouds.

CLP—Contract Laboratory Program
- Certified Lease Professional www.clpfoundation.org

CLS—Community Liaison Officer

CLTD—The Cooling Load Temperature Difference.°

Clunkers—Also known as gross-polluting or super-emitting vehicles, i.e., vehicles that emit far in excess of the emission standards by which the vehicle was certified when it was new.

CLUP—Comprehensive Land Use Plan

CM—Central Maintenance
- Corrective Measure
- Centimeters

cm/s—Centimeters per Second

CMA—Chemical Manufacturers Association www.socma.com

CMAS—Cross-Media Analysis Staff

CMB—Chemical Mass Balance

CMC—The 2001 California Mechanical Code.

CMD—Contracts Management Division

CME—Comprehensive Monitoring Evaluation

CMEP—Critical Mass Energy Project

CMHP—Contaminated Materials Handling Plan

CMI—Corrective Measures Implementation
- Corrective Measures Investigation

CMO—Contract Management Office

CMPBS—Center for Maximum Potential Building Systems www.cmpbs.org

CMR—Chemical Monitoring Reform

CMRA—Construction Materials Recycling Association, www.cdrecycling.org

CMS—Case Management System
- Continuous Monitoring System
- Corrective Measures Study

CMTW—Committee to Minimize Toxic Waste

CNEL—Community Noise Equivalent Level

CNF—Cost No Fee

CNG—Compressed Natural Gas

CNR—Composite Noise Rating

CO—Carbon Monoxide
- Change Order
- Commissioned Officer
- Consent Order
- Contracting Officer
- Custodial Officer

CO Control Period ("seasons")—The portion of the year in which a CO non-attainment area is prone to high ambient levels of carbon monoxide. This portion of the year is to be specified by the Environmental Protection Agency but is to be not less than 4 months in length.

CO Non-attainment Area (EPA)—Areas with carbon monoxide design values of 9.5 parts per million or more, generally based on data for 1988 and 1989.

CO_2—Carbon Dioxide—a greenhouse gas

CO_2 Equivalent (CO_2 e)—The quantity of a given greenhouse gas multiplied by its total global warming potential. This is the standard unit for comparing the degree of harm which can be caused by different greenhouse gases.

COA—Construction Quality Assurance

COAL—Black or brown rock, formed under pressure from organic fossils in prehistoric times, that is mined and burned to produce heat energy. It consists of more than 50 percent by weight and more than 70 percent by volume of carbonaceous material. It is formed from plant remains that have been compacted, hardened, chemically altered, and metamorphosed by heat and pressure over geologic time.

Coal Analysis—Determines the composition and properties of coal so it can be ranked and used most effectively.

- Proximate analysis determines, on an as-received basis, the moisture content, volatile matter (gases released when coal is heated), fixed carbon (solid fuel left after the volatile matter is driven off), and ash (impurities consisting of silica, iron, alumina, and other incombustible matter). The moisture content affects the ease with which coal can be handled and burned. The amount of volatile matter and fixed carbon provides guidelines for determining the intensity of the heat produced. Ash increases the weight of coal, adds to the cost of handling, and can cause problems such as clinkering and slagging in boilers and furnaces.

- Ultimate analysis determines the amount of carbon, hydrogen, oxygen, nitrogen, and sulfur. Heating value is determined in terms of Btu, both on an as received basis (including moisture) and on a dry basis.

- Agglomerating refers to coal that softens when heated and forms a hard gray coke; this coal is called caking coal. Not all caking coals are coking coals. The agglomerating value is used to differentiate between coal ranks and also is a guide to determine how a particular coal reacts in a furnace.

- Other tests include the determination of the ash softening temperature, the ash fusion temperature (the temperature at which the ash forms clinkers or slag), the free swelling index (a guide to a coal's coking characteristics), the Gray King test (which determines the suitability of coal for making coke), and the Hardgrove grindability index (a measure of the ease with which coal can be pulverized). In a petrographic analysis, thin sections of coal or highly polished blocks of coal are studied with a microscope to determine the physical composition, both for scientific purposes and for estimating the rank and coking potential.

Coal Bed—A bed or stratum of coal. Also called a coal seam.

Coal Bed Degasification—This refers to the removal of methane or coal bed gas from a coal mine before or during mining.

Coal Bed Methane—Methane is generated during coal formation and is contained in the coal microstructure. Typical recovery entails pumping water out of the coal to allow the gas to escape. Methane is the principal component of natural gas. Coal bed methane can be added to natural gas pipelines without any special treatment.

Coal Bed Methane Recovery—A Coal Bed Methane emission reduction project captures methane released from coal bed seams during the mining process for flaring or energy use.

Coal Briquets—Anthracite, bituminous, and lignite briquets comprise the secondary solid fuels manufactured from coal by a process in which the coal is partly dried, warmed to expel excess moisture, and then compressed into briquets, usually without the use of a binding substance. In the reduction of briquets to coal equivalent, different conversion factors are applied according to their origin from hard coal, peat, brown coal, or lignite.

Coal Carbonized—The amount of coal decomposed into solid coke and gaseous products by heating in a coke oven in a limited air supply or in the absence of air.

Coal Chemicals—Coal chemicals are obtained from the gases and vapor recovered from the manufacturing of coke. Generally, crude tar, ammonia, crude light oil, and gas are the basic products recovered. They are refined or processed to yield a variety of chemical materials.

Coal Coke—See *Coke (coal)*.

Coal Consumption—The quantity of coal burned for the generation of electric power (in short tons), including fuel used for maintenance of standby service.

Coal Conversion—Changing coal into synthetic gas or liquid fuels. See *Gasification*.

Coal Delivered—Coal which has been delivered from the coal supplier to any site belonging to the electric power company.

Coal Equivalent of Fuels Burned—The quantity of coal (tons) of a stated kind and heat value which would be required to supply the Btu equivalent of all fuels burned. In determining this coal equivalent, the Btu content of other fuels is generally divided by the representative heat value per ton of coal burned.

Coal Exports—Amount of U.S. coal shipped to foreign destinations, as reported in the U.S. Department of Commerce, Bureau of Census, "Monthly Report EM 545."

Coal Face—This is the exposed area from which coal is extracted.

Coal Financial Reporting Regions—A geographic classification of areas with coal resources which is used for financial reporting of coal statistics.

- Eastern Region. Consists of the Appalachian Coal Basin. The following comprise the Eastern Region: Alabama, eastern Kentucky, Georgia, Maryland, Mississippi, Ohio, Pennsylvania, Virginia, Tennessee, North Carolina, and West Virginia.
- Midwest Region. Consists of the Illinois and Michigan Coal Basins. The following comprise the Midwest Region: Illinois, Indiana, Michigan, and western Kentucky.
- Western Region. Consists of the Northern Rocky, Southern Rocky, West Coast Coal Basins and Western Interior. The following comprise the Western Region: Alaska, Arizona, Arkansas, California, Colorado, Idaho, Iowa, Kansas, Louisiana, Missouri, Montana, New Mexico, North Dakota, Oklahoma, Oregon, Texas, South Dakota, Utah, Washington, and Wyoming.

Coal Fines—Coal with a maximum particle size usually less than one-sixteenth inch and rarely above one-eighth inch.

Coal Gas—Substitute natural gas produced synthetically by the chemical reduction of coal at a coal gasification facility.

Coal Gasification—The process of converting coal into gas. The basic process involves crushing coal to a powder, which is then heated in the presence of steam and oxygen to produce a gas. The gas is then refined to reduce sulfur and other impurities. The gas can be used as a fuel or processed further and concentrated into chemical or liquid fuel.

Coal Grade—This classification refers to coal quality and use.

- Briquettes are made from compressed coal dust, with or without a binding agent such as asphalt.
- Cleaned coal or prepared coal has been processed to reduce the amount of impurities present and improve the burning characteristics.
- Compliance coal is a coal, or a blend of coal, that meets sulfur dioxide emission standards for air quality without the need for flue-gas de-sulfurization.
- Culm and silt are waste materials from preparation plants. In the anthracite region, culm consists of coarse rock fragments containing as much as 30 percent small-sized coal. Silt is a mixture of very fine coal particles (approximately 40 percent) and rock dust that has settled out from waste water from the plants. The terms culm and silt are sometimes used interchangeably and are sometimes called refuse. Culm and silt have a heat value ranging from 8 to 17 million Btu per ton.
- Low-sulfur coal generally contains 1 percent or less sulfur by weight. For air quality standards, "low sulfur coal" contains 0.6 pounds or less sulfur per million Btu, which is equivalent to 1.2 pounds of sulfur dioxide per million Btu.
- Metallurgical coal (or coking coal) meets the requirements for making coke. It must have a low ash and sulfur content and form a coke that is capable of supporting the charge of iron ore and limestone in a blast furnace. A blend of two or more bituminous coals is usually required to make coke.
- Pulverized coal is a coal that has been crushed to a fine dust in a grinding mill. It is blown into the combustion zone of a furnace and burns very rapidly and efficiently.
- Slack coal usually refers to bituminous coal one-half inch or smaller in size.
- Steam coal refers to coal used in boilers to generate steam to produce electricity or for other purposes.
- Stoker coal refers to coal that has been crushed to specific sizes (but not powdered) for burning on a grate in automatic firing equipment.

Coal Imports—Amount of foreign coal shipped to the United States, as reported in the U.S. Department of Commerce, Bureau of the Census, "Monthly Report IM 145."

Coal Liquefaction—A chemical process that converts coal into clean-burning liquid hydrocarbons, such as synthetic crude oil and methanol.

Coal Mining Productivity—Coal mining productivity is calculated by dividing total coal production by the total direct labor hours worked by all mine employees.

Coal Oil—Oil that can be obtained by distilling bituminous coal.

Coal Preparation—The process of sizing and cleaning coal to meet market specifications by removing impurities such as rock, sulfur, etc. It may include crushing, screening, or mechanical cleaning.

Coal Producing Districts—A classification of coal fields defined in the Bituminous Coal Act of 1937. The districts were originally established to aid in formulating minimum prices of bituminous and sub-bituminous coal and lignite. Because much statistical information was compiled in terms of these districts, their use for statistical purposes has continued since the abandonment of that legislation in 1943. District 24 was added for the anthracite-producing district in Pennsylvania.

Coal Production—The sum of sales, mine consumption, issues to miners, and issues to coke, briquetting, and other ancillary plants at mines. Production data include quantities extracted from surface and underground mines, and normally exclude wastes removed at mines or associated reparation plants.

Coal Rank—The classification of coals according to their degree of progressive alteration from lignite to anthracite. In the United States, the standard ranks of coal include lignite, sub-bituminous coal, bituminous coal, and anthracite and are based on fixed carbon, volatile matter, heating value, and agglomerating (or caking) properties.

Coal Regions—The following regional definitions are used to report domestic coal reserves, production, and other operating statistics. Eastern Region. Consists of the Northern Appalachian Coal Basin. The following States comprise the Eastern Region: Alabama, Georgia, Ohio, Maryland, Mississippi, Pennsylvania, Virginia, Tennessee, North Carolina, West Virginia, and Eastern Kentucky. Midwest Region. Consists of the Illinois and Michigan Coal Basins. The following States comprise the Midwest Region: Illinois, Indiana, Michigan, and Western Kentucky. Western Region. Consists of the Northern Rocky, Southern Rocky, Western Interior, and West Coast Coal Basins. The following States comprise the Central Western Region: Alaska, Arizona, Arkansas, California, Colorado, Idaho, Iowa, Kansas, Louisiana, Missouri, Montana, New Mexico, North Dakota, Oklahoma, Oregon, Texas, South Dakota, Utah, Washington, and Wyoming.

Coal Sampling—The collection and proper storage and handling of a relatively small quantity of coal for laboratory analysis. Sampling may be done for a wide range of purposes, such as: coal resource exploration and assessment, characterization of the reserves or production of a mine, to characterize the results of coal cleaning processes, to monitor coal shipments or receipts for adherence to coal quality contract specifications, or to subject a coal to specific combustion or reactivity tests related to the customer's intended use. During pre-development phases, such as exploration and resource assessment, sampling typically is from natural outcrops, test pits, old or existing mines in the region, drill cuttings, or drilled cores. Characterization of a mine's reserves or production may use sample collection in the mine, representative cuts from coal conveyors or from handling and loading equipment, or directly from stockpiles or shipments (coal rail cars or barges). Contract specifications rely on sampling from the production flow at the mining or coal handling facility or at the loadout, or from the incoming shipments at the receiver's facility. In all cases, the value of a sample taken depends on its being representative of the coal under consideration, which in turn requires that appropriate sampling procedures be carefully followed.

For coal resource and estimated reserve characterization, appropriate types of samples include:

- Face channel or channel sample: a sample taken at the exposed coal in a mine by cutting away any loose or weathered coal then collecting on a clean surface a sample of the coal seam by chopping out a channel of uniform width and depth; a face channel or face sample is taken at or near the working face, the most freshly exposed coal where actual removal and loading of mined coal is taking place. Any partings greater than 3/8 inch and/or mineral concretions greater than 1/2 inch thick and 2 inches in maximum diameter are normally discarded from a channel sample so as better to represent coal that has been mined, crushed, and screened to remove at least gross non-coal materials.
- Column sample: a channel or drill core sample taken to represent the entire geologic coalbed; it includes all partings and impurities that may exist in the coalbed.

- Bench sample: a face or channel sample taken of just that contiguous portion of a coalbed that is considered practical to mine, also known as a "bench;" For example, bench samples may be taken of minable coal where impure coal that makes up part of the geologic coalbed is likely to be left in the mine, or where thick partings split the coal into two or more distinct minable seams, or where extremely thick coalbeds cannot be recovered by normal mining equipment, so that the coal is mined in multiple passes, or benches, usually defined along natural bedding planes.
- Composite sample: a recombined coalbed sample produced by averaging together thickness-weighted coal analyses from partial samples of the coalbed, such as from one or more bench samples, from one or more mine exposures or outcrops where the entire bed could not be accessed in one sample, or from multiple drill cores that were required to retrieve all local sections of a coal seam.

Coal Seam—A mass of coal, occurring naturally at a particular location that can be commercially mined.

Coal Slurry Pipeline—A pipe system that transports pulverized coal suspended in water.

Coal Stocks—Coal quantities that are held in storage for future use and disposition. Note: When coal data are collected for a particular reporting period (month, quarter, or year), coal stocks are commonly measured as of the last day of this period.

Coal Sulfur—Coal sulfur occurs in three forms: organic, sulfate, and pyritic. Organic sulfur is an integral part of the coal matrix and cannot be removed by conventional physical separation. Sulfate sulfur is usually negligible. Pyritic sulfur occurs as the minerals pyrite and marcasite; larger sizes generally can be removed by cleaning the coal.

Coal Type—The classification is based on physical characteristics or microscopic constituents. Examples of coal types are banded coal, bright coal, cannel coal, and splint coal. The term is also used to classify coal according to heat and sulfur content. See *Coal Grade* above.

Coal Zone—A series of laterally extensive and (or) lenticular coal beds and associated strata that arbitrarily can be viewed as a unit. Generally, the coal beds in a coal zone are assigned to the same geologic member or formation.

Coal-derived Fuel—Any fuel, whether is a solid, liquid, or gaseous state, produced by the mechanical, thermal or chemical processing of coal (e.g., pulverized coal, coal refuse, liquefied or gasified coal, washed coal, chemically cleaned coal, coal-oil mistures, and coke).

Coal-Producing Regions—Appalachian Region. Consists of Alabama, Georgia, Eastern Kentucky, Maryland, North Carolina, Ohio, Pennsylvania, Tennessee, Virginia, and West Virginia.
- Interior Region (with Gulf Coast). Consists of Arkansas, Illinois, Indiana, Iowa, Kansas, Louisiana, Michigan, Mississippi, Missouri, Oklahoma, Texas, and Western Kentucky.
- Western Region. Consists of Alaska, Arizona, Colorado, Montana, New Mexico, North Dakota, Utah, Washington, and Wyoming.

Note: Some States discontinue producing coal as reserves are depleted or as production becomes uneconomic.

Coastal Zone Management Act—National law encouraging states to develop coastal management programs. The act establishes a method for the federal government to approve states' plans, establishes funding for approved plans, and requires federal actions to be consistent with states' plans.

COB—Close of Business

Cob Construction—A traditional building technique using hand formed lumps of earth mixed with sand and straw.

COC—Chain of Custody
- Chemical of Concern
- Certificate of Compliance
- Cost of Capital: The cost to a company of acquiring funds to finance its operations, including borrowed money, preferred stock dividends, etc.

COCO—Contractor-Owned/Contractor-Operated

Co-Control Benefit—The additional benefits derived from an environmental policy that is designed to control one type of pollution, while reducing the emissions of other pollutants as well. For example, a policy to reduce carbon dioxide emissions might reduce the combustion of coal, but when coal combustion is reduced, so too are the emissions of particulates and sulfur dioxide. The benefits associated with reductions in emissions of particulates and sulfur dioxide are the co-control benefits of reductions in carbon dioxide.

COD—See *Chemical Oxygen Demand*.

Code of Federal Regulations—A compilation of the general and permanent rules of the executive departments and agencies of the Federal Government as published in the Federal Register. The code is divided into 50 titles that represent broad areas subject to Federal regulation. Title 18 contains the FERC regulations.

Codes—Legal documents that regulate construction to protect the health, safety, and welfare of people. Codes establish minimum standards but do not guarantee efficiency or quality.

Coefficient of Expansion—The change in length per unit length or the change in volume per unit volume, per degree change in temperature.

Coefficient of Heat Transmission (U-Value)—A value that describes the ability of a material to conduct heat. The number of Btu's that flow through 1 square foot of material, in one hour. It is the reciprocal of the R-Value (U-Value = 1/R-Value).

Coefficient of Performance (COP)—A ratio of the work or useful energy output of a system versus the amount of work or energy inputted into the system as determined by using the same energy equivalents for energy in and out. Is used as a measure of the steady state performance or energy efficiency of heating, cooling, and refrigeration appliances. The COP is equal to the Energy Efficiency Ratio (EER) divided by 3.412. The higher the COP, the more efficient the device.

Coefficient of Performance (COP), Cooling—Is the ratio of the rate of net heat removal to the rate of total energy input, calculated under designated operating conditions and expressed in consistent units, as determined using the applicable test method in the Appliance Efficiency Regulations (in California) or Section 112.

Coefficient of Performance (COP), Heating—The ratio of the rate of net heat output to the rate of total energy input, calculated under designated operating conditions and expressed in consistent units, as determined using the applicable test method in the Appliance Efficiency Regulations (in California) or Section 112.

Coefficient of Utilization (CU)—A term used for lighting appliances; the ratio of lumens received on a flat surface to the light output, in lumens, from a lamp; used to evaluate the effectiveness of luminaries in delivering light.

• Coefficient of Utilization (Lighting) is the ratio of Lumens on the work surface to total Lumens emitted by the lamps.

Coefficients—Coefficients are the constants in Baseline Models. They are the values that are multiplied by the Independent Variables to get the model results, and are determined during the baseline modeling process.

Cofiring—The use of two or more different fuels (e.g. wood and coal) simultaneously in the same combustion chamber of a power plant.

COG—Compliance Order Guidance
• Cost of Goods
• Council of Governors

Cogeneration—Simultaneous on-site production of electric energy and process steam or heat from the same power source. Capturing the exhaust heat created by the power generation process, then using it to heat water, create steam, heat spaces, or as a second source of energy.

Cogeneration means the sequential use of energy for the production of electrical and useful thermal energy. The sequence can be thermal use followed by power production or the reverse, subject to the following standards:
• At least 5 percent of the cogeneration project's total annual energy output shall be in the form of useful thermal energy.
• Where useful thermal energy follows power production, the useful annual power output plus one-half the useful annual thermal energy output equals not less than 42.5 percent of any natural gas and oil energy input.

Cogeneration Facility—An industrial structure, installation, plant, building, or self-generating facility that has sequential or simultaneous generation of multiple forms of useful energy (usually mechanical and thermal) in a single, integrated system.

Cogeneration System—A system using a common energy source to produce both electricity and steam for other uses, resulting in increased fuel efficiency.

Cogenerator—A generating facility that produces electricity and another form of useful thermal energy (such as heat or steam), used for industrial, commercial, heating, or cooling purposes. To receive status as a qualifying facility (QF) under the Public Utility Regulatory Policies Act (PURPA), the facility must produce electric energy and "another form of useful thermal energy through the sequential use of energy" and meet certain ownership, operating, and efficiency criteria established by the Federal Energy Regulatory Commission (FERC). (See the Code of Federal Regulations, Title 18, Part 292.)

COH—Coefficient of Haze

Coil—As a component of a heating or cooling appliance, rows of tubing or pipe with fins attached through which a heat transfer fluid is circulated and to deliver heat or cooling energy to a building.

Coincidence Factor—The ratio of the coincident, maximum demand or two or more loads to the sum of their non-coincident maximum demand for a given period; the reciprocal of the diversity factor, and is always less than or equal to one.

Coincident Demand—The demand of a consumer of electricity at the time of a power supplier's peak system demand.

Coincidental Demand—The sum of two or more demands that occur in the same time interval.

Coincidental Peak Load—The sum of two or more peak loads that occur in the same time interval.

Coke (Coal)—A solid carbonaceous residue derived from low-ash, low-sulfur bituminous coal from which the volatile constituents are driven off by baking in an oven at temperatures as high as 2,000 degrees Fahrenheit so that the fixed carbon and residual ash are fused together. Coke is used as a fuel and as a reducing agent in smelting iron ore in a blast furnace. Coke from coal is grey, hard, and porous and has a heating value of 24.8 million Btu per ton.

Coke (Petroleum)—A residue high in carbon content and low in hydrogen that is the final product of thermal decomposition in the condensation process in cracking. This product is reported as marketable coke or catalyst coke. The conversion is 5 barrels (of 42 U.S. gallons each) per short ton. Coke from petroleum has a heating value of 6.024 million Btu per barrel. Includes catalyst coke deposited on a catalyst during the refining process which must be burned off in order to regenerate the catalyst.

Coke Breeze—The term refers to the fine sizes of coke, usually less than one-half inch, that are recovered from coke plants. It is commonly used for sintering iron ore.

Coke Burn-off—The coke removed from the surface of a catalyst by combustion in the catalyst regenerator.

Coke Button—A button-shaped piece of coke resulting from standard laboratory tests that indicates the coking or free-swelling characteristics of a coal; expressed in numbers and compared with a standard.

Coke Oven Gas—The mixture of permanent gases produced by the carbonization of coal in a coke oven at temperatures in excess of 1,000 degrees Celsius.

Coke Plants—Plants where coal is carbonized for the manufacture of coke in slot or beehive ovens.

Coking—Thermal refining processes used to produce fuel gas, gasoline blendstocks, distillates, and petroleum coke from the heavier products of atmospheric and vacuum distillation. Includes:

- Delayed Coking. A process by which heavier crude oil fractions can be thermally decomposed under conditions of elevated temperatures and pressure to produce a mixture of lighter oils and petroleum coke. The light oils can be processed further in other refinery units to meet product specifications. The coke can be used either as a fuel or in other applications such as the manufacturing of steel or aluminum.

- Flexicoking. A thermal cracking process which converts heavy hydrocarbons such as crude oil, tar sands bitumen, and distillation residues into light hydrocarbons. Feedstocks can be any pumpable hydrocarbons including those containing high concentrations of sulfur and metals.

- Fluid Coking. A thermal cracking process utilizing the fluidized-solids technique to remove carbon (coke) for continuous conversion of heavy, low-grade oils into lighter products.

Coking Coal—Bituminous coal suitable for making coke. See *Coke (coal) above.*

COLA—Cost of Living Adjustment

Cold Night Sky—The low effective temperature of the sky on a clear night.

Cold-Deck Imputation—A statistical procedure that replaces a missing value of an item with a constant value from an external source such as a value from a previous survey. See *Imputation.*

Coliform Bacteria—Bacteria whose presence in waste water is an indicator of pollution and of potentially dangerous contamination.

Collar—A contract between a buyer and seller for a commodity that assures the buyer of a set maximum price and assures that the seller will receive a set minimum price.

Collar Agreement—Agreed-upon adjustments in the number of shares offered in a stock-for-stock exchange to account for price fluctuations before the completion of the deal.

Collateral—The asset(s), such as securities, cash accounts, real property, or equipment, which is offered as security for a financial contact.

Collateral Assignment—The Assignment of a debt and the security therefore, such as a Note and Deed of Trust to secure the performance of an obligation by the Assignor. The Assignee holds Title for security purposes.

Collector—The component of a solar energy heating system that collects solar radiation, and that contains components to absorb solar radiation and (with solar thermal collectors) transfer the heat to a heat transfer fluid (air or liquid).

Collector Efficiency—The ratio of solar radiation captured and transferred to the collector (heat transfer) fluid.

Collector Fluid—The fluid, liquid (water or water/antifreeze solution) or air, used to absorb solar energy and transfer it for direct use, indirect heating of interior air or domestic water, and/or to a heat storage medium.

Collector Loop—The plumbing loop in a solar hot water system that includes the solar collectors. The collectors heat the fluid in the collector, and the heated fluid can be used directly (if water) or the heat can be exchanged to a potable water loop.

Collector Tilt—The angle that a solar collector is positioned from horizontal.

Color Rendering or Rendition—A measure of the ability of a light source to show colors, based on a color rendering index.

Color Rendition (Rendering) Index (CRI)—A measure of light quality. The maximum CRI value of 100 is given to natural daylight and incandescent lighting. The closer a lamp's CRI rating is to 100, the better its ability to show true colors to the human eye.

Color Temperature—A measure of the quality of a light source by expressing the color appearance correlated with a black body.

COM—Continuous Opacity Monitor

Comb—Combustion

Combination Space-Heating and Water-Heating Appliance—An appliance that is designed to provide both space heating and water heating from a single primary energy source.

Combined Accounts—When two or more meters are combined for billing purposes under the following conditions: Where combinations of meter readings are specifically provided for in rate schedules. Where the maintenance of adequate service and/or where a company's operating convenience shall require the installation

Combined Collector—A photovoltaic device or module that provides useful heat energy in addition to electricity.

Combined Cycle—An electric generating technology in which electricity is produced from otherwise lost waste heat exiting from one or more gas (combustion) turbines. The exiting heat is routed to a conventional boiler or to a heat recovery steam generator for utilization by a steam turbine in the production of electricity. This process increases the efficiency of the electric generating unit.

Combined Cycle Unit—An electric generating unit that consists of one or more combustion turbines and one or more boilers with a portion of the required energy input to the boiler(s) provided by the exhaust gas of the combustion turbine(s).

Combined Heat & Power Association—Trade association: www.chpassociation.org

Combined Heat and Power (CHP)—See *Cogeneration*.

Combined Heat and Power (CHP) Plant—A plant designed to produce both heat and electricity from a single heat source. Note: This term is being used in place of the term "cogenerator" that was used by EIA in the past. CHP better describes the facilities because some of the plants included do not produce heat and power in a sequential fashion and, as a result, do not meet the legal definition of cogeneration specified in the Public Utility Regulatory Policies Act (PURPA).

Combined Household Energy Expenditures—The total amount of funds spent for energy consumed in, or delivered to, a housing unit during a given period of time and for fuel used to operate the motor vehicles that are owned or used on a regular basis by the household. The total dollar amount for energy consumed in a housing unit includes state and local taxes but excludes merchandise repairs or special service charges. Electricity, and natural gas expenditures are for the amount of those energy sources consumed. Fuel oil, kerosene, and LPG expenditures are for the amount of fuel purchased, which may differ from the amount of fuel consumed. The total dollar amount of fuel spent for vehicles is the product of fuel consumption and price.

Combined Hydroelectric Plant—A hydroelectric plant that uses both pumped water and natural stream flow for the production of power.

Combined Hydronic Space/Water Heating—A system in which both space heating and domestic water heating are provided by the same water heater(s).

Combined Hydronic Space/Water Heating System—A system which both domestic hot water and space heating is supplied from the same water heating equipment. Combined hydronic space heating may include both radiant floor systems and convective or fan coil systems.

Combined Pumped-Storage Plant—A pumped-storage hydroelectric power plant that uses both pumped water and natural stream flow to produce electricity.

Combined-Cycle Power Plant—The combination of a gas turbine and a steam turbine in an electric generation plant. The waste heat from the gas turbine provides the heat energy for the steam turbine.

Combiner Box—A box where wires from individual PV modules or strings are combined into larger wires to run to the battery bank. Can also contain overcurrent protection devices.

Combuster Basket, Can or Chamber—That part of a gas turbine into which fuel is injected and burned.

Combustible Constituents—The components of a fuel that will burn. In natural gas, this is mostly methane.

Combustible Material—Combustible material, as pertaining to material adjacent to or in contact with heat producing appliances, chimney connectors and vent connectors, steam and hot water pipes, and warm air ducts, means material made of or surfaced with wood, compressed paper, plant fibers, or other material that will ignite and burn. Such material shall be considered as combustible even though flameproofed, fire retardant treated, or plastered.

Combustion—Burning. The transformation of biomass fuel into heat, chemicals, and gases through chemical combination of hydrogen and carbon in the fuel with oxygen in the air. Chemical oxidation accompanied by the generation of light and heat.

Combustion Air—The air fed to a fire to provide oxygen for combustion of fuel. It may be preheated before injection into a furnace.

Combustion Analysis—The determination of combustion characteristics, such as exhaust gas composition and temperature, air-fuel ratio, the relation of these to perfect combustion.

Combustion Burning—Rapid oxidation, with the release of energy in the form of heat and light.

Combustion Chamber—Any wholly or partially enclosed space in which combustion takes place.

Combustion Efficiency—The actual heat produced by combustion divided by the total heat potential of the fuel consumed.

Combustion Emissions—Criteria pollutants such NO_X, SO_X, VOCs, etc. or greenhouse gas emissions occurring during the exothermic reaction of a fuel with oxygen.

Combustion Gases—The gases released from a combustion process. The composition will depend on, among other things, the fuel; the temperature of burning; and whether air, oxygen or another oxidizer is used. In simple cases the combustion gases are carbon dioxide and water. In some other cases, nitrogen and sulfur oxides may be produced as well. Incinerators and boilers must be controlled carefully to be sure that they do not emit more than the allowable amounts of more complex, hazardous compounds. This often requires use of emission-control devices.

Combustion Source—A stationary fuel fired boiler, turbine, or internal combustion engine.

Combustion Turbine—A fossil-fuel-fired power plant that uses the conversion process known as the Brayton cycle. The fuel, oil, or gas is combusted and drives a turbine-generator. See *Gas Turbine*.

Combustion Vapor Mixture—The composition range over which air containing vapor of an organic compound will burn or even explode when set off by a flame or spark. Outside that range the reaction does not occur, but the mixture may nevertheless be hazardous because it does not contain enough oxygen to support life, or because the vapor is toxic.

Comfort Conditioning—The process of treating air to simultaneously control its temperature, humidity, cleanliness, and distribution to meet the comfort requirements of the occupants of the conditioned space.

Comfort Zone—A frequently used room or area that is maintained at a more comfortable level than the rest of the house; also known as a "warm room."

Command and Control—This is the policy approach common to many areas of environmental regulation. The essential features of it include:

- Polluters are given highly specific regulations, often including specific technologies to adopt.
- Polluters need to satisfy rigid regulatory requirements often based on ad-hoc judgments of regulators.

This approach is highly unlikely to be cost-effective unless the regulatory authority knows the MAC curves of all polluters, and designs the regulations accordingly.

Commencement Date of the Lease—The date on which a lessor makes an underlying asset available for use by a lessee.

Commercial—A utility's consumers that are non-manufacturing business establishments, including hotels, motels, restaurants, wholesale businesses, retail stores, and health, social, and educational institutions. The utility may classify commercial service as all consumers whose demand or annual use exceeds some specified limit. The limit may be set by the utility based on the rate schedule of the utility.

Commercial Building—A building with more than 50 percent of its floor space used for commercial activities. Commercial buildings include, but are not limited to, stores, offices, schools, churches, gymnasiums, libraries, museums, hospitals, clinics, warehouses, and jails. Government buildings are included except for buildings on military bases or reservations.

Commercial Facility—An economic unit that is owned or operated by one person or organization and that occupies two or more commercial buildings at a single location. A university and a large hospital complex are examples of a commercial multi-building facility.

Commercial Forest Land—Forested land which is capable of producing new growth at a minimum rate of 20 cubic feet per acre per year, excluding lands withdrawn from timber production by statute or administrative regulation.

Commercial Loan—A formal agreement in which a lender provides a borrower with funds for a stated purpose, and which is backed by the full faith and credit of the borrower. Loans may be secured by the particular asset being purchased, and/or with additional assets of the company (i.e., accounts receivables, property, other tangible assets, etc.).

Commercial Mortgage-backed Securities (CMBS)—A bond or other investment instrument backed by loans secured with commercial rather than residential property.

Commercial Operation—Commercial operation occurs when control of the generator is turned over to the system dispatcher.

Commercial Operation (Nuclear)—The phase of reactor operation that begins when power ascension ends and the operating utility formally declares the nuclear power plant to be available for the regular production of electricity. This declaration is usually related to the satisfactory completion of qualification tests on critical components of the unit.

Commercial Paper (CP)—Bearer debentures with a maturity of seven days to two years (generally 30-90 days), which are not publicly quoted. Their tradeability in comparison to other money market investments makes them attractive.

Commercial Sector—An energy-consuming sector that consists of service-providing facilities and equipment of: businesses; Federal, State, and local governments; and other private and public organizations, such as religious, social, or fraternal groups. The commercial sector includes institutional living quarters. It also includes sewage treatment facilities. Common uses of energy associated with this sector include space heating, water heating, air conditioning, lighting, refrigeration, cooking, and running a wide variety of other equipment. Note: This sector includes generators that produce electricity and/or useful thermal output primarily to support the activities of the above-mentioned commercial establishments.

Commercialization—Programs or activities that increase the value or decrease the cost of integrating new products or services into the electricity sector. (See *Sustained Orderly Development*.)

Commingled Gas—A homogeneous mix of gas obtained from various physical and contractual supply sources.

Commingling—The mixing of one utility's generated supply of electric energy with another utility's generated supply within a transmission system.

Commission—Federal Energy Regulatory Commission (FERC), or local public utility regulatory commission (PUC).

Commission (California)—The California State Energy Resources Conservation and Development Commission, also known as the California Energy Commission.

Commissioned Agent—An agent who wholesales or retails a refined petroleum product under a commission arrangement. The agent does not take title to the product or establish the selling price, but receives a percentage of fixed fee for serving as an agent.

Commissioner—Most often refers to the Commissioner of Internal Revenue Service as it relates to tax laws and issues.

Commissioning (Cx)—The process by which a power plant, apparatus, or building is approved for operation based on observed or measured operation that meets design specifications.

• The process of documenting and verifying through adjusting/remedying the performance of building facility systems so that they operate in conformity with the design intent. An independent party often performs the function.

• Elements to be commissioned are identified, installation is observed, sampling is conducted, test procedures are devised and executed, staff training is verified, and operations and maintenance manuals are reviewed.

Commitment Fee—A fee required by the lender or underwriter, at the time a proposal or commitment is accepted by the credit applicant, to lock in a specific finance rate and/or other finance terms.

Commitment Letter—A document prepared by a lender or lessor that sets forth its commitment, including rate and term, to provide financing to the borrower. This document, if utilized, precedes final documentation, and may or may not be subject final credit approval by the funding source.

Commitment Period—The commitment period, sometimes referred to as the "compliance period" or the "budget period," is the time frame given to Parties to the Kyoto Protocol to meet their quantified emission limitation and reduction commitments (QELRCs) established in Annex B. Under the Kyoto Protocol the first commitment period is 2008-2012, during which the assigned amount (of emissions) for each Party (on average, 5% below 1990 emission levels) included in Annex I must be equal to or lower than the percentage listed for it in Annex B multiplied by five.

Commitment Period Reserve—The minimum number of Kyoto Units as calculated and required to be held in a Registry pursuant to the Kyoto Protocol.

Commodity Charge—Charge for the actual energy that flows through the physical assets. Typically measured in MCF for natural gas and kWh for electricity.

Commodity Costs (Rate)—That part of the total cost of service which must be recovered through use of a commodity rate; i.e., a rate for each Mcf of gas sold. Revenue from a commodity rate varies with throughput.

Commodity Future—A futures contract on a commodity, which differs from financial futures because prices of commodity futures are determined by supply and demand instead of cost-of-carry of the underlying commodity.

Common Area (Buildings)—The total area within a property that is not designed for sale or rental but is available for common use by all owners, tenants, or their invitees, e.g., parking and its appurtenances, malls, sidewalks, landscaped areas, recreation areas, public toilets, truck and service facilities.

Common Area Maintenance (CAM)—The expense of operating and maintaining common areas; may or may not include management charges and expenditures or tenant improvements or other improvements to the property.

Common Carrier—An individual, partnership, or corporation that transports persons or goods for compensation.

Common Equity (Book Value)—The retained earnings and common stock earnings plus the balances in common equity reserves and all other common stock accounts. This also includes the capital surplus, the paid-in surplus, and the premium on common stocks, except those balances specifically related to preferred or preference stocks; less any common stocks held in the treasury.

Common Stock—A unit of ownership of a corporation. In the case of a public company, the stock is traded between investors on various exchanges. Owners of common stock are typically entitled to vote on the selection of directors and other important events and in some cases receive dividends on their holdings. Investors who purchase common stock hope that the stock price will increase so the value of their investment will appreciate. Common stock offers no performance guarantees. Additionally, in the event that a corporation is liquidated, the claims of secured and unsecured creditors and owners of bonds and preferred stock take precedence over the claims of those who own common stock.

Common Stock Equity—The funds (including retained earnings) invested in the business by the residual owners whose claims to income and assets are subordinate to all other claims.

Common Stock Equivalent—A security which is not, in form, a common stock but which usually contains provisions to enable its holder to become a common stockholder and which, because of its terms and circumstances under which it was issued, is in substance equivalent to a common stock. Convertible debt, convertible preferred stock, stock options and stock warrants meeting certain criteria are considered common stock equivalents.

Community Independent Transaction Log—The transaction log which will be established under the EU emissions Trading Scheme, through which all Transactions will be communicated and recorded, checked, and completed or rejected as appropriate.

Compact Fluorescent—A smaller version of standard fluorescent lamps which can directly replace standard incandescent lights. These lights consist of a gas filled tube, and a magnetic or electronic ballast.

Compact Fluorescent Light (CFL)—These are also known as "screw-in fluorescent replacements for incandescent" or "screw-ins." Compact fluorescent bulbs combine the efficiency of fluorescent lighting with the convenience of a standard incandescent bulb. There are many styles of compact fluorescent, including exit light fixtures and floodlights (lamps containing reflectors). Many screw into a standard light socket, and most produce a similar color of light as a standard incandescent bulb. Compact fluorescent bulbs come with ballasts that are electronic (lightweight, instant, no-flicker starting, and 10 to 15% more efficient) or magnetic (much heavier and slower starting). Other types of compact fluorescent bulbs include adaptive circulation and PL and SL lamps and ballasts. Compact fluorescent bulbs are designed for residential uses; they are also used in table lamps, wall sconces, and hall and ceiling fixtures of hotels, motels, hospitals, and other types of commercial buildings with residential-type applications.

Company Automotive (Retail) Outlet—Any retail outlet selling motor fuel under the brand name of a company reporting in the EIA Financial Reporting System.

Company Operated—A company 'retail' outlet operated by personnel paid by the reporting company. Lessee. An independent marketer who leases the station and land and has use of tanks, pumps, signs, etc. A lessee dealer typically has a supply agreement with a refiner or a distributor, and purchases products at dealer tank wagon prices. The term 'lessee dealer' is limited to those dealers who are supplied directly by a refiner or any affiliate or subsidiary company of a refiner. 'Direct supply' includes use of commission agent or common carrier delivery. Open. An independent marketer who owns or leases (from a third party—not a refiner) the station or land of a retail outlet and has use of tanks, pumps, signs, etc. An open dealer typically has a supply agreement with a refiner or a distributor, and purchases products at or below dealer tank wagon prices.

Company-Lessee Automotive Outlet—One of three types of company automotive (retail) outlets. This type of outlet is operated by an independent marketer who leases the station and land and has use of tanks, pumps, signs, etc. A lessee dealer typically has a supply agreement with a refiner or a distributor and purchases products at dealer tank wagon prices. The term includes outlets operated by commissioned agents and is limited to those dealers who are supplied directly by a refiner or any affiliate or subsidiary company of a refiner.

Company-Open Automotive Outlet—One of three types of company automotive (retail) outlets. This type of outlet is operated by an independent marketer who owns or leases (from a third party that is not a refiner) the station or land of a retail outlet and has use of tanks, pumps, signs, etc. An open dealer typically has a supply agreement with a refiner or a distributor and purchases products based on either rack or dealer tank wagon prices.

Company-Operated Automotive Outlet—One of three types of company automotive (retail) outlets. This type of outlet is operated by salaried or commissioned personnel paid by the reporting company.

Company-Operated Retail Outlet—Any retail outlet (i.e., service station) which sells motor vehicle fuels and is under the direct control of a firm that sets the retail product price and directly collects all or part of the retail margin. The category includes retail outlets operated by (1) salaried employees of the firm and/or its subsidiaries and affiliates, (2) licensed or commissioned agents, and/or (3) personnel services contracted by the firm.

Comparability—When a transmission owner provides access to transmission services at rates, terms and conditions equal to those the owner incurs for its own use.

Comparable Sales—As part of the appraisal process, those relatively recently sold properties which will be compared to the subject property (the property being appraised) for the purpose of forming an opinion of value for the subject property.

Comparison Group—A selected group of customers that do not participate in a DSM program, but otherwise have the same characteristics as the participating group. The comparison group is used to isolate program effects from other factors that affect demand. Also known as a "Control Group."

Compensating Balances—The amount of funds that a bank requires a borrower to keep on deposit during the term of a loan or other financial contract. The about of this non-interest earning deposit is typically based upon some percentage of the loan and effectively increases the borrower's interest cost since those funds are not available to the borrower for other financial purposes in the business.

Competition—Allowing two or more entities to sell similar goods and services in the same market. Electric competition means that consumers have a choice of which company they may purchase their electricity from.

Competitive Bidding (Utilities)—This is a procedure that utilities use to select suppliers of new electric capacity and energy. Under competitive bidding, an electric utility solicits bids from prospective power generators to meet current or future power demands.

Competitive Franchise—A process whereby a municipality (or group of municipalities) issues a franchise to supply electricity in the community to the winner of a competitive bid process. Such franchises can be for bundled electricity and transmission/distribution, or there can be separate franchises for the supply of electricity services and the transmission and distribution function. Franchises can be, but typically are not, exclusive licenses.

Competitive Transition Charge—A non-bypassable charge levied on each customer of the distribution utility, including those who are served under contracts with non-utility suppliers, for recovery of the utility's stranded costs that develop because of competition.

Competitive Transmission Charge (CTC)—A non-bypassable charge that customers pay to a utility for the recovery of its Stranded Costs.

Complete Building—An entire building with one occupancy making up 90 percent of the conditioned floor area. See *Entire Building*.

Complete Mix Digester—A type of anaerobic digester that has a mechanical mixing system and where temperature and volume are controlled to maximize the anaerobic digestion process for biological waste treatment, methane production, and odor control.

Completed Contract Method—In accounting, a method of revenue recognition used in long-term construction contracts when the date of completion cannot be forecast. Revenue and related costs are recognized when the project is completed.

Completion (oil/gas production)—The term refers to the installation of permanent equipment for the production of oil or gas. If a well is equipped to produce only oil or gas from one zone or reservoir, the definition of a "well" (classified as an oil well or gas well) and the definition of a "completion" are identical. However, if a well is equipped to produce oil and/or gas separately from more than one reservoir, a "well" is not synonymous with a "completion." (See *Well*.)

Completion Date (oil/gas production)—The date on which the installation of permanent equipment has been completed as reported to the appropriate regulatory agency. The date of completion of a dry hole is the date of abandonment as reported to the appropriate agency. The date of completion of a service well is the date on which the well is equipped to perform the service for which it was intended.

Compliance—A state whereby a source is achieving all that is required of it by the applicable rules and regulatory agencies.

Compliance Account—The place in the NO_X or SO_2 Allowance Tracking System where Allowances are recorded and held by a NO_X affected source or an SO_2 source.

Compliance Approach (California)—(in California, Title 24) Any one of the allowable methods by which the design and construction of a building may be demonstrated to be in compliance with Part 6. The compliance approaches are the performance compliance approach and the prescriptive compliance approach. The requirements for each compliance approach are set forth in Section 100 d) 2, of Part 6.

Compliance Coal—A coal or a blend of coals that meets sulfur dioxide emission standards for air quality without the need for flue gas de-sulfurization.

Compliance Documentation—(in California) The set of forms and other data prepared in order to demonstrate to the building official that a building complies with the Standards. The compliance forms for the residential and nonresidential standards are contained in the Residential Manual and the Nonresidential Manual (Title 24).

Compliance Period or Year—In states with RPS requirements, facilities subject to the requirement have to ensure that they have sufficient certificates in their account for each compliance period. For example, in Texas by March 31, each competitive retailer must submit credits to the program administrator from its account equivalent to its REC requirement for the previous compliance period (calendar year).

Compliance Year (California)—In the RECLAIM (South Coast Air Quality Management District) program, the 12 month period beginning on January 1 and ending on December 31 for cycle 1 facilities, and beginning on July 1 and ending on June 30 for cycle 2 facilities.

Component Depreciation—Allocating the cost of a building to its various structural components and computing the depreciation in each component based on its useful life.

Composite Book Depreciation—A method of determining an allowance for depreciation to be included as an element of cost in a cost of service study. The method looks to the service life of the total plant investment for determining depreciation rates, rather than the individual plant components. Under a composite method, an item of plant is not considered fully depreciated until that item is retired from service.

Composting—The process of degrading organic material (biomass) by microorganisms in aerobic conditions.

Composting Toilet—A self-contained toilet that use the process of aerobic decomposition (composting) to break down feces into humus and odorless gases.

Compound Annual Growth Rate (CAGR)—The year-over-year growth rate applied during a multiple-year period.

Compound Interest—An interest method that calculates interest on interest earned in prior periods.

Compound Paraboloid Collector—A form of solar concentrating collector that does not track the sun.

Comprehensive Environmental Response, Compensation and Liability Act of 1980 (CERCLA)—Also known as Superfund, this Federal law authorizes U.S. EPA to respond directly to releases of hazardous substances that may endanger public health or the environment. The superfund Amendments and Reauthorization Act of 1986 (SARA), amended and reauthorized CERCLA for five years at a total funding level of $8.5 billion. SARA also strengthened state involvement in the cleanup process, and encouraged the use of new treatment technologies and permanent solutions. CERCLA has since been extended by other laws. In particular, SARA Title III is known as the Emergency Planning and Community Right-to-Know Act of 1986 (EPCRA). It requires each state to have an emergency response plan as described, and any company that produces, uses or stores more than certain amounts of listed chemicals must meet emergency planning requirements, including release reporting.

Comprehensive National Energy Policy Act—Federal legislation in 1992 that opened the U.S. electric utility industry to increase competition at the wholesale level and left authority for retail competition to the states.

Compressed Air Storage—The storage of compressed air in a container for use to operate a prime mover for electricity generation.

Compressed Natural Gas (CNG)—Natural gas that has been compressed under high pressure, typically between 2,000 and 3,600 pounds per square inch, held in a container. The gas expands when released for use as a fuel. It is used as a fuel for natural gas powered vehicles.

Compression Chiller—A cooling device that uses mechanical energy to produce chilled water.

Compression Cycles—Adiabatic (isentropic) compression takes place when there is not heat added to or removed from the system. Compression follows the formula $p1V1k=p2V2k$, where exponent k is the ratio of the specific heat capacities. Although an adiabatic cycle is never totally obtained in practice, it is approached typically with most positive-displacement machines and is generally the base to which they are referred. Isothermal compression takes place when the temperature is kept constant as the pressure increases, requiring continuous removal of heat generated during compression. Compression follows the formula $p1V1=p2V2$. However, in practice it is never possible to remove the heat of compression as rapidly as it is generated. Polytropic compression is a compromise between the two basic processes, the adiabatic and the isothermal. It is primarily applicable to dynamic continuous-flow machines such as centrifugal or axial compressors. Compression follows the formula $p1V1n=p2V2n$, where exponent n is experimentally determined for a particular type of machine. It may be lower or higher than the exponent k used in adiabatic cycle calculations.

Compression Efficiency—The ratio of the theoretical work requirement (using a stated process) to the actual work required to compress a given quantity of gas. It accounts for the gas friction losses, internal leakage and other variations from the idealized thermodynamic process.

Compression Ratio—The relationship of absolute outlet pressure at a compressor to absolute inlet pressure.

Compressor—A device used to compress air for mechanical or electrical power production, and in air conditioners, heat pumps, and refrigerators to pressurize the refrigerant and enabling it to flow through the system.

Compressor Station—Any combination of facilities that supply the energy to move gas in transmission or distribution lines or into storage by increasing the pressure.

COMPTER—Multiple Source Air Quality Model

Concentrating (Solar) Collector—A solar collector that uses reflective surfaces to concentrate sunlight onto a small area, where it is absorbed and converted to heat or, in the case of solar photovoltaic (PV) devices, into electricity. Concentrators can increase the power flux of sunlight hundreds of times. The principal types of concentrating collectors include: compound parabolic, parabolic trough, fixed reflector moving receiver, fixed receiver moving reflector, Fresnel lens, and central receiver. A PV concentrating module uses optical elements (Fresnel lens) to increase the amount of sunlight incident onto a PV cell. Concentrating PV modules/arrays must track the sun and use only the direct sunlight because the diffuse portion cannot be focused onto the PV cells. Concentrating collectors for home or small business solar water heating applications are usually parabolic troughs that concentrate the sun's energy on an absorber tube (called a receiver), which contains a heat-transfer fluid.

Concentrating Solar or Solar Thermal Power System—A solar energy conversion system characterized by the optical concentration of solar rays through an arrangement of mirrors to generate a high temperature working fluid. Also see *Solar Rough, Solar Power Tower, or Solar Dish*. Concentrating solar power (but not Solar thermal power) may also refer to a system that focuses solar rays on a photovoltaic cell to increase conversion efficiency. Concentrators can increase the power flux of sunlight hundreds of times.

Concentration—A number expressing the percent of the specified constituent in a mixture to the total quantity of the mixture, as pounds of salt per pound of brine.

Concentrator (Solar)—A reflective or refractive device that focuses incident insolation onto an area smaller than the reflective or refractive surface, resulting in increased insolation at the point of focus. A photovoltaic module that includes optical components, such as lenses, to direct and concentrate sunlight onto a solar cell of smaller area. Most concentrator arrays must directly face or track the sun.

Concession—The operating right to explore for and develop petroleum fields in consideration for a share of production in kind (equity oil).

• An inducement for a tenant to lease space, usually in the form of free rent, an additional tenant improvement allowance, moving costs, etc.

Concessionary Purchases—The quantity of crude oil exported during a reporting period, which was acquired from the producing government under terms that arise from the firm's participation in a concession. It includes preferential crude where the reporting firm's access to such crude is derived from a former concessionary relationship.

Cond—Condition

Condemnation—The exercise of the power of Eminent Domain, i.e., the taking of property for a public use upon payment of just compensation; also refers to Condemnation of unsafe structures under the Police Power.

Condensate—The liquid resulting when water vapor contacts a cool surface; also the liquid resulting when a vaporized working fluid (such as a refrigerant) is cooled or depressurized.

Condensate (Lease Condensate)—A natural gas liquid recovered from associated and non-associated gas wells from lease separators or field facilities, reported in barrels of 42 U.S. gallons at atmospheric pressure and 60 degrees Fahrenheit.

Condensation—The process by which water in air changes from a vapor to a liquid due to a change in temperature or pressure; occurs when water vapor reaches its dew point (condensation point); also used to express the existence of liquid water on a surface.

Condenser—Heat exchanger in which the refrigerant is compressed to hot air and condensed to liquid by eliminating heat. A heat-transfer device that reduces a fluid from a vapor phase to a liquid phase.

Condenser Coil—The device in an air conditioner or heat pump through which the refrigerant is circulated and releases heat to the surroundings when a fan blows outside air over the coils. This will return the hot vapor that entered the coil into a hot liquid upon exiting the coil.

Condenser Cooling Water—A source of water external to a boiler's feed system is passed through the steam leaving the turbine in order to cool and condense the steam. This reduces the steam's exit pressure and recaptures its heat, which is then used to preheat fluid entering the boiler, thereby increasing the plant's thermodynamic efficiency.

Condensing Furnace—A type of heating appliance that extracts so much of the available heat content from a combusted fuel that the moisture in the combustion gases condenses before it leaves the furnace. Also this furnace circulates a liquid to cool the furnace's heat exchanger. The heated liquid may either circulate through a liquid-to-air heat exchanger to warm room air, or it may circulate through a coil inside a separate indirect-fired water heater.

Condensing Power—Power generated through a final steam turbine stage where the steam is exhausted into a condenser and cooled to a liquid to be recycled back into a boiler.

Condensing Turbine—A turbine used for electrical power generation from a minimum amount of steam. To increase plant efficiency, these units can have multiple uncontrolled extraction openings for feed-water heating.

Condensing Unit—The component of a central air conditioner that is designed to remove heat absorbed by the refrigerant and transfer it outside the conditioned space.

Condensing, Controlled Extraction Turbines—A controlled turbine that bleeds off (condenses) part of the main stream flow at one (single extraction) or two (double extraction) points. Used when process steam is required at pressures below the inlet pressure and above the exhaust pressure.

Conditional Demand Analysis—A method that is used to estimate equipment-specific energy consumption, without requiring end-use metered data for the appliances. Instead, it relies on the statistical analysis of consumption data, appliance saturation data, and other data such as demographic, household, weather, economic and market data.

Conditional Energy Intensity—Total consumption of a particular energy source(s) or fuel(s) divided by the total floor space of buildings that use the energy source(s) or fuel(s); i.e., the ratio of consumption to energy source-specific floor space.

Conditional Sales Agreement—An agreement for the purchase of an asset in which the borrower is treated as the owner of the asset for federal income tax purposes (thereby being entitled to the tax benefits of ownership, such as depreciation or energy tax credits), but does not become the legal owner of the asset until the terms and conditions of the agreement have been satisfied. Also known as a security agreement.

Conditional Use Permit—A permit, with conditions, allowing an approved use on a site outside the appropriate zoning class.

Conditionally Effective Rates—An electric rate schedule that has been put into effect by the FERC subject to refund pending final disposition or refiling.

Conditioned Floor Area (CFA)—The floor area of enclosed conditioned spaces on all floors measured from the interior surfaces of exterior partitions for nonresidential buildings and from the exterior surfaces of exterior partitions for residential buildings. [See California Code of Regulations, Title 24, Section 2-5302]

Conditioned Footprint—A projection of all conditioned space on all floors to a vertical plane. The conditioned footprint area may be equal to the first floor area, or it may be greater, if upper floors project over lower floors. One way to think of the conditioned footprint area is as the area of the largest conditioned floor in the building plus the conditioned floor area of any projections from other stories that extend beyond the outline of that largest floor

Conditioned Space—Enclosed space that is either directly conditioned space or indirectly conditioned space. [See California Code of Regulations, Title 24, Section 2-5302]

Conditioned Space, Directly—An enclosed space that is provided with heating equipment that has a capacity exceeding 10 Btu's/(hr-ft2), or with cooling equipment that has a capacity exceeding 10 Btu's/(hr-ft2). An exception is if the heating and cooling equipment is designed and thermostatically controlled to maintain a process environment temperature less than 65 degrees Fahrenheit or greater than 85 degrees Fahrenheit for the whole space the equipment serves. [See California Code of Regulations, Title 24, Section 2- 5302]

Conditioned Space, Indirectly—Enclosed space that: (1) has a greater area weighted heat transfer coefficient (u-value) between it and directly conditioned spaces than between it and the outdoors or unconditioned space; (2) has air transferred from directly conditioned space moving through it at a rate exceeding three air changes per hour.

Conditioned Volume—The total volume (in cubic feet) of the conditioned space within a building.

Conductance—The quantity of heat, in Btu's, that will flow through one square foot of material in one hour, when there is a 1 degree F temperature difference between both surfaces. Conductance values are given for a specific thickness of material, not per inch thickness.

Conduction—The transfer of heat through a material by the transfer of kinetic energy from particle to particle; the flow of heat between two materials of different temperatures that are in direct physical contact.

Conduction Band—An energy band in a semiconductor in which electrons can move freely in a solid, producing a net transport of charge.

Conductivity (k)—The quantity of heat that will flow through one square foot of homogeneous material, one inch thick, in one hour, when there is a temperature difference of one degree Fahrenheit between its surfaces.

Conductivity (Thermal)—This is a positive constant, k, that is a property of a substance and is used in the calculation of heat transfer rates for materials. It is the amount of heat that flows through a specified area and thickness of a material over a specified period of time when there is a temperature difference of one degree between the surfaces of the material.

Conductor—Metal wires, cables, and bus-bar used for carrying electric current. Conductors may be solid or stranded, that is, built up by an assembly of smaller solid conductors.

Conduit—A tubular material used to encase and protect one or more electrical conductors.

• An alliance between mortgage originators and an unaffiliated organization that acts as a funding source by regularly purchasing loans, usually with a goal of pooling and securitizing them.

Conduit Financing—The financial intermediary that sponsors the conduit between the lender(s) originating loans and the ultimate investor. The conduit makes or purchases loans from third-party correspondents under standardized terms, underwriting, and documents and then, when sufficient volume has been obtained, pools the loans for sale to investors in the commercial mortgage-backed securities (CMBS) market.

Conference of the Parties (COP)—The collection of nations that have ratified the Framework Convention on Climate Change (FCCC). The primary role of the COP is to keep implementation of the FCCC under review and make the decisions necessary for its effective implementation. The role of the COP, which consists of more than 170 nations that ratified or acceded to the Framework Convention on Climate Change, is to promote and review the implementation of the convention.

Confined Space—Any space not intended for continuous employee occupancy, having a limited means of egress.

Confirmed Letter of Credit—A letter of credit, issued by a foreign bank, with validity confirmed by a U.S. bank. When confirmed, the U.S. bank undertakes responsibility for payment even if the foreign buyer or bank defaults.

Confiscatory Rates—Approved rates which yield a rate of return insufficient to attract new capital.

Conflict of Interest—A situation in which, because of other activities or relationships with other persons or organizations, a person or firm is unable or potentially unable to render an impartial verification or opinion of a potential client's GHG emissions, or criteria pollutants, or the person or firm's objectivity in performing verification activities is or might be otherwise compromised.

CONG—Congressional Committee

Congestion—A condition that occurs when insufficient transfer capacity is available to implement all of the preferred schedules simultaneously.

Congestion Management—Alleviation of congestion by the ISO.

Congressional (Energy) Committees—House Subcommittee on Energy and Environment—This committee has legislative jurisdiction and general and special oversight and investigative authority on all matters relating to energy and environmental research and development and demonstration.

• House Water and Power Committee—This committee has oversight over the generation and marketing of electric power from federal water projects by federally charted or Federal RPM authorities, measures and matters concerning water resources planning, compacts relating to use and apportionment of interstate waters, water rights or power movement programs, measures and matters pertaining to irrigation and reclamation projects and other water resources development programs.

• House Water and Power Committee—This committee has oversight over the generation and marketing of electric power from federal water projects by federally charted or Federal RPM authorities, measures and matters concerning water resources planning, compacts relating to use and apportionment of interstate waters, water rights or power movement programs, measures and matters pertaining to irrigation and reclamation projects and other water resources development programs.

• Senate Committee on Energy and Natural Resources—This committee has jurisdiction on: coal production, distribution and utilization; energy policy; energy research, conservation, and development; hydroelectric power; irrigation; mineral conservation; nonmilitary development of nuclear energy; solar energy systems; and over territorial possessions, including trusteeships of the United States.

• Senate Subcommittee on Energy Research, Development, Production and Regulation—This committee has jurisdiction on the oversight and legislative responsibilities for: coal, nuclear, and non-nuclear energy commercialization projects;

DOE National Laboratories; global climate change; new technologies research and development; commercialization of new technologies including, solar energy systems; Federal energy conservation programs; energy information; and power provider policy.

Connected Load—The sum of the continuous ratings or the capacities for a system, part of a system, or a customer's electric power consuming apparatus.

Connection—The physical connection (e.g., transmission lines, transformers, switch gear, etc.) between two electric systems permitting the transfer of electric energy in one or both directions.

Connection Charge—An amount paid by a customer for being connected to an electricity supplier's transmission and distribution system.

Consent Degree—A legal document, approved and issued by a judge, formalizing an agreement between an environmental regulator and the parties potentially responsible for site contamination. The decree describes cleanup and other actions that the potentially responsible parties are required to perform and the costs incurred by the government that they will reimburse, together with the roles, responsibilities and enforcement options that the government may exercise in the event of non-compliance.

Consenting Party—A party (the company, staff, a customer, or other interested party) that supports a stipulation and agreement or settlement in a rate or other proceeding.

Conservation—Steps taken to cause less energy to be used than would otherwise be the case. These steps may involve improved efficiency, avoidance of waste, reduced consumption, etc. They may involve installing equipment (such as a computer to ensure efficient energy use), modifying equipment (such as making a boiler more efficient), adding insulation, changing behavior patterns, etc.
- Proactive measures taken to reduce energy usage including for example improved efficiency, reduced waste, reduced consumption, enhanced building envelop, equipment modifications, etc.
- Efficiency of energy use, production, transmission, or distribution that results in a decrease of energy consumption while providing the same level of service.

Conservation and Other DSM—This Demand-Side Management category represents the amount of consumer load reduction at the time of system peak due to utility programs that reduce consumer load during many hours of the year. Examples include utility rebate and shared savings activities for the installation of energy efficient appliances, lighting and electrical machinery, and weatherization materials. In addition, this category includes all other Demand-side Management activities, such as thermal storage, time-of-use rates, fuel substitution, measurement and evaluation, and any other utility-administered Demand-Side Management activity designed to reduce demand and/or electricity use.

Conservation Cost Adjustment—A means of billing electric power consumers to pay for the costs of demand side management/energy conservation measures and programs. (See also *Benefits Charge*.)

Conservation Feature—A feature in the building designed to reduce the usage of energy.

Conservation Pricing—Pricing that provides an incentive to reduce average or peak use, or both.

Conservation Program—A program in which a utility company furnishes home weatherization services free or at reduced cost or provides free or low cost devices for saving energy, such as energy efficient light bulbs, flow restrictors, weather stripping, and water heater insulation.

Conservation Supply Curve—A graph showing the quantity of energy savings of individual efficiency measures on the x-axis and the total cost per unit of energy saved on the y-axis.

Conservation-Energy—Means of reducing the energy resources required to do a task such as heating a house, transporting freight between two points, or producing steel.

Consideration—The inducement for entering into a contract. It consists of either a benefit to the Promisor, or a loss or detriment to the Promisee.

Consignee—The person or firm to whom something is sold or shipped. Buyer or importer.

Consignor—The person or firm from whom the goods have been received for shipment—the seller, shipper, or exporter.

Consolidated Metropolitan Statistical Area (CMSA)—An area that meets the requirements of a metropolitan statistical area, has a population of one million or more, and consists of two or more component parts that are recognized as primary metropolitan statistical areas.

Consolidated Parent Company—The parent company (the manufacturer or dealer in a leasing context) combined with its many subsidiaries. A parent company can be combined with its subsidiaries for tax and/or financial reporting purposes.

Constant Dollars—The value or purchasing power of a dollar in a specified year carried forward or backward.

Constant-Speed Wind Turbines—Wind turbines that operate at a constant rotor revolutions per minute (RPM) and are optimized for energy capture at a given rotor diameter at a particular speed in the wind power curve.

Constraint (Financial)—A limit placed on the financial solution of a transaction, especially when optimizing.

Construction & Demolition Recycling Association (CDRA)—The Construction & Demolition Recycling Association (CDRA) promotes and defends the environmentally sound recycling of the more than 325 million tons of recoverable construction and demolition (C&D) materials that are generated in the United States annually. These materials include aggregates such as concrete, asphalt, asphalt shingles, gypsum wallboard, wood and metals. www.cdrecycling.org

Construction (Energy Projects)—An energy-consuming sub sector of the industrial sector that consists of all facilities and equipment used to perform land preparation and construct, renovate, alter, install, maintain, or repair major infrastructure or individual systems therein. Infrastructure includes buildings; industrial plants; and other major structures, such as tanks, towers, monuments, roadways, tunnels, bridges, dams, pipelines, and transmission lines.

Construction Costs (of the electric power industry)—All direct and indirect costs incurred in acquiring and constructing electric utility plant and equipment and proportionate shares of common utility plants. Included are the cost of land and improvements, nuclear fuel and spare parts, allowance for funds used during construction, and general overheads capitalized, less the cost of acquiring plant and equipment previously operated in utility service.

Construction Documents—Construction Documents is a set of drawings (plans) and written specifications that describe construction requirements for a building. Detailed floor plans, elevations, sections and drawings of specific areas (such as window, door, and staircase details; detailed engineering plans (structural, mechanical, plumbing, electrical, electronic); and detailed written specifications. Construction documents are reviewed after the de-sign development stage.

Construction Expenditures (Utilities)—The gross expenditures for construction costs (including the cost of replacing worn out plants), and electric construction costs, and land held for future use.

• Cost of construction, including work-in-progress (CWIP), overhead or contributions in aid of construction and allowances for funds used during construction (AFUDC), for additions to, renewals and replacements of plant facilities, but excluding the purchase cost of an acquired operating unit or system of utility plant, accounting transfers and adjustments to utility plant, and cost to remove plant facilities from service.

Construction Layers—Roof, wall and floor constructions which represent an assembly of layers. Some layers are homogeneous, such as gypsum board and plywood sheathing, while other layers are non-homogeneous such as the combination of wood framing and cavity insulation typical in many buildings.

Construction Loan—Financing arranged for the construction of personal or real property; generally a short-term, floating-rate debt repaid with the proceeds from permanent financing. Construction loans with a permanent financing commitment are designated as loans "with a takeout commitment" and those that do not have permanent financings are loans "without a takeout commitment." May include an interest carry and/or a very small permanent component to see the property from the post-construction phase through full commercial operation or tenancy.

Construction Pipeline (of a nuclear reactor)—The various stages involved in the acquisition of a nuclear reactor by a utility. The events that define these stages are the ordering of a reactor, the licensing process, and the physical construction of the nuclear generating unit. A reactor is said to be "in the pipeline" when the reactor is ordered and "out of the pipeline" when it completes low power testing and begins operation toward full power.

Construction Specifications Institute (CSI)—The mission of CSI is to advance building information management and education of project teams to improve facility performance. www.csinet.org

Construction Work in Progress (CWIP)—The balance shown on a utility's balance sheet for construction work not yet completed but in process. This balance line item may or may not be included in the rate base.

Constructive Notice—Notice given by the public records of a claim of Ownership or interest in property. Generally, the law presumes that one has the same knowledge of instruments properly recorded as if one were actually acquainted with them.

Constructive Surplus or Deficit—The amounts representing the exchange of services, supplies, etc., between the utility department and the municipality and its other departments without charge or at a reduced charge. Charges to this account include utility and other services, supplies, etc., furnished by the utility department to the municipality or its other departments without charge, or the amount of the reduction, if furnished at a reduced charge. Credits to the account consist of services, supplies, office space, etc., furnished by the municipality to the utility department without charge on the amount of the reduction, if furnished at a reduced charge.

Consumer (Energy)—Any individually metered dwelling, building, establishment, or location using natural gas, synthetic natural gas, and/or mixtures of natural and supplemental gas for feedstock or as fuel for any purpose other than in oil or gas lease operations; natural gas treating or processing plants; or pipeline, distribution, or storage compressors.

Consumer Charge—An amount charged periodically to a consumer for such utility costs as billing and meter reading, without regard to demand or energy consumption.

Consumer Price Index (CPI)—These prices are collected in 85 urban areas selected to represent all urban consumers about 80 percent of the total U.S. population.

Consumption (Fuel)—Amount of fuel used for gross generation.

Consumption Charge—The part of a power provider's charge based on actual energy consumed by the customer; the product of the kilowatt-hour rate and the total kilowatt-hours consumed.

Consumption per Square Foot—The aggregate ratio of total consumption for a particular set of buildings to the total floor space of those buildings.

Contact Resistance—The resistance between metallic contacts and the semiconductor.

Container—A uniform, sealed, reusable metal "box" (generally 40 feet in length, able to hold about 40,000 pounds) in which goods are shipped by vessel, truck or rail. The use of containers (or containerization) in trade is generally thought to require less labor and reduce losses due to breakage, spoil-

age, and pilferage than more traditional shipment methods.

Containment—Enclosing or containing hazardous substances in a structure to prevent the migration of contaminants into the environment.

Content of Fuel—The heat value per unit of fuel expressed in Btu as determined from tests of fuel samples. Examples: Btu per pound of coal, per gallon of oil, per cubic foot of gas.

Contiguous—Being in actual contact; adjoining or touching.

Continental Shelf—The portion of the sea bottom that slopes gradually from the edge of a continent. Usually defined as areas where water is less than 200 meters or 600 feet deep.

Contingency Planning—The Energy Commission's strategy to respond to impending energy emergencies such as curtailment or shortage of fuel or power because of natural disasters or the result of human or political causes, or a clear threat to public health, safety or welfare. The contingency plan specifies state actions to alleviate the impacts of a possible shortage or disruption of petroleum, natural gas or electricity. The plan is reviewed and updated at least every five years. Legislative authority for the California Energy Shortage Contingency Plan is found in Public Resources Code, Section 25216.5. All states have contingency plans.

Contingent—Dependent upon an uncertain event.

Contingent Rentals—Rentals in which the amount of the rents is dependent upon some future event such as a price index or borrowing rate, rather than the passage of time.

• Rental payments which depend on events occurring subsequent to the start of the lease, such as prevailing interest rates or revenue generated through the use of the leased asset

Contingent Valuation Method—This is a survey-based method for putting a dollar value on the benefits from environmental cleanup. It is subject to a number of biases that make it imprecise. The most serious of these is hypothetical bias.

Continuous Delivery Energy Sources—Those energy sources provided continuously to a building.

Continuous Dimming—A lighting control method that is capable of varying the light output of lamps over a continuous range from full light output to minimum light output.

Continuous Emissions Monitoring System (CEMS)—The total equipment required to determine a continuous measurement of a gas concentration or emission rate from combustion or industrial process.

Continuous Fermentation—A steady-state fermentation process.

Continuous Mining—A form of room pillar mining in which a continuous mining machine extracts and removes coal or other mineral from the working face in one operation; no blasting is required.

Continuous Monitoring—The taking and recording of measurement at regular and frequent intervals during operation of a facility or item of equipment.

Continuous Output Rating—The maximum amount of power an inverter may deliver to a load (or loads) for a sustained period of time.

Contract—An agreement between two or more parties that creates enforceable rights and obligations.

Contract Path—The most direct physical transmission tie between two interconnected entities. When utility systems interchange power, the transfer is presumed to take place across the "contract path," notwithstanding the electrical fact that power flow in the network will distribute in accordance with network flow conditions. This term can also mean to arrange for power transfer between systems. (See also *Parallel Path Flow*)

Contract Pressure—The maximum or minimum required operating pressure at a receipt or delivery point as specified in the Service Agreement.

Contract Price—Price of fuels marketed on a contract basis covering a period of one or more years. Contract prices reflect market conditions at the time the contract was negotiated and therefore remain constant throughout the life of the contract or are adjusted through escalation clauses. Generally, contract prices do not fluctuate widely.

Contract Quantity Method—A method to allocate demand costs by function to customer classes based on the customer classes' contract quantity or a company's obligation to serve the customer class.

Contract Receipts—Purchases based on a negotiated agreement that generally covers a period of 1 or more years.

Contracted Gas—Any gas for which Interstate Pipeline has a contract to purchase from any domestic or foreign source that cannot be identified to a specific field or group. This includes tailgate plant purchases, single meter point purchases, pipeline purchases, natural gas imports, SNG purchases, and LNG purchases.

Contracted Reserves (Gas)—Natural gas reserves ded-

icated to the fulfillment of gas purchase contracts.

Contractor's Liability Insurance—Insurance purchased and maintained by the contractor to protect the contractor from specified claims which may arise out of, or result from, the contractor's operations under the contract, whether such operations are by the contractor or by any subcontractor, or by any one directly or indirectly employed by any of them, or by anyone who acts for any of them and may be liable.

Contracts for Differences—A type of bilateral contract where the electrical generation seller is paid a fixed amount over time which is a combination of the short-term market price and an adjustment with the purchaser for the difference.

Contracts for Differences (CFD)—A type of bilateral contract where the electric generation seller is paid a fixed amount over time which is a combination of the short-term market price and an adjustment with the purchaser for the difference. For example, a generator may sell a distribution company power for ten years at 6-cents/kilowatt-hour (kWh). That power is bid into Poolco at some low/kWh value (to ensure it is always taken). The seller then gets the market clearing price from the pool and the purchaser pays the producer the difference between the Poolco selling price and 6-cents/kWh (or vice versa if the pool price should go above the contract price).

Contrast—The difference between the brightness of an object compared to that of its immediate background.

Contribution to Net Income—The FRS (Financial Reporting System survey) segment equivalent to net income. However, some consolidated items of revenue and expense are not allocated to the segments, and therefore they are not equivalent in a strict sense. The largest item not allocated to the segments is interest expense since this is regarded as a corporate level item for FRS purposes.

Control—Including the terms "controlling," "controlled by," and "under common control with," means the possession, direct or indirect, of the power to direct or cause the direction of the management and policies of a person, whether through the ownership of voting shares, by contract, or otherwise.

Control Area—An electric power system, or a combination of electric power systems, to which a common automatic generation control (AGC) is applied to match the power output of generating units within the area to demand.

Control Plot—A plot in which no vegetation will be cut and natural succession will occur. A control plot serves as a baseline to compare other treatments (Early, Mid and Late Seral).

Control Total—The number of elements in the population or a subset of the population. The sample weights for the observed elements in a survey are adjusted so that they add up to the control total. The value of a control total is obtained from an outside source. The control totals are given by the number of households in one of the 12 cells by categorizing households by the four Census regions and by three categories of metropolitan status (Metropolitan Statistical Area central city, Metropolitan Statistical Area outside central city, and non-Metropolitan Statistical Area). The control totals are obtained from the Current Population Survey.

Controlled Ventilation Crawl Space (CVC)—A crawl space in a residential building where the side walls of the crawlspace are insulated rather than the floor above the crawlspace. A CVC has automatically controlled crawl space vents. Credit for a CVC is permitted for low-rise residential buildings that use the performance approach to compliance.

CONUS—Contiguous or Continental United States

Convection—Transferring heat by moving air, or transferring heat by means of upward motion of particles of liquid or gas heat from beneath.

Convection, Forced—Heat transfer through a fluid such as air or water by currents caused by a device powered by an external energy source, such as an electric fan or pump.

Convection, Natural—A method of heat transfer where a fluid (liquid, gas or molten metal) picks up heat from one object and carries it to another by currents that result from the rising of lighter, warm fluid and the sinking of heavier, cool fluid. The two objects exchanging energy don't have to physically touch. Convection is the source of the "wind chill factor" and the "cooling breeze" one experiences outdoors and the "draft" a person often feels while indoors.

Convector—An agency of convection. In heat transfer, a surface designed to transfer its heat to a surrounding fluid largely or wholly by convection. The heated fluid may be moved mechanically or by gravity (gravity convector). Such a surface may or may not be enclosed or concealed.

Conventional Energy Resource—Electric generation facilities or technologies that have been in practical use for a long time (i.e., hydroelectricity) or which

represent the majority of generation resources in use (i.e., coal, natural-gas, nuclear). Although some utilities may still consider renewable energy technologies as "alternative resources" they have reached the status of mainstream, if not conventional resources.

Conventional Forest Products—Any commercial roundwood product (boards, dimension lumber, pulp and paper products) except fuel wood.

Conventional Fuel—The fossil fuels: coal, oil, and natural gas.

Conventional Gasoline—Finished motor gasoline not included in the oxygenated or reformulated gasoline categories. Note: This category excludes reformulated gasoline blendstock for oxygenate blending (RBOB) as well as other blendstock.

Conventional Heat Pump—This type of heat pump is known as an air-to air system.

Conventional Hydroelectric Plant—A plant in which all of the power is produced from natural stream flow as regulated by available storage.

Conventional Mill (Uranium)—A facility engineered and built principally for processing of uraniferous ore materials mined from the earth and the recovery, by chemical treatment in the mill's circuits, of uranium and/or other valued co-product components from the processed one.

Conventional Mining—The oldest form of room pillar mining, which consists of a series of operations that involve cutting the coal bed, so it breaks easily when blasted with explosives or high pressure air, and then loading the broken coal.

Conventional Power—Power generation from sources such as petroleum, natural gas, or coal. In some cases, large-scale hydropower and nuclear power generation are considered conventional sources.

Conventional Thermal Electricity Generation—Electricity generated by an electric power plant using coal, petroleum, or gas as its source of energy.

Conventionally Fueled Vehicle—A vehicle that runs on petroleum-based fuels such as motor gasoline or diesel fuel.

Conversion—Device or kit by which a conventional fuel vehicle is changed to an alternative fuel vehicle.

Conversion Company—An organization that performs vehicle conversions on a commercial basis.

Conversion Efficiency—The amount of energy produced as a percentage of the amount of energy consumed. The ratio of the electrical energy generated by a solar PV cell to the solar energy impacting the cell. The balance is lost as heat or reflected light.

Conversion Factor—A number that translates units of one measurement system into corresponding values of another measurement system. Conversion factors can be used to translate physical units of measure for various fuels into Btu equivalents. Note: For specific conversion factors, see *EIA Data Products*.

Conversion Ratio—The number of shares of stock into which a convertible security may be converted. The conversion ratio equals the par value of the convertible security divided by the conversion price.

Converted (Alternative-fuel) Vehicle—A vehicle originally designed to operate on gasoline/diesel that was modified or altered to run on an alternative fuel after its initial delivery to an end-user.

Converter—Any technology that changes the potential energy in a fuel into a different from of energy such as heat or motion. The term also is used to mean an apparatus that changes the quantity or quality of electrical energy. An electronic device for DC power that steps up voltage and steps down current proportionally (or vice-versa).

Convertible Bonds—Bonds that may be exchanged for other securities of the corporation, usually common stock. The buyer of a convertible bond or preferred stock has the security of the promised interest or preferred dividend yet can enjoy profits from the rise in price of the stock into which the convertible security can be converted once that stock's price exceeds the stipulated conversion price.

Convertible Currency—The currency of a nation that may be exchanged for that of another nation without restriction. Also referred to as hard currency.

Convertible Debenture—Debt instrument that automatically or voluntarily converts to some other security, either debt or equity.

Convertible Preferred Stock—Preferred stock that may be converted into common stock or another class of preferred stock, either voluntarily or mandatory.

Convertible Securities—Securities which are convertible into other classes of securities (usually common stock) of the same corporation at the option of the security holder, but only in accordance with prescribed conditions.

Convertible Security—A bond, debenture or preferred stock that is exchangeable for another type of security (usually common stock) at a pre-stated price. Convertibles are appropriate for investors who want higher income, or liquidation-preference protection, than is available from common stock,

together with greater appreciation potential than regular bonds offer.

Conveyance—A written instrument transferring the Title to land or personal property or an interest therein from one Party to another.

Conveying System—A device for transporting materials from one piece of equipment or location to another location within a facility. Conveying systems include but are not limited to the following: feeders, belt conveyors, bucket elevators and pneumatic systems.

Conveyor—A mechanical apparatus for carrying bulk material from place to place; for example, an endless moving belt or a chain of receptacles.

Cool Roof—A roofing material with high thermal emittance and high solar reflectance, or lower thermal emittance and exceptionally high solar reflectance, that reduces heat gain through the roof.

Cool Roof Rating Council (CRRC)—A not-for-profit organization designated by the Commission (in California) as the Supervisory Entity with responsibility to rate and label the reflectance and emittance of roof products. www.coolroofs.org

Cooling—Conditioning of room air for human comfort by a refrigeration unit (such as an air conditioner or heat pump) or by circulating chilled water through a central cooling or district cooling system. Use of fans or blowers by themselves, without chilled air or water, is not included in this definition of cooling.

Cooling Balance Point—The Cooling Balance Point is the outdoor temperature above which additional energy will be consumed at a meter to satisfy cooling loads.

Cooling Capacity—The quantity of heat that a cooling appliance is capable of removing from a room in one hour.

Cooling Capacity, Latent—Available refrigerating capacity of an air conditioning unit for removing latent heat from the space to be conditioned.

Cooling Capacity, Sensible—Available refrigerating capacity of an air conditioning unit for removing sensible heat from the space to be conditioned.

Cooling Capacity, Total—Available refrigerating capacity of an air conditioner for removing sensible heat and latent heat from the space to be conditioned.

Cooling Degree Day (CDD)—A value used to estimate interior air cooling requirements (load) calculated as the number of degrees per day (over a specified period) that the daily average temperature is above 65 degrees Fahrenheit (or some other, specified base temperature). The daily average temperature is the mean of the maximum and minimum temperatures recorded for a specific location for a 24 hour period.

Cooling Degree-Days—A measure of how warm a location is over a period of time relative to a base temperature, most commonly specified as 65 degrees Fahrenheit. The measure is computed for each day by subtracting the base temperature (65 degrees) from the average of the day's high and low temperatures, with negative values set equal to zero. Each day's cooling degree-days are summed to create a cooling degree-day measure for a specified reference period. Cooling degree-days are used in energy analysis as an indicator of air conditioning energy requirements or use.

Cooling Equipment—Equipment used to provide mechanical cooling for a room or room in a building.

Cooling Load—That amount of cooling energy to be supplied (or heat and humidity removed) based on the sensible and latent loads.

Cooling Load Temperature Difference (CLTD)—An equivalent temperature difference used for calculating the instantaneous external cooling loads across a wall or roof. The cooling load is the CLTD x U-factor x Area.

Cooling Pond—A natural or manmade body of water that is used for dissipating waste heat from power plants.

Cooling System—An equipment system that provides water to the condensers and includes water intakes and outlets; cooling towers; and ponds, pumps, and pipes.

Cooling Tower—A structure used to cool power plant water; water is pumped to the top of the tubular tower and sprayed out into the center, and is cooled by evaporation as it falls, and then is either recycled within the plant or is discharged.

Co-Op—This is the commonly used term for a rural electric cooperative. Rural electric cooperatives generate and purchase wholesale power, arrange for the transmission of that power, and then distribute the power to serve the demand of rural customers. Co-ops typically become involved in ancillary services such as energy conservation, load management and other demand-side management programs in order to serve their customers at least cost.

Cooperative Electric Utility—An electric utility legally established to be owned by and operated for the benefit of those using its service. The utility company will generate, transmit, and/or distribute supplies of electric energy to a specified area not being serviced by another utility. Such ventures are generally exempt from Federal income tax laws. Most electric cooperatives have been initially financed by the Rural Utilities Service (prior Rural Electrification Administration), U.S. Department of Agriculture.

Coordination Service—The sale, exchange, or transmission of electricity between two or more electric utilities that typically have sufficient generation and transmission capacity to supply their load requirements under normal conditions.

Coordination Service Pricing—The typical price components of a bulk power coordination sale are an energy charge, a capacity, or reservation charge, and an adder. The price for a particular sale may embody some or all of these components. The energy charge is made on a per-kilowatt basis and is intended to recover the seller's system incremental variable costs of making a sale. Because the non-fuel expenses are usually hard to quantify, and small relative to fuel expense, energy charges quoted are usually based on fuel cost. A capacity charge is set at a certain level per kilowatt and is normally paid whether or not energy is taken by the buyer. An adder is added to that energy charge to recover the hard quantify non-fuel variable costs.

There are three types of adders: percentage, fixed, and split savings. A percentage adder increases the energy charge by a certain percentage. A fixed adder adds a fixed amount per kilowatt hour to the energy charge. Split savings adders are used only in economy energy transactions. They split production costs savings between the seller and the buyer by adding one half of the savings to the energy cost.

COP (Coefficient of Performance)—Used to rate the performance of a heat pump, the COP is the ratio of the rate of useful heat output delivered by the complete heat pump unit (exclusive of supplementary heating) to the corresponding rate of energy input, in consistent units and under specific conditions. [See California Code of Regulations, Title 24, Section 2-1602(c)(4)]

Copper Indium Disenenide (CuInSe2, or CIS)—A polycrystalline thin-film photovoltaic material (sometimes incorporating gallium (CIGS) and/or sulfur).

Coppice Regeneration—The ability of certain hardwood species to regenerate by producing multiple new shoots from a stump left after harvest.

Coproducts—The potentially useful byproducts of ethanol fermentation process.

Cord—A measure of volume, 4 by 4 by 8 feet, used to define amounts of stacked wood available for use as fuel. Burned, a cord of wood produces about 5 million calories of energy. One cord contains about 1.2 bone dry tons.

Core/Noncore Customers (Utilities)—Two basic customer classes. Core includes residential and small business users. Noncore includes large industrial users who generally also have the ability to switch to other fuels. Some large companies prefer to be treated like core customers, have the utility purchase and deliver gas or electricity for them, but pay a handling fee to the utility for special treatment.

Corn Stover—Residue materials from harvesting corn consisting of the cob, leaves and stalk.

Corporate Average Fuel Economy (CAFE)—A sales-weighted average fuel mileage calculation, in terms of miles per gallon, based on city and highway fuel economy measurements performed as part of the federal emissions test procedures. CAFE requirements were instituted by the Energy Policy and Conservation Act of 1975 (89 Statute. 902) and modified by the Automobile Fuel Efficiency Act of 1980 (94 Statute. 1821). For major manufacturers, CAFE levels in 1996 are 27.5 miles per gallon for light-duty automobiles. CAFE standards also apply to some light trucks. The Alternative Motor Fuels Act of 1988 allows for an adjusted calculation of the fuel economy of vehicles that can use alternative fuels, including fuel-flexible and dual-fuel vehicles.

Corporate Charter—The document prepared when a corporation is formed. The Charter sets forth the objectives and goals of the corporation, as well as a complete statement of what the corporation can and cannot do while pursuing these goals.

Corporate Citizenship—A company's responsible involvement with the wider community in which it is situated.

Corporate Health—The idea that companies, especially commercial businesses, have a duty to care for all of their stakeholders in all aspects of their operations.

Corporate Resolution—A document signed by a registered corporate officer, designating company representatives authorized to sign financial contracts which obligate the company.

Corporate Responsibility Report—A periodically-published report of a company's corporate responsibility practices, goals, and progress toward achieving those goals that may be included with the company's annual report or as a separate publication that focuses on the company's social and environmental impact. The process of creating this report is meant to uncover strengths and weaknesses as well as enhance transparency for all company stakeholders.

Corporate Social Responsibility—The continuing commitment by businesses to behave ethically and contribute to economic development while improving the quality of life of the workplace as well as the local community and society at large. A company's obligation to be accountable to all of its stakeholders in all its operations and activities (including financial stakeholders as well as suppliers, customers, and employees) with the aim of achieving sustainable development not only in the economic dimension but also in the social and environmental dimensions.

Corporate Sustainability Report—A periodic report published by a company to outline its progress toward meeting its financial, environmental, and social sustainability goals. Often published in compliance with third-party standards such as the UN Global Compact or Global Initiative.

Corporation—The term "corporation" includes associations, joint-stock companies, and insurance companies.

Correlation (Statistical Term)—In its most general sense, correlation denotes the interdependence between quantitative or qualitative data. It would include the association of dichotomized attributes and the contingency of multiple classified attributes. The concept is quite general and may be extended to more than two variates. The word is most frequently used in a somewhat narrower sense to denote the relationship between measurable variates or ranks.

Corrosion—Destruction of a metal by chemical or electrochemical reaction with its environment.

COS—Cost Accounting System

• Cost of Service: A method of using utility costs in rate design; a cost of service study measures a utility's costs incurred in serving each customer class, including a reasonable return on investment.

Cost—The amount of money actually paid for property, material or services. When the consideration given is other than cash in a purchase and sale transaction, as distinguished from a transaction involving the issuance of common stock in a merger or a pooling of interest, the value of such consideration shall be determined on a cash basis.

Cost Approach (Appraisal)—A set of procedures through which a value indication is derived for the fee simple interest in a property by estimating the current cost to construct a reproduction of (or replacement for) the existing structure, including an entrepreneurial incentive, deducting depreciation from the total cost, and adding the estimated land value. Adjustment may then be made to the indicated fee simple value of the subject property to reflect the value of the property interest being appraised.

Cost Effectiveness (Environmental)—A policy is cost-effective if it achieves a given environmental quality target at minimum cost. Equally, the cost effective policy will get the maximum abatement (environmental quality) for a given cost. Thus, cost-effectiveness analysis is the process of choosing the least cost method of attaining a given environmental target.

The cost per unit of reduced emissions of greenhouse gases adjusted for its global warming potential.

Cost Model for Undiscovered Resources—A computerized algorithm that uses the uranium endowment estimated for a given geological area and selected industry economic indexes to develop random variables that describe the undiscovered resources ultimately expected to be discovered in that area at chosen forward cost categories.

Cost of Acquisition—Cost of acquisition, account: actual expenditure made to acquire an economic asset. The cost of acquisition is the basis for capitalizing/preparing accounts at a leasing or PPA provider company.

Cost of Capital (Utility)—The rate of return a utility must offer to obtain additional funds. The cost of capital varies with the leverage ratio, the effective income tax rate, conditions in the bond and stock markets, growth rate of the utility, its dividend strategy, stability of net income, the amount of new capital required, and other factors dealing with business and financial risks. It is a composite of the cost for debt interest, preferred stock dividends, and common stockholders' earnings that provide the facilities used in supplying utility service.

Cost of Cary—Generally, the costs associated with holding an investment over time—opportunity costs, fees and other expenses. In carbon markets, where the bulk of trade in emission allowances and credits is in forward contracts for their future delivery, it's largely the time value of money, or the interest rate. Here, 'cost of carry' represents the investment return foregone by the seller (who does not receive the sale proceeds until delivery) over the contract period, and is reflected in a premium built into forward carbon prices over spot prices.

Cost of Compliance—For firms, this is the total cost of complying with environmental regulations. In the case of an effluent fee, this is the abatement costs plus the fees to be paid. In the case of permits, it's the abatement cost plus the cost of additional permits acquired (or less the number of permits sold).

Cost of Debt—The interest rate paid on new increments of debt capital multiplied by 1 minus the tax rate. The costs incurred by a firm to fund the acquisition of assets through the use of borrowings. A firm's component cost of debt is used in calculating the firm's overall weighted-average cost of capital.

Cost of Equity—The return on investment required by the equity holders of a firm. Cost of equity can be calculated using any number of different theoretical approaches and must take into consideration the current and long-term yield requirements of a firm's investors. A firm's component cost of equity is used in calculating the firm's overall weighted-average cost of capital.

Cost of Goods Sold—The sum of all costs required to acquire and prepare goods for sale.

Cost of Preferred Stock—The preferred stock dividends divided by the net price of the preferred stock.

Cost of Removal—The cost of demolishing, dismantling, tearing down or otherwise removing plant, including the cost of transportation and handling incidental thereto.

Cost of Retained Earnings—The residual of an entity's earnings over expenditures, including taxes and dividends, that are reinvested in its business. The cost of these funds is always lower than the cost of new equity capital, due to taxes and transactions costs. Therefore, the cost of retained earnings is the yield that retained earnings accrue upon reinvestment.

Cost of Service (Utilities)—A ratemaking concept used for the design and development of rate schedules to ensure that the filed rate schedules recover only the cost of providing the electric service at issue. This concept attempts to correlate the utility's costs and revenue with the service provided to each of the various customer classes.

• A term used in public utility regulation to mean the total number of dollars required to supply any total utility service (i.e., revenue requirements); it must include all of the supplier's costs, an amount to cover operation and maintenance expenses, and other necessary costs such as taxes, including income taxes, depreciation, depletion, and amortization of the property not covered by ordinary maintenance. Included also is a fair return in order that the utility can maintain its financial integrity, attract new capital, and compensate the owners of the property for the risks involved. A "cost of service study" is made in order to assist in determining the total revenue requirements to be recovered from each of the various classes of service. The amounts to be recovered from each of the classes of service are determined by the management or a commission after study of the various factors involved in rate design. Cost analysis or cost allocation is an important factor in rate design but only one of several important factors. Cost analysis does not produce a precise inflexible "cost of service" for any individual class of service because cost analysis involves judgement in certain cost areas. Its principal value is in determining the minimum costs attributable to each class of service. Other factors that must be considered in rate design are the value of the service, the cost of competitive services, the volume and load factor of the service and their relation to system load equalization and stabilization of revenue, promotional factors and their relation to the social and economic growth of the service area, political factors such as the sizes of minimum bills, and regulatory factors.

Cost, Insurance, Freight (CIF)—A type of sale in which the buyer of the product agrees to pay a unit price that includes the f.o.b. value of the product at the point of origin plus all costs of insurance and transportation. This type of transaction differs from a "delivered" purchase in that the buyer accepts the quantity as determined at the loading port (as certified by the Bill of Loading and Quality Report) rather than pay on the basis of the quantity and quality ascertained at the unloading port. It is similar to the terms of an f.o.b. sale except that the seller, as a service for which he is compensated, arranges for transportation and insurance.

Cost-Based Pricing—A method of setting rates so that a utility can recover the costs of providing that particular service.

Cost-Effective—A term describing a resource that is available within the time it is needed and is able to meet or reduce electrical power demand at an estimated incremental system cost no greater than that of the least-costly, similarly reliable and available alternative.

Cost-of-Service Regulation—A traditional electric utility regulation under which a utility is allowed to set rates based on the cost of providing service to customers and the right to earn a limited profit.

Cost-Plus Contract—Government contracts, popular during World War II, to control manufacturer profits on federal work. Often used in conjunction with leasing contracts.

Costs (Imports of Natural Gas)—All expenses incurred by an importer up to the U.S. point of delivery for the reported quantity {of natural gas} imported.

COTE—AIA Committee on the Environment

COTR—Contracting Officer's Technical Representative

COTS—Commercially available off-the-shelf

COU—Consumer-owned Utility: Independent, usually electric, utility established to provide at-cost electricity to its members. Also known as a "cooperative." The utility will generate, transmit, and/or distribute supplies of electric energy to a specified area not being serviced by another utility. Such ventures are generally exempt from Federal income tax laws. Most electric cooperatives have been initially financed by the Rural Electrification Administration, U.S. Department of Agriculture.

Coulomb—A unit for the quantity of electricity transported in 1 second by a current of 1 ampere.

Council of Petroleum Accountants Societies (CO-PAS)—An organization that publishes industry guidelines for oil and gas accounting and shipping.

Council on Environmental Quality (CEQ)—An advisory council to the President established by the National Environmental Policy Act of 1969. The CEQ reviews federal programs for their effect on the environment, conducts environmental studies, and advises the President on environmental matters.

Counterflow Heat Exchanger—A heat exchanger in which two fluids flow in opposite directions for transfer heat energy from one to the other.

Countries with Economies in Transition (EIT)—Under the Kyoto Protocol, the Central and East European countries, Russia, and the former republics of the Soviet Union that are in transition from centrally-planned economies to market-based economies.

Courtyard—An open space through one or more floor levels surrounded by walls within a building.

Covenant & Condition—A formal and binding agreement or promise between two or more parties for the performance of some action. Usually contained in a finance agreements.

Covenants—Restrictions on the use of a property

- A legally binding commitment by the issuer of municipal bonds to the bondholder. An impairment of a covenant can lead to a Technical Default.
- A protective clause in an agreement

COWPS—Council on Wage and Price Stability

CPA—Certified Public Accountant
- Contract Property Administrator

CPAF—Cost Plus Award Fee

CPBD—Certified Professional Building Designer

CPC—Chemical Protective Clothing

CPCN—Certificate of Public Convenience and Necessity: Prior to beginning operations or making significant additions to its plants, a public utility must apply to the Commission for permission. This procedure helps ensure that no plant facilities are built, or transportation services begun, that the public does not need. The meaning of this term is not spelled out by statute, however, and the Commission has broad discretion in granting or denying these applications.

CPDA—Chemical Producers and Distributors Association www.cpda.com

CPF—Cancer Potency Factor
- Carcinogenic Potency Factor

CPFF—Cost Plus Fixed Fee

CPG—Comprehensive Procurement Guidelines

CPI—Customer Price Index: An index measuring the change in the cost of typical wage earner purchases of goods and services in some base period; also, cost-of-living index.
- Chemical Process Industries

CPIF—Cost Plus Incentive Fee

CPM—Continuous Particle Monitor
- Compliance Project Manager
- Certified Property Manager

CPO—Certified Project Officer

CPP—Compliance Policy and Planning
- Critical Peak Pricing: A time-of-use pricing that effectively reduces a customer's electricity consumption.

CPS—Contract Payment System
- Compliance Program and Schedule

CPSA—Consumer Product Safety Act

CPSC—Consumer Product Safety Commission

CPT—Comprehensive Performance Test

CPUC—California Public Utilities Commission (United States) www.cpuc.ca.gov

CPV—Concentrating Photovoltaics

CQA—Construction Quality Assurance

CQAP—Construction Quality Assurance Plan

CR—Community Relations

- Cost Reimbursable
- Conversion Ratio

CRA—Community Reinvestment Act
- Civil Rights Act
- Classification Review Area

Cracking—The process of breaking down larger molecules into smaller molecules, utilizing catalysts and/or elevated temperatures and pressures.

Cradle to Grave—A design philosophy put forth by architect William McDonough that considers the life-cycle of a material or product, and ensures that the product is completely recycled at the end of its defined lifetime. A procedure in which hazardous materials are identified and followed as they are produced, treated, transported, and disposed of by a series of permanent, linkable, descriptive documents (e.g., manifests).

Cramming—Refers to the placement of charges on a consumer's bill that the consumer has not authorized.

CRAVE—Cancer Risk Assessment Verification Endeavor

Crawl Space—A space immediately under the first floor of a building adjacent to grade.

Crawlspace—The unoccupied, and usually unfinished and unconditioned space between the floor, foundation walls, and the slab or ground of a building.

CRC—Community Relations Coordinator
- Contamination Reduction Corridor

CRD—Community Relations Division

CRDL—Contract-Required Detection Limit

CREB—*Clean Renewable Energy Bond* CREBs are part of the Energy Policy Act of 2005 signed into law on August 8, 2005. It is a program designed to give an incentive to develop clean, renewable energy sources by providing very low-cost capital. It is designed to provide a similar incentive to the production tax credit (PTC) program currently offered to private investors.

Under the energy Policy Act, a qualified issuer can issue CREBs. Then, instead of paying interest to the bondholder, the federal government provides a tax credit to the bond purchaser. The proceeds from these bonds are then available to finance new renewable energy projects.

The same projects that qualify under the production tax credit program are eligible under the program. This includes wind, closed-loop biomass, solar, refined coal production, small irrigation power, landfill gas and qualified hydropower.

Credit Enhancement—A technique to reduce historical loss risk and concentration risk of a loan portfolio in order to achieve a better credit rating and lower borrowing costs.

Credit for Early Action—Within the Kyoto Protocol, Annex B governments cannot receive credits before the first commitment period (2008-2012) towards their emission obligation, except under the Clean Development Mechanism. However, some governments have suggested giving credit for early action taken before 2008 with the intent to stimulate investment in their emission abatement projects.

Credit Portfolio Management (CPM)—The purchase of short, medium and long-term receivables. As the seller of his products and services, the customer retains complete contact to his own customers while simultaneously transferring the credit risk to CPM.

Credit Scoring—An objective method of quantifying credit worthiness by assigning numerical values based on meeting established credit criteria.

Credit Standing—Creditworthiness and ability to pay off a debtor/security of a cash receivable.

Creosote—A liquid byproduct of wood combustion (or distillation) that condenses on the internal surfaces of vents and chimneys, which if not removed regularly, can corrode the surfaces and fuel a chimney fire.

CRF—Combustion Research Facility

CRGS—Chemical Regulations and Guideline system

CRIB—Criteria Reference Information Bank

CRIMDOCK—Criminal Docket System

CRISP—Comprehensive Risk Information Structure Project

Criteria Pollutant—A pollutant determined to be hazardous to human health and regulated under EPA's National Ambient Air Quality Standards. The 1970 amendments to the Clean Air Act require EPA to describe the health and welfare impacts of a pollutant as the "criteria" for inclusion in the regulatory regime. U.S. EPA has identified six "criteria pollutants," ozone, carbon monoxide, nitrogen dioxide, sulfur dioxide, particulate matter, and lead as indicators of air quality, and for each is an established maximum concentration above which adverse effects on human health may occur.

Critical Compression Pressure—The highest possible pressure in a fuel-air mixture before spontaneous ignition occurs.

Critical Habitat—Under the Endangered Species Act, critical habitat is defined as "the specific areas within the geographic area occupied by a species on which are found those physical and biological features essential to the conservation of the species, and that may require special management considerations or protection; and specific areas outside the geographic area occupied by a species at the time it is listed, upon determination that such areas are essential for the conservation of the species."

Critical Path Method (CPM)—A process used to estimate project duration by sequencing individual project components based on which has the least amount of scheduling flexibility.

CRL—Certified Reporting Limit

CROP—Consolidated Rules of Practice

Crop Residue—Organic residue remaining after the harvesting and processing of a crop.

Crop Tree—Usually a conifer tree grown to provide wood products.

Cross Border—In contracts, the debtor and creditor are residents of different countries.

Cross Collateralization—A grouping of personal or real properties that serves to jointly secure one debt obligation.

Cross Country Pipeline—A physical pipe that starts at the well head (production zone) that transports large volumes of gas to a state or local distribution company.

Cross Flow Turbine—A turbine where the flow of water is at right angles to the axis of rotation of the turbine.

Cross-Border Lease—A lease structured to take advantage of the tax laws in two different countries. Typically, the lessor is located in one country and the lessee is located in a different country.

CRP—Child Resistant Packaging
• Community Relations Plan
• Conservation Reserve Program

CRQL—Contract-Required Quantitation Limit

CRR—Center for Renewable Resources www.rredc.nrel.gov.

CRRC—See *Cool Roof Rating Council*

CRS—Community Relations Staff
• Congressional Research Service

CRSTER—Single Source Dispersion Model

CRTK—Community Right-to-Know

Crude Oil—A mixture of hydrocarbons that exists in liquid phase in natural underground reservoirs and remains liquid at atmospheric pressure after passing through surface separating facilities. Depending upon the characteristics of the crude stream, it may also include:

1. Small amounts of hydrocarbons that exist in

gaseous phase in natural underground reservoirs but are liquid at atmospheric pressure after being recovered from oil well (casinghead) gas in lease separators and are subsequently commingled with the crude stream without being separately measured. Lease condensate recovered as a liquid from natural gas wells in lease or field separation facilities and later mixed into the crude stream is also included;

2. Small amounts of non-hydrocarbons produced with the oil, such as sulfur and various metals;

3. Drip gases, and liquid hydrocarbons produced from tar sands, gilsonite, and oil shale.

Liquids produced at natural gas processing plants are excluded. Crude oil is refined to produce a wide array of petroleum products, including heating oils; gasoline, diesel and jet fuels; lubricants; asphalt; ethane, propane, and butane; and many other products used for their energy or chemical content.

Crude Oil Acquisitions (Unfinished Oil Acquisitions)—The volume of crude oil either:

• acquired by the respondent for processing for his own account in accordance with accounting procedures generally accepted and consistently and historically applied by the refiner concerned, or

• in the case of a processing agreement, delivered to another refinery for processing for the respondent's own account.

Crude oil that has not been added by a refiner to inventory and that is thereafter sold or otherwise disposed of without processing for the account of that refiner is deducted from its crude oil purchases at the time when the related cost is deducted from refinery inventory in accordance with accounting procedures generally applied by the refiner concerned.

Crude oil processed by the respondent for the account of another is not a crude oil acquisition.

Crude Oil F.O.B. Price—The crude oil price actually charged at the oil producing country's port of loading. Includes deductions for any rebates and discounts or additions of premiums, where applicable. It is the actual price paid with no adjustment for credit terms.

Crude Oil Input—The total crude oil put into processing units at refineries.

Crude Oil Landed Cost—The price of crude oil at the port of discharge, including charges associated with purchasing, transporting, and insuring a cargo from the purchase point to the port of discharge. The cost does not include charges incurred at the discharge port (e.g., import tariffs or fees, wharfage charges, and demurrage).

Crude Oil Less Lease Condensate—A mixture of hydrocarbons that exists in liquid phase in natural underground reservoirs and remains liquid at atmospheric pressure after passing through surface separating facilities. Such hydrocarbons as lease condensate and natural gasoline recovered as liquids from natural gas wells in lease or field separation facilities and later mixed into the crude stream are excluded. Depending upon the characteristics of the crude stream, crude oil may also include:

1. Small amounts of hydrocarbons that exist in gaseous phase in natural underground reservoirs but are liquid at atmospheric pressure after being recovered from oil well (casinghead) gas in lease separators and are subsequently commingled with the crude stream without being separately measured;

2. Small amounts of non-hydrocarbons produced with the oil, such as sulfur and various metals.

Crude Oil Losses—Represents the volume of crude oil reported by petroleum refineries as being lost in their operations. These losses are due to spills, contamination, fires, etc., as opposed to refining processing losses.

Crude Oil Production—The volume of crude oil produced from oil reservoirs during given periods of time. The amount of such production for a given period is measured as volumes delivered from lease storage tanks (i.e., the point of custody transfer) to pipelines, trucks, or other media for transport to refineries or terminals with adjustments for (1) net differences between opening and closing lease inventories, and (2) basic sediment and water (BS&W).

Crude Oil Qualities—Refers to two properties of crude oil, the sulfur content, and API gravity, which affect processing complexity and product characteristics.

Crude Oil Refinery Input—The total crude oil put into processing units at refineries.

Crude Oil Stocks—Stocks held at refineries and at pipeline terminals. Does not include stocks held on leases (storage facilities adjacent to the wells). In California, crude oil stocks in 1990 are approximately 18 million barrels on any given day.

Crude Oil Used Directly—Crude oil consumed as fuel by crude oil pipelines and on crude oil leases.

Crude Oil, Refinery Receipts—Receipts of domestic and foreign crude oil at a refinery. Includes all crude oil in transit except crude oil in transit by pipeline. Foreign crude oil is reported as a receipt only after entry through customs. Crude oil of foreign origin held in bonded storage is excluded.

Cryogen—A material that is a gas at ambient conditions but can be liquefied at below-ambient temperatures. This includes all of the ambient temperature gases.

Cryogenic—The science of producing very low temperatures such as natural gas liquefaction.

Crystalline Fully Refined Wax—A light colored paraffin wax having the following characteristics: viscosity at 210 degrees Fahrenheit (D88)-59.9 SUS (10.18 centistokes) maximum; oil content (D721)-0.5 percent maximum; other +20 color, Saybolt minimum.

Crystalline Other Wax—A paraffin wax having the following characteristics: viscosity at 210 deg. F(D88)-59.9 SUS (10.18 centistokes) maximum; oil content (D721)-0.51 percent minimum to 15 percent maximum.

Crystalline Silicon Photovoltaic Cell—A type of photovoltaic cell made from a single crystal or a polycrystalline slice of silicon. Crystalline silicon cells can be joined together to form a module (or panel).

CRZ—Contamination Reduction Zone

CS—Caustic Scrubber
- Compliance Staff
- Contract Specialist
- Cost Share

CSA—Chemical Safety Audit

CSA (Canadian Standards Association)—CSA is a not-for-profit membership-based association serving business, industry, government and consumers in Canada and the world marketplace. To date, CSA has published more than 3,000 standards, codes and related products. The standards developed by CSA committees are voluntary. Many CSA standards are referenced in legislation by governments or other regulatory bodies in jurisdictions throughout North America.

CSA develops, administers and distributes standards, guidance documents and related products in a wide variety of areas including construction products and materials such as building materials and masonry, concrete, forest products, plumbing products and materials, and national construction codes; electrical and electronics; environment including environmental technology, sustainable forestry, and environmental management systems; and energy including energy efficient and renewable energy.

The Canadian Standards Association is a division of CSA Group, which also consists of CSA International, which provides testing and certification services for electrical, mechanical, plumbing, gas and a variety of other products.

CSB—Chemical Safety and Hazard Investigation Board

CSCT—Committee for Site Characterization

CSD—Criteria and Standards Division

CSE (California Seasonal Efficiency)—See *Seasonal Efficiency*.

CSG—Council of State Governments www.csg.org

CSGWPP—Comprehensive State Ground Water Protection Program

CSHO—Compliance Safety and Health Officer

CSI—California Solar Initiative—a rebate program for solar in California
- Chemical Substances Inventory
- Common Sense Initiative
- Compliance Sampling Inspection
- Construction Specification Institute www.csi-net.org

CSIN—Chemical Substances Information Network

CSLB—Contractors State License Board (in California)

CSM—Conceptual Site Model

CSO—Combined Sewer Overflow

CSP—Competitive Sealed Proposal
- Concentrating Solar Photovoltaics

CSPA—Council of State Planning Agencies

CSPD—Chemicals and Statistical Policy Division

CSPI—Center for Science in the Public Interest. Center for Science in the Public Interest is a Washington, D.C.-based non-profit watchdog and consumer advocacy group fighting for safer, more nutritious food. www.cspinet.org

CSRL—Center for the Study of Responsive Law www.csrl.org

CSS—Clerical Support Staff

CSSD—Computer Services and Systems Division

CST—Combined Statistical and Technology Based MACT Floor Procedure

CT—Contact Time
- Combustion turbine
- Current transformer

CTARC—Chemical Testing and Assessment Research Commission

CTC—Control Technology Center

CTD—Control Technology Document

CTG—Control Techniques Guidelines
- Combustion turbine generator

CTI—Cooling Tower Institute www.cti.org

CTI STD-201—The Cooling Tower Institute document entitled "Certification Standard for Commercial Water Cooling Towers," 2002 (CTI STD-201-02).

CTL—Coal-to-liquid conversion

CTM—Complex Terrain Data Base

CTO—Control Technology Office

CTS—Correspondence Tracking System

CTSA—Cleaner Technologies Substitutes Assessment

CTT—Chronic Toxicity Test

Cube Law—In reference to wind energy, for any given instant, the power available in the wind is proportional to the cube of the wind velocity; when wind speed doubles, the power availability increases eight times.

Cubic Foot (cf)—The most common unit of measurement of natural gas volume. It equals the amount of gas required to fill a volume of one cubic foot under stated conditions of temperature, pressure and water vapor. One cubic foot of natural gas has an energy content of approximately 1,000 Btu's. One hundred (100) cubic feet equals one therm (100 ft3 = 1 therm).

Cubic Foot (cf), Natural Gas—The amount of natural gas contained at standard temperature and pressure (60 degrees Fahrenheit and 14.73 pounds standard per square inch) in a cube whose edges are one foot long.

CUBP—Coincidental Unavoidable By-Product

Cull (Lumber)—Any item of production picked out for rejection because it does not meet certain specifications. Chip culls and utility culls are specifically defined for purposes of log grading by percentage of sound wood content.

Cull Section—A log cut from a tree that is rejected because of defects making it unsuitable for conventional forest products.

Cull Trees—Live saw-timber and pole-timber size trees which do not contain a merchantable sawlog due to poor form, quality, or undesirable species.

Cull Wood—Wood logs, chips, or wood products that are burned.

Culm—Waste from Pennsylvania anthracite preparation plants, consisting of coarse rock fragments containing as much as 30 percent small sized coal; sometimes defined as including very fine coal particles called silt. Its heat value ranges from 8 to 17 million Btu per short ton.

Cultivar—A horticulturally or agriculturally derived variety of a plant.

Cumulative Depletion—The sum in tons of coal ex-tracted and lost in mining as of a stated date for a specified area or a specified coal bed.

Cumulative Effects—Effects on the environment resulting from actions that are individually minor but that add up to a greater total effect as they take place over a period of time.

Cumulative Emission—This is a pollutant that accumulates in the environment once released. A good example is lead or mercury. Since the environment's assimilative capacity is essentially non-existent, they tend to accumulate in the environment.

Cumulative Impact—The term cumulative impact is used in several ways. As the effect of exposure to more than one compound; as the effect of exposure to emissions from more than one facility; the combined effects of a facility and surrounding facilities or projects on the environment; or some combination of these.

Cumulative Preferred Stock—A stock having a provision that if one or more dividend payments are omitted, the omitted dividends (arrearage) must be paid before dividends may be paid on the company's common stock.

Cumulative Voting Rights—When shareholders have the right to pool their votes to concentrate them on an election of one or more directors rather than apply their votes to the election of all directors. For example, if the company has 12 openings to the Board of Directors, in statutory voting, a shareholder with 10 shares casts 10 votes for each opening (10x12 = 120 votes). Under the cumulative voting method however, the shareholder may opt to cast all 120 votes for one nominee (or any other distribution they might choose).

CUR—Carbon Usage Rate

Curable Depreciation—Items of physical deterioration or functional obsolescence that is economically feasible to cure. Economic feasibility is indicated if the cost to cure is equal to or less than the anticipated increase in the value of the property.

CURE—Chemical Unit Record Estimates Data Base
* California Unions for Reliable Energy

Curie—A measure of radioactivity.

Current—The movement of electrical charge (electrons or ions) through a circuit.

Current (Electric)—A flow of electrons in an electrical conductor. The strength or rate of movement of the electricity is measured in amperes.

Current Assets—Cash and other assets that are expected to be turned into cash, sold, or exchanged within the normal operating cycle of the company, usually one year. Current assets include cash, marketable securities, receivables, inventory, and current prepayments.

Current at Maximum Power (Imp)—The current at which maximum power is available from a module. [UL 1703]

Current Dollars—The value or purchasing power of a dollar that has not been reduced to a common basis of constant purchasing power, but instead reflects anticipated future inflation; when used in computations the assumed inflation rate must be stated.

Current Liabilities—A debt or other obligation that must be discharged within one year or the normal operating cycle of the company by expending a current asset or the incurrence of another short-term obligation. Current liabilities include accounts payable, short-term notes payable, and accrued expenses payable such as taxes payable and salaries payable.

Current Maturity—Current time to maturity on an outstanding note, bond or other money market instrument.

Current Ratio—The ratio of current assets divided by current liabilities that shows the ability of a company to pay its current obligations from its current assets. A measure of liquidity, the higher the ratio, the more assurance that current liabilities can be paid.

Current Transducer (CT)—Used to measure the amperage through a wire or bus bar.

Current Value—In accounting, synonymous with "market value."

Current Value Accounting—The practice of showing both the book value and the current market value of assets on the annual financial statement of publicly owned real estate corporations.

Curtailable Rate Schedule—Also referred to as an interruptible rate schedule (electricity). A type of rate schedule that allows the transmission provider to interrupt all or part of a transmission service under specified terms due to constraints that reduce the capability of the transmission network to provide that service.

Curtailment—When the natural gas supply is reduced or shut off due to system problems or for economical reasons.

Curtain Wall—Non-load-bearing wall placed as a weather-proof membrane around a structure; usually made of glass or metal.

CUS—Chemical Update System

Customer Charge—The customer's monthly payment to the local gas company (or electric utility) for providing natural gas delivery. This charge typically will include metering, billing and customer services provided by the local distribution company.

Customer Choice—The right of customers to purchase energy from a supplier other than their traditional supplier or from more than one seller in the retail market.

Customer Class—Categories of energy consumers, as defined by consumption or demand levels, patterns, and conditions, and generally included residential, commercial, industrial, agricultural.

Customs District (Coal)—Customs districts, as defined by the Bureau of the Census, U.S. Department of Commerce, " Monthly Report EM 545," are as follows:

- Eastern: Bridgeport, CT, Washington, DC, Boston, MA, Baltimore, MD, Portland, ME, Buffalo, NY, New York City, NY, Ogdensburg, NY, Philadelphia, PA, Providence, RI, Norfolk, VA, St. Albans, VT.
- Southern: Mobile, AL, Savannah, GA, Miami, FL, Tampa, FL, New Orleans, LA, Wilmington, NC, San Juan, PR, Charleston, SC, Dallas-Fort Worth, TX, El Paso, TX, Houston-Galveston, TX, Laredo, TX, Virgin Islands.
- Western: Anchorage, AK, Nogales, AZ, Los Angeles, CA, San Diego, CA, San Francisco, CA, Honolulu, HI, Great Falls, MT, Portland, OR, Seattle, WA.
- Northern: Chicago, IL, Detroit, MI, Duluth, MN, Minneapolis, MN, St. Louis, MO, Pembina, ND, Cleveland, OH, Milwaukee, WI.

Cut-In-Speed—The lowest wind speed at which a wind turbine begins producing usable power.

Cut-Off Grade (Uranium)—The lowest grade, in percent U_3O_8, of uranium ore at a minimum specified thickness that can be mined at a specified cost.

Cut-off Voltage (Solar)—The voltage levels at which the charge controller (regulator) disconnects the PV array from the battery, or the load from the battery.

Cut-Out-Speed—The highest wind speed at which a wind turbine stops producing power.

CUWA—California Urban Water Agencies www.cuwa.org

CUWCC—California Urban Water Conservation Council www.cuwcc.org

CV—Chemical Vocabulary

CVAA—Cold Vapor Atomic Absorption

C-Value—(also known as C-factor) is the time rate of heat flow through unit area of a body induced by a unit temperature difference between the body surfaces, in Btu/(hr. x ft.^{2} x °F). It is not the same as K-value or K-factor.

CVM—Contingent Valuation Method

CVS—Constant Volume Sampler

CW—Continuous Working-level Monitoring

CWA—Clean Water Act: The law governing health of the nation's rivers, lakes, estuaries and coastal waters. It was originally enacted in 1948 as the Federal Water Pollution Control Act. In 1972 it was revised and renamed the Clean Water Act.

CWAP—Clean Water Action project

CWFT—Chilled Water From Tower

CWIP—Construction Work in Progress: A bookkeeping account that accumulates all costs in building new plant facilities until they begin to serve the public, at which time they are transferred to the appropriate plant accounts and included in rate base.

CWMB—California Waste Management Board www.calrecycle.ca.gov

CWRT—Center Waste Reduction Technologies

CWS—Community Water System
- Compressed Work Schedule

CWT—Centralized Waste Treatment

CWTT—Condenser Water to Tower

C$_x$A—Commissioning Agent or Commissioning Authority are equivalent terms used to describe the designated individual in charge of leading the commissioning process.

CY—Calendar Year

Cycle—In the RECLAIM emissions trading program (Southern California), Cycle 1 commences on January 1 and ends on December 31. Cycle 2 commences on July 1 and ends on the following June 30.
- In alternating current, the current goes from zero potential or voltage to a maximum in one direction, back to zero, and then to a maximum potential or voltage in the other direction. The number of complete cycles per second determines the current frequency; in the U.S. the standard for alternating current is 60 cycles.
- The discharge and re-charge of a battery, one complete charge/discharge cycle of the battery

Cycle Life—Number of charge-discharge cycles a battery can perform under specified conditions before it fails to meet its specified performance (e.g. capacity decreases to 80% of nominal capacity).

Cycle/Reactor History—A group of assemblies that have been irradiated in the same cycles in an individual reactor and are said to have the same cycle/

reactor history.

Cycling (Natural Gas)—The practice of producing natural gas for the extraction of natural gas liquids, returning the dry residue to the producing reservoir to maintain reservoir pressure and increase the ultimate recovery of natural gas liquids. The re-injected gas is produced for disposition after cycling operations are completed.

Cycling Losses—The loss of heat as the water circulates through a water heater tank and inlet and outlet pipes. The actual efficiency of a heating or cooling system is reduced because of start-up and shutdown losses. Oversizing a heating or cooling system increases cycling losses.

Cyclone Burner—A furnace/combustion chamber in which finely ground fuel is blown in spirals in the combustion chamber to maximize combustion efficiency.

Cyclone Separator—A device used to remove particulate matter suspended in exhaust gases.

CZM—Coastal Zone Management

CZMA—Coastal Zone Management Act (1972)

CZMP—Coastal Zone Management Plan

Czochralski Process—A method of growing large size, high quality semiconductor crystal by slowly lifting a seed crystal from a molten bath of the material under careful cooling conditions.

D

D&F—Determination and Findings

D/E—Debt-to-equity Ratio

DA—Deputy Administrator
- Designated Agent
- Dilution Air
- Direct Access: The ability of a retail customer to purchase commodity electricity directly from the wholesale market rather than through a local distribution utility.

DAA—Deputy Assistant Administrator

DACA—Days After Contract Award

DAIG—Deputy Assistant Inspector General

Dam—A physical barrier constructed across a river or waterway to control the flow of or raise the level of water. The purpose of construction may be for flood control, irrigation needs, hydroelectric power production, and/or recreation usage. The reservoir creates, in effect, stored energy.

Damper—A movable plate used to control air flow; in a wood stove or fireplace, used to control the amount and direction of air going to the fire.

Dangling Bonds—A chemical bond associated with an

atom on the surface layer of a crystal. The bond does not join with another atom of the crystal, but extends in the direction of exterior of the surface.

DAR—Direct Assistance Request

Darcy—A measure of permeability. A permeability of one darcy means that the material will pass a fluid of one centipoise viscosity through a section of one cubic centimeter at a rate of one cubic centimeter per second with a drop in pressure of one standard atmosphere.

Darrius (Wind) Machine—A type of vertical-axis wind machine that has long, thin blades in the shape of loops connected to the top and bottom of the axle; often called an "eggbeater windmill."

DART—Damage Assessment Regulations Team

DAS—Data Analysis System

Day-Ahead and Hour-ahead Markets—Forward markets where electricity quantities and market clearing prices are calculated individually for each hour of the day on the basis of participant bids for energy sales and purchases.

Day-Ahead Market—The forward market for energy and ancillary services to be supplied during the settlement period of a particular trading day that is conducted by the ISO, the PX, and other Scheduling Coordinators. This market closes with the ISO's acceptance of the final day-ahead schedule.
  • Also known as the Integrated Forward Market, the Day Ahead Market co-optimizes energy and ancillary services (AS) to assure a feasible, secure, and least cost operating plan for the next day.

Day-Ahead Schedule—A schedule prepared by a scheduling coordinator or the independent system operator before the beginning of a trading day. This schedule indicates the levels of generation and demand scheduled for each settlement period that trading day.

Day-Lighting—The use of direct, diffuse, or reflected sunlight to provide supplemental lighting for building interiors. An energy-saving strategy that provides interior building lighting using indirect sunlight often provided by clerestory windows, skylights, and light shelves. Natural day-lighting has been proven to improve productivity, attendance, and occupant health and happiness.

Day-Lighting Controls—A system of sensors that assesses the amount of daylight and controls lighting or shading devices to maintain a specified lighting level. The sensors are sometimes referred to as "photocells."

Daylit Area—The floor area that is illuminated by day-light through vertical glazing or skylights as specified in Section 131(c).

Days of Autonomy—The number of consecutive days that a stand-alone renewable energy system will meet a defined load without additional energy input.

Days of Storage—The number of days that a stand-alone system will power a specified load without solar energy input. A measure of system autonomy.

dB—Decibel

DB—Dry Bult

DBH—See *Diameter at Breast Height*.

DBMS—Data Base Management System

DC—Direct Current Electrical current that flows only in one direction. The most common form of electricity used in boats and RVs. The type of electricity produced by solar photovoltaic systems.

DC Motor, Brushless—High-technology motor used in centrifugal-type DC submersible pumps and other applications. The motor is filled with oil to keep water out. An electronic system is used to precisely alternate the current, causing the motor to spin.

DC Motor, Brush-type—The traditional DC motor, in which small carbon blocks called "brushes" conduct current into the spinning portion of the motor. They are used in many applications, including DC surface pumps and also in DC submersible diaphragm pumps. Brushes naturally wear down after years of use, and may be replaced.

DC Motor, Permanent Magnet—A variable speed motor that uses permanent magnets instead of wound coils. Reduced voltage (in low sun) produces proportionally reduced speed, and causes no harm to the motor.

DC to DC Converter—Electronic circuit to convert dc voltages (e.g., PV module voltage) into other levels (e.g., load voltage). Can be part of a maximum power point tracker (MPPT).

DCA—Document Control Assistant

DCAA—Defense Contract Audit Agency

DCF—Discounted Cash Flow. A method of evaluating a company or receivable by estimating future free cash flows that will be available for disbursement or investment.

DCN—Document Control Number

DCO—Delayed Compliance Order
  • Document Control Officer

DCR—Document Control Register

DCS—Distributed Control System

DCTL—Double Circuit Transmission Line

DD—Deputy Director

• Double Duct

DD&A—Abbreviation for depreciation, depletion and amortization.

DE—Destruction Efficiency

De Minimus Risk—A level of risk that the scientific and regulatory community asserts is too insignificant to regulate.

Deadweight Tons—The lifting capacity of a ship expressed in long tons (2,240 lbs.), including cargo, commodities, and crew.

Deaeration—Removal of gases from a liquid.

Deaerator—The apparatus used to separate the dissolved gases from the condensate.

Dealer Tank Wagon (DTW) Sales—Wholesale sales of gasoline priced on a delivered basis to a retail outlet.

Debenture—A company's long-term IOU (bond) backed by the general credit of the firm, rather than by a lien on any specific asset.
 • A debt instrument; basically the same as a Promissory Note.

Debt Capacity—The total amount of debt a company can prudently support, given its earnings expectations and equity base.

Debt Capitalization Ratio—The ratio of a company's debt to its capitalization. The higher this ratio, the greater the financial leverage and the risk.

Debt Coverage Principle—A method to determine the cost of common equity component of return based on cost of the fixed components, debt and preferred stock.

Debt Coverage Ratio (DCR)—The ratio of net operating income to annual debt service; measures the ability of a property (real or personal) to meet its debt service out of net operating income; also called debt service coverage ration (DSCR).

Debt Equity Ratio—The ratio of a company's debt to its equity. The higher this ratio, the greater the financial leverage of the company.

Debt Leverage—The amplification in the return earned on equity funds when an investment is financed partly with borrowed money.

Debt Optimization—A method of borrowing funds in a leveraged lease in which the equity participants borrow and repay the debt in such a manner as to maximize their return on equity, maintain a constant return and offer a lower lease payment, maximize cash flow or maximize a combination of factors.

Debt Participant—A long-term lender in a leveraged lease or PPA transaction. Frequently, these transactions have more than one debt participant.

Debt Service—The money needed to pay the amount due on a loan.
 • Payment of principal and interest due lenders under a leveraged lease agreement.

Debt Service Coverage Ratio (DSCR)—In corporate finance, it is the amount of cash flow available to meet annual interest and principal payments on debt, including sinking fund payments
 • In government finance, it is the amount of export earnings needed to meet annual interest and principal payments on a country's external debts.
 • In general, it is calculated by dividing *Net Operating Income by Total Debt Service*.
 • A DSCR of less than 1 would mean a negative cash flow. A DSCR of .95 would mean that there is only enough net operating income to cover 95% of annual debt payments.

Debt-to-Equity Ratio—Financial term comparing total liabilities to equity consisting of retained earnings and contributed capital.

DEC—Department of Environmental Control
 • Department of Environmental Conservation
 • Direct Embedded Cost: The total current cost of owning, operating and maintaining a company's existing system. Under an embedded cost structure, a substantial portion reflects historical capital investments, including depreciation and return on investment.

Decatherm—Ten therms or 1,000,000 Btu.

Decentralized (Energy) System—Energy systems supply individual, or small-groups, of energy loads.

Decibels (dBA)—A scale to measure sound levels.

Decision Notice—The written record of a federal agency decision after an environmental assessment. The decision notice chooses one of the alternatives, or a blend of the alternatives. A decision notice is subject to appeal.

Declination—The angular position of the sun at solar noon with respect to the plane of the equator.

Declining Balance Depreciation—A type of accelerated depreciation in which a constant percentage of an asset's declining remaining basis is depreciated each year. The constant percentage amount is often calculated at 125%, 150% or 200% (double declining balance) of the straight-line percentage over the same recovery period.

Declining Block Rate—An electricity supplier rate structure in which the per unit price of electricity decreases as the amount of energy increases. Normally only available to very large consumers.

Decommissioning—To remove from service
- The process of dismantling power plants after they have stopped producing power, including disposal of radioactive waste materials, the destruction or resale of plant equipment, and the return of the power plant site to its original state.

Decomposition—The process of breaking down organic material; reduction of the net energy level and change in physical and chemical composition of organic material.

Decontamination—Removal of unwanted radioactive or hazardous contamination by a chemical or mechanical process.

Decorative Gas Appliance—A gas appliance that is designed or installed for visual effect only, cannot burn solid wood, and simulates a fire in a fireplace.

Decoupling—A regulatory design that breaks the link between utility revenues and energy sales to encourage utility investment in conservation.

Dedicated Reserves—The volume of recoverable, salable gas reserves committed to, controlled by, or possessed by the reporting pipeline company and used for acts and services for which both the seller and the company have received certificate authorization from the Federal Energy Regulatory Commission (FERC). Reserves include both company-owned reserves (including owned gas in underground storage), reserves under contract from independent producers, and short-term and emergency supplies from the intrastate market. Gas volumes under contract from other interstate pipelines are not included as reserves, but may constitute part or all of a company's gas supply.

Dedicated Vehicle—A vehicle that operates only on an alternative fuel, as when a vehicle is configured to operate on compressed natural gas. Note: A vehicle powered by an electric motor is not to be treated as dedicated.

Deductible (Tax)—Able to be deducted from taxable income. Common examples are depreciation, interest paid, or fees amortized.

Deed of Trust—A legal instrument similar to a mortgage, which, when executed and delivered, conveys or transfers property title to a trustee.

De-Energize (d)—To disconnect a transmission and/or distribution line; a power line that is not carrying a current; to open a circuit.

Deep Cycle Battery—A battery designed to regularly discharge 80% of its capacity before recharging.

Deep Discharge—Discharging a battery to 20 percent or less of its full charge capacity.

Deep Gas—Gas found at depths greater than the average for a particular area; for FERC purposes, it is gas found at depths of more than 15,000 feet.

Deep Mining—Extraction of coal or minerals at depths greater than 1,000 feet. Coal usually is deep-mined at not more than 1,500 feet.

Deepest Total Depth—The deepest total depth of a given well is the distance from a surface reference point (usually the Kelly bushing) to the point of deepest penetration measured along the well bore. If a well is drilled from a platform or barge over water, the depth of the water is included in the total length of the well bore.

Default—The debtor is in default when they fail to pay after the debt matures despite reminders. If a specific calendar date is fixed for payment, they are in default even without a reminder if they do not pay by this date. Failure to perform a duty or to discharge an obligation.

Default Premium—The increased return on a security required to compensate investors for the risk the company will default on its obligation(s).

Default Service—Electricity service available to consumers who choose not to select an alternative electricity service provider.

Defeasance—In loans, leases and swaps, a substitution of a lump sum payment for the present value of a stream of payments.
- Up-front settlement of an obligation. Often used in reference to a lessee's obligation to pay rent, which can be defeased by a single payment to a trustee, who then makes periodic payments to the lessor.

Defeased Lease—A lease in which the rents, purchase option and other obligations of the lessee have been defeased by the lessee depositing funds with a third party who agrees to pay such obligations when due (normally only from the funds they hold).

Deferred Cost—An expenditure not recognized as a cost of operation of the period in which incurred, but carried forward to be written off in future periods.

Deferred Credits—Accounts carried on the liability side of the balance sheet in which are recorded items being amortized as credits to income over a period of time (such as Unamortized Premium on Debt) and items held in suspense pending final transfer or disposition (such as Customer Advances for Construction, etc.)

Deferred Debits—Accounts carried on the asset side of the balance sheet in which are recorded items being amortized as charges against income over a period time (such as Unamortized Debt Discount and Expense) and items held in suspense pending final transfer or disposition (such as Extraordinary Property Losses, Clearing Accounts (Net Debit), etc.).

Deferred Fuel Costs—An expenditure for fuel that is not recognized for bookkeeping practices as a cost in the operating period incurred, but carried forward to be written off in future periods.

Deferred Income Tax (Liability)—A liability in the balance sheet representing the additional Federal income taxes that would have been due if a utility had not been allowed to compute tax expenses differently for income tax reporting purposes than for ratemaking purposes.

Deferred Maintenance—Curable, physical deterioration that should be corrected immediately, although work has not commenced. Denotes the need for immediate expenditures, but does not necessarily suggest inadequate maintenance in the past.

Deferred Taxes—Taxes accrued and reflected as an expense in a company's income statement, but not payable to the taxing authority in that time period. These taxes are accrued to compensate for an understatement of income tax expense that would occur if only the tax currently due to the taxing authority was reflected as the total income tax expense

Deficiency Agreement—An agreement to guarantee revenues will be received or expenses paid to make up a shortfall needed to pay a debt.

Deficiency Letter—A letter sent by the SEC to the issuer of a new issue regarding omissions of material fact in the registration statement.

Deforestation—The net removal of trees from forested land. Examples include cutting or burning to provide land for agricultural purposes, residential or industrial building sites, roads etc., or harvesting the trees for building materials or fuel.

Degasification System—The methods employed for removing methane from a coal seam that could not otherwise be removed by standard ventilation fans and thus would pose a substantial hazard to coal miners. These systems may be used prior to mining or during mining activities.

Degradable Organic Carbon—The portion of organic carbon present in such solid waste as paper, food waste, and yard waste that is susceptible to biochemical decomposition.

Degree Day—A unit for measuring the extent that the outdoor daily average temperature (the mean of the maximum and minimum daily dry-bulb temperatures) falls below (in the case of heating, see *Heating Degree Day*), or falls above (in the case of cooling, see *Cooling Degree Day*) an assumed base temperature, normally taken as 65 degrees Fahrenheit, unless otherwise stated. One degree day is counted for each degree below (for heating) or above (in the case of cooling) the base, for each calendar day on which the temperature goes below or above the base.

Degree Day, Heating—A unit, based upon temperature difference and time, used in estimating fuel consumption and specifying nominal annual heating load of a building. For any one day, when the mean temperature is less than 65°F, there exist as many degree days as there are Fahrenheit degrees difference in temperature between the mean temperature for the day and 65°F.

Degree Hour—The product of 1 hour, and usually the number of degrees Fahrenheit the hourly mean temperature is above a base point (usually 65 degrees Fahrenheit); used in roughly estimating or measuring the cooling load in cases where processes heat, heat from building occupants, and humidity are relatively unimportant compared to the dry-bulb temperature.

Dehumidifier—A device that cools air by removing moisture from it.

Dehumidify—To reduce by any process the quantity of water vapor contained in a solid or gas.

Dehydration—The process of removing liquids and moisture content from gas or other matter.

Deintegration—(See *Disaggregation*)

Deionized Water—Water which has been specifically treated to remove minerals.

DEIR—Draft Environmental Impact Report: Either created by the PUC, or a contractor hired by the PUC, to discuss the environmental effects of a project. The DEIR is usually issued for comment and then the Final Environmental Impact Report (FEIR) is issued. Usually done to comply with the California Environmental Quality Act (CEQA).

DEIS—See *Draft Environmental Impact Statement*.

Dekatherm—One million Btu of natural gas.

Delayed Coking—A process by which heavier crude oil fractions can be thermally decomposed under conditions of elevated temperatures and pressure to produce a mixture of lighter oils and petroleum coke. The light oils can be processed further in other refinery units to meet product specifications. The coke can be used either as a fuel or in other applications such as the manufacturing of steel or aluminum.

Deliverability—Represents the number of future years during which a pipeline company can meet its annual requirements for its presently certificated delivery capacity from presently committed sources of supply. The availability of gas from these sources of supply are governed by the physical capabilities of these sources to deliver gas by the terms of existing gas-purchase contracts, and by limitations imposed by State or Federal regulatory agencies.

Delivered (Gas)—The physical transfer of natural, synthetic, and/or supplemental gas from facilities operated by the responding company to facilities operated by others or to consumers.

Delivered Cost—The cost of fuel, including the invoice price of fuel, transportation charges, taxes, commissions, insurance, and expenses associated with leased or owned equipment used to transport the fuel.

Delivered Energy—The amount of energy delivered to the site (building); no adjustment is made for the fuels consumed to produce electricity or district sources. This is also referred to as net energy.

Deliveries (Electric)—Energy generated by one system and delivered to another system through one or more transmission lines.

Delphi Survey—A survey of professionals who are especially knowledgeable about market conditions, economic trends, and governmental policies. Such a survey is used to develop a prognostication for a specific industry.

Delta-T—A difference in temperature. Often used in the context of the difference between the design indoor temperature and the outdoor temperature.

DEM—Department of Environmental Management

Demand—The rate at which electricity is delivered to or by a system, part of a system, or piece of equipment expressed in kilowatts, kilovoltamperes, or other suitable unit, at a given instant or averaged over a specified period of time.

Demand (Tankless) Water Heater—A type of water heater that has no storage tank thus eliminating storage tank stand-by losses. Cold water travels through a pipe into the unit, and either a gas burner or an electric element heats the water only when needed.

Demand (Utility)—The level at which electricity or natural gas is delivered to users at a given point in time.

Electric demand is expressed in kilowatts.

Demand Bid—A bid into the power exchange indicating a quantity of energy or an ancillary service that an eligible customer is willing to purchase and, if relevant, the maximum price that the customer is willing to pay.

Demand Billing—The electric capacity requirement for which a large user pays. It may be based on the customer's peak demand during the contract year, on a previous maximum or on an agreed minimum. Measured in kilowatts.

Demand Charge—A charge for the maximum rate at which energy is used during peak hours of a billing period. That part of a power provider service charged for on the basis of the possible demand as distinguished from the energy actually consumed.

Demand Charge Credit—Compensation received by the buyer when the delivery terms of the contract cannot be met by the seller.

Demand Charge Management—The ability to reduce or eliminate specified utility service charges associated with use of energy during high-demand periods.

Demand Deposit—Money placed with a financial institution that must be returned upon demand by its owner; checking account is the most common form.

Demand Indicator—A measure of the number of energy-consuming units, or the amount of service or output, for which energy inputs are required.

Demand Interval—The time period during which flow of electricity is measured (usually in 15-, 30-, or 60-minute increments.)

Demand Power—See *Peak Power*

Demand Registration—Resale registration that gives the investor the right to require the Company to file a Registration Statement registering the resale of the securities issued to the investor in a private offering.

Demand Response—Demand response refers to the reduction of customer energy usage at times of peak usage. Demand response programs may include dynamic pricing/tariffs, price-responsive demand bidding, contractually obligated and voluntary curtailment, and direct load control/cycling.

Demand Side Bidding—Process in which a utility issues a request for proposals to acquire DSM resources from energy service companies and customers, reviews proposals, and negotiates contracts with winning bidders for a specified amount of energy savings.

Demand Side Management (DSM)—The planning, implementation, and monitoring of utility activities designed to encourage consumers to modify patterns of electricity usage, including the timing and level of electricity demand. It refers to only energy and load-shape modifying activities that are undertaken in response to utility-administered programs. It does not refer to energy and load-shaped changes arising from the normal operation of the marketplace or from government-mandated energy-efficiency standards. Demand-Side Management covers the complete range of load-shape objectives, including strategic conservation and load management, as well as strategic load growth.

Demand Side Resource—Resources obtained through the implementation of DSM that may be used as an alternative to traditional supply-side resources.

Demand Side Resource Portfolio—Comprehensive collection of DSM resources, both viable and non-viable, that are available, both currently and in the future to the utility.

Demand(ed) Factor—The ratio of the maximum demand on an electricity generating and distribution system to the total connected load on the system; usually expressed as a percentage.

Demand, Integrated—The demand averaged over a specified period, usually determined by an integrating demand meter or by the integration of a load curve. It is the average of the instantaneous demands during a specified demand interval.

Demand-Metered—Having a meter to measure peak demand (in addition to total consumption) during a billing period. Demand is not usually metered for other energy sources.

Demising Partitions—Barriers that separate conditioned space from enclosed unconditioned space.

Demising Wall—A wall that is a demising partition.

Demonstrated Reserve Base (coal)—A collective term for the sum of coal in both measured and indicated resource categories of reliability, representing 100 percent of the in-place coal in those categories as of a certain date. Includes beds of bituminous coal and anthracite 28 or more inches thick and beds of sub-bituminous coal 60 or more inches thick that can occur at depths of up to 1,000 feet. Includes beds of lignite 60 or more inches thick that can be surface mined. Includes also thinner and/or deeper beds that presently are being mined or for which there is evidence that they could be mined commercially at a given time. Represents that portion of the identified coal resource from which reserves are calculated.

Demonstrated Resources—Same qualifications as identified resources, but include measured and indicated degrees of geologic assurance and excludes the inferred.

Demonstration—The application and integration of a new product or service into an existing or new system. Most commonly, demonstration involves the construction and operation of a new electric technology interconnected with the electric utility system to demonstrate how it interacts with the system. This includes the impacts the technology may have on the system and the impacts that the larger utility system might have on the functioning of the technology.

Demonstration and Test Vehicles—Vehicles operated by a motor vehicle dealer solely for the purpose of promoting motor vehicle sales or permitting potential purchasers to drive the vehicle for pre-purchase or pre-lease evaluation; or a vehicle that is owned and operated by a motor vehicle manufacturer or motor vehicle component manufacturer, or owned or held by a university research department, independent testing laboratory, or other such evaluation facility, solely for the purpose of evaluating the performance of such vehicles for engineering, research and development, or quality control reasons.

Demonstration Scale Plant—A plant between pilot and commercial size built to demonstrate the commercial feasibility of a process.

Demurrage—The charge paid to the vessel owner or operator for detention of a vessel at the port(s) beyond the time allowed, usually 72 hours, for loading and unloading.

Denatured—Ethanol that has had a substance added to make it unfit for human consumption.

Dendrite—A slender threadlike spike of pure crystalline material, such as silicon.

Dendritic Web Technique—A method for making sheets of polycrystalline silicon in which silicon dendrites are slowly withdrawn from a melt of silicon whereupon a web of silicon forms between the dendrites and solidifies as it rises from the melt and cools.

Densification—A mechanical process to compress biomass (usually wood waste) into pellets, briquettes, cubes, or densified logs.

Densitometer—An instrument used to determine the ratio of the density of a substance to the density of a given substance (as water or hydrogen) taken as a standard, when both densities are obtained by weighing in air.

Density—The mass per unit volume of a construction material as documented in an ASHRAE handbook, a comparably reliable reference or manufacturer's literature.

DEP—Department of Environmental Protection

Department of Agriculture (USDA)—A federal government agency involved in rural development, marketing and regulatory programs, food safety, research, education and economics, food, nutrition and consumer service, farm and foreign agricultural services, and natural resources and environment programs.

Department of Energy (US DOE)—The federal department established by the Department of Energy Organization Act to consolidate the major federal energy functions into one cabinet-level department that would formulate a comprehensive, balanced national energy policy. DOE's main headquarters are in Washington, D.C.

Dependable Capacity—The load-carrying ability of an electric power plant during a specific time interval and period when related to the characteristics of the load to be/being supplied; determined by capability, operating power factor, and the portion of the load the station is to supply.

Depletable Energy Sources (California)—(1) electricity purchased from a public utility, (2) energy obtained from burning coal, oil, natural gas or liquefied petroleum gases. [See California Code of Regulations, Title 24, Section 2-5302]

Depletable Sources—Energy obtained from electricity purchased from a public utility, or energy obtained from burning coal, oil, natural gas, or liquefied petroleum gases.

Depleted Resources—Resources that have been mined; include coal recovered, coal lost in mining, and coal reclassified as sub-economic because of mining.

Depleted Storage Field—A sub-surface natural geological reservoir, usually a depleted gas or oil field, used for storing natural gas.

Depletion (Coal)—The subtraction of both tonnage produced and the tonnage lost to mining from identified resources to determine the remaining tonnage as of a certain time.

Depletion Allowance—A term for either (1) a periodic assignment to expense of recorded amounts or (2) an allowable income tax deduction that is related to the exhaustion of mineral reserves. Depletion is included as one of the elements of amortization. When used in that manner, depletion refers only to book depletion.

Book. The portion of the carrying value (other than the portion associated with tangible assets) prorated in each accounting period, for financial reporting purposes, to the extracted portion of an economic interest in wasting natural resource.

Tax-cost. A deduction (allowance) under U.S. Federal income taxation normally calculated under a formula whereby the adjusted basis of the mineral property is multiplied by a fraction, the numerator of which is the number of units of minerals sold during the tax year and the denominator of which is the estimated number of units of un-extracted minerals remaining at the end of the tax year plus the number of units of minerals sold during the tax year.

Tax-percentage (for Statutory). A deduction (allowance) allowed to certain mineral producers under U.S. Federal income taxation calculated on the basis of a specified percentage of gross revenue from the sale of minerals from each mineral property not to exceed the lesser of 50 percent of the taxable income from the property computed without allowance for depletion. (There are also other limits on percentage depletion on oil and gas production.) The taxpayer is entitled to a deduction representing the amount of tax-cost depletion or percentage (statutory) depletion, whichever is higher.

Excess statutory depletion. The excess of estimated statutory depletion allowable as an income tax deduction over the amount of cost depletion otherwise allowable as a tax deduction, determined on a total enterprise basis.

Depletion Factor—The multiplier applied to the tonnage produced to compute depletion. This multiplier takes into account both the tonnage recovered and the tonnage lost due to mining. The depletion factor is the reciprocal of the recovery factor in relation to a given quantity of production.

Depreciation—The process of allocating the actual cost of a tangible fixed asset, less salvage value, over its estimated useful life in a rational and systematic manner. Land cannot be depreciated. Also see definition for *Amortization*. Depreciation is used in both a financial reporting and tax context, and is considered a tax benefit because the depreciation deductions cause a reduction in taxable income, thereby lowering a firm's tax liability.

Depreciation and Amortization of Property, Plant, and Equipment—The monthly provision for depreciation and amortization (applicable to utility property other than electric plant, electric plant in service, and equipment).

Depreciation Reserve Ratio—The ratio of the accumulated depreciation to the recorded cost of surviving plant at any given date.

Depreciation, Flow-Through—An accounting procedure under which current Net Income reflects decreases or increases in current taxes on income, arising from the use of liberalized depreciation or accelerated amortization for tax purposes instead of the straightline method.

Depreciation, Normalized—An accounting method under which Net Income includes charges or credits equal to the decreases or increases in current taxes on income, arising from the use of liberalized depreciation or accelerated amortization for tax purposes instead of the straight-line method. The contra entries for such charges to Net Income are suspended in Balance Sheet accounts. In future years, there is a feedback of these suspended amounts to Net Income when increases in the then current taxes on income occur because liberalized depreciation or accelerated amortization was used for tax purposes in prior years.

Depreciation, Unit of Production—A method of depreciation whereby the asset is depreciated over an estimated life expressed in units of output rather than over an estimated life expressed as a period of time.

Depth of Deepest Production—The depth of the deepest production is the length of the well bore measured (in feet) from the surface reference point to the bottom of the open hole or the deepest perforation in the casing of a producing well.

Depth of Discharge (DOD)—The amount of energy withdrawn from a battery or cell expressed as a percentage of its rated capacity. For example, the removal of 25 ampere-hours from a fully charged 100 ampere-hour rated cell results in a 25 percent depth of discharge. Depth of discharge is the opposite of state of charge (SOC).

DEQ—Department of Environmental Quality.

DER—Department of Environmental Resources
  • Distributed Energy Resources

Derating—The production of energy by a system or appliance at a level less than its design or nominal capacity.

Deregulation—The elimination of some or all regulations from a previously regulated industry or sector of an industry.
  • The process of changing the laws and regulations that control the electric industry to allow competition of electricity service and retail sales. This results in customer choice of an electricity provider. See also "Restructuring"

Deriming—Adding heat to remove accumulated solid water or carbon dioxide constituents from low-temperature process equipment.

Derivatives—A specialized security or contract that has no intrinsic overall value, but whose value is based on an underlying security or factor as an index. A generic term that, in the energy field, may include options, futures, forwards, etc.
  • A financial instrument derived from a cash market commodity, futures contract or other financial instrument, which can be traded on regulated exchange markets or over-the-counter.

DERs—Data Evaluation Records

Desertification—A process whereby the productivity of the land is reduced through deforestation, water logging and salinization, chemical degradation by nutrient leaching, range mismanagement such as overgrazing, soil erosion and aridity and semi-aridity.

Desiccant—A material used to desiccate (dry) or dehumidify air.
  • Any absorbent or adsorbent, liquid or solid, that will remove water or water vapor from a material. In a refrigeration circuit, the desiccant must be insoluble in the refrigerant.

Desiccant Cooling—To condition/cool air by desiccation.

Desiccation—The process of removing moisture; involves evaporation.

Design Conditions—The parameters and conditions used to determine the performance requirements of space-conditioning systems. Design conditions for determining design heating and cooling loads are specified in Section 144 (b) for nonresidential, high-rise residential, and hotel/motel buildings and in Section 150 (h) for low-rise residential buildings (in California).

Design Cooling Load—The amount of conditioned air to be supplied by a cooling system; usually the maximum amount to be delivered based on a specified number of cooling degree days or design temperature.

Design Day—A 24-hour period of demand which is used as a basis for planning energy capacity requirements.

Design Electrical Rating (Capacity) Net—The nominal net electrical output of a nuclear unit, as specified by the utility for the purpose of plant design.

Design Factor—The Design Factor is a ratio comparing a proposed system's expected generation output with that of a baseline system.

Design for Environment—An engineering perspective in which the environmentally related characteristics of a product, process, or facility design are optimized.

Design Head—The achieved river, pondage, or reservoir surface height (forebay elevation) that provides the water level to produce the full flow at the gate of the turbine in order to attain the manufacturer's installed nameplate rating for generation capacity.

Design Heat Gain Rate—The total calculated heat gain through the building envelope under design conditions.

Design Heat Loss Rate—The total calculated heat loss through the building envelope under design conditions.

Design Heating Load—The amount of heated air, or heating capacity, to be supplied by a heating system; usually the maximum amount to be delivered based on a specified number of heating degree days or design outside temperature.

Design Life—Period of time a system or appliance (or component of) is expected to function at its nominal or design capacity without major repair.

Design Month (Solar)—The month in which the combination of insolation and load requires the maximum energy from the array.

Design Temperature—The temperature that a system is designed to maintain (inside) or operate against (outside) under the most extreme conditions.

Design Tip Speed Ratio—For a wind turbine, the ratio of the speed of the tip of a turbine blade for which the power coefficient is at maximum.

Design Voltage—The nominal voltage for which a conductor or electrical appliance is designed; the reference voltage for identification and not necessarily the precise voltage at which it operates.

Design-Build—A method of project delivery in which the owner contracts directly with a single entity that is responsible for both design and construction services for a construction project. Proceeds in a quasi-parallel fashion construction starts before design is completed, and pricing occurs before/during design.

Desired Future Condition—A vision of the desired future state of a specific area. A description of land and resource conditions expected to result if management goals and objectives are achieved.

Destruction and Removal Efficiency (DRE)—A percentage that represents the number of molecules of compound removed or destroyed in an incinerator or other device relative to the number of molecules that entered the incinerator or device. A DRE of 99.99 percent means that 9,999 molecules of a compound are destroyed for every 10,000 molecules that

enter the system.

Desulphurization—The removal of sulfur, as from molten metals, petroleum oil, or flue gases.

Desulphurizer—A component part that removes sulfur from a fuel.

Desuperheater—An energy saving device in a heat pump that, during the cooling cycle, recycles some of the waste heat from the house to heat domestic water.

DET—Determination of Equivalent Treatment

Detention—In a stormwater management, ponding of runoff in pools and basins for water-quality improvement and flood prevention.

Developed Countries—Under the Kyoto Protocol, industrialized countries (identified in Annex I and Annex B).

Developer's Fee—A term subject to various interpretations. Many appraisers associate a developer's fee with payment for overseeing the development of a project from inception to completion and include it among the direct and indirect costs of development. Others use the term interchangeably with "entrepreneurial profit," equating it with compensation for the time, energy, and experience a developer invests in a project as well as a reward for the risk the developer takes.

Developing Countries—Under the Kyoto Protocol, countries which are in the process of industrialization which have constrained resources to address their economic and environmental problems. Developing countries also referred to as Less Developed Countries (LDC).

Development (Mineral)—The preparation of a specific mineral deposit for commercial production; this preparation includes construction of access to the deposit and of facilities to extract the minerals. The development process is sometimes further distinguished between a pre-production stage and a current stage, with the distinction being made on the basis of whether the development work is performed before or after production from the mineral deposit has commenced on a commercial scale

Development Costs (Oil and Gas)—Costs incurred to obtain access to proven reserves and to provide facilities for extracting, treating, gathering, and storing the oil and gas. More specifically, development costs, depreciation and applicable operating costs of support equipment and facilities, and other costs of development activities, are costs incurred to:

- Gain access to and prepare well locations for drilling, including surveying well locations for the purpose of determining specific development drilling sites; clearing ground; draining; road building; and relocating public roads, gas lines, and power lines to the extent necessary in developing the proved reserves.

- Drill and equip development wells, development-type stratigraphic test wells, and service wells, including the costs of platforms and of well equipment such as casing, tubing, pumping equipment, and the wellhead assembly.
- Acquire, construct, and install production facilities such as lease flow lines, separators, treaters, heaters, manifolds, measuring devices, production storage tanks, natural gas cycling and processing plants, and utility waste disposal systems.
- Provide improved recovery systems.

Development Drilling—Drilling done to determine more precisely the size, grade, and configuration of an ore deposit subsequent to when the determination is made that the deposit can be commercially developed. Not included are: (1) secondary development drilling, (2) solution-mining drilling for production, or (3) production-related underground and open-pit drilling done for control of mining operations.

Development Well—A well drilled within the proved area of an oil or gas reservoir to the depth of a stratigraphic horizon known to be productive. Also see *Well*.

Developmental Drilling—To delineate the boundaries of a known mineral deposit to enhance the productive capacity of the producing mineral property.

Devonian Shale—Geological formations, typically hundreds of feet thick, that underlie much of the Appalachian Basin. It is known to contain much natural gas, but usually lacks sufficient natural permeability for ordinary production.

Dewpoint—The temperature to which air must be cooled, at constant pressure and water vapor content, in order for saturation or condensation to occur; the temperature at which the saturation pressure is the same as the existing vapor pressure; also called saturation point.

DFE—Design for the Environment

DFG—Department of Fish & Game

DG—Distributed Generation: Also called Self-Generation, whereby consumers use small-scale power generation technologies (typically in the range of 3 to 10,000 kW) located close to where electricity is used (e.g., a home or business) to provide an alternative to or an enhancement of the traditional electric power system.

DHCP (Dynamic Host Control Protocol)—A device IP address assignment protocol that automatically generates an IP address once a device is connected to a local server/router. With DHCP, a gateway may get a different IP address within a fixed range each time it reboots. Used for devices that send data from a remote site to a central station.

DHS—Department of Hazardous Substances
- Designated Hazardous Substances
- Department of Health Services

DHW—Domestic Hot Water. Refers to any system that provides hot water for domestic use.

Diameter at Breast Height (DBH)—The diameter of a tree measured 4 feet 6 inches above the ground.

Diesel Engine—A compression-ignition piston engine in which fuel is ignited by injecting it into air that has been heated (unlike a spark-ignition engine).

Diesel Fuel—A fuel composed of distillates obtained in petroleum refining operation or blends of such distillates with residual oil used in motor vehicles. The boiling point and specific gravity are higher for diesel fuels than for gasoline.

Diesel Fuel System—Diesel engines are internal combustion engines that burn diesel oil rather than gasoline. Injectors are used to spray droplets of diesel oil into the combustion chambers, at or near the top of the compression stroke. Ignition follows due to the very high temperature of the compressed intake air, or to the use of "glow plugs," which retain heat from previous ignitions (spark plugs are not used). Diesel engines are generally more fuel-efficient than gasoline engines but must be stronger and heavier because of high compression ratios.

Diesel Oil—Fuel for diesel engines obtained from the distillation of petroleum. It is composed chiefly of aliphatic hydrocarbons. Its volatility is similar to that of gas oil. Its efficiency is measured by cetane number.

Diesel-Electric Plant—A generating station that uses diesel engines to drive its electric generators.

Difference in Life-Cycle Costs—Present-Value Savings. The collar value of the energy savings over the life of the system.

Difference of Potential—The difference in electrical pressure (voltage) between any two points in an electrical system or between any point in an electrical system and the earth.

Differential Controller—An electronic switch that turns off or on based on the difference between two temperatures. In a solar hot water system, the controller measures the temperature at the collector and compares it to the water temperature in a storage tank to turn the pump on or off.

Differential Thermostat—A type of automatic thermostat (used on solar heating systems) that responds to temperature differences (between collectors and the storage components) so as to regulate the functioning of appliances (to switch transfer fluid pumps on and off).

Differentiation—Differing national circumstances that might require differing emission reduction obligations in the Kyoto Protocol.

Diffuse Isolation—Incident sunlight received indirectly because of scattering due to clouds, fog, particulates, or other obstructions in the atmosphere. The other component of sunlight is Direct. Opposite of direct insolation.

Diffuse Radiation—Solar radiation, scattered by water vapor, dust and other particles as it passes through the atmosphere, so that it appears to come from the entire sky. Diffuse radiation is higher on hazy or overcast days than on clear days.

Diffuse Solar Radiation—Sunlight scattered by atmospheric particles and gases so that it arrives at the earth's surface from all directions and cannot be focused.

Diffusion—The movement of individual molecules through a material; permeation of water vapor through a material.

Diffusion Furnace—Furnace used to make junctions in semiconductors by diffusing dopant atoms into the surface of the material.

Diffusion Length—The mean distance a free electron or hole moves before recombining with another hole or electron.

Diffusive Transport—The process by which particles of liquids or gases move from an area of higher concentration to an area of lower concentration.

DIG—Deputy Inspector General

Digester—An airtight vessel or enclosure in which bacteria decomposes biomass in water to produce biogas.

Digester (Anaerobic)—A device in which organic material is bio-chemically decomposed (digested) by anaerobic bacteria to treat the material and/or to produce biogas.

Digirail—Modbus device used to count pulses, such as for an anemometer or KYZ meter.

Diluent—A neutral fluid added to another fluid to reduce the concentration of the second fluid in a mixture.

Dilution—A reduction in the percentage ownership of a given shareholder in a company caused by the issuance of new shares.

Dilution Protection—Mainly applies to convertible securities. Standard provision whereby the conversion ratio is changed accordingly in the case of a stock dividend or extraordinary distribution to avoid dilution of a convertible bondholder's potential equity position. Adjustment usually requires a split or stock dividend in excess of 5% or issuance of stock below book value. Share Purchase Agreements also typically contain anti-dilution provisions to protect investors in the event that a future round of financing occurs at a valuation that is below the valuation of the current round.

Dimmer—A light control device that allows light levels to be manually adjusted. A dimmer can save energy by reducing the amount of power delivered to the light while consuming very little themselves.

Diode—An electronic device that allows current to flow in one direction only.

Dioxin—A family of compounds, some of which are hazardous, that result from combustion of carbon materials. The most toxic of these compounds is 2, 3, 7, 8-tetrachlorodibenzo-p-dioxin.

Dip Tube—A tube inside a domestic water heater that distributes the cold water from the cold water supply line into the lower area of the water heater where heating occurs.

DIR—Department of Industrial Relations

Direct Access—The ability of an electric power consumer to purchase electricity from a supplier of their choice without being physically inhibited by the owner of the electric distribution and transmission system to which the consumer is connected to. (See also *Open Access*.)

Direct Beam Radiation—Solar radiation that arrives in a straight line from the sun. Measured by a pyrheliometer with a solar aperture of 5.7° to transcribe the solar disc.

Direct Capitalization—A method used to convert an estimate of a single year's income expectancy into an indication of value in one direct step, either by dividing the income estimate by an appropriate rate or by multiplying the income estimate by an appropriate factor.

Direct Costs—Expenditures for the labor and materials used in the construction of improvements or capital assets. Also called hard costs.

Direct Current (DC)—A type of electricity transmission and distribution by which electricity flows in one direction through the conductor; usually relatively low voltage and high current; typically abbreviated as dc.

Direct Digital Control (DCC)—The automated control of a condition or process by a digital device.

Direct Electricity Load Control—The utility installs a radio-controlled device on the HVAC equipment. During periods of particularly heavy use of electricity, the utility will send a radio signal to the building in its service territory with this device and turn off the HVAC for a certain period.

Direct Emissions—Emissions from sources that are owned or controlled by the reporting entity.

Direct Energy Conversion—Production of electricity from an energy source without transferring the energy to a working fluid or steam. For example, photovoltaic cells transform light directly into electricity. Direct conversion systems have no moving parts and usually produce direct current.

Direct Expansion (Refrigeration)—Any system that, in operation between an environment where heat is absorbed (heat source), and an environment into which unwanted heat is directed (heat sink) at two different temperatures, is able to absorb heat from the heat source at the lower temperature and reject heat to the heat sink at the higher temperature. The cooling effect is obtained directly from a fluid called a refrigerant that absorbs heat at a low temperature and pressure, and transfers heat at a higher temperature and higher pressure.

Direct Finance Lease—A lessor capital lease (per FASB 13) that does not give rise to manufacturers or dealer's profit (or loss) to the lessor. A lease classification for a financing lessor in which the lease meets any of the criteria defining a capital lease and for which (a) collectability of the minimum lease payments is reasonably assured; and (b) there exists no important uncertainties as to costs yet to be incurred by the lessor under the lease agreement.

Direct Gain—In passive solar heating, a direct gain system relies on the sunshine to directly hit the substance or mass being heated. Direct gain systems used today usually rely on a layer(s) of glass to assist in holding the heat within a space where the heat is desirable. In direct-gain buildings, sunlight directly enters the building through the windows and is absorbed and stored in massive floors or walls. These buildings are elongated in the east-west direction, and most of their windows are on the south side. The area devoted to south windows varies throughout the country. It could be as much as 20% of the floor area in sunny cold climates, where advanced glazing or moveable insulation are recommended to prevent heat loss at night. These buildings have high insulation levels and added thermal mass for heat storage.

Direct Insolation—Sunlight falling directly upon a collector. Opposite of diffuse insolation.

Direct Installation Program—DSM program in which the utility directly installs DSM measures within customers homes or businesses.

Direct Labor Hours (Mining)—Direct labor hours worked by all mining employees at a mining operation during the year. Includes hours worked by those employees engaged in production, preparation, development, maintenance, repair, shop or yard work management, and technical or engineering work. Excludes office workers. Excludes vacation and leave hours.

Direct Lease—A lease in which the lessor provides the entire purchase price for the leased asset from the lessor's own funds.

Direct Load Control—This Demand-Side Management category represents the consumer load that can be interrupted at the time of annual peak load by direct control of the utility system operator. Direct Load Control does not include Interruptible Load. This type of control usually involves residential consumers.

Direct Methanol Fuel Cell—A type of fuel cell that uses a polymer membrane as the electrolyte, however, the catalyst itself draws the hydrogen from the liquid methanol, eliminating the need for a fuel reformer.

Direct Milling Cost—Operating costs directly attributable to the processing of ores or other feed materials, including labor, supervision, engineering, power, fuel, supplies, reagents, and maintenance.

Direct Non-process End Use—Those end uses that may be found on commercial, residential, or other sites, as well as at manufacturing establishments. They include heating, ventilation, and air conditioning (HVAC), facility lighting, facility support, onsite transportation, conventional electricity generation, and other non-process uses. "Direct" denotes that only the quantities of electricity or fossil fuel used in their original state (i.e., not transformed) are included in the estimates.

Direct Process End Use—Those end uses that are specific to the carrying out of manufacturing. They include process heating, process cooling and refrigeration, machine drive, electrochemical processes, and other process uses. "Direct" denotes that only the quantities of electricity or fossil fuel used in their original state (i.e., not transformed) are included in the estimates.

Direct Radiation—Light that has traveled in a straight path from the sun (also referred to as beam radiation). An object in the path of direct radiation casts a shadow on a clear day.

Direct Solar Gain—Solar energy collected from the sun (as heat) in a building through windows, walls, skylights, etc.

Direct Solar Water Heater—These systems use water as the fluid that is circulated through the collector to the storage tank. Also known as "open-loop" systems.

Direct Sunlight—That portion of daylight arriving at a specified location directly from the sun, without diffusion.

Direct Use—Use of electricity that (1) is self-generated, (2) is produced by either the same entity that consumes the power or an affiliate, and (3) is used in direct support of a service or industrial process located within the same facility or group of facilities that house the generating equipment. Direct use is exclusive of station use.

Direct Vent Heater—A type of combustion heating system in which combustion air is drawn directly from outside and the products of combustion are vented directly outside. These features are beneficial in tight, energy-efficient homes because they will not depressurize a home and cause air infiltration, and backdrafting of other combustion appliances.

Direct Water Heater—A type of water heater in which heated water is stored within the tank. Hot water is released from the top of the tank when a hot water faucet is turned. This water is replaced with cold water that flows into the tank and down to just above the bottom plate under which are the burners.

Direct-Fired—A heating unit in which the combustion products are mixed with the air or liquid being heated.

Direct-Gain—The process by which sunlight directly enters a building through the windows and is absorbed and stored in massive floors or walls.

Directional (Deviated) Well—A well purposely deviated from the vertical, using controlled angles to reach an objective location other than directly below the surface location. A directional well may be the original hole or a directional "sidetrack" hole that deviates from the original bore at some point below the surface. The new footage associated with directional "sidetrack" holes should not be confused with footage resulting from remedial sidetrack drilling. If there is a common bore from which two or more wells are drilled, the first complete bore from the surface to the original objective is classified and reported as a well drilled. Each of the deviations from the common bore is reported as a separate well.

Directional Drilling—That is deliberately made to depart significantly from the vertical.

Directly Conditioned Space—An enclosed space that is provided with wood heating, is provided with mechanical heating that has a capacity exceeding 10 Btu/(hr.xft.^{2}), or is provided with mechanical cooling that has a capacity exceeding 5 Btu/(hr. xft.^{2}), unless the space-conditioning system is designed and thermostatically controlled to maintain a process environment temperature less than 55°F or to maintain a process environment temperature greater than 90°F for the whole space that the system serves, or unless the space-conditioning system is designed and controlled to be incapable of operating at temperatures above 55°F or incapable of operating at temperatures below 90°F at design conditions.

Director—Person elected by shareholders to serve on the board of directors. The directors appoint the president, vice president and all other operating officers, and decide when dividends should be paid (among other matters).

Disaggregation—The functional separation of the vertically integrated utility into smaller, individually owned business units (i.e., generation, dispatch/control, transmission, distribution). The terms "deintegration," "disintegration" and "delamination" are sometimes used to mean the same thing. (See also *Divestiture*.)

Discharge—Withdrawal of electrical energy from a battery.

Discharge Factor—A number equivalent to the time in hours during which a battery is discharged at constant current usually expressed as a percentage of the total battery capacity, i.e., C/5 indicates a discharge factor of 5 hours.

Discharge Rate—A measure of the current withdrawn from a battery over time, expressed as a percentage of battery capacity. A C/5 discharge rate indicates a current of one-fifth of the rated capacity of the battery. Usually expressed in amperes or time, at which electrical current is taken from the battery.

Discharged Fuel—Irradiated fuel removed from a nuclear reactor during refueling. Also see *Spent Fuel*.

Disclosure Statement—A booklet outlining the risk factors associated with an investment.

DISCO—Distribution Company

Disconnect—Switch used to connect or disconnect components in a PV system.

Discount on Capital Stock—The excess of par or stated value over the price paid to the company by the shareholders for original issue shares of its capital stock.

Discount Rate—The interest rate at which the Federal Reserve System stands ready to lend reserves to commercial banks. The rate is proposed by the 12 Federal Reserve banks and determined with the approval of the Board of Governors.

• A certain interest rate that is used to bring a series of future cash flows to their present value in order to state them in current, or today's, dollars. Use of a discount rate removes the time value of money from future cash flows.

Discounted Cash Flow (DCF)—A cash flow occurring in the future which has been discounted by a given discount factor on a compounded basis; the present value of a future cash flow.

Discounted Cash Flow Analysis—The procedure in which a discount rate is applied to a set of projected income and expense streams and a reversion. The analyst specifies the quantity, variability, timing, and duration of the income/expense streams as well as the quantity and timing of the reversion and discounts each to its present value at a specified yield rate. The analysis can be applied with any yield capitalization technique and may be performed on either a transactional or aggregate basis.

The analysis of cash flow projections, period by period over a presumed term of ownership, to commute the present value for a given rate of return or to compute the internal rate of return indicated by a series of cash flows.

Discounted Payback—The number of years it will take for a project to pay back the full costs of the assets, taking into consideration the value of money over time.

Discounting—A method of converting future dollars into present values, accounting for interest costs or forgone investment income. Used to convert a future payment into a value that is equivalent to a payment now.

A method used by economists to determine the dollar value today of a project's future costs and benefits. This is done by weighting money values that occur in the future by a value less than 1, or "discounting" them. Because environmental decision makers are increasingly forced to evaluate policies with costs and benefits that will be spread out over tens—perhaps hundreds—of years, discounting is

used to help evaluate the value of measures that deal with problems such as stratospheric ozone depletion, global climate change, and the disposal of low- and high-level radioactive wastes.

Discrete Early Action—Greenhouse gas reduction measures enforceable by January 1, 2010.

Discrete Emission Reduction Credits (DER)—DERs, DERCs, or open market credits, are reductions in emissions that occur over a specified time period and do not continue on into the future. Generally, unlike ERCs, DERs are not evaluated and verified by the relevant local or state government air agency. Mass-based ERCs are one type of DER.

Discrete-Delivery Energy Sources—Energy sources that must be delivered to a site.

Dispatchability—The ability of a generating unit or other source of electric power to vary output.

• Operational control over the periods when a storage resource is employed to generate, supply or charge electrical power.

Dispatchability of DSM—The ability of the utility to schedule and control, directly or indirectly, manually or automatically, the timing and volumes of DSM measures.

Dispatchable Power—Energy output that can be planned on and typically provides a continuous power output. Solar power and Wind power is not dispatchable without configuration with some other power or storage mechanism. Hydrocarbon based power plants or nuclear plants are dispatchable.

Dispatch—Centralized—The ability to control which plants are operating at which levels from a single location. There are three types of electrical generators: Baseload, cycling and peaking facilities. Baseload units usually operate in the range of 40-85% of the year; cycling facilities operate 15-40% of the time; and peaking facilities generally run less than 15% of the time.

"Economic Dispatch" (also known as cost based dispatch) refers to the process by which facilities are chosen for operation—those with the lowest price power are called on first. Compliance with a strict definition of the next most expensive kilowatt being chosen for generation to meet demand is sometimes criticized since system or capacity constraints regularly preclude such truly economic choices. No consideration is usually given to the lowest pollution emitting plants in this model.

Bid based dispatch is described as a true market. Also called a "spot market," the power sold would be the least expensive power that is purchased, not necessarily the lease expensive produced. As in any market, if a seller wishes to sell below their costs, that would be their choice.

Dispatching—The operating control of an integrated electric system involving operations such as (1) the assignment of load to specific generating stations and other sources of supply to effect the most economical supply as the total or the significant area loads rise or fall; (2) the control of operations and maintenance of high-voltage lines, substations, and equipment; (3) the operation of principal tie lines and switching; (4) the scheduling of energy transactions with connecting electric utilities.

Displacement Power—A source of power (electricity) that can displace power from another source so that source's power can be transmitted to more distant loads.

Displacement Ventilation—Ventilation that uses natural convection processes to move warm air up and out of a volume. Displacement ventilation tends to use less energy than conventional forced air ventilation, as it works with natural convection processes. Often used extensively in Net Zero Energy Buildings.

Disposition, Natural Gas—The removal of natural, synthetic, and/or supplemental gas, or any components or gaseous mixtures contained therein, from the responding company's facilities within the report State by any means or for any purpose, including the transportation of such gas out of the report State.

Disposition, Petroleum—A set of categories used to account for how crude oil and petroleum products are transferred, distributed, or removed from the supply stream. The categories include stock change, crude oil losses, refinery inputs, exports, and products supplied for domestic consumption.

Dissolved Gas—Natural gas that can be developed for commercial use, and which is found mixed with oil in naturally occurring underground formations.

Dissolved Oxygen—Oxygen found in water and required by organisms for survival. As the amount of sewage increases in water, bacteria multiply to feed on the sewage and consume more oxygen, thereby decreasing the amount in the water available for use by other animals living there.

Distillate—A general classification for one of the petroleum fractions produced in conventional distillation operations. Included are kerosene and products known as heating oils and diesel fuels, specifically No. 1, No. 2, and No. 4 Fuel Oils and No. 1, No. 2, and No. 4 Diesel Fuels.

Distillate Fuel Oil—A general classification for one of the petroleum fractions produced in conventional distillation operations. It includes diesel fuels and fuel oils. Products known as No. 1, No. 2, and No. 4 diesel fuel are used in on-highway diesel engines, such as those in trucks and automobiles, as well as off-highway engines, such as those in railroad locomotives and agricultural machinery. Products known as No. 1, No. 2, and No. 4 fuel oils are used primarily for space heating and electric power generation.

No. 1 Distillate: A light petroleum distillate that can be used as either a diesel fuel (See *No. 1 Diesel Fuel*) or a fuel oil. See No. 1 Fuel Oil.

No. 1 Diesel Fuel: A light distillate fuel oil that has distillation temperatures of 550 degrees Fahrenheit at the 90-percent point and meets the specifications defined in ASTM Specification D 975. It is used in high-speed diesel engines, such as those in city buses and similar vehicles. See No. 1 Distillate above.

No. 1 Fuel Oil: A light distillate fuel oil that has distillation temperatures of 400 degrees Fahrenheit at the 10-percent recovery point and 550 degrees Fahrenheit at the 90-percent point and meets the specifications defined in ASTM Specification D 396. It is used primarily as fuel for portable outdoor stoves and portable outdoor heaters. See No. 1 Distillate above.

No. 2 Distillate: A petroleum distillate that can be used as either a diesel fuel (See *No. 2 Diesel* Fuel definition below) or a fuel oil. See *No. 2 Fuel Oil* below.

No. 2 Diesel Fuel: A fuel that has distillation temperatures of 500 degrees Fahrenheit at the 10-percent recovery point and 640 degrees Fahrenheit at the 90-percent recovery point and meets the specifications defined in ASTM Specification D 975. It is used in high-speed diesel engines, such as those in railroad locomotives, trucks, and automobiles. See *No. 2 Distillate* above.

Low Sulfur No. 2 Diesel Fuel: No. 2 diesel fuel that has a sulfur level no higher than 0.05 percent by weight. It is used primarily in motor vehicle diesel engines for on-highway use.

High Sulfur No. 2 Diesel Fuel: No. 2 diesel fuel that has a sulfur level above 0.05 percent by weight.

No. 2 Fuel oil (Heating Oil): A distillate fuel oil that has distillation temperatures of 400 degrees Fahrenheit at the 10-percent recovery point and 640 degrees Fahrenheit at the 90-percent recovery point and meets the specifications defined in ASTM Specification D 396. It is used in atomizing type burners for domestic heating or for moderate capacity commercial/industrial burner units. See *No. 2 Distillate* above.

No. 4 Fuel: A distillate fuel oil made by blending distillate fuel oil and residual fuel oil stocks. It conforms with ASTM Specification D 396 or Federal Specification VV-F-815C and is used extensively in industrial plants and in commercial burner installations that are not equipped with preheating facilities. It also includes No. 4 diesel fuel used for low- and medium-speed diesel engines and conforms to ASTM Specification D 975.

Distillate Oil—Any distilled product of crude oil. A light petroleum product used for home heating and most machinery.

Distillation—The process to separate the components of a liquid mixture by boiling the liquid and then recondensing the resulting vapor.

Distillation Unit (Atmospheric)—The primary distillation unit that processes crude oil (including mixtures of other hydrocarbons) at approximately atmospheric conditions. It includes a pipe still for vaporizing the crude oil and a fractionation tower for separating the vaporized hydrocarbon components in the crude oil into fractions with different boiling ranges. This is done by continuously vaporizing and condensing the components to separate higher oiling point material. The selected boiling ranges are set by the processing scheme, the properties of the crude oil, and the product specifications.

Distillers' Dried Grains (DDG)—The dried byproduct of the grain fermentation process. Typically used as a high-protein animal feed.

Distributed Emissions—Annual air pollutant (GHG or other) from fuel combustion at co-generation facilities distributed between energy stream outputs including thermal energy, electricity generation, and where applicable other product outputs such as hydrogen.

Distributed Energy Resources (DER)—A variety of small, modular power-generating technologies that can be combined with energy management and storage systems and used to improve the operation of the electricity delivery system, whether or not those technologies are connected to an electricity grid.

Distributed Generation (DG)—This type of system involves small amounts of generation located on the utility's distribution system for the purpose of meeting substation level (local) peak loads and/or eliminating the need to upgrade local distribution lines. Installed systems that are installed at or near the location where the electricity or thermal load is used, as opposed to central systems that supply electricity to grids.

Distributed Power Generation—A distributed generation system involves small amounts of generation located on a utility's distribution system for the purpose of meeting local (substation level) peak loads and/or displacing the need to build additional (or upgrade) local distribution lines.

• The concept of generating power onsite—at the point of use. Distributed generation means moving power generation from central, large-scale, power plants to smaller units located close to the end-users, thereby enabling significant environmental benefits to be derived from the use of sustainable energy sources.

• These decentralized resources are often managed using remote, web-enabled monitoring and control systems, thereby cutting operating costs. Distributed generation can be used to meet local demands for electricity, heating and cooling; or it can be connected to existing regional or national electricity grids for the purpose of buying and selling, according to need.

Distributed System—Systems that are installed at or near the location where the electricity is used, as opposed to central systems that supply electricity to grids. A residential photovoltaic system is a distributed system if it is installed on the residence where the occupants reside and use the power directly.

Distributed/Point-of-use Water-heating System—A system for heating hot water, for other than space heating purposes, which is located at more than one space within a building. A point-of-use water heater is located at the faucet and heats water only as required for immediate use. Because water is not heated until it is required, this equipment is more energy-efficient.

Distribution—The process of distributing electricity; usually defines that portion of a power provider's power lines between a power provider's power pole and transformer and a customer's point of connection/meter.

Distribution Feeder—(See *Feeder*)

Distribution Line—One or more circuits of a distribution system on the same line or poles or supporting structures' usually operating at a lower voltage relative to the transmission line.

Distribution System—Distribution System refers to the conveying means, such as ducts, pipes and wires, to bring substances or energy from a source to the point of use.
- Electric Grid
- The portion of the transmission and facilities of an electric system that is dedicated to delivering electric energy to an end-user.
- The substations, transformers and lines that convey electricity from high-power transmission lines to ultimate consumers. See *Grid*.

Distribution Upgrade Deferral—The avoided cost of deferred infrastructure on the distribution system.

Distribution Use—Natural gas used as fuel in the respondent's operations.

Distribution Utility/Company (Disco)—The regulated electric utility entity that constructs and maintains the distribution wires connecting the transmission grid to the final customer. The Disco can also perform other services such as aggregating customers, purchasing power supply and transmission services for customers, billing customers and reimbursing suppliers, and offering other regulated or non-regulated energy services to retail customers. The "wires" and "customer service" functions provided by a distribution utility could be split so that two totally separate entities are used to supply these two types of distribution services.

Distributor—A company primarily engaged in the sale and delivery of natural and/or supplemental gas directly to consumers through a system of mains.

District Chilled Water—Chilled water from an outside source used as an energy source for cooling in a building. The water is chilled in a central plant and piped into the building. Chilled water may be purchased from a utility or provided by a central physical plant in a separate building that is part of the same multi-building facility (for example, a hospital complex or university).

District Heat—Steam or hot water from an outside source used as an energy source in a building The steam or hot water is produced in a central plant and piped into the building. The district heat may be purchased from a utility or provided by a physical plant in a separate building that is part of the same facility (for example, a hospital complex or university).

District Heating—A heating system in which steam or hot water for space heating or hot water is piped from a central boiler plant or electric power/heating plant to a cluster of buildings.

District Heating or Cooling—A system that involves the central production of hot water, steam, or chilled water and the distribution of these transfer media to heat or cool buildings.

Distrigas Method—A formula used to allocate overhead costs of a parent to its affiliates. This method deviates from the "Mass Formula" by replacing the gross revenue factor with a factor that is computed based on net operating revenues (operating income before interest and federal taxes). This modification to the "Mass Formula" was the result of the Commission's concern, that the gross revenue factor would be distorted when allocating costs from an unregulated entity because of the inclusion of purchase gas costs in gross revenue for regulated pipelines.

Diversification—The process of spreading investments among various types of securities and various companies in different fields.

Diversion Rate—The share of solid waste diverted from landfill as a percentage of the total waste stream. (Recycling usually achieves about 20%, whereas the wet-dry schemes can achieve 50-80% diversion rates).

Diversity—The electric utility system's load is made up of many individual loads that make demands upon the system usually at different times of the day. The individual loads within the customer classes follow similar usage patterns, but these classes of service place different demands upon the facilities and the system grid. The service requirements of one electrical system can differ from another by time-of-day usage, facility usage, and/or demands placed upon the system grid.

Diversity Exchange—An exchange of capacity or energy, or both, between systems whose peak loads occur at different times.

Diversity Factor—The ratio of the sum of the non-coincidental maximum demands of two or more loads to their coincidental maximum demands for the same period.

Divest—To deprive of a right or Title.

Divestiture—The stripping off of one utility function from the others by selling (spinning-off) or in some other way changing the ownership of the assets related to that function. Most commonly associated with spinning-off generation assets so they are no longer owned by the shareholders that own the transmission and distribution assets. (See also *Dis-*

aggregation.)
Removal of utility functions from one another by selling-off or in some other way changing the ownership of the related assets. For example, this is most commonly associated with generation assets that are 'spun-off,' so these assets are no longer associated with the shareholders that own the transmission and distribution (T&D) assets.

Dividend—The payments designated by the Board of Directors to be distributed pro-rata among the shares outstanding. On preferred shares, it is generally a fixed amount. On common shares, the dividend varies with the fortune of the company and the amount of cash on hand and may be omitted if business is poor or if the Directors determine to withhold earnings to invest in capital expenditures or research and development.

Dividend Appropriations—Amounts declared payable out of unappropriated retained earnings as dividends on outstanding preferred or common stock.

Dividers—Are wood, aluminum or vinyl glazing dividers including mullions, muntins, munnions and grilles. Dividers may truly divide lights, be between the panes, or be applied to the exterior or interior of the glazing.

dL—Decaliter

DL—Detection Limit

DMIS—Duns Marketing Identification System (Dunn & Bradstreet)

DMS—Data Management System

DNC—Declared Net Capacity

DNO—Distribution Network Operator

DNR—Department of Natural Resources

DOB—Date of Birth

DOC—Department of Commerce
- Dissolved Organic Carbon
- Documentation of Compliance

Docket—A formal record of a Federal Energy Regulatory Commission proceeding. These records are available for inspection and copying by the public. Each individual case proceeding is identified by an assigned number.

Documentation Author—(in California) The person completing the compliance documentation that demonstrates whether a building complies with the standards. Compliance documentation requirements are defined in the Residential Manual.

DOD—Department of Defense
- Depth of Discharge—from 100% state of charge (SOC), in a battery or battery system

DOE—Department of Energy. A cabinet-level, executive department of the federal government responsible for a variety of regulatory, research, and marketing programs related to energy production and use.
- Department of Ecology

DOE-2.1—A computer software program that simulates energy consumption of commercial buildings; used for design and auditing purposes.

DOH—Department of Health

DOI—Department of the Interior

DOJ—Department of Justice

DOL—Department of Labor

Dollar-Years—The aggregate years of service for plant dollars during the life of the plant. Expired dollar-years are the aggregate years of service realized as of any given date. Future dollar-years are the aggregate years of service remaining. The area under the survivor curve represents the total dollar-years of service.

Dome (Geodesic)—An architectural design invented by Buckminster Fuller with a regular polygonal structure based on radial symmetry.

Domestic (legal)—The term "domestic" when applied to a corporation or partnership means created or organized in the United States or under the law of the United States or of any State unless, in the case of a partnership, the Secretary provides otherwise by regulations.

Domestic Hot Water—Water heated for residential washing, bathing, etc.

Domestic Inland Consumption—Domestic inland consumption is the sum of all refined petroleum products supplied for domestic use (excludes international marine bunkers). Consumption is calculated by product by adding production, imports, crude oil burned directly, and refinery fuel and losses, and then subtracting exports and charges in primary stocks (net withdrawals is a plus quantity and net additions is a minus quantity).

Domestic Operations—Domestic operations are those operations located in the United States. The U.S. is defined as the 50 states, including their offshore territorial waters, the District of Columbia, U.S. commonwealth territories, and protectorates.

Domestic Uranium Industry—Collectively, those businesses (whether U.S. or foreign-based) that operate under the laws and regulations pertaining to the conduct of commerce within the United States and its territories and possessions and that engage in activities within the United States, its territories, and possessions specifically directed toward uranium exploration, development, mining, and milling;

marketing of uranium materials; enrichment; fabrication; or acquisition and management of uranium materials for use in commercial nuclear power plants.

Domestic Vehicle Producer—An Original Vehicle Manufacturer that assembles vehicles in the United States for domestic use. The term "domestic" pertains to the fifty states, the District of Columbia, commonwealths, territories, and possessions of the United States.

Dominant Occupancy—The occupancy type in mixed occupancy buildings with the greatest percentage of total conditioned floor area.

Donor—In a solar photovoltaic device, an n-type dopant, such as phosphorus, that puts an additional electron into an energy level very near the conduction band; this electron is easily exited into the conduction band where it increases the electrical conductivity over that of an undoped semiconductor.

Dopant—A chemical element (impurity) added in small amounts to an otherwise pure semiconductor material to modify the electrical properties of the material. An n-dopant introduces more electrons. A p-dopant creates electron vacancies (holes).

Doping—The addition of dopants to a semiconductor.

Dormitory—A building consisting of multiple sleeping quarters and having interior common areas such as dining rooms, reading rooms, exercise rooms, toilet rooms, study rooms, hallways, lobbies, corridors, and stairwells, other than high-rise residential, low-rise residential, and hotel/motel occupancies.

DOS—Department of State

Dose—The amount of ionizing radiation energy absorbed per unit mass of irradiated material at a specific location, such as a part of a human body.

DOT—Department of Transportation

Double Bundle Chiller—A Double Bundle Chiller is a condenser, usually located in a refrigeration machine, that contains two separate tube bundles, allowing the option of rejecting heat to a cooling tower or to another building system requiring heat input.

Double Circuit Line—A transmission line having two separate circuits. Because each carries three-phase power, at least six conductors, three per circuit, are required.

Double Dividend—Refers to the notion that environmental taxes can both reduce pollution (the first dividend) and reduce the overall economic costs associated with the tax system by using the revenue generated to displace other more distortionary taxes that slow economic grow at the same time (the second dividend).

Double Glazing—Windows having two sheets of glass with an airspace between.

Double Leverage—Concept used in developing a company's proper capital structure. It occurs when the company participates in a project financed joint venture through a wholly owned subsidiary.

Double Wall Heat Exchanger—A heat exchanger in a solar water heating system that has two distinct walls between the heat transfer fluid and the domestic water, to ensure that there is no mixing of the two.

Double-dip Lease—A series of lease transactions between parties in two countries. The tax rules for leases are different to the extent that both parties, the lessee and the lessor, receive favorable tax treatment.

Double-pane or Glazed Window—A type of window having two layers (panes or glazing) of glass separated by an air space. Each layer of glass and surrounding air space reradiates and traps some of the heat that passes through thereby increasing the windows resistance to heat loss (R-value).

Down Draft (Back Draft)—A flow of air down the chimney or flue because of adverse draft conditions.

Downdraft Gasifier—A gasifier in which the product gases pass through a combustion zone at the bottom of the gasifier.

Downgradient—The direction in which groundwater flows

Downstream—Any point in the economy, and in particular, at the level of energy consumers rather than suppliers. It is commonly interpreted to be industrial boilers, electric utilities and other major energy users, but also applies, in theory, to all consumers of gasoline, coal, electricity etc. Conversely, upstream refers to the point (or close to it) where fossil fuels enter the economy. In the U.S., it means at the input to oil refineries, at coal processing plants and where natural gas enters pipelines.

Downstream Impacts—Environmental impacts caused by consumer use and product disposal.

Downtime—Time when the PV system cannot provide power to the load, expressed either in hours per year or as a percentage.

Downwind—In relation to a wind turbine, the direction away from the source of wind. A downwind turbine has its blades on the downwind side of the tower.

Downwind Wind Turbine—A horizontal axis wind turbine in which the rotor is downwind of the tower.

DPA—Deepwater Ports Act

DPAS—Defense Priority Allocation System

DPC—Domestic Policy Council

DPS—Dividends Per Share

DQA—Data Quality Assessment

DRA—Deputy Regional Administrator

Draft—A column of burning combustion gases that are so hot and strong that the heat is lost up the chimney before it can be transferred to the house. A draft brings air to the fire to help keep it burning.

Draft Diverter—A door-like device located at the mouth of a fireplace chimney flue for controlling the direction and flow of the draft in the fireplace as well as the amount of oxygen that the fire receives.

Draft Environmental Impact Statement—(DEIS) A draft statement of environmental effects. Section 102 of the National Environmental Policy Act requires a DEIS for all major federal actions. The DEIS is released to the public and other agencies for comment and review.

Draft Hood—A device built into or installed above a combustion appliance to assure the escape of combustion byproducts, to prevent backdrafting of the appliance, or to neutralize the effects of the stack action of the chimney or vent on the operation of the appliance.

Draft Trades—Under emission trading, it is a deal which the buyer is in the process of compiling, and is saved within draft trades. The buyer can edit and add other trades to this saved document. Once decided on the trade, it can be submitted to the broker.

Draft Tube—A tube added to the outfall of a hydro turbine to increase energy production by taking advantage of the drop in the tailrace.

Drag—Resistance caused by friction in the direction opposite to that of movement (i.e., motion) of components such as wind turbine blades.

Drag-Along Rights—A majority shareholders' right, obligating shareholders whose shares are bound into the shareholders' agreement to sell their shares into an offer the majority wishes to execute.

Drainage—See *Watershed*.

Drainage Basin—The land drained by a river system.

Drainback (Solar) Systems—A closed-loop solar heating system in which the heat transfer fluid in the collector loop drains into a tank or reservoir whenever the booster pump stops to protect the collector loop from freezing.

Draindown (Solar) Systems—An open-loop solar heating system in which the heat transfer fluid from the collector loop and the piping drain into a drain whenever freezing conditions occur.

Drawdown (Maximum)—The distance that the water surface of the reservoir is lowered from the normal full elevation to the lowest allowable elevation as the result of the withdrawal of water for the purposes of generating electricity.

DRC—Deputy Regional Counsel

DRE—Destruction/Removal Efficiency

Dredge Mining—A method of recovering coal from rivers or streams.

Drift Mine—A mine that opens horizontally into the coal bed or coal outcrop.

Drilling—The act of boring a hole (1) to determine whether minerals are present in commercially recoverable quantities and (2) to accomplish production of the minerals (including drilling to inject fluids).

- Exploratory. Drilling to locate probable mineral deposits or to establish the nature of geological structures; such wells may not be capable of production if minerals are discovered.

- Developmental. Drilling to delineate the boundaries of a known mineral deposit to enhance the productive capacity of the producing mineral property.

- Directional. Drilling that is deliberately made to depart significantly from the vertical.

Drilling and Equipping of Wells—The drilling and equipping of wells through completion of the 'christmas tree'.

Drilling Arrangement—A contractual agreement under which a working interest owner (assignor) assigns a part of a working interest in a property to another party (the assignee) in exchange for which the assignee agrees to develop the property. The term may also be applied to an agreement under which an operator assigns fractional shares in production from a property to participants for cash considerations as a means of acquiring cash for developing the property. Under a "disproportionate cost" drilling arrangement, the participants normally pay a greater total share of costs than the total value of the fractional shares of the property received in the arrangement.

DRIP—Dividend Reinvestment Plan

Drought Tolerance—The capacity of a landscape plant to function well in drought conditions. Used extensively for landscaping in arid regions.

DRP—Default Risk Premium

DRR—Data Review Record

Dry (Coal) Basis—Coal quality data calculated to a theoretical basis in which no moisture is associated with the sample. This basis is determined by measuring the weight loss of a sample when its inherent moisture is driven off under controlled conditions of low temperature air-drying followed by heating to just above the boiling point of water (104 to 110 degrees Centigrade).

Dry Bottom Boiler—No slag tanks at furnace throat area. The throat area is clear. Bottom ash drops through the throat to the bottom ash water hoppers. This design is used where the ash melting temperature is greater than the temperature on the furnace wall, allowing for relatively dry furnace wall conditions.

Dry Bulb Temperature—A measure of the sensible temperature of air. Dry-Bulb Temperature is the temperature of air indicated on an ordinary thermometer. It does not account for the effects of humidity.

Dry Cell Battery—A battery that uses a solid paste for an electrolyte.

Dry Gas—See *Dry Natural Gas* below.

Dry Hole—A drilled well that does not yield gas and/or oil quantities or condition to support commercial production; also applied to gas that has been produced and from which liquid components have been removed.

• An exploratory or development well found to be incapable of producing either oil or gas in sufficient quantities to justify completion as an oil or gas well. Also See *Well*.

Dry Hole Charge—The charge-off to expense of a previously capitalized cost upon the conclusion of an unsuccessful drilling effort.

Dry Hole Contribution—A payment (either in cash or acreage) that is required by agreement only if a test well is unsuccessful and that is made in exchange for well test and evaluation data.

Dry Lease—A net lease. This term traditionally is used in aircraft and marine leasing to describe a lease agreement that provides financing only and, therefore, requires the lessee to separately procure personnel, fuel and provisions necessary to operate the craft.

Dry Natural Gas—Natural gas which remains after: 1) the liquefiable hydrocarbon portion has been removed from the gas stream (i.e., gas after lease, field, and/or plant separation); and 2) any volumes of non-hydrocarbon gases have been removed where they occur in sufficient quantity to render the gas unmarketable. Note: Dry natural gas is also known as consumer-grade natural gas. The parameters for measurement are cubic feet at 60 degrees Fahrenheit and 14.73 pounds per square inch absolute.

Dry Natural Gas Production—The process of producing consumer-grade natural gas. Natural gas withdrawn from reservoirs is reduced by volumes used at the production (lease) site and by processing losses. Volumes used at the production site include (1) the volume returned to reservoirs in cycling, repressuring of oil reservoirs, and conservation operations; and (2) gas vented and flared. Processing losses include (1) non-hydrocarbon gases (e.g., water vapor, carbon dioxide, helium, hydrogen sulfide, and nitrogen) removed from the gas stream; and (2) gas converted to liquid form, such as lease condensate and plant liquids. Volumes of dry gas withdrawn from gas storage reservoirs are not considered part of production. Dry natural gas production equals marketed production less extraction loss.

Dry Production—See *Dry Natural Gas Production* above.

Dry Steam—The conventional type of geothermal energy used for electricity production in California. Dry steam captured at the earth's surface is used to run electric turbines. The principal dry steam resource area is the Geysers in Northern California; one of only two known areas in the world for dry steam—the other being Larderello, Italy.

Dry Steam Geothermal Plants—Conventional turbine generators are used with the dry steam resources. The steam is used directly, eliminating the need for boilers and boiler fuel that characterizes other steam-power-generating technologies. This technology is limited because dry-steam hydrothermal resources are extremely rare. The Geysers, in California, is the nation's only dry steam field.

Dry Ton—2,000 pounds of material dried to a constant weight.

dscf—Dry Standard Cubic Feet

dscfm—Dry Standard Cubic Feet per Minute

dscm—Dry Standard Cubic Meter

DSIRE—Database of State Incentives for Renewables & Efficiency www.dsireusa.org

DSM—Demand side management is a method used to manage energy demand including energy efficiency, load management, alternate fuels, and load build.

DSM Costs, Administrative—Costs incurred by a utility for DSM program planning, design, marketing, implementation, and evaluation. Included are labor-related costs, office supplies and expenses, data processing, and other such costs. Excluded are costs of marketing materials and advertising, purchases of equipment for specific programs, and rebates or other incentives.

DSM Costs, Equipment—The price of all equipment that a utility directly purchases for a DSM program, whether for its own use or for distribution to program participants.

DSM Costs, Marketing—All DSM costs directly associated with the preparation and implementation of the strategies designed to encourage participation in a program.

DSM Costs, Monitoring and Evaluation—DSM expenditures associated with the collection and analysis of data used to assess program operation and effects.

DSO—Days Sales Outstanding

DSS—Data Systems Staff
- Decision Support System
- Domestic Sewage Study

Dth—Decatherm (equals one MMBtu): The approximate heat content of 1,000 cubic feet of gas; 10 therms.

DTSC—Department of Toxic Substances Control. A department within the California Environmental Protection Agency charged with the regulation of hazardous waste from generation to final disposal, and for overseeing the investigation and clean-up of hazardous waste sites. www.dtsc.ca.gov

Dual Duct System—An air conditioning system that has two ducts, one is heated and the other is cooled, so that air of the correct temperature is provided by mixing varying amounts of air from each duct.

Dual Fuel (or Flex Fuel) Vehicle—A vehicle with an engine capable of operating on two different types of fuels.

Dual Fuel Vehicle—A motor vehicle that is capable of operating on an alternative fuel and on gasoline or diesel fuel. These vehicles have at least two separate fuel systems which inject each fuel simultaneously into the engine combustion chamber.
- A motor vehicle that is capable of operating on an alternative fuel and on gasoline or diesel fuel. This term is meant to represent all such vehicles whether they operate on the alternative fuel and gasoline/diesel simultaneously (e.g., flexible-fuel vehicles) or can be switched to operate on gasoline/diesel or an alternative fuel (e.g., bi-fuel vehicles).

Dual-Duct System—A central plant heating, ventilation and air conditioning (HVAC) system that produces conditioned air at two temperatures and humidity levels. The air is then supplied through two independent duct systems to the points of usage where mixing occurs.

Dual-Fired Unit—A generating unit that can produce electricity using two or more input fuels. In some of these units, only the primary fuel can be used continuously; the alternate fuel(s) can be used only as a start-up fuel or in emergencies.

Dual-Fuel or Bi-Fuel Vehicle—Refers to a vehicle capable of operating on two different fuels, in distinct fueling systems, such as compressed natural gas and gasoline.

Dual-Glazed Greenhouse Windows—A type of dual-glazed fenestration product which adds conditioned volume but not conditioned floor area to a building.

Dual-Paned (Double-Glazed)—Two panes of glass or other transparent material, separated by a space.

Duct Fan—An axial flow fan mounted in a section of duct to move conditioned air.

Duct Losses—Heat transfer into or out of a space conditioning system duct through conduction or leakage.

Duct Sealing—A procedure for installing a space conditioning distribution system that minimizes leakage of air from or to the distribution system. Minimum specifications for installation procedures, materials, diagnostic testing and field verification are contained in the Residential and Nonresidential ACM Approval Manuals (in California).

Duct(s)—The round or rectangular tube(s), generally constructed of sheet metal, fiberglass board, or a flexible plastic-and-wire composite, located within a wall, floor, and ceiling that distributes heated or cooled air in buildings.

Due Diligence—The detailed, multifaceted investigation of physical, legal, economic, and other aspects of real estate transactions (especially investment properties), M&A deals, and project finance arrangements.

Duff—The layer of forest litter.

Dump—Excess hydropower that cannot be stored or conserved. Also known as Spill Energy.

Dump Energy—Energy generated in a hydroelectric plant by water that cannot be stored or conserved and which energy is in excess of the needs of the system producing the energy.

Duration—A measure of the speed of amortization, often used to gauge the exposure to interest rate changes. Calculated as the PV of the debt service.

Duration-Energy Storage—A measure of how long a storage device can discharge, or supply electrical energy; may be measured in a range from milliseconds to hours.

Dutch Oven Furnace—One of the earliest types of furnaces, having a large, rectangular box lined with firebrick (refractory) on the sides and top. Commonly used for burning wood. Heat is stored in the refractory and radiated to a conical fuel pile in the center of the furnace.

Duty Cycle—The duration and periodicity of the operation of a device. The ratio of active to total time, used to describe the operating regime of loads in PV systems.

Duty Rating—The amount of time an inverter can operate at full rated power. Some inverters can operate at their rated power for only a short time without overheating.

DWR—Department of Water Resources (California) www.water.ca.gov

Dynamic Baseline—Dynamic baseline is a forecast baseline which adjusts to the changes in the business environment over time.

Dynamic Head—The pressure equivalent of the velocity of a fluid.

Dynamo—A machine for converting mechanical energy into electrical energy by magneto-electric induction; may be used as a motor.

Dynamometer—An apparatus for measuring force or power, especially the power developed by a motor.

Dyne—The absolute centimeter-gram-second unit of force; that force that will impart to a free mass of one gram an acceleration of one centimeter per second per second.

E

E&P—Exploration and Production

E85—A fuel containing a mixture of 85 percent ethanol and 15 percent gasoline.

E95—A fuel containing a mixture of 95 percent ethanol and 5 percent gasoline

EA—Effective Aperture
- Enforcement Agreement
- Environmental Action
- Environmental Assessment
- Environmental Audit

EAC—Electricity Authority of Cambodia (Cambodia)

EAD—Economic Analysis Division
- Energy and Air Division

EAP—Environmental Action Plan

EAR—Effective Annual Rate

Early Action—The action of reducing emissions, investing in Clean Development Mechanism projects, Joint Implementation, or trading emissions before the start for the Kyoto Commitment Period.

Early Buy Out (EBO)—Early buyout. A provision in a lease allowing but not requiring the lessee to purchase the asset for a certain amount at an agreed-upon date prior to the end of the term.

Early Crediting—A provision that allows crediting of emission reductions achieved prior to the start of a legally imposed emission control period. These credits can then be used to assist in achieving compliance once a legally imposed system begins.

Early Seral Species—Shrubs (such as ceanothus) and hardwoods (usually in tree form, such as red alder, bitter cherry and big leaf maple) that start growing in natural succession soon after a disturbance (fire or logging).

Earnest Money—A good faith deposit given to bind a contract between Buyer and Seller.

Earnings/Price Ratio (E/P)—A method to determine the cost of common equity component of return using the ratio of earnings per share to the stock price.

Earth—Refers to physically connecting a part of an electrical system to the ground, done as a safety measure, by means of a conductor embedded in suitable soil.

Earth Berm—A mound of dirt next to exterior walls to provide wind protection and insulation. The daily and seasonal temperature of the earth is less variable than the atmosphere. The earth prevents convection and therefore eliminates the wind chill factor.

Earth Cooling Tube—A long, underground metal or plastic pipe through which air is drawn. As air travels through the pipe it gives up some of its heat to the soil, and enters the house as cooler air.

Earth Leakage Circuit Breaker (ELCB)—A device used to prevent electrical shock hazards in mains voltage power systems, including independent power systems. Also known as residual current devices (RCDs).

Earth Sheltered Houses—Houses that have earth berms around exterior walls.

Earth-Coupled Ground Source (Geothermal) Heat Pump—A type of heat pump that uses sealed horizontal or vertical pipes, buried in the ground, as heat exchangers through which a fluid is circulated to transfer heat.

Earth-Ship—A registered trademark name for houses built with tires, aluminum cans, and earth.

Easement—An incorporated right, liberty, privilege, or use of another entity's property, distinct from ownership, without profit or compensation; a right-of-way.

East-Facing—Means that a surface is oriented such that its normal is within 45 degrees of true east, including 45°0'0" south of east (SE), but excluding 45°0'0" north of east (NE).

EB—Emissions Balancing

Ebike—Electric bicycle

EBIT—Earnings Before Interest and Taxes.

EBITDA—Earnings Before Interest, Taxes, Depreciation, and Amortization. A measure of cash flow calculated as: Revenue—Expenses (excluding tax, interest, depreciation, and amortization).

EBITDA looks at the cash flow of a company. By not including interest, taxes, depreciation, and amortization, we can clearly see the amount of money a company brings in. This is especially useful when one company is considering a takeover of another because the EBITDA would cover any loan payments needed to finance the takeover.

EC—Electrical Conductivity: A measurement of how well a material accommodates the transport of electric charge.

ECA—Enforceable Consent Order

ECABF—Energy Cost Adjustment Billing Factor: The positive or negative surcharge on customers' bills used to recover energy costs in rates.

ECAO—Environmental Criteria and Assessment Office

Eccentric—A device for converting continuous circular motion into reciprocating rectilinear motion.

ECDB—Emissions Certification Data Base

ECM—Energy conservation measure

Eco-Efficiency—The creation of more goods and services while using fewer resources and creating less waste and pollution.

Eco-Friendly—A product, practice or process that is "green" or good for the environment, creating no unnecessary or hazardous waste and minimizing use of non-renewable, natural resources.

Ecojustice—The concept that all components of an ecosystem (such as plant and animal life as well as natural resources) have a right to be free from human exploitation and free from destruction, discrimination, bias, or extinction.

Ecological Footprint—The total amount of land, food, water, and other resources used by, or the total ecological impact of, a person or organization's subsistence. Usually measured in acres or hectares of productive land.

Ecology—The study of the interrelationships between organisms and their environment. From the Greek word *oikos*, meaning "house."

Economic and Technology Advancement Advisory Committee (ETAAC)—A committee which advises California Air Resources Board on activities that will facilitate investment in and implementation of technological research and development opportunities including, but not limited to, identifying new technologies, research, demonstration projects, funding opportunities, developing state, national, and international partnerships and technology transfer opportunities, and identifying and assessing research and advanced technology investment and incentive opportunities that will assist in the reduction of greenhouse gas emissions.

Economic Efficiency—A term that refers to the optimal production and consumption of goods and services. This generally occurs when prices of products and services reflect their marginal costs. Economic efficiency gains can be achieved through cost reduction, but it is better to think of the concept as actions that promote an increase in overall net value (which includes, but is not limited to, cost reductions).

Economic Energy Intensity—The energy intensity per unit of economic output.

Economic Feasibility—The ability of a project or enterprise to meet defined investment objectives. An investment's ability to produce sufficient revenue to pay all expenses and charges and to provide a reasonable return on and recapture of the money invested.

Economic Impact Statement—A report detailing a major real estate or energy project's potential impact on the local economy, which may include estimates of the project's market value and potential gross sales as well as indications of its business, occupational, labor and tax impact on the community.

Economic Life—The estimated period during which the property is expected to be economically usable by one or more users (or the number of production or similar units expected to be obtained from an asset by one or more users), with normal repairs and maintenance, for the purpose for which it was intended at the inception.

Economic Potential (Utility)—In DSM, an estimate of energy savings based on the assumption that all energy-efficient options will be adopted and all existing equipment will be replaced with the most efficient measure possible whenever it is cost-effective to do so, without regard to market acceptance.

Economic Sector—A subdivision of economic activities based on major purpose (for example, "commercial sector" or "private sector").

Economic Value Added (EVA)—EVA describes a value margin (key figure) that is multiplied with the capital a company invests during a defined period.

Economic/Market Clauses—Contract provisions which allow price redetermination at specified times or conditions at prices prevailing in the area, or at market prices.

Economies of Scale—Economies of scale exist where the industry exhibits decreasing average long-run costs with size. Economic principle that, as the volume of production increases, the cost of producing each unit decreases.

Economizer—A heat exchanger for recovering heat from flue gases for heating water or air.

  • An arrangement of tubes through which the feed water passes before entering boiler drum and flue gases leave burners. Economizers are invariably counter flow; meaning the water flows opposite to the gases, and heat of gases is transferred to the water.

Economizer-Air—A ducting arrangement and automatic control system that allows a heating, ventilation and air conditioning (HVAC) system to supply up to 100 percent outside air to satisfy cooling demands, even if additional mechanical cooling is required. An Air Economizer is a duct-and-damper arrangement and automatic control system that together allow a cooling system to supply outside air. This lowers energy usage by reducing or eliminating the need for mechanical cooling during mild or cold weather. Quite often, interior heat loads (from people, equipment, lighting) in a building will create a cooling requirement even though the ambient outside air temperature is below room temperature. Air brought in for ventilation will provide some cooling, but an Air Economizer will bring in more outside air than the ventilation minimum to take advantage of the available free cooling.

Economizer-Water—A system that uses either direct evaporative cooling, or a secondary evaporatively cooled water loop and cooling coil to satisfy cooling loads, even if additional mechanical cooling is required. A Water Economizer (sometimes called a water-side economizer) is a system by which the supply air of a cooling system is cooled indirectly with water that is itself cooled by heat or mass transfer to the environment without the use of mechanical cooling. Typically this involves using a cooling plant cooling tower during the winter to provide chilled water via a heat exchanger. Water Economizer systems are appropriate only in buildings that require chilled water through the winter (i.e., buildings that have significant concentrated heat loads, such as computer rooms). Many buildings can satisfy any winter cooling requirement more efficiently using standard Air Economizer systems.

Economy Energy—Electricity purchased by one utility from another to take the place of electricity that would have cost more to produce on the utility's own system.

Economy of Scale—The principle that larger production facilities have lower unit costs than smaller facilities.

ECOS—Environmental Council of the States www.ecos.org

Ecosystem—The system of interactions between living organisms and their environment.

ECP—External Compliance Programs

ECR—Enforcement Case Review

ECRA—Economic Cleanup Responsibilities Act

ECU—Environmental Crimes Unit

EDA—Economic Development Administration

  • Emergency Declaration Area

EDD—Acronym for "Enforcement Decision Document" which means a document that provides an explanation to the public of EPA's selection of the cleanup alternative at enforcement sites on the National Priorities List. Similar to a Record of Decision.

  • Economic Development Department

Edge of Glass—The portion of fenestration glazing that is within two and one half inches of the spacer.

Edge-Defined Film-Fed Growth (EFG)—A method for making sheets of polycrystalline silicon (for solar photovoltaic devices) in which molten silicon is drawn upward by capillary action through a mold.

EDI—Electronic Data Interchange

EDR—Energy Development Report

EDS—Energy Data System

Edwards Balance—An instrument for determining the specific gravity of gases.

EE—Energy efficiency

EEA—Energy and Environmental Analysis

EEC—Estimated Environmental Concentration

EECBG—Energy Efficiency and Conservation Block Grants

EEI—The Edison Electric Institute is an association of electric companies born in 1933 to exchange information concerning industry developments and to function as an advocate in behalf of the utilities on subjects of national interest.

EER—Excess Emission Report

  • Energy Efficiency Ratio

EER (Energy Efficiency Ratio)—The ratio of cooling capacity of an air conditioning unit in Btu's per hour to the total electrical input in watts under specified test conditions. [See California Code of Regulations, Title 20, Section 1602(c)(6)] The EER is the ratio of net equipment cooling capacity in Btu/hour to total rate of electric input in watts (W) under designated operating conditions. If the output capacity in Btu/hour is converted to watts (to create consistent units), the result is equal to the cooling COP (EER 3.41 = COP 1.0.)

EERE—US Department of Energy: Office of Energy Efficiency and Renewable Energy

EERU—Environmental Emergency Response Unit

EESA—Emergency Economic Stabilization Act of 2008 (H.R. 1424, "Bailout Bill")

EESI—Environment and Energy Study Institute www.eesi.org

EETD—Lawrence Berkeley National Labs—Environmental Energy Technology Division

EF—Emission Factor
- Enrichment Factor

Effective Aperture (EA)—Is the extent that vertical glazing or skylights are effective for providing daylighting. The effective aperture for vertical glazing is specified in Exception 1 to Section 131(c) of Title 24 in California, The effective aperture for skylights is specified in Section 146 (a) 4 F.

Effective Capacity—The maximum load that a device is capable of carrying.

Effective Date—The date at which the analysis, opinions, and advice in an appraisal, review, or consulting service apply.

Effective Full-power Days—The number of effective full-power days produced by a unit is a measure of the unit's energy generation. It is determined using the following ratio: Heat generation (planned or actual) in megawatt days thermal (MWdt)(divided by) Licensed thermal power in megawatts thermal (MWt)

Effective Rate—The interest rate in a financial contract stated on an annual basis. The rate includes the compounding effect of interest during the year.
- Yields are annual. However, yields are usually compounded monthly, or whenever there is a cash flow, and yield interest is applied to each period, rather than to the whole year at once. The effective yield (as opposed to nominal) is an annual expression of the yield interest rate for each period that attempts to take into consideration the effect of compounding.

Efficacy—The amount of energy service or useful energy delivered per unit of energy input. Often used in reference to lighting systems, where the visible light output of a luminary is relative to power input; expressed in lumens per Watt; the higher the efficacy value, the higher the energy efficiency.

Efficacy Lighting—Ratio of light from a lamp to the electricity consumed, including ballast loss in terms of lumens per watt.
- The ratio of light from a lamp to the electrical power consumed, including ballast losses, expressed as lumens per watt. [See California Code of Regulations, Title 24, Section 2-5302]

Efficacy, Lamp—The quotient of rated initial lamp lumens divided by the rated lamp power (watts), without including auxiliaries such as ballasts, measured at 25°C according to IESNA and ANSI Standards.

Efficacy, Lighting System—The quotient of rated initial lamp lumens measured at 25°C according to IESNA and ANSI Standards, times the ballast factor, divided by the input power (watts) to the ballast or other auxiliary device (e.g. transformer), expressed in lumens per watt.

Efficiency—Under the First Law of Thermodynamics, efficiency is the ratio of work or energy output to work or energy input, and cannot exceed 100 percent. Efficiency under the Second Law of Thermodynamics is determined by the ratio of the theoretical minimum energy that is required to accomplish a task relative to the energy actually consumed to accomplish the task. Generally, the measured efficiency of a device, as defined by the First Law, will be higher than that defined by the Second Law.
- The effectiveness of a device to convert energy from one form to another, or to transfer energy from one body to another. An electric pump that is 60 percent efficient converts 60 percent of the input energy into work—pumping water. The remaining 40 percent becomes waste heat.

Efficiency (Appliance) Ratings—A measure of the efficiency of an appliance's energy efficiency.

Efficiency (Solar)—The ratio of power output of a photovoltaic cell to the incident power from the sun or simulated sun sources under specified standard insolation conditions. A solar cell that converts 1/10 of the sun's energy that strikes its surface to electricity has an efficiency of 10 percent.

Effluent—The products of combustion and air being discharged from gas utilization equipment.
- Wastewater, treated or untreated, that flows out of a treatment plant, sewer or industrial outfall. Generally refers to wastes discharge into surface waters.

Effluent Fee—An effluent fee is a fixed tax rate per unit (litre or kilogram) of emissions. They are also referred to as emission charges or emission taxes.

EFIN—Environmental Financing Information Network www.sustainable.org

EFL—Electricity Feed Law

EFO—Emergency Flow Order: A mandate that end-use customers' usage must be less than or equal to supply. The order is used when actual or forecast supply and/or capacity shortages threaten delivery of power to end-use customers.

EFS—Energy Facilities Siting

EFSC—See *Energy Facility Siting Council*.

EFVR—Estimated Fair Value Range

EG—Emission Guidelines

EHS—Extremely Hazardous Substance

EI—Emissions Inventory

EIA—*The Energy Information Administration*. An independent agency within the U.S. Department of Energy that develops surveys, collects energy data, and analyzes and models energy issues. The Agency must meet the requests of Congress, other elements within the Department of Energy, Federal Energy Regulatory Commission, the Executive Branch, its own independent needs, and assist the general public, or other interest groups, without taking a policy position

Economic Impact Assessment

EIN—Employer Identification Number

EIP—Economic Incentive Program

EIR—Environmental Impact Report: A report prepared on the potential effects of a project proposed by a utility on the environment.

• Exposure Information Report

EIS—See *Environmental Impact Statement*.

EISA 2007—Energy Independence and Security Act of 2007

EITF—Emerging Issues Task Force (EPA)

EITs—Economies In Transition. Those nations in Annex I of the Kyoto Protocol considered developed but currently in transition to a market economy. Generally the nations and former republics of the old Soviet bloc.

EJ—Environmental Justice

ekWh—A standard unit of energy consumption used to compare energy sources.

EL—Exposure Level

Elasticity of Demand—The ratio of the percentage change in the quantity of a good or service demanded to the percentage change in the price.

Elastomer—A material which at room temperature can be stretched repeatedly to at least twice its original length and upon immediate release of the stress, will return to its approximate original length and shape.

ELCON—Electricity Consumers Resources Council. ELCON is an association of 28 large industrial consumers of electricity. ELCON members account for over five percent of all electricity consumed in the United States. ELCON was formed in 1976 "to enable member companies to "work cooperatively for

the development of coordinated, rational and consistent policies affecting electric energy supply and pricing at the federal, state, and local levels."

ELCR—Excess Lifetime Cancer risk

Electric Baseboard—An individual space heater with electric resistance coils mounted behind shallow panels along baseboards. Electric baseboards rely on passive convection to distribute heated air to the space.

Electric Circuit—Path followed by electrons from a power source (generator or battery) through an external line (including devices that use the electricity) and returning through another line to the source.

Electric Current—The flow of electrons measured in Amps.

Electric Energy—The amount of work accomplished by electrical power, usually measured in kilowatt-hours (kWh). One kWh is 1,000 Watts and is equal to 3,413 Btu.

Electric Expenses—The cost of labor, material, and expenses incurred in operating a facility's prime movers, generators, auxiliary apparatus, switching gear, and other electric equipment for each of the points where electricity enters the transmission or distribution grid.

Electric Furnace—An air heater in which air is blown over electric resistance heating coils.

Electric Generation Industry—Stationary and mobile generating units that are connected to the electric power grid and can generate electricity. The electric generation industry includes the "electric power sector" (utility generators and independent power producers) and industrial and commercial power generators, including combined-heat-and-power producers, but excludes units at single-family dwellings.

Electric Generator—A system that converts heat, chemical, or mechanical energy into electricity. A facility that produces only electricity, commonly expressed in kilowatt-hours (kWh) or megawatt hours (MWh). Electric generators include electric utilities and independent power producers.

Electric Grid—A network for electricity distribution across a large area.

Electric Heating—An electrically powered heating source, such as electric resistance, heat pumps with no auxiliary heat or with electric auxiliary heat, solar with electric backup, etc.

Electric Heating Pump and Air-Conditioning Efficiency—Energy Efficiency Ratio (EER)—a ratio calculated by dividing the cooling capacity in Btu per hour by the power input in watts at any given set of rating conditions. Heating Seasonal Performance Factor (HSPF)—the total heating output of a heat pump during its normal annual usage period for heating divided by the total electric power input in watt-hours during the same period. Seasonal Energy Efficiency Ratio (SEER)—the total cooling capacity of a central unitary air conditioner or unitary heat pump in Btu's during its normal annual usage period for cooling divided by the total electric energy input in watt-hours during the same period.

Electric Hybrid Vehicle—An electric vehicle that either (1) operates solely on electricity, but contains an internal combustion motor that generates additional electricity (series hybrid); or (2) contains an electric system and an internal combustion system and is capable of operating on either system (parallel hybrid).

Electric Industry Re-regulation—The design and implementation of regulatory practices to be applied to the remaining traditional utilities after the electric power industry has been restructured. Re-regulation applies to those entities that continue to exhibit characteristics of a natural monopoly. Re-regulation could employ the same or different regulatory practices as those used before restructuring.

Electric Industry Restructuring—The process of replacing a monopolistic system of electric utility suppliers with competing sellers, allowing individual retail customers to choose their supplier but still receive delivery over the power lines of the local utility. It includes the reconfiguration of vertically-integrated electric utilities.

Electric Motor Vehicle—A motor vehicle powered by an electric motor that draws current from rechargeable storage batteries, fuel cells, photovoltaic arrays, or other sources of electric current.

Electric Operating Expenses—Summation of electric operation-related expenses, such as operation expenses, maintenance expenses, depreciation expenses, amortization, taxes other than income taxes, Federal income taxes, other income taxes, provision for deferred income taxes, provision for deferred income-credit, and investment tax credit adjustment.

Electric Plant (Physical)—A facility containing prime movers, electric generators, and auxiliary equipment for converting mechanical, chemical, and/or fission energy into electric energy.

Electric Plant Acquisition Adjustment—The difference between (a) the cost to the respondent utility of an electric plant acquired as an operating unit or system by purchase and (b) the depreciated original cost, estimated if not known, of such property.

Electric Power—The rate at which electric energy is transferred. Electric power is measured by capacity and is commonly expressed in megawatts (MW).

Electric Power Grid—A system of synchronized power providers and consumers connected by transmission and distribution lines and operated by one or more control centers. In the continental United States, the electric power grid consists of three systems: the Eastern Interconnect, the Western Interconnect, and the Texas Interconnect. In Alaska and Hawaii, several systems encompass areas smaller than the State (e.g., the interconnect serving Anchorage, Fairbanks, and the Kenai Peninsula; individual islands).

Electric Power Plant—A station containing prime movers, electric generators, and auxiliary equipment for converting mechanical, chemical, and/or fission energy into electric energy.

Electric Power Sector—An energy-consuming sector that consists of electricity only and combined heat and power (CHP) plants whose primary business is to sell electricity, or electricity and heat, to the public—i.e., North American Industry Classification System 22 plants.

Electric Power System—An individual electric power entity—a company; an electric cooperative; a public electric supply corporation as the Tennessee Valley Authority; a similar Federal department or agency such as the Bonneville Power Administration; the Bureau of Reclamation or the Corps of Engineers; a municipally owned electric department offering service to the public; or an electric public utility district (a "PUD"); also a jointly owned electric supply project such as the Keystone.

Electric Power Transmission—The transmission of electricity through power lines.

Electric Pump for Well Water—This pump forces the water from a well below ground level up into the water pipes that circulate through the house. When this pump is not working, there is a limited supply of running water in the house.

Electric Radiant Heating—A heating system in which electric resistance is used to produce heat that radiates to nearby surfaces. There is no fan component to a radiant heating system.

Electric Rate—The price set for a specified amount and type of electricity by class of service in an electric rate schedule or sales contract.

Electric Rate Schedule—A statement of the electric rate and the terms and conditions governing its application, including attendant contract terms and conditions that have been accepted by a regulatory body with appropriate oversight authority.

Electric Reliability Council of Texas, or ERCOT—The Electric Reliability Council of Texas, Inc. is the corporation that administers the Texas's power grid. ERCOT serves approximately 85 percent of Texas's electric load and oversees the operation of approximately 70,000 megawatts of generation and over 37,000 miles of transmission lines.

Electric Resistance Heater—A device that produces heat through electric resistance. For example, an electric current is run through a wire coil with a relatively high electric resistance, thereby converting the electric energy into heat that can be transferred to the space by fans.

Electric Resistance Heating—A heating system that converts electric energy directly into heat energy by passing a current through an electric resistance. Electric resistance heat is inherently less efficient than gas as a heating energy source because it must account for losses associated with generation frown depletable fossil fuels and transmission to the building site.

Electric Service Provider (ESP)—Also known as competitive power supplier or power marketer, an ESP sells electricity in the retail market. Some suppliers own generation units, while others buy from outside generators and then resell it. In any case, your distribution company (in most cases your local electric utility) delivers the electricity sold by an electric service provider to your home.

Electric System—The physically connected generation, transmission, and distribution facilities and components operated as a unit.

Electric System Loss(es)—The total amount of electric energy loss in an electric system between the generation source and points of delivery.

Electric System Reliability—The degree to which the performance of the elements of the electrical system results in power being delivered to consumers within accepted standards and in the amount desired. Reliability encompasses two concepts, adequacy and security. Adequacy implies that there are sufficient generation and transmission resources installed and available to meet projected electrical demand plus reserves for contingencies. Security implies that the system will remain intact operationally (i.e., will have sufficient available operating capacity) even after outages or other equipment failure. The degree of reliability may be measured by the frequency, duration, and magnitude of adverse effects on consumer service.

Electric Utility—A corporation, person, agency, authority, or other legal entity or instrumentality aligned with distribution facilities for delivery of electric energy for use primarily by the public. Included are investor-owned electric utilities, municipal and State utilities, Federal electric utilities, and rural electric cooperatives. A few entities that are tariff based and corporately aligned with companies that own distribution facilities are also included. Note: Due to the issuance of FERC Order 888 that required traditional electric utilities to functionally un-bundle their generation, transmission, and distribution operations, "electric utility" currently has inconsistent interpretations from State to State.

Electric Utility Divestiture—The separation of one electric utility function from others through the selling of the management and ownership of the assets related to that function. It is most commonly associated with selling generation assets so they are no longer owned or controlled by the shareholders that own the company's transmission and distribution assets.

Electric Utility Restructuring—The introduction of competition into at least the generation phase of electricity production, with a corresponding decrease in regulatory control.

Electric Utility Sector—The electric utility sector consists of privately and publicly owned establishments that generate, transmit, distribute, or sell electricity primarily for use by the public and that meet the definition of an electric utility. Non-utility power producers are not included in the electric sector.

Electric Vehicles—A battery-powered electrically driven vehicle.

Electric Zone—A portion of the grid controlled by the independent system operator.

Electrical Charge—A condition that results from an imbalance between the number of protons and the number of electrons in a substance.

Electrical Grid—An integrated system of electricity distribution, usually covering a large area.

Electrical Horsepower—See *Horsepower*.

Electrical System—All the conductors and electricity using devices that are connected to a source of electromotive force (or generator).

Electrical System Energy Losses—The amount of energy lost during generation, transmission, and distribution of electricity, including plant and unaccounted for use.

Electricity—A property of the basic particles of matter. A form of energy having magnetic, radiant and chemical effects. Electric current is created by a flow of charged particles (electrons).

Electricity Broker—An entity that arranges the sale and purchase of electric energy, the transmission of electricity, and/or other related services between buyers and sellers but does not take title to any of the power sold.

Electricity Congestion—A condition that occurs when insufficient transmission capacity is available to implement all of the desired transactions simultaneously.

Electricity Demand—The rate at which energy is delivered to loads and scheduling points by generation, transmission, and distribution facilities.

Electricity Demand Bid—A bid into the power exchange indicating a quantity of energy or an ancillary service that an eligible customer is willing to purchase and, if relevant, the maximum price that the customer is willing to pay.

Electricity Generation—The process of producing electric energy or the amount of electric energy produced by transforming other forms of energy, commonly expressed in kilowatt-hours (kWh) or megawatt hours (MWh).

Electricity Grid—A common term referring to an electricity transmission and distribution system.

Electricity Industry Restructuring—The process of changing the structure of the electric power industry from one of guaranteed monopoly over service territories, as established by the Public Utility Holding Company Act of 1935, to one of open competition between power suppliers for customers in any area.

Electricity Only Plant—A plant designed to produce electricity only. See also *Combined heat and power (CHP)* plant.

Electricity Paid by Household—The household paid the electric utility company directly for all household uses of electricity (such as water heating, space heating, air-conditioning, cooking, lighting, and operating appliances.) Bills paid by a third party are not counted as paid by the household.

Electricity Sales—The amount of kilowatt-hours sold in a given period of time; usually grouped by classes of service, such as residential, commercial, industrial, and other. "Other" sales include sales for public street and highway lighting and other sales to public authorities, sales to railroads and railways, and interdepartmental sales.

Electrochemical—Refers to the process or device in which chemical reactions take place at electrodes, resulting in the separate transfer of electrons and ions to or from reactants, which, in a fuel cell, are hydrogen and oxygen gases. This differs from combustion where the fuel and oxygen combine directly.

Electrochemical Cell—A device containing two conducting electrodes, one positive and the other negative, made of dissimilar materials (usually metals) that are immersed in a chemical solution (electrolyte) that transmits positive ions from the negative to the positive electrode and thus forms an electrical charge. One or more cells constitute a battery.

Electrochemical Process—The direct process end use in which electricity is used to cause a chemical transformation. Major uses of electrochemical process occur in the aluminum industry in which alumina is reduced to molten aluminum metal and oxygen, and in the alkalies and chlorine industry, in which brine is separated into caustic soda, chlorine, and hydrogen.

Electrode—An electrically conductive material, forming part of an electrical device, often used to lead current into or out of a liquid or gas. In a battery, the electrodes are also known as plates.

Electrodeposition—Electrolytic process in which a metal is deposited at the cathode from a solution of its ions.

Electrolysis—A chemical change in a substance that results from the passage of an electric current through an electrolyte. The production of commercial hydrogen by separating the elements of water, hydrogen, and oxygen, by charging the water with an electrical current.

Electrolyte—A liquid conductor of electricity
  • The medium in a fuel cell which provides the ion transport mechanism between the anode and cathode necessary to sustain the electrochemical process and also insulates against electron transfer. In a PEM fuel cell, the electrolyte allows the transport of positively charged hydrogen ions (protons) from the anode, where they are produced, to the cathode where they react with oxygen molecules and electrons to produce water.

Electrolyte (Battery)—The medium that provides ionic transport between the electrodes of a battery. All common batteries contain an electrolyte, such as the sulfuric acid used in lead-acid batteries.

Electromagnetic Energy—Energy generated from an electromagnetic field produced by an electric current flowing through a superconducting wire kept at a specific low temperature.

Electromagnetic Fields (EMF)—Ordinary every day use of electricity produces magnetic and electric fields. These 60 Hertz fields (fields that go back and forth 60 times a second) are associated with electrical appliances, power lines and wiring in buildings.

Electromagnetic Radiation—Magnetic radiation produced by a changing electrical current, such as alternating current (AC).

Electromotive Force—The amount of energy derived from an electrical source per unit quantity of electricity passing through the source.

Electron—A negatively charged particle. The movement of electrons in an electrical conductor constitutes an electric current.

Electron—An elementary particle of an atom with a negative electrical charge and a mass of $1/1837$ of a proton; electrons surround the positively charged nucleus of an atom and determine the chemical properties of an atom.

Electron Volt—The amount of kinetic energy gained by an electron when accelerated through an electric potential difference of 1 Volt; equivalent to 1.603×10^{-19} volt, a unit of energy or work; abbreviated as eV.

Electronic Ballast—A device that uses electronic components to regulate the voltage of fluorescent lamps.

Electronic Data Interchange (EDI)—The computer-to-computer exchange of business documents and information through the use of standard document formats.

Electronic Gas Measurement (EGM)—"Real time" monitoring of natural gas quantities, and characteristics, as it passes through a specific location.

Electronic Metering—A system that measures energy usage and sends daily electronic data by telephone or fixed network systems to the energy company.

Electronically Commutated Motor—A brushless DC motor with a permanent magnet rotor that is surrounded by stationary motor windings, and an electronic controller that varies rotor speed and direction by sequentially supplying DC current to the windings.

Electro-Osmotic Drag—In a PEM fuel cell, water molecules are attracted to the proton, and, move with it from anode to cathode.

Electrostatic Precipitator—A device used to remove particulate matter from the waste gasses of a combustion power plant or other source of emission gases.

Element—A substance consisting entirely of atoms of the same atomic number.

Elevation—(1) The height above sea level (altitude); (2) A geometrical projection, such as a building, on a plane perpendicular to the horizon.

Elevator Presentation—An extremely concise presentation of an entrepreneur's idea, business model, company solution, marketing strategy, and competition delivered to potential investors. Should not last more than a few minutes, or the duration of an elevator ride.

ELFIN—Electric Utility Financial and Production Simulation Model

ELI—Environmental Law Institute www.eli.org

Eligibility Criteria—The Kyoto Protocol and jurisdictional criteria that must be met by an emissions reduction project to produce reductions which can be banked, traded or offset against emissions.

Eligible Renewable Generator—Renewable generators who have legal documentation verifying the name, address, capacity, fuel type and operational data used and whose generation types fall within the relevant Green-e definition of renewable resources.

Eligible Renewable Resource Facility—A facility generating electricity from renewable resources that meet the Green-e Standard.

Eligible Renewable Resource Product—An electricity product that meets the criteria set forth in the Green-e Standard, thus being eligible for Green-e certification.

Elkor Meter—Stand-alone power meter. Specific subtypes can be used with millivolt, milliamp, and 5 amp current transducers, though the meters are not universal for any CT type. While Elkor meters are of the highest accuracy, not all Elkor meter and CT configurations meet state public utility commission PBI (Performance Based Incentive) requirements.

Ellipsoidal Reflector Lamp—A lamp where the light beam is focused 2 inches ahead of the lamp reducing the amount of light trapped in the fixture.

ELR—Environmental Law Reporter

Elution—Activities of removing "elutes" a material (uranium) adsorbed on ion exchange resin from the "eluant" solution.

EMA—Emergency Management Agency

Embedded Costs Exceeding Market Prices (ECEMP)—Embedded costs of utility investments exceeding market prices are: 1) costs incurred pursuant to a regulatory or contractual obligation; 2) costs that are reflected in cost-based rates; and 3) cost-based rates that exceed the price of alternatives in the marketplace. ECEMPS may become "stranded costs" where they exceed the amount that can be recovered through the asset's sale. Regulatory questions involve whether such costs should be recovered by utility shareholders and if so, how they should be recovered. "Transition costs" are stranded costs which are charged to utility customers through some type of fee or surcharge after the assets are sold or separated from the vertically-integrated utility. "Stranded assets" are assets that cannot be sold for some reason. The British nuclear plants are an example of stranded assets that no one would buy. (Also referred to as Transition Costs.)

Embedded Derivative—A derivative is any financial instrument whose value depends on an underlying asset, price or index. An embedded derivative is the same as a traditional derivative; its placement, however, is different. Traditional derivatives stand alone and are traded independently. Embedded derivatives are incorporated into a contract, called the host contract. Together, the host contract and the embedded derivative form an entity known as a hybrid instrument.A component of a hybrid security that is embedded in a non-derivative instrument. An embedded derivative can modify the cash flows of the host contract because the derivative can be related to an exchange rate, commodity price or some other variable which frequently changes. For example, a French company might enter into a sales contract with a Mexican company, creating a host contract. If the contract is denominated in a foreign currency, such as the U.S. dollar, an embedded foreign currency derivative is created. According to the International Financial Reporting Standards (IFRS), the embedded derivative has to be separated from the host contract and accounted for separately unless the economic and risk characteristics of both the embedded derivative and host contract are closely related. Combining derivatives with traditional contracts, or embedding derivatives, changes the way that risk is distributed among the parties to the contracts.

Embodied Energy—The energy consumed by all of the processes associated with the production of a material. This includes the energy required in mining, transport, manufacturing, administration, use, disposal, etc.

• The total amount of energy used to create a product or project, including energy expended in raw materials extraction, processing, manufacturing and transportation. Embodied energy is often used as a rough measure of the environmental impact of a product or project.

EMCS—Energy Management Control System

Emergency—The failure of an electric power system to generate or deliver electric power as normally intended, resulting in the cutoff or curtailment of service.

Emergency Backup Generation—The use of electric generators only during interruptions of normal power supply.

Emergency Core Cooling System (ECCS)—Equipment designed to cool the core of a nuclear reactor in the event of a complete loss of the coolant.

Emergency Energy—Electric energy provided for a limited duration, intended only for use during emergency conditions.

EMF—Electric and magnetic fields

Eminent Domain—The power to take private property for a public purpose upon payment of just compensation.

• The right of government to take private property for public use upon the payment of just compensation. The fifth amendment of the U.S. Constitution, also known as the takings clause, guarantees payment of just compensation upon appropriation of private property.

Emission Allowance—Emission allowances are the total emissions allowed to be released by an emission source (often a net emitting firm) within a given period of time. Emission Allowance are created by a regulating entity and distributed to emitters by grant, auction, or a combination of the two.

Emission Cap—A regulatory device that sets a ceiling on emissions that can be released into the atmosphere within a designated timeframe. Caps are effectively the same as 'Allowances' however caps more often refer to national emission limitations and allowances to individual emitters. Pollutant released into either air or waterways from industrial processes, households or transportation vehicles.

Emission Factor—A measure of the average amount of a specified pollutant or material emitted for a specific type of fuel or process. Also means a unique value for determining an amount of a GHG emitted for a given quantity of activity data (e.g., million metric tons of carbon dioxide emitted per barrel of fossil fuel burned).

Emission Forecast—An emission forecast refers to the forecasts of emissions produced by an emitter for its internal management purposes. Forecasts are hypothetical and incorporate knowledge about the firm's future operational, regulatory and economic impacts to determine emission projections. This process is to baseline forecasting except that baselines are used to quantify emission reductions and are subject to far more scrutiny.

Emission Inventory—Emission Inventory is an archive of historical emissions. An emission inventory can begin once systems boundaries are defined.

Emission Limitation—A requirement established by a State, local government, or the EPA Administrator which limits the quantity, rate, or concentration of emissions of air pollutants on a continuous basis, including any requirements which limit the level of opacity, prescribe equipment, set fuel specifications, or prescribe operation or maintenance procedures for a source to assure continuous emission reduction.

Emission Offset—The use of an ERC (emission reduction credit) to offset, or mitigate, an emission increase governed by New Source Review Rules.
• A reduction in the air pollution emissions of existing sources to compensate for emissions from new sources.

Emission Permit—A non-transferable, non-tradable allocation of entitlement by a government to an individual firm to emit a specified amount of a substance.

Emission Quota—The portion or share of total allowable emissions assigned to a country or group of countries within a framework of maximum total emissions and mandatory allocations of resources or assessments.

Emission Reduction Credit (ERC)—ERCs are reductions in emission that have been recognized by the relevant local or state government air agency as being real, permanent, surplus, and enforceable. ERCs are usually measured as a weight over time (e.g., pounds per day or tons per year). Such rate-based ERCs can be used to satisfy emission offset requirements of new major sources and new major modifications of existing major sources. Mass-based ERCs, more akin to DERs, are issued with the weight and without reference to time.

Emission Reduction Unit (ERU)—Emissions reduction units (ERUs) are units of Greenhouse Gas reductions (or, portion of a country's Assigned Amount) that have been generated via Joint Implementation under Article 6 of the Kyoto Protocol—as opposed to Certified Emission Reduction units (CERs)—which have been generated and certified under the provisions of Article 12 of the Kyoto Protocol, the Clean Development Mechanism..

Emission Reporting Boundaries—The scope of emission sources included in an emission inventory or forecast for a particular firm. This scope can be defined according to jurisdictional reporting requirements or it can be broader which may allow greater opportunities for reductions. For example, national requirements may only require a business to report on emissions from the production cycle but the firm's own internal reporting boundaries may include emissions from waste etc. Also known as "System Boundaries."

Emission Standard—The maximum amount of a pollutant legally permitted to be discharged from a single source.

Emission Target—Emission targets are emission limits imposed on emitters by a regulatory body.

Emission Taxes—Surcharge or levy placed on emissions sources, usually on a per ton basis. Emission taxes are designed to provide incentives to firms and households to reduce their emissions as a means to control pollution (carbon tax is a subset of an emissions tax).

Emissions—Waste substances released into the air or water from sources and processes in a facility. In the context of global climate change, they consist of greenhouse gases (e.g., the release of carbon dioxide during fuel combustion).

Emissions Coefficient—A unique value for scaling emissions to activity data in terms of a standard rate of emissions per unit of activity (e.g., pounds of carbon dioxide emitted per Btu of fossil fuel consumed).

Emissions Data Report—The report, prepared by an operator or retail provider each year, that provides the information required by a regulator and is submitted using a written or electronic reporting tool and using formats approved by the regulator.

Emissions Excursion—In the Illinois ERMS program, an event that occurs when a Participating Source does not hold sufficient ATUs at the end of the Reconciliation Period to account for its volatile organic material emissions from the preceding Seasonal Allotment Period.

Emissions Leakage—A concept often used by policymakers in reference to the problem that emissions abatement achieved in one location may be offset by increased emissions in unregulated locations. Such leakage can arise, for example, in the short term as emissions abaters reduce energy demand or timber supply, influencing world prices for these commodities and increasing the quantity emitted elsewhere; and it can arise in the longer term, for example, as industries relocate to avoid controls.

Emissions Reduction Market System—The ERMS is a cap and trade regulatory program for stationary sources emitting volatile organic material in the ozone nonattainment area located in Northeastern Illinois.

Emissions Reduction Unit (ERU)—Emissions reductions generated by projects in Annex B countries that can be used by another Annex B country to help meet its commitments under the Kyoto Protocol. Reductions must be additional to those that would otherwise occur.

Emissions Trading—Emissions trading is a regulatory program that allows firms the flexibility to select cost-effective solutions to achieve established environmental goals. With emissions trading, firms can meet established emissions goals by: (a) reducing emissions from a discrete emissions unit; (b) reducing emissions from another place within the facility; or (c) securing emission reductions from another facility. Emissions trading encourages compliance and financial managers to pursue cost-effective emission reduction strategies and incentives emitting entrepreneurs to develop the means by which emissions can inexpensively be reduced.

Emissions(s)—Anthropogenic releases of gases to the atmosphere. In the context of global climate change, they consist of radiatively important greenhouse gases (e.g., the release of carbon dioxide during fuel combustion).

Emissivity—The ratio of the radiant energy (heat) leaving (being emitted by) a surface to that of a black body at the same temperature and with the same area; expressed as a number between 0 and 1. The measure of the ability of a material to radiate heat. Good glazing material should have a low emissivity (low-e).

Emittance—The emissivity of a material, expressed as a fraction. Emittance values range from 0.05 for brightly polished metals to 0.96 for flat black paint.

Emittance, Thermal—Is the ratio of the radiant heat flux emitted by a sample to that emitted by a blackbody radiator at the same temperature.

EMM—Electricity Market Model

EMPC—Estimated Maximum Possible Concentration

Employee Stock Option Plan (ESOP)—A plan established by a company whereby a certain number of shares is reserved for purchase and issuance to key employees. Such shares usually vest over a certain period of time to serve as an incentive for employees to build long-term value for the company.

EMR—Environmental Management Report

EMS—Enforcement Management Systems
• Environmental Management System
• Environmental Mutagen Society www.ems-us.org

Emulsifier—Substance that helps in mixing liquids that don't normally mix; e.g., oil and water.

Enclosed Space—Space that is substantially surrounded by sold surfaces.

Enclosure—The housing around a motor that supports the active parts and protects them. They come in different varieties (open, protected) depending on the degree of protection required.

Encroachment—The extension of an improvement onto the property of another.

Encumbrance—A lien or charge on land or personal property. Any right to or interest in land or personal property that affects its value, including loans, unpaid taxes, easements, junior liens or (in the case of real property), deed restrictions.

End of Term Lease Options—Options stated in the lease agreement that give the lessee flexibility in its treatment of the leased equipment at the end of the lease term. Common end-of-term options include purchasing the equipment, renewing the lease or returning the equipment to the lessor.

End Taker—The user taking the product produced by a project. The term is often used in connection with a take-or-pay contract. For example, electricity produced by a solar PV project under a power purchase agreement.

End Use—The purpose for which useful energy or work is consumed.

End User—A firm or individual that purchases products for its own consumption and not for resale (i.e., an ultimate consumer).

Endangered Species—See *Threatened, Endangered, and Sensitive Species*.

Endemic—Naturally existing at low levels in the environment.

Ending Stocks—Primary stocks of crude oil and petroleum products held in storage as of 12 midnight on the last day of the month. Primary stocks include crude oil or petroleum products held in storage at (or in) leases, refineries, natural gas processing plants, pipelines, tank farms, and bulk terminals that can store at least 50,000 barrels of petroleum products or that can receive petroleum products by tanker, barge, or pipeline. Crude oil that is in-transit by water from Alaska or that is stored on Federal leases or in the Strategic Petroleum Reserve is included. Primary Stocks exclude stocks of foreign origin that are held in bonded warehouse storage.

Endorsement—To express approval or support of; to write (something) on the back of a document or paper.

Endothermic—A heat absorbing reaction or a reaction that requires heat.

End-Use Sectors—The residential, commercial, industrial, and transportation sectors of the economy.

Energize(d)—To send electricity through an electricity transmission and distribution network; a conductor or power line that is carrying current.

Energy—The capacity for doing work. Forms of energy include: thermal, mechanical, electrical and chemical. Energy may be transformed from one form into another. Most of the world's convertible energy comes from fossil fuels that are burned to produce heat that is then used as a transfer medium to mechanical or other means in order to accomplish tasks. Electrical energy is measured in kilowatt-hours, while heat energy is usually measured in British thermal units.

• The capacity for doing work as measured by the capability of doing work (potential energy) or the conversion of this capability to motion (kinetic energy). Energy has several forms, some of which are easily convertible and can be changed to another form useful for work. Most of the world's convertible energy comes from fossil fuels that are burned to produce heat that is then used as a transfer medium to mechanical or other means to accomplish tasks.

• Power consumed multiplied by the duration of use. For example, 1000 Watts used for four hours is 4000 Watt hours.

Energy 2020—An economy-wide energy use model that predicts the investment behavior of both energy suppliers and consumers.

Energy Audit—A program carried out by a utility company or other licensed professional in which an auditor inspects a facility and suggests ways energy can be saved (generally referred to as an "energy conservation measure" or ECM). Determines the amount of energy used at a facility.

Energy Auditor—A person or organization specializing in identifying and assessing energy conservation opportunities in existing buildings. Energy Auditors require a broad knowledge range, including energy billing analysis, HVAC engineering, heating and cooling systems, lighting design, plumbing fixtures, technical writing and financial modeling.

Energy Broker System—Introduced into Florida by the Public Service Commission, the energy broker system is a system for exchanging information that allows utilities to efficiently exchange hourly quotations of prices at which each is willing to buy and sell electric energy. For the broker system to operate, utility systems must have in place bilateral agreements between all potential parties, must have transmission arrangements between all potential parties, and must have transmission arrangements that allow the exchanges to take place.

Energy Budget—A requirement in the Building Energy Efficiency Standards that a proposed building be designed to consume no more than a specified number of British thermal units (Btu's) per year per square foot of conditioned floor area.

• The maximum amount of Time Dependent Valuation (TDV) energy that a proposed building, or portion of a building, can be designed to consume, calculated with the approved procedures specified in Part 6 of Title 24 in California.

Energy Charge—That portion of the charge for electric service based upon the electric energy (kWh) consumed or billed.

Energy Charter Treaty (ECT)—International treaty that provides for a choice of dispute resolution mechanisms including arbitration at the election of the investor

• Provides protections for foreign investment against discrimination, expropriation and nationalization, unjustified restrictions on the transfer of funds and more

• Only multilateral investment treaty dealing with intergovernmental cooperation on energy—covers whole energy value chain

• Increases confidence by investors and financial community and promotes investment and trade flow among members

• Signed 1994; entered into force 1998

• Main areas of coverage are trade, transit, investment, energy efficiency and dispute resolution on an international level

Energy Conservation Features—This includes building shell conservation features, HVAC conservation features, lighting conservation features, any conservation features, and other conservation features incorporated by the building. However, this category does not include any demand-side management (DSM) program participation by the building. Any DSM program participation is included in the DSM Programs.

Energy Conservation Maintenance and Operating Procedure—Modification or modifications in maintenance and operations of a facility, and any installations within the facility, which are designed to reduce energy consumption in the facility and which require no significant expenditure of funds.

Energy Conservation Measure (ECM)—An installation or modification of an installation in a facility which is primarily intended to reduce energy consumption or allow use of an alternative energy source. It could be replacement of a component or system, installation of new equipment or changes in operating practices that result in energy conservation.

Energy Conservation Service—A service which provides pre-established levels of heating, cooling, lighting, and equipment use at reduced energy consumption levels. The services may include, but are not limited to, providing financing, design, installation, repair, maintenance, management, technical advice, and/or training.

Energy Consumption—The amount of electrical energy and demand, natural gas, oil, propane or other fuel consumed in a facility in any billing period. It also applies to utility services, such as water and sewer, which require energy to be consumed to supply the services to the facility.

Energy Contribution Potential—Recombination occurring in the emitter region of a photovoltaic cell.

Energy Conversion Factors—

To Convert	Into	Multiply By
kilowatt-hours per square meter	megajoules per square meter	3.60
kilowatt-hours per square meter	Btus per square foot	317.2
kilowatt-hours per square meter	Langleys	86.04
kilowatt-hours per square meter	calories per square centimeter	86.04
meters	feet	3.281
meters per second	miles per hour	2.237
millibars	pascals	100.0
millibars	atmospheres	0.0009869
millibars	kilograms per square meter	10.20
millibars	pounds per square inch	0.0145
degrees Centigrade	degrees Fahrenheit	°C x 1.8 + 32
degrees (angle)	radians	0.017453
degree days (base 18.3°C)	degree days (base 65 °F)	1.8

Energy Cost Savings—Energy savings converted into dollar savings.

Energy Credit—A Federal investment tax credit for energy property calculated as either 30 percent or 10 percent of the eligible basis of energy property placed in service during a taxable year. Energy property includes solar energy equipment, equipment to produce energy from a geothermal deposit, qualified fuel cell property, qualified microturbine property, combined heat and power (CHP) system property, qualified small wind energy property, or equipment using ground or ground water as a thermal energy source.

Energy Crops—Crops grown specifically for their fuel value. These include food crops such as corn and sugarcane and nonfood crops such as poplar trees and switch grass. Currently, two energy crops are under development in the United States: short-rotation woody crops, which are fast-growing hardwood trees harvested in 5 to 8 years, and herbaceous energy crops, such as perennial grasses, which are harvested annually after taking 2 to 3 years to reach full productivity.

Energy Deliveries—Energy generated by one electric utility system and delivered to another system through one or more transmission lines.

Energy Demand—The requirement for energy as an input to provide products and/or services.

Energy Density—The ratio of the energy available from an energy storage device such as a battery to its volume (Wh/liter) or weight (Wh/kg).

Energy Effects—The changes in aggregate electricity use (measured in megawatt hours) for consumers that participate in a utility DSM (demand-side management) program. Energy effects represent changes at the consumer's meter (i.e., exclude transmission and distribution effects) and reflect only activities that are undertaken specifically in response to utility-administered programs, including those activities implemented by third parties under contract to the utility. To the extent possible, Energy effects should exclude non-program related effects such as changes in energy usage attributable to non-participants, government-mandated energy-efficiency standards that legislate improvements in building and appliance energy usage, changes in consumer behavior that result in greater energy use after initiation in a DSM program, the natural operations of the marketplace, and weather and business-cycle adjustments.

Energy Efficiency—Using less energy/electricity to perform the same function. Programs designed to use electricity more efficiently—doing the same with less. Energy efficiency is distinguished from DSM programs in that the latter are utility-sponsored and -financed, while the former is a broader term not limited to any particular sponsor or funding source. "Energy conservation" is a term which has also been used but it has the connotation of doing without in order to save energy rather than using less energy to do the something and so is not used as much today. Many people use these terms interchangeably.

Refers to programs that are aimed at reducing the energy used by specific end-use devices and systems, typically without affecting the services provided. These programs reduce overall electricity consumption (reported in megawatt hours), often without explicit consideration for the timing of program-induced savings. Such savings are generally achieved by substituting technically more advanced equipment to produce the same level of end-use services (e.g. lighting, heating, motor drive) with less electricity. Examples include high-efficiency appliances, efficient lighting programs, high-efficiency heating, ventilating and air conditioning (HVAC) systems or control modifications, efficient building design, advanced electric motor drives, and heat recovery systems.

• Energy Efficiency occurs when you use less energy to accomplish the same task, for example heating your home or washing clothes. Using less energy means less air pollution and lower costs. To save energy in your home, you can use weather stripping, a water heater blanket or compact fluorescent light bulbs. Also when shopping for household appliances, look for the Energy Star to find appliances that use less energy and lower your electricity costs.

Energy Efficiency Ratio (EER)—The measure of the instantaneous energy efficiency of room air conditioners; the cooling capacity in Btu/hr divided by the watts of power consumed at a specific outdoor temperature (usually 95 degrees Fahrenheit).

• The ratio of net cooling capacity (in Btu/hr.) to total rate of electrical energy (in watts).of a cooling system under designated operating conditions, as determined using the applicable test method in the Appliance Efficiency Regulations or Section 112 of Title 24 (in California).

Energy Efficiency Resource Standard (EERS)—A simple, market-based mechanism to encourage more efficient generation, transmission, and use of electricity and natural gas. An EERS consists of electric and/or gas energy savings targets for utilities, often with flexibility to achieve the target through a market-based trading system. All EERS's include end-user energy saving improvements that are aided and documented by utilities or other program operators. Often used in conjunction with a Renewable Portfolio Standard (RPS).

Energy Efficient Mortgages (EEM)—A type of home mortgage that takes into account the energy savings of a home that has cost-effective energy saving improvements that will reduce energy costs thereby allowing the homeowner to more income to the mortgage payment. A borrower can qualify for a larger loan amount than otherwise would be possible.

Energy Efficient Motors—Are also known as "high-efficiency motors" and "premium motors." They are virtually interchangeable with standard motors, but differences in construction make them more energy efficient.

Energy End-Use Sectors—Major energy consuming sectors of the economy. The Commercial Sector includes commercial buildings and private companies. The Industrial Sector includes manufacturers and processors. The Residential Sector includes private homes. The Transportation Sector includes automobiles, trucks, rail, ships, and aircraft.

Energy Exchange—Any transaction in which quantities of energy are received or given up in return for similar energy products. See exchange, electricity; exchange, petroleum; and exchange, natural gas (see definitions further below).

Energy Expenditures—The money directly spent by consumers to purchase energy. Expenditures equal the amount of energy used by the consumer multiplied by the price per unit paid by the consumer.

Energy Factor (EF)—The measure of overall efficiency for a variety of appliances. For water heaters, the energy factor is based on three factors: 1) the recovery efficiency, or how efficiently the heat from the energy source is transferred to the water; 2) standby losses, or the percentage of heat lost per hour from the stored water compared to the content of the water: and 3) cycling losses. For dishwashers, the energy factor is defined as the number of cycles per kWh of input power. For clothes washers, the energy factor is defined as the cubic foot capacity per kWh of input power per cycle. For clothes dryers, the energy factor is defined as the number of pounds of clothes dried per kWh of power consumed.

Energy Guide Labels—The labels placed on appliances to enable consumers to compare appliance energy efficiency and energy consumption under specified test conditions as required by the Federal Trade Commission.

Energy Independence and Security Act of 2007 (EISA 2007)—Law covering issues from fuel economy standards for cars and trucks to renewable fuel and electricity to training programs for a "green collar" workforce to the first federal mandatory efficiency standards for appliances and lighting.

Energy Information—Includes (A) all information in whatever form on fuel reserves, extraction, and energy resources (including petrochemical feedstocks) wherever located; production, distribution, and consumption of energy and fuels wherever carried on; and (B) matters relating to energy and fuels, such as corporate structure and proprietary relationships, costs, prices, capital investment, and assets, and other matters directly related thereto, wherever they exist.

Energy Information Administration (EIA)—An independent agency within the U.S. Department of Energy that develops surveys, collects energy data, and does analytical and modeling analyses of energy issues. The Agency must satisfy the requests of Congress, other elements within the Department of Energy, Federal Energy Regulatory Commission, the Executive Branch, its own independent needs, and assist the general public, or other interest groups, without taking a policy position. www.eia.gov.

Energy Intensity—The relative extent that energy is required for a process. Energy Intensity is the energy consumption per unit of output.

Energy Intensity (Commercial Buildings Energy Consumption Survey)—The ratio of consumption to floor space.

Energy Intensity Indicator—The Energy Intensity Indicator is a dimensionless ratio equal to the Energy Intensity in a particular year divided by the Energy Intensity of a reference year. The Energy Intensity Indicator for the reference year equals 1.0.

Energy Management and Control System (EMCS)—An energy conservation feature that uses mini/microcomputers, instrumentation, control equipment, and software to manage a building's use of energy for heating, ventilation, air conditioning, lighting, and/or business-related processes. These systems can also manage fire control, safety, and security. Not included as EMCS are time-clock thermostats.

Energy Management Practices—Involvement, as a part of the building's normal operations, in energy efficiency programs that are designed to reduce the energy used by specific end-use systems. This includes the following: EMCS, DSM Program Participation, Energy Audit, and a Building Energy Manager.

Energy Management System—A control system (often computerized) designed to regulate the energy consumption of a building by controlling the operation of energy consuming systems, such as the heating, ventilation and air conditioning (HVAC), lighting and water heating systems.

Energy Modeling—Process to determine the energy use of a building based on software analysis. Also called building energy simulation.

Energy Obtained from Depletable Sources—Electricity purchased from a public utility, or any energy obtained from coal, oil, natural gas, or liquefied petroleum gases.

Energy Obtained from Non-Depletable Sources—Energy that is not energy obtained from depletable sources.

Energy Payback Time—The time required for any energy producing system or device to produce as much energy as was required in its manufacture. For solar electric panels, this is normally in the range 6-36 months.

Energy Policy Act of 1992 (EPACT)—This legislation creates a new class of power generators, exempt wholesale generators, that are exempt from the provisions of the Public Holding Company Act of 1935 and grants the authority to the Federal Energy Regulatory Commission to order and condition access by eligible parties to the interconnected transmission grid. Major revisions were passed in 2005 and 2007.

Energy Policy Act of 2005—Also known as the Domenici-Barton Energy Policy Act of 2005. Major act enacted in 2005 that provides more than $14.5 Billion in tax breaks to oil, gas and renewable energy companies over a ten-year period. Increased the federal Energy Business Tax Credit from 10% to 30% for tax years 2006 and 2007.

Energy Production—See production terms associated with specific energy types.

Energy Receipts—Energy brought into a site from another location.

Energy Reserves—Estimated quantities of energy sources that are demonstrated to exist with reasonable certainty on the basis of geologic and engineering data (proved reserves) or that can reasonably be expected to exist on the basis of geologic evidence that supports projections from proved reserves (probable/indicated reserves). Knowledge of the location, quantity, and grade of probable/indicated reserves is generally incomplete or much less certain than it is for proved energy reserves. Note: This term is equivalent to "Demonstrated Reserves" as defined in the resource/reserve classification contained in the U.S. Geological Survey Circular 831, 1980. Demonstrated reserves include measured and indicated reserves but exclude inferred reserves.

Energy Resources—Everything that could be used by society as a source of energy. The available supply and price of fossil and alternative resources will play a huge role in estimating how much a greenhouse gas constraint will cost. In the U.S. context, natural gas supply (and thus price) is particularly important, as it is expected to be a transition fuel to a lower carbon economy.

Energy Resources Program Account (ERPA)—The state law that directs California electric utility companies to gather a state energy surcharge of two-tenths of one mil ($0.0002) per kilowatt hour of electricity consumed by a customer. These funds are used for operation of the California Energy Commission.

Energy Risk Management—Using futures or offsetting contracts to hedge the risk of fluctuating energy prices.

Energy Sale(s)—The transfer of title to an energy commodity from a seller to a buyer for a price or the quantity transferred during a specified period.

Energy Saving Performance Contract (ESPC)—An ESPC is usually entered into with an Energy Service Company (ESCO) and allows you to leverage your operating budget to finance capital improvements. You can make significant improvements to your heating, ventilation, air conditioning, lighting, water systems, and building envelope.

With an ESPC, the improvements are self-funding, paid for over time from the energy savings delivered by an ESCO's solutions. The ESCO makes the up-front investment and assumes the risk of performance. The ESCO can also provide maintenance for the life of the contract with. Your energy savings are guaranteed by the ESCO under the ESPC.

Energy Savings—The amount of energy expressed in standard units (e.g., therms, gallons, kilowatt hours) of energy saved by an energy conservation measure or service.

• A reduction in the amount of electricity used by end users as a result of participation in energy efficiency programs and load management programs.

Energy Security Act of 1980—Legislation authorizing a U.S. biomass and alcohol fuel program, and that authorized loan guarantees and price guarantees and purchase agreements for alcohol fuel production.

Energy Security/Fuel Security—Policy that considers the risk of dependence on fuel sources located in remote and unstable regions of the world and the benefits of domestic and diverse fuel sources.

Energy Service Company (ESCO)—Energy Service Company that provides customers with various price options and non-traditional energy services. ESCO refers to a business operator that guarantees a certain amount of energy savings from the planning stages of an energy improvement project and that, working on the basis of a performance contract, provides its clients with integrated services to improve the client's energy efficiency that range from proposing plans to performing repairs and providing managerial expertise. These services can be applied to public buildings, private-sector office buildings, or factories, with costs and investments recovered in the form of lower energy bills. In this way, clients can receive the benefits of reduced energy costs, while the ESCO involved can receive appropriate compensation for the expertise and services it provides. What sets this type of business activity off from all others is the fact that all of the costs associated with making improvements in energy conservation are covered by the savings achieved through lower energy consumption.

Energy Service Provider—An energy entity that provides service to a retail or end-use customer. Also referred to as an Electric Service Provider.

Energy Source—Any substance or natural phenomenon that can be consumed or transformed to supply heat or power. Examples include petroleum, coal, natural gas, nuclear, biomass, electricity, wind, sunlight, geothermal, water movement, and hydrogen in fuel cells.

Energy Star®—A rating system for appliances established by the Federal government. Includes appliance efficiency standards and new building codes. Administered by the Environmental Protection Agency.

Energy Storage—The process of storing, or converting energy from one form to another, for later use;

storage devices and systems include batteries, conventional and pumped storage hydroelectric, flywheels, compressed gas, and thermal mass.

Energy Supplier—Fuel companies supplying electricity, natural gas, fuel oil, kerosene, or LPG (liquefied petroleum gas) to the household.

Energy Supply—Energy made available for future disposition. Supply can be considered and measured from the point of view of the energy provider or the receiver.

Energy Time Shift—The differential value derived by using energy during off-peak periods to charge an energy storage device that can be discharged during a peak or other period of higher prices (a.k.a., Energy Arbitrage).

Energy Used in the Home—For electricity or natural gas, the quantity is the amount used by the household during the 365- or 366-day period. For fuel oil, kerosene, and liquefied petroleum gas (LPG), the quantity consists of fuel purchased, not fuel consumed. If the level of fuel in the storage tank was the same at the beginning and end of the annual period, then the quantity consumed would be the same as the quantity purchased.

Energy/Fuel Diversity—Policy that encourages the development of energy technologies to diversify energy supply sources, thus reducing reliance on conventional (petroleum) fuels; applies to all energy sectors.

Energy-Use Sectors—A group of major energy-consuming components of U.S. society developed to measure and analyze energy use. The sectors most commonly referred to in EIA are: residential, commercial, industrial, transportation, and electric power.

Energy-Weighted Industrial Output—The weighted sum of real output for all two-digit Standard Industrial Classification (SIC) manufacturing industries plus agriculture, construction, and mining. The weight for each industry is the ratio between the quantity of end-use energy consumption to the value of real output.

Enforcement Agency—The city, county, or state agency responsible for issuing a building permit.

Engine Size—The total volume within all cylinders of an engine when pistons are at their lowest positions. The engine is usually measured in "liters" or "cubic inches of displacement (CID)." Generally, larger engines result in greater engine power, but less fuel efficiency. There are 61.024 cubic inches in a liter.

Enhanced Greenhouse Effect—The increase in the natural greenhouse effect resulting from increases in atmospheric concentrations of GHGs due to emissions from human activities.

ENPA—Environmental Performance Agreement

Enriched Uranium—Uranium in which the U-235 isotope concentration has been increased to greater than the 0.711 percent U-235 (by weight) present in natural uranium.

Enrichment Feed Deliveries—Uranium that is shipped under contract to a supplier of enrichment services for use in preparing enriched uranium product to a specified U-235 concentration and that ultimately will be used as fuel in a nuclear reactor.

Enrichment Tails Assay—A measure of the amount of fissile uranium (U-235) remaining in the waste stream from the uranium enrichment process. The natural uranium "feed" that enters the enrichment process generally contains 0.711 percent (by weight) U-235. The "product stream" contains enriched uranium (more than 0.711 percent U-235) and the "waste" or "tails" stream contains depleted uranium (less than 0.711 percent U-235). At the historical enrichment tails assay of 0.2 percent, the waste stream would contain 0.2 percent U-235. A higher enrichment tails assay requires more uranium feed (thus permitting natural uranium stockpiles to be decreased), while increasing the output of enriched material for the same energy expenditure.

Enthalpy—A thermodynamic property of a substance, defined as the sum of its internal energy plus the pressure of the substance times its volume, divided by the mechanical equivalent of heat. The total heat content of air; the sum of the enthalpies of dry air and water vapor, per unit weight of dry air; measured in Btu per pound (or calories per kilogram).

Entire Building—The ensemble of all enclosed space in a building, including the space for which a permit is sought, plus all existing conditioned and unconditioned space within the structure.

Entitlement—Electric energy or generating capacity that a utility has a right to access under power exchange or sales agreements.

Entrained Bed Gasifier—A gasifier in which the feedstock (fuel) is suspended by the movement of gas to move it through the gasifier.

Entropy—A measure of the unavailable or unusable energy in a system; energy that cannot be converted to another form.

Entry Into Force—The point at which international

climate change agreements become binding. The United Nations Framework Convention on Climate Change (UNFCCC) has entered into force. In order for the Kyoto Protocol to do so as well, 55 Parties to the Convention must ratify (approve, accept, or accede to) the Protocol, including Annex I Parties accounting for 55 percent of that group's carbon dioxide emissions in 1990. As of June 2003, 110 countries had ratified the Protocol, representing 43.9 percent of Annex I emissions.

Environment—All the natural and living things around us. The earth, air, weather, plants, and animals all make up our environment.

Environmental Assessment (EA)—A public document that analyzes a proposed federal action for the possibility of significant environmental impacts. The analysis is required by the National Environmental Policy Act. If the environmental impacts will be significant, the federal agency must then prepare an environmental impact statement.

Environmental Dynamic Revenue Assessment Model (E-DRAM)—A dynamic general equilibrium forecasting model that simulates the way that changes in energy investment, price and use affect how citizens live their lives.

Environmental Economics—Questions of the social costs and benefits that accompany issues relating to pollution, resource depletion, and environmental degradation fall within the area of environmental economics. Few people today would disagree that a factory emitting large amounts of smoke causes air pollution, which affects the health of local residents. A fundamental issue in environmental economics is the assessment of the costs of that pollution to the residents, in terms of illness; to society, in terms of health-care costs and lost work time; and, ultimately, to the world, as the facility smoke contributes to the formation of acid rain and may increase the likelihood of global warming. Equally important, however, is the inclusion of the cost equation, of the values created by the activities of the pollution facility: the usefulness of its products and the worth of the jobs it creates.

Environmental Equity—*Or* environmental justice refers to the environmental protection for all citizens so that no segment of the population, regardless of race, ethnicity, culture, or income, bears a disproportionate burden of the consequences of environmental pollution.

Environmental Footprint—For an industrial setting, this is a company's environmental impact determined by the amount of depletable raw materials and nonrenewable resources it consumes to make its products, and the quantity of wastes and emissions that are generated in the process. Traditionally, for a company to grow, the footprint had to get larger. Today, finding ways to reduce the environmental footprint is a priority for leading companies. An environmental footprint can be determined for a building, city, or nation as well, and gives an indication of the sustainability of the unit.

Environmental Impact Assessment (EIA)—An assessment of potential environmental effects of development projects. Required by the National Environmental Policy Act (NEPA) for any proposed major federal action with significant environmental impact.

Environmental Impact Statement (EIS)—A report that documents the information required to evaluate the environmental impact of a project. It informs decision makers and the public of the reasonable alternatives that would avoid or minimize adverse impacts or enhance the quality of the environment. It is a document required by federal agencies under the National Environmental Policy Act (NEPA) of 1969 for major projects or legislative proposals significantly affecting the environment. The EIS describes the positive and negative effects of the undertaking and cites alternative actions that could be taken.

Environmental Impact Study—An investigation to assess the comprehensive, long-range environmental impact of a proposed land use, including both direct and indirect effects over all phases of use.

Environmental Justice—The concept of equal access to environmental resources and protection from environmental hazards regardless of race, ethnicity, national origin, or income.

Environmental Justice Advisory Committee (EJAC)—A committee created by the legislation of AB 32 in California whose mission is to advise The California Air Resources Board in developing the Scoping Plan and any other pertinent matter in implementing AB 32.

Environmental Protection Agency (EPA)—A federal agency created in 1970 to permit coordinated governmental action for protection of the environment by systematic abatement and control of pollution through integration or research, monitoring, standards setting and enforcement activities.

Environmental Protection Agency (EPA) Certification Files—Computer files produced by EPA for analysis purposes. For each vehicle make, model and year, the files contain the EPA test MPGs (city, highway, and 55/45 composite). These MPG's are associated with various combinations of engine and drive-train technologies (e.g., number of cylinders, engine size, gasoline or diesel fuel, and automatic or manual transmission). These files also contain information similar to that in the DOE/EPA Gas Mileage Guide, although the MPG's in that publication are adjusted for shortfall.

Environmental Restoration—Although usually described as "cleanup," this function encompasses a wide range of activities, such as stabilizing contaminated soil; treating ground water; decommissioning process buildings, nuclear reactors, chemical separations plants, and many other facilities; and exhuming sludge and buried drums of waste.

Environmental Restrictions—In reference to coal accessibility, land-use restrictions that constrain, postpone, or prohibit mining in order to protect environmental resources of an area; for example, surface- or groundwater quality, air quality affected by mining, or plants or animals or their habitats.

Environmental Risk Assessment (ERA)—The tracking and rating of environmental risks, such as emissions, associated with a product and its manufacturing.

Environmental Valuation—The inclusion of environmental costs and benefits into accounting practices using such mechanisms as taxes, tax incentives, and subsidies by quantifying environmentally-related costs and revenues. Better management decisions and increased investment in environmental protection and improvement are encouraged.

Environmental, Social and Government (ESG)—An acronym commonly used by investment firms to refer to the types of issues or factors considered in measuring a company's "responsible practices." These issues or factors include the environmental effects of a company's business practices, social metrics such as fair pay and treatment of labor and community involvement and ethical corporate governance practices that are both transparent and anti-corruption.

Environmentally Superior Product—A product that reflects 1) a greater proportion of renewable energy and 2) lower emissions per kilowatt-hour of SO_x, NO_x, and greenhouse gases than the default system power.

Enzymatic Hydrolysis—A process by which enzymes (biological catalysts) are used to break down starch or cellulose into sugar.

EO—Executive Officer
- Executive Order

EOJ—End of Job

EOP—Emergency Operations Plan
- End of Pipe Treatment

EOR—Enhanced Oil Recovery

EOT—Emergency Operations Team

EOY—End of Year

EP Toxicity—A test defined by the federal Environmental Protection Agency to check a substance for the presence of arsenic, barium, cadmium, chromium, lead, mercury, selenium, or silver. 40 CFR 261.24 defines the concentrations constituting hazardous waste and the test procedure.

EPA—The Environmental Protection Agency is a federal agency born in 1970 -charged with protecting the environment.
- Economic Price Adjustment

EPA Certification—A permanent label on fireplace inserts and freestanding wood stoves manufactured after July 1, 1988, indicating that the equipment meets EPA standards for clean burning.

EPA Composite MPG—The harmonic mean of the EPA city and highway MPG (miles per gallon), weighted under the assumption of 55 percent city driving and 45 percent highway driving.

EPAA—Environmental Programs Assistance Act

EPAC—Emergency Preparedness Advisory Committee

EPAct—The Energy Policy Act of 1992 addresses a wide variety of energy issues. The legislation creates a new class of power generators, exempt wholesale generators (EWG's), that are exempt from the provisions of the Public Utilities Holding Company Act of 1935 and grants the authority to FERC to order and condition access by eligible parties to the interconnected transmission grid.

EPAct 2005—Energy Policy Act of 2005

EPBB—Expected Performance-Based Buydown. A solar rebate program found in California. Cash is provided that is based upon the expected performance of a new installed solar system (both residential and commercial) and is paid on an installed watt basis.

EPC—Emergency Preparedness Coordinator

EPC Contract—Engineering, Procurement and Construction Contract. An agreement between the owner or expected owner of an energy facility and the contractor or installer of the project. The EPC outlines the contractor's responsibilities, compensation, timing, etc. in completing the installation of the facility.

EPCA—Energy Policy and Conservation Act (1975)

EPCRA—Emergency Planning and Community Right-to-Know Act (1986)—also known as Title III-Superfund Amendments and Reauthorization Act

EPI—Environmental Policy Institute
  • Environmental Priorities Initiative

EPIA—European Photovoltaic Industries Association

Epitaxial Growth—In reference to solar photovoltaic devices, the growth of one crystal on the surface of another crystal. The growth of the deposited crystal is oriented by the lattice structure of the original crystal.

Epoxy Resins—Resins made by the reaction of epoxides or oxiranes with other materials such as amines, alcohols, phenols, carboxylic acids, acid anhydrides, and unsaturated compounds.

EPR—Energy Profit Ratio

EPRG—Environmental Policy Review Group

EPRI—Electric Power Research Institute www.epri.com

EPS—Earnings Per Share

Equal Rate Treatment (Utility)—Term used to designate a test of the reasonableness of an allocation of costs. In this test the rates designed to recoup the costs allocated to jurisdictional business are applied to the billing units of non-jurisdictional business to determine whether such rates will produce more or less revenues than the costs which have been assigned to the non-jurisdictional business. The "equal rate treatment" is also used as an allocation methodology. In this instance rates are designed which will recoup the total cost of service and are applied equally to jurisdictional and non-jurisdictional business.

Equalization Charge—Periodical overcharging the batteries for a short time to mix the electrolyte solution in batteries.

Equilibrium Cycle—An analytical term that refers to fuel cycles that occur after the initial one or two cycles of a reactor's operation. For a given type of reactor, equilibrium cycles have similar fuel characteristics.

Equinox—The two times of the year when the sun crosses the equator and night and day are of equal length; usually occurs on March 21st (spring equinox) and September 23 (fall equinox).

Equipment—Any article, machine, or other contrivance, or combination thereof, which may cause the issuance or control the issuance of air or water contaminants.

Equipment Schedule—A document, incorporated by reference into the contract that describes in detail the equipment being financed or used. The schedule may state the contract terms, commencement date, repayment schedule and location of the equipment.

Equitable Title—Title of the Purchaser under a Contract of Sale or the right to acquire the Legal Title.

Equity—The value of the unencumbered interest in real or personal property as determined by subtracting the total amount of the debt balances plus the sum of any current liens from the property's fair market value.

Equity (Financial)—Ownership of shareholders in a corporation represented by stock. Ownership in the capital of a Company. In corporations, it is called "stock"; in limited partnerships or LLCs, it is called "interests" or "units." Net worth—assets minus liabilities.

Equity Capital—The sum of capital from retained earnings and the issuance of stock.

Equity Crude Oil—The proportion of production that a concession owner has the legal and contractual right to retain.

Equity Funding (Lease Contract)—The equity monies which are used in a lease transaction by the equity investor to partially pay for the leased equipment. The balance will be paid through some form of debt.

Equity in Earnings of Unconsolidated Affiliates—A company's proportional share (based on ownership) of the net earnings or losses of an unconsolidated affiliate.

Equity Investments—The development and financing of infrastructure projects for which the company also uses its own capital. The goal of such investments is to earn annual dividends and participate in the long-term value growth of the investment.

Equity Investor or Participant—An entity that provides equity funding in a leveraged financial transaction and thereby becomes the owner and ultimate owner of the equipment.

Equity Kicker—Option for private equity investors to purchase shares at a discount. Typically associated with mezzanine financings where a small number of shares or warrants are added to what is primarily a debt financing.

Equity Warrants—Equity warrants may be included with bonds or equity issues. An equity warrant gives the holder the right to buy shares in the company at a fixed price at a point of time in the future. The lower the exercise price for the shares, the more valuable the warrants will be, because the holder has a greater chance of exercising their rights profitably.

Equivalent Direct Radiation—Heat expressed in terms of a square foot of steam radiator surface emitting 240 Btu per hour. (Btu per hour divided by 240).

ER—Energy Report

ERA—Economic Regulatory Agency
* Environmental Risk Assessment

ERC (Emission Reduction Credit)—An ERC (or Emission Reduction Credit) is effectively the same as a tradable emission permit. Under Title IV of the US Clean Air Act, ERCs are created when a polluter reduces pollution below their target level. Such ERCs can then be sold to other firms whose emissions exceed their target.
* ERCs are reductions in emissions that have been recognized by the relevant local or state government air agency as being real, permanent, surplus, and enforceable. ERCs are usually measured as a weight over time (e.g., pounds per day or tons per year). Such rate-based ERCs can be used to satisfy emission offset requirements of new major sources and new major modifications of existing major sources. Mass-based ERCs, more akin to DERs, are issued with the weight and without reference to time.

ERCOT—Electric Reliability Council of Texas

ERDA—Energy Research and Development Administration

Erg—A unit of work done by the force of one dyne acting through a distance of one centimeter.

ERNS—Emergency Response Notification System

EROEI—Energy Return on Energy Invested

ERP—Enforcement Response Policy
* Emerging Renewables Program (in California). The ERP is an Energy Commission program offering cash rebates on eligible grid-connected renewable energy electric-generating systems.

ERP—Equity Risk Premium

ERPA—Emissions Reduction Purchase Agreement. Contracts governing the sale of CER carbon credits from UN CDM and JI projects. Heavily used for forward sales of CERs not yet issued, in projects under development, as a means of project financing. The price of such primary CERs is discounted in ERPAs to reflect the risks of non-delivery.

ERPG—Emergency Response Planning Guidelines

ERT—Emergency Response Team
* Environmental Response Team

ERTA—Economic Recovery Tax Act (1981)

ERU—Emission Reduction Unit. Tradable credits generated from activities to reduce greenhouse emissions in industrialized countries, particularly those of the former Soviet-bloc, under the Kyoto Protocol's Joint Implementation (JI) mechanism.

ES—Enforcement Strategy
* Engineering Staff
* Entrainment Separator
* Expert System

ES&H—Environmental Safety and Health

ESA—Energy Service Agreement
* Endangered Species Act
* Environmentally Sensitive Area
* Environmental Site Assessment

Escalation Clause—A lease provision that allows the lessor to increase the rents based on an increase in the CPI, bank prime rate or other index rate.

Escheat—Forfeiture of Title to property to the State—either Real or Personal.

ESCO—Energy Service Company. A business entity that designs, builds, develops, owns, operates or any combination thereof, self-generation Projects for the sake of providing energy or energy services to a Host Customer (including both electrical and thermal).

Escrow—Placing money or assets in a special and separate account under the control of another party, usually a financial institution or trustee, to be held in trust until the completion of conditions set forth in an agreement.
* A transaction in which an impartial third party acts as Agent for both Seller and Buyer, or for both Borrower and Lender, in carrying out Instructions, delivering papers and documents, and disbursing funds.

ESD—Environmental Services Department

ESE—Environmental Science and Engineering

ESECA—Energy Supply and Environmental Coordination Act (1974)

ESP—Energy Service Provider. An ESP or Energy Services Provider is a company that offers all the same services as an ESCO and has the additional expertise to provide energy supply through the development and implementation of build/own/operate distributed generation, cogeneration, or combined heat and power (CHP) projects; the arrangement of supply on a consulting basis; or the firm contracting of energy supply.

ESPC (Energy Service Performance Contract)—An agreement with a private energy service company (ESCO). The ESCO will identify and evaluate energy-saving opportunities and then recommend a package of improvements to be paid for through savings. ESPCs come in many variations. In general, performance contracts contain three component parts: a) a project development agreement, b) an energy services agreement, and c) a financing agreement. Performance contracts come in many varieties and may differ greatly in their content and coverage.

Establishment—An economic unit, generally, at a single physical location where business is conducted or where services or industrial operations are performed. However, "establishment" is not synonymous with "building."

Estimated Additional Resources (EAR)—The uranium in addition to reasonable assured resources (RAR) that is expected to occur, mostly on the basis of direct geological evidence, in extensions of well-explored deposits, little-explored deposits, and undiscovered deposits believed to exist along a well-defined geologic trend with known deposits, such that the uranium can subsequently be recovered within the given cost ranges. Estimates of tonnage and grade are based on available sampling data and on knowledge of the deposit characteristics as determined in the best known parts of the deposit or in similar deposits. EAR correspond to DOE's Probable Potential Resource Category.

Estimation (Tax)—The process of predicting the taxes due at each tax payment date to meet the full year's liability.

Estoppel—A doctrine which bars one from asserting rights which are inconsistent with a previous position or representation.

Estoppel Certificate—A statement of material facts or conditions on which another person can rely because it cannot be denied at a later date.

ET—Emissions Trading
  • Ecotox Thresholds

Et Al—And others.

ETA—Energy Tax Act

ETBE (Ethyl Tertiary Butyl Ether)—An oxygenate blend stock formed by the catalytic etherification of isobutylene with ethanol.

Ethane (C_2H_6)—A normally gaseous straight-chain hydrocarbon. It is a colorless paraffinic gas that boils at a temperature of -127.48 degrees Fahrenheit. It is extracted from natural gas and refinery gas streams.

Ethanol (CH_3-CH_2OH)—A clear, colorless, flammable oxygenated hydrocarbon. Ethanol is typically produced chemically from ethylene, or biologically from fermentation of various sugars from carbohydrates found in agricultural crops and cellulosic residues from crops or wood. It is used in the United States as a gasoline octane enhancer and oxygenate (blended up to 10 percent concentration). Ethanol can also be used in high concentrations (E85) in vehicles designed for its use. Note: The lower heating value, equal to 76,000 Btu per gallon, is assumed for estimates in the Renewables Energy Annual report.

Ether—A generic term applied to a group of organic chemical compounds composed of carbon, hydrogen, and oxygen, characterized by an oxygen atom attached to two carbon atoms (e.g., methyl tertiary butyl ether).

Ethyl Tertiary Butyl Ether (ETBE)—An aliphatic ether similar to MTBE. This fuel oxygenate is manufactured by reacting isobutylene with ethanol. Having high octane and low volatility characteristics, ETBE can be added to gasoline up to a level of approximately 17 percent by volume. ETBE is used as an oxygenate in some reformulated gasolines.

Ethylene (C_2H_4)—An olefinic hydrocarbon recovered from refinery processes or petrochemical processes. Ethylene is used as a petrochemical feedstock for numerous chemical applications and the production of consumer goods.

Ethylene Dichloride—A colorless, oily liquid used as a solvent and fumigant for organic synthesis, and for ore flotation.

ETOR—Estimated Time of Restoration: The time when a utility company estimates service will be restored to customers after an outage.

ETP—Emissions Trading Policy

ETPS—Emissions Trading Policy Statement

ETS—Emissions Tracking System
  • Emissions Trading Scheme

ETSR—Energy Technologies Status Report

EU—European Union (1994)

EU Bubble—Under the Kyoto Protocol, the individual countries that comprise the European Union have aggregated their emissions and accepted an aggregated emissions reduction target. This has been reallocated back to the individual countries to allow differentiation of national reduction programs. The arrangement allows the target to be shared among all countries within the bubble.

EU ETS—European Union Emissions Trading Scheme.

EUA—European Union Allowances. Tradable emission credits from the EU Emissions Trading Scheme.

Each allowance carries the right to emit one ton of carbon dioxide.

EUC—End Use Consumption

Eudiometer—An instrument for the volumetric measurement and analysis of gases.

EUL—Enhanced use lease

EUP—Environmental Use Permit
- Experimental Use Permit

European Community—As a regional economic integration organization, the European Community can be and is a Party to the UNFCCC; however, it does not have a separate vote from its members (Austria, Belgium, Denmark, Finland, France, Germany, Greece, Ireland, Italy, Luxemburg, the Netherlands, Portugal, Spain, Sweden, and the United Kingdom).

Eutectic—A mixture of substances that has a melting point lower than that of any mixture of the same substances in other proportions.

Eutectic Salts—Salt mixtures with potential applications as solar thermal energy storage materials.

EV—Enterprise Value. Equal to the market value of equity, preferred equity, outstanding net debt and minorities.

EV (Electric Vehicle)—A vehicle powered by electricity, usually provided by batteries but may also be provided by photovoltaic (solar) cells or a fuel cell.

EV Charging Station—The facility that provides battery charging for EVs. Many new installations provide electricity from wind and solar sources.

EVA—Economic Value Added

EVA (Ethylene Vinyl Acetate)—An encapsulant used between the glass cover and the solar cells in PV modules. It is durable, transparent, resistant to corrosion, and flame retardant.

Evacuated-Tube Collector—A collector is the mechanism in which fluid (water or diluted antifreeze, for example) is heated by the sun in a solar hot water system. Evacuated-tube collectors are made up of rows of parallel, transparent glass tubes. Each tube consists of a glass outer tube and an inner tube, or absorber. The absorber is covered with a selective coating that absorbs solar energy well but inhibits radiative heat loss. The air is withdrawn ("evacuated") from the space between the tubes to form a vacuum, which eliminates conductive and convective heat loss. Evacuated-tube collectors are used for active solar hot water systems and usually produce water at temperatures of greater than 160° F.

Evaporation—The conversion of a liquid to a vapor (gas), usually by means of heat.

Evaporation Pond—A containment pond (that preferably has an impermeable lining of clay or synthetic material such as hypalon) to hold liquid wastes and to concentrate the waste through evaporation.

Evaporative Cooler—Provides cooling to a building by either direct contact with water (direct evaporative cooler), no direct contact with water (indirect evaporative cooler), or a combination of direct and indirect cooling (indirect/direct evaporative cooler). The credit offered for evaporative coolers depends on building type and climate.

Evaporative Cooling—Cooling by exchange of latent heat from water sprays, jets of water, or wetted material.

Evaporator Coil—The inner coil in a heat pump that, during the cooling mode, absorbs heat from the inside air and boils the liquid refrigerant to a vapor, which cools the house.

Evapotranspiration—The water released from plants as they grow. The evaporation of water from plant surfaces and adjacent soil.

Evergreen Lease—This is a lease that automatically renews itself each year unless the lessee gives notice of its termination within a specified period of time.

EWEA—European Wind Energy Association www.ewea.org

EWG—Exempt Wholesale Generator: Created under the 1992 Energy Policy Act, these wholesale generators are exempt from certain financial and legal restrictions stipulated in the Public Utilities Holding Company Act of 1935.

Exaction—A requirement imposed by a local government that a developer contribute to the community by providing an amenity, paying an impact fee, or making some other monetary contribution as a condition for the right to construct the development. Typical exactions are road rights of way, park land or cash in lieu of park land dedication, and utility or drainage easements. These donations must be roughly proportional to the value of the governmental services that will be required for the new development. For example, a residential developer may be required to dedicate a park to serve the development's residents or to improve the roadway that provides access to the development.

Exception—A deduction, subtraction, or exclusion.

Exceptional Method—An approved alternative calculation method that analyzes designs, materials, or devices that cannot be adequately modeled using public domain computer programs. Exceptional methods must be submitted to and approved by the California Energy Commission. [See California Code of Regulations, Title 20, Section 1409(b)3] Two examples of exceptional methods are the controlled ventilation crawl space (CVC) credit and the combined hydronic space and water heating method.

Excess Annual Growth—The amount by which new forest growth exceeds removal in a year. The annual quantity of wood produced in a forest in excess of market demand.

Excess Statutory Depletion—The excess of estimated statutory depletion allowable as an income tax deduction over the amount of cost depletion otherwise allowable as a tax deduction, determined on a total enterprise basis.

Exchange (Electric utility)—Agreements between utilities providing for purchase, sale and trading of power. Usually relates to capacity (kilowatts) but sometimes energy (kilowatt-hours).

Exchange Act—["34 Act"] Regulates periodic reporting by companies with publicly traded securities, companies with more than 500 shareholders, and brokers and dealers in securities.

Exchange Agreement—A contractual agreement in which quantities of crude oil, petroleum products, natural gas, or electricity are delivered, either directly or through intermediaries, from one company to another company, in exchange for the delivery by the second company to the first company of an equivalent volume or heat content. The exchange may take place at the same time and location or at different times and/or locations. Such agreements may also involve the payment of cash. Note: EIA excludes volumes sold through exchange agreements to avoid double counting of data. See *Energy Exchange* above.

Exchange Energy—See *Exchange, Electricity* below.

Exchange Rate—The ratio of prices at which the currencies of nations are exchanged.

Exchange, Electricity—A type of energy exchange in which one electric utility agrees to supply electricity to another. Electricity received is returned in kind at a later time or is accumulated as an energy balance until the end of a specified period, after which settlement may be made by monetary payment. Note: This term is also referred to as Exchange Energy.

Exchange, Natural Gas—A type of energy exchange in which one company agrees to deliver gas, either directly or through intermediaries, to another company at one location or in one time period in exchange for the delivery by the second company to the first company of an equivalent volume or heat content at a different location or time period. Note: Such agreements may or may not include the payment of fees in dollar or volumetric amounts.

Exchange, Petroleum—A type of energy exchange in which quantities of crude oil or any petroleum product(s) are received or given up in return for other crude oil or petroleum products. It includes reciprocal sales and purchases.

Excitation—The power required to energize the magnetic field of a generator.

Executive Officer—A term generally used to refer to the senior officer of a regulatory agency.

Executive Order Number 6—A provision under the California Emergency Services Act permits the Governor to establish, by Executive Order Number 6, a state Petroleum Fuels Set-Aside Program after proclamation of an energy emergency.

Executor—The executor or administrator of the decedent, or, if there is no executor or administrator appointed, qualified, and acting within the United States, then any person in actual or constructive possession of any property of the decedent.

Executory Costs—Recurring costs in a financial contract, such as insurance, maintenance and taxes on the assets, whether paid by the lender or the borrower. Executory costs also include amounts paid by the borrower in consideration for a third-party residual guarantee, as well as any profits realized by the lender on any executory costs paid by the lender and passed on to the borrower.

Exempt Wholesale Generator (EWG)—Wholesale generators created under the 1992 Energy Policy Act that are exempt from certain financial and legal restrictions stipulated in the Public Utilities Holding Company Act of 1935.

Exemption—An immunity from some burden or obligation.

Exercise Right—The price at which an option or warrant can be exercised.

EXEX—Expected Exceedence

Exfiltration—Uncontrolled outward air leakage from inside a building, including leakage through cracks and interstices, around windows and doors, and through any other exterior partition or duct penetration.

Exhaust—Air removed deliberately from a space, by a fan or other means, usually to remove contaminants from a location near their source.

Exhaust Fan—Small fans located in the wall or ceiling

that exhaust air, odors, and moisture from the bathroom, kitchen, or basement to the outside.

Exit Strategy—A fund's or a company's intended method for liquidating its holdings while achieving the maximum possible return. These strategies depend on the exit climates, including market conditions and industry trends. Exit strategies can include selling or distributing the portfolio company's shares after an initial public offering (IPO), a sale of the portfolio company, or a recapitalization.

Exothermic—A reaction or process that produces heat; a combustion reaction.

Expanded Polystyrene—A type of insulation that is molded or expanded to produce coarse, closed cells containing air. The rigid cellular structure provides thermal and acoustical insulation, strength with low weight, and coverage with few heat loss paths. Often used to insulate the interior of masonry basement walls.

Expansion Ratio—The ratio of gas volume after expansion to the gas volume before expansion.

Expansion Tank—A tank used in a closed-loop solar heating system that provides space for the expansion of the heat transfer fluid in the pressurized collector loop.

Expansion Valve—The device that reduces the pressure of liquid refrigerant thereby cooling it before it enters the evaporator coil in a heat pump.

Expected Savings—Expected savings are those reported in the Post-installation Report. They are based on as-built conditions and post-installation verification activities, and are the savings expected for year 1 of the project.

Expected Value—A measure of investment risk that employs weighted probabilities. An event is weighted based on the probability of the occurrence.

Expenditure—The incurrence of a liability to obtain an asset or service.

Expenditures per Million Btu—The aggregate ratio of a group of buildings' total expenditures for a given fuel to the total consumption of that fuel.

Expenditures per Square Foot—The aggregate ratio of a group of buildings' total expenditures for a given fuel to the total floor space in those buildings.

Expense Stop—A clause in a lease that limits the landlord's expense obligation because the lessee assumes any expenses above an established level.

Exploration—The identification of areas that may warrant examination and to examine specific areas that are considered to have prospects of containing oil and gas reserves, including drilling exploratory wells and exploratory-type stratigraphic test wells. Exploration costs may be incurred both before acquiring the related property (sometimes referred to in part as prospecting costs) and after acquiring the property.

Exploration Costs—Costs, including depreciation and applicable operating costs, of support equipment and facilities and other costs directly identifiable with exploration activities, such as:

- Costs of topographical, geological, and geophysical studies, rights of access to properties to conduct those studies, and salaries and other expenses of geologists, geophysical crews, and others conducting those studies. Collectively, these costs are sometimes referred to as geological and geophysical, or 'G&G' costs.
- Costs of carrying and retaining undeveloped properties, such as delay rentals, ad valorem taxes on the properties, legal costs for title defense, and the maintenance of land and lease records.
- Dry hole contributions and bottom hole contributions.
- Costs of drilling and equipping exploratory wells.
- Costs of drilling exploratory-type stratigraphic test wells.

Exploration Drilling—Drilling done in search of new mineral deposits, on extensions of known ore deposits, or at the location of a discovery up to the time when the company decides that sufficient ore reserves are present to justify commercial exploration. Assessment drilling is reported as exploration drilling.

Exploratory Well—A hole drilled: a) to find and produce oil or gas in an area previously considered unproductive area; b) to find a new reservoir in a known field, i.e., one previously producing oil and gas from another reservoir, or c) to extend the limit of a known oil or gas reservoir.

Exports—Shipments of goods from within the 50 States and the District of Columbia to U.S. possessions and territories or to foreign countries.

Exports (Electric Utility)—Power capacity or energy that a utility is required by contract to supply outside of its own service area and not covered by general rate schedules.

Exposed Thermal Mass—Mass that is directly exposed (uncovered) to the conditioned space of the building. Concrete floors that are covered by carpet are not considered exposed thermal mass.

Extension Agreement—Agreement granting further time for performance.

Extensions—Any new reserves credited to a previously producing reservoir because of enlargement of its proved area. This enlargement in proved area is usually due to new well drilling outside of the previously known productive limits of the reservoir.

Extensions, Discoveries, and Other Additions—Additions to an enterprise's proved reserves that result from (1) extension of the proved acreage of previously discovered (old) reserves through additional drilling in periods subsequent to discovery and (2) discovery of new fields with proved reserves or of new reservoirs of proved reserves in old fields.

Exterior Door—A door through an exterior partition that is opaque or has a glazed area that is less than or equal to one-half of the door area. Doors with a glazed area of more than one half of the door area are treated as a fenestration product.

Exterior Floor/Soffit—A horizontal exterior partition, or a horizontal demising partition, under conditioned space. For low-rise residential occupancies, exterior floors also include those on grade.

Exterior Partition—An opaque, translucent, or transparent solid barrier that separates conditioned space from ambient air or space that is not enclosed. For low-rise residential occupancies, exterior partitions also include barriers that separate conditioned space from unconditioned space, or the ground.

Exterior Roof/Ceiling—An exterior partition, or a demising partition, that has a slope less than 60 degrees from horizontal, that has conditioned space below, and that is not an exterior door or skylight.

Exterior Wall—Any wall or element of a wall, or any member or group of members, which defines the exterior boundaries or courts of a building and which has a slope of 60 degrees or greater with the horizontal plane. An exterior wall or partition is not an exterior floor/soffit, exterior door, exterior roof/ceiling, window, skylight, or demising wall.

Exterior Zones—The portions of a building, with significant amounts of exterior walls, windows, roofs, or exposed floors. Such zones have heating or cooling needs largely dependent upon weather conditions.

External Combustion Engine—An engine in which fuel is burned (or heat is applied) to the outside of a cylinder; a Stirling engine.

External Rate of Return (ERR)—A method of yield calculation. ERR is a modified internal rate of return (IRR) that allows for the incorporation of specific reinvestment, borrowing and sinking fund assumptions.

Externalities—Benefits or costs, generated as a byproduct of an economic activity, that do not accrue to the parties involved in the activity. Environmental externalities are benefits or costs that manifest themselves through changes in the physical or biological environment. For example the impact of environmental degradation resulting from an activity that is not incorporated into the economics of the activity.

Externality—The environmental, social, and economic impacts of producing a good or service that are not directly reflected in the market price of the good or service.

Extra High Voltage (EHV)—Voltage levels higher than those normally used on transmission lines. Generally EHV is considered to be 345,000 volts or higher.

Extraction Loss—The reduction in volume of natural gas due to the removal of natural gas liquid constituents such as ethane, propane, and butane at natural gas processing plants.

Extraction Well—Well that is used primarily to remove contaminated groundwater from the ground. Water level measurements and water samples can also be collected from extraction wells.

Extractive Industries—Industries involved in the activities of (1) prospecting and exploring for wasting (non-regenerative) natural resources, (2) acquiring them, (3) further exploring them, (4) developing them, and (5) producing (extracting) them from the earth. The term does not encompass the industries of forestry, fishing, agriculture, animal husbandry, or any others that might be involved with resources of a regenerative nature.

Extraordinary Income Deductions (Electric Utility)—Those items related to transactions of a nonrecurring nature that are not typical or customary business activities of the utility and that would significantly distort the current year's net income if reported other than as extraordinary items.

Extraordinary Item—Income and expense items associated with events and transactions that possess a high degree of abnormality and are of a type that would not reasonably be expected to recur in the foreseeable future.

Extraordinary Property Losses—An amortizable (Deferred Debit) account, which includes the depreciated value of property abandoned or damaged by circumstances that could not have been reasonably anticipated and which is not covered by insurance.

Extrinsic Semiconductor—The product of doping a pure semiconductor.

Extruded Polystyrene—A type of insulation material with fine, closed cells, containing a mixture of air and refrigerant gas. This insulation has a high R-value, good moisture resistance, and high structural strength compared to other rigid insulation materials.

F

F—Fahrenheit (temperature measurement increment)

F.A.S. Value—Free alongside ship value. The value of a commodity at the port of exportation, generally including the purchase price plus all charges incurred in placing the commodity alongside the carrier at the port of exportation in the country of exportation.

F.O.B. Price—The price actually charged at the producing country's port of loading. The reported price should be after deducting any rebates and discounts or adding premiums where applicable and should be the actual price paid with no adjustment for credit terms.

F.O.B. Value (Coal)—Free-on-board value. This is the value of coal at the coal mine or of coke and breeze at the coke plant without any insurance or freight transportation charges added.

FAA—Federal Aviation Administration

Fabricated Fuel—Fuel assemblies composed of an array of fuel rods loaded with pellets of enriched uranium dioxide.

FAC—Federal Advisory Committee
- Fully Allocated Costs: Embedded costs that include both direct and indirect costs.

FACA—Federal Advisory Committee (ACT)
- Financial Assurance for Corrective Action

Face Value—The maturity value of a bond or other debt instrument. Sometimes referred to as the bond's par value of nominal value.
- The stated worth of a note, insurance policy, mortgage, etc. Synonymous to "par value" in capital stocks.

Facilities Charge—An amount to be paid by the customer in a lump sum, or periodically as reimbursement for facilities furnished. The charge may include operation and maintenance as well as fixed costs.

Facility—An existing or planned location or site at which prime movers, plants, building, structure, source, electric generators, and/or equipment for converting mechanical, chemical, and/or nuclear energy into electric energy are situated or will be situated. Usually located on one or more contiguous or adjacent properties, in actual physical contact or separated solely by a public roadway or other public right-of way, and under common operational control, that emits or may emit any GHG or other pollutant and is considered a single major industrial grouping. A facility may contain more than one generator of either the same or different prime mover type. For a co generator, the facility includes the industrial or commercial process.

Facility Group—A group of any number of Facilities. It is a logical group used for reporting purposes (i.e., for consolidated reporting on groups of buildings within the entire portfolio). Group assignments are usually made according to geographic regions or internal cost centers. They have no impact on engineering analysis or Opportunity identification and therefore do not necessarily have to be defined or assigned as part of those activities.

Facility Variable—An Independent Variable (an external factor affecting energy usage that is changing regularly), other than Billing Period or weather-related heating and cooling load. Examples of Facility Variables include production volume, arena refrigeration days, occupied suites, hot meals served and production shifts.

Facsimile—An exact and precise copy.

Factory Assembled Cooling Towers—Are cooling towers constructed from factory assembled modules either shipped to the site in one piece or put together in the field.

Facts & Circumstances—A tax depreciation method based upon the historical accuracy of a company's experience.

Facultative Ponds—Ponds having an aerobic zone on the top and an anaerobic zone on the bottom.

FAF—Fuel adjustment factor. An ancillary charge on some ocean freight shipments to account for fluctuation in fuel costs. Also referred to as BAF or bunker (a fuel) adjustment factor.

Fahrenheit—A temperature scale in which the boiling point of water is 212 degrees and its freezing point is 32 degrees. To convert Fahrenheit to Celsius, subtract 32, multiply by 5, and divide the product by 9. For example: 100 degrees Fahrenheit - 32 = 68; 68 x 5 = 340; 340/9 = 37.77 degrees Celsius.

Failure or Hazard—Any electric power supply equipment or facility failure or other event that, in the judgment of the reporting entity, constitutes a hazard to maintaining the continuity of the bulk electric power supply system such that a load reduction action may become necessary and reportable outage may occur. Types of abnormal conditions that should be reported include the imposition of a special operating procedure, the extended purchase of emergency power, other bulk power system actions that may be caused by a natural disaster, a major equipment failure that would impact the bulk power supply, and an environmental and/or regulatory action requiring equipment outages.

Fair—As in "fair" rate of return. In ratemaking "fair" is a subjective term requiring significant study to support the proposed level.

Fair Market Value—The value of a piece of equipment if the equipment were to be sold in a transaction determined at arm's length, between a willing buyer and a willing seller, for equivalent property and under similar terms and conditions.

Fair Market Value Cap—A high end limit on a FMV lease that protects the lessee upside risk for executing the residual at the end of the lease.

Fair Market Value Purchase Option—An option to purchase leased property at the end of the lease term at its then fair market value. The lessor does not have the ability to retain title to the equipment if the lessee chooses to exercise the purchase option.

Fair Rental Value—The theoretical amount of periodic rental that should be paid for an asset. Used by the IRS as a guideline in Revenue Ruling 55-540.

Fair Trade—An international trading partnership that seeks to help marginalized producers and workers achieve financial self-sufficiency by establishing direct lines of trade between producers and consumers, guaranteeing producers fair prices for goods, restricting exploitative labor processes, and favoring environmentally-sustainable production processes through a system of labeling products as "fair trade."

FAN—Final Acceptance Notice

Fan Coil—A component of a heating, ventilation and air conditioning (HVAC) system containing a fan and heating or cooling coil, used to distribute heated or cooled air.

Fan Velocity Pressure—The pressure corresponding to the outlet velocity of a fan; the kinetic energy per unit volume of flowing air.

FAR—Federal Acquisitions Regulation

Farad—A unit of electrical capacitance; the capacitance of a capacitor between the plates of which there appears a difference of 1 Volt when it is charged by one coulomb of electricity.

Farm Out (in) Arrangement—An arrangement, used primarily in the oil and gas industry, in which the owner or lessee of mineral rights (the first party) assigns a working interest to an operator (the second party), the consideration for which is specified exploration and/or development activities. The first party retains an overriding royalty or other type of economic interest in the mineral production. The arrangement from the viewpoint of the second party is termed a "farm-in arrangement."

Farm Use—Energy use at establishments where the primary activity is growing crops and/or raising animals. Energy use by all facilities and equipment at these establishments is included, whether or not it is directly associated with growing crops and/or raising animals. Common types of energy-using equipment include tractors, irrigation pumps, crop dryers, smudge pots, and milking machines. Facility energy use encompasses all structures at the establishment, including the farm house.

FASB—Financial Accounting Standards Board www.fasb.org

FASB 13—Statement issued by the Financial Accounting Standards Board establishing financial accounting standards for reporting leases for both lessees and lessors.

Fast Breeder Reactor (FBR)—A reactor in which the fission chain reaction is sustained primarily by fast neutrons rather than by thermal or intermediate neutrons. Fast reactors require little or no moderator to slow down the neutrons from the speeds at which they are ejected from fissioning nuclei. This type of reactor produces more fissile material than it consumes.

Fast Pyrolysis—Thermal conversion of biomass by rapid heating to between 450ø to 600øC in the absence of oxygen.

Fault—An event generated by a manufacturer of a particular device which is installed on a remote site.

FBC—Fluidized Bed Combustion

FBE—Functional Basis Earthquake

FBPOC—Facility Boundary Point of Compliance

FBR—Fluidized Bed Reactor

FBS—Federal Base System

FCA—Full Cost Accounting

FCAA—Federal Clean Air Act

FCC—Fluid Catalytic Converter
- Federal Communications Commission
- Final Cost Certification. A report issued by an independent certified public accountant certifying the total and eligible costs incurred in an energy

project, typically in the form of an agreed-upon procedures or examination report.

FCCC—Framework Convention on Climate Change

FCCU—Fluid Catalytic Cracking Unit

FCF—Free Cash Flow

FCIA—Foreign Credit Insurance Association

FCQAS—Financial Compliance and Quality Assurance Staff

FDA—Food and Drug Administration

FDCA—Food, Drug and Cosmetic Act

FDI—Foreign Direct Investment

FDIC—Federal Deposit Insurance Corporation

FDL—Final Determination Letter

FE—Fugitive Emissions

FEA—Federal Energy Administration

Feasibility Factor—A factor used to adjust potential energy savings to account for cases where it is impractical to install new equipment. For example, certain types of fluorescent lighting require room temperature conditions. They are not feasible for outdoor or unheated space applications. Some commercial applications, such as color-coded warehouses, require good color rendition, so color distortions could also make certain types of lighting infeasible. The feasibility factor equals 100 percent minus the percent of infeasible applications.

Feather—In a wind energy conversion system, to pitch the turbine blades so as to reduce their lift capacity as a method of shutting down the turbine during high wind speeds.

Federal Coal Lease—A lease granted to a mining company to produce coal from land owned and administered by the Federal Government in exchange for royalties and other revenues.

Federal Electric Utility—A utility that is either owned or financed by the Federal Government.

Federal Emergency Management Agency (FEMA)—FEMA is the federal agency in charge of disaster recovery in locations declared 'official' disaster areas.

Federal Energy Management Agency—The federal agency in charge of disaster recovery in locations that have been declared disaster areas by a state's Governor and the President of the United States.

Federal Energy Management Program (FEMP)—A program of the U.S. Department of Energy (DOE) that implements energy legislation and presidential directives. FEMP provides project financing, technical guidance and assistance, coordination and reporting, and new initiatives for the federal government. It also helps federal agencies identify the best technologies and technology demonstrations for their use.

Federal Energy Regulatory Commission (FERC)—An independent regulatory commission within the U.S. Department of Energy that has jurisdiction over energy producers that sell or transport fuels for resale in interstate commerce; the authority to set oil and gas pipeline transportation rates and to set the value of oil and gas pipelines for ratemaking purposes; and regulates wholesale electric rates and hydroelectric plant licenses. FERC is the successor to the Federal Power Commission.

Federal Network for Sustainability (FNS)—FNS promotes cost-effective, energy- and resource-efficient operations across all branches of government. www.federalsustainability.org

Federal Power Act—Enacted in 1920, and amended in 1935, the Act consists of three parts. The first part incorporated the Federal Water Power Act administered by the former Federal Power Commission, whose activities were confined almost entirely to licensing non-Federal hydroelectric projects. Parts II and III were added with the passage of the Public Utility Act. These parts extended the Act's jurisdiction to include regulating the interstate transmission of electrical energy and rates for its sale as wholesale in interstate commerce. The Federal Energy Regulatory Commission is now charged with the administration of this law.

Federal Power Commission (FPC)—The predecessor agency of the Federal Energy Regulatory Commission. The Federal Power Commission was created by an Act of Congress under the Federal Water Power Act on June 10, 1920. It was charged originally with regulating the electric power and natural gas industries. It was abolished on September 30, 1977, when the Department of Energy was created. Its functions were divided between the Department of Energy and the Federal Energy Regulatory Commission, an independent regulatory agency.

Federal Power Marketing Administrations (PMA)—These are separate and distinct organizational agencies within the U.S. DOE that market power at federal multipurpose water projects at lowest possible rates to consumers consistent with sound business principles. There are five PMA's:

- Alaska Power Administration
- Bonneville Power Administration
- Southeastern Power Administration
- Southwestern Power Administration

•Western Area Power Administration.

Federal Region—In a Presidential directive issued in 1969, various Federal agencies (among them the currently designated Department of Health and Human Services, the Department of Labor, the Office of Economic Opportunity, and the Small Business Administration) were instructed to adopt a uniform field system of 10 geographic regions with common boundaries and headquarters cities. The action was taken to correct the evolution of fragmented Federal field organization structures that each agency or component created independently, usually with little reference to other agencies' arrangements. Most Federal domestic agencies or their components have completed realignments and relocations to conform to the Standard Federal Administration Regions (SFARs).

Federal Water Pollution Control Act—A federal regulatory law administered by the states. The act created the National Pollution Discharge Elimination System.

FEDS—Federal Energy Data System

Fee Interest—The absolute, legal possession and ownership of land, property, or rights, including mineral rights. A fee interest can be sold (in its entirety or in part) or passed on to heirs or successors.

Feed—The prepared and mixed materials, which include but are not limited to materials such as limestone, coal, clay, shale, sand, iron ore, mill scale, cement kiln dust and fly ash, that are fed into a process at a facility. Feed does not include the fuels used in the process.

Feeder—An electrical supply line, either overhead or underground, which runs from the substation, through various paths, ending with the transformers. It is a distribution circuit, usually less than 69,000 volts, which carries power.

Feeder Line—An electrical line that extends radially from a distribution substation to supply electrical energy within an electric area or sub-area.

Feeder Lockout—This happens when a main circuit is interrupted at the substation by automatic protective devices and cannot be restored until crews investigate. This indicates a serious problem on the circuit, usually equipment failure of a broken conductor.

Feed-in Tariff—A policy that sets a fixed price at which power producers can sell renewable power into the electric power network. Some policies provide a fixed tariff while others provide fixed premiums added to market- or cost-related tariffs. Some pro-

vide both.

Feedstock—Any material that can be converted to another form of fuel or energy product. It is the raw material supplied to a process.

FEIR—Final Environmental Impact Report

FEIS—Final Environmental Impact Statement
• Fugitive Emissions Information System

Fell—To cut down a tree. Cutting down trees and sawing them to manageable lengths is referred to as "felling and bucking" or "falling and bucking."

Feller-Buncher—A self-propelled machine that cuts trees with giant shears near ground level and then stacks the trees into piles to await skidding.

FEMA—Federal Emergency Management Administration: FEMA's continuing mission is to lead the effort to prepare the nation for all hazards and effectively manage federal response and recovery efforts following any national incident. FEMA also initiates proactive mitigation activities, trains first responders, and manages the National Flood Insurance Program and the U.S. Fire Administration.

FEMP—Federal Energy Management Program

Fenestration—In simplest terms, windows or glass doors. Technically fenestration is described as any transparent or translucent material plus any sash, frame, mullion or divider. This includes windows, sliding glass doors, French doors, skylights, curtain walls and garden windows.

Fenestration Product—Any transparent or translucent material plus any sash, frame, mullions and dividers, in the envelope of a building, including, but not limited to, windows, sliding glass doors, French doors, skylights, curtain walls, garden windows, and other doors with a glazed area of more than one half of the door area.

Fenestration Products—A collection of fenestration products included in the design of a building.

FEPCA—Federal Environmental Pesticides Control Act (amendments to FIFRA)

FERC—Federal Energy Regulatory Commission: A Federal independent regulatory body within the Department of Energy that regulates interstate gas and electric rates and facilities, as well as hydroelectric plant licenses.

FERC Guidelines—A compilation of the Federal Energy Regulatory Commission's enabling statutes; procedural and program regulations; and orders, opinions, and decisions.

Fermentation—The decomposition of organic material to alcohol, methane, etc., by organisms, such as yeast or bacteria, usually in the absence of oxygen.

Fermi Level—Energy level at which the probability of finding an electron is one-half. In a metal, the Fermi level is very near the top of the filled levels in the partially filled valence band. In a semiconductor, the Fermi level is in the band gap.

Fertile Material—Material that is not itself fissionable by thermal neutrons but can be converted to fissile material by irradiation. The two principal fertile materials are uranium-238 and thorium-232.

FF&E—Furniture, fixtures and equipment

FFA—Flammable Fabrics Act

FFAR—Fuel and Fuel Additive Registration

FFC—Federal Facility Coordinator

FFDCA—Federal Food, Drug and Cosmetic Act

FFF—Firm Financial Facility

FFL—Fossil Fuel Levy

FFP—Firm Fixed Price

FFS—Federal financing specialist

FGD—Flue Gas Desulfurization

FHA—Farmers Home Administration
- Federal Housing Authority

FHLBB—Federal Home Loan Bank Board

FHSA—Federal Hazardous Substances Act

FHSLA—Federal Hazardous Substance Labeling Act

FHWA—Federal Highway Administration

FIA—Federal Insurance Administration

FIATS—Freedom of Information Action Tracking System

Fiberglass Insulation—A type of insulation, composed of small diameter pink, yellow, or white glass fibers, formed into blankets or batts, or used in loose-fill and blown-in applications.

FICA—Federal Insurance Contributions Act

Fiduciary—One who holds something of value in trust for another—a guardian, trustee, executor, administrator, receiver, conservator, or any person acting in any fiduciary capacity for any person.

Field—An area consisting of a single reservoir or multiple reservoirs all grouped on or related to the same individual geological structural feature and/or stratigraphic condition. There may be two or more reservoirs in a field, which are separated vertically by intervening impervious strata, or laterally by local geologic barriers, or by both.

Field Area—A geographic area encompassing two or more pools that have a common gathering and metering system, the reserves of which are reported as a single unit. This concept applies primarily to the Appalachian region.

Field Discovery Year—The calendar year in which a field was first recognized as containing economically recoverable accumulations of oil and/or gas.

Field Erected Cooling Towers—Cooling towers which are custom designed for a specific application and which cannot be delivered to a project site in the form of factory assembled modules due to their size, configuration, or materials of construction.

Field Production—Represents crude oil production on leases, natural gas liquids production at natural gas processing plants, new supply of other hydrocarbons/oxygenates and motor gasoline blending components, and fuel ethanol blended into finished motor gasoline.

Field Separation Facility—A surface installation designed to recover lease condensate from a produced natural gas stream usually originating from more than one lease and managed by the operator of one or more of these leases.

Field-Fabricated Fenestration Product or Exterior Door—A fenestration product or exterior door whose frame is made at the construction site of standard dimensional lumber or other materials that were not previously cut, or otherwise formed with the specific intention of being used to fabricate a fenestration product or exterior door. Field fabricated does not include site-built fenestration with a label certificate or products required to have temporary or permanent labels.

FIFO—First In/First Out (an accounting term as it regards inventory)

FIFRA—Federal Insecticide, Fungicide and Rodenticide Act (1972)

Filament—A coil of tungsten wire suspended in a vacuum or inert gas-filled bulb. When heated by electricity the tungsten "filament" glows.

File Fate Schedule—The rate for a particular electric service, including attendant contract terms and conditions, accepted for filing by a regulatory body with appropriate oversight authority.

Filing—Any written application, complaint, declaration, petition, protest, answer, motion, brief, exception, rate schedule, or other pleading, amendment to a pleading, document, or similar paper that is submitted to a utility commission.

Fill Factor—The ratio of a photovoltaic cell's actual power to its power if both current and voltage were at their maxima. A key characteristic in evaluating cell performance. On an I-V (current-voltage) curve characterizing the output of a solar cell or module, the ratio of the maximum power to the product of the open-circuit voltage and the short-circuit current. The higher the fill factor (FF) the "squarer" the shape of the I-V curve.

Filter (Air)—A device that removes contaminants, by mechanical filtration, from the fresh air stream before the air enters the living space. Filters can be installed as part of a heating/cooling system through which air flows for the purpose of removing particulates before or after the air enters the mechanical components.

Fin—A thin sheet of material (metal) of a heat exchanger that conducts heat to a fluid.

Final Deal—Fully matched and negotiated bids/offers that are under contract for delivery.

Final Order—A final ruling by FERC that terminates an action, decides some matter litigated by the petitioning parties, operates to some right, or completely disposes of the subject matter.

• An order issued by the EPA Administrator after an appeal of an initial decision, accelerated decision, decision to dismiss or default order, disposing of a matter in controversy between the parties, or

• An initial decision which becomes a final order

Finance Lease—An expression oftentimes used in the industry to refer to a capital lease or a nontax lease. It is also a type of tax-oriented lease that was introduced by the Tax Equity and Fiscal Responsibility Act of 1982, to be effective in 1984, but later repealed by the Tax Reform Act of 1986.

Financial Accounting Standards Board (FASB)—An independent board responsible, since 1973, for establishing generally accepted accounting principles. Its official pronouncements are called "Statements of Financial Accounting Standards" and "Interpretations of Financial Accounting Standards."

Financial Attributes—Financial attributes measure the financial health of the company. Utility management, security analysts, investors, and regulators use these attributes to evaluate a utility's performance against its historic records and industry averages. Key financial attributes include capital requirements, earnings per share of common equity, capitalization ratios, and interest coverage ratios.

Financial Value Chain—A company's financial flows from data entry through to accounts (balance sheet, profit-and-loss statement, cash flow statement), reports (controlling) and company evaluations (see *Discounted Cash Flow*) to peer group comparisons (rating).

Financing Statement—A Personal Property Security Instrument which replaced a Chattel Mortgage upon adoption of the Uniform Commercial Code.

• A notice of a security interest filed under the Uniform Commercial Code (UCC).

Finding of No Significant Impact (FONSI)—A document describing the reasons why the impacts of a proposed federal action are not significant. Required by the National Environmental Policy Act after an environmental assessment when a federal agency is not preparing an environmental impact statement.

Fine—A very small particle of material such as very fine sander dust or very small pieces of bark.

Finish—Both a noun and a verb to describe the exterior surface of building elements (walls, floors, ceilings, etc.) and furniture, and the process of applying it.

Finish Charge—The final stage of battery charging, when the battery is charged at a slow rate over a long period of time.

Finished Leaded Gasoline—Contains more than 0.05 gram of lead per gallon or more than 0.005 gram of phosphorus per gallon. Premium and regular grades are included, depending on the octane rating. Includes leaded gasohol. Blendstock is excluded until blending has been completed. Alcohol that is to be used in the blending of gasohol is also excluded.

Finished Motor Gasoline—A complex mixture of relatively volatile hydrocarbons, with or without small quantities of additives, blended to form a fuel suitable for use in spark-ignition engines. Specification for motor gasoline, as given in ASTM Specification D439-88 or Federal Specification VV-G-1690B, include a boiling range of 122 degrees to 158 degrees Fahrenheit at the 10-percent point to 365 degrees to 374 degrees Fahrenheit at the 90-percent point and a Reid vapor pressure range from 9 to 15 psi. "Motor gasoline" includes finished leaded gasoline, finished unleaded gasoline, and gasohol. Blendstock is excluded until blending has been completed. (Alcohol that is to be used in the blending of gasohol is also excluded.)

Finished Unleaded Gasoline—Contains not more than 0.05 gram of lead per gallon and not more than 0.005 gram of phosphorus per gallon. Premium and regular grades are included, depending on the octane rating. Includes unleaded gasohol. Blendstock is excluded until blending has been completed. Alcohol that is to be used in the blending of gasohol is also excluded.

FIP—Federal Implementation Plan
• Federal Information Plan
• Final Implementation Plan

- Fields in Production

Fire Brick—Heat resistant refractory ceramic material formed into bricks and used to line fire boxes of boilers, furnaces, or other combustion chambers.

Fire Classification—Classifications of fires developed by the National Fire Protection Association.

Fire Clay—A special kind of clay that will not melt or fuse at high temperatures.

Fire Point—Minimum temperature at which a substance will continue to burn after being ignited.

Fireplace—A wood or gas burning appliance that is primarily used to provide ambiance to a room. Conventional, masonry fireplaces without energy saving features, often take more heat from a space than they put into it.

Fireplace Insert—A wood or gas burning heating appliance that fits into the opening or protrudes on to the hearth of a conventional fireplace.

Fire-Rating—The ability of a building construction assembly (partition, wall, floor, etc.) to resist the passage of fire. The rating is expressed in hours.

Firewall—A wall to prevent the spread of fire; usually made of non-combustible material.

Firing Rate—The rate at which fuel is fed to a burner, expressed as volume, heat units, or weight per unit time.

The amount of Btus/hour or kWs produced by a heating system from the burning of a fuel.

Firm—An association, company, corporation, estate, individual, joint venture, partnership, or sole proprietorship, or any other entity, however organized, including: (a) charitable or educational institutions; (b) the Federal Government, including corporations, departments, Federal agencies, and other instrumentalities; and State and local governments. A firm may consist of (1) a parent entity, including the consolidated and unconsolidated entities (if any) that it directly or indirectly controls; (2) a parent and its consolidated entities only; (3) an unconsolidated entity; or (4) any part or combination of the above.

Firm Capacity—When referring to interstate pipeline, backbone transmission, or storage, firm capacity is capacity that is available under virtually all operating conditions 365 days a year.

Firm Energy—Power supplies that are guaranteed to be delivered under terms defined by contract.

Firm Gas—This is the most stable source of supply. During system emergencies and high demand periods, this type of gas is the last to be curtailed.

Firm Power—Power which is guaranteed by the supplier to be available at all times during a period covered by a commitment. That portion of a customer's energy load for which service is assured by the utility provider.

Firm Service Level—Power supplies that are guaranteed to be delivered under terms defined by contract. For electric utility customers who are on an interruptible or curtailable rate, only generation that serves the portion of their electric load that is designated as firm services is eligible for California Solar Initiatives (CSI). Under the CSI program, customers must agree to maintain the firm service level at or above capacity of the proposed generating system for the duration of the required applicable warranty period. Customers may submit a letter requesting an exemption to the firm service rule if they plan to terminate or reduce a portion of their available load.

Firm Transmission Service—Service that is reserved for at least one year.

First Law of Thermodynamics—States that energy cannot be created or destroyed, but only changed from one form to another. First Law efficiency measures the fraction of energy supplied to a device or process that it delivers in its output. Also called the law of conservation of energy.

First Loss Provision—A guarantee that is measured by some percentage of the total liability. The guarantor suffers the first loss up to that amount.

First Purchase (of Crude Oil)—An equity (not custody) transaction commonly associated with a transfer of ownership of crude oil associated with the physical removal of the crude oil from a property for the first time (also referred to as a lease sale). A first purchase normally occurs at the time and place of ownership transfer where the crude oil volume sold is measured and recorded on a run ticket or other similar physical evidence of purchase. The volume purchased and the cost of such transaction shall not be measured farther from the wellhead than the point at which the value for landowner royalties is established, if there was a separate landowner.

First Purchase Price—The marketed first sales price of domestic crude oil, consistent with the removal price defined by the provisions of the Windfall Profits Tax on Domestic Crude Oil (Public Law 96-223, Sec. 4998 (c)).

Fiscal (vs. Calendar)—Referring to the business year of an organization, in contrast to the calendar year.

Fiscal Funding Clause—A provision by which the debt

obligation is cancelable if the legislature or other funding authority does not appropriate the funds necessary for the governmental unit to fulfill its obligations under the debt agreement.

Fiscal Year (FY)—The U.S. Government's fiscal year runs from October 1 through September 30. The fiscal year is designated by the calendar year in which it ends; e.g., fiscal year 2002 begins on October 1, 2001 and ends on September 30, 2002. In non-public corporate entities, the year end of the corporation that is other than December 31 of each year.

Fissile Material—Material that can be caused to undergo atomic fission when bombarded by neutrons. The most important fissionable materials are uranium-235, plutonium-239, and uranium-233.

Fission—The process whereby an atomic nucleus of appropriate type, after capturing a neutron, splits into (generally) two nuclei of lighter elements, with the release of substantial amounts of energy and two or more neutrons.

Fissionable Material—A substance whose atoms can be split by slow neutrons. Uranium-235, plutonium-239 and uranium-233 are fissionable materials.

FIT—Field Investigation Team
- Feed-in Tariff. An incentive used to encourage the use of renewable energy through legislation. Under a feed-in tariff system, utilities are obligated to purchase electricity from renewable energy facilities at specific rates set by the government that are based on the cost of renewable energy generation.

Fixed Asset Turnover—A ratio of revenue to fixed assets which is a measure of the productivity and efficiency of property, plant, and equipment in generating revenue. A high turnover reflects positively on the entity's ability to utilize properly its fixed assets in business operations.

Fixed Assets—Tangible property used in the operations of an entity, but not expected to be consumed or converted into cash in the ordinary course of events. With a life in excess of one year, not intended for resale to customers, and subject to depreciation (with the exception of land), they are usually referred to as property, plant, and equipment.

Fixed Assets to Total Debt Ratio—This ratio is a measure of the amount of fixed assets owned by a company which are available to reduce the company's debt is a possible liquidation. The higher this ratio, the higher the protection to creditors.

Fixed Carbon—The nonvolatile matter in coal minus the ash. Fixed carbon is the solid residue other than ash obtained by prescribed methods of destructive distillation of a coal. Fixed carbon is the part of the total carbon that remains when coal is heated in a closed vessel until all matter is driven off.

Fixed Charge Coverage—The ratio of earnings available to pay so-called fixed charges to such fixed charges. Fixed charges include interest on funded debt, including leases, plus the related amortizations of debt discount, premium, and expense. Earnings available for fixed charges may be computed before or after deducting income taxes. Occasionally credits for the "allowance for funds used during construction" are excluded from the earnings figures. The precise procedures followed in calculating fixed charge or interest coverages vary widely.

Fixed Cost (Expense)—An expenditure or expense that does not vary with volume level of activity.

Fixed Operating Costs—Costs other than those associated with capital investment that do not vary with the operation, such as maintenance and payroll.

Fixed Purchase Option—An option contained in the lease agreement allowing the lessee (but not requiring) to purchase the equipment at a predetermined price at end of the lease term.

Fixed Tilt Array—A solar PV array set at a fixed angle to the horizontal.

Fixture—A thing which was originally Personal Property but which has become attached to and is considered as part of the Real Property.

Fixture-Lighting—A Lighting Fixture is the component of a Luminaire that houses the lamp(s), positions the lamp, shields it from view and distributes the light. The fixture also provides for connection to the power supply, which may require the use of a Ballast.

Flagging—Noticeable deformation of trees from prevailing winds. Flagging is an indication of an effective wind site. Lack of flagging is not necessarily an indication of a poor wind site.

Flame Spread Classification—A measure of the surface burning characteristics of a material.

Flame Spread Rating—A measure of the relative flame spread, and smoke development, from a material being tested. The flame spread rating is a single number comparing the flame spread of a material with red oak, arbitrarily given the number 100 and asbestos cement board with a flame spread of 0. Building codes require a maximum flame spread of 25 for insulation installed in exposed locations.

Flare—A combustion device that uses an open flame to burn combustible gases with combustion air provided by uncontrolled ambient air around the flame.

Flare Gas—Unwanted natural gas that is disposed of by burning as it is released from an oil field or refinery.

Flared—Gas disposed of by burning in flares usually at the production sites or at gas processing plants.

Flared Natural Gas—See *Flared* above.

Flaring—Burning of gas for the purpose of safe disposal.

Flashing—Metal, usually galvanized sheet metal, used to provide protection against infiltration of precipitation into a roof or exterior wall; usually placed around roof penetrations such as chimneys.

Flashpoint—The minimum temperature at which sufficient vapor is released by a liquid or solid (fuel) to form a flammable vapor-air mixture at atmospheric pressure.

Flash-Steam Geothermal Plants—When the temperature of the hydrothermal liquids is over 350 F (177 C), flash-steam technology is generally employed. In these systems, most of the liquid is flashed to steam. The steam is separated from the remaining liquid and used to drive a turbine generator. While the water is returned to the geothermal reservoir, the economics of most hydrothermal flash plants are improved by using a dual-flash cycle, which separates the steam at two different pressures. The dual-flash cycle produces 20% to 30% more power than a single-flash system at the same fluid flow.

Flat and Meter Rate Schedule—An electric rate schedule consisting of two components, the first of which is a service charge and the second a price for the energy consumed.

Flat Demand Rate Schedule—An electric rate schedule based on billing demand that provides no charge for energy.

Flat Plate—A device used to collect solar energy. It is a piece of metal painted black on the side facing the sun, to absorb the sun's heat. A solar PV array or module that does not contain concentrating devices and so responds to both direct and diffuse sunlight.

Flat Plate Collector—A device used to trap the sun's heat and transfer it to a fluid. Collectors for low temperature applications (pool heating) can be made of plastic. For higher temperature applications (water and space heating) the collectors are usually made of copper and aluminum and the collector surfaces have a selective surface coating.

They are typically enclosed in a glass-covered insulated box facing south.

Flat Plate Photovoltaics—Refers to a PV array or module that consists of non-concentrating elements. Flat-late arrays and modules use direct and diffuse sunlight, but if the array is fixed in position, some portion of the direct sunlight is lost because of oblique sun-angles in relation to the array.

Flat Plate Pumped—A medium-temperature solar thermal collector that typically consists of a metal frame, glazing, absorbers (usually metal), and insulation and that uses a pumped liquid as the heat-transfer medium: predominant use is in water-heating applications.

Flat Plate Solar Photovoltaic Module—An arrangement of photovoltaic cells or material mounted on a rigid flat surface with the cells exposed freely to incoming sunlight.

Flat Plate Solar Thermal/Heating Collectors—Large, flat boxes with glass covers and dark-colored metal plates inside that absorb and transfer solar energy to a heat transfer fluid. This is the most common type of collector used in solar hot water systems for homes or small businesses.

Flat Rate—A fixed charge for goods and services that does not vary with changes in the amount used, volume consumed, or units purchased.

Flat Rate Service—A rate payment plan by which customer pays a single monthly, quarterly, or annual amount for unlimited use of a utility service.

Flat Roof—A slightly sloped roof, usually with a tar and gravel cover. Most commercial buildings use this kind of roof.

Flat-Black Paint—Non-glossy paint with a relatively high absorptance.

Fleet Vehicle—Any motor vehicle a company owns or leases that is in the normal operations of a company. Vehicles which are used in the normal operation of a company, but are owned by company employees are not fleet vehicles. If a company provides services in addition to providing natural gas, only those vehicles that are used by the natural gas provider portion of a company should be counted as fleet vehicles. Vehicles that are considered "off-road" (e.g., farm or construction vehicles) or demonstration vehicles are not to be counted as fleet vehicles. Fleet vehicles include gasoline/diesel powered vehicles and alternative-fuel vehicles.

Flex Space—Industrial space designed to allow flexible conversion of warehouse or manufacturing space to a higher percentage of office space. Alternatively known as a service center or tech space.

Flexibility Mechanisms—The Kyoto Protocol has provisions that allow for flexibility in how, where, and when emissions reductions are made via three mechanisms: the Clean Development Mechanism, International Emission Trading and Joint Implementation. These mechanisms have been established to increase flexibility and hence reduce the costs of reducing emissions.

Flexible Fuel Vehicle—A vehicle that can operate on:

1. alternative fuels (such as M85 or E85)
2. 100-percent petroleum-based fuels
3. any mixture of an alternative fuel (or fuels) and a petroleum-based fuel.

Flexible fuel vehicles have a single fuel system to handle alternative and petroleum-based fuels. Flexible fuel vehicle and variable fuel vehicle are synonymous terms.

Flexible Load Shape—The ability to modify a utility's load shape on short notice. When resources are insufficient to meet load requirements, load shifting or peak clipping may be appropriate.

Flexible Retail PoolCo—This provides a model for the restructured electric industry that features an Independent System Operator (ISO) operating in parallel with a commercial Power Exchange, which allows end-use customers to buy from a spot market or "pool" or to contract directly.

Flexicoking—A thermal cracking process which converts heavy hydrocarbons such as crude oil, tar sands bitumen, and distillation residues into light hydrocarbons. Feedstocks can be any pumpable hydrocarbons, including those containing high concentrations of sulfur and metals.

Flipping—The act of buying shares in an IPO and selling them immediately for a profit. Brokerage firms underwriting new stock issues tend to discourage flipping and will often try to allocate shares to investors who intend to hold on to the shares for some time. However, the temptation to flip a new issue once it has risen in price sharply is too irresistible for many investors who have been allocated shares in a hot issue.

FLM—Federal Land Manager

Float Charge—A battery charge current that is equal too, or slightly greater than, the self-discharge rate. A low rate of charge that will maintain a battery at a full state of change without overcharging the battery.

Float Life—The number of years that a battery can keep its stated capacity when it is kept at float charge.

Float Service—A battery operation in which the battery is normally connected to an external current source; for instance, a battery charger which supplies the battery load< under normal conditions, while also providing enough energy input to the battery to make up for its internal quiescent losses, thus keeping the battery always up to full power and ready for service.

Floating Rental Rate—Rental that is subject to upward or downward adjustments during the lease term. Floating rents sometimes are adjusted in proportion to prime interest rate, commercial paper rate or other changes in the cost of money during the term of the lease.

Float-Zone Process—In reference to solar photovoltaic cell manufacture, a method of growing a large-size, high-quality crystal whereby coils heat a polycrystalline ingot placed atop a single-crystal seed. As the coils are slowly raised the molten interface beneath the coils becomes a single crystal.

Flooded Cell Battery—A form of rechargeable battery where the plates are completely immersed in a liquid electrolyte. Most cars use flooded-cell batteries. Flooded cell batteries are the most commonly used type for independent and remote area power supplies.

Floodplain—Relatively flat surfaces adjacent to active stream or river channels, formed by deposition of sediments during major floods. The floodplain may be covered by water during floods:

- 100-year floodplain: That area that would be covered by water during the 100-year flood event.
- Historic floodplain: An area larger than the 100-year floodplain.

Floor—A contract between a buyer and seller that assures the seller will receive a minimum price.

- The upward facing structure of a building.

Floor (Coal)—The upper surface of the stratum underlying a coal seam. In coals that were formed in persistent swamp environments, the floor is typically a bed of clay, known as "underclay," representing the soil in which the trees or other coal-forming swamp vegetation was rooted.

Floor Area—The floor area (in square feet) of enclosed conditioned or unconditioned space on all floors of a building, as measured at the floor level of the exterior surfaces of exterior walls enclosing the conditioned or unconditioned space. See *Conditioned Floor Area*.

Floor Price—A price specified in a market-price contract as the lowest purchase price of the uranium, even if the market price falls below the specified price. The floor price may be related to the seller's production costs.

Floor Space—The area enclosed by exterior walls of a building, including parking areas, basements, or other floors below ground level. It is measured in square feet.

Floor, Wall, or Pipeless Furnace—Space-heating equipment consisting of a ductless combustor or resistance unit, having an enclosed chamber where fuel is burned or where electrical-resistance heat is generated to warm the rooms of a building. A floor furnace is located below the floor and delivers heated air to the room immediately above or (if under a partition) to the room on each side. A wall furnace is installed in a partition or in an outside wall and delivers heated air to the rooms on one or both sides of the wall. A pipeless furnace is installed in a basement and delivers heated air through a large register in the floor of the room or hallway immediately above.

Floor/Soffit Type—A type of floor/soffit assembly having a specific heat capacity, framing type, and U-value.

Flow—In hydro-electric terms, flow refers to the quantity of water supplied to a water source or exiting a nozzle per unit of time. Commonly measured in gallons per minute.

Flow Condition—In reference to solar thermal collectors, the condition where the heat transfer fluid is flowing through the collector loop under normal operating conditions.

Flow Control—The laws, regulations, and economic incentives or disincentives used by waste managers to direct waste generated in a specific geographic area to a designated landfill, recycling, or waste-to-energy facility.

Flow Rate—The amount of water that moves through an area (usually pipe) in a given period of time. Usually measured in volume/minute or volume/hour. (1 litre/min = 0.264 US gallons/minute)

Flow Restrictor—A water and energy conserving device that limits the amount of water that a faucet or shower head can deliver.

Flow-Through Method—An accounting method under which decreases or increases in state or federal income taxes resulting from the use of liberalized depreciation and the Investment Tax Credit for income tax purposes are carried down to net income in the year in which they are realized. For rate-making purposes, the flow-through method passes on savings from liberalized depreciation and investment credit to the benefit of rate payers through lower rates.

FLP—Flash Point

FLPMA—Federal Land Policy and Management Act (1976)

Flue—The structure (in a residential heating appliance, industrial furnace, or power plant) into which combustion gases flow and are contained until they are emitted to the atmosphere.

Flue Gas—Gas that is left over after fuel is burned and which is disposed of through a pipe or stack to the outer air.

Flue Gas Desulphurization—Equipment used to remove sulfur oxides from the combustion gases of a boiler plant before discharge to the atmosphere. Also referred to as scrubbers. Chemicals such as lime are used as scrubbing media.

Flue Gas Desulphurization Unit (Scrubber)—Equipment used to remove sulfur oxides from the combustion gases of a boiler plant before discharge to the atmosphere. Chemicals such as lime are used as the scrubbing media.

Flue Gas Particulate Collector—Equipment used to remove fly ash from the combustion gases of a boiler plant before discharge to the atmosphere. Particulate collectors include electrostatic precipitators, mechanical collectors (cyclones), fabric filters (baghouses), and wet scrubbers.

Fluffing—The practice of installing blow-in, loose-fill insulation at a lower density than is recommended to meet a specified R-Value.

Fluid Catalytic Cracking—The refining process of breaking down the larger, heavier, and more complex hydrocarbon molecules into simpler and lighter molecules. Catalytic cracking is accomplished by the use of a catalytic agent and is an effective process for increasing the yield of gasoline from crude oil. Catalytic cracking processes fresh feeds and recycled feeds.

Fluid Catalytic Cracking Unit (FCCU)—A process unit in a refinery in which petroleum derivative feedstock is charged and fractured into smaller molecules in the presence of a catalyst; or reacts with a contact material to improve feedstock quality for additional process; and the catalyst or contact material is regenerated by burning off coke and other deposits. The unit includes, but is not limited to, the riser, reactor, regenerator, air blowers, spent catalyst, and all equipment for controlling air pollutant emissions and recovering heat.

Fluid Catalytic Cracking Unit Regenerator—The portion of the fluid catalytic cracking unit (*FCCU above*) in which coke burn-off and catalysts regeneration occurs, and includes the regenerator combustion air blower(s).

Fluid Coking—A thermal cracking process utilizing the fluidized-solids technique to remove carbon (coke) for continuous conversion of heavy, low-grade oils into lighter products.

Fluidized Bed Combustion (FBC)—A process for burning powdered coal that is poured in a liquid-like stream with air or gases. The process reduces sulfur dioxide emissions from coal combustion.

Fluidized-Bed Boiler—A large, refractory-lined vessel with an air distribution member or plate in the bottom, a hot gas outlet in or near the top, and some provisions for introducing fuel. The fluidized bed is formed by blowing air up through a layer of inert particles (such as sand or limestone) at a rate that causes the particles to go into suspension and continuous motion. The super-hot bed material increased combustion efficiency by its direct contact with the fuel.

Fluidized-Bed Combustion—A method of burning particulate fuel, such as coal, in which the amount of air required for combustion far exceeds that found in conventional burners. The fuel particles are continually fed into a bed of mineral ash in the proportions of 1 part fuel to 200 parts ash, while a flow of air passes up through the bed, causing it to act like a turbulent fluid.

Fluorescent Lamp—A tubular electric lamp that is coated on its inner surface with a phosphor and that contains mercury vapor whose bombardment by electrons from the cathode provides ultraviolet light which causes the phosphor to emit visible light either of a selected color or closely approximating daylight.

Fluorescent Light—A form of lighting that uses long thin tubes of glass which contain mercury vapor and various phosphor powders (chemicals based on phosphorus) to produce white light. Generally considered to be the most efficient form of home lighting.

The conversion of electric power to visible light by using an electric charge to excite gaseous atoms in a glass tube. These atoms emit ultraviolet radiation that is absorbed by a phosphor coating on the walls of the lamp tube. The phosphor coating produces visible light.

Fluorescent Lighting (other than compact fluorescent bulbs)—In fluorescent lamps, energy is converted to light by using an electric charge to "excite" gaseous atoms within a fluorescent tube. Common types are "cool white," "warm white," etc. Special energy efficient fluorescent lights have been developed that produce the same amount of light while consuming less energy.

Fluorocarbon—Carbon-fluorine compounds that often contain other elements such as hydrogen, chlorine, or bromine. Common fluorocarbons include chlorofluorocarbons and related compounds (ozone depleting substances), hydrofluorocarbons (HFCs), and perfluorocarbons (PFCs).

Fluorocarbon Gases—Propellants used in aerosol products and refrigerants that are believed to be causing depletion of the earth's ozone shield. See *CFCs*.

Flushout—A process used to remove VOCs from a building by operating the building's HVAC system at 100 percent outside air for a specific period of time.

Fluvial—Pertaining to, or produced by, stream action.

Flux—The rate of the energy flow per unit area.

Flux Material—A substance used to promote fusion, e.g., of metals or minerals.

Fly Ash—Particulate matter mainly from coal ash in which the particle diameter is less than 1×10^4 meter and which may also have metals attached to them, are carried up the stack of a combustion unit with gases during combustion. This ash is removed from the flue gas using flue gas particulate collectors such as fabric filters and electrostatic precipitators.

Flywheel Effect—The damping of interior temperature fluctuations by massive construction.

FMC—Federal Maritime Commission

FME—Free Market Economies. Countries that are members of the Council for Mutual Economic Assistance (CMEA) are not included.

FMFIA—Federal Managers Financial Integrity Act (1982)

FML—Flexible Membrane Liner

FMP—Facility Management Plan
  • Financial Management Plan

FMS—Financial Management System

FMV—Fair market value

FNS—Federal Network for Sustainability

FO—Facilities Office

Foam (Insulation)—A high R-value insulation product usually made from urethane that can be injected into wall cavities, or sprayed onto roofs or floors, where it expands and sets quickly.

Foam Board—A plastic foam insulation product, pressed or extruded into board-like forms, used as sheathing and insulation for interior basement or crawl space walls or beneath a basement slab; can also be used for exterior applications inside or outside foundations, crawl spaces, and slab-on-grade foundation walls.

Foam Core Panels—A type of structural, insulated product with foam insulation contained between two facings of drywall, or structural wood composition boards such as plywood, waferboard, and oriented strand board.

FOB—Freight on Board

FOI—Freedom of Information

FOIA—Freedom of Information Act

FOM—Field Operations Manual

FONSI—See *Finding of No Significant Impact*.

Food Miles—Refers to the distance foodstuffs travel through the various stages of production and processing to the point at which they reach the consumer. A measure of both distance traveled and mode of transportation allows comparisons of the energy use and the contribution to greenhouse emissions associated with various food products and their origin.

Food Preparation Equipment—Cooking equipment intended for commercial use, including coffee machines, espresso coffee makers, conductive cookers, food warmers including heated food servers, fryers, griddles, nut warmers, ovens, popcorn makers, steam kettles, ranges, and cooking appliances for use in commercial kitchens, restaurants, or other business establishments where food is dispensed.

Foot Candle—A Foot Candle is a standard I-P reference unit used when measuring intensity of light at a location. One Foot Candle equals the total intensity of light that falls upon a one square foot surface that is placed one foot away from a point source of light.

Foot Pound—The amount of work done in raising one pound one foot.

Footage Drilled—Total footage for wells in various categories, as reported for any specified period. Includes: (1) the deepest total depth (length of well bores) of all wells drilled from the surface, (2) the total of all bypassed footage drilled in connection with reported wells, and (3) all new footage drilled for directional 'sidetrack' wells. Footage reported for directional 'sidetrack' wells do not include footage in the common bore; reported as footage for the original well. In the case of old wells drilled deeper, the reported footage is that which was drilled below the total depth of the old well. Deepest Total Depth. The deepest total depth of a given well is the distance from a surface reference point (usually the Kelly bushing) to the point of deepest penetration measured along the well bore. If a well is drilled from a platform or barge over water, the depth of the water is included in the total length of the well bore. Sidetrack Drilling. This is a remedial operation that results in the creation of a new section of well bore for the purpose of (1) detouring around junk, (2) re-drilling lost hole, or (3) straightening key seats and crooked holes. Directional 'sidetrack' wells do not include footage in the common bore that is reported as footage for the original well.

FOP—Fuel Oil Price

Force—The push or pull that alters the motion of a moving body or moves a stationary body; the unit of force is the dyne or poundal; force is equal to mass time velocity divided by time.

Force Majeure—The title of a standard clause in almost all contracts exempting the parties for non-fulfillment of their obligations as a result of conditions beyond their control, such as earthquakes, floods, or war.

Forced Air System or Furnace—A type of heating system in which heated air is blown by a fan through air channels or ducts to rooms.

Forced Air Unit (FAU)—A central furnace equipped with a fan or blower that provides the primary means for circulation of air.

Forced Outage—The shutdown of a generating unit, transmission line, or other facility for emergency reasons or a condition in which the generating equipment is unavailable for load due to unanticipated breakdown.

Forced Ventilation—A type of building ventilation system that uses fans or blowers to provide fresh air to rooms when the forces of air pressure and gravity are not enough to circulate air through a building.

Foreclosure—A proceeding to enforce a lien by a sale of real or personal property in order to satisfy the debt.

Foreign (legal)—The term "foreign" when applied to a corporation or partnership means a corporation or partnership which is not domestic. *(see "Domestic")*

Foreign Access—Refers to proven reserves of crude,

condensate, and natural gas liquids applicable to long-term supply agreements with foreign governments or authorities in which the company or one of its affiliates acts as producer.

Foreign Currency Transaction Gains and Losses—Gains or losses resulting from the effect of exchange rate changes on transactions denominated in currencies other than the functional currency (for example, a U.S. enterprise may borrow Swiss francs or a French subsidiary may have a receivable denominated in kroner from a Danish customer). Gains and losses on those foreign currency transactions are generally included in determining net income for the period in which exchange rates change unless the transaction hedges a foreign currency commitment or a net investment in a foreign entity. Inter-company transactions of a long-term investment nature are considered part of a parent's net investment and hence do not give rise to gains or losses.

Foreign Currency Translation Effects—Gains or losses resulting from the process of expressing amounts denominated or measured in one currency in terms of another currency by use of the exchange rate between the two currencies. This process is generally required to consolidate the financial statements of foreign affiliates into the total company financial statements and to recognize the conversion of foreign currency or the settlement of a receivable or payable denominated in foreign currency at a rate different from that at which the item is recorded. Translation adjustments are not included in determining net income, but are disclosed as separate components of consolidated equity.

Foreign Operations—These are operations that are located outside the United States. Determination of whether an enterprise's mobile assets, such as off-shore drilling rigs or ocean-going vessels, constitute foreign operations should depend on whether such assets are normally identified with operations located outside the United States. Foreign operations are segregated into the following areas for FRS reporting purposes: OECD Europe. Includes Austria, Belgium, Denmark, Finland, France, the Federal Republic of Germany, Greece, Iceland, Ireland, Italy, Luxembourg, the Netherlands, Norway, Portugal, Spain, Sweden, Switzerland, Turkey, and the United Kingdom. Former Soviet Union (FSU) and East Europe. The Baltic States of Estonia, Latvia, and Lithuania, as well as Armenia, Azerbaijan, Belarus, Georgia, Kazakhstan, Kyr-gystan, Moldova, Russia, Tajikistan, Turkmenistan, Ukraine, Uzbekistan, Albania, Bulgaria, Czech Republic, Hungary, Poland, Romania, Slovakia, and Yugoslavia. Middle East. Includes Saudi Arabia, the United Arab Emirates, Iraq, Iran, Kuwait, the Iraq- Saudi Arabia Neutral Zone, Qatar, Dubai, Bahrain, Oman, Yemen, Syria, Jordan, and Israel. Canada. Africa (the African continent). Other Eastern Hemisphere. Areas eastward of the Greenwich prime meridian to 180 degrees longitude and not included in other specified domestic or foreign classifications. Other Western Hemisphere. Areas westward of the Greenwich prime meridian to 180 degrees longitude not included in other domestic or foreign classifications.

Foreign Sales Corporation (FSC)—A legal entity created to export U.S. manufactured goods to a foreign country. If a FSC meets the qualifications for FSC status, a portion of its income is exempt from current U.S. taxation. There are two types of FSC structures: Commission FSCs and Ownership FSCs.

Foreign Source Income—Net income earned overseas that is reported to the IRS for federal income tax purposes.

Foreign Tax Credit—Also called FTC. Under § 901 of the U.S. tax code, a foreign tax credit is allowed for taxes paid or accrued to a foreign country or U.S. possession during the taxable year.

Foreign-Controlled Firms (Coal)—Foreign-controlled firms are U.S. coal producers with more than 50 percent of their stock or assets owned by a foreign firm.

Forest Health—A condition of ecosystem sustainability and attainment of management objectives for a given forest area. Usually considered to include green trees, snags, resilient stands growing at a moderate rate and endemic levels of insects and disease. Natural processes still function or are duplicated through management intervention. A more fire-tolerant forest condition and the elimination of unnatural woody biomass accumulations that have resulted from past fire suppression.

Forest Plan—The document that sets goals, objectives, desired future condition, standards and guidelines, and overall programmatic direction for a National Forest. Required by the National Forest Management Act of 1976.

Forest Residues—Material not harvested or removed from logging sites in commercial hardwood and softwood stands as well as material resulting from forest management operations such as pre-commercial thinnings and removal of dead and dying trees.

Forested Areas or Land—Any land that is capable of producing or has produced forest growth or, if lacking forest growth, has evidence of a former forest and is not now in other use.

Forfeiture—A loss of some right, title, estate, or interest in consequence of Default under an obligation.

Form 10-K—This is the annual report that most reporting companies file with the Securities & Exchange Commission. It provides a comprehensive overview of the registrant's business.

Form 10-KSB—This is the annual report filed by reporting "small business issuers." It provides a comprehensive overview of the company's business, although its requirements call for slightly less detailed information than required by Form 10-K.

Form S-4—Type of Registration Statement under which public company mergers and security exchange offers may be registered with the SEC.

Form SB-2—This form may be used by "small business issuers" to register securities to be sold for cash. This form requires less detailed information about the issuer's business than Form S-1.

Formaldehyde—A chemical used as a preservative and in bonding agents. It is found in household products such as plywood, furniture, carpets, and some types of foam insulation. It is also a by-product of combustion and is a strong-smelling, colorless gas that is an eye irritant and can cause sneezing, coughing, and other health problems.

Forward Contract—Purchase or sale of a specific quantity of reductions, offsets, or allowances at the current or spot price, with delivery and settlement scheduled for a specified future date. The seller of a forward stream of emission reductions must make physical delivery and the buyer must pay for contracted reductions.

Forward Cost (Uranium)—1.   Forward costs are those operating and capital costs yet to be incurred at the time an estimate of reserves is made. Profits and "sunk" costs, such as past expenditures for property acquisition, exploration, and mine development, are not included. Therefore, the various forward-cost categories are independent of the market price at which uranium produced from the reserves would be sold.

2.   The operating and capital costs still to be incurred in the production of uranium from in-place reserves. By using forward costing, estimates for reserves for ore deposits in differing geological settings and status of development can be aggregated and reported for selected cost categories. Included are costs for labor, materials, power and fuel, royalties, payroll taxes, insurance, and applicable general and administrative costs. Excluded from forward cost estimates are prior expenditures, if any, incurred for property acquisition, exploration, mine development, and mill construction, as well as income taxes, profit, and the cost of money. Forward costs are neither the full costs of production nor the market price at which the uranium, when produced, might be sold.

Forward Coverage—Amount of uranium required to assure uninterrupted operation of nuclear power plants.

Forward Market—A forward market deals in forward contracts which are agreements to buy or sell an asset at a certain time in the future for a certain price. They generally constitute a private agreement between two entities including a mutually agreed delivery date. Forward contracts are not marked to market daily like futures contracts. As a comparison, most futures contracts are marked to market daily on an exchange and typically have a range of delivery dates. Whereas most futures contracts are closed out prior to delivery, most forward contracts do lead to delivery of the physical asset or to final settlement in cash. Entities can reverse their forward position by entering into an offsetting transaction. However, forward contracts generally cannot be unsold.

Forward Settlement—Purchase or sale of a specific quantity of reductions, offsets, or allowances at the current or spot price, with delivery and settlement scheduled for a specified future date.

Forwarder—A self-propelled vehicle to transport harvested material from the stump area to the landing. Trees, logs, or bolts are carried off the ground on a stake-bunk, or are held by hydraulic jaws of a clam-bunk. Chips are hauled in a dumpable or open-top bin or chip-box.

Fossil Dismantlement—The dismantlement and disposal of all buildings, structures, equipment, tanks and stacks at the site and restoration of the site to a usable condition.

Fossil Fuel—Solid, liquid or gaseous fuels formed in the ground after millions of years by chemical and physical changes in plant and animal residues under high temperature and pressure. Oil, natural gas and coal are fossil fuels. All are carbon based.

Fossil Fuel Electric Generation—Electric generation

in which the prime mover is a turbine rotated by high-pressure steam produced in a boiler by heat from burning fossil fuels.

Fossil Fuel Plant—A plant using coal, petroleum, or gas as its source of energy.

Fossil Fuel Steam-electric Power Plant—An electricity generation plant in which the prime mover is a turbine rotated by high-pressure steam produced in a boiler by heat from burning fossil fuels.

Fossil Fuels—Fuels formed in the ground from the remains of dead plants and animals. It takes millions of years to form fossil fuels. Oil, natural gas, and coal are fossil fuels.

Foundation—The supportive structure of a building.

Founders Shares—Shares owned by a company's founders upon its establishment.

Foundry—An operation where metal castings are produced, using coke as a fuel.

Foundry Coke—This is a special coke that is used in furnaces to produce cast and ductile iron products. It is a source of heat and also helps maintain the required carbon content of the metal product. Foundry coke production requires lower temperatures and longer times than blast furnace coke.

FP—Fine Particulate

FPA—Federal Pesticide Act (1978)
- Federal Power Act: Enacted in 1920, and amended in 1935, the Act consists of three parts. The first part incorporated the Federal Water Power Act administered by the former Federal Power Commission, whose activities were confined almost entirely to licensing non-Federal hydroelectric projects. Parts II and III were added with the passage of the Public Utility Act. These parts extended the Act's jurisdiction to include regulating the interstate transmission of electrical energy and rates for its sale as wholesale in interstate commerce. The Federal Energy Regulatory Commission is now charged with the administration of this law.

FPAS—Foreign Purchase Acknowledge Statements

FPC—Federal Power Commission www.ferc.gov

FPPA—Federal Pollution Prevention Act (1990)

FPR—Federal Procurement Regulation
- Final proposal revision

FQPA—Food Quality Protection Act (1996)

FR—Federal Register
- Final Rule Making

FRA—Federal Register Act

FRAC—Gas industry term used to refer to the method used to increase the deliverability of a production or underground storage well by pumping a liquid or other substance into a well under pressure to crack (fracture) and prop open the gas-bearing formation.

Fractional Distillation—The process of refining crude oil into various oil products. The various products are separated out in the order of their boiling points.

Fractional Horse Power Motor—An electric motor rated at less than one horse power (hp).

Fractionation—The process by which saturated hydrocarbons are removed from natural gas and separated into distinct products, or "fractions," such as propane, butane, and ethane.

Fracturing—A process of opening up underground channels in hydrocarbon-bearing formations by force rather than by chemical action such as in acidizing. High pressure is hydraulically or explosively directed at the rock, causing it to fracture.

Frame (Window)—The outer casing of a window that sits in a designated opening of a structure and holds the window panes in place.

Framed Partition or Assembly—A partition or assembly constructed using separate structural members spaced not more than 32 inches on center.

Framework Convention on Climate Change (FCCC)—An agreement opened for signature at the "Earth Summit" in Rio de Janeiro, Brazil, on June 4, 1992, which has the goal of stabilizing greenhouse gas concentrations in the atmosphere at a level that would prevent significant anthropogenically forced climate change.

Framing—The structural materials and elements used to construct a wall.

Framing Effects—The effect of framing (wood or metal studs, joists, beams, etc.) on the overall U-value of a wall, roof, floor, window or other building surface. Framing generally increases the U-Value and decreases the R-Value of insulated surfaces.

Framing Percentage—The area of actual framing in an envelope assembly divided by the overall area of the envelope assembly. This percentage is used to calculate the overall U-value of an assembly.

Franchise—A right or privilege conferred by Law or Contract.

Franchise Area—This is the territory in which a utility system supplies service to customers.

Franchise Monopoly—Under this system, a utility has the right to be the sole or principal supplier of electric power at a retail level in a specific region or area knows as the franchise service territory. In return for its sole supplier privilege, the utility has an obligation to serve anyone who requests service, and agrees to be accountable to state and/or federal regulatory bodies that regulate the utility's performance, accounting procedures, pricing structures, and plant planning and siting.

Franchising—The right or license granted to an individual or group to market a company's goods or services in a particular territory.

Francis Turbine—A type of hydropower turbine that contains a runner that has water passages through it formed by curved vanes or blades. As the water passes through the runner and over the curved surfaces, it causes rotation of the runner. The rotational motion is transmitted by a shaft to a generator.

FRB—Federal Reserve Board www.frb.gov

FRC—Federal Records Center
- Functional Residual Capacity

FREE—Fund for Renewable Energy and the Environment

Free Alongside Ship (F.A.S.)—The value of a commodity at the port of exportation, generally including the purchase price plus all charges incurred in placing the commodity alongside the carrier at the port of exportation.

Free Cash Flow—The cash flow of a company available to service the capital structure of the firm. Typically measured as operating cash flow less capital expenditures and tax obligations.

Free on Board (F.O.B.)—A sales transaction in which the seller makes the product available for pick up at a specified port or terminal at a specified price and the buyer pays for the subsequent transportation and insurance.

Free Well—A well drilled and equipped by an assignee as consideration for the assignment of a fractional share of the working interest, commonly under a farm-out agreement.

Freestanding Tower—A wind generator tower with no guy wires. This can be either a lattice tower or a monopole. Freestanding towers are the most expensive type of tower, requiring large excavations and large amounts of concrete.

French Drain System—A pit or trench filled with crushed rock and used to collect and divert stormwater or wastewater. Most often, perforated piping at the bottom provides easy drainage.

Freon—A registered trademark for a chlorofluorocarbon (CFC) gas that is highly stable and that has been historically used as a refrigerant.

Frequency—The number of cycles or repetitions per unit time of a complete waveform (AC current), in electrical applications usually expressed in cycles per second or Hertz (Hz). Electrical equipment in the United States requires 60 Hz, in Europe 50Hz.

Frequency Regulation—An ancillary service category that provides support for maintaining grid stability within a defined range above or below 60 Hertz (a.k.a., 60 cycles per second).

Fresh Feed Input—Represents input of material (crude oil, unfinished oils, natural gas liquids, other hydrocarbons and oxygenates or finished products) to processing units at a refinery that is being processed (input) into a particular unit for the first time. Examples:
1. Unfinished oils coming out of a crude oil distillation unit that are put into a catalytic cracking unit are considered fresh feed to the catalytic cracking unit.
2. Unfinished oils coming out of a catalytic cracking unit being looped back into the same catalytic cracking unit to be reprocessed are not considered fresh feed.

Fresh Feeds—Crude oil or petroleum distillates that are being fed to processing units for the first time.

Fresnel Lens—An optical device for concentrating light that is made of concentric rings that are faced at different angles so that light falling on any ring is focused to the same point. A concentrating lens, positioned above and concave to a PV material to concentrate light on the material.

Friction Head—The energy lost from the movement of a fluid in a conduit (pipe) due to the disturbances created by the contact of the moving fluid with the surfaces of the conduit, or the additional pressure that a pump must provide to overcome the resistance to fluid flow created by or in a conduit.

FRM—Final Rule Making

FRN—Federal Register Notice
- Final Rulemaking Notice

From S-1—The form can be used to register securities for which no other form is authorized or prescribed, except securities of foreign governments or political sub-divisions thereof.

Front—The primary entry side of the building (front façade) used as a reference in defining the orientation of the building or unit plan. The orientation of the front facade may not always be the same as that for the front door itself.

FRS—Financial Reporting System Survey (EIA survey)
- Formal Reporting System

FS—Factor of Safety
- Feasibility Study

- Forest Service

FSA—Food Security Act (1985)
- Final Staff Assessment

FSC—Foreign Sales Corporation. A foreign subsidiary of a U.S. corporation which allows exemption of a portion of the income generated by the subsidiary for U.S. tax purposes.

FSC (Forest Stewardship Council)—FSC-certified wood is wood from a certified, well-managed forest. Any FSC product can be traced back to its original certified source. FSC has 12 accredited independent third-party certifiers worldwide that evaluate forest management activities and tracking of forest product. Many green building programs including LEED offer credit for using FSC-certified wood.

FSC Pure products are made completely of wood or fiber that originated from FSC-certified forests. FSC Mixed products are a combination of FSC-certified material, controlled wood and/or reclaimed wood.

FT—Full time
- Fischer-Tropsch process of converting methane, biomass, or coal to liquid fuels

FTA—Federal Transit Administration

FTC—Federal Trade Commission

FTE—Full Time Equivalent

FTP—Federal Test Procedure

FTS Gas—Firm Transportation Gas

FTT—Full Time Temporary

FUA—Fuel Use Act (1987)
- Fields Under Appraisal

FUD—Fields Under Development

Fuel—Any material substance that can be consumed to supply heat or power. Included are petroleum, coal, and natural gas (the fossil fuels), and other consumable materials, such as uranium, biomass, and hydrogen. Can be solid, liquid or gaseous combustible material.

Fuel Adjustment—A clause in the rate schedule that provides for adjustment of the amount of a bill as the cost of fuel varies from a specified base amount per unit. The specified base amount is determined when rates are approved. This item is shown on all customer bills and indicates the current rate for any adjustment in the cost of fuel used by the company. It can be a credit or a debit. The fuel adjustment lags two months behind the actual price of the fuel. For example, the cost of oil in January will be reflected in March's fuel adjustment.

Fuel Cell—A device or an electrochemical engine with no moving parts that converts the chemical energy of a fuel, such as hydrogen, and an oxidant, such as oxygen, directly into electricity. The principal components of a fuel cell are catalytically activated electrodes for the fuel (anode) and the oxidant (cathode) and an electrolyte to conduct ions between the two electrodes, thus producing electricity.

Fuel Cycle—The series of steps required to produce electricity. The fuel cycle includes mining or otherwise acquiring the raw fuel source, processing and cleaning the fuel, transport, electricity generation, waste management and plant decommissioning.
- Refers to the total life of a fuel in all of its uses and forms. The stages of a fuel cycle may include extraction or generation, transportation, combustion, air emissions, byproduct removal, further transportation, and/or disposal.

Fuel Diversity—A utility or power supplier that has power stations using several different types of fuel. Avoiding over-reliance on one fuel helps avoid the risk of supply interruption and price spikes.

Fuel Efficiency—The ratio of heat produced by a fuel for doing work to the available heat in the fuel. See *Miles per Gallon*

Fuel Emergencies—An emergency that exists when supplies of fuels or hydroelectric storage for generation are at a level or estimated to be at a level that would threaten the reliability or adequacy of bulk electric power supply. The following factors should be taken into account to determine that a fuel emergency exists:
1. Fuel stock or hydroelectric project water storage levels are 50 percent or less of normal for that particular time of the year and a continued downward trend in fuel stock or hydroelectric project water storage level is estimated; or
2. Unscheduled dispatch or emergency generation is causing an abnormal use of a particular fuel type, such that the future supply of stocks of that fuel could reach a level that threatens the reliability or adequacy of bulk electric power supply.

Fuel Escalation—The annual rate of increase of the cost of fuel, including inflation and real escalation, resulting from resource depletion, increased demand, etc.

Fuel Ethanol (CH_3-CH_2OH)—An anhydrous denatured aliphatic alcohol intended for gasoline blending as described in the definition of oxygenates. It is also used in high concentrations to produce E85.

Fuel Expenses—These costs include the fuel used in

the production of steam or driving another prime mover for the generation of electricity. Other associated expenses include unloading the shipped fuel and all handling of the fuel up to the point where it enters the first bunker, hopper, bucket, tank, or holder in the boiler-house structure.

Fuel Gas—Synthetic gas used for heating or cooling. It has less energy content than pipeline-quality gas.

Fuel Grade Alcohol—Usually refers to ethanol to 160 to 200 proof.

Fuel Handling System—A system for unloading wood fuel from vans or trucks, transporting the fuel to a storage pile or bin, and conveying the fuel from storage to the boiler or other energy conversion equipment.

Fuel Injection—A fuel delivery system whereby gasoline is pumped to one or more fuel injectors under high pressure. The fuel injectors are valves that, at the appropriate times, open to allow fuel to be sprayed or atomized into a throttle bore or into the intake manifold ports. The fuel injectors are usually solenoid operated valves under the control of the vehicle's on-board computer (thus the term "electronic fuel injection"). The fuel efficiency of fuel injection systems is less temperature-dependent than carburetor systems. Diesel engines always use injectors.

Fuel Mix—The proportions of each fuel type (e.g. nuclear, coal, solar electric, oil, wind, hydro, etc.) used by a power plant to generate electricity. The fuel mix is displayed on the Power Content Label (Green-e program)

Fuel Oil—A liquid petroleum product less volatile than gasoline, used as an energy source. Fuel oil includes distillate fuel oil (No. 1, No. 2, and No. 4), and residual fuel oil (No. 5 and No. 6).

Fuel Purchase Agreement—An agreement between a company and a fuel provider which stipulates that the company agrees to purchase its fuel from the fuel provider. If the company has a credit card for use at a fuel provider's locations, but is not bound by an additional agreement to purchase fuel from that provider, the credit card agreement alone is not considered a fuel purchase agreement.

Fuel Rate—The amount of fuel necessary to generate one kilowatt-hour of electricity.

Fuel Ratio—The ratio of fixed carbon to volatile matter in coal.

Fuel Reprocessing (Nuclear)—The means for obtaining usable, fissionable material from spent reactor fuel.

Fuel Rod (Nuclear)—A long slender tube that holds fissionable material (fuel) for nuclear reactor use. Fuel rods are assembled into bundles called fuel elements or assemblies, which are loaded individually into the reactor core.

Fuel Security—See *Energy Security*

Fuel Switching—Fuel switching is the substitution of conventional and existing technologies for more efficient and less carbon-intensive fuel technologies including repowering, upgrading instrumentation, controls, and/or equipment, more efficient utilization of fuel and fuel switching.

Fuel Switching Capability—The short-term capability of a manufacturing establishment to have used substitute energy sources in place of those actually consumed. Capability to use substitute energy sources means that the establishment's combustors (for example, boilers, furnaces, ovens, and blast furnaces) had the machinery or equipment either in place or available for installation so that substitutions could actually have been introduced within 30 days without extensive modifications. Fuel-switching capability does not depend on the relative prices of energy sources; it depends only on the characteristics of the equipment and certain legal constraints.

Fuel Wood—Wood and wood products, possibly including scrubs and branches, etc., bought or gathered, and used by direct combustion.

Fuel/Fabricator Assembly Identifier—Individual assembly identifier based on a numbering scheme developed by individual fuel fabricators. Most fuel fabricator assembly identifiers schemes closely match the scheme developed by the American National Standards Institute (ANSI) and are therefore unique.

Fuel-Cell Furnace—A variation of the Dutch oven design, that usually incorporates a primary and secondary combustion chamber (cell). The primary chamber is a vertical refractory-lined cylinder with a grate at the bottom in which combustion is partially completed. Combustion is completed in the secondary chamber.

Fuels Solvent De-asphalting—A refining process for removing asphalt compounds from petroleum fractions, such as reduced crude oil. The recovered stream from this process is used to produce fuel products.

Fuel-Switching DSM Program Assistance—DSM program assistance where the sponsor encourages consumers to change from one fuel to another for a particular end-use service. For example, utilities might encourage consumers to replace electric water heaters with gas units or encourage industrial consumers to use electric microwave heaters instead of natural gas-heaters.

Fuel-Use Attributes—Fuel-use attributes are important to utilities concerned about reliance on a single fuel or reduction in usage of a particular fuel. These attributes include annual fuel consumption by type and percent energy generation by fuel.

Fugitive Emissions—Unintended leaks of gases into the atmosphere from the extraction, processing or transportation of fossil fuels. For example, gas emissions from leaking pipelines or methane escaping from the ground during the mining of coal.
• Fugitive emissions are intentional or unintentional releases of gases from anthropogenic activities such as the processing, transmission or transportation of gas or petroleum. In particular, they may arise from the production, processing, transmission, storage and use of fuels, and include emissions from combustion only where it does not support a productive activity (e.g., flaring of natural gases at oil and gas production facilities).

Full Forced Outage—The net capability of main generating units that are unavailable for load for emergency reasons.

Full Hybrid—A hybrid electric vehicle capable of running on battery power only.

Full Power Day—The equivalent of 24 hours of full power operation by a reactor. The number of full power days in a specific cycle is the product of the reactor's capacity factor and the length of the cycle.

Full Power Operation—Operation of a unit at 100 percent of its design capacity. Full-power operation precedes commercial operation.

Full Ratchet Antidilution—The sale of a single share at a price less than the favored investors paid reduces the conversion price of the favored investors' convertible preferred stock "to the penny." For example, from $1.00 to 50 cents, regardless of the number of lower-priced shares sold.

Full Requirements Consumer—A wholesale consumer without other generating resources whose electric energy seller is the sole source of long-term firm power for the consumer's service area. The terms and conditions of sale are equivalent to the seller's obligations to its own retail service, if any.

Full Service Rent—An all-inclusive rental rate that includes operating expenses and real estate taxes for the first year. The tenant is generally still responsible for any increase in operating expenses over the base year amount.

Full Sun—The amount of power density in sunlight received at the earth's surface at noon on a clear day (about 1,000 Watts/square meter). Lower levels of sunlight are often expressed as 0.5 sun or 0.1 sun.

Full Term Rate—The rate that discounts the interim and base rents to the equipment cost at delivery date.

Full-Payout Lease—A lease in which the lessor recovers, through the lease payments, all costs incurred in the lease plus an acceptable rate of return, without any reliance upon the leased equipment's future residual value.

Full-Service Lease—A lease that includes additional services such as maintenance, insurance and property taxes that are paid for by the lessor, the cost of which is built into the lease payments.

Fully Diluted Earnings Per Share—Earnings per share expressed as if all outstanding convertible securities and warrants have been exercised.

Fully Diluted Outstanding Shares—The number of shares representing total company ownership, including common shares and current conversion or exercised value of the preferred shares, options, warrants, and other convertible securities.

Fumarole—A vent from which gas or steam issue; a geyser or spring that emits gases.

Functional Obsolescence—An element of depreciation resulting from deficiencies or superadequacies in the structure.

Functional Unbundling—The functional separation of generation, transmission, and distribution transactions within a vertically integrated utility without selling of "spinning off" these functions into separate companies.

Funds from Operations—Calculated by adding non-cash charges back to net income or contribution to net income. Deferred taxes and depreciation, depletion, and amortization (DD&A) are the largest non-cash charges.

Funds, Total Sources of—The total source of funds including net income plus non-cash charges such as DD&A and deferred taxes, issuance of stocks and bonds, and proceeds from the sale or property, plant, and equipment. The concept is similar to cash flow generated, but does not attempt to account for changes in working capital items. Thus, for example, an inventory buildup or drawdown would not be accounted for under the 'funds' concept since both cash and inventory are items of working capital.

Fungi—Plant-like organisms with cells with distinct nuclei surrounded by nuclear membranes, incapable of photosynthesis. Fungi are decomposers of waste organisms and exist as yeast, mold, or mildew.

Furling—The process of forcing, either manually or automatically, a wind turbines blades out of the direction of the wind in order to stop the blades from turning.

Furnace—An enclosed chamber or container used to burn biomass in a controlled manner to produce heat for space or process heating.

Furnace (Residential)—A combustion heating appliance in which heat is captured from the burning of a fuel for distribution, comprised mainly of a combustion chamber and heat exchanger.

Furnace Coke Plant—A coke plant whose coke production is used primarily by the producing company.

Furnaces That Heat Directly, Without Using Steam or Hot Water (similar to a residential furnace)—Furnaces burn natural gas, fuel oil, propane/butane (bottled gas), or electricity to warm the air. The warmed air is then distributed throughout the building through ducts. Many people use the words "boilers" and "furnaces" interchangeably. They are not the same. We mean that warm air is produced directly by burning some fuel.

• Warm-air furnaces typically rely on air ducts to carry the warm air throughout the building. Warm-air furnaces are often built in combination with central air-conditioning systems, so that they can use the same air ducts for both heating or air-conditioning (depending on the season).

• Other terms for describing this type of equipment include: "central system," "split system," and "forced air/forces air furnace."

Fuse—A safety device consisting of a short length of relatively fine wire, mounted in a holder or contained in a cartridge and connected as part of an electrical circuit. If the circuit source current exceeds a predetermined value, the fuse wire melts (i.e. the fuse 'blows') breaking the circuit and preventing damage to the circuit protected by the fuse.

Fusion Energy—A power source, now under development, based on the release of energy that occurs when atoms are combined under the most extreme heat and pressure. It is the energy process of the sun and the stars.

Futures Contract—This exchange-traded supply contract between a buyer and seller obligates the buyer to take delivery and obligates the seller to provide delivery of a fixed amount of a commodity at an established price at a specific location.

• Futures Contract is technically and functionally different from a Forward contract. It is an agreement to buy or sell a specific amount of a commodity or financial instrument at a certain time in the future for a particular price. The price is established between the buyer and seller on a commodity exchange via a standardized contract defined by the exchange. Futures Contracts typically have a range of delivery dates and are marked to market daily. Most Futures Contracts close out their position before maturity, either through an offsetting transaction or by selling the futures contract i.e. a Futures Contract is tradable in its own right. Futures Contracts are highly defined instruments usually based upon a strong cash market for the underlying commodity. At this stage, a greenhouse gas emissions futures market does not exist under the Kyoto Protocol—most transactions are forward contracts.

Futures Market—A trade center for quoting prices on contracts for the delivery of a specified quantity of a commodity at a specified time and place in the future.

FVMP—Federal Visibility Monitoring Program

FWCA—Fish and Wildlife Coordination Act

FWPCA—Federal Water Pollution Control Act (1972)

FWS—Fish and Wildlife Service

FY—Fiscal Year (financial year)

G

g/day—Grams per Day

g/m³—Micro grams (10^{-6} grams) per cubic meter

G/MI—Grams per Mile

GAAP—See *Generally Accepted Accounting Principles* below

GACT—Generally Available Control Technology

gal/d—Gallons per day

Gallium Arsenide—A compound used to make certain types of solar photovoltaic cells.

Gallon—A volumetric measure equal to 4 quarts (231 cubic inches) used to measure fuel oil. One barrel equals 42 gallons.

Gamma Radiation—A high-energy photon emitted from the nucleus of certain radioactive atoms. Gamma rays are the most penetrating of the three common types of radiation (the other two are alpha particles and beta particles) and are best stopped by dense materials such as lead.

GAO—General Accounting Office

Gap Width—The distance between glazings in multi-glazed systems. This is typically measured from inside surface to inside surface, though some manufacturers may report "overall" IG width, which is measured from outside surface to outside surface.

GAQM—Guideline on Air Quality Models

Garnishment—A Statutory Proceeding whereby property, money, or credits of a Debtor in possession of another are seized and applied to payment of the debt.

Gas—Gaseous fuel (usually natural gas) that is burned to produce heat energy. The word also is used, colloquially, to refer to gasoline.

Gas Broker—A company that finds customers who want to purchase gas and completes transactions with suppliers who want to sell gas.

Gas Bubble—An excess of natural gas deliverability relative to demand requirements at current prices.

Gas Cooled Fast Breeder Reactor (GCFB)—A fast breeder reactor that is cooled by a gas (usually helium) under pressure.

Gas Cooling Equipment—Cooling equipment that produces chilled water or cold air using natural gas or liquefied petroleum gas as the primary energy source.

Gas Day—A period of twenty-four (24) consecutive hours commencing at a specified hour on a given calendar day and ending at the same specified hour on the next succeeding calendar day.

Gas Engine—A piston engine that uses gaseous fuel rather than gasoline. Fuel and air are mixed before they enter cylinders; ignition occurs with a spark.

Gas Heating System—A natural gas or liquefied petroleum gas heating system.

Gas Imbalance—This occurs when the customer uses either more or less gas than was scheduled for a certain time period.

Gas Impurities—Undesirable matter in gas, such as dust, excessive water vapor, hydrogen sulphide, tar, and ammonia.

Gas Infills—Are air, argon, krypton, CO_2, SF_6, or a mixture of these gasses between the panes of glass in insulated glass units.

Gas Log—A self-contained, free-standing, open-flame, gas-burning appliance consisting of a metal frame or base supporting simulated logs, and designed for installation only in a vented fireplace.

Gas Oil—European and Asian designation for No. 2 heating oil and No. 2 diesel fuel.

Gas Plant—Any plant which performs one of the following functions: removing liquefiable hydrocarbons from wet gas or casinghead gas (gas processing); removing undesirable gaseous and particulate elements from natural gas (gas treatment); removing water or moisture from the gas stream (dehydration). Also, the original cost of property, plant and equipment owned and used by the utility in its gas operations and having an expectation of life in service of more than one year from the date of installation.

Gas Plant Operator—Any firm, including a gas plant owner, which operates a gas plant and keeps the gas plant records. A gas plant is a facility in which natural gas liquids are separated from natural gas or in which natural gas liquids are fractionated or otherwise separated into natural gas liquid products or both.

Gas Processing Unit—A facility designed to recover natural gas liquids from a stream of natural gas that may or may not have passed through lease separators and/or field separation facilities. Another function of natural gas processing plants is to control the quality of the processed natural gas stream. Cycling plants are considered natural gas processing plants.

Gas Producer—An individual or business that owns gas wells and sends the gas into the interstate or intrastate market.

Gas Research Institute (GRI)—An organization sponsored by a number of U.S. gas companies to investigate new sources of supply and new uses (applications) for natural gas. www.gastechnology.org

Gas Shift Process—A process in which carbon monoxide and hydrogen react in the presence of a catalyst to form methane and water.

Gas Supply Coordinator—A representative of a company assigned the task of managing the operations under Transportation, Sales or Purchase Service agreements. Responsibilities typically include scheduling activity, imbalance management and volume confirmation.

Gas Surcharge—An unbundled rate component included on gas customer bills to fund public purpose programs including energy efficiency, low-income services, and research and development.

Gas Synthesis—A method producing synthetic gas from coal. Also called the FISCHER-TROPSCH PROCESS.

Gas To Liquids (GTL)—A process that combines the carbon and hydrogen elements in natural gas molecules to make synthetic liquid petroleum products, such as diesel fuel.

Gas Turbine—A type of turbine in which combusted, pressurized gas is directed against a series of blades connected to a shaft, which forces the shaft to turn to produce mechanical energy.

Gas Turbine Plant—A plant in which the prime mover is a gas turbine. A gas turbine consists typically of an axial-flow air compressor and one or more combustion chambers where liquid or gaseous fuel is burned and the hot gases are passed to the turbine and where the hot gases expand to drive the generator and are then used to run the compressor.

Gas Utility—Any person engaged in, or authorized to engage in, distributing or transporting natural gas, including, but not limited to, any such person who is subject to the regulation of the Public Utilities Commission.

Gas Well—A well completed for production of natural gas from one or more gas zones or reservoirs. Such wells contain no completions for the production of crude oil.

Gas Well Productivity—Derived annually by dividing gross natural gas withdrawals from gas wells by the number of producing gas wells on December 31 and then dividing the quotient by the number of days in the year.

Gas, Conventional—Gas that can be produced with current technology at a cost that is no higher than its current market value.

Gas, Liquefied Petroleum (LPG)—A gas containing certain specific hydrocarbons which are gaseous under normal atmospheric conditions but can be liquefied under moderate pressure at normal temperatures. Propane and butane are the principal examples.

Gas, Natural—A naturally occurring mixture of hydrocarbon and nonhydrocarbon gases found in porous geologic formations beneath the earth's surface, often in association with petroleum. The principal constituent is methane.

Gas, Unconventional—Gas that can not be economically produced using current technology.

GASB—Government Accounting Standards Board. GASB was organized in 1984 by the Financial Accounting Foundation (FAF) to establish standards of financial accounting and reporting for state and local governmental entities. GASB standards guide the preparation of external financial reports of those entities. www.gasb.org

Gasification—The conversion of carbonaceous material into gas or the extraction of gas from another fuel.
- The process during which liquified natural gas (LNG) is returned to its vapor or gaseous state through an increase in temperature and a decrease in pressure.

Gasification—The process where biomass fuel is reacted with sub-stoichiometric quantities of air and oxygen usually under high pressure and temperature along with moisture to produce gas which contains hydrogen, methane, carbon monoxide, nitrogen, water and carbon dioxide. The gas can be burned directly in a boiler, or scrubbed and combusted in an engine-generator to produce electricity. The three types of gasification technologies available for biomass fuels are the fixed bed updraft, fixed bed downdraft and fluidized bed gasifiers. Gasification is also the production of synthetic gas from coal.

Gasifier—A device for converting solid fuel into gaseous fuel. In biomass systems, the process is also referred to as pyrolitic distillation. See *Pyrolysis*.

Gasket/Seal—A seal used to prevent the leakage of fluids, and also maintain the pressure in an enclosure.

Gasohol—In the United States, gasohol (E10) refers to gasoline that contains 10 percent ethanol by volume. This term was used in the late 1970s and early 1980s but has been replaced in some areas of the country by terms such as E-10, Super Unleaded Plus Ethanol, or Unleaded Plus.

Gas-oil Ratio—The quantity of gas produced with oil from an oil well, usually expressed as the number of cubic feet of gas produced per barrel of oil produced.

Gasoline—A light petroleum product obtained by refining oil, and used as motor vehicle fuel.

Gasoline Blending Components—Naphthas which will be used for blending or compounding into finished aviation or motor gasoline (e.g., straight-run gasoline, alkylate, reformate, benzene, toluene, and xylene). Excludes oxygenates (alcohols, ethers), butane, and pentanes plus.

Gasoline Grades—The classification of gasoline by octane ratings. Each type of gasoline (conventional, oxygenated, and reformulated) is classified by three grades—Regular, Midgrade, and Premium. Note: Gasoline sales are reported by grade in accordance with their classification at the time of sale. In general, automotive octane requirements are lower at high altitudes. Therefore, in some areas of the United States, such as the Rocky Mountain States, the octane ratings for the gasoline grades may be 2 or more octane points lower.
- Regular gasoline: Gasoline having an antiknock index, i.e., octane rating, greater than or equal to 85 and less than 88. Note: Octane

requirements may vary by altitude.

- Midgrade gasoline: Gasoline having an anti-knock index, i.e., octane rating, greater than or equal to 88 and less than or equal to 90. Note: Octane requirements may vary by altitude.
- Premium gasoline: Gasoline having an anti-knock index, i.e., octane rating, greater than 90. Note: Octane requirements may vary by altitude.

Gasoline Motor, (Leaded)—Contains more than 0.05 grams of lead per gallon or more than 0.005 grams of phosphorus per gallon. The actual lead content of any given gallon may vary. Premium and regular grades are included, depending on the octane rating. Includes leaded gasohol. Blendstock is excluded until blending has been completed. Alcohol that is to be used in the blending of gasohol is also excluded.

Gasoline Motor, (Unleaded)—Contains not more than 0.05 grams of lead per gallon and not more than 0.005 grams of phosphorus per gallon. Premium and regular grades are included, depending on the octane rating. Includes unleaded gasohol. Blendstock is excluded until blending has been completed. Alcohol that is to be used in the blending of gasohol is also excluded.

Gassing—Gaseous by-products when charging a battery, e.g. hydrogen from a lead acid battery. This gas is very explosive in concentrations of more than 2 PPM (parts per million).

- The evolution of gas from one or more of the electrodes in the cells of a battery. Gassing commonly results from local action self-discharge or from the electrolysis of water in the electrolyte during charging.

Gassing Current—The portion of charge current that goes into electrolytical production of hydrogen and oxygen from the electrolytic liquid. This current increases with increasing voltage and temperature.

Gas-Turbine Electric Power Plant—A plant in which the prime mover is a gas turbine. A gas turbine typically consists of an axial-flow air compressor and one or more combustion chambers where liquid or gaseous fuel is burned. The hot gases expand to drive the generator and then are used to run the compressor.

Gate Station—Location where the pressure of natural gas being transferred from the transmission system to the distribution system is lowered for transport through small diameter, low pressure pipelines.

- Generally a location at which gas changes own-ership, from one party to another, neither of which is the ultimate consumer. It should be noted, however, that the gas may change from one system to another at this point without changing ownership. Also referred to as city gate station, town border station, or delivery point.

Gatherer—A company primarily engaged in the gathering of natural gas from well or field lines for delivery, for a fee, to a natural gas processing plant or central point. Gathering companies may also provide compression, dehydration, and/or treating services.

GATT—General Agreement on Tariffs and Trade

Gauss—The unit of magnetic field intensity equal to 1 dyne per unit pole.

GB—Giga-barrels = 1 billion barrels

GBA—Green Building Alliance www.gbapgh.org

GBI (Green Building Initiative)—The Green Building Initiative was formed to help local Home Builder Associations (HBAs) develop green building programs modeled after the National Association of Home Builders (NAHB) Model Green Home Building Guidelines. In 2004, GBI acquired the rights to promote and distribute the Green Globes green building rating and assessment tool in the U.S. In 2005, GBI became the first green building organization to be accredited as a standards developer by the American National Standards Institute (ANSI), and began the process of establishing Green Globes as an official ANSI standard. The GBI is governed by a multi-stakeholder board of 15 directors featuring representatives from industry, NGOs, construction companies, architectural firms and academic institutions.

GBL—Government Bill of Lading

GBP—Gravity Based Penalty

GC—General Counsel

GCC—Global Climate Convention

GCIM—Gas Cost Incentive Mechanism: Replaces the Reasonableness Review as a means of reviewing natural gas purchasing activities for retail core customers. The purpose of the GCIM is to provide market-based incentives to reduce the cost of gas to core customers and to provide appropriate objective standards against which to measure a utility's performance in gas procurement and transportation functions on behalf of core customers.

GCP—Good Combustion Practices

GCVTC—Grand Canyon Visibility Transport Commission

GDP—Gross Domestic Product, a measure of overall

economic activity.

Gearing—Debt to equity ratio

GEI (GREENGUARD Environmental Institute)—Founded in 2001, this organization establishes acceptable indoor air quality standards for indoor products, environments and buildings and oversees the GREENGUARD Certification Program.

Gel-type Battery—Lead-acid battery in which the electrolyte is immobilized in a gel. Usually used for mobile installations and when batteries will be subject to high levels of shock or vibration.

General Business Credit—The term given to a group of credits listed in IRC Section 38. The ITC and PTC for renewable energy are included in this group of credits.

General Circulation Models (GCMs)—Complex computer simulations of climate and its various components used by researchers and policy analysts to predict climate change. Typically run on "super computers," these models can approximate future climates and give some clues to how climate has changed or might change over time.

• A global, three-dimensional computer model of the climate system, which can be used to simulate human-induced climate change. GCMs are highly complex and they represent the effects of such factors as reflective and absorptive properties of atmospheric water vapor, greenhouse gas concentrations, clouds, annual and daily solar heating, ocean temperatures and ice boundaries. The most recent GCMs include global representations of the atmosphere, oceans, and land surface.

General Lighting—Lighting designed to provide a substantially uniform level of illumination throughout an area, exclusive of any provision for special visual tasks or decorative effect. When designed for lower-than-task illuminance used in conjunction with other specific task lighting systems, it is also called "ambient" lighting.

General Obligation Bond—A bond secured by a pledge of the issuer's taxing powers (limited or unlimited). More commonly the general obligation bonds of local governments are paid from ad valorem property taxes and other general revenues. Considered the most secure of all municipal debt.

General Partner (GP)—The partner in a limited partnership responsible for all management decisions of the partnership. The GP has a fiduciary responsibility to act for the benefit of the limited partners (LPs) and is fully liable for its actions.

General Rate Case—Every couple of years the traditional, investor-owned utilities revisit their operating budgets, expenses, liabilities, etc. and make determinations on how to adequately recover their costs. In the interest of their shareholders and ratepayers the utilities work with the Public Utilities Commission and various intervenors to decide which customer classes will pay for the services offered and how rates will be adjusted to meet those revenue goals.

Generally Accepted Accounting Principles (GAAP)—Defined by the FASB as the conventions, rules, and procedures necessary to define accepted accounting practice at a particular time, includes both broad guidelines and relatively detailed practices and procedures.

Generating Facility—An existing or planned location or site at which electricity is or will be produced and includes one or more generating units at the same location.

Generating Station—A station that consists of electric generators and auxiliary equipment for converting mechanical, chemical, or nuclear energy into electric energy.

Generating Unit—Any combination of physically connected generators, reactors, boilers, combustion turbines, and other prime movers operated together to produce electric power.

Generation (Electricity)—The process of producing electric energy by transforming other forms of energy; also, the amount of electric energy produced, expressed in kilowatt-hours.

Generation Attribute—A non-price characteristic of electrical energy output of a Generation Unit including, but not limited to, the Unit's fuel type, emissions, vintage and RPS eligibility.

Generation Charges—Part of the basic service charges on every customer's bill for producing electricity. Generation service is competitively priced and is not regulated by Public Utility Commissions. This charge depends on the terms of service between the customer and the supplier.

Generation Circulation Models (GCMs)—Computer programs that attempt to mathematically simulate the global climate. The complex and large computer programs are based on mathematical equations derived from knowledge of the physics and chemistry that govern the earth atmosphere system.

Generation Company (GENCO)—A regulated or non-regulated entity (depending upon the industry structure) that operates and maintains existing generating plants. The Genco may own the generation plants or interact with the short-term market on behalf of plant owners. In the context of restructuring the market for electricity, Genco is sometimes used to describe a specialized "marketer" for the generating plants formerly owned by a vertically-integrated utility.

Generation Curtailment—A forced limitation of electrical energy output from a facility due to lack of demand, insufficient transmission capacity, or the sufficient availability of economically superior resources.

Generation Unit—A facility that converts a fuel or an energy resource into electrical energy.

Generation, Dispatch and Control—Aggregating and dispatching (sending off to some location) generation from various generating facilities, providing backups and reliability services. Ancillary services include the provision of reactive power, frequency control, and load following. (Also see "*Power Pool*" and "Poolco" below.)

Generation, Non-Utility—Generation by producers having generating plants for the purpose of supplying electric power required in the conduct of their industrial and commercial operations. Generation by mining, manufacturing, and commercial establishments and by stationary plants of railroads and railways for active power is included.

Generation-Sited Storage—A category of energy storage solutions that are co-located with large-scale generation (vs. distributed generation); includes molten salt or other media (co-located with concentrated solar thermal), and storage co-located with natural gas combustion turbines.

Generator—A mechanical device used to produce DC electricity. Power is produced by coils of wire passing through magnetic fields inside the generator. Most alternating current generating sets are also referred to as generators.

Generator (Green-e program)—The facility that physically generates the electricity.

Generator Capacity—The maximum output, commonly expressed in megawatts (MW), that generating equipment can supply to system load, adjusted for ambient conditions.

Generator Nameplate Capacity (Installed)—The maximum rated output of a generator, prime mover, or other electric power production equipment under specific conditions designated by the manufacturer. Installed generator nameplate capacity is commonly expressed in megawatts (MW) and is usually indicated on a nameplate physically attached to the generator.

Geographical Information System (GIS)—A GIS is a research tool that allows analysts to view geographically referenced information (maps, charts and diagrams) to perform trend and spatial analyses with indicators.

Geologic Assurance—State of sureness, confidence, or certainty of the existence of a quantity of resources based on the distance from points where coal is measured or sampled and on the abundance and quality of geologic data as related to thickness of overburden, rank, quality, thickness of coal, area extent, geologic history, structure, and correlations of coal beds and enclosing rocks. The degree of assurance increases as the nearness to points of control, abundance, and quality of geologic data increases.

Geologic Considerations—Conditions in the coal deposit or in the rocks in which it occurs that may complicate or preclude mining. Geologic considerations are evaluated in the context of the current state of technology and regulations, so the impact on mining may change with time.

Geological and Geophysical (G&G) Costs—Costs incurred in making geological and geophysical studies, including, but not limited to, costs incurred for salaries, equipment, obtaining rights of access, and supplies for scouts, geologists, and geophysical crews.

Geological Repository—A mined facility for disposal of radioactive waste that uses waste packages and the natural geology as barriers to provide waste isolation.

Geomorphic—Pertaining to those processes that affect the form or shape of the surface of the earth.

Geophysical Logging—A general term that encompasses all techniques for determining whether a subsurface geological formation may be sufficiently porous or permeable to serve as an aquifer. These techniques typically involve lowering a sensing device into a borehole to measure properties of the subsurface formation.

Geophysical Survey—Searching and mapping of the subsurface structure of the earth's crust by use of geophysical methods, to locate probable reservoir structures capable of containing gas or oil.

Geophysics—A study of subsurface geological conditions of structure or material through the interpretation of measurement variations in density, magnetics, elasticity, electrical conductivity, temperature, and/or radioactivity.

Geopressured—A type of geothermal resource occurring in deep basins in which the fluid is under very high pressure.

Geopressurized Brines—These brines are hot (300 F to 400 F) (149 C to 204 C) pressurized waters that contain dissolved methane and lie at depths of 10,000 ft (3048 m) to more than 20,000 ft (6096 m) below the earth's surface. The best known geopressured reservoirs lie along the Texas and Louisiana Gulf Coast. At least three types of energy could be obtained: thermal energy from high-temperature fluids; hydraulic energy from the high pressure; and chemical energy from burning the dissolved methane gas.

Geothermal—An electric generating station in which steam tapped from the earth drives a turbine-generator, generating electricity.

Geothermal Element—An element of a county general plan consisting of a statement of geothermal development policies, including a diagram or diagrams and text setting forth objectives, principles, standards, and plan proposals, including a discussion of environmental damages and identification of sensitive environmental areas, including unique wildlife habitat, scenic, residential, and recreational areas, adopted pursuant to Section 65303 of the Government Code.

Geothermal Energy—Energy produced by the internal heat of the earth; geothermal heat sources include: hydrothermal convective systems; pressurized water reservoirs; hot dry rocks; manual gradients; and magma. Geothermal energy can be used directly for heating or to produce electric power.

Geothermal Gradient—The change in the earth's temperature with depth. As one goes deeper, the earth becomes hotter.

Geothermal Heat Pump—A type of heat pump that uses the ground, ground water, or ponds as a heat source and heat sink, rather than outside air. Ground or water temperatures are more constant and are warmer in winter and cooler in summer than air temperatures. Geothermal heat pumps operate more efficiently than "conventional" or "air source" heat pumps.

Geothermal Plant—A plant in which the prime mover is a steam turbine. The turbine is driven either by steam produced from hot water or by natural steam that derives its energy from heat found in rocks or fluids at various depths beneath the surface of the earth. The energy is extracted by drilling and/or pumping.

Geothermal Power Station—An electricity generating facility that uses geothermal energy.

Geothermal Steam—Steam drawn from deep within the earth.

GEP—Good Engineering Practice

Geyser—A special type of thermal spring that periodically ejects water with great force.

GHG—Greenhouse Gas

GHPC—Geothermal Heat Pump Consortium www.ghpc.org

Giga—One billion

Gigawatt (GW)—One thousand megawatts (1,000 MW) or, one million kilowatts (1,000,000 kW) or one billion watts (1,000,000,000 watts) of electricity. One gigawatt is enough to supply the electric demand of about one million average homes.

Gigawatt Hour—A measurement of energy. One Gigawatt-hour is equal to one Gigawatt being used for a period of one hour, or one Megawatt being used for 1000 hours.

Gigawatt-Electric (GWe)—One billion watts of electric capacity.

Gilsonite—Trademark name for uintaite (or uintahite), a black, brilliantly lustrous natural variety of asphalt found in parts of Utah and western Colorado.

Gin Pole—A pole used to assist in raising a tower. Either of two different types of devices used with wind generator towers. With a tilt-up tower, it describes the lever that helps tilt the tower up. With a fixed tower, it describes a temporary crane used to raise tower sections or the wind generator.

Girdling—Killing a tree by removing a strip of bark from around its trunk.

GIS—Geographic Information System
  • Gas Insulated Switchgear
  • GIS—Green investment scheme. An arrangement whereby Annex 1 industrialized countries buy the surplus Kyoto carbon emissions credits, AAUs, of eastern European countries on the condition they invest the proceeds in low-emissions technology. GISs came about due to pressure on former Soviet bloc countries to use these surpluses, known as "hot air," responsibly, ie. to build environmentally-sustainable industry. Credits traded through GISs are termed 'greened AAUs'.

Glare—The discomfort or interference with visual perception when viewing a bright object against a dark background.

Glass-Steagall Act—A 1933 Federal Act requiring the separation of commercial and investment banking operations.

Glauber's Salt—A salt, sodium sulfate decahydrate, that melts at 90 degrees Fahrenheit; a component of eutectic salts that can be used for storing heat.

Glazing—Transparent or translucent material (glass or plastic) used to admit light and/or to reduce heat loss; used for building windows, skylights, or greenhouses, or for covering the aperture of a solar collector. The term can refer to a fiberglass or plastic covering as well.

Global Climate Change—Gradual changing of global climates due to buildup of carbon dioxide and other greenhouse gases in the earth's atmosphere. Carbon dioxide produced by burning fossil fuels has reached levels greater than what can be absorbed by green plants and the seas.

Global Environmental Facility (GEF)—The multi-billion-dollar GEF was established by the World Bank, the UN Development Programme, and the UN Environment Programme in 1990. It operates the Convention's 'financial mechanism' on an interim basis and funds developing country projects that have global climate change benefits.

Global Insolation (or Solar Radiation)—The total diffuse and direct insolation on a horizontal surface, averaged over a specified period of time.

Global Reporting Initiative (GRI)—A multi-stakeholder process and independent institution whose mission is to develop and disseminate globally applicable Sustainability Reporting Guidelines. A reporting standard generally accepted to be the leading international standard for reporting social, environmental and economic performance. www.globalreporting.org

Global Warming—An increase in the near surface temperature of the Earth. Global warming has occurred in the distant past as the result of natural influences, but the term is today most often used to refer to the warming some scientists predict will occur as a result of increased anthropogenic emissions of greenhouse gases.

Global Warming Potential (GWP)—An index used to compare the relative radiative forcing of different gases without directly calculating the changes in atmospheric concentrations. GWPs are calculated as the ratio of the radiative forcing that would result from the emission of one kilogram of a greenhouse gas to that from the emission of one kilogram of carbon dioxide over a fixed period of time, such as 100 years.

• In the greenhouse gas program, the GWP is an index found in the Kyoto Protocol that allows for the comparison of greenhouse gases with each other in the context of their relative potential to contribute to global warming.

• A system of multipliers devised to enable warming effects of different gases to be compared. The cumulative warming effect, over a specified time period, of an emission of a mass unit of CO_2 is assigned the value of 1. Effects of emissions of a mass unit of non-CO_2 greenhouse gases are estimated as multiples. For example, over the next 100 years, a gram of methane (CH_4) in the atmosphere is currently estimated as having 23 times the warming effect as a gram of carbon dioxide; methane's 100-year GWP is thus 23. Estimates of GWP vary depending on the time-scale considered (e.g., 20-, 50-, or 100-year GWP), because the effects of some GHGs are more persistent than others.

The GWP is used for converting other types of greenhouse gas emissions into CO_2 equivalents.

Glycol—An antifreeze, heat transfer fluid that is circulated through closed loop solar hot water collectors. (Propylene Glycol)

GM—Genetically Modified

GMCC—Global Monitoring for Climatic Change

GMO—Genetically Modified Organism

GMT—Greenwich Mean Time

GNP—Gross National Product

GO—General Order: A PUC order that sets standards, procedures, or guidelines applicable to a class of utilities, as distinguished from a decision affecting only a single utility.

GOCO—Government-Owned/Contractor-Operated

GOGO—Government-Owned/Government-Operated

Going Concern—An operating business enterprise that is expected to continue.

Going Concern Value—The value of a business enterprise that is expected to continue to operate into the future. The intangible elements of going-concern value result from factors such as having a trained work force, an operational plant, and the necessary licenses, systems, and procedures in place.

Golden Handcuff—This occurs when an employee is required to relinquish unvested stock when terminating their employment contract early.

Golden Parachute—Employment contract of upper management that provides a large payout upon the occurrence of certain control transactions, such as a certain percentage share purchase by an outside entity or when there is a tender offer for a certain percentage of a company's shares. This is discussed in more detail in the Executive Employment Agreement.

Good Utility Practice—Methods and practices that are approved by a significant portion of the industry.

Goodwill—The intangible assets of a firm, calculated at the excess purchase price paid over the book value of the firm.

GOP—General Operating Procedures
• Good Operating Practices

GOPO—Government-Owned/Privately-Operated

GOSP—Gas Oil Separation Plant

Government-Owned Stocks—Oil stocks owned by the national government and held for national security. In the United States, these stocks are known as the Strategic Petroleum Reserve.

Governor—A device used to regulate motor speed, or, in a wind energy conversion system, to control the rotational speed of the rotor.

GPAD—Gallons per Acre per Day

GPCD—Gallons per Capita per Day: The amount of water used on average by an individual each day. Total gpcd is calculated by dividing total water use in the area, including industrial and commercial uses, by the number of users. Residential gpcd is the number resulting from only considering domestic water use.

GPD—Gallons per Day

GPG—Grams per Gallon

GPM—Gallons per Minute

GPO—Government Printing Office

GPS—Global Positioning System. A system that using one or more receivers one could determine accurate positioning in the x, y or z axes by means of signals from satellites.

gr—Grains (7,000 grains per pound)

Gr/dscf—Grains per dry standard cubic feet

gr/SCF—Grains of pollutant per standard cubic foot of gas. A measure of dust particles in a gas stream.

Grain Alcohol—Ethanol

Grain Boundaries—The boundaries where crystallites in a multi-crystalline material meet.

Grandfather Clause—The continuation of a former rule, clause, or policy (usually in a contractual agreement) where a change to a new rule or policy would be patently unfair to those covered by the former.

Grandfathering—Grandfathering of emissions permits is a method by which permits for greenhouse gas emissions may be allocated among emitters and firms in a domestic emissions trading regime according to their historical emissions. Supporters of this method of emissions trading assert that this would be administratively simple but some critics argue that this method would reward firms with high historical emissions and unfairly complicate entry into markets by new firms and emitters.

Grant—A transfer of real or personal property
• Funds received from a private foundation or charitable group, federal, state of local government that do not have to be repaid

Grantee—The person to whom a grant is made.

Grantor—The person who makes a grant.

Grantor Trust—A trust used as the owner trust in a leveraged lease transaction or as a special purpose vehicle on a securitized transaction. Usually funded by the equity participant(s) or investors.

Granular Activated Carbon (GAC)—A form of crushed and hardened charcoal. GAC has a strong potential to attract and absorb volatile organic compounds from extracted groundwater and gases.

Graywater—The used water discharged by sinks, showers, bathtubs and clothes washing machines.

GRC—General Rate Case: A proceeding in which the Commission takes a broad, in-depth look at a utility's revenues, expenses, and financial outlook and considers quality of service and other factors to arrive at just and reasonable rates. These are the major regulatory proceedings that come before the Commission.

Green Accounting—The incorporation of the amount of natural resources used and pollutants expelled into conventional economic accounting in order to provide a detailed measure of all environmental consequences of any and all economic activities.

Green Building—A comprehensive process of design and construction that employs techniques to minimize adverse environmental impacts and reduce the energy consumption of a building, while contributing to the health and productivity of its occupants. A common metric for evaluating green buildings is the LEED (Leadership in Energy and Environmental Design) certification.

Green Certificates—Green certificates represent the environmental attributes of power produced from renewable resources. By separating the environmental attributes from the power, clean power generators are able to sell the electricity they produce to power providers at a competitive market value. The additional revenue generated by the sale of the green certificates covers the above-market costs associated with producing power made from renewable energy sources. Also known as green tags, renewable energy certificates, or tradable renewable certificates.

Green Design—The design of products, services, buildings, or experiences that are sensitive to environmental issues and achieve greater efficiency and effectiveness in terms of energy and materials use.

Green Globes—Green Globes is an online rating system for new and existing buildings. Builders and operators can use Green Globes to assess how green a building is, and they can also have the building results verified by a third party and then use the official "Green Globes" symbol for the green level garnered by the structure. Buildings are rated in eight areas: energy, water, resources, site, emissions/effluents and other impacts, indoor environment, project management, and site. Depending on the number of points awarded, buildings can achieve from rating one to four globes, with four being the top rating.

In the U.S., Green Globes is operated by the Green Building Initiative while in Canada it is operated by BOMA Canada under the name "Go Green" (Visez vert).

Green House Effect—The increasing mean global surface temperature of the earth caused by gases in the atmosphere (including carbon dioxide, methane, nitrous oxide, ozone, and chlorofluorocarbon). The greenhouse effect allows solar radiation to penetrate but absorbs the infrared radiation returning to space.

Green Label—In 1992, CRI launched its Green Label testing and approval program, which sets limits for the level of VOC emissions from carpet, adhesives and cushion that can be released into the indoor air. Carpet systems that meet or exceed CRI's Green Label or Green Label Plus programs can contribute one full Indoor Environmental Quality Credit to the LEED ratings of the U.S. Green Building Council, according to CRI. Also, Green Label carpet is also used as a specification standard for the American Lung Association's Healthy Home program and the Collaborative for High Performance Schools (CHPS) in conjunction with the state of California, and the Green Guide for Health Care awards one point to healthcare facilities that install Green Label Plus carpet.

Green Logging—The logging of timber that is still alive.

Green Power—A popular term for energy produced from clean, renewable energy resources. As defined by Green-e, this term is synonymous with "eligible renewable electricity product."

Green Power Board—The Governing Board for the Green-e Program. The Board meets twice annually as a full board, and may convene meetings in subcommittee as necessary.

Green Power Purchasing—Voluntary purchases of green power by residential, commercial, government, or industrial customers, from utility companies, a third-party renewable energy generator, or with "renewable energy certificates." With utility green pricing or competitive sales, a customer's electricity demand is matched by an equivalent amount of renewable energy generation feeding into the power grid. Green certificates allow the renewable energy production to be located anywhere.

Green Pricing—In the case of renewable electricity, green pricing represents a market solution to the various problems associated with regulatory valuation of the non-market benefits of renewables. Green pricing programs allow electricity customers to express their willingness to pay for renewable energy development through direct payments on their monthly utility bills.

Green Roof—The roof of a building that is partially or completely covered with vegetation and planted over a water-proof membrane. Green roofs reduce rooftop and building temperatures, filter pollution, lessen pressure on sewer systems, and reduce the heat island effect.

Green Roof—Contained green space on, or integrated with, a building roof. Green roofs maintain living plants in a growing medium on top of a membrane and drainage system. Green roofs are considered a sustainable building strategy in that they have the capacity to reduce stormwater runoff from a site, they modulate temperature in and around the building, have thermal insulating properties, can provide habitat for wildlife and open space for humans, and other benefits.

Green Seal—Founded in 1989, Green Seal is a nonprofit organization that issues environmental standards in more than 40 major product categories including paints and coatings and windows and doors. Product standards are developed with the input of the public and industry member, academia and government agencies. The standards need to meet U.S. Environmental Protection Agency (EPA) requirements, International Standards Organization (ISO) requirements and third-party certifier requirements.

Green Ton—2,000 pounds of undried biomass material. Moisture content must be specified if green tons are used as a measure of fuel energy.

Green-e—A voluntary certification program for renewable energy. The Green-e logo indicates energy options that meet strict standards set through a collaborative process with environmentalists, consumer advocates, and energy experts. The Green-e Program verifies that participating suppliers are purchasing enough renewable electricity or certificates to meet their customer's needs.

Greenfield Plant—A new electric power generating facility built from the ground up.

Greenfield Site—Greenfield sites are those that are not previously developed or graded and remain in a natural state. Previously developed sites are those previously containing buildings, roadways, parking lots, or were graded or altered by direct human activities.

Greenguard—The GREENGUARD Certification Program is an industry-independent, third-party testing program for low-emitting products and materials including building products such as flooring, paint and furniture. GREENGUARD products are regularly tested to make sure that their chemical and particle emissions meet acceptable indoor air quality pollutant standards and guidelines.

Since awarding its first GREENGUARD certification in 2002, GEI has also introduced the GREENGUARD Children & Schools standard and the GREENGUARD for Building Construction Program, which is a mold risk reduction program that certifies the design, construction and ongoing operations of newly constructed multifamily and commercial properties. (Buildings with the GREENGUARD for Building Construction certification are monitored for ongoing compliance throughout the term of the loan or building life.)

Greenhouse Effect—The presence of trace atmospheric gases make the earth warmer than would direct sunlight alone. These gases (carbon dioxide [CO_2], methane [CH_4], nitrous oxide [N_2O], tropospheric ozone [O_3], and water vapor [H_2O]) allow visible light and ultraviolet light (shortwave radiation) to pass through the atmosphere and heat the earth's surface. This heat is re-radiated from the earth in form of infrared energy (longwave radiation). The greenhouse gases absorb part of that energy before it escapes into space. This process of trapping the longwave radiation is known as the greenhouse effect. Scientists estimate that without the greenhouse effect, the earth's surface would be roughly 54 degrees Fahrenheit colder than it is today—too

cold to support life as we know it. See *Global Climate Change*.

• The effect produced as greenhouse gases allow incoming solar radiation to pass through the Earth's atmosphere, but prevent most of the outgoing long-wave infra-red radiation from the surface and lower atmosphere from escaping into outer space. This envelope of heat-trapping gases keeps the Earth about 30°C warmer than if these gases did not exist.

Greenhouse Effect (Relating to Buildings)—The characteristic tendency of some transparent materials (such as glass) to transmit radiation with relatively short wavelengths (such as sunlight) and block radiation of longer wavelengths (such as heat). This tendency leads to a heat build-up within the space enclosed by such a material.

Greenhouse Gas— Any gas that contributes to the "greenhouse effect," whereby heat is trapped within the Earth's atmosphere, including: carbon dioxide, methane, nitrous oxide, hydrofluorocarbons, perfluorocarbons and sulfur hexafluoride.

Greenhouse Gas Effect—In the greenhouse gas program, a concept that refers to the effect that releasing greenhouse gas emissions has on the relative warming of the earth's atmosphere. The release of too much greenhouse gas over a period of time results in a gradual warming of the earth's atmosphere.

Greenhouse Gas Protocol Initiative—Produced the most widely used international accounting tool for government and business leaders to understand, quantify, and manage greenhouse gas emissions. www.ghgprotocol.org

Greenhouse Gas Reduction—A greenhouse gas reduction is a reduction in emissions that is recognized to contribute to climate change, e.g. carbon dioxide, methane, nitrous oxide, hydrofluorocarbons, perfluorocarbons, and sulfur hexofluoride. Greenhouse gas reductions are often measured in tons of carbon dioxide-equivalent. For example, 1 ton of methane has the same global warming potential as 20.9 tons of carbon dioxide.

Greenhouse Gas Source—Any physical unit or process that releases a GHG into the atmosphere.

Greenhouse Gases—A gas that absorbs and re-emits infrared radiation, warming the earth's surface and contributing to climate change. Those gases, such as water vapor, carbon dioxide, nitrous oxide, methane, hydrofluorocarbons (HFCs), perfluorocarbons (PFCs) and sulfur hexafluoride, that are transparent to solar (short-wave) radiation but opaque to long-wave (infrared) radiation, thus preventing long-wave radiant energy from leaving Earth's atmosphere. The net effect is a trapping of absorbed radiation and a tendency to warm the planet's surface. The three principal greenhouse gases are carbon dioxide, methane and nitrous oxide.

Greenhouse Intensity—Refers to the ratio of a nation's greenhouse gas emissions to its GDP, or the volume of emissions per unit of economic output. A country's greenhouse intensity may often be falling yet overall emissions are rising due to an expanding economy. Greenhouse intensity measures are also used at a company, plant or industry sector level.

Greenhouse Window—A type of fenestration product which adds conditioned volume but no conditioned floor area to a building.

Greenwashing—The process by which a company publicly and misleadingly declares itself to be environmentally-friendly but internally participates in environmentally- or socially- unfriendly practices.

Greenwood—Freshly cut, unseasoned, wood.

Grey Water—Waste water from a household source other than a toilet. This water can be used for landscape irrigation depending upon the source of the grey water.

GRHC—Green Roofs for Healthy Cities www.greenroofs.org

GRI Reporting Framework—The GRI Reporting Framework is intended to provide a generally accepted framework for reporting on an organization's economic, environmental, and social performance. The Framework consists of the Sustainability Reporting Guidelines, the Indicator Protocols, Technical Protocols, and the Sector Supplements.

Grid—The electric utility companies' transmission and distribution system that links power plants to customers through high power transmission line service (110 kilovolt [kv] to 765 kv); high voltage primary service for industrial applications and street rail and bus systems (23 kv-138 kv); medium voltage primary service for commercial and industrial applications (4 kv to 35 kv); and secondary service for commercial and residential customers (120 v to 480 v). Grid can also refer to the layout of a gas distribution system of a city or town in which pipes are laid in both directions in the streets and connected at intersections.

Grid Connected—A power delivery connection system consisting of an independent power source that normally operates in parallel with a utility power system.
  • A PV system in which the PV array acts like a central generating plant, supplying power to the grid.

Grid Connection—Joining a plant that generates electric power to a utility system so that electricity can flow in either direction between the utility system and the plant.

Grid Line (Solar)—Metallic contacts fused to the surface of a solar cell to provide a low resistance path for electrons to flow out to the cell interconnect wires.

Grid Parity—The point at which renewable energy facilities are able to produce renewable electricity at a rate equal to or cheaper than convention power sources.

Grid-Connected System—Independent power systems that are connected to an electricity transmission and distribution system (referred to as the electricity grid) such that the systems can draw on the grid's reserve capacity in times of need, and feed electricity back into the grid during times of excess production.

Grid-Tie System (Solar)—A renewable energy system that is connected to the utility grid, selling excess energy back to the utility. Also called a utility-interactive system.

GRIM—Government Regulatory Impact Model

Gross Additions to Construction Work in Progress for the Month—This amount should include the monthly gross additions for an electric plant in the process of construction.

Gross Area—The area of a surface including areas not belonging to that surface (such as windows and doors in a wall). The total surface area of a solar collector including the frame, manifold, absorber or other componentry.

Gross Building Demand—A configuration that uses current and voltage sensors placed after the grid connection to the building and after the inverter connection to the building. The voltage and current sensors measure the total energy usage of the building, regardless of the amount of electricity generated by the renewable energy system.

Gross Calorific Value—The heat produced by combusting a specific quantity and volume of fuel in an oxygen-bomb colorimeter under specific conditions.

Gross Capacity—The full-load continuous rating of a generator, prime mover, or other electric equipment under specified conditions as designated by the manufacturer. It is usually indicated on a nameplate attached to the equipment.

Gross Company-Operated Production—Total production from all company-operated properties, including all working and nonworking interests.

Gross Domestic Product (GDP)—The total value of goods and services produced by labor and property located in the United States. As long as the labor and property are located in the United States, the supplier (that is, the workers and, for property, the owners) may be either U.S. residents or residents of foreign countries.

Gross Domestic Product (GDP) Implicit Price Deflator—The implicit price deflator, published by the U.S. Department of Commerce, Bureau of Economic Analysis, is used to convert nominal figures to real figures.

Gross Energy Intensity—Total consumption of a particular energy source(s) or fuel(s) by a group of buildings, divided by the total floor space of those buildings, including buildings and floor space where the energy source or fuel is not used, i.e., the ratio of consumption to gross floor space.

Gross Exterior Roof Area—The sum of the skylight area and the exterior roof/ceiling area.

Gross Exterior Wall Area—The sum of the window area, door area, and exterior wall area.

Gross Gas Withdrawal—The full-volume of compounds extracted at the wellhead, including non-hydrocarbon gases and natural gas plant liquids.

Gross Generation—The total amount of electric energy produced by generating units and measured at the generating terminal in kilowatt-hours (kWh) or megawatt hours (MWh) per year.

Gross Head—A dam's maximum allowed vertical distance between the upstream's surface water (headwater) forebay elevation and the downstream's surface water (tailwater) elevation at the tail-race for reaction wheel dams or the elevation of the jet at impulse wheel dams during specified operation and water conditions.

Gross Heating Value—(GHV) The maximum potential energy in the fuel as received, considering moisture content (MC). It reflects the heat used to evaporate moisture. Compare Higher Heating Value (HHV). Expressed as:

$$GHV = HHV\,((1 - MC)/100)$$

Gross Input to Atmospheric Crude Oil Distillation Units—Total input to atmospheric crude oil distillation units. Includes all crude oil, lease condensate, natural gas plant liquids, unfinished oils, liquefied refinery gases, slop oils, and other liquid hydrocarbons produced from tar sands, gilsonite, and oil shale.

Gross Inputs—The crude oil, unfinished oils, and natural gas plant liquids put into atmospheric crude oil distillation units.

Gross Lease—Opposite of a net lease. The lessor pays property taxes, insurance and maintenance costs on the assets on the lease agreement.

Gross Margin (%)—Gross profit divided by revenues

Gross National Product (GNP)—The total value of goods and services produced by the nation's economy before deduction of depreciation charges and other allowances for capital consumption. It includes the total purchases of goods and services by private consumers and government, gross private domestic capital investment, and net foreign trade.

Gross Pre-Tax Yield—The yield calculated in a lease before considering tax benefits and costs of doing business such as bad debt and general and administrative expenses.

Gross Profit Margin (GPM)—Used in modeling the financial situation of captive financing subsidiaries with financial modeling software such as Super-Trump®.

Gross Rent Multiplier (GRM)—The relationship or ratio between the sale price or value of a property and its gross rental income.

Gross Residual Asset—The amount a lessor expect to derive from an underlying asset following the end of the lease term, measured on a discounted basis.

Gross Vehicle Weight Rating (GVWR)—Vehicle weight plus carrying capacity.

Gross Withdrawals—Full well stream volume, including all natural gas plant liquid and non-hydrocarbon gases, but excluding lease condensate. Also includes amounts delivered as royalty payments or consumed in field operations.

Gross Working Interest Ownership Basis—Gross working interest ownership is the respondent's working interest in a given property plus the proportionate share of any royalty interest, including overriding royalty interest, associated with the working interest.

Ground—A device used to protect the user of any electrical system or appliance from shock. The connection of electrical components to the earth and/or each other for the purposes of dissipating static charge or protecting against a short circuit or lightning.

Ground Cover—Low-growing plants often grown to keep soil from eroding and to discourage weeds.

Ground Fault—Unwanted current path to ground.

Ground Floor Area—Defined as the slab-on-grade area of a slab-on-grade building and the conditioned footprint area of a raised floor building (for compliance with the low-rise residential standards in California).

Ground Lease—A lease that grants the right to use and occupy land. Improvements made by the ground lease typically revert to the ground lessor.

Ground Loop—In geothermal heat pump systems, a series of fluid-filled plastic pipes buried in the shallow ground, or placed in a body of water, near a building. The fluid within the pipes is used to transfer heat between the building and the shallow ground (or water) in order to heat and cool the building.

• An undesirable feedback condition caused by two or more circuits sharing a common electrical line, usually a grounded conductor.

Ground Mount—A photovoltaic (PV) rack designed to be installed on the ground or other flat surface.

Ground Reflection—Solar radiation reflected from the ground onto a solar collector.

Ground Rod—A metal rod (typically 5/8 inch diameter) that is driven into the earth (typically 8 feet deep) and is electrically connected to the negative conductor and/or any metal parts, wiring enclosures, or conduit of an electrical circuit.

Ground Source Heat Pump—A heat pump that uses the earth as a source of energy for heating and a sink for energy when cooling. Some systems pump water from an aquifer in the ground and return the water to the ground after transferring heat from or to the water. A few systems use refrigerant directly in a loop of piping buried in the ground. Those heat pumps that use either a water loop or pump water from an aquifer have efficiency test methods that are accepted by the Energy Commission (in California). These efficiency values are certified to the Energy Commission by the manufacturer and are expressed in terms of heating Coefficient of Performance (COP) and cooling Energy Efficiency Ratio (EER).

Groundwater Contamination—The introduction of hazardous or toxic material into the underground water supply or aquifers.

Group—A group is a logical grouping of assemblies with similar characteristics. All assemblies in a group have the same initial average enrichment, the same cycle/reactor history, the same current location, the same burn up, the same owner, and the same assembly type.

Group Name—The DOE/EIA-assigned name identifying a composite supply source (i.e., commonly metered gas streams from more than one field), which is often the case in contract areas, field areas, and plants. A group name can also be a pipeline purchase (i.e., FERC Gas Tariff, Canadian Gas, Mexican Gas, and Algerian LNG). Emergency purchases and short-term purchases are also group names. Group Code—The DOE/EIA-assigned code identifying a composite supply source.

Group of 77 and China—An international organization established in 1964 by 77 developing countries; membership has now increased to 133 countries. The group acts as a major negotiating bloc on some issues including climate change.

Group Quarters—Living arrangement for institutional groups containing ten or more unrelated persons. Group quarters are typically found in hospitals, nursing or rest homes, military barracks, ships, halfway houses, college dormitories, fraternity and sorority houses, convents, monasteries, shelters, jails, and correctional institutions. Group quarters may also be found in houses or apartments shared by ten or more unrelated persons. Group quarters are often equipped with a dining area for residents.

GS—Geological Survey

GSA—General Services Administration

GT—Gas Turbine

GTL—Gas-to-liquids conversion

Guaranteed Residual Value—A situation in which the lender or an unrelated third party (e.g., equipment manufacturer, insurance company) guarantees to the lender that the financed equipment will be worth a certain fixed amount at the end of the lease term. The guarantor agrees to reimburse the lender for any deficiency realized if the financed equipment is subsequently salvaged at an amount below the guaranteed residual value.

Guaranteed Savings—A program in which a company guarantees a user a predetermined reduction in energy costs. The company guarantees that energy costs plus all costs of the energy conservation measures or services provided will be less than the user's normal energy costs.

Guarantor—The party that promises to pay the debt payments to the lender in the event the principal debtor defaults.

Guideline Lease—A tax lease that meets or follows the IRS guidelines, as established by Revenue Ruling 75-21, for a leveraged lease.

Guy Wire—Cable use to secure a wind turbine tower to the ground in a safe, stable manner.

GVP—Gasoline Vapor Pressure

GVW—Gross Vehicle Weight

GW—Groundwater

GWh—Gigawatt-hours: One billion watt-hours.

GWP—Global warming potential. This refers to the potency of greenhouse gases, that is, their ability to trap heat in the atmosphere. The GWP is a numerical measure relative to carbon dioxide, the most abundant greenhouse gas. So carbon dioxide itself has a GWP of 1.

GWTR—Groundwater Treatment Rule

Gypsum—Calcium sulfate dihydrate (C_aSO_4 $2H_2O$) a sludge constituent from the conventional lime scrubber process, obtained as a byproduct of the dewatering operation and sold for commercial use.

H

Habitable Story—A story that contains space in which humans may work or live in reasonable comfort, and that has at least 50 percent of its volume above grade.

Habitat—The area where a plant or animal lives and grows under natural conditions. Habitat includes living and non-living attributes and provides all requirements for food and shelter.

Hadronic Heating—A system that heats a space using hot water which may be circulated through a convection or fan coil system or through a radiant baseboard or floor system.

Haircut—A discount.

Half-Life—The time it takes for an isotope to lose half of its radioactivity. It is also used to describe: The time for a pollutant to lose one half of its concentration, as through biological action; and the time for elimination of one half of a total dose of a drug from a body.

Half-Year Convention—A tax depreciation convention that assumes all equipment is purchase or sold at the midpoint of a taxpayer's tax year. The half-year convention allows an equipment owner to claim a half-year of depreciation deductions in the year of acquisition, as well as in the year of disposition, regardless of the actual date within the year that the equipment was placed in service, or disposed of.

Halogen Lamp—A type of incandescent lamp that lasts much longer and is more efficient than the common incandescent lamp. The lamp uses a halogen gas, usually iodine or bromine, that causes the evaporating tungsten to be re-deposited on the filament, thus prolonging its life. Efficiency is better than a normal incandescent, but not as good as a fluorescent light. Also see *Incandescent Lamp*.

Halogenated Substances—A volatile compound containing halogens, such as chlorine, fluorine or bromine.

Halogens—The family of elements that includes fluorine, chlorine, bromine and iodine. Halogens are very reactive and have many industrial uses. They are also commonly used in disinfectants and insecticides.

Hammermill—A device consisting of a rotating head with free-swinging hammers which reduce chips or hogged fuel to a predetermined particle size through a perforated screen.

Hand Loading—An underground loading method by which coal is removed from the working face by manual labor through the use of a shovel for conveyance to the surface. Though rapidly disappearing, it is still used in small-tonnage mines.

Hand Pile—A pile of slash constructed by a crew, not by machine. Hand piles are typically less than 10' high and less than 12' in diameter.

HAP—Hazardous Air Pollutant

HAPPS—Hazardous Air Pollutant Prioritization System

Hard Coat—A low emissivity metallic coating applied to the glass, which will be installed in a fenestration product, through a pyrolytic process (at or near the melting point of the glass so that it bonds with the surface layer of glass). Hard coatings are less susceptible to oxidation and scratching as compared to soft coats. Hard coatings generally do not have as low emissivity as soft coats.

Hardwoods—Usually broad-leaved and deciduous trees.

Harmonic Content—Frequencies in the output waveform in addition to the primary frequency (usually 50 or 60 Hz.) Energy in these harmonics is lost and can cause undue heating of the load.

Harmonic(s)—A sinusoidal quantity having a frequency that is an integral multiple of the frequency of a periodic quantity to which it is related.

Harmonized System (HS)—The international classification system for goods used by most major trading countries for tariff classification, trade statistics, and transport documentation. Officially known as the Harmonized Commodity Description and Coding system.

HASWOPER—Hazardous Waste Operations and Emergency Response

Haulage Cost—Cost of loading ore at a mine site and

transporting it to a processing plant.

Hazardous Waste—Any waste or combination of wastes which pose a substantial present or potential hazard to human health or living organisms because such wastes are non-degradable or persistent in nature or because they can be biologically magnified, or because they can be lethal, or because they may otherwise cause or tend to cause detrimental cumulative effects.

HBL—Health Based Level

H-Coal Process—A means of making coal cleaner so it will produce less ash and less sulfur emissions.

HCP—Habitat Conservation Plan

HDD—Heavy Duty Diesel
  • Heating Degree Day: A unit of measurement derived from the variance between the average temperature during a given time period (month, season, year) and a reference point, usually 65 degrees Fahrenheit.

HDT—Heavy Duty Truck

HEA—Health Effects Assessment

Head—The product of the water's weight and a usable difference in elevation gives a measurement of the potential energy possessed by water.
  • The vertical distance water drops from the highest level to the level of the receiving body of water.
  • The differential or pressure, usually expressed in terms of the height of a liquid column that the pressure will support. Also, the differential across a primary measuring device in feet of flowing fluid.

Header—A pipe from which two or more tributary pipes run.

Headlease—A lease to a single entity, which is intended to be the holder of subsequent leases to sub-lessees that will be the tenants in possession of the leased premises. Also called a "master lease."

Headrace—A flume or channel that feeds water into a hydro turbine.

Headstation—A point at which gas enters the pipeline's main transmission line, either at the interconnection of the gathering system or of a third-party transporter.

Health-based Remediation Targets—Levels to which hazardous substances on the site will be cleaned up. These target levels are health-based, meaning that exposure to the hazardous substances at or below the target is not expected to present a significant health risk.

Healthy Building Network (HBN)—HBN is a national network of green building professionals, environmental and health activists, socially responsible investment advocates, and others interested in promoting healthier building materials as a means of improving public health and preserving the global environment. www.healthybuilding.net

Heap Leach Solutions—The separation, or dissolving-out from mined rock of the soluble uranium constituents by the natural action of percolating a prepared chemical solution through mounded (heaped) rock material. The mounded material usually contains low grade mineralized material and/or waste rock produced from open pit or underground mines. The solutions are collected after percolation is completed and processed to recover the valued components.

Heat—A form of thermal energy resulting from combustion, chemical reaction, friction, or movement of electricity. As a thermodynamic condition, heat, at a constant pressure, is equal to internal or intrinsic energy plus pressure times volume.

Heat Absorbing Window Glass—A type of window glass that contains special tints that cause the window to absorb as much as 45% of incoming solar energy, to reduce heat gain in an interior space. Part of the absorbed heat will continue to be passed through the window by conduction and re-radiation.

Heat Balance—The outdoor temperature at which a building's internal heat gain (from people, lights and machines) is equal to the heat loss through windows, roof and walls.
  • The accounting of the energy output and losses from a system to equal the energy input.

Heat Capacity (HC)—The amount of heat necessary to raise the temperature of all the components of a unit area in an assembly by VF.—It is calculated as the sum of the average thickness times the density times the specific heat for each component, and is expressed in Btu per square foot per °F.

Heat Content—Measurement: The gross heat content (or heating value), is the number of British thermal units (Btu) produced by the combustion, at constant pressure, of the amount of the gas that would occupy a volume of one cubic foot at a temperature of 60 degrees Fahrenheit, if saturated with water vapor and under a pressure equivalent to 30 inches of mercury at 32 degrees Fahrenheit and under standard gravitational force (980.665 cm per sec.^{2}), with air of the same temperature and pressure as the gas, when the products of combustion are cooled to the initial temperature of gas and air and when the water formed by combustion is condensed to the liquid state.

Heat Engine—A device that produces mechanical energy directly from two heat reservoirs of different temperatures. A machine that converts thermal energy to mechanical energy, such as a steam engine or turbine.

Heat Exchanger—A device used to transfer heat from a fluid (liquid or gas) to another fluid where the two fluids are physically separated.

Heat Gain—An increase in the amount of heat contained in a space, resulting from direct solar radiation, heat flow through walls, windows, and other building surfaces, and the heat given off by people, lights, equipment, and other sources.

Heat Island Effect—A "dome" of elevated temperatures over an urban area caused by structural and pavement heat fluxes, and pollutant emissions.

Heat Loss—A decrease in the amount of heat contained in a space, resulting from heat flow through walls, windows, roof and other building surfaces and from exfiltration of warm air. It is the loss of heat from a building when the outdoor temperature is lower than the desired indoor temperature. It represents the amount of heat that must be provided to a space to maintain indoor comfort conditions.

Heat Pipe—A device that transfers heat by the continuous evaporation and condensation of an internal fluid as found in evacuated tube solar thermal units.

Heat Pump—Heating and/or cooling equipment that, during the heating season, draws heat into a building from outside and, during the cooling season, ejects heat from the building to the outside. Heat pumps are vapor-compression refrigeration systems whose indoor/outdoor coils are used reversibly as condensers or evaporators, depending on the need for heating or cooling.

Heat Pump (Air Source)—An air-source heat pump is the most common type of heat pump. The heat pump absorbs heat from the outside air and transfers the heat to the space to be heated in the heating mode. In the cooling mode the heat pump absorbs heat from the space to be cooled and rejects the heat to the outside air. In the heating mode when the outside air approaches 32°F or less, air-source heat pumps loose efficiency and generally require a back-up (resistance) heating system.

Heat Pump (Geothermal)—A heat pump in which the refrigerant exchanges heat (in a heat exchanger) with a fluid circulating through an earth connection medium (ground or ground water). The fluid is contained in a variety of loop (pipe) configurations depending on the temperature of the ground and the ground area available. Loops may be installed horizontally or vertically in the ground or submersed in a body of water.

Heat Pump Efficiency—The efficiency of a heat pump, that is, the electrical energy to operate it, is directly related to temperatures between which it operates. Geothermal heat pumps are more efficient than conventional heat pumps or air conditioners that use the outdoor air since the ground or ground water a few feet below the earth's surface remains relatively constant throughout the year. It is more efficient in the winter to draw heat from the relatively warm ground than from the atmosphere where the air temperature is much colder, and in summer transfer waste heat to the relatively cool ground than to hotter air. Geothermal heat pumps are generally more expensive ($2,000-$5,000) to install than outside air heat pumps. However, depending on the location geothermal heat pumps can reduce energy consumption (operating cost) and correspondingly, emissions by more than 20 percent compared to high-efficiency outside air heat pumps. Geothermal heat pumps also use the waste heat from air-conditioning to provide free hot water heating in the summer.

Heat Pump Water Heaters—A water heater that uses electricity to move heat from one place to another instead of generating heat directly.

Heat Rate—The amount of fuel energy required by a power plant to produce one kilowatt-hour of electrical output. A measure of generating station thermal efficiency, generally expressed in Btu per net kWh. It is computed by dividing the total Btu content of fuel burned for electric generation by the resulting net kWh generation.

Heat Recovery—Use of byproduct heat as a source of energy.

Heat Recovery Ventilator—A device that captures the heat from the exhaust air from a building and transfers it to the supply/fresh air entering the building to preheat the air and increase overall heating efficiency.

Heat Register—The grilled opening into a room by which the amount of warm air from a furnace can be directed or controlled; may include a damper.

Heat Sink—A structure or media that absorbs heat. The medium—air, water or earth—which receives heat rejected from a heat pump. Could also act as a heat source.

Heat Source—A structure or media from which heat can be absorbed or extracted.

Heat Storage—A device or media that absorbs heat for storage for later use.

Heat Storage Capacity—The amount of heat that a material can absorb and store.

Heat Transfer—The flow of heat from one area to another by conduction, convection, and/or radiation. Heat flows naturally from a warmer to a cooler material or space.

Heat Transfer Coefficient—The quantity of heat transferred through a unit area of a material in a unit time per unit of temperature difference between the two sides of the material.

Heat Transfer Efficiency—The useful heat output released to a room divided by the actual heat produced in the firebox.

Heat Transfer Fluid—A gas or liquid used to move heat energy from one place to another; a refrigerant.

Heat Transmission Coefficient—Any coefficient used to calculate heat transmission by conduction, convection, or radiation through materials or structures.

Heat, Latent—Change in heat content of a substance when its physical state is changed without a change in temperature; i.e., boiling or melting.

Heat, Sensible—That heat which, when added or subtracted, results in a change of temperature, as distinguished from latent heat.

Heat, Specific—The heat required to raise a unit mass of a substance through a degree of temperature difference. Also, the ratio of the thermal capacity of a substance to that of water at 60 degrees F (15.6 degrees C). Interchangeable with "heat capacity" in common usage.

Heated Floor Space—The area within a building that is space heated.

Heated Slab Floor—A concrete slab floor or a lightweight concrete topping slab laid over a raised floor, with embedded space heating hot water pipes. The heating system using the heated slab is sometimes referred to as radiant slab floors or radiant heating.

Heater, Infrared Radiant—A self-contained, vented, or unvented heater used to convert the combustion energy to radiant energy, a substantial portion of which is in the infrared spectrum, for the purpose of direct heat transfer.

Heating Balance Point—The outdoor temperature below which additional energy will be consumed at a meter to satisfy heating loads.

Heating Capacity (Also specific heat)—The quantity of heat necessary to raise the temperature of a specific mass of a substance by one degree.

Heating Degree Day(s) (HDD)—A measure of how cold a location is over a period of time relative to a base temperature, most commonly specified as 65 degrees Fahrenheit. The measure is computed for each day by subtracting the average of the day's high and low temperatures from the base temperature (65 degrees), with negative values set equal to zero. Each day's heating degree-days are summed to create a heating degree-day measure for a specified reference period. Heating degree-days are used in energy analysis as an indicator of space heating energy requirements or use.

Heating Equipment—Any equipment designed and/or specifically used for heating ambient air in an enclosed space. Common types of heating equipment include: central warm air furnace, heat pump, plug-in or built-in room heater, boiler for steam or hot water heating system, heating stove, and fireplace. Note: A cooking stove in a housing unit is sometimes reported as heating equipment, even though it was built for preparing food.

Heating Fuel Units—Standardized weights or volumes for heating fuels.

Heating Fuels—Any gaseous, liquid, or solid fuel used for indoor space heating.

Heating Intensity—The ratio of space-heating consumption or expenditures to square footage of heated floor space and heating degree-days (base 65 degrees Fahrenheit). This ratio provides a way of comparing different types of housing units and households by controlling for differences in housing unit size and weather conditions. The square footage of heated floor space is based on the measurements of the floor space that is heated. The ratio is calculated on a weighted, aggregate basis according to the following formula: Heating Intensity = Btu for Space Heating/(Heated Square Feet * Heating Degree-Days).

Heating Load—The rate of heat flow required to maintain a specific indoor temperature; usually measured in Btu per hour.

Heating Season—The coldest months of the year; months where average daily temperatures fall below 65 degrees Fahrenheit creating demand for indoor space heating.

Heating Seasonal Performance Factor (HSPF)—A representation of the total heating output of a central air-conditioning heat pump in Btu's during its normal usage period for heating, divided by the total electrical energy input in watt-hours during the same period, as determined using the test procedure specified in the California Code of Regulations, Title 20, Section 1603(c).

Heating Stove Burning Wood, Coal, or Coke—Any free-standing box or controlled-draft stove; or a stove installed in a fireplace opening, using the chimney of the fireplace. Stoves are made of cast iron, sheet metal, or plate steel. Free-standing fireplaces that can be detached from their chimneys are considered heating stoves.

Heating Value—The amount of heat produced from the complete combustion of a unit of fuel. The higher (or gross) heating value is that when all products of combustion are cooled to the pre-combustion temperature, water vapor formed during combustion is condensed, and necessary corrections have been made. Lower (or net) heating value is obtained by subtracting from the gross heating value the latent heat of vaporization of the water vapor formed by the combustion of the hydrogen in the fuel.

Heating, Ventilating and Air Conditioning (HVAC) System—Is the mechanical heating, ventilating and air conditioning system of the building, also known as the HVAC system. The standards use various measures of equipment efficiency defined according to the type of equipment installed.

Gas (fossil fuel) heating equipment is rated by the Annual Fuel Utilization Efficiency (AFUE). The heating efficiency of electric heat pumps with less than 65,000 Btu/h cooling capacity is rated by the Heating Seasonal Performance Factor (HSPF). The heating efficiency of heat pumps with cooling capacity of 65,000 Btu/h or more is rated by the Coefficient of Performance (COP). Electric resistance heating is rated by HSPF or COP.

All electric cooling equipment (including heat pump cooling equipment) with less than 65,000 Btu/h output capacity is rated by the Seasonal Energy Efficiency Ratio (SEER) (equipment of this size may also be rated by the EER). Electric cooling equipment (including heat pump cooling equipment) with an output capacity of 65,000 Btu/h or more is rated by the Energy Efficiency Ratio (EER).

Heavy Gas Oil—Petroleum distillates with an approximate boiling range from 651 degrees to 1000 degrees Fahrenheit.

Heavy Metals—Metallic elements, including those required for plant and animal nutrition, in trace concentration but which become toxic at higher concentrations. Examples are mercury, chromium, cadmium, and lead.

Heavy Oil—The fuel oils remaining after the lighter oils have been distilled off during the refining process. Except for start-up and flame stabilization, virtually all petroleum used in steam plants is heavy oil. Includes fuel oil numbers 4, 5, and 6; crude; and topped crude.

Heavy Rail—An electric railway with the capacity for a "heavy volume" of traffic and characterized by exclusive rights-of-way, multi-car trains, high speed and rapid acceleration, sophisticated signaling, and high platform loading. Also known as "subway," elevated (railway), "metropolitan railway (metro)."

Heavy Water—Water containing a significantly greater proportion of heavy hydrogen (deuterium) atoms to ordinary hydrogen atoms than is found in ordinary (light) water. Heavy water is used as a moderator in some reactors because it slows neutrons effectively and also has a low cross section for absorption of neutrons.

Heavy-Water-Moderated Reactor—A reactor that uses heavy water as its moderator. Heavy water is an excellent moderator and thus permits the use of inexpensive natural (unenriched) uranium as fuel.

Hectare—An area equal to 2.47 acres. There are 100 hectares in 1 square kilometer.

Hedging—The buying and selling of futures contracts so as to protect energy traders from unexpected or adverse price fluctuations.

• Any method of minimizing the risk of price change. Since the movement of cash prices is usually in the same direction and about in the same degree as the movement of the present prices of futures contracts, any loss (or gain) resulting from carrying the actual merchandise is approximately offset by a corresponding gain (or loss) when the contract is liquidated.

Hedging Contracts—Contracts which establish future prices and quantities of electricity independent of the short-term market. Derivatives may be used for this purpose. (See *Contracts for Differences, Forwards, Futures Market, and Options.*)

HEI—Health Effects Institute www.healtheffects.org

Helio Chemical—Using solar radiation to cause chemical reactions.

Helio Thermal—A process that uses the sun's rays to produce heat.

Heliochemical Process—The utilization of solar energy through photosynthesis.

Heliodon—A device used to simulate the angle of the sun for assessing shading potentials of building structures or landscape features.

Heliostat—A mirror that reflects solar rays onto a central receiver. A heliostat automatically adjusts its position to track daily or seasonal changes in the sun's position. The arrangement of heliostats around a central receiver is also called a solar collector field.

Heliothermal—Any process that uses solar radiation to produce useful heat.

Heliothermic—Site planning that accounts for natural solar heating and cooling processes and their relationship to building shape, orientation, and siting.

Heliothermometer—An instrument for measuring solar radiation.

Heliotropic—Any device (or plant) that follows the sun's apparent movement across the sky.

Hell-or-high-water Clause—A clause in a lease that states the unconditional obligation of the lessee to pay rent for the entire term of the lease, regardless of any event affecting the equipment or any change in the circumstances of the lessee.

HELM—Hourly Electric Load Model

Hemispherical Bowl Technology—A solar energy concentrating technology that uses a linear receiver that tracks the focal area of a reflector or array of reflectors.

Henry Hub—A pipeline hub on the Louisiana Gulf coast. It is the delivery point for the natural gas futures contract on the New York Mercantile Exchange (NYMEX).

Heptachlor—An organochloride insecticide once widely used on food crops, especially corn, but has not been in use since 1988. It is listed as a cancer-causing chemical under Proposition 65 in California.

Herbicide—A chemical used to kill unwanted vegetation.

Herfindahl-Hirschman Index (HHI)—A measure of market concentration. The index is frequently used by the Department of Justice and the Federal Trade Commission to analyze mergers and acquisitions.

Hertz (HZ)—A measure of the number of cycles or wavelengths of electrical energy per second; U.S. electricity supply has a standard frequency of 60 hertz.

Heterogeneity—The condition or state of being different in kind or nature.

Heterojunction—A region of electrical contact between two different materials.

HEV—Hybrid electric vehicle with two or more sources of power, with one source electric.

HFC—Hydrofluorocarbon

Hg—Mercury

HGWP (High Global Warming Potential)—Some industrially produced gases such as sulfur hexafluoride (SF_6), perfluorocarbons (PFCs), and hydrofluorocarbons (HFCs) have extremely high GWPs. Emissions of these gases have a much greater effect on global warming than an equal emission (by weight) of the naturally occurring gases. Most of these gases have GWPs of 1,300-23,900 times that of CO_2. These GWPs can be compared to the GWPs of CO_2, CH_4, and N_2O which are presently estimated to be 1, 23 and 296, respectively.

HHV—See *Higher Heating Value*.

HI—The Hydronics Institute of the Gas Appliance Manufacturers Association (GAMA).

HI HTG Boiler Standard—The Hydronics Institute document entitled "Testing and Rating Standard for Rating Boilers," 1989.

Hibernacula—Caves or other structures used by bats for hibernation.

High Efficiency Ballast—A lighting conservation feature consisting of an energy-efficient version of a conventional electromagnetic ballast. The ballast is the transformer for fluorescent and high-intensity discharge (HID) lamps, which provides the necessary current, voltage, and wave-form conditions to operate the lamp. A high-efficiency ballast requires lower power input than a conventional ballast to operate HID and fluorescent lamps.

High Efficiency Lighting—Lighting provided by high-intensity discharge (HID) lamps and/or fluorescent lamps.

High Gearing—A high debt to equity ratio.

High Heat Value (HHV)—The high or gross heat content of the fuel with the heat of vaporization included. The water vapor is assumed to be in a liquid state.

High Voltage Disconnect—Voltage at which the charge controller will disconnect the array to prevent overcharging the batteries.

High Voltage Disconnect Hysteresis—The voltage difference between the high voltage disconnect set point and the voltage at which the full photovoltaic array current will be reapplied.

Higher Heating Value (HHV)—The maximum heating value of a fuel sample, which includes the calorific value of the fuel (bone dry) and the latent heat of vaporization of the water in the fuel. (See *Moisture Content* and *Net (lower) Heating Value*, below.)

Highest and Best Use—The reasonably probable and

legal use of vacant land or an improved property, which is physically possible, appropriately supported, financially feasible, and that results in the highest value. The four criteria the highest and best use must meet are legal permissibility, physical possibility, financial feasibility, and maximum productivity.

High-Intensity Discharge (HID) Lamp—A lamp that produces light by passing electricity through gas, which causes the gas to glow. Examples of HID lamps are mercury vapor lamps, metal halide lamps, and high-pressure sodium lamps. HID lamps have extremely long life and emit far more lumens per fixture than do fluorescent lights.

High-low Debt—Debt with higher payments early in the term of the agreement.

High-Mileage Households—Households with estimated aggregate annual vehicle mileage that exceeds 12,500 miles.

High-Priority Customers—Customers with priority in use in utility curtailment

High-Rise Residential Building—(in California) A building, other than a hotel/motel, of Occupancy Group R, Division 1 with four or more habitable stories.

High-Sulfur Coal—Coal whose weight is more than one percent sulfur.

High-Temperature Collector—A solar thermal collector designed to operate at a temperature of 180 degrees Fahrenheit or higher.

Highwall—The unexcavated face of exposed over-burden and coal in a surface mine.

Hinshaw Amendment—An amendment to the Natural Gas Act which exempts from Federal Energy Regulatory Commission regulation the transportation and sale for resale of natural gas received within the boundaries of a state, provided (1) all such gas is ultimately consumed within the state, and (2) the facilities and rates are regulated by the state. Pipelines qualifying under this amendment are called Hinshaw Pipelines (see below).

Hinshaw Pipeline—A pipeline or local distribution company that has received exemptions from regulations pursuant to the Natural Gas Act (NGA). These companies transport interstate natural gas not subject to regulations under NGA.

Historical Cost—The actual cost of land, buildings, pipelines and other plant items to the company, when used in ratemaking it assumes the company's acquisition costs are prudent. The difference with original cost is the acquisition adjustment.

HL—Hubbert Linearization

HLBV—*Hypothetical Liquidation at Book Value* An accounting term used in partnership accounting of alternative energy projects. Structuring options can be used by tax investors to mitigate post-flip GAAP (generally accepted accounting principles) losses calculated by the Hypothetical Liquidation at Book Value (HLBV) accounting method for flipping partnerships. In some circumstances, these GAAP losses can dramatically affect the reported earnings for publicly traded companies and influence investment decisions in renewable energy projects.

HMTA—Hazardous Materials Transportation Act (1975)

HMTR—Hazardous Materials Transportation Regulations

HN—Host Nations

HO—Headquarters Office

Hog Fuel or Hogged Fuel—Wood residues processed through a chipper or mill to produce coarse chips normally used for fuel. Bark, sawdust, planer shavings, wood chunks, dirt, and fines may be included.

Holdback—A portion of a loan commitment that is not funded until an additional requirement is met, such as completion of construction.

Holding Company—A company that confines its activities to owning stock in and supervising management of other companies. The Securities and Exchange Commission, as administrator of the Public Utility Holding Company Act of 1935, defines a holding company as "a company which directly or indirectly owns, controls or holds 10 percent or more of the outstanding voting securities of a holding company" (15 USC 79b, par. a (7)).

Holding Period—The amount of time an investor has held an investment. The period begins on the date of purchase and ends on the date of sale, and determines whether a gain or loss is considered short-term or long-term, for capital-gains-tax purposes.

Holding Pond—A structure built to contain large volumes of liquid waste to ensure that it meets environmental requirements prior to release.

Hole—The vacancy where an electron would normally exist in a solid; behaves like a positively charged particle.

Home Energy Assistance Program (HEAP)—A centrally operated direct payment program that assists eligible households in offsetting the cost of heating and cooling their homes. Payments are generally made in the form of dual party warrants (checks)

made payable to the applicant and their designated utility company. The program is administered by the California Department of Economic Opportunity using federal and state funds. The toll-free number for the HEAP Program is (800) 433-4327. For more information about your utility bills.

Home Energy Rating System Provider—An organization that the Commission has approved to PROVIDER—administer a home energy rating system program, certify raters and maintain quality control over field verification and diagnostic testing required for compliance with the Energy Efficiency Standards.

Home Energy Rating System Rater—(in California) A person certified by a Commission approved HERS Provider to perform the field verification and diagnostic testing required for demonstrating compliance with the Energy Efficiency Standards.

Home Energy Rating Systems (HERS)—A nationally recognized energy rating program that gives builders, mortgage lenders, secondary lending markets, homeowners, sellers, and buyers a precise evaluation of energy losing deficiencies in homes. Builders can use this system to gauge the energy quality in their home and also to have a star rating on their home to compare to other similarly built homes.

Homojunction—The region between an n-layer and a p-layer in a single material, photovoltaic cell.

Horizontal Axis Wind Turbine—The most common type of wind turbine where the axis of rotation is oriented horizontally. Also see *Wind Turbine*.

Horizontal Ground Loop—In this type of closed-loop geothermal heat pump installation, the fluid-filled plastic heat exchanger pipes are laid out in a plane parallel to the ground surface. The most common layouts either use two pipes, one buried at six feet, and the other at four feet, or two pipes placed side-by-side at five feet in the ground in a two-foot wide trench. The trenches must be at least four feet deep. Horizontal ground loops are generally most cost-effective for residential installations, particularly for new construction where sufficient land is available.

Horizontal Wells—Extraction and monitoring wells are typically drilled vertically. A horizontal well has the advantage of providing a large area of groundwater capture for a lower overall cost.

Horizontal-Axis Wind Turbines—Turbines in which the axis of the rotor's rotation is parallel to the wind stream and the ground.

Horsepower (HP)—A unit for measuring the rate of mechanical energy output. The term is usually applied to engines or electric motors to describe maximum output. 1 hp = 745.7 Watts = 0.746 kW = 2,545 Btu/hr.

Horsepower Hour (hph)—One horsepower provided over one hour; equal to 0.745 kilowatt-hour or 2,545 Btu.

Horsepower, Boiler (Bhp)—The equivalent evaporation of 34.5 lbs. of water per hour at 212 degrees F and above. This is equal to a heat output of 33,475 Btu per hour.

Horsepower, Brake (bhp)—The power developed by the engine, as measured at the crank shaft or flywheel by the Prony brake or other device.

Host Country—The country where the reduction, avoidance or sequestration of greenhouse gas takes place.

Host Customer—An individual or entity that meets all of the following criteria:
- Has legal rights to occupy the site;
- Receives retail level electric service from a utility;
- Is the utility customer of record at the site;
- Is connected to the electric grid; and
- Is the recipient of the net electricity generated from the energy generation equipment

Purchaser of electricity, heat or other type of energy. The host allows energy production equipment to be installed on its site or facility and enters into a power purchase agreement or similar agreement under which it agrees to purchase the energy from the owner of the energy system.

Host Government—The government (including any government-controlled firm engaged in the production, refining, or marketing of crude oil or petroleum products) of the foreign country in which the crude oil is produced.

HOT (Colloquial)—The word is sometimes used to describe electric utility lines that are carrying electric currently. It also is used to refer to anything that is highly radioactive.

Hot Air—A situation in which emissions (of a country, sector, company or facility) are well below a target due to the target being above emissions that materialized under the normal course of events (i.e. without deliberate emission reduction efforts). Hot air can result from over-optimistic projections of growth. Emissions are often projected to grow roughly in proportion to GDP, and GDP is often projected to grow at historic rates. If a recession occurs and fuel use declines, emissions may be well below targets since targets are generally set in relation to emission projections. If emission trading is allowed, an emitter could sell the difference between actual emissions and emission targets. Such emissions are considered hot air because they do not represent reductions from what would have occurred in the normal course of events.

Hot Air Furnace—A heating unit where heat is distributed by means of convection or fans.

Hot Dry Rock—A geothermal energy resource that consists of high temperature rocks above 300°F (150°C) that may be fractured and have little or no water. To extract the heat, the rock must first be fractured, then water is injected into the rock and pumped out to extract the heat. In the western United States, as much as 95,000 square miles (246,050 square km) have hot dry rock potential.

Hot Spot (Solar)—A phenomenon where one or more cells within a PV module or array act as a resistive load, resulting in local overheating or melting of the cells.

Hot Spot Criteria—Cleanup levels for small areas on the site that have particularly high concentrations of hazardous substances.

Hot Tub—Water-filled wood, plastic, or ceramic container in which up to 12 people can lounge. Normally equipped with a heater that heats the water from 80 degrees to 106 degrees Fahrenheit. It may also have jets to bubble the water. The water is not drained after each use. An average-size hot tub holds 200 to 400 gallons of water. All reported hot tubs are assumed to include an electric pump. These are also called spas or jacuzzis.

Hot Water Heating Systems—(See *Hydronic*)

Hotel/Motel—A building or buildings incorporating six or more guest rooms or a lobby serving six or more guest rooms, where the guest rooms are intended or designed to be used, or which are used, rented, or hired out to be occupied, or which are occupied for sleeping purposes by guests, and all conditioned spaces within the same building envelope. Hotel/motel also includes all conditioned spaces which are (1) on the same property as the hotel/motel, (2) served by the same central heating, ventilation, and air-conditioning system as the hotel/motel, and (3) integrally related to the functioning of the hotel/motel as such, including, but not limited to, exhibition facilities, meeting and conference facilities, food service facilities, lobbies, and laundries.

Hourly Metering—Tracking or recording a customer's consumption during specific periods of time that can be tied to the price of energy.

Hourly Non-Firm Transmission Service—Transmission scheduled and paid for on an as-available basis and subject to interruption.

Hours Under Load—The hours the boiler is operating to drive the generator producing electricity.

Household—A family, an individual, or a group of up to nine unrelated persons occupying the same housing unit. "Occupy" means that the housing unit is the person's usual or permanent place of residence.

Household Energy Expenditures—The total amount of funds spent for energy consumed in, or delivered to, a housing unit during a given period of time.

Housing Unit—A house, an apartment, a group of rooms, or a single room if it is either occupied or intended for occupancy as separate living quarters by a family, an individual, or a group of one to nine unrelated persons. Separate living quarters means the occupants (1) live and eat separately from other persons in the house or apartment and (2) have direct access from the outside of the buildings or through a common hall—that is, they can get to it without going through someone else's living quarters. Housing units do not include group quarters such as prisons or nursing homes where ten or more unrelated persons live. A common dining area used by residents is an indication of group quarters. Hotel and motel rooms are considered housing units if occupied as the usual or permanent place of residence.

HOV—High Occupancy Vehicle

Hp—See *Horsepower*
  • HEPA Filter

HQ—Headquarters

HRA—Hourly Rolling Average
  • Health Risk Assessment

HRSG—Heat Recovery Steam Generator

HSIA—Halogenated Solvent Industry Alliance. www.hsia.org.

HSPF (Heating Seasonal Performance Factor)—A measure of heating efficiency for the total heating output of a central air-conditioning heat pump. Efficiency is derived according to federal test methods by using the total Btu's during its normal usage period for heating divided by the total electrical energy input in watt-hours during the same period. California Code of Regulations, Section 2-1602(c)(7).

HTC—Historic Tax Credits. Federal investment tax credits provided to taxpayers that rehabilitate a certified historic building.

HTF—Heat Transfer Fluid

HTHE—High Temperature Heat Exchanger

HTP—High Temperature and Pressure

HTS—Harmonized Tariff Schedule
- High-temperature superconducting

Hub Height—In a horizontal-axis wind turbine, the distance from the turbine platform to the rotor shaft.

Humidifier—A humidifier adds moisture to the air (often needed in winter when indoor air is very dry). It may be a portable unit or attached to the heating system.

Humidistat—A regulating device, actuated by changes in humidity, used for the automatic control of relative humidity.

Humidity—The moisture content of air. Relative humidity is the ratio of the amount of water vapor actually present in the air to the greatest amount possible at the same temperature.

Humidity, Relative—The ratio of the weight of water vapor in the atmosphere to the weight the air would hold if completely saturated at that temperature, expressed as a percentage.

Hurdle Rate—The minimum return or reward a company will accept in order to fund a project. In leasing, the hurdle rate is the lowest return a lease company would accept on a given lease.
- The internal rate of return that a fund or company must achieve before its general partners or managers may receive an increased interest in the proceeds of the fund. Often, if the expected rate of return on an investment is below the hurdle rate, the project is not undertaken.

HV—High Voltage

HVAC (Heating Ventilation and Air Conditioning)—An abbreviation for the heating, ventilation, and air-conditioning system; the system or systems that condition air in a building.

HVAC Conservation Feature—A building feature designed to reduce the amount of energy consumed by the heating, cooling, and ventilating equipment.

HVAC DSM Program—A DSM (demand-side management) program designed to promote the efficiency of the heating or cooling delivery system, including replacement. Includes ventilation (economizers; heat recovery from exhaust air), cooling (evaporative cooling, cool storage; heat recovery from chillers; high-efficiency air conditioning), heating, and automatic energy management systems.

HVDC—High Voltage Direct Current

HW—Hazardous Waste

HWC—Hazardous Waste Combustor

HWCL—Hazardous Waste Control Law

HWM—Hazardous Waste Management

HX—Heat Exchanger

Hybrid Solar Lighting System—Systems that use solar energy to illuminate the inside of a structure using fiber-optic distributed sunlight.

Hybrid System—A renewable energy system that includes two different types of technologies that produce the same type of energy; for e.g., a wind turbine and a solar photovoltaic array combined to meet a power demand. A self-generation system that combines more than one type of distributed generation technology and is located behind a single Electric Utility service meter.

Hybrid Transmission Line—A double-circuit line that has one alternating current and one direct circuit. The AC circuit usually serves local loads along the line.

Hybrid Vehicle—Usually a hybrid EV, a vehicle that employs a combustion engine system together with an electric propulsion system. Hybrid technologies expand the usable range of EV's beyond what an all-electric-vehicle can achieve with batteries only.

Hydraulic Fracturing—Fracturing of rock at depth with fluid pressure. Hydraulic fracturing at depth may be accomplished by pumping water into a well at very high pressures. Under natural conditions, vapor pressure may rise high enough to cause fracturing in a process known as hydrothermal brecciation.

Hydraulic Head—The distance between the respective elevations of the upstream water surface (headwater) above and the downstream surface water (tailwater) below a hydroelectric power plant.

Hydraulic Load—Amount of liquid going into a system.

Hydro Thermal Systems—Underground reservoirs that produce either dry steam or a mixture of steam and water.

Hydrocarbon—An organic chemical compound of hydrogen and carbon in either the gaseous, liquid, or solid phase. The molecular structure of hydrocarbon compounds varies from the simplest (e.g., methane, a constituent of natural gas) to the very heavy and very complex.

Hydrochloric Acid—Clear, colorless and acidic solution of hydrogen chloride in water often used in metal cleaning and electroplating. Many hazardous wastes contain chlorine compounds which create small amounts of hydrogen chloride when they are burned. This can contribute to the formation of acid rain. Regulations require that air pollution control equipment remove either 99% of the hydrochloric acid, or that the emissions contain less than four pounds per hour.

Hydrochlorofluorocarbons (HCFCs)—Chemicals composed of one or more carbon atoms and varying numbers of hydrogen, chlorine, and fluorine atoms. It is one of the primary GHGs primarily used as refrigerants, consisting of a class of gases containing hydrogen, fluorine, and carbon.

• Compounds containing carbon, hydrogen, and fluorine that are emitted in various refrigeration processes. They do not destroy ozone. The market for HFCs is expanding as CFCs are being phased out.

Hydroelectric Plant—A plant in which the turbine generators are driven by falling water.

Hydroelectric Power—Electricity produced by falling water that turns a turbine generator. Also referred to as HYDRO.

Hydroelectric Power Plant—A power plant that produces electricity by the force of water falling through a hydro turbine that spins a generator.

Hydroelectric Spill Generation—Hydroelectric generation in existence prior to January 1, 1998, that has no storage capacity and that, if backed down, would spill. This term also refers to a hydro resource that has exceeded or has inadequate storage capacity and is spilling, even though generators are operating at full capacity.

Hydrofluorocarbons (HFCs)—HFCs are synthetic industrial gases, primarily used in refrigeration and semi-conductor manufacturing as commercial substitutes for chlorofluorocarbons (CFCs). There are no natural sources of HFCs. The atmospheric lifetime of HFCs is decades to centuries, and they have 100-year "global warming potentials" thousands of times that of CO_2, depending on the gas. HFCs are among the six greenhouse gases to be curbed under the Kyoto Protocol.

Hydrogen—A colorless, odorless, highly flammable gaseous element. It is the lightest of all gases and the most abundant element in the universe, occurring chiefly in combination with oxygen in water and also in acids, bases, alcohols, petroleum, and other hydrocarbons.

Hydrogen Fuel Cell—A device that converts hydrogen to DC electricity.

Hydrogen Plant—A facility that produces hydrogen with steam hydrocarbon reforming, partial oxidation of hydrocarbons, or other processes.

Hydrogen Sulfide (H_2S)—A poisonous, corrosive compound consisting of two atoms of hydrogen and one of sulfur, gaseous in its natural state. It is found in manufactured gas made from coals or oils containing sulphur and must be removed. It is also found to some extent in some natural gas. It is characterized by the odor of rotten eggs.

Hydrogenated Amorphous Silicon—Amorphous silicon with a small amount of incorporated hydrogen. The hydrogen neutralizes dangling bonds in the amorphous silicon, allowing charge carriers to flow more freely.

Hydrogeology—The geology of groundwater, with particular emphasis on the chemistry and movement of water.

Hydrolysis—A process of breaking chemical bonds of a compound by adding water to the bonds.

Hydrometer—A hydrometer is an instrument for measuring the density of liquids in relation to the density of water. The hydrometer is used to indicate the state of charge in lead-acid cells by measuring the specific gravity of the electrolyte.

Hydronic Cooling System—Any cooling system which uses water or a water solution as a source of cooling or heat rejection, including chilled water systems (both air and water-cooled) as well as water-cooled or evaporatively cooled direct expansion systems, such as water source (water-to-air) heat pumps.

Hydronic Heating Systems—A type of heating system where water is heated in a boiler and either moves by natural convection or is pumped to heat exchangers or radiators in rooms; radiant floor systems have a grid of tubing laid out in the floor for distributing heat. The temperature in each room is controlled by regulating the flow of hot water through the radiators or tubing.

Hydronic Space Heating System—A system that uses water-heating equipment, such as a storage tank water heater or a boiler, to provide space heating. Hydronic space heating systems include both radiant floor systems and convective or fan coil systems.

See *Combined Hydronic Space/Water Heating System*

Hydronics—Heating and/or cooling with circulated water.

Hydropower—Electrical energy produced by falling or flowing water.

Hydrothermal Fluids—These fluids can be either water or steam trapped in fractured or porous rocks; they are found from several hundred feet to several miles below the Earth's surface. The temperatures vary from about 90 F to 680 F (32 C to 360 C) but roughly 2/3 range in temperature from 150 F to 250 F (65.5 C to 121.1 C). The latter are the easiest to access and, therefore, the only forms being used commercially.

Hydroxyl Radical (OH)—An important chemical scavenger of many trace gases in the atmosphere that are greenhouse gases. Atmospheric concentrations of OH affect the atmospheric lifetimes of greenhouse gases, their abundance, and, ultimately, the effect they have on climate.

Hygas—A process that uses water to help produce pipeline-quality gas from coal.

Hygrometer—An instrument for determining the relative humidity of air or other gases.

Hypothecate—To give a thing as security without parting with possession.

Hypothesis—A proposition tentatively assumed. Science tests the logical consequences of a hypothesis against facts that are known or that may be determined.

Hypothetical Resources (Coal)—Undiscovered coal resources in beds that may reasonably be expected to exist in known mining districts under known geologic conditions. In general, hypothetical resources are in broad areas of coalfields where points of observation are absent and evidence is from distant outcrops, drill holes, or wells. Exploration that confirms their existence and better defines their quantity and quality would permit their reclassification as identified resources. Quantitative estimates are based on a broad knowledge of the geologic character of coalbed or region. Measurements of coal thickness are more than 6 miles apart. The assumption of continuity of coalbed is supported only by geologic evidence.

Hz—Hertz

I

I&M—Inspection & Maintenance

IA—Inter-Agency Agreement

IAEA—International Atomic Energy Agency www.iaea.org

IAG—Inter-Agency Agreement

IAP—Indoor Air Pollution

IAQ—Indoor Air Quality

IAR—Issues and Alternatives Report

IAS—International Accounting Standards. Regulations governing the submission of accounts with a numerical designation (IAS 1 to IAS 32) that were issued by the IASC (International Accounting Standards Committee). The key aim of these accounting regulations is to provide decision-related information for a large circle of parties involved in the submission of annual accounts while maintaining the basic principles of comprehensibility, relevance to decision-making, comparability and reliability.

IBEW—International Brotherhood of Electrical Workers: Represents approximately 750,000 members who work in a wide variety of fields, including utilities, construction, telecommunications, broadcasting, manufacturing, railroads and government. www.ibew.org

IBR—Incorporation by Reference

IBRD—International Bank for Reconstruction and Development

IC—Internal Combustion
  • Ion Chromatography

ICC—Interstate Commerce Commission

ICCA—International Council of Chemical Associations www.icca-chem.org

ICCP—International Climate Change Partnership www.iccp.net

ICE—Internal Combustion Engine

ICV—Initial Calibration Verification

ID—Inside Diameter

Ideal Gas Law—The ideal gas law is the combination of the volume, temperature, and pressure relationships of Boyle's and Charles' laws resulting in the relationship PV=RT. Real gases deviate by varying amounts from the ideal gas law.

Identified Resources (Coal)—Coal deposits whose location, rank, quality, and quantity are known from geologic evidence supported by engineering measurements. Included are beds of bituminous coal and anthracite (14 or more inches thick) and beds of sub-bituminous coal and lignite (30 or more inches thick) that occur at depths to 6,000 feet. The existence and quantity of these beds have been delineated within specified degrees of geologic assurance as measured, indicated, or inferred. Also included are thinner and/or deeper beds that presently are being mined or for which there is evidence that they could be mined commercially.

IDIQ—Indefinite delivery, indefinite quantity

Idle Capacity—The component of operable capacity that is not in operation and not under active repair, but capable of being placed in operation within 30 days; and capacity not in operation but under active repair that can be completed within 90 days.

IDR—Interval Data Recorder

IEA—International Energy Agency www.iea.org

IEC (International Electrotechnical Commission)—The IEC is an international organization that prepares and publishes international standards for all electrical, electronic and related technologies. IEC standards cover a number of areas as they impact electricity and electronics including the environment, electrical energy efficiency, renewable energies and safety and performance. The IEC also manages conformity assessment schemes that certify that equipment, systems or components conform to its international standards. www.iec.ch.

IEEE—Institute of Electrical and Electronics Engineers, Inc. www.ieee.usa.org.

IEP—Independent Energy Producers Association: California's oldest and leading trade association representing the interests of developers and operators of independent energy facilities and independent power marketers. Independent energy producers include producers of renewable products derived from biomass, geothermal, small hydro, solar, and wind; producers of highly efficient cogeneration; and owners/operators of gas-fired merchant facilities. www.iepa.com

IEPR—Integrated Energy Policy Report

IER—Incremental Energy Rate: A measure of the efficiency of an electric utility's power production.

IESNA Lighting Handbook—The Illuminating Engineering Society National Association document entitled "The IESNA Lighting Handbook: Reference and Applications, Ninth Edition." (2000)

IFB—Invitation for Bid

IFR—Interim Final Rule

IG—Inspector General

IGA—Investment grade audit

IGCC—Integrated Gasification Combined Cycle

Ignitability—A characteristic of hazardous waste. If a liquid (containing less than 24% alcohol) has a flash point less than 140° F, it is a hazardous waste in the United States.

Ignite—To heat a gaseous mixture to the temperature at which combustion takes place.

Ignition Point—The minimum temperature at which combustion of a solid or fluid can occur.

Ignition Temperature—The temperature at which a substance, such as gas, will ignite and continue burning with adequate air supply.

IIR—Iswsues Identification Report

ILEV (Inherently Low Emission Vehicle)—Term used by federal government for any vehicle that is certified to meet the California Air Resources Board's Low Emission Vehicle (LEV) standards for non-methane organic gases and carbon monoxide, ULEV standards for nitrogen oxides and does not emit any evaporative emissions.

Illuminance—A measure of the amount of light incident on a surface; measured in foot-candles or Lux.

Illuminants—The group of unsaturated or heavy hydrocarbons in a manufactured gas, such as ethylene and benzene, which burn with a luminous flame.

Imbalance Energy—The real-time change in generation output or demand requested by the ISO to maintain reliability of the ISO-controlled grid. Sources of imbalance energy include regulation, spinning and non-spinning reserves, replacement reserve, and energy from other generating units that are able to respond to the ISO's request for more or less energy.

IMBY—In my backyard

Imhoff Treatment System—In wastewater treatment, a tank without aeration or oxygenation where solids settle out. The solids are digested in a separate compartment in the bottom.

Impact Fee—A fee charged to developers to cover part or all of the costs of providing service.

Impedance—The opposition to power flow in an AC circuit. Also, any device that introduces such opposition in the form of resistance, reactance, or both. The impedance of a circuit or device is measured as the ratio of voltage to current, where a sinusoidal voltage and current of the same frequency are used for the measurement; it is measured in ohms.

Impermeable—Any formation that prohibits the passage of fluid or gas through it.

Implicit Price Deflator—The implicit price deflator, published by the U.S. Department of Commerce, Bureau of Economic Analysis, is used to convert nominal figures to real figures.

Implicit Rate—The discount rate that, when applied to the minimum lease payments (excluding executory costs) together with any unguaranteed residual, caused the aggregate present value at the inception of the lease to be equal to the fair market value (reduced by any lessor retained Investment Tax Credits) of the leased property. This rate is a critical component when determining whether a lease transaction is classified as an operating lease or a capital lease on the lessee's balance sheet.

• The rate which discounts all guaranteed and unguaranteed lease payments ("rent and residual") to the equipment cost at delivery date. Used in calculating the PV of the rents for the lessor's capital lease test.

Implied—Presumed or inferred, rather than expressed.

Imported Crude Oil Burned as Fuel—The amount of foreign crude oil burned as a fuel oil, usually as residual fuel oil, without being processed as such. Imported crude oil burned as fuel includes lease condensate and liquid hydrocarbons produced from tar sands, gilsonite, and oil shale.

Imported Refiners' Acquisition Cost (IRAC)—The average price for imported oil paid by U.S. refiners.

Imports—Receipts of goods into the 50 States and the District of Columbia from U.S. possessions and territories or from foreign countries.

Imports (Electric utility)—Power capacity or energy obtained by one utility from others under purchase or exchange agreement.

Impound Account—Funds collected by a Lender from a Borrower to guarantee payment of such items as Taxes, maintenance costs, and Hazard Insurance Premiums when due.

Impoundment—A body of water confined by a dam, dike, floodgate or other artificial barrier.

Improved Recovery—The operation whereby crude oil or natural gas is recovered using any method other than those that rely primarily on the use of natural reservoir pressure, gas lift, or a pump.

Impulse Turbine—A turbine that is driven by high velocity jets of water or steam from a nozzle directed to vanes or buckets attached to a wheel. (A pelton wheel is an impulse hydro turbine).

Imputed Capitalization—A method to adjust a projected capital structure for accumulated deferred income taxes.

IMS—Information Management Staff

In Situ Leach Mining (ISL)—The recovery, by chemical leaching, of the valuable components of a mineral deposit without physical extraction of the mineralized rock from the ground. Also referred to as "solution mining."

Inadvertent Power Exchange—An unintended power exchange among utilities that is either not previously agreed upon or in an amount different from the amount agreed upon.

Incandescent Lamp—A glass enclosure in which light is produced when a tungsten filament is electrically heated so that it glows. Much of the energy is converted into heat; therefore, this class of lamp is a relatively inefficient source of light. Included in this category are the familiar screw-in light bulbs, as well as somewhat more efficient lamps, such as tungsten halogen lamps, reflector or r-lamps, par-

abolic aluminized reflector (PAR) lamps, and ellipsoidal reflector (ER) lamps.

Incandescent Light—An electric lamp which is evacuated or filled with an inert gas and contains a filament (commonly tungsten). The filament emits visible light when heated to extreme temperatures by passage of electric current through it.

Incandescent Light Bulbs, Including Regular or Energy-efficient Light Bulbs—An incandescent bulb is a type of electric light in which light is produced by a filament heated by electric current. The most common example is the type you find in most table and floor lamps. In commercial buildings, incandescent lights are used for display lights in retail stores, hotels and motels. This includes the very small, high-intensity track lights used to display merchandise or provide spot illumination in restaurants. Energy efficient light bulbs, known as "watt-savers," use less energy than a standard incandescent bulb. "Long-life" bulbs, bulbs that last longer than standard incandescent but produce considerably less light, are not considered energy-efficient bulbs. This category also includes halogen lamps. Halogen lamps are a special type of incandescent lamp containing halogen gas to produce a brighter, whiter light than standard incandescent. Halogen lamps come in three styles: bulbs, models with reflectors, and infrared models with reflectors. Halogen lamps are especially suited to recessed or "canned fixtures," track lights, and outdoor lights.

Incentive—A rebate or some form of payment used to encourage people to implement a given demand-side management (DSM) technology. The incentive is calculated as the amount of the technology costs that must be paid by the utility for the participant test to equal one and achieve the desired benefit/cost ratio to drive the market.

Incentive Agreement—An agreement executed between the program participant and the Utility that documents the estimated electric and gas savings and the estimated incentive amount for the project. Funds are reserved for a period of 48 months upon execution of this agreement (in California).

Incentive Rate of Return (IROR)—A variable regulatory rate that reduces the allowed return in the event of cost overruns.

Incentive-Based Regulation—Uses the economic behavior of firms and households to attain desired environmental goals. Incentive-based programs involve taxes on emissions or tradable emission permits. The primary strength of incentive-based regulation is the flexibility it provides the polluter to find the least-cost way to reduce emissions.

Incentives Demand-Side Management (DSM) Program Assistance—This DSM program assistance offers monetary or non-monetary awards to encourage consumers to buy energy-efficient equipment and to participate in programs designed to reduce energy usage. Examples of incentives are zero or low-interest loans, rebates, and direct installation of low cost measures, such as water heater wraps or duct work for distributing the cool air; the units condition air only in the room or areas where they are located.

Inch of Mercury—A pressure unit representing the pressure required to support a column of mercury one inch high at a specified temperature; 2.036 inches of mercury (at 32 degrees F and standard gravity of 32.174 ft/sec^2) is equal to a gauge pressure of one pound per square inch.

Inch of Water—A pressure unit representing the pressure required to support a column of water one inch high. Usually reported as inches W.C. (water column) at a specified temperature; 27.707 inches of water (at 60° and standard gravity of 32.174 ft/sec^2) is equal to a gauge pressure of one pound per square inch.

Incidence Angle Modifier (IAM)—Refers to the change in performance on a solar system as the sun's angle in relation to the collector surface changes. Perpendicular to the collector (usually midday) is expressed as 0°, with negative angles in the morning and positive angles in the afternoon.

Incident Light—Light that shines on to the surface of a PV cell or module.

Incident Solar Radiation—The amount of solar radiation striking a surface per unit of time and area.

Incineration—The process of reducing refuse material to ash.

Incinerator—Any device used to burn solid or liquid residues or wastes as a method of disposal. In some incinerators, provisions are made for recovering the heat produced.

Inclined Grate—A type of furnace in which fuel enters at the top part of a grate in a continuous ribbon, passes over the upper drying section where moisture is removed, and descends into the lower burning section. Ash is removed at the lower part of the grate.

Income Capitalization Approach (Appraisal)—A set of procedures through which an appraiser derives a value indication for an income-producing property by converting its anticipated benefits (cash lows and reversion) into property value. This conversion can be accomplished in two ways.

One year's income expectancy can be capitalized at a market-derived capitalization rate or at a capitalization rate that reflects a specified income pattern (see *Cap Rate*), return on investment, and change in the value of the investment. Alternatively, the annual cash flows for the holding period and the reversion can be discounted at a specified yield rate.

Income Fund—An investment vehicle sold to investors. The income fund generates its income by investing in financing transactions. Dividends, which are derived from the investment activities, are declared and paid to investors.

Income Statement—A financial statement that shows the amount of income (revenue) earned by a business over s specific accounting period. All costs (expenses) are subtracted from the gross revenues (sales) to determine net income, which outlines the profit-and-loss financial statement (P & L).

Incompatible Wastes—Wastes which create a hazard of some form when mixed together. This could be intense heat or toxic gases, for example.

Inconvertible Currency—A currency that may not be exchanged without restrictions for another currency.

Incremental Borrowing Rate—Is the interest rate that, at the inception of the financial contract, debtor would have incurred to borrow the funds necessary to purchase the asset over a similar term. Used when determining whether a lease qualifies as an operating or capital lease.

Incremental Cost—The additional costs incurred from the production or delivery of an additional number of units, usually the minimum capacity or production that can be added. The additional cost divided by the additional capacity or output is defined as the incremental cost.

Incremental Effects—The annual changes in energy use (measured in megawatt hours) and peak load (measured in kilowatts) caused by new participants in existing DSM (Demand-Side Management) programs and all participants in new DSM programs during a given year. Reported Incremental Effects are annualized to indicate the program effects that would have occurred had these participants been initiated into the program on January 1 of the given year. Incremental effects are not simply the Annual Effects of a given year minus the Annual Effects of the prior year, since these net effects would fail to account for program attrition, equipment degradation, building demolition, and participant dropouts. Please note that Incremental Effects are not a monthly disaggregate of the Annual Effects, but are the total year's effects of only the new participants and programs for that year.

Incremental Energy Costs—The cost of producing and transporting the next available unit of electrical energy. Short run incremental costs (SRIC) include only incremental operating costs. Long run incremental costs (LRIC) include the capital cost of new resources or capital equipment.

Incurable Functional Obsolescence—An element of depreciation. A defect caused by a deficiency or superadequacy in the structure, materials, or design, which cannot be practically or economically corrected.

Indemnification (Lease Contract)—A clause in a master lease agreement or other financial contract that requires lessees to indemnify lessors against any and all claims, suits, actions, damages, liabilities, expenses, cost, including attorney fees, whether or not suit is instituted, arising out of or incurred in connection with the equipment.

Indemnitee—A term used to describe the class of persons entitled to indemnification under the general indemnity and general tax indemnity provisions of an agreement.

Indemnity Agreement—An agreement to compensate another Party for a potential loss. A "Hold-Harmless" agreement.

Indemnity (or other contract)—The indemnity provisions in a lease: general indemnity, general tax indemnity and special tax indemnity. In leveraged leases, the tax indemnity clauses can be quite lengthy and sometimes are contained in a special supplement to the lease agreement or in a separate agreement.

Indenture Trust—An agreement between the owner trustee and the indenture trustee. Similar to a mortgage.

Indenture Trustee—The party who holds the security interest in the leased equipment for the benefit of the lender.

Independent Identity—Having no financial interest in, and not advocating or recommending the use of any product or service as a means of gaining increased business with, firms or persons specified in Section 1673(i) of the California Home Energy Rat-

ing System Program regulations (California Code of Regulations, Title 20, Division 2, Chapter 4, Article 8). (Financial Interest is an ownership interest, debt agreement, or employer/employee relationship. Financial interest does not include ownership of less than 5% of the outstanding equity securities of a publicly traded corporation.)
NOTE: The definitions of "independent entity" and "financial interest," together with Title 20, Section 1673(i), prohibit conflicts of interest between HERS Providers and HERS Raters, or between Providers/Raters and builders/subcontractors.

Independent Petroleum Association of America (IPAA)—A trade group representing independent oil and gas producers. www.ipaa.org

Independent Power Producer—An Independent Power Producer (IPP) generates power that is purchased by an electric utility at wholesale prices. The utility then resells this power to end-use customers. Although IPP's generate power, they are not franchised utilities, government agencies or QF's. IPP's usually do not own transmission lines to transmit the power that they generate and do not sell power in any retail service territory where they have a franchise.

Independent System Operator (ISO)—An independent, Federally regulated entity established to coordinate regional transmission in a non-discriminatory manner and ensure the safety and reliability of the electric system.

Independent Variable—A parameter that is expected to change regularly in an energy project and have a measurable impact on the energy use of a building or system. An Independent Variable is a factor affecting energy usage that is essentially outside the control of building operations and energy management staff. Examples include Billing Period, weather-related Independent Variables (such as HDDs and CDDs), and user variables that are related to specific building use.

Index Price Gas—Price based on the monthly spot market sales. Several gas contracts are averaged to create a system or location wide average gas price.

Indexing—Tying the commodity price in a contract to other published prices, such as spot prices for gas or alternate fuels, or general indexes like the Consumer Price Index or Producer Price Index.

Indian Coal—Coal that is produced from coal reserves that on June 14, 2005 was owned by an Indian tribe or held in trust by the United States for the benefit of an Indian tribe or its members.

Indian Coal Lease—A lease granted to a mining company to produce coal from Indian lands in exchange for royalties and other revenues; obtained by direct negotiation with Indian tribal authorities, but subject to approval and administration by the U.S. Department of the Interior.

Indian Tribal Government—The governing body of any tribe, bank, community, village, or group of Indians, or (if applicable) Alaska Natives, which is determined by the Secretary of Treasury, after consultation with the Secretary of the Interior, to exercise governmental functions.

Indicated Resources, Coal—Coal for which estimates of the rank, quality, and quantity are based partly on sample analyses and measurements and partly on reasonable geologic projections. Indicated resources are computed partly from specified measurements and partly from projection of visible data for a reasonable distance on the basis of geologic evidence. The points of observation are 1/2 to 1-1/2 miles apart. Indicated coal is projected to extend as a 1/2-mile-wide belt that lies more than 1/4 mile from the outcrop, points of observation, or measurement.

Indicator Protocol—(GRI Reporting Framework) An Indicator Protocol provides definitions, compilation guidance, and other information to assist report preparers, and to ensure consistency in the interpretation of the Performance Indicators. An Indicator Protocol exists for each of the Performance Indicators (see Performance Indicator) contained in the Guidelines.

Indigenous Energy Resources—Power and heat derived from sources native to a particular location. These include geothermal, hydro, biomass, solar and wind energy. The term usually is understood to include cogeneration facilities.

Indirect Cost—Expenditures or allowances for items other than labor and materials that are necessary for construction, but are not typically part of the construction contract. Indirect costs may include administrative costs, professional fees, financing costs and the interest paid on construction loans, taxes and the builder's or developer's all-risk insurance during construction, and marketing, sales and lease-up costs incurred to achieve occupancy or sales. Also called "soft costs."

• Costs not directly related to production operations, such as overhead, insurance, security, office expenses, property taxes, and similar administrative expenses.

Indirect Emissions—Emissions that are a consequence of actions of the reporting entity, but are produced by sources owned or controlled by another entity.

Indirect Energy—Electricity, thermal, or other energy sources provided by a retail provider or facility not owned or operated by the user of the energy.

Indirect Liquefaction—Conversion of biomass to a liquid fuel through a synthesis gas intermediate step.

Indirect Solar Gain System—A passive solar heating system in which the sun warms a heat storage element, and the heat is distributed to the interior space by convection, conduction, and radiation.

Indirect Solar Water Heater—These systems circulate fluids other than water (such as diluted antifreeze) through the collector. The collected heat is transferred to the household water supply using a heat exchanger. Also known as "closed-loop" systems.

Indirect Uses (End-Use Category)—The end-use category that handles boiler fuel. Fuel in boilers is transformed into another useful energy source, steam or hot water, which is in turn used in other end uses, such as process or space heating or electricity generation. Manufacturers find measuring quantities of steam as it passes through to various end uses especially difficult because variations in both temperature and pressure affect energy content. Thus, the MECS (an EIA survey) does not present end-use estimates of steam or hot water and shows only the amount of the fuel used in the boiler to produce those secondary energy sources.

Indirect Utility Cost—Any cost that is not identified with a specific DSM category such as administration, marketing, etc.

Indirectly Conditioned Space—Enclosed space, including, but not limited to, unconditioned volume in atria, that (1) is not directly conditioned space; and (2) either (a) has a thermal transmittance area product (UA) to directly conditioned space exceeding that to the outdoors or to unconditioned space and does not have fixed vents or openings to the outdoors or to unconditioned space, or (b) is a space through which air from directly conditioned spaces is transferred at a rate exceeding three air changes per hour.

Indium Oxide—A wide band gap semiconductor that can be heavily doped with tin to make a highly conductive, transparent thin film. Often used as a front contact or one component of a heterojunction solar cell.

Indoor Air Quality (IAQ)—Good indoor air quality includes the introduction and distribution of adequate ventilation air, control of airborne contaminants, and maintenance of acceptable temperature and relative humidity. According to ASHRAE Standard 62-1989, indoor air quality is defined as "air in which there are no known contaminants at harmful concentrations as determined by cognizant authorities and with which a substantial majority (80 percent or more) of the people exposed do not express dissatisfaction."

Indoor Environmental Quality (IEQ)—LEED Rating System category. Prerequisites and credits in this category focus on the strategies and systems that result in a healthy indoor environment for building occupants.

Induction—The production of an electric current in a conductor by the variation of a magnetic field in its vicinity.

Induction Generator—A device that converts the mechanical energy of rotation into electricity based on electromagnetic induction. An electric voltage (electromotive force) is induced in a conducting loop (or coil) when there is a change in the number of magnetic field lines (or magnetic flux) passing through the loop. When the loop is closed by connecting the ends through an external load, the induced voltage will cause an electric current to flow through the loop and load. Thus rotational energy is converted into electrical energy.

Induction Generator—A variable speed multi-pole electric generator.

Induction Motor—A motor in which a three phase (or any multiphase) alternating current (i.e. the working current) is supplied to iron-cored coils (or windings) within the stator. As a result, a rotating magnetic field is set up, which induces a magnetizing current in the rotor coils (or windings). Interaction of the magnetic field produced in this manner with the rotating field causes rotational motion to occur.

Industrial—This utility customer class is generally defined as manufacturing, construction, mining agriculture, fishing and forestry establishments Standard Industrial Classification (SIC) codes 01-39. The utility may classify industrial service using the SIC codes, or based on demand or annual usage exceeding some specified limit. The limit may be set by the utility based on the rate schedule of the utility.

Industrial Equipment—Manufactured equipment used in industrial processes.

Industrial Process Heat—The thermal energy used in an industrial process.

Industrial Production—The Federal Reserve Board calculates this index by compiling indices of physical output from a variety of agencies and trade groups, weighting each index by the Census' value added, and adding it to the cost of materials. When physical measures are not available, the Federal Reserve Board uses the number of production workers or amount of electricity consumed as the basis for the index. To convert industrial production into dollars, multiply by the "real value added" estimate used by the Federal Reserve Board.

Industrial Restrictions (Coal)—Land-use restrictions that constrain, postpone, or prohibit mining in order to meet other industrial needs or goals; for example, resources not mined due to safety concerns or due to industrial or societal priorities, such as to preserve oil or gas wells that penetrate the coal reserves; to protect surface features such as pipelines, power lines, or company facilities; or to preserve public or private assets, such as highways, railroads, parks, or buildings.

Industrial Sector—An energy-consuming sector that consists of all facilities and equipment used for producing, processing, or assembling goods The industrial sector encompasses the following types of activity: manufacturing (NAICS codes 31-33); agriculture, forestry, fishing and hunting (NAICS code 11); mining, including oil and gas extraction (NAICS code 21); natural gas distribution (NAICS code 2212); and construction (NAICS code 23). Overall energy use in this sector is largely for process heat and cooling and powering machinery, with lesser amounts used for facility heating, air conditioning, and lighting. Fossil fuels are also used as raw material inputs to manufactured products. Note: This sector includes generators that produce electricity and/or useful thermal output primarily to support the above-mentioned industrial activities.

Inert—A material not acted upon chemically by the surrounding environment. Nitrogen and carbon dioxide are examples of inert constituents of natural gases; they dilute the gas and do not burn, and thus add no heating value.

Inert Gas—A gas that does not react with other substances; e.g. argon or krypton; sealed between two sheets of glazing to decrease the U-value (increase the R-Value) of windows.

Inferred Resources—Coal in unexplored extensions of demonstrated resources for which estimates of the quality and size are based on geologic evidence and projection. Quantitative estimates are based largely on broad knowledge of the geologic character of the bed or region and where few measurements of bed thickness are available. The estimates are based primarily on an assumed continuation from demonstrated coal for which there is geologic evidence. The points of observation are 1-1/2 to 6 miles apart. Inferred coal is projected to extend as a 2-1/4-mile wide belt that lies more than 3/4 mile from the outcrop, points of observation, or measurement.

Infiltration—Uncontrolled inward air leakage from outside a building or unconditioned space, including leakage through cracks and interstices, around windows and doors, and through any other exterior or demising partition or pipe or duct penetration.

Infiltration Barrier—A material placed on the outside or the inside of exterior wall framing to restrict inward air leakage, while permitting the outward escape of water vapor from the wall cavity. [See California Code of Regulations, Title 24, Section 2-5302]

Infiltration Controls—Measures taken to control the infiltration of air. Mandatory Infiltration control measures include weather-stripping, caulking, and sealing in and around all exterior joints and openings.

Inflow—Water (and pollutants) that enter sewage systems through street inlets, roof drains, and similar sources.

Influent—Wastewater going into the sewage treatment plant.

Infrared Radiation—Electromagnetic radiation whose wavelengths lie in the range from 0.75 micrometer to 1000 micrometers; invisible long wavelength radiation (heat) capable of producing a thermal or photovoltaic effect, though less effective than visible light.

Infrastructure (vehicle)—Generally refers to the recharging and refueling network necessary to successful development, production, commercialization and operation of alternative fuel vehicles, including fuel supply, public and private recharging and refueling facilities, standard specifications for refueling outlets, customer service, education and training, and building code regulations.
• The basic facilities serving an area: transportation and communication systems, power plants, and roads.

In-House Demand-Side Management (DSM) Program

Sponsor—The building's owner or management encourages consumers in the building to improve energy efficiency, reduce energy costs, change timing or energy usage, or promote the use of a different energy source by sponsoring its own DSM programs.

Initial Direct Costs (Lease Contract)—Costs incurred by the lessor that are directly associated with negotiating and consummating a lease and which would not have been occurred without entering into the lease. These costs include, but are not necessarily limited to, commissions, legal fees, costs of credit investigations, the cost of preparing and processing documents for new leases acquired and so forth.

Initial Enrichment—Average enrichment for a fresh fuel assembly as specified and ordered in fuel cycle planning. This average should include axial blankets and axially and radially zoned enrichments.

Initial Operation—First availability of a newly constructed unit to provide power to the grid. For a nuclear unit, this time is when the Full Power Operating License for the unit is received.

Initial Public Offering (IPO)—First-time public offering of a company in which investors at large may purchase stock in that company.
• The sale or distribution of a stock of a portfolio company to the public for the first time. IPOs are often an opportunity for the existing investors (often venture capitalists) to receive significant returns on their original investment. During periods of market downturns or corrections, the opposite is true.

Injection (Petroleum)—Forcing gas or water into an oil well to increase pressure and cause more oil to come to the surface. See *Thermally Enhanced Oil Recovery*.

Injections—Natural gas injected into storage reservoirs.

Inland Bill of Lading—A bill of lading used in transporting goods overland to the exporter's international carrier. Although a through bill of lading can sometimes be used, it is usually necessary to prepare both an inland bill of lading and an ocean bill of lading for export shipments.

Inorganic Compounds—Those compounds lacking carbon but including carbonates and cyanides. Compounds not having the organized anatomical structure of animal or vegetable life.

In-Situ Combustion—An experimental means of recovering hard-to-get petroleum by burning some of the oil in its natural underground reservoir. Also called Fireflooding.

In-Situ Gasification—Converting coal into synthetic gas at the place where the coal is found in nature.

In-Situ Soil Aeration—Applying a vacuum to vapor extraction wells to draw air through the soil so that chemicals in the soil are brought to the surface where they can be treated.

Insolation—The total amount of solar radiation (direct, diffuse, and reflected) striking a surface exposed to the sky. Usually expressed in Watt hours per square meter per day. Could also be expressed in a "Langley"—a unit of heat energy equivalent to one calorie falling on one square centimeter of surface. One Btu per square foot is the equivalent of 0.27125 Langleys.

• The solar power density incident on a surface of stated area and orientation, usually expressed as Watts per square meter or Btu per square foot per hour.

Installation Certificate (CF-6R)—(in California) A document with information required by the Commission that is prepared by the builder or installer verifying that the measure was installed to meet the requirements of the standards.

Installation Period—That time period after the Baseline Period but before the M&V Period, when audit results are prepared, investment decisions are made and Energy Conservation Measures are implemented. Installation start and end dates can be defined separately for each Project Opportunity.

Installed Capacity—The total capacity of electrical generation devices in a power station or system.

Installment Note—A Promissory Note providing for payment of the principal in two or more certain amounts at different stated times.

Instantaneous Efficiency (of a Solar Collector)—The amount of energy absorbed (or converted) by a solar collector (or photovoltaic cell or module) over a 15-minute period.

Instantaneous Peak Demand—The maximum demand at the instant of greatest load.

Instantaneous Water Heater—Also called a "tankless" or "point-of-use" water heater. The water is heated at the point of use as it is needed.

Institute of Electrical and Electronics Engineers (IEEE)—A professional association for the advancement of technology. Particularly interested in emphasizing standards to be applied universally across all electronics manufacturers. Designers of Interconnection Standard 1547, which is used as a basic interconnection model by a number of states and utilities. See www.ieee.org

Institutional Investor—Organizations that professionally invest, including insurance companies, depository institutions, pension funds, investment companies, mutual funds, and endowment funds.

Institutional Living Quarters—Space provided by a business or organization for long-term housing of individuals whose reason for shared residence is their association with the business or organization. Such quarters commonly have both individual and group living spaces, and the business or organization is responsible for some aspects of resident life beyond the simple provision of living quarters. Examples include prisons; nursing homes and other long-term medical care facilities; military barracks; college dormitories; and convents and monasteries.

Instrument—A writing or document, such as a Deed, made and executed as the expression of some act, contract, or proceeding.

Insulating Glass Unit—A self-contained unit, including the glazings, spacer(s), films (if any), gas infills, and edge caulking, that is installed in fenestration products. It does not include the frame.

Insulation—Any material or substance that provides a high resistance to the flow of heat from one surface to another. The different types include blanket or batt, foam, or loose fill, which are used to reduce heat transfer by conduction. Dead air space is an insulating medium in storm windows and storms as it reduces passage of heat through conduction and convection. Reflective materials are used to reduce heat transfer by radiation.

Insulation is a material that limits heat transfer.

Insulating material of the types and forms listed in Section 118(a) of the Standards (in California), may be installed only if the manufacturer has certified that the insulation complies with the Standards for Insulating Material, Title 24, Part 12, Chapter 1213 of the California Code of Regulations.

Insulation must be placed within or contiguous with a wall, ceiling or floor, or over the surface of any appliance or its intake or outtake mechanism for the purpose of reducing heat transfer or reducing adverse temperature fluctuations of the building, room or appliance.

Insulation may be installed in wall, ceiling/roof and raised floor assemblies and at the edge of a slab-on-grade. Movable insulation is designed to cover windows and other glazed openings part of the time to reduce heat loss and, heat gain.

Insulation Around Heating and/or Cooling Ducts—Extra insulation around the heating and/or cooling ducts intended to reduce the loss of hot or cold air as it travels to different parts of the residence.

Insulation Around Hot-water Pipes—Wrapping of insulating material around hot-water pipes to reduce the loss of heat through the pipes.

Insulation Around Water Heater—Blanket insulation wrapped around the water heater to reduce loss of heat. To qualify under this definition, this wrapping must be in addition to any insulation provided by the manufacturer.

Insulation Blanket—A pre-cut layer of insulation applied around a water heater storage tank to reduce stand-by heat loss from the tank.

Insulation, Thermal—A material having a relatively high resistance of heat flow and used principally to retard heat flow by limiting conduction, convection, radiation or evaporation. See *R-Value*.

Insulator—A material that is a very poor conductor of electricity. The insulating material is usually a ceramic or fiberglass when used in the transmission line and is designed to support a conductor physically and to separate it electrically from other conductors and supporting material.

Insured Value—An agreed-upon value that the insurance company will pay the beneficiary if the equipment is destroyed while being financed.

Intake—In a hydro system, the structure that receives the water and feeds it into the penstock (pipeline). Usually incorporates screening or filtering to keep debris and aquatic life out of the system.

Intangible Asset—Intangible assets include such items or accounts as: goodwill, patents and patent rights, deferred charges and unamortized bond premium. This can include franchises, trademarks, patents, copyrights, good will, equities, mineral rights, securities and contracts, as distinguished from physical assets such as facilities and equipment.

Intangible Drilling and Development Costs (IDC)—Costs incurred in preparing well locations, drilling and deepening wells, and preparing wells for initial production up through the point of installing control valves. None of these functions, because of their nature, have salvage value. Such costs would include labor, transportation, consumable supplies, drilling tool rentals, site clearance, and similar costs.

Intangible Transmission Charge—The amounts on all customer bills, collected by the electric utility to recover transition bond expenses.

Integral Collector Storage (ICS)—A solar thermal collector in which incident solar radiation is absorbed directly by the storage medium.

Integral Collector Storage System—This simple passive solar hot water system consists of one or more storage tanks placed in an insulated box that has a glazed side facing the sun. An integral collector storage system is mounted on the ground or on the roof. Some systems use "selective" surfaces on the tank(s). These surfaces absorb sun well but inhibit radiative loss. Also known as bread box systems or batch heaters.

Integrated Demand—The summation of the continuously varying instantaneous demand averaged over a specified interval of time. The information is usually determined by examining a demand meter.

Integrated Design—An inclusive process (sometimes called Whole systems Design) that consider the many disparate parts of a building project, and examines the interaction between design, construction, and operations, to optimize the resources (natural, spiritual, social, and technical) used. The strength of this process is that all relevant issues are considered simultaneously during the design process by the whole team of designers, builders, and maintenance people in order to "solve for pattern: or solve many problems with one solution. The goal of integrated design is to create developments that have the potential to heal damaged environments and become net producers of energy, healthy food, clean water and air, and healthy human and biological communities.

Integrated Gasification-Combined Cycle Technology (IGCC)—Coal, water, and oxygen are fed to a gasifier, which produces syngas. This medium-Btu gas is cleaned (particulates and sulfur compounds removed) and is fed to a gas turbine. The hot exhaust of the gas turbine and heat recovered from the gasification process are routed through a heat-recovery routed through a heat-recovery generator to produce steam, which drives a steam turbine to produce electricity.

Integrated Heating Systems—A type of heating appliance that performs more than one function, for example space and water heating.

Integrated Part Load Value (IPLV)—A single number figure of merit based on part load EER or COP expressing part load efficiency for air-conditioning and heat pump equipment on the basis of weighted operation at various load capacities for the equipment as determined using the applicable test method in the Appliance Efficiency Regulations or Section 112 (in California).

Integrated Resource Planning (IRP)—A public planning process and framework within which the costs and benefits of both demand- and supply-side resources are evaluated to develop the least-total-cost mix of utility resource options. In many states, IRP includes a means for considering environmental damages caused by electricity supply/transmission and identifying cost-effective energy efficiency and renewable energy alternatives. IRP has become a formal process prescribed by law in some states and under some provisions of the Clean Air Act amendments of 1992. See *Least Cost Planning*.

Integrated Resource Planning Principles—The underlying principles of IRP can be distinguished from the formal process of developing an approved utility resource plan for utility investments in supply- and demand-side resources. A primary principle is to provide a framework for comparing a variety of supply- and demand-side and transmission resource costs and attributes outside of the basic provision (or reduction) of electric capacity and energy. These resources may be owned or constructed by any entity and may be acquired through contracts as well as through direct investments. Another principle is the incorporation of risk and uncertainty into the planning analysis. The public participation aspects of IRP allow public and regulatory involvement in the planning rather than the siting stage of project development.

Integrated Waste Management—The complementary use of a variety of practices to handle solid waste safely and effectively. Techniques include source reduction, recycling, composing, combustion and landfilling.

Intensity—The amount of a quantity per unit floor space. This method adjusts either the amount of energy consumed or expenditures spent, for the effects of various building characteristics, such as size of the building, number of workers, or number of operating hours, to facilitate comparisons of energy across time, fuels, and buildings.

Intensity per Hour—Total consumption of a particular fuel(s) divided by the total floor space of buildings that use the fuel(s) divided by total annual hours of operation.

Intensive Management—Planned, active treatment to improve the quality and quantity of timber within a stand. A general term that distinguishes active forest management from passive forest management.

Interactive Effects—Energy impacts to one system resulting from changes made to another building system.

Interchange (Electric Utility)—The agreement among interconnected utilities under which they buy, sell and exchange power among themselves. This can, for example, provide for economy energy and emergency power supplies.

Interchange Energy—Kilowatt-hours delivered to or received by one electric utility or pooling system from another. Settlement may be payment, returned in kind at a later time, or accumulated as energy balances until the end of the stated period.

Intercity Bus—A bus designed for high speed, long distance travel; equipped with front doors only, high backed seats, and usually restroom facilities.

Interconnect—A conductor within a module or other means of connection which provides an electrical interconnection between the solar cells. [UL 1703]

Interconnected System—A system consisting of two or more individual power systems normally operating with connecting tie lines.

Interconnection—Two or more electric systems having a common transmission line that permits a flow of energy between them. The physical connection of the electric power transmission facilities allows for the sale or exchange of energy.

Interconnection (Electric Utility)—The linkage of transmission lines between two utilities, enabling power to be moved in either direction. Interconnections allow the utilities to help contain costs while enhancing system reliability.

Interconnection Agreement—A legal document authorizing the flow of electricity between the facilities of two electric systems. Usually between a utility customer who has an energy generating system and the utility.

Inter-creditor Agreement—An agreement between the lenders to a company or project as to the rights of credits in the event of default, covering such topics as collateral, waiver, security and set-offs.

Interdepartmental Sales—Includes amounts charged by the electric department at tariff or other specified rates for electricity supplied by it to other utility departments.

Interdepartmental Service (Electric)—Interdepartmental service includes amounts charged by the electric department at tariff or other specified rates for electricity supplied by it to other utility departments.

Interest Coverage Ratio—The number of times that fixed interest charges were earned. It indicates the margin of safety of interest on fixed debt. The times-interest-earned ratio is calculated using net income before and after income taxes; and the credits of interest charged to construction being treated as other income. The interest charges include interest on long-term debt, interest on debt of associated companies, and other interest expenses.

Interest Expense—An amount paid to a lender in return for a loan. Typically the interest is paid out over time, accompanied by a reduction in loan principal.

Interest Rate Implicit in a Lease—The discount rate which, when applied to minimum lease payments (excluding executory costs paid by the lessor) and unguaranteed residual value, causes the aggregate present value at the beginning of the lease term to be equal to the fair value of the leased property at the inception of the lease.

Interested Party—Any person whom the commission finds and acknowledges as having a real and direct interest in any proceeding or action carried on, under, or as a result of the operation of, this division.

Intergenerational Equity—In the context of environmental policy refers to the fairness of the distribution of the costs and benefits of a long-lived policy when those costs and benefits are borne by different generations. In the case of a climate change policy designed to reduce greenhouse gas emissions, the costs of the emissions reductions will be borne by the current and near term generations, while the benefits of an unchanged climate will be enjoyed by far distant generations.

Intergovernmental Panel on Climate Change (IPCC)—The World Meteorological Organization (WMO) and the United Nations Environment Program (UNEP) formed the Intergovernmental Panel on Climate Change (IPCC) in 1988. The IPCC represents the collective work of over 2,000 scientists, principally in the atmospheric sciences, but also comprising social, economic and other environmental components potentially impacted by climate change. Between its three Working Groups, the IPCC assesses the scientific and socio-economic aspects of human-induced climate change, as well as options for greenhouse gas reduction and other forms of climate change mitigation. Its Task Force on National Greenhouse Gas Inventories is responsible for overseeing the National Greenhouse Gas Inventories Program (NGGIP).

The IPCC neither conducts original research nor monitors climate-related data, but its periodic assessment reports and technical papers play a very important role in the creation of climate change policies worldwide. The IPCC was instrumental in establishing the Intergovernmental Negotiating Committee for the United Nations Framework Convention on Climate Change (UNFCCC or the Convention) in 1992.

Interim Loan—A short-term loan under circumstances anticipating a subsequent long-term loan. Sometimes referred to as a construction loan when a project is being built.

Interim Remedial Actions (IRAs)—Also known as Interim Remedial Measures. Cleanup actions taken to protect public health and the environment while long-term solutions are being developed.

Interim Rent—A charge for the use of a piece of equipment from its in-service date, or delivery date, until the date on which the base term of the lease commences. The daily interim rent charge is typically equal to the daily equivalent of the base rental payment. The use of interim rent allows the lessor to have one common base term commencement date for a lease agreement having multiple deliveries of equipment.

Interior Partition—An interior wall or floor/ceiling that separates one area of conditioned space from another within the building envelope.

Interior Zones—The portions of a building which do not have significant amounts of exterior surfaces. Such zones have heating or cooling needs largely dependent upon internal factors such as lighting.

Interlocking Directorates—The holding of a significant position in management or a position on the corporate board of a utility while simultaneously holding a comparable position with another utility, or with a firm doing business with the utility.

Interlocutory Decree—A Decree that does not finally dispose of a Cause of Action but requires that some further steps be taken.

Intermediate Grade Gasoline—A grade of unleaded gasoline with an octane rating intermediate between "regular" and "premium." Octane boosters are added to gasolines to control engine pre-ignition or "knocking" by slowing combustion rates.

Intermediate Load (Electric System)—The range from base load to a point between base load and peak. This point may be the midpoint, a percent of the peak load, or the load over a specified time period.

Intermittent Electric Generator or Intermittent Resource—An electric generating plant with output controlled by the natural variability of the energy resource rather than dispatched based on system requirements. Intermittent output usually results from the direct, non-stored conversion of naturally occurring energy fluxes such as solar energy, wind energy, or the energy of free-flowing rivers (that is, run-of-river hydroelectricity).

Intermittent Generators—Power plants, whose output depends on a factor(s) that cannot be controlled by the power generator because they utilize intermittent resources such as solar energy or the wind.

Intermodal—Pertaining to transportation involving more than one form of carrier, such as truck, ship and rail.

Internal Collector Storage (ICS)—A solar thermal collector in which incident solar radiation is absorbed by the storage medium.

Internal Combustion Engine (I/C)—An engine in which fuel is burned inside the engine. A car's gasoline engine or rotary engine is an example of a internal combustion engine. It differs from engines having an external furnace, such as a steam engine.

Internal Combustion Plant—A plant in which the prime mover is an internal combustion engine. An internal combustion engine has one or more cylinders in which the process of combustion takes place, converting energy released from the rapid burning of a fuel-air mixture into mechanical energy. Diesel or gas-fired engines are the principal types used in electric plants. The plant is usually operated during periods of high demand for electricity.

Internal Gain—The heat produced by sources of heat in a building (occupants, appliances, lighting, etc).

Internal Loan—The portion of the equity funded through internal debt (as opposed to true equity). Often used in the calculations of return on (internal) equity.

Internal Mass—Materials with high thermal energy storage capacity contained in or part of a building's walls, floors, or freestanding elements.

Internal Rate of Return (IRR)—The unique discount rate that equates the present value of a series of cash inflows to the present value of the cash outflows. IRR is the most common method used to compute yields.

• The rate which discounts a stream of positive and negative cash flows to 0.

Internal Trading—An intra-company emissions trading system allowing the trade of emission permits among a firm's own business units with the objective of maximizing cost effective internal emission abatement opportunities.

International Emissions Trading (IET)—IET is a flexibility mechanism of the Kyoto Protocol which allows the trade of Assigned Amount Units (AAUs) among Annex B countries. It is expected that this activity will be delegated by national governments to entities within their jurisdictions so that international trading between entities will occur. This will adjust each nations 'pool' of AAUs.

International Energy Agency (IEA)—An organization formed in 1973 by major oil-consuming nations to manage future oil supply shortfalls. www.iea.org

International Organization—The term "international organization" means a public international organization entitled to enjoy privileges, exemptions, and immunities as an international organization under the International Organizations Immunities Act (22 U.S.C. 288-288f).

Interpollutant Trading—The use of reductions of one type of pollutant to offset the increases of another. In new source review programs, ERCs of pollutants considered to be precursors to a second pollutant can be used to offset the increases of the second pollutant. For example, SO_x is often considered to be a precursor to PM. As such, in some areas SO_x reductions can be used to offset PM emission increases (though perhaps at a ratio of greater than 1:1).

Interruptible Capacity—Interstate pipeline, backbone transmission or storage capacity, which may be available from time to time, but cannot be assured under all operating conditions.

Interruptible Gas—Gas sold to customers with a provision that permits curtailment or cessation of service at the discretion of the distributing company under certain circumstances, as specified in the service contract.

Interruptible Load—This Demand-Side Management category represents the consumer load that, in accordance with contractual arrangements, can be interrupted at the time of annual peak load by the action of the consumer at the direct request of the system operator. This type of control usually involves large-volume commercial and industrial consumers. Interruptible Load does not include Direct Load Control.

Interruptible or Curtailable Rate—A special electricity or natural gas arrangement under which, in return for lower rates, the customer must either reduce energy demand on short notice or allow the electric or natural gas utility to temporarily cut off the energy supply for the utility to maintain service for higher priority users. This interruption or reduction in demand typically occurs during periods of high demand for the energy (summer for electricity and winter for natural gas).

Interruptible Power—Power and usually the associated energy made available by one utility to another. This transaction is subject to curtailment or cessation of delivery by the supplier in accordance with a prior agreement with the other party or under specified conditions.

Interruptible Service (Electric Utility)—Electricity service which is offered at a lower level of cost and can be interrupted on short notice.

Interstate Companies—Natural gas pipeline companies subject to Federal Energy Regulatory Commission (FERC) jurisdiction.

Interstate Gas—Gas that is moved through the pipeline from one state to another state.

Interstate Pipeline—Any person engaged in natural gas transportation subject to the jurisdiction of Federal Energy Regulatory Commission (FERC) under the Natural Gas Act.

Interstate Pipeline Purchase—Any gas supply contracted from and volumes purchased from other interstate pipelines, overland natural gas import purchases, and LNG, SNG, or coal gas purchases from domestic or foreign sources. Purchases from intrastate pipelines to section 311 (b) of the NGPA of 1978 and from independent producers are not included with interstate pipelines purchase.

Intertie—A transmission line that links two or more regional electric power systems.

Interval Data Recorder (IDR)—IDR is a metering device capable of recording minimum data required.

Interval Metering—The process by which power consumption is measured at regular intervals in order that specific load usage for a set period of time can be determined.

Intervenor—A person, business entity, or public body that is granted the right to participate in a rate case or hearing.

Intransit Deliveries—Redeliveries to a foreign country of foreign gas received for transportation across U.S. territory, and deliveries of U.S. gas to a foreign country for transportation across its territory and redelivery to the United States.

Intransit Receipts—Receipts of foreign gas for transpor-

tation across U.S. territory and redelivery to a foreign country, and redeliveries to the United States of U.S. gas transported across foreign territory.

Intrastate Companies—Companies not subject to Federal Energy Regulatory Commission (FERC) jurisdiction.

Intrastate Gas—Gas that is moved from one location to another location within the same state.

Intrastate Pipeline—Any person engaged in natural gas transportation (not including gathering) that is not subject to the jurisdiction of the Commission under the Natural Gas Act (other than any such pipeline that is not subject to the jurisdiction of the Commission solely by reason of Section 1(c) of the Natural Gas Act).

Intrinsic Layer—A layer of semiconductor material (as used in a solar photovoltaic device) whose properties are essentially those of the pure, undoped, material.

Inure—To accrue to the benefit of a person or entity.

In-Use (Vehicles)—Implies that a vehicle is:
- Registered with the Government of one or more States, the District of Columbia, the Commonwealth of Puerto Rico, or the Virgin Islands; or
- The vehicle is owned or operated by a Government or military organization within the United States that is not required to register vehicles with the Government agencies listed under 1 above. For example, civilian Federal vehicles are generally not required to register with the State Government in which they are assigned.

Inventory Turnover Ratio—A measure of management's control of its investment in inventory, usually defined as cost of goods sold divided by ending inventory.

Inverted Lease—A term used to describe a form of ownership structure that is used to facilitate investment that qualify for the investment tax credit under IRC section 46 (the ITC) and which involve using two limited partnerships—one which owns the property that earns the ITCs (the owner or lessor) and another (the tenant or lessee) that leases the property from the owner. Under this structure the owner is able to pass-through the ITCs to the tenant. The investor limited partner(s) usually own a limited partnership interest in the tenant, which in turn invests in the owner. This ownership structure is also known as the lease pass-through and/or the master-tenant.

Inverter—A device that converts direct current electricity (from for example a solar photovoltaic module or array) to alternating current for use directly to operate appliances or to supply power to an electricity grid.

Inverter Efficiency—The AC power output of the inverter divided by the DC power input. Inverter efficiency is lowest when operating at low loads.

Investment and Advances to Unconsolidated Affiliates—The balance sheet account representing the cost of investments and advances to unconsolidated affiliates. Generally, affiliates that are less than 50 percent owned by a company may not be consolidated into the company's financial statements.

Investment Banker—Serves as a middleman between the suppliers of capital and the users of capital; also known as an underwriter.

Investment Company Act—Investment Company Act means the Investment Company Act of 1940, as amended, including the rules and regulations promulgated thereunder.

Investment Grade—Bond issuers that the three major bond rating agencies (Moody's, Standard & Poor's, and Fitch) rate BBB or Baa or better. Many fiduciaries, trustees, and some mutual fund managers can only invest in securities with an "investment grade rating."

Investment Grade Audit—A comprehensive assessment of a facility's energy and water usage characteristics, identifying and analyzing energy conservation measures. It is an analysis of energy usage history, along with a thorough examination of building systems and operating practices and a feasibility study of contemplated energy efficiency opportunities. It differs from a traditional energy audit in three ways. First, it addresses the operating implications of the opportunities being considered, not just conservation. It also looks at the opportunities in the context of asset management and renewal, so it makes use of accounting measures like internal rate of return and life-cycle costing. Finally, it considers how implemented measures will behave over time.

Investment Letter—A letter signed by an investor purchasing unregistered long securities under Regulation D, in which the investor attests to the long-term investment nature of the purchase. These securities must be held for a minimum of one year before they can be sold.

Investment of Municipality—The investment of the municipality in its utility department, when such investment is not subject to cash settlement on demand or at a fixed future time. Include the cost of

debt-free utility plant constructed or acquired by the municipality and made available for the use of the utility department, cash transferred to the utility department for working capital, and other expenditures of an investment nature.

Investment Tax Credit (ITC)—A specified percentage of the dollar amount of certain new investments that a company can deduct as a credit against its income tax bill. A credit allowed under IRC Section 48 and is equal to 30 percent of the cost of certain qualified renewable energy property or a lesser amount on other qualified property. It is a credit directly applied to any taxes the taxpayer has and not as a "pre-tax" deduction (like depreciation).

Investment Value—The specific value of an investment to a particular investor or class of investors based on individual investment requirements. Distinguished from market value, which is impersonal and detached.

Investments and Advances to Unconsolidated Affiliates—The balance sheet account representing the cost of investments and advances to unconsolidated affiliates. Generally, affiliates that are less than 50-percent owned by a company may not be consolidated into the company's financial statements.

Investor-Owned Utility (IOU)—A private company that provides a utility, such as water, natural gas or electricity, to a specific service area. The investor-owned utility is regulated by a Public Utilities Commission. In California the investor owned utilities supplying energy are:

- Canadian Pacific National Corporation
- Pacific Gas and Electric Company
- Pacific Power and Light Company
- San Diego Gas & Electric
- Sierra Pacific Power Company
- Southern California Edison Company
- Southern California Gas Company (The Gas Company)
- Southwest Gas Corporation

• (IOU) A private power company owned by and responsible to its shareholders and regulated by a public service commission.

• A privately-owned electric utility whose stock is publicly traded. It is rate regulated and authorized to achieve an allowed rate of return.

Ion—An electrically charged atom or group of atoms that has lost or gained electrons; a loss makes the resulting particle positively charged; a gain makes the particle negatively charged.

Ion Exchange—Reversible exchange of ions adsorbed

on a mineral or synthetic polymer surface with ions in solution in contact with the surface. A chemical process used for recovery of uranium from solution by the interchange of ions between a solution and a solid, commonly a resin.

Ionizer—A device that removes airborne particles from breathable air. Negative ions are produced and give up their negative charge to the particles. These new negative particles are then attracted to the positive particles surrounding them. This accumulation process continues until the particles become heavy enough to fall to the ground.

IOU—An investor owned utility. A company, owned by stockholders for profit, that provides utility services. A designation used to differentiate a utility owned and operated for the benefit of shareholders from municipally owned and operated utilities and rural electric cooperatives.

IP—Inhalable Particles (Particulates)
- Intellectual Property
- Ingress Protection—describes with two figures the protection level against mechanical impact and water penetration.
- Inflation Premium

IPIECA—International Petroleum Industry Environmental Conservation Association www.ipieca.org

IPLV—See *Integrated Part Load Value*

IPM—Inhalable Particulate Matter
- Integrated Pest Management

IPMPV—The International Performance Measurement and Verification Protocol (IPMVP) provides a framework for 'measuring' energy or water savings. It presents common terminology and defines full disclosure, to support rational discussion of M&V issues.

IPMVP is published in several languages by the Efficiency Valuation Organization, a non-profit group of experts from around the world.

It is published in three Volumes:
- Volume I—Concepts and Options for Determining Energy and Water Savings
- Volume II—Indoor Environmental Quality (IEQ) Issues
- Volume III—Applications

IPO—Initial Public Offering

IPP—Independent Power Producer. A private entity that operates a generation facility and sells power to electric utilities for resale to retail customers.

IR—Infrared

IRA Rollover—The reinvestment of assets received as a lump-sum distribution from a qualified tax-deferred retirement plan. Reinvestment may be the entire lump sum or a portion thereof. If reinvestment is done within 60 days, there are no tax consequences.

IRC—Internal Revenue Code

IREC—Interstate Renewable Energy Council

IRIS—Integrated Risk Information System

Iron and Steel Industry—Steel Works, Blast Furnaces (Including Coke Ovens), and Rolling Mills: Establishments primarily engaged in manufacturing hot metal, pig iron, and silvery pig iron from iron ore and iron and steel scrap; converting pig iron, scrap iron, and scrap steel into steel; and in hot-rolling iron and steel into basic shapes, such as plates, sheets, strips, rods, bars, and tubing.

IRPTC—International Register of Potentially Toxic Chemicals

IRR—*Internal Rate of Return*. A typical measure of how VC Funds measure performance. IRR is technically a discount rate: the rate at which the present value of a series of investments is equal to the present value of the returns on those investments.

Irradiance—The direct, diffuse, and reflected solar radiation that strikes a surface. The solar power incident on a surface, usually expressed in kilowatts per square meter. Irradiance multiplied by time gives insolation.
- A measurement of sunlight strength, typically in Watts per square meter or "suns" (1000 W/m^2), considered to represent total available energy for usage by a PV panel. It will vary with cloud cover, time of year, and time of day. Readings may exceed 1000 W/m^2 under normal conditions. Readings above 1300 W/m^2 are typically due to reflection.

Irradiated Nuclear Fuel—Nuclear fuel that has been exposed to radiation in the reactor core at any power level.

Irrevocable—Not to be revoked or withdrawn.

Irritant—A chemical that can cause temporary irritation at the site of contact.

IRS—Internal Revenue Service

IRTA—Institute for Research and Technical Assistance

IS—Initial Study

ISCC—Integrated Solar Combined Cycle

ISD—Information Services Division
- Information Systems Division
- Interim Status Document

ISO—Independent System Operator. A neutral operator responsible for maintaining instantaneous balance of the grid system. The ISO performs its function by controlling the dispatch of flexible plants to ensure that loads match resources available to the system.

ISO (International Organization for Standardization)—ISO is a developer and publisher of proprietary industrial and commercial international standards. Founded in 1947, it is a network of the national standards institutes of 157 countries (one member per country) and has its headquarters in Geneva, Switzerland.

ISO 13256-1—The International Organization for Standardization document entitled "Water-source heat pumps—Testing and rating for performance—Part 1: Water-to-air and brine-to-air heat pumps," 1998.

ISO 14001—International standard for Environmental Management Systems. The principles underlying ISO 14001—"plan, act, check, review"—are similar to those underlying the Quality Management standards ISO9001/2. So are the principles for certification.

ISO 14021—International standard providing guidance for declarations made by an organization about the environmental aspects of its products or services (these "self-declared" environmental claims are sometimes called Type II Declarations). ISO 14021 is one of a series of ISO standards covering product-related environmental declarations—the ISO 14020 series.

ISO 14031—International standard providing guidance on environmental performance evaluation. Provides a method for identifying performance indicators that an organization can use to monitor its progress towards environmental performance objectives. The mechanism shares many features of the mechanism for establishing an environmental management system described by ISO14001.

ISO 50001—International standard providing guidance on energy management. ISO 50001 is based on the management system model of continual improvement also used for other well-known standards such as ISO 9001 or ISO 14001. This makes it easier for organizations to integrate energy management into their overall efforts to improve quality and environmental management.

Isobutane (C_4H_{10})—A normally gaseous branch-chain hydrocarbon. It is a colorless paraffinic gas that boils at a temperature of 10.9 degrees Fahrenheit. It is extracted from natural gas or refinery gas streams.

Isobutylene (C_4H_8)—An olefinic hydrocarbon recovered from refinery processes or petrochemical processes.

Isohexane (C_6H_{14})—A saturated branch-chain hydrocarbon. It is a colorless liquid that boils at a temperature of 156.2 degrees Fahrenheit.

Isolated Solar Gain System—A type of passive solar heating system where heat is collected in one area for use in another.

Isolation Device—A device that prevents the conditioning of a zone or group of zones in a building while other zones of the building are being conditioned.

Isomerization—A refining process that alters the fundamental arrangement of atoms in the molecule without adding or removing anything from the original material. Used to convert normal butane into isobutane (C_4), an alkylation process feedstock, and normal pentane and hexane into isopentane (C_5) and isohexane (C_6), high-octane gasoline components.

Isopach—A line on a map drawn through points of equal thickness of a designated unit (such as a coal bed).

Isopentane (C_5H_{12})—A saturated branched-chain hydrocarbon (C_5H_{12}) obtained by fractionation of natural gasoline or isomerization of normal pentane.

Isotopes—Forms of the same chemical element that differ only by the number of neutrons in their nucleus. Most elements have more than one naturally occurring isotope. Many isotopes have been produced in reactors and scientific laboratories.

ISPRA Guidelines—Guidelines for the assessment of photovoltaic power plants, published by the Joint Research Centre of the Commission of the European Communities, Ispra, Italy.

ISS—Interim Status Standards

Issuer—A state or local unit of government that borrows money through the sale of bonds and/or notes.
- Refers to the organization issuing or proposing to issue a security.

Issues—Unresolved conflicts regarding alternative uses of available resources. Subjects of public interest.

ISTEA—Intermodal Surface Transportation Efficiency Act (1991)

ISWMP—Integrated Solid Waste Management Plan

ITC—International Trade Commission
- Investment Tax Credit. One of six federal tax credits for businesses. Investment tax credits are typically calculated based on the amount of eligible costs incurred upon installation of a facility or rehabilitation of a historic building.

ITC Recapture Period—The five-year period over which a project must continue to satisfy the various requirements of IRC Section 48 in order to avoid tax credit recapture on a qualified energy facility that has claimed the ITC. The recapture period begins with the first taxable year of the credit period.

ITCS—Interstate Transition Cost Surcharge: A balancing account that records the difference between the full rate for an energy utility's interstate pipeline capacity reserved for the noncore and brokering revenues received for that capacity. The balance is allocated to all customers on an equal cents-per-therm basis, except that core customers' liability is limited to 10 percent of the cost of the core capacity reservation.

ITL—International Transaction Log. The means by which carbon allowances and credits generated under the mechanisms of the Kyoto Protocol—AAUs, CERs and ERUs—are traded between countries. An online IT platform that connects UN and national greenhouse emissions registries, facilitating the emerging global carbon market.

I-Type Semiconductor—A semiconductor material that is left intrinsic, or undoped so that the concentration of charge carriers is characteristic of the material itself rather than of added impurities.

I-V Curve—A graphical plot or representation the current and voltage output of a solar photovoltaic cell or module as a load on the device is increased from short circuit (no load) condition to the open circuit condition; used to characterize cell/module performance.

• A graph that plots the current versus the voltage from a PV cell as the electrical load (or resistance) is increased from short circuit (no load) to open circuit (maximum voltage). The shape of the curve characterizing cell performance. Three important points on the I-V curve are the open-circuit voltage, short-circuit current, and peak or maximum power (operating) point.

I-V Data—The relationship between current and voltage of a photovoltaic device in the power-producing quadrant, as a set of ordered pairs of current and voltage readings in a table, or as a curve plotted in a suitable coordinate system [ASTM E 1036]

IV&V—Independent Validation and Verification

IWC—In-Stream Waste Concentration

J

J&A—Justification and approval

Jack Pump—A submerged pump mechanically activated by a rod extending above the well head to a reciprocating engine, motor or any other rotating device.

Jacket—The enclosure on a water heater, furnace, or boiler.

JAPCA—Journal of Air Pollution Control Association

JEC—Joint Economic Committee

JES—Joint Environmental Statement

Jet Fuel—A refined petroleum product used in jet aircraft engines. It includes kerosene-type jet fuel and naphtha-type jet fuel.

JI—Joint Implementation. A Kyoto Protocol mechanism which allows developed countries, particularly those in transition to a market economy, to host carbon-reducing projects funded by another developed country. The arrangement sees the credits generated, called ERUs, go to the investor country while the emission allowances (AAUs) of the host country are reduced by the same amount. See *Joint Implementation* below.

JICA—Japanese International Cooperation Agency

JNCP—Justification for Noncompetitive Procurement

JOFOC—Justification for Other Than Full and Open Competition

Joint Implementation (JI)—The Kyoto Protocol establishes a mechanism whereby a developed country can receive "emissions reductions units" when it helps to finance projects that reduce net emissions in another developed country (including countries with economies in transition). Some aspects of this approach are being tested as Activities Implemented Jointly.

• Agreements made between two or more nations under the auspices of the Framework Convention on Climate Change (FCCC) whereby a developed country can receive "emissions reduction units" when it helps to finance projects that reduce net emissions in another developed country (including countries with economies in transition).

• This is a method of achieving reductions in CO_2 emissions whereby rich countries (which will probably have made binding commitments to cut emissions) can get partial credit for emission reductions projects which are funded by them, but which are undertaken in poor countries. This is now referred to as the Clean Development Mechanism (CDM) to denote respect for developing countries' right to develop.

• This mechanism is an attempt to make a system of marketable permits more equitable, by recognizing that wealthy countries were the major cause of the CO_2 buildup. It also makes carbon reduction more politically possible. It also has the advantage of being more cost effective than trying to achieve emissions reductions solely in the rich countries.

Joint Venture—A legal entity formed by two or more parties to conduct a specific business transaction.

Joint-Use Facility—A multiple-purpose hydroelectric plant. An example is a dam that stores water for both flood control and power production.

Joist—A structural, load-carrying building member with an open web system that supports floors and roofs utilizing wood or specific steels and is designed as a simple span member.

Joule (J)—A unit of work or energy equal to the amount of work done when the point of application of force of 1 newton is displaced 1 meter in the direction of the force. It takes 1,055 joules to equal a British thermal unit. It takes about 1 million joules to make a pot of coffee.

Joule's Law—The rate of heat production by a steady current in any part of an electrical circuit that is proportional to the resistance and to the square of the current, or, the internal energy of an ideal gas depends only on its temperature.

Joule-Thomson Effect—The cooling which occurs when a compressed gas is allowed to expand in such a way that no external work is done. The effect is approximately 7 degrees Fahrenheit per 100 psi for natural gas.

JPA—Joint Permitting Agreement
  • Joint Powers Authority

Junction—A region of transition between semiconductor layers, such as a p/n junction, which goes from a region that has a high concentration of acceptors (p-type) to one that has a high concentration of donors (n-type).

Junction Box—A PV junction box is a protective enclosure on a PV module where PV strings are electrically connected and where electrical protection devices such as diodes can be fitted.

Junction Diode—A semiconductor device, having a junction and a built-in potential, that passes current better in one direction than the other. All solar cells are junction diodes.

Junior Lien—A subordinate lien.

Junk Bonds—Bonds of a speculative grade which represent a higher risk to investors but offer the opportunity for higher interest. Looked upon as undervalued assets and frequently used as a money source in takeover attempts.

Jurat—A certificate evidencing the fact that an Affidavit was properly made before an Authorized Officer.

Jurisdiction—Portion of the company's activities that are subject to the rules and regulations of the particular government entity which regulates it.

Jurisdictional Utilities—Utilities regulated by public laws.

JUSSCANNZ—The JUSSCANNZ is a group of non-European Union industrialized nations in the Kyoto Protocol negotiations including Japan, United States, Switzerland, Canada, Australia, Norway, and New Zealand. Iceland, Mexico, and the Republic of Korea.

K

K—Kelvin

Kaplan Turbine—A type of turbine that that has two blades whose pitch is adjustable. The turbine may have gates to control the angle of the fluid flow into the blades.

KBtu—One-thousand (1,000) Btu's.

KCM—Thousand circular mils (also KCmil) (electricity conductor)

Keelage—A duty charged for permitting a ship to enter and anchor in a port or harbor.

Kerosene—A light petroleum distillate that is used in space heaters, cook stoves, and water heaters and is suitable for use as a light source when burned in wick-fed lamps. Kerosene has a maximum distillation temperature of 400 degrees Fahrenheit at the 10-percent recovery point, a final boiling point of 572 degrees Fahrenheit, and a minimum flash point of 100 degrees Fahrenheit. Included are No. 1-K and No. 2-K, the two grades recognized by ASTM Specification D 3699 as well as all other grades of kerosene called range or stove oil, which have properties similar to those of No. 1 fuel oil.

Kerosene-type Jet Fuel—A kerosene-based product having a maximum distillation temperature of 400 degrees Fahrenheit at the 10-percent recovery point and a final maximum boiling point of 572 degrees Fahrenheit and meeting ASTM Specification D 1655 and Military Specifications MIL-T-5624P and MIL-T-83133D (Grades JP-5 and JP-8). It is used for commercial and military turbojet and turboprop aircraft engines.

Ketone-Alcohol (cyclohexanol)—An oily, colorless, hygroscopic liquid with a camphor-like odor. Used in soap making, dry cleaning, plasticizers, insecticides, and germicides.

Kg—Kilogram

KGRA—Known Geothermal Resource Area

Kiln—A device, including any associated preheater or precalciner devices, that produces clinker by heating limestone and other materials for subsequent production of Portland cement.

Kilovolt (kv)—One-thousand volts (1,000). Distribution lines in residential areas usually are 12 kv (12,000 volts).

Kilovolt-Ampere (kVa)—A unit of apparent power, equal to 1,000 volt-amperes; the mathematical product of the volts and amperes in an electrical circuit.

Kilowatt (kW)—One thousand (1,000) watts, which is equivalent to approximately 1.34 horsepower. A unit of measure of the amount of electricity needed to operate given equipment. On a hot summer afternoon a typical home, with central air conditioning and other equipment in use, might have a demand of four kW each hour.

Kilowatt Hour (kWh)—A measure of energy equivalent to the expenditure of one kilowatt for one hour. For example, 1 kWh will light a 100-watt light bulb for 10 hours. 1 kWh = 3,413 Btu.

• The most commonly-used unit of measure telling the amount of electricity consumed over time. It means one kilowatt of electricity supplied for one hour. 1 kWh=3600 kJ. In 1989, a typical California household consumed 534 kWh in an average month.

• This is the standard unit of billing for electrical energy.

Kilowatt-Electric (kWe)—One thousand watts of electric capacity.

Kinetic Energy—Energy available as a result of motion that varies directly in proportion to an object's mass and the square of its velocity.

Kitchen—A low-rise residential building is a room or area used for cooking, food storage and preparation and washing dishes, including associated counter tops and cabinets, refrigerator, stove, ovens, and floor area. Adjacent areas are considered kitchen if the lighting for the adjacent areas is on the same -circuit as the lighting for the kitchen.

Km—Kilometer

Kneewall—A wall usually about 3 to 4 feet high located that is placed in the attic of a home, anchored with plates between the attic floor joists and the roof joist. Sheathing can be attached to these walls to enclose an attic space.

A sidewall separating conditioned space from attic space under a pitched roof. Knee walls should be insulated as an exterior wall as specified by the chosen method of compliance.

Knutson-Vandenberg Act—(KV) Federal law that allows the U.S. Forest Service to collect money from a timber sale for resource enhancement, protection, and improvement work in the timber sale vicinity.

KOP—Key Observation Point

kV—Kilovolt: Unit of measurement of electromotive force equal to 1,000 volts. A volt is the force required to produce a current of one ampere through a resistance of one ohm.

KVA—Kilovolt-ampere (transformer size rating)

KVAR—Kilovolt-ampere reactive

kW—Kilowatt: A kilowatt is 1,000 watts

kWe—Kilowatt, electric

kWh—Kilowatt-hour: A kilowatt hour is the amount of kilowatts (1,000 watts) of electricity used in one hour of operation.

kWp—Peak kilowatt

Kyoto Commitment Period—The Kyoto commitment period is the period in which Annex B countries have committed to reduce their collective emissions of greenhouse gases by an average of 5.2%. There are currently no emissions targets after the commitment period specified in the Kyoto Protocol from 2008 to 2012. These targets, if the United Nations Framework Convention on Climate Change (UNFCCC or the Convention) process continues in its present form, will be negotiated closer to the expiration of the first commitment period. It is expected that the current model of five-year periods of commitment will be maintained. Major questions regarding future commitment periods include the level of allowed emissions among capped (Annex I) countries and the extent to which additional countries take on caps (that is, developing country participation).

Kyoto Mechanisms—The Kyoto Mechanisms (commonly referred to as Emissions Trading) allow for the creation and transfer of emissions permits between countries. Based on economic market principals, they are designed to minimize the cost of reducing global greenhouse emissions and include: Joint Implementation (Article 6), the Clean Development Mechanisms (Article 12), and International Emissions Trading (Article 17).

Kyoto Protocol—The result of negotiations at the third Conference of the Parties (COP-3) in Kyoto, Japan, in December of 1997. The Kyoto Protocol sets binding greenhouse gas emissions targets for countries that sign and ratify the agreement. The gases covered under the Protocol include carbon dioxide, methane, nitrous oxide, hydrofluorocarbons (HFCs), perfluorocarbons (PFCs) and sulfur hexafluoride. 175 parties have so far ratified the

Protocol and are legally bound to adhere to its principles.

L

L&I—License and Inspection

L&MP—Lease and management plan

LAA—Lead Agency Attorney

LADD—Lifetime Average Daily Dose

LADWP—Los Angeles Department of Water and Power

LAER—See *Lowest Achievable Emission Rate.*

LAFCO—Local Agency Formation Commission: State mandated local agency that oversees boundary changes to cities and special districts, the formation of new agencies including incorporation of new cities, and the consolidation of existing agencies.

Lagoon—In wastewater treatment or livestock facilities, a shallow pond used to store wastewater where sunlight and biological activity decompose the waste.

Lamp—A light source composed of a metal base, a glass tube filled with an inert gas or a vapor, and base pins to attach to a fixture. Lamps are consumable items with a known life expectancy. Types include fluorescent, incandescent PAR, compact fluorescent and high-intensity discharge lamps.

Land Use—The ultimate uses to be permitted for currently contaminated lands, waters, and structures at each Department of Energy installation. Land-use decisions will strongly influence the cost of environmental management.

Land Use, Land-Use Change and Forestry (LULUCF)—Land uses and land-use changes can act either as sinks or as emission sources. It is estimated that approximately one-fifth of global emissions result from LULUCF activities. The Kyoto Protocol allows Parties to receive emissions credit for certain LULUCF activities that reduce net emissions.

Landfill Gas—Gas generated by the natural degrading and decomposition of municipal solid waste by anaerobic microorganisms in sanitary landfills. The gases produced, carbon dioxide and methane, can be collected by a series of low-level pressure wells and can be processed into a medium Btu gas that can be burned to generate steam or electricity.

Landing—A cleared working area on or near a timber harvest site at which processing steps are carried out.

Landlord Waiver—A document required by a lender when a debtor is placing the underlying equipment on a property leased from another party, to secure lender's rights.

Landscape—A region consisting of interacting ecosystems determined by geology, soils, climate, biota, and human influences. A landscape is made up of watersheds and smaller ecosystems.

Landscaping—Features and vegetation on the outside of or surrounding a building for aesthetics and energy conservation.

Land-Use Restrictions—Constraints placed upon mining by societal policies to protect surface features or entities that could be affected by mining. Because laws and regulations may be modified or repealed, the restrictions, including industrial and environmental restrictions, are subject to change.

Langley—A unit or measure of solar radiation; 1 calorie per square centimeter or 3.69 Btu per square foot. 1 L = 41.84 kJ/m2.

LAO—Local Area Operations
- Local Area Office

Large Passenger Car—A passenger car with more than 120 cubic feet of interior passenger and luggage volume.

Large Pickup Truck—A pickup truck weighing between 4,500-8,500 lbs gross vehicle weight (GVW).

Large Woody Debris—Dead woody material greater than 20″ in diameter on the ground or in a stream or river. It may consist of logs, trees, or parts of trees. Large woody debris contributes to long-term site productivity and health in several ways. It supplies nutrients to the soil, supports symbiotic fungi that are beneficial to conifers, and provides habitat for beneficial rodents and insects.

Laser—A very intense, uniform beam of electromagnetic radiation. Acronym for Light Amplification by Simulated Emission of Radiation.

Late Seral Species—Shade tolerant species, primarily vine maple shrubs and western red cedar and western hemlock trees. These species follow the mid seral species in natural succession.

Late Seral Treatment—A treatment in which late seral species will be established after thinning.

Latency—Latency refers to the process by which some diseases develop long after an individual is exposed to the causative agent. Some diseases develop many years after individuals are exposed to toxic chemicals. In such cases, it takes longer testing to prove that a compound is safe.

Latent Cooling—The removal of the heat load created by moisture in the air, including from outside air infiltration and from indoor sources such as occupants, plants and everyday activities such as cooking and showering. Latent Cooling is the same as dehumidification. Latent Cooling differs from Sensible Cooling in that the air temperature does not change under pure Latent Cooling. Most air conditioning processes provide both Latent and Sensible Cooling.

Latent Cooling Load—The load created by moisture in the air, including from outside air infiltration and that from indoor sources such as occupants, plants, cooking, showering, etc.

Latent Heat—A change in the heat content that occurs without a corresponding change in temperature, usually accompanied by a change of state (as from liquid to vapor during evaporation).

Latent Heat of Vaporization—The quantity of heat produced to change a unit weight of a liquid to vapor with no change in temperature.

Latent Load—The cooling load caused by moisture in the air.

Later Stage—A stage of company growth characterized by viable products, a developed market, significant customers, sustained revenue growth, and both profits and positive cash flow from operations. Later-stage companies would generally be candidates for an IPO. Investments in the C round or after qualify as later stage.

Late-Successional Forest—Forest seral stages which include mature and old-growth age classes.

Late-Successional Reserve—An area of forest where the management objective is to protect and enhance conditions of late successional and old-growth forest ecosystems.

Latitude—The angular distance north or south of the equator, measured in degrees of arc.

Latitude and Longitude—The distance on the earth's surface measured, respectively, north or south of the equator and east or west of the standard meridian, expressed in angular degrees, minutes, and seconds.

Lattice—The regular periodic arrangement of atoms or molecules in a crystal of semiconductor material.

Law(s) of Thermodynamics—The first law states that energy cannot be created or destroyed; the second law states that when a free exchange of heat occurs between two materials, the heat always moves from the warmer to the cooler material.

Lay Up—Lay up is another term for cold storage and describes the status of equipment (such as a power plant) that has been placed in storage ("mothballed") for later use.

Layoff (Electric Utility)—Excess capacity of a generating unit, available for a limited time under the terms of a power sales agreement.

LBI—Limited Background Investigation

LBNL—Lawrence Berkeley National Laboratory

LBO (Leveraged Buyout)—The purchase of a company financed by borrowing on its assets. Takeover of a company by external or internal investors. This type of corporate acquisition is characterized by the fact that little equity is used.

lbs/hr—Pounds per hour

LC—Lethal Concentration
- Liquid Chromatography

LCA—Life Cycle Assessment

LCC—Life-cycle Cost

LCCA—Lead Contamination Control Act of 1988, which deals with the recall of lead-lined drinking water coolers

LCD—Local Climatological Data

LCDC—See *Land Conservation and Development Commission.*

LCFS—Low Carbon Fuel Standard

LCL—Lower Control Limit

LCM—Life Cycle Management

LCOE—Levelized Cost of Energy

LD—Liquidated Damages
- Land Disposal
- Lethal Dose
- Light Duty

LDAR—Leak Detection and Repair

LDC—Local Distribution Company: A public utility that delivers natural gas to end-use customers through its own distribution system.
- London Dumping convention

LDD—Light Duty Diesel

LDPE—Low Density Polyethylene

LDS—Light Duty Vehicle/Truck Certification
- Leak Detection System

Leachate—The liquid that has percolated through the soil or other medium. Typically, water that has come in contact with hazardous wastes.

Leachates—Liquids percolated through waste piles. Leachate can include various minerals, organic matter, or other contaminants and can contaminate surface water or ground water.

Leaching—A solution mining process to remove salt and form gas storage caverns in salt domes.

Lead—A heavy metal present in small amounts everywhere in the human environment. Lead can get into the body from drinking contaminated water, eating vegetables grown in contaminated soil, or breathing dust when children play or adults work in lead-contaminated areas or eating lead-based paint. It can cause damage to the nervous system or blood cells.

Lead Acid Battery—A type of battery that consists of plates made of lead, lead-antimony, or lead-calcium and lead-oxide, surrounded by a sulfuric acid electrolyte. The most common type of battery used in RAPS systems.

Lead Agency—A public agency which has the principal responsibility for ordering and overseeing site investigation and cleanup.

Lead Bank—The bank which negotiates a large loan with a borrower and solicits other lenders to join the syndicate making the loan.

Lead Investor—Also known as a bell cow investor. Member of a syndicate of private equity investors holding the largest stake, in charge of arranging the financing and most actively involved in the overall project.

Leaded Gasoline—A fuel that contains more than 0.05 gram of lead per gallon or more than 0.005 gram of phosphorus per gallon. Gasoline containing tetraethyl lead, an important constituent in antiknock gasoline. Leaded gasoline is no longer sold in the United States.

Leaded Premium Gasoline—Gasoline having an antiknock index (R+M/2) greater than 90 and containing more than 0.05 grams of lead or 0.005 grams of phosphorus per gallon.

Leaded Regular Gasoline—Gasoline having an antiknock index (R+M/2) greater than or equal to 87 and less than or equal to 90 and containing more than 0.05 grams of lead or 0.005 grams of phosphorus per gallon.

Leading Edge—In reference to a wind energy conversion system, the area of a turbine blade surface that first comes into contact with the wind.

Leakage—Occurs when laws or activities designed to cut greenhouse gas emissions implemented in one jurisdiction or project area lead to the shifting of the targeted emitting activities elsewhere, thus undermining the attempt to reduce emissions.

• Leakage is the indirect effect of emission reduction policies or activities that lead to a rise in emissions elsewhere (e.g. fossil fuel substitution leads to a decline in fuel prices and a rise in fuel use elsewhere). For land use change and forestry activities, leakage can be defined as the unexpected loss of estimated net carbon sequestered. Specific to CDM/AIJ/JI projects in both forestry and energy sectors, leakage can be a result of unexpected effects including unforeseen circumstances, improperly defined baseline, improperly defined project lifetime or project boundaries, and inappropriate project design.

Leaking Electricity—Related to stand-by power, leaking electricity is the power needed for electrical equipment to remain ready for use while in a dormant mode or operation. Electricity is still used by many electrical devices, such as TVs, stereos, and computers, even when you think they are turned "off."

Lease—A contract that conveys the right to use an asset (the underlying asset) for a period of time in exchange for consideration.

Lease and Plant Fuel—Natural gas used in well, field, and lease operations (such as gas used in drilling operations, heaters, dehydrators, and field compressors) and as fuel in natural gas processing plants.

Lease Bonus—An amount paid by a lessee to a Lessor as consideration for granting a lease, usually as a lump sum; this payment is in addition to any rental or royalty payments.

Lease Condensate—A mixture consisting primarily of pentanes and heavier hydrocarbons which is recovered as a liquid from natural gas in lease separation facilities. This category excludes natural gas plant liquids, such as butane and propane, which are recovered at downstream natural gas processing plants or facilities.

Lease Equipment—All equipment located on the lease except the well to the point of the "Christmas tree."

Lease Fuel—Natural gas used in well, field, and lease operations, such as gas used in drilling operations, heaters, dehydrators, and field compressors.

Lease Liability—A lessee's obligation to make lease payments arising from a lease, measured on a discounted basis.

Lease Line—A lease line of credit. A lessee can add equipment without having to renegotiate a new lease each time. Also known as a Master Lease.

Lease Operations—Any well, lease, or field operations related to the exploration for or production of natural gas prior to delivery for processing or transportation out of the field. Gas used in lease operations includes usage such as for drilling operations, heaters, dehydraters, field compressors, and net used for gas lift.

Lease Pass-Through—A term used to describe a form of ownership structure that is used to facilitate investments that qualify for the investment tax credit (ITC) under section 46 and which involve using two limited partnerships.

Lease Payments—Payments made by a lessee to a lessor relating to the right to use an underlying asset during the lease term, consisting of the following:

- Fixed payments, less any lease incentives received or receivable from the lessor
- Variable lease payments that depend on an index or a rate or are in-substance fixed payments
- The exercise price of a purchase option if the lessee has a significant economic incentive to exercise that option
- Payments for penalties for terminating the lease, if the lease term reflects the lessee exercising an option to terminate the lease

For the lessee, lease payments also include amounts expected to be payable by the lessee under residual value guarantees. Lease payments do not include payments allocated to non-lease components of a contract except when the lessee is required to combine non-lease and lease components and account for them as a single lease components.

For the lessor, lease payments also include lease payments structured as residual value guarantees. Lease payments do not include payments allocated to non-lease components.

Lease Receivable—A lessor's right to receive lease payments arising from a lease, measured on a discounted basis.

Lease Rollover—The expiration of a lease and the subsequent re-leasing of the space or physical asset.

Lease Separation Facility (Lease Separator)—A facility installed at the surface for the purpose of (a) separating gases from produced crude oil and water at the temperature and pressure conditions set by the separator and/or (b) separating gases from that portion of the produced natural gas stream that liquefies at the temperature and pressure conditions set by the separator.

Lease vs Buy—A comparison of the costs incurred in obtaining the use of an asset for a specific period of time through either leasing or purchasing. Costs are typically compared on an after-tax, present value basis

Lease-buy Analysis—A financial analysis comparing the cost of leasing equipment with the cost of direct purchasing of equipment.

Leasehold Improvement—An improvement to leased property, considered an intangible asset to the lessee that becomes the property of the lessor at the end of the lease.

Leasehold Interest—The interest held by the lessee through a lease transferring the rights of use or occupancy for a stated term under certain conditions.

Leasehold Reserves—Natural gas liquid reserves corresponding to the leasehold production defined above.

Leasing—Municipal or Capital—A tax exempt lease where the cost of equipment is amortized over the lease term. At the end of the lease period ownership passes to the lessee. This is also known as a lease purchase.

Leasing—Operating—Using a piece of property without transferring ownership. Leasing is an alternative to direct ownership of energy saving equipment. This is also known as an operating lease.

Least Cost Planning—(Integrated resource planning) A method of power planning that recognizes load uncertainty, embodies an emphasis on risk management, and reviews all available and reliable resources to meet future loads. It takes into consideration all costs of a resource, including capital, labor, fuel, maintenance, decommissioning, known environmental impacts, and the difficulty in quantifying the consequences of selecting one resource over another. Least cost planning seeks to minimize total energy costs.

Least Developed County (LDC)—Forty-nine countries are currently designated by the United Nations as "least developed countries" (LDCs). The list is reviewed every three years by the Economic and Social Council (ECOSOC). The criteria underlying the current list of LDCs are: low income, weak human resources, and a low level of economic diversification.

LED—Light Emitting Diode

LEED—The Leadership in Energy and Environmental Design (LEED) Green Building Rating System™ encourages and accelerates global adoption of sustainable green building and development practices through the creation and implementation of universally understood and accepted tools and performance criteria. LEED is a third-party certification program and the nationally accepted benchmark for the design, construction and operation of high performance green buildings. LEED gives building owners and operators the tools they need to have an immediate and measurable impact on their buildings' performance. LEED promotes a whole-building approach to sustainability by recognizing performance in five key areas of human and environmental health: sustainable site development, water savings, energy efficiency, materials selection and indoor environmental quality. The program is sponsored by the United States Green Building Council (USGBC).

LEED AP (LEED Accredited Professional)—LEED Accredited Professionals are building industry practitioners who have passed the USGBC's exam that tests their understanding of green building practices and principles, and familiarity with LEED requirements.

LEED for Commercial Interiors—Addresses the specifics of tenant spaces mainly in office, retail and institutional buildings. It provides a way for tenants who lease their space or do not occupy the entire building to LEED certify their space as a green interior.

LEED for Core and Shell—A green building system designed to provide a set of performance criteria for certifying the sustainable design and construction of speculative developments and core and shell buildings. According to the USGBC, core and shell construction covers base building elements, such as the structure, envelope and building-level systems, such as central HVAC, LEED for Core and Shell complements both the Commercial Interiors and Existing Buildings rating systems.

LEED for Existing Buildings—O&M is a revised tool for the ongoing operations and maintenance of existing buildings. The certification system identifies and rewards current best practices and provides an outline for buildings to use less energy, water and natural resources; improve the indoor environment; and uncover operating inefficiencies, the USGBC says.

LEED for Homes—A green home rating system to help ensure that homes are designed and built to be energy- and resource-efficient and healthy for occupants. It can be applied to single- and multi-family homes.

LEED for Neighborhood Development—Integrates the principles of smart growth, new urbanism, and green building into a standard for neighborhood design. Being developed by USGBC in partnership with the Congress for the New Urbanism (CNU) and the Natural Resources Defense Council (NRDC), LEED for Neighborhood Development recognizes development projects that successfully protect and enhance the overall health, natural environment, and quality of life of communities. This pilot program began in 2007 and is currently underway.

LEED for New Construction and Major Renovations—A rating system for buildings created to guide and distinguish high performance buildings that have less of an impact on the environment, are healthier for those who work and/or live in the building, and are more profitable than their conventional counterparts, according to the USGBC. It can be applied to commercial, institutional and high-rise residential projects, with a focus on office buildings, and also has also been applied to K-12 schools, multi-unit residential buildings, laboratories, and manufacturing facilities.

Left Side—The left side of the building as one faces the front facade from the outside. This designation is used on the Certificate of Compliance (in California) and other compliance documentation.

Legacy—A gift of Personal Property by will, usually money.

Legal Opinion—A written opinion from counsel that an issue of bonds and/or leases was duly authorized and issued. The opinion usually includes the statement, "interest received thereon is exempt from federal taxes and, in certain circumstances, from state and local taxes."

Legatee—One to whom Personal Property is given by will.

LEL—Lower Explosive Limit
  • Lower Exposure Limit
  • Lowest Effect Level

LEP—Laboratory Evaluation Program

LEPC—Local Emergency Planning Committee

LEPD—Legal Enforcement Policy Division

LERC—Local Emergency Response Committee

Less Intensive Verification Activities—The verification activities done in interim years between full verifications that only require data checks on a facility's emissions based on the most current sampling plan developed as part of the most current full verification activities.

Lessee—An entity that enters into a contract to obtain the right to use an underlying asset for a period of time in exchange for consideration.
  • An independent marketer who leases the station and land and has use of tanks, pumps, signs, etc. A lessee dealer typically has a supply agreement with a refiner or distributor and purchases products at dealer tank-wagon prices. The term "lessee dealer" is limited to those dealers who are supplied directly by a refiner or any affiliate or subsidiary of the reporting company. "Direct supply" includes use

of commission agent or common carrier delivery.

Lessor—An entity that enteres into a contract to provide the right to use an underlying asset for a period of time in exchange for consideration.

Lethe—A measure of air purity that is equal to one complete air change (in an interior space).

Letter of Credit (L/C)—A bank's written guarantee of funds available of drafts written on it.

LEV—Low Emissions Vehicle *see Low Emission Vehicle*

Levelized—A lump sum that has been divided into equal amounts over period of time.

Levelized Cost—The present value of the total cost of building and operating a generating plant over its economic life, converted to equal annual payments. Costs are levelized in real dollars (i.e., adjusted to remove the impact of inflation).

Levelized Life-Cycle Cost—The present value of the cost of a resource, including capital, financing and operating costs, expressed as a stream of equal annual payments. This stream of payments can be converted to a unit cost of energy by dividing the annual payment amount by the annual kilowatt-hours produced or saved. By levelizing costs, resources with different lifetimes and generating capabilities can be compared.

Leverage (Lease)—The funds in a leveraged lease provided by third-party debt. The leverage is combined with equity to provide the funds for the initial purchase of the asset to be leased.

Leverage Buyout (LBO)—A takeover of a company, using a combination of equity and borrowed funds. Generally, the target company's assets act as the collateral for the loans taken out by the acquiring group. The acquiring group then repays the loan from the cash flow of the acquired company. For example, a group of investors may borrow funds, using the assets of the company as collateral, in order to take over a company. Or the management of the company may use this vehicle as a means to regain control of the company by converting a company from public to private. In most LBOs, public shareholders receive a premium to the market price of the shares.

Leverage Ratio—A measure that indicates the financial ability to meet debt service requirements and increase the value of the investment to the stockholders. (i.e. the ratio of total debt to total assets).

Leveraged—Indebted. Refers to the amount of debt in a transaction or lease company. A firm becomes more leveraged as it uses more borrowed funds, relative to equity infusions, to finance its operations.

Leveraged Lease—A specific form of lease involving at least three parties: a lessor, lessee and funding source. The lessor borrows a significant portion of the equipment cost on a non-recourse basis by assigning the future lease payment stream to the lender in return for up-front funds (the borrowing). The lessor puts up a minimal amount of its own equity funds (the difference between the equipment cost and the present value of the assigned lease payments) and generally is entitled to the full tax benefits of equipment ownership.

Levy—A seizure of property by Judicial Process. Includes the power of distraint and seizure by any means. To impose a tax, fine or other penalty.

LFG—Landfill Gas

LFL—Lower Flammability Limit

LGP—Low Ground Pressure

LGR—Local Governments Reimbursement Program

LH$_2$O—The heat (Btu) needed to vaporize and superheat one pound of water.

LHDDV—Light Heavy Duty Diesel Vehicle

LHMA—Labeling of Hazardous Materials Act. Federal act that requires that all chronically hazardous materials be labeled as inappropriate for children's use.

LHW—Liquid Hazardous Waste

Liability—An amount payable in dollars or by future services to be rendered.

LIBOR—LIBOR is an abbreviation for "London Interbank Offered Rate," and is the interest rate offered by a specific group of London banks for U.S. dollar deposits of a stated maturity. LIBOR is used as a base index for setting rates of some adjustable rate financial instruments.

Licensed Site Capacity—Capacity (number of assemblies) for which the site is currently licensed.

Licensees—Entity that has been granted permission to engage in an activity otherwise unlawful (i.e., hydropower project).

LIEE—Low-Income Energy Efficiency: A PUC program that provides no-cost weatherization services to low-income households who meet the California Alternative Rates for Energy income guidelines. Seniors who are 60 years old and over, and disabled persons qualify for these services if their incomes are at or below 200% of the federal poverty guidelines. Services provided include attic insulation, energy efficient refrigerators, energy efficient furnaces, weather-stripping, caulking, low-flow showerheads, water heater blankets, and door and building envelope repairs that reduce air infiltration.

Lien—A security interest on property to protect the lender in the event of debtor default.

• A charge upon property for the payment of a debt or performance of an obligation. A form of encumbrance. Taxes, Special Assessments, and Judgments, as well as Mortgages, are liens. In addition, there are Mechanic's and Materialmen's Liens for furnishing labor or materials.

Life—The period during which a system can operate above a specified performance level.

Life Cycle Assessment (LCA)—A process of evaluating the effects of a product or its designated function on the environment over the entire period of the product's life in order to increase resource-use efficiency and decrease liabilities. Commonly referred to as "cradle-to-grave" analysis.

Life Cycle Cost—The sum of all the costs both recurring and nonrecurring, related to a product, structure, system, or service during its life span or specified time period. Includes operating costs, maintenance costs and initial cost.

Life Extension—Restoration or refurbishment of a plant to its original performance without the installation of new combustion technologies. Life extension results in 10 to 20 years of plant life beyond the anticipated retirement date, but usually does not result in larger capacity.

• A term used to describe capital expenses that reduce operating and maintenance costs associated with continued operation of electric utility boilers. Such boilers usually have a 40-year operating life under normal circumstances.

Life-Cycle Costing—A method of comparing costs of equipment or buildings based on original costs plus all operating and maintenance costs over the useful life of the equipment. Future costs are discounted.

Lifeline Rates—Rates charged by a utility company for the low income, the disadvantaged and senior citizens. The rates provide a discount for minimum necessary utilities, such as electricity requirements of typically 300 to 400 kilowatt/hours per month.

LIFO—Last In/First Out. A method using in accounting for inventory valuation and cash management.

Lift—The force that pulls a wind turbine blade, as opposed to drag.

Lifting Costs—The costs associated with the extraction of a mineral reserve from a producing property. Refer to Production Cost.

Light—A range of wavelengths of electromagnetic radiation given off by hot objects and certain chemical or physical reactions. It includes the visible, infrared, ultraviolet, and x-ray spectra. The speed of light in a vacuum is a constant 186,281 miles per second.

Light Bulbs—A term generally used to describe a man-made source of light. The term is often used when referring to a "bulb" or "tube."

Light Emitting Diode—A semi conductor device composed of a p-n junction designed such that electrons emit visible light during their migration across the junction.

Light Gas Oils—Liquid petroleum distillates heavier than naphtha, with an approximate boiling range from 401 degrees to 650 degrees Fahrenheit.

Light Oil—Lighter fuel oils distilled off during the refining process. Virtually all petroleum used in internal combustion and gas-turbine engines is light oil. Includes fuel oil numbers 1 and 2, kerosene, and jet fuel.

Light Quality—A description of how well people in a lighted space can see to do visual tasks and how visually comfortable they feel in that space.

Light Rail—An electric railway with a "light volume" traffic capacity compared to "heavy rail." Light rail may use exclusive or shared rights-of-way, high or low platform loading, and multi-car trains or single cars. Also known as "street car," "trolley car," and "tramway."

Light Shelf—A light-colored horizontal surface placed inside or outside and just below a clerestory window. It is designed to reflect sunlight up to the ceiling and deeper into the building while reducing glare.

Light Trapping—The trapping of light inside a semiconductor material by refracting and reflecting the light at critical angles; trapped light will travel further in the material, greatly increasing the probability of absorption and hence of producing charge carriers.

Light Trucks—All single unit two-axle, four-tire trucks, including pickup trucks, sports utility vehicles, vans, motor homes, etc. This is the Department of Transportation definition. The Energy Information defined light truck as all trucks weighing 8,500 pounds or less.

Light Water—Ordinary water (H_2O), as distinguished from heavy water or deuterium oxide (D_2O).

Light Water Reactor (LWR)—A nuclear power unit that uses ordinary water to cool its core. The LWR may be a boiling water reactor or a pressurized water reactor.

Light-Duty Vehicles—Vehicles weighing less than 8,500 lbs (include automobiles, motorcycles, and light trucks).

Light-Induced Defects—Defects, such as dangling bonds, induced in an amorphous silicon semiconductor upon initial exposure to light.

Lighting Arrestor—Devices that protect electronics from lightning-induced surges by carrying the charge to ground.

Lighting Conservation Feature—A building feature or practice designed to reduce the amount of energy consumed by the lighting system.

Lighting Demand-Side Management (DSM) Program—A DSM program designed to promote efficient lighting systems in new construction or existing facilities. Lighting DSM programs can include: certain types of high-efficiency fluorescent fixtures including T-8 lamp technology, solid state electronic ballasts, specular reflectors, compact fluorescent fixtures, LED and electro-luminescent Emergency Exist Signs, High Pressure Sodium with switchable ballasts, Compact Metal Halide, occupancy sensors, and daylighting controllers.

Lighting Zone—A section of a Facility within which all parts can be considered to have similar light scheduling requirements. Lighting zones are identified to compare occupancy patterns with lighting control patterns.

Lights—All of the light bulbs controlled by one switch are counted as one light. For example, a chandelier with multiple lights controlled by one switch is counted as one light. A floor lamp with two separate globes or bulbs controlled by two separate switches would be counted as two lights. Indoor and outdoor lights were counted if they were under the control of the householder. This would exclude lights in the hallway of multifamily buildings.

Lignin—An amorphous polymer that, together with cellulose, forms the cell walls of woody plants. Lignin acts as the bonding agent between cells.

Lignite—The lowest rank of coal, often referred to as brown coal, used almost exclusively as fuel for steam-electric power generation. It is brownish-black and has a high inherent moisture content, sometimes as high as 45 percent The heat content of lignite ranges from 9 to 17 million Btu per ton on a moist, mineral-matter-free basis. The heat content of lignite consumed in the United States averages 13 million Btu per ton, on the as-received basis (i.e., containing both inherent moisture and mineral matter).

LIHEAP (Low-Income Home Energy Assistance Program)—See definition further below.

LIHTC—Low-Income Housing Tax Credit. Federal tax credit generated from the construction or rehabilitation of affordable residential rental housing.

Limited Partner (LP)—An investor in a limited partnership who has no voice in the management of the partnership. LPs have limited liability and usually have priority over GPs upon liquidation of the partnership.

Limited Partnership—Tax entity formed by investors to shelter income. Often used to finance energy operations.

• An organization comprised of a general partner, who manages a fund, and limited partners, who invest money but have limited liability and are not involved with the day-to-day management of the fund. In the typical venture capital fund, the general partner receives a management fee and a percentage of the profits (or carried interest). The limited partners receive income, capital gains, and tax benefits.

• A Partnership composed of one or more general partners and one or more limited partners, whose contribution and liability are limited.

Line—A line is a system of poles, conduits, wires, cables, transformers, fixtures, and accessory equipment used for the distribution of electricity to the public.

Line Loss (or Drop)—Electrical energy lost due to inherent inefficiencies in an electrical transmission and distribution system under specific conditions.

Line of Credit—Short-term financing usually granted by a bank up to a predetermined limit; debtor borrows as needed up to the limit of credit without need to renegotiate the loan.

Linear Current Booster—An electronic circuit that matches PV output directly to a motor. Used in array direct water pumping.

Linear Regression—A statistical technique in which a straight line is fitted to a set of data points (energy usage values) to define the effect of one or more Independent Variables such as Billing Period, heating load or production volume. During the baseline modeling process, energy auditors try to find the best Linear Regression fit for all Independent Variables by varying Heating and Cooling Balance Point temperatures and observing the resulting Linear Regression. Generally, the goal is to maximize the overall R^2 (R squared) while also maximizing the T-statistic for each Independent Variable.

Line-Commutated Inverter—An inverter that is tied to a power grid or line. The commutation of power (conversion from direct current to alternating current) is controlled by the power line, so that, if there is a failure in the power grid, the photovoltaic system cannot feed power into the line.

Line-Miles of Seismic Exploration—The distance along the Earth's surface that is covered by seismic surveying.

Liquefaction—The process of making synthetic liquid fuel from coal. The term also is used to mean a method for making large amounts of gasoline and heating oil from petroleum.
  • The sudden temporary loss of shear strength in saturated, loose to medium dense, granular sediments subjected to ground shaking, as would typically occur with an earthquake.

Liquefied Gases—Gases that have been or can be changed into liquid form. These include butane, butylene, ethane, ethylene, propane and propylene.

Liquefied Natural Gas (LNG)—Natural gas (primarily methane) that has been liquefied by reducing its temperature to -260 degrees Fahrenheit at atmospheric pressure.

Liquefied Petroleum Gases (LPG)—A group of hydrocarbon-based gases derived from crude oil refining or natural gas fractionation. They include ethane, ethylene, propane, propylene, normal butane, butylene, isobutane, and isobutylene. For convenience of transportation, these gases are liquefied through pressurization.

Liquefied Refinery Gases (LRG)—Liquefied petroleum gases fractionated from refinery or still gases. Through compression and/or refrigeration, they are retained in the liquid state. The reported categories are ethane/ethylene, propane/propylene, normal butane/butylene, and isobutane/isobutylene. Excludes still gas.

Liquid Asset—A liquid asset is one that can be converted easily and rapidly into cash without a substantial loss of value.

Liquid Brine—A type of geothermal energy resource that depends on naturally occurring hot water solution found within the earth. Technology for this novel energy source is being developed in the Salton Sea area in Southern California.

Liquid Collector—A medium-temperature solar thermal collector, employed predominantly in water heating, which uses pumped liquid as the heat-transfer medium.

Liquid Electrolyte Battery—A battery containing a liquid solution of an electrolyte in a solvent (e.g. sulfuric acid in water). Also called a flooded battery because the plates are covered with the electrolyte solution.

Liquid Hydrocarbon—One of a very large group of chemical compounds composed only of carbon and hydrogen. The largest source of hydrocarbons is petroleum.

Liquid Line—The refrigerant line that leads from the condenser to the evaporator in a split system air conditioner or heat pump. The refrigerant in this line is in a liquid state and is at an elevated temperature. This line should not be insulated.

Liquid Metal Fast Breeder Reactor—A nuclear breeder reactor, cooled by molten sodium, in which fission is caused by fast neutrons.

Liquid Petroleum Gas—See *LPG*

Liquidation—The process of closing down a company, selling its assets, paying off its creditors, and distributing any remaining cash to owners of the company.

Liquid-based Solar Heating System—A solar heating system that uses a liquid as the heat transfer fluid.

Liquidity—The ability to convert assets into cash. A measure of how easily assets can be converted into cash.

Liquid-to-Air Heat Exchanger—A heat exchanger that transfers the heat contained in a liquid heat transfer fluid to air.

Liquid-to-Liquid Heat Exchanger—A heat exchanger that transfers heat contained in a liquid heat transfer fluid to another liquid.

Lis Pendens—A recorded notice of the filing of an Action.

List Price—The price at which an asset is usually sold to a customer and from which a discount is computed by a dealer. Various items, such as the residual and fees can be entered as a percentage of the list price. Usually it is the same as the cost.

Listed—A project is considered "listed" once the Climate Action Reserve has satisfactorily reviewed all project submitted forms and tentatively accepted the project in the Reserve. The project will now appear in the public interface of the Reserve system.

Lithium-Ion Battery—A popular lightweight, high capacity battery. Different versions have differing chemistries, but they are found in everything from EVs to cell phones.

Lithium-Sulfur Battery—A battery that uses lithium in the negative electrode and a metal sulfide in the positive electrode, and the electrolyte is molten salt; can store large amounts of energy per unit weight.

Littoral—Pertaining to the shore of a lake, sea, or ocean.

Live Bottom—A material storage bin or truck with a floor which incorporates a device for removing or unloading the material contained in the bin.

Live Steam—Steam available directly from a boiler under full pressure.

LL—Load Loss

LLP—Loan Loss Provisions

LLRW—Low Level Radioactive Waste

LLRWPA—Low Level Radioactive Waste Policy Act. The LLRWPA requires each state to provide disposal facilities for commercial low-level waste generated within its borders. It also encourages states to work together to develop regional disposal facilities.

LMFBR—Liquid Metal Fast Breeder Reactor

LNEP—Low Noise Emission Product

LNG (Liquefied Natural Gas)—Natural gas that has been condensed to a liquid, typically by cryogenically cooling the gas to minus 327.2 degrees Fahrenheit (below zero).

Load (Electric)—The amount of electric power delivered or required at any specific point or points on a system. The requirement originates at the energy-consuming equipment of the consumers. Anything in an electrical circuit that, when the circuit is turned on, draws power from that circuit. An end-use device or an end-use customer that consumes power. Load should not be confused with demand, which is the measure of power that a load receives or requires.

Load Analysis—Assessing and quantifying the discrete components that comprise a load. This analysis often includes time of day or season as a variable.

Load Building—Programs aimed at increasing use of existing electric equipment or the addition of new equipment.

Load Centers—A geographical area where large amounts of power are drawn by end-users.

Load Circuit—The wiring including switches and fuses that connects the load to the power source.

Load Control Program—A program in which the utility company offers a lower rate in return for having permission to turn off the air conditioner or water heater for short periods of time by remote control. This control allows the utility to reduce peak demand.

Load Current—The current required to power the electrical device.

Load Curve—The relationship of power supplied to the time of occurrence. Illustrates the varying magnitude of the load during the period covered.

Load Diversity—The condition that exists when the peak demands of a variety of electric customers occur at different times. This is the objective of "load molding" strategies, ultimately curbing the total capacity requirements of a utility.

Load Duration Curve—A curve that displays load values on the horizontal axis in descending order of magnitude against the percent of time (on the vertical axis) that the load values are exceeded.

Load Factor—A percent telling the difference between the amount of electricity a consumer used during a given time span and the amount that would have been used if the usage had stayed at the consumer's highest demand level during the whole time. The term also is used to mean the percentage of capacity of an energy facility—such as power plant or gas pipeline—that is utilized in a given period of time.

Load Following—Regulation of the power output of electric generators within a prescribed area in response to changes in system frequency, tie-line loading, or the relation of these to each other, so as to maintain the scheduled system frequency and/or established interchange with other areas within predetermined limits.

Load Forecast—An estimate of power demand at some future period.

Load Leveling—Any load control technique that dampens the cyclical daily load flows and increases base-load generation. Peak load pricing and time-of-day charges are two techniques that electric utilities use to reduce peak load and to maximize efficient generation of electricity.

Load Loss (3 hours)—Any significant incident on an electric utility system that results in a continuous outage of 3 hours or longer to more than 50,000 customers or more than one half of the total customers being served immediately prior to the incident, whichever is less.

Load Management—Steps taken to reduce power demand at peak load times or to shift some of it to off-peak times. This may be with reference to peak hours, peak days or peak seasons. The main thing affecting electric peaks is air-conditioning usage, which is therefore a prime target for load management efforts. Load management may be pursued by persuading consumers to modify behavior or by using equipment that regulates some electric consumption.

Load Management Technique—Utility demand management practices directed at reducing the maximum kilowatt demand on an electric system and/or modifying the coincident peak demand of one or more classes of service to better meet the utility system capability for a given hour, day, week, season, or year.

Load on Equipment—One hundred percent load is the maximum continuous net output of the unit at normal operating conditions during the annual peak load month. For example, if the equipment is capable of operating at 5% overpressure continuously, use this condition for 100% load.

Load Profile—This is information on a customer's electricity usage over a period of time.

Load Profile or Shape—A curve on a chart showing power (kW) supplied (on the horizontal axis) plotted against time of occurrence (on the vertical axis) to illustrate the variance in a load in a specified time period.

Load Ratio Share—Ratio of a transmission customer's network load to the provider's total load calculated on a rolling twelve-month basis.

Load Reduction Request—The issuance of any public or private request to any customer or the general public to reduce the use of electricity for the reasons of maintaining the continuity of service of the reporting entity's bulk electric power supply system. Requests to a customer(s) served under provisions of an interruptible contract are not a reportable action unless the request is made for reasons of maintaining the continuity of service of the reporting entity's bulk electric power supply.

Load Resistance—The electrical resistance of the load.

Load Serving Entity (LSE)—A retail electricity provider. Utilities, marketers or aggregators who provide electric power to a large number of end-use customers.

Load Shape—A method of describing peak load demand and the relationship of power supplied to the time of occurrence.

Load Shedding—Intentional action by a utility that results in the reduction of more than 100 megawatts (MW) of firm customer load for reasons of maintaining the continuity of service of the reporting entity's bulk electric power supply system. The routine use of load control equipment that reduces firm customer load is not considered to be a reportable action.

Load Shifting—A load management objective that moves loads from on-peak periods to off-peak periods.

Load Voltage Cut-off—The voltage at which a controller will disconnect the load from the battery.

LOAEL—Lowest Observed Adverse Effect Level

Loan to Value Ratio—The ratio between a mortgage loan and the value of the property pledged as security, usually expressed as a percentage. Also called "loan ratio."

Local Capacity Requirement (LCR)—The California Independent System Operator (CAISO) performs annual studies to identify the minimum local resource capacity required in each local area to meet established reliability criteria. Based on the study results, load serving entities receive a proportional allocation of the minimum required local resource capacity by transmission access charge area, and submit resource adequacy plans to show that they have procured the necessary capacity.

Local Distribution Company (LDC)—A legal entity engaged primarily in the retail sale and/or delivery of natural gas through a distribution system that includes mainlines (that is, pipelines designed to carry large volumes of gas, usually located under roads or other major right-of-ways) and laterals (that is, pipelines of smaller diameter that connect the end user to the mainline). Since the restructuring of the gas industry, the sale of gas and/or delivery arrangements may be handled by other agents, such as producers, brokers, and marketers that are referred to as "non-LDC."

Local Emission—A pollutant whose primary impacts occur close to the emissions. Examples: garbage deposited in landfills, particulates.

Local Publicly Owned Electric Utility—A municipal corporation, a municipal utility district, an irrigation district, or a joint power authority furnishing electric services over its own transmission facilities, or furnishing electric service over its own or its members' distribution system.

Local Solar Time—A system of astronomical time in which the sun crosses the true north-south meridian at 12 noon, and which differs from local time according to longitude, time zone, and equation of time.

Location Normalization—A type of weather adjustment used to compare the energy performance of facilities in different climatic zones. It simulates energy use for each of the facilities as if they were all in the same location. This is necessary for valid comparisons, since facilities in harsh climates will naturally use more energy than similar buildings in mild climates.

This is how it works. First, a weather sensitive statistical model of energy consumption over a chosen time period is produced for each utility meter in each facility, using the local weather for that facility. Next, a common weather set is chosen. This would reasonably be Typical Meteorological Year weather for a station somewhere in the geographic center of the group of facilities. Finally the single weather set is applied to each meter model to generate location-normalized consumption values that can be used for comparison.

Lock-up Period—The period of time that certain stockholders have agreed to waive their right to sell their shares of a public company. Investment banks that underwrite initial public offerings generally insist upon lockups for a set period of time, typically 180 days from large shareholders (such as 1% ownership or more) in order to allow an orderly market to develop in the shares. The shareholders that are subject to lockup usually include the management and directors of the company, strategic partners, and such large investors. These shareholders have typically invested prior to the IPO at a significantly lower price to that offered to the public and therefore stand to gain considerable profits. If a shareholder attempts to sell shares that are subject to lockup during the lockup period, the transfer agent will not permit the sale to be completed.

LOD—Limit of Detection

LOEL—Lowest Observed Effect Level

Log Choker—A length of cable or chain that is wrapped around a log or harvested tree to secure the log to the winch cable of a skidder or to an overhead cable yarding line.

Log Law—In reference to a wind energy conversion system, the wind speed profile in which wind speeds increase with the logarithmic of the height of the wind turbine above the ground.

Logging Residues—The unused portion of wood and bark left on the ground after harvesting merchantable wood. The material may include tops, broken pieces, and un-merchantable species.

LOHAS Market—An acronym for Lifestyles of Health and Sustainability. A market that consist of mindful consumers passionate about the environment, sustainability, social issues and health.

Long-term Procurement Proceeding (LTPP)—The biennial LTPP proceeding evaluates utilities' need for new fossil-fired resources and establishes rules for rate recovery of procurement transactions. It also serves as the "umbrella" proceeding to consider, in an integrated fashion, all loading-order resource procurement policies and programs.

Long Ton—A unit that equals 20 long hundredweight or 2,240 pounds. Used mainly in England.

Long-Term Debt—Debt securities or borrowings having a maturity of more than one year.

Long-Term Productivity—The capacity of a site to support forest ecosystems over generations of humans and trees as measured against some defined reference.

Long-Term Purchase—A purchase contract under which at least one delivery of material is scheduled to occur during the second calendar year after the contract-signing year. Deliveries also can occur during the contract-signing year, during the first calendar year thereafter, or during any subsequent calendar year.

Longwall Mining—An automated form of underground coal mining characterized by high recovery and extraction rates, feasible only in relatively flat-lying, thick, and uniform coalbeds. A high-powered cutting machine is passed across the exposed face of coal, shearing away broken coal, which is continuously hauled away by a floor-level conveyor system. Longwall mining extracts all machine-minable coal between the floor and ceiling within a contiguous block of coal, known as a panel, leaving no support pillars within the panel area. Panel dimensions vary over time and with mining conditions but currently average about 900 feet wide (coal face width) and more than 8,000 feet long (the minable extent of the panel, measured in direction of mining). Longwall mining is done under movable roof supports that are advanced as the bed is cut. The roof in the mined-out area is allowed to fall as the mining advances.

Long-Wave Radiation—Infrared or radiant heat.

Loop Flow—The movement of electric power from generator to load by dividing along multiple parallel paths; it especially refers to power flow along an unintended path that loops away from the most direct geographic path or contract path.

Loose Fill Insulation—Insulation made from rockwool fibers, fiberglass, cellulose fiber, vermiculite or perlite minerals, and composed of loose fibers or granules can be applied by pouring directly from the bag or with a blower.

LORS—Laws, ordinances, regulations and standards

Loss of Load Probability (LOLP)—A measure of the probability that a system demand will exceed capacity during a given period; often expressed as the estimated number of days over a long period, frequently 10 years or the life of the system.

Loss of Load Risk—The evaluation of the risk of a system not adequately meeting the load demand of firm customers under normal operating conditions. It is based upon the evaluation of supply and capacity reliabilities and the uncertainty of demand forecast, weather variability, and other uncertainties.

Loss of Service (15 minutes)—Any loss in service for greater than 15 minutes by an electric utility of firm loads totaling more than 200 MW, or 50 percent of the total load being supplied immediately prior to the incident, whichever is less. However, utilities with a peak load in the prior year of more than 3000 MW are only to report losses of service to firm loads totaling more than 300 MW for greater than 15 minutes. (The DOE shall be notified with service restoration and in any event, within three hours after the beginning of the interruption.)

Loss Payable Clause—An Endorsement to an Insurance Policy (Hazard) specifying parties (Lenders) entitled to participate in proceeds in the event of a loss.

Losses (Electric Utility)—Electric energy or capacity that is wasted in the normal operation of a power system. Some kilowatt-hours are lost in the form of waste heat in electrical apparatus such as substation conductors. LINE LOSSES are kilowatts or kilowatt-hours lost in transmission and distribution lines under certain conditions.

Losses (Energy)—A general term applied to the energy that is converted to a form that cannot be effectively used (lost) during the operation of an energy producing, conducting, or consuming system.

Low Btu Gas—A fuel gas with a heating value between 90 and 200 Btu per cubic foot.

Low Emission Vehicle (LEV)—A vehicle certified by the California Air Resources Board to have emissions from zero to 50,000 miles no higher than 0.075 grams/mile (g/mi) of non-methane organic gases, 3.4 g/mi of carbon monoxide, and 0.2 g/mi of nitrogen oxides. Emissions from 50,000 to 100,000 miles may be slightly higher.

Low Flow Showerheads—Reduce the amount of water flow through the showerhead from 5 to 6 gallons a minute to 3 gallons a minute.

Low Flush Toilet—A toilet that uses less water than a standard one during flushing, for the purpose of conserving water resources.

Low Head—Vertical difference of 100 feet or less in the upstream surface water elevation (headwater) and the downstream surface water elevation (tailwater) at a dam.

Low Heat Value (LHV)—The low or net heat of combustion for a fuel assumes that all products of combustion, including water vapor, are in a gaseous state.

Low Income Home Energy Assistance Program (LIHEAP)—The purpose of LIHEAP is to assist eligible households to meet the cost of heating or cooling in residential dwellings. The Federal government provides the funds to the States that administer the program.

Low Power Testing—The period of time between a plant's nuclear generating unit's initial fuel loading date and the issuance of its operating (full-power) license. The maximum level of operation during this period is 5 percent of the unit's thermal rating.

Low Sloped Roof—A roof that has a ratio of rise to run of 2:12 or less.

Low Temperature Solar Thermal Collector—Metallic or nonmetallic collectors that generally operate at temperatures below 110° Fahrenheit and use pumped liquid or air as the heat transfer medium. They usually contain no glazing and no insulation, and they are often made of plastic or rubber, although some are made of metal.

Low Voltage Disconnect (LVD)—The voltage at which the charge controller will disconnect the load from the batteries to prevent over-discharging.

Low Voltage Disconnect Hysteresis—The voltage difference between the low voltage disconnect set point and the voltage at which the load will be reconnected.

Low Voltage Warning—A warning buzzer or light that indicates the low battery voltage set-point has been reached.

Low-E—A special coating, usually a very thin layer of metal that reduces the emissivity of a window assembly by reflecting low frequency radiation (heat) and thereby reducing the heat transfer through the assembly.

Low-E Ceiling—A lightweight barrier installed between the roof and ice sheet in an arena to increase energy efficiency by reducing the refrigeration heat load due to radiant heat from the roof. A Low-E Ceiling is constructed of aluminum foil, fiberglass and fire-resistant material. It may be installed close to the ceiling or suspended from wires as a horizontal sheet. Low-E Ceiling manufacturers claim that radiant heat transfer from the roof to the ice sheet is reduced by 90% to 95%, that condensation problems are reduced and that light levels are increased due to the reflective underside of the material.

Low-E Coating—A low emissivity metallic coating applied to glazing in fenestration products.

Low-E Coatings & (Window) Films—A coating applied to the surface of the glazing of a window to reduce heat transfer through the window.

Low-E Glass—Low-emission glass reflects up to 90% of long-wave radiation, which is heat, but let's in short-wave radiation, which is light. Windows are glazed with a coating that bonds a microscopic, transparent, metallic substance to the inside surface of the double-pane or triple-pane windows.

Low-E Window—Low-Emissivity Windows are energy-efficient windows that have a window film or coating applied to the surface of the glass to reduce heat transfer through the window.

Low-Emissivity Windows & (Window) Films—Energy-efficient windows that have a coating or film applied to the surface of the glass to reduce heat transfer through the window.

Lower (Net) Heating Value—The lower or net heat of combustion for a fuel that assumes that all products of combustion are in a gaseous state. (See *Net Heating Value* below.)

Lower Heating Value (LHV)—The potential energy in a fuel if the water vapor from combustion of hydrogen is not condensed. The Lower Heating Value (also known as net calorific value, net CV or LHV) of a fuel is the amount of heat released by burning a specified quantity and returning the temperature of the combustion products to 302° F (150° C). It is used in energy calculations where flue gas condensation is not practical or heat below 302° F cannot be put to good use.

Lowest Achievable Emissions Rate (LAER)—Used to describe air emissions control technology. A rate of emissions defined by the permitting agency. LEAR sets emission limits for non-attainment areas.

Low-Flow Solar Water Heating Systems—The flow rate in these systems is 1/8 to 1/5 the rate of most solar water heating systems. The low-flow systems take advantage of stratification in the storage tank and theoretically allows for the use of smaller diameter piping to and from the collector and a smaller pump.

Low-high Debt—Debt with lower payments early in the term.

Low-Income Housing Tax Credit (LIHTC)—A credit for providing affordable rental housing for low-income individuals.

Low-Pressure Sodium Lamp—A type of lamp that produces light from sodium gas contained in a bulb operating at a partial pressure of 0.13 to 1.3 Pascal.

The yellow light and large size make them applicable to lighting streets and parking lots.

Low-Rise Enclosed Space—An enclosed space located in a building with 3 or fewer stories.

Low-Rise Residential Building—(in California) A building, other than a hotel/motel that is of Occupancy Group R, Division 1, and is three stories or less, or that is of Occupancy Group R, Division 3.

Low-Sulfur Coal—Coal having one percent or less of sulfur by weight.

Low-Sulfur Oil—Oil having one percent or less of sulfur by weight.

LP—Limited Partnership
• Liquidity Premium

LPG (Liquefied Petroleum Gas)—A mixture of gaseous hydrocarbons, mainly propane and butane that change into liquid form under moderate pressure. LPG or propane is commonly used as a fuel for rural homes for space and water heating, as a fuel for barbecues and recreational vehicles, and as a transportation fuel. It is normally created as a by-product of petroleum refining and from natural gas production.

LPTA—Low price, technically acceptable

LQER—Lesser Quantity Emission Rates

LQG—Large Quantity Generator

LQHC—Low Quality Hydrocarbons

LRTAP—Long Range Transboundary Air Pollution

LRTP—Long Range Transportation Plan

LSE—Load Serving Entities: Utilities that serve consumers.

LST—Low-Solvent Technology

LT—Lifetime

LTC—Long-Term Concentration

LTHE—Low Temperature Heat Exchanger

LTOP—Lease to Purchase

LTU—Land Treatment Unit

Lubricants—Substances used to reduce friction between bearing surfaces, or incorporated into other materials used as processing aids in the manufacture of other products, or used as carriers of other materials. Petroleum lubricants may be produced either from distillates or residues. Lubricants include all grades of lubricating oils, from spindle oil to cylinder oil to those used in greases.

LUFT—Leaking Underground Fuel Tank

LULUCF—Land use, land use change and forestry. The term given to the sector covering reforestation and afforestation, land clearing and agriculture. Each of these activities can make significant contributions to atmospheric carbon emissions and/or removals.

Lumen—An empirical measure of the quantity of light. It is based upon the spectral sensitivity of the photo sensors in the human eye under high (daytime) light levels. Photometrically it is the luminous flux emitted with a solid angle (1 steradian) by a point source having a uniform luminous intensity of 1 candela.

Lumen Depreciation—The decrease in Lumen output of a light source over time; every lamp type has a unique Lumen depreciation curve (sometimes called Lumen maintenance curve) depicting the pattern of decreasing light output.

Lumen Maintenance Control—An electrical control device designed to vary the electrical consumption of a lighting system in order to maintain a specified illumination level.

Lumens/Watt (LPW)—A measure of the efficacy (efficiency) of lamps. It indicates the amount of light (lumens) emitted by the lamp for each unit of electrical power (Watts) used.

• The amount of light available from a given light source (lumens) divided by the power requirement for that light source (watts). The more usable light that a light source provides per watt, the greater its efficacy. (See *Efficacy)*

Luminaire—A complete lighting unit consisting of a lamp and the parts designed to distribute the light, to position and protect the lamp, and to connect the lamp to the power supply; commonly referred to as "lighting fixtures" or "instruments."

Luminance—The physical measure of the subjective sensation of brightness; measured in lumens.

Luminaries—A complete lighting unit consisting of a lamp or lamps together with the parts designed to distribute the light, to position and protect the lamps and to connect the lamps to the power supply. California Code of Regulations, Section 2-1602(h)].

LUST—Leaking Underground Storage Tank

Lux—A unit of illumination equal to the direct illumination on a surface that is everywhere one meter from a uniform point source of one candle; a unit of illumination that is equal to one lumen per square meter. The unit of luminance in the International System of Units (SI). It is equal to one Lumen per meter squared (lm/m^2), which is a very small unit. Converting the imperial measurement system: 1 lx = 0.0929 fc (foot-candles) Also see *Footcandle.*

LV—Low voltage

LVE—Low Volume Exemption

LVM—Low Volatility Metals

LWA—Light Weight Aggregate

LWAK—Light Weight Aggregate Kiln

LWOP—Lease With Option to Purchase

M

m—Meter

M&A—Mergers and Acquisitions. Brokerage of acquisitions and divestments of companies or company divisions. The phrase describes a division of banks (investment bankers) that, among other things, consults companies on mergers and takeovers.

M&E—Measurement and Evaluation. A process or protocol to evaluate the performance of an energy system.

M&V—Measurement and Verification. A process or protocol to confirm the actual energy savings realized from a project once the project is implemented and operating.

M&V Plan—The Measurement and Verification (M&V) Plan is a document that defines project-specific M&V methods and techniques that will be used to determine savings resulting from a specific performance contracting project.

M/B—Market to Book Ratio

M100—100 percent (neat) methanol used as a motor fuel in dedicated methanol vehicles, such as some heavy-duty truck engines.

M85—A blend of 85 percent methanol and 15 percent unleaded regular gasoline, used as a motor fuel.

MAC—Management Advisory Committee

• Multiple-award Contract

Machine Drive (Motors)—The direct process end use in which thermal or electric energy is converted into mechanical energy. Motors are found in almost every process in manufacturing. Therefore, when motors are found in equipment that is wholly contained in another end use (such as process cooling and refrigeration), the energy is classified there rather than in machine drive.

Macroclimate—The climate of a defined region, such as a valley, plain, desert, coastline, or mountain range, created by a unique combination of wind, topography, solar exposure, soil, and vegetation of the region. One macroclimate region will have many local microclimates, which may be significantly different.

MACRS—*"Modified Accelerated Cost Recovery System"* (tax depreciation method) The current tax depreciation system as introduced by the Tax Reform Act of 1986, effective for equipment placed in service after December 31, 1986.

MACRS Class Life—The specific tax cost recovery (depreciation) period for a class of assets as defined by (Modified Accelerated Cost Recovery System (MACRS.) Asset class lives (ADR midpoint lives) are used to determine an asset's MACRS class life and, hence, its recovery period.

MACT—Maximum Available (or Achievable) Control Technology

Made Available (Vehicle)—A vehicle is considered "Made available" if it is available for delivery to dealers or users, whether or not it was actually delivered to them. To be "Made available," the vehicle must be completed and available for delivery; thus, any conversion to be performed by an original equipment manufacturer (OEM) Vehicle Converter or Aftermarket Vehicle Converter must have been completed.

MAER—Maximum Allowable Emission Rate

MAG—Management Advisory Group

Magma—The molten rock and elements that lie below the earth's crust. The heat energy can approach 1,000 degrees Fahrenheit and is generated directly from a shallow molten magma resource and stored in adjacent rock structures. To extract energy from magma resources requires drilling near or directly into a magma chamber and circulating water down the well in a convection- type system. California has two areas that may be magma resource sites: the Mono- Long Valley Caldera and Coso Hot Springs Known Geothermal Resource Areas.

Magnetic Ballast—A type of florescent light ballast that uses a magnetic core to regulate the voltage of a florescent lamp.

Magnetic Declination—The number of degrees east or west of true south from magnetic south.

Magneto Hydro Dynamics (MHD)—A means of producing electricity directly by moving liquids or gases through a magnetic field.

MAI—Member of the Appraisal Institute

Main Heating Equipment—Equipment primarily used for heating ambient air in the housing unit.

Main Heating Fuel—The form of energy used most frequently to heat the largest portion of the floor space of a structure. The energy source designated as the main heating fuel is the source delivered to the site for that purpose, not any subsequent form into which it is transformed on site to deliver the heat energy (e.g., for buildings heated by a steam boiler, the main heating fuel is the main input fuel to the boiler, not the steam or hot water circulated through the building.) Note: In commercial buildings, the heating must be to at least 50 degrees Fahrenheit.

Mains—A system of pipes for transporting gas within a distributing gas utility's retail service area to points of connection with consumer service pipes.

Maintenance Expenses—That portion of operating expenses consisting of labor, materials, and other direct and indirect expenses incurred for preserving the operating efficiency and/or physical condition of utility plants used for power production, transmission, and distribution of energy.

Maintenance Free Battery—A sealed battery to which water cannot be added to maintain the level of the electrolyte solution.

Maintenance of Boiler Plant (Expenses)—The cost of labor, material, and expenses incurred in the maintenance of a steam plant. Includes furnaces; boilers; coal, ash-handling, and coal-preparation equipment; steam and feed water piping; and boiler apparatus and accessories used in the production of steam, mercury, or other vapor to be used primarily for generating electricity. The point at which an electric steam plant is distinguished from an electric plant is defined as follows:

- Inlet flange of throttle valve on prime mover.
- Flange of all steam extraction lines on prime mover.
- Hotwell pump outlet on condensate lines.
- Inlet flange of all turbine-room auxiliaries.
- Connection to line side of motor starter for all boiler-plant equipment.

Maintenance of Structures (expenses)—The cost of labor, materials, and expenses incurred in maintenance of power production structures. Structures include all buildings and facilities to house, support, or safeguard property or persons.

Maintenance Supervision and Engineering Expenses—The cost of labor and expenses incurred in the general supervision and direction of the maintenance of power generation stations. The supervision and engineering included consists of the pay and expenses of superintendents, engineers, clerks, other employees, and consultants engaged in supervising and directing the maintenance of each utility function. Direct supervision and engineering of specific activities, such as fuel handling, boiler room operations, generator operations, etc., are charged to the appropriate accounts.

Major Electric Utility—A utility that, in the last 3 consecutive calendar years, had sales or transmission services exceeding one of the following: (1) 1 million megawatt hours of total annual sales; (2) 100 megawatt hours of annual sales for resale; (3) 500 megawatt hours of annual gross interchange out; or (4) 500 megawatt hours of wheeling (deliveries plus losses) for others.

Major Energy Sources—Fuels or energy sources such as electricity, fuel oil, natural gas, district steam, district hot water, and district chilled water. District chilled water is not included in any totals for the sum of major energy sources or fuels; all other major fuels are included in these totals.

Major Fuels—Fuels or energy sources such as: electricity, fuel oil, liquefied petroleum gases, natural gas, district steam, district hot water, and district chilled water.

Major Interstate Pipeline Company—A company whose combined sales for resale, including gas transported interstate or stored for a fee, exceeded 50 million thousand cubic feet in the previous year.

Major Marketer—Any person who sells natural gas or oil in amounts determined by the commission as having a major effect on energy supplies.

Major Natural Gas Producer—Any person who produces natural gas in amounts determined by the commission as having a major effect on energy supplies.

Major Oil Producer—Means any person who produces oil in amount determined by the commission as having a major effect on energy supplies.

Major Source—A source that emits, or has the potential to emit, a pollutant regulated under the Clean Air Act in excess of a specified rate in a non-attainment area.

Majority Carrier—Current carriers (either free electrons or holes) that are in excess in a specific layer of a semiconductor material (electrons in the n-layer, holes in the p-layer) of a cell.

Make-Up Air—Air brought into a building from outside to replace exhaust air.

Malathion—Malathion is an insecticide that, at high doses, affects the human nervous system.

Mall Building—A single building enclosing a number of tenants and occupants wherein two or more tenants have a main entrance into one or more malls.

Management Activities—Planned activities initiated by land managers to meet the desired future condition for an area. Management activities may include thinning, timber harvest, prescribed burning, tree planting, and other activities.

Management Area—Management areas are specific geographical areas defined by a forest plan. Each management area has a set of objectives and a management prescription unique to it.

Management Indicator Species—Species selected as ecological indicators. The welfare of a management indicator species is presumed to be an indicator of the welfare of other species using the same habitat. The condition and welfare of these species can be used to assess the impacts of management actions on particular areas or habitats.

Management Plan—A plan guiding overall management of an area administered by a federal or state agency. A management plan usually includes objectives, goals, standards and guidelines, management actions, and monitoring plans.

Mandatory Measures Checklist (MF-1R)—(in California) a form used by the building plan checker and field inspector to verify compliance of the building with the prescribed list of mandatory features, equipment efficiencies and product certification requirements. The documentation author indicates compliance by initialing, checking, or marking NIA (for features not applicable) in the boxes or spaces provided for the designer.

Manhattan Project—The U.S. Government project that produced the first nuclear weapons during World War II. Started in 1942, the Manhattan Project formally ended in 1946. The Hanford Site, Oak Ridge Reservation, and Los Alamos National Laboratory were created for this effort. The project was named for the Manhattan Engineer District of the U.S. Army Corps of Engineers.

Manifest—A list of passengers or an invoice of cargo.

Manual—Capable of being operated by personal intervention.

Manual Dimmer Switches—These are like residential-style dimmer switches. They are not generally used with fluorescent and high-intensity discharge (HID) lamps.

Manual J—The standard method for calculating residential cooling loads developed by the Air-Conditioning and Refrigeration Institute (ARI) and the Air Conditioning Contractors of America (ACCA) based largely on the American Society of Heating, Refrigeration, and Air-Conditioning Engineer's (ASHRAE) "Handbook of Fundamentals."

Manufactured Device—Any heating, cooling, ventilation, lighting, water heating, refrigeration, cooking, plumbing fitting, insulation, door, fenestration product, or any other appliance, device, equipment, or system subject to Sections 1-10 through 119 of Title 24, Part 6 (in California).

Manufactured Fenestration Product—A fenestration product constructed of materials which are factory cut or otherwise factory formed with the specific intention of being used to fabricate a fenestration product. A manufactured fenestration product is typically assembled before delivery to a job site. However a "knocked-down" or partially assembled product sold as a fenestration product is also a manufactured fenestration product when provided with temporary and permanent labels as described in Section 10-111; otherwise it is a site-built fenestration product.

Manufactured Gas—A gas obtained by destructive distillation of coal or by the thermal decomposition of oil, or by the reaction of steam passing through a bed of heated coal or coke. Examples are coal gases, coke oven gases, producer gas, blast furnace gas, blue (water) gas, carbureted water gas. Btu content varies widely.

Manufacturing—An energy-consuming sub sector of the industrial sector that consists of all facilities and equipment engaged in the mechanical, physical, chemical, or electronic transformation of materials, substances, or components into new products. Assembly of component parts of products is included, except for that which is included in construction.

Manufacturing Division—One of 10 fields of economic activity defined by the Standard Industrial Classification Manual. The manufacturing division includes all establishments engaged in the mechanical or chemical transformation of materials or substances into new products. The other divisions of the U.S. economy are agriculture, forestry, fishing, hunting, and trapping; mining; construction; transportation, communications, electric, gas, and sanitary services; wholesale trade; retail trade; finance, insurance, and real estate; personal, business, professional, repair, recreation, and other services; and public administration. The establishments in the manufacturing division constitute the universe for the MECS (an EIA survey).

Manufacturing Establishment—An economic unit at a single physical location where mechanical or chemical transformation of materials or substances into new products are performed.

MAPPS Model—A global biological and geographical model which simulates the potential natural vegetation that can be supported at any site in the world under a long-term steady-state climate. Its acronym stands for mapped atmosphere-plant-soil.

MAR—Management System Review

Marginal Abatement Cost (MAC)—The cost of reducing emissions by one ton of CO_2 e. An aggregation of these costs against total tons abated creates a firm's marginal abatement cost curve. The lower the MAC curve, the more effective the firm's emission reduction strategies.

Marginal Cost—The sum that has to be paid the next increment of product of service. The marginal cost of electricity is the price to be paid for kilowatt-hours above and beyond those supplied by presently available generating capacity. In the utility context, the cost to the utility of providing the next (marginal) kilowatt-hour of electricity, irrespective of sunk costs.

Marginal Cost of Capital—The incremental cost of financing above a previous level.

Marginal Rate—The Marginal Rate for a utility meter is a unit rate representing the bill increase that would be incurred for an additional unit of usage. In tiered rate structures, the Marginal Rate is lower than the average rate because customers are given volume discounts at various levels as they use more. Generally, the Marginal Rate at a meter will be the unit rate in the highest tier normally billed for that meter.

Marginal Tax Rate—The tax rate that would have to be paid on an additional dollar of taxable income earned.

Marine Freight—Freight transported over rivers, canals, the Great Lakes, and domestic ocean waterways.

Market Benefits—Benefits of a climate policy that can be measured in terms of avoided market impacts such as changes in resource productivity (e.g., lower agricultural yields, scarcer water resources) and damages to human-built environment (e.g., coastal flooding due to sea-level rise).

Market Capitalization—The total dollar value of all outstanding shares. Computed as shares multiplied by current price per share. Prior to an IPO, market capitalization is arrived at by estimating a company's future growth and by comparing a company with similar public or private corporations.

Market Clearing Price—The price at which supply equals demand. The Day Ahead and Hour Ahead Markets.

Market Eligibility—The percentage of equipment still available for retrofit to the demand-side management measure. For example, if 20 percent of customers where demand controllers are feasible have already purchased demand controllers, then the eligible market eligibility factor is 80 percent.

Market Participant—An entity, including a Scheduling Coordinator, who participates in the energy marketplace through the buying, selling, transmission, or distribution of energy or ancillary services into, out of, or through the ISO-controlled grid.

Market Price Contract—A contract in which the price of a product or commodity is not specifically determined at the time the contract is signed but is based instead on the prevailing market price at the time of delivery. A market price contract may include a floor price, that is, a lower limit on the eventual settled price. The floor price and the method of price escalation generally are determined when the contract is signed. The contract may also include a price ceiling or a discount from the agreed-upon market price reference.

Market Price Settlement—The price paid for product delivery under a market-price contract. The price is commonly (but not always) determined at or sometime before delivery and may be related to a floor price, ceiling price, or discount.

Market Value—The price that property would reasonably be expected to bring if it were offered for sale with a reasonable sales effort over a reasonable period of time.

Marketable Coke—Those grades of coke produced in delayed or fluid cokers that may be recovered as relatively pure carbon. This "green" coke may be sold as is or further purified by calcining.

Marketable Discharge Permit—A permit that allows a certain source to emit a specified volume of some pollutant. (Also called transferable discharge permits)

Market-Based Incentives—In the context of climate change, this refers to measures (such as subsidies, taxes, emissions trading) intended to directly change relative prices of "climate—friendly" technologies in order to overcome market barriers.

Market-Based Price—A price set by the mutual decisions of many buyers and sellers in a competitive market.

Market-Based Pricing—Prices of electric power or other forms of energy determined in an open market system of supply and demand under which prices are set solely by agreement as to what buyers will pay and sellers will accept. Such prices could recover less or more than full costs, depending upon what the buyers and sellers see as their relevant opportunities and risks.

Marketed Production—Gross withdrawals less gas used for repressuring, quantities vented and flared, and non-hydrocarbon gases removed in treating or processing operations. Includes all quantities of gas used in field and processing plant operations.

Marketer—An agent for generation projects who markets power on behalf of the generator. The marketer may also arrange transmission, firming or other ancillary services as needed. Though a marketer may perform many of the same functions as a broker, the difference is that a marketer represents the generator while a broker acts as a middleman.

Marsh Gas—A common term for gas that bubbles to the surface of the water in a marsh or swamp. It is colorless, odorless and can be explosive.

MASH—Multifamily Affordable Solar Housing Program. A California program offering incentives for solar installations on existing multifamily affordable housing.

Masonry—A general term covering wall construction using masonry materials such as brick, concrete block, stone, and tile that are set in mortar; also included is stucco. The category does not include concrete panels because concrete panels represent a different method of constructing buildings.

Masonry Stove—A type of heating appliance similar to a fireplace, but much more efficient and clean burning. They are made of masonry and have long channels through which combustion gases give up their heat to the heavy mass of the stove, which releases the heat slowly into a room. Often called Russian or Finnish fireplaces.

Mass Balance Condition—The mass balance condition is that the mass of all the inputs used to produce goods and services (output) must equal the mass of the resulting output(s) plus the mass of the wastes.

Mass Burn Facility—A type of municipal solid waste (MSW) incineration facility in which MSW is burned with only minor presorting to remove oversize, hazardous, or explosive materials. Mass burn facilities can be large, with capacities of 3000 tons (2.7 million kg) of MSW per day or more. They can be scaled down to handle the waste from smaller communities, and modular plants with capacities as low as 25 tons (22.7 thousand kg) per day have been built. Mass burn technologies represent over 75% of all the MSW-to-energy facilities constructed in the United States to date. The major components of a mass burn facility include refuse receiving and handling, combustion and steam generation, flue gas cleaning, power generation (optional), condenser cooling water, residue ash hauling and landfilling.

Master File—A file maintained by the PX for use in bidding and bid evaluation protocol that contains information on generating units, loads, and other resources eligible to bid into the PX.

Master Lease—A lease line of credit provided by a lessor that allows a lessee to obtain additional leased equipment under the same basic lease terms and conditions as originally agreed to, without having to renegotiate and execute a new lease contract with the lessor. The actual lease rate for a specific piece of equipment will generally be set upon equipment delivery to the lessee.

Master-Metered—The system by which multi-unit buildings or mobile home parks are connected to a single meter. The master-meter holder receives a single bill from the utility and collects from tenants individually. The utility has no liability for repairs or maintenance beyond the master meter.

Master-Metering—Measurement of electricity or natural gas consumption of several tenants or housing units using a single meter. That is, one meter measures the energy usage for several households collectively.

Master-Tenant—A term used to describe a form of ownership structure that is used to facilitate investments that qualify for the investment tax credit (ITC) under section 46 of the revenue code in the U.S. The structure involves using two limited partnerships—one which owns the property that earns the ITCs (the owner or lessor) and another partnership (the tenant or lessee) which leases the property from the owner. Under this structure the owner is able to pass-through the ITCs to the tenant. The investor limited partner(s) usually own a limited partnership interest in the tenant, which in turn invests in the owner. This ownership structure is also known as the inverted lease and/or the lease pass-through.

MATC—Maximum Allowable Toxicant Concentration

Match Funded Debt—Debt incurred by the lessor in a lease agreement (or energy provider in a PPA agreement) to fund a specific piece of capital equipment, the terms and repayment of which are structured to correspond to the repayment of the contract obligation by the debtor.

Material Misstatement—An error or errors in reporting emissions that result in the total reported emissions being outside the 95% accuracy required to receive a positive verification opinion.

Materials Recovery Facility—A recycling facility for municipal solid waste.

Mauna Loa Record—The record of measurement of atmospheric CO_2 concentrations taken at Mauna Loa Observatory, Mauna Loa, Hawaii, since March 1958. This record shows the continuing increase in average annual atmospheric CO_2 concentrations.

MAWP—Maximum Allowable Working Pressure

Maximum Contaminant Level (MCL)—A contaminant level for drinking water, established by the California Department of Health Services, Division of Drinking Water and Environmental Management, or by the U.S. Environmental Protection Agency. These levels are legally-enforceable standards based on health risk (primary standards) or non-health concerns such as odor or taste (secondary standards).

Maximum Deliverability—The maximum deliverability rate (Mcf/d) estimated at the present developed maximum operating capacity.

Maximum Demand—The greatest of all demands of the load that has occurred within a specified period of time.

Maximum Dependable Capacity, Net—The gross electrical output measured at the output terminals of the turbine generator(s) during the most restrictive seasonal conditions, less the station service load.

Maximum Established Site Capacity (Reactors)—The maximum established spent fuel capacity for the site is defined by DOE as the maximum number of intact assemblies that will be able to be stored at some point in the future (between the reporting date and the reactor's end of life) taking into account any established or current studies or engineering evaluations at the time of submittal for licensing approval from the NRC.

Maximum Hourly Load—This is determined by the interval in which the 60-minute integrated demand is the greatest.

Maximum Power Point (MPP)—Operating a PV array at that voltage will produce maximum power. The point on the current-voltage (I-V) curve of a module under illumination, where the product of current and voltage is maximum. [UL 1703] This corresponds to the point on an I-V curve that represents the largest area rectangle that can be drawn under the curve. For a typical silicon cell panel, this is about 17 volts for a 36 cell configuration. For a typical silicon cell, this is at about 0.45 volts.

Maximum Power Point Tracker (MPPT)—A power conditioning unit that automatically operates the PV generator at its MPP under all conditions. An

MPPT will typically increase power delivered to the system by 10% to 40%, depending on climate conditions and battery state of charge.

Maximum Power Tracking—Operating a photovoltaic array at the peak power point of the array's I-V curve where maximum power is obtained. Also called peak power tracking.

Maximum Stream flow—The maximum rate of water flow past a given point during a specified period.

MB—Mass Balance
• Millions of barrels

MBD—Millions of barrels per day

MBDA—Minority Business Development Agency

MBE—Minority Business Enterprise

MBE/WBE—Certification Questionnaire—Minority/Woman Business Enterprise

MBF—One thousand board feet of lumber.

MBI—Management Buy In. Takeover of a company by external managers.

MBO—Management Buy Out. Takeover of a company by its own managers.

MBO—Management By Objectives

MC—See *Moisture Content*.

MCE—Maximum credible earthquake

MCF—One thousand cubic feet or natural gas, having an energy value of one million Btu. A typical home might use six MCF in a month.
• Methyl Chloroform

MCL—Maximum Contaminant Level

MCM—Thousand circular mil (electricity conductor)

MCP—Market Clearing Price: The price at which supply equals demand for the Day Ahead and/or Hour Ahead Markets.

MDL—Method Detection Limit (Level)

Mean Annual Increment—The annual average growth rate for a tree, computed over its entire life cycle.

Mean Indoor Temperature—The "usual" temperature. If different sections of the house are kept at different temperatures, the reported temperature is for the section where the people are. A thermostat setting is accepted if the temperature is not known.

Mean Operating Hours—The arithmetic average number of operating hours per building is the lighted sum of the number of operating hours divided by the weighted sum of the number of buildings.

Mean Power Output (of a Wind Turbine)—The average power output of a wind energy conversion system at a given mean wind speed based on a Raleigh frequency distribution.

Mean Square Feet per Building—The arithmetic average square feet per building is the weighted sum of the total square feet divided by the weighted sum of the number of buildings.

Mean Wind Speed—The arithmetic wind speed over a specified time period and height above the ground (the majority of U.S. National Weather Service anemometers are at 20 feet (6.1 meters).

Measure Impact—The expected effect of an Energy Conservation Measure on a meter component. It is defined as either a +/- % of the physical quantity or as a fixed value (in physical units) for each month.

Measure Life—The length of time that the demand-side management technology will last before requiring replacement. The measure life equals the technology life. These terms are used synonymously.

Measured Heated Area of Residence—The floor area of the housing unit that is enclosed from the weather and heated. Basements are included whether or not they contain finished space. Garages are included if they have a wall in common with the house. Attics that have finished space and attics that have some heated space are included. Crawl spaces are not included even if they are enclosed from the weather. Sheds and other buildings that are not attached to the house are not included. "Measured" area means the measurement of the dimensions of the home, using a metallic, retractable, 50-foot tape measure. "Heated area" is that portion of the measured area that is heated during most of the season. Rooms that are shut off during the heating season to save on fuel are not counted. Attached garages that are unheated and unheated areas in the attics and basements are also not counted.

Measured Reserves—See *Proved Energy Reserves*.

Measured Resources, Coal—Coal resources for which estimates of the rank, quality, and quantity have been computed, within a margin of error of less than 20 percent, from sample analyses and measurements from closely spaced and geologically well known sample sites. Measured resources are computed from dimensions revealed in outcrops, trenches, mine workings, and drill holes. The points of observation and measurement are so closely spaced and the thickness and extent of coals are so well defined that the tonnage is judged to be accurate within 20 percent. Although the spacing of the points of observation necessary to demonstrate continuity of the coal differs from region to region, according to the character of the coalbeds, the point of observation are no greater than 1/2 mile apart. Measured coal is projected to extend as a belt 1/4 mile wide from the outcrop or points of observation or measurement.

Measurement and Evaluation (M&E)—A process or protocol to evaluate the performance of an energy system. As a condition of receiving incentive payments under the CSI (California) program, System Owners and Host Customers agree to participate in Measurement and Evaluation (M&E) activities as required by the CPUC. M&E activities will be performed by the Program Administrator or the Program Administrator's independent third-party consultant and include but are not limited to, periodic telephone interviews, on-site visits, development of a M&E monitoring Plan, access for installation of metering equipment, collection and transfer of data from installed system monitoring equipment, whether installed by Host Customer, System Owner, a third party, or the Program Administrator.

Measurement and Verification (M&V)—A process or protocol to confirm the actual energy savings realized from a project once the project is implemented and operating.

Measurement and Verification Approach—An evaluation procedure for determining energy and cost savings. M&V techniques include engineering calculations, metering, utility bill analysis, and computer simulations.

Measurement, Continuous—Measurements repeated at regular intervals over the baseline period or contract term.

Measurement, Long-Term—Measurements taken over a period of several years.

Measurement, Short-Term—Measurements taken for several hours, weeks or months.

Measurement, Spot—Measurements taken one-time. Snap-shot measurement.

Measures—Actions that can be taken by a government or a group of governments, often in conjunction with the private sector, to accelerate the use of technologies or other practices that reduce GHG emissions.

Measuring, Monitoring and Reporting (MM&R)—The process where emissions are measured, monitored and reported for record keeping purposes. Normally done to satisfy the requirement of a regulatory agency. Accomplished with data instrumentation using protocols established by various regulatory and licensing agencies.

MEC—Model Energy Code

MECA—Manufacturers of Emission Controls Association www.meca.org

Mechanic's Lien—A Statutory Lien in favor of laborers and materialmen who have contributed to a work of improvement.

Mechanical Cooling—Lowering the temperature within a space using refrigerant compressors or absorbers, desiccant dehumidifiers, or other systems that require energy from depletable sources to directly condition the space. In nonresidential, high-rise residential, and hotel/motel buildings cooling of a space by direct or indirect evaporation of water alone is not considered mechanical cooling.

Mechanical Equivalent of Heat—The conversion factor for transforming heat units into mechanical units of work. One Btu equals 778 foot-pounds.

Mechanical Heating—Raising the temperature within a space using electric resistance heaters, fossil fuel burners, heat pumps, or other systems that require energy from depletable sources to directly condition the space.

Mechanical Systems—Those elements of building used to control the interior climate. See *HVAC System.*

Mechanical Ventilation—The active process of supplying or removing air to or from an indoor space by powered equipment such as motor-driven fans and blowers, but not by passive devices like wind-driven turbine ventilators or operable windows.

MED—Minimum Effective Dose

Median—The middle number of a data set when the measurements are arranged in ascending (or descending) order.

Median Stream flow—The middle rate of flow of water past a given point for which there have been several greater and lesser rates of flow occurring during a specified period.

Median Water Condition—The middle precipitation and run-off condition for a distribution of water conditions that have happened over a long period of time. Usually determined by examining the water supply record of the period in question.

Median Wind Speed—The wind speed with 50 percent probability of occurring.

Medium Btu Gas—Fuel gas with a heating value of between 200 and 300 Btu per cubic foot.

Medium Pressure—For valves and fittings, implies that they are suitable for working pressures between 125 to 175 pounds per square inch.

Medium-Temperature Collector—A collector designed to operate in the temperature range of 140 degrees to 180 degrees Fahrenheit, but that can also operate at a temperature as low as 110 degrees Fahrenheit. The collector typically consists of a metal frame, metal absorption panels with integral flow channels (attached tubing for liquid collectors or integral ducting for air collectors), and glazing and insulation on the sides and back.

Megawatt (MgW) (MW)—One thousand kilowatts (1,000 kW) or one million (1,000,000) watts. One megawatt is enough energy to power 1,000 average California homes.

Megawatt Electric (MWe)—One million watts of electric capacity.

Megawatt Hour (MWh)—One thousand kilowatt-hours, or an amount of electricity that would supply the monthly power needs of a typical home having an electric hot water system.

Megawatt Thermal (MWth)—A unit of heat-supply capacity used to measure the potential output from a heating plant, such as might supply a building or neighborhood. More recently used to measure the capacity of solar hot water/heating installations. Represents an instantaneous heat flow and should not be confused with units of produced heat, or megawatt-hours-thermal.

MEI—Maximum Exposed Individual

MEK—Methyl Ethyl Ketone

Member System—An eligible customer operating as part of an agency composed exclusively of other eligible customers.

Membrane—A thin sheet of natural or synthetic material that is permeable to substances in solution.

Membrane Electrode Assembly—The core of a fuel cell, consisting of two electrodes and the proton exchange membrane electrolyte bonded to form a single structure.

MEP—Maximum Extent Practicable
  • Multiple Extraction Procedure

MER—Maximum effective rate (of production)

Mercalli—A standard scale of relative measurement of earthquake intensity.

Mercaptan—An organic chemical compound that has a sulfur like odor that is added to natural gas before distribution to the consumer, to give it a distinct, unpleasant odor (smells like rotten eggs). This serves as a safety device by allowing it to be detected in the atmosphere, in cases where leaks occur.

Merchant Coke Plant—A coke plant where coke is produced primarily for sale on the commercial (open) market.

Merchant Facilities—High-risk, high-profit facilities that operate, at least partially, at the whims of the market, as opposed to those facilities that are constructed with close cooperation of municipalities and have significant amounts of waste supply guaranteed.

Merchant MTBE Plants—MTBE (methyl tertiary butyl ether) production facilities primarily located within petrochemical plants rather than refineries. Production from these units is sold under contract or on the spot market to refiners or other gasoline blenders.

Merchant Oxygenate Plants—Oxygenate production facilities that are not associated with a petroleum refinery. Production from these facilities is sold under contract or on the spot market to refiners or other gasoline blenders.

Merchantable—Logs from which at least part of the volume can be converted into sound grades of lumber ("standard and better" framing lumber).

Mercury—Also known as "quicksilver," this metal is used in the paper pulp and chemical industries, in the manufacture of thermometers, thermostats, and in fungicides. Mercury exists in three biologically important forms: elemental, inorganic and organic. It is highly toxic and affects the nervous system, kidneys and other organs. It also accumulates in animals that are high in the food chain (predators). Organic mercury compounds are the most toxic, and transformations between the three forms of mercury do occur in nature.

Mercury Vapor Lamp—A high-intensity discharge lamp that uses mercury as the primary light-producing element. Includes clear, phosphor coated, and self-ballasted lamps.

Merger—A combining of companies or corporations into one, often by issuing stock of the controlling corporation to replace the greater part of that of the other.

Mesne—Intermediate; intervening.

Mesophilic—An optimum temperature for bacterial growth in an enclosed digester (25° to 40°C).

Met—An approximate unit of heat produced by a resting person, equal to about 18.5 Btu per square foot per hour.

Metal Building—A complete integrated set of mutually dependent components and assemblies that form a building, which consists of a steel framed superstructure and metal skin. This does not include structural glass or metal panels such as in a curtain wall system.

Metal Halide Lamp—A high-intensity discharge lamp type that uses mercury and several halide additives as light-producing elements. These lights have the best Color Rendition Index (CRI) of the

High-Intensity Discharge lamps. They can be used for commercial interior lighting or for stadium lights.

Metallic—The metallic material composition of the collector's absorber system.

Metallurgical Coal—Coking coal and pulverized coal consumed in making steel.

Meta-Trend—A global and overarching force that will affect many multidimensional changes. For example, environmental impacts on business, individual and countires.

Meter—A device for measuring levels and volumes of a customer's gas and electricity use.

Meter Constant—This represents the ratio between instrument transformers (CTs, PTs) and the meter. It is used as a multiplier of the difference between meter readings to determine the kWh used. The meter constant is also used as a multiplier of the demand reading to determine the actual demand.

Meter Modeling Data—The set of numerical parameters established during the baseline modeling process, to mathematically define energy usage at a meter.

Metered Data—End-use data obtained through the direct measurement of the total energy consumed for specific uses within the individual household. Individual appliances can be sub-metered by connecting the recording meters directly to individual appliances.

Metered Peak Demand—The presence of a device to measure the maximum rate of electricity consumption per unit of time. This device allows electric utility companies to bill their customers for maximum consumption, as well as for total consumption.

Metes and Bounds—Measurements and boundaries.

METH—Methylene Chloride

Methanation—Catalytic upgrading of synthetic fuel gas to high Btu. Hydrogen and carbon monoxide react to form methane.

Methane—A light hydrocarbon that is the main component of natural gas and marsh gas. It is the product of the anaerobic decomposition of organic matter, enteric fermentation in animals and is one of the greenhouse gases. Pure methane has a heating value of 1,1012 Btu per standard cubic foot. Chemical formula is CH_4.

CH_4 is among the six greenhouse gases to be curbed under the Kyoto Protocol. Atmospheric CH_4 is produced by natural processes, but there are also substantial emissions from human activities such as landfills, livestock and livestock wastes, natural gas and petroleum systems, coalmines, rice fields, and wastewater treatment. CH_4 has a relatively short atmospheric lifetime of approximately 10 years, but its 100-year GWP is currently estimated to be approximately 23 times that of CO_2.

Methanogens—Bacteria that synthesize methane, requiring completely anaerobic conditions for growth.

Methanol (also known as Methyl Alcohol, Wood Alcohol, CH_3OH)—A colorless, flammable alcohol fuel made from natural gas or coal and used as an antifreeze, general solvent, fuel, and denaturant for ethyl alcohol. A liquid formed by catalytically combining carbon monoxide (CO) with hydrogen (H_2) in a 1:2 ratio, under high temperature and pressure. Commercially it is typically made by steam reforming natural gas. Also formed in the destructive distillation of wood.

Methanol Blend—Mixtures containing 85 percent or more (or such other percentage, but not less than 70 percent) by volume of methanol with gasoline. Pure methanol is considered an "other alternative fuel."

Methanotrophs—Bacteria that use methane as food and oxidize it into carbon dioxide.

Methyl Chloroform (trichloroethane)—An industrial chemical (CH_3CCl_3) used as a solvent, aerosol propellant, and pesticide and for metal degreasing.

Methyl Tertiary Butyl Ether (MTBE)—An ether manufactured by reacting methanol and isobutylene. The resulting ether has a high octane and low volatility. MTBE is a fuel oxygenate and is permitted in unleaded gasoline up to a level of 15 percent. It is one of the primary ingredients in reformulated gasolines. Added to unleaded gasoline to reduce carbon monoxide emissions.

Methylene Chloride—A colorless liquid, non-explosive and practically nonflammable. Used as a refrigerant in centrifugal compressors, a solvent for organic materials, and a component in nonflammable paint removers.

Metric Conversion Factors (for floor space)—Floor space estimates may be converted to metric units by using the relationship, 1 square foot is approximately equal to .0929 square meters. Energy estimates may be converted to metric units by using the relationship, 1 Btu is approximately equal to 1,055 joules. One kilowatt hour is exactly 3,600,000 joules. One gigajoule is approximately 278 kilowatt-hours (kWh).

Metric Ton or Tonne—1000 kilograms. 1 metric ton =

2,204.62 lb = 1.023 short tons.

Metropolitan—Located within the boundaries of a metropolitan area.

Metropolitan Area—A geographic area that is a metropolitan statistical area or a consolidated metropolitan statistical area as defined by the U.S. Office of Management and Budget.

Metropolitan Statistical Area (MSA)—A county or group of contiguous counties (towns and cities in New England) that has (1) at least one city with 50,000 or more inhabitants; or (2) an urbanized area of 50,000 inhabitants and a total population of 100,000 or more inhabitants (75,000 in New England). These areas are defined by the U.S. Office of Management and Budget. The contiguous counties or other jurisdictions to be included in an MSA are those that, according to certain criteria, are essentially metropolitan in character and are socially and economically integrated with the central city or urbanized area.

Mezzanine Financing—Refers to the stage of venture financing for a company immediately prior to its IPO. Investors entering in this round have lower risk of loss than those investors who have invested in an earlier round. Mezzanine-level financing can take the structure of preferred stock, convertible bonds, or subordinated debt.

MFBI—Major Fuel Burning Installation

Mg—Megagram

MG—Milli gauss

mg/dscm—Milligrams per Dry Standard Cubic Meter

mg/hr—Megagrams per Year

mg/kg—Milligrams per Kilogram

mg/l—Milligrams per Liter

mg/l—Milligrams per liter

mg/m3—Milligrams per Cubic Meter

mgd—Million gallons per day

MGD—Millions of Gallons per Day

MG-Si—Metallurgical-grade Silicon

MHDDV—Medium Heavy Duty Diesel Vehicle

MHRA—Maximum Hourly Rolling Average

MIA—Manufacturer Impact Analysis

Microalgae—Unicellular, photosynthetic aquatic plants.

Microclimate—The local climate of specific place or habitat, as influenced by landscape features (wind, topography, solar exposure, soil, vegetation, etc.). All microclimates are a subset of larger regional macroclimates, which may be significantly different.

Microcrystalline Wax—Wax extracted from certain petroleum residues having a finer and less apparent crystalline structure than paraffin wax and having the following physical characteristics: penetration at 77 degrees Fahrenheit (D1321)-60 maximum; viscosity at 210 degrees Fahrenheit in Saybolt Universal Seconds (SUS); (D88)-60 SUS (10.22 centistokes) minimum to 150 SUS (31.8 centistokes) maximum; oil content (D721)-5 percent minimum.

Microgrid—A defined geographic area, set of buildings or campus facilities capable of operating autonomously from the electrical grid by supplying all of its own generation.

Microgroove—A small groove scribed into the surface of a solar photovoltaic cell which is filled with metal for contacts.

Micrometer (or Micron)—One-millionth of a meter. It can also be expressed as 10^{-6} meter.

MICROMORT—A One-in-a-Million Chance of Death From an Environmental Hazard

MicroTurbine—Small combustion turbines approximately the size of a large refrigerator with outputs of 25kW to 500kW. They consist of a compressor, combustor, turbine, alternator, recuperator and generator. They are typically classified by the physical arrangement of the component parts: single shaft or two-shaft, simple cycle or recuperated, inter-cooled, and reheat.

Microwave—Electromagnetic radiation with wavelengths of a few centimeters. It falls between infrared and radio wavelengths on the electromagnetic spectrum. The radio wave beam can deliver electrical energy over long distances.

Microwave Oven—A household cooking appliance consisting of a compartment designed to cook or heat food by means of microwave energy. It may also have a browning coil and convection heating as additional features.

Microwave Sounding Units (MSU)— Sensors carried aboard Earth orbiting satellites that have been used since 1979 to monitor tropospheric temperatures.

Mid Seral Species—Shade intolerant species, primarily Douglas-fir trees and vine maple shrubs. These species typically follow the early seral species in natural succession.

Mid Seral Treatment—A treatment in which a stand of predominately mid seral species will be established.

Mid-American Interconnected Network (MAIN)—One of the ten regional reliability councils that make up the North American Electric Reliability Council (NERC).

Mid-Atlantic Area Council (MAAC)—One of the ten regional reliability councils that make up the North American Electric Reliability Council (NERC).

Mid-Continent Area Power Pool (MAPP)—One of the ten regional reliability councils that make up the North American Electric Reliability Council (NERC).

Middle Distillates—A general classification of refined petroleum products that includes distillate fuel oil and kerosene.

Middlings—In coal preparation, this material called mid-coal is neither clean nor refuse; due to their intermediate specific gravity, middlings sink only partway in the washing vessels and are removed by auxiliary means.

Midgrade Gasoline—Gasoline having an antiknock index, i.e., octane rating, greater than or equal to 88 and less than or equal to 90. Note: Octane requirements may vary by altitude.

Mid-Quarter Convention—A depreciation convention (replacing half-year convention for certain taxpayers in certain years) that assumes all equipment is placed in service halfway through the quarter in which it was actually placed in service. Allowable acquisition and disposition year depreciation deductions are prorated based upon the mid-quarter date of the quarter in which the asset was placed in service.

Mid-Size Passenger Car—A passenger car with between 110 and 119 cubic feet of interior passenger and luggage volume.

Migration—The movement of chemical contaminants through soils or groundwater.

Mil—One-tenth of one cent $0.001.

Mild Hybrid—An HEV in which the electric motor requires an additional source of power, such as a gasoline engine.

Mileage—A term denoting payment for providing fast-regulation services, defined in units of "MW – miles" as the regulation provided in an hour and is calculated as the sum of the absolute value of positive and negative movements requested by the grid operator to provide regulation.

Miles per Gallon (MPG)—A measure of vehicle fuel efficiency. Miles per gallon or MPG represents "Fleet Miles per Gallon." For each subgroup or "table cell," MPG is computed as the ratio of the total number of miles traveled by all vehicles in the subgroup to the total number of gallons consumed. MPG's are assigned to each vehicle using the EPA certification files and adjusted for on-road driving.

Military Use—Includes sales to the Armed Forces, in-

cluding volumes sold to the Defense Fuel Supply Center (DFSC) for use by all branches of the Department of Defense (DOD).

Mill—A monetary cost and billing unit used by utilities; it is equal to 1/1000 of the U.S. dollar (equivalent to 1/10 of 1 cent). A tenth of a cent ($0.001).

Mill Capital—Cost for transportation and equipping a plant for processing ore or other feed materials.

Mill Feed—Uranium ore supplied to a crusher or grinding mill in an ore-dressing process.

Mill Residue—Wood and bark residues produced in processing logs into lumber, plywood, and paper.

Mill/kWh—A common method of pricing electricity. Tenths of a cent per kilowatt hour.

Milling—The grinding or crushing of ore, concentration, and other benefication, including the removal of valueless or harmful constituents and preparation for market.

Milling Capacity—The maximum rate at which a mill is capable of treating ore or producing concentrate.

Milling of Uranium—The processing of uranium from ore mined by conventional methods, such as underground or open pit methods, to separate the uranium from the undesired material in the ore.

Minable—Capable of being mined under current mining technology and environmental and legal restrictions, rules, and regulations.

Mine Capital—Cost for exploration and development, pre-mining stripping, shaft sinking, and mine development (including in situ leaching), as well as the mine plant and its equipment.

Mine Count—The number of mines, or mines collocated with preparation plants or tipples, located in a particular geographic area (state or region). If a mine is mining coal across two counties within a state, or across two states, then it is counted as two operations. This is done so that EIA can separate production by state and county.

Mineral—Any of the various naturally occurring inorganic substances, such as metals, salt, sand, stone, sulfur, and water, usually obtained from the earth. Note: For reporting on the Financial Reporting System the term also includes organic non-renewable substances that are extracted from the earth such as coal, crude oil, and natural gas.

Mineral Interests in Properties (referred to as Properties)——These include fee ownership or a lease, concession, or other contractual interest representing the right to extract minerals subject to such terms as may be imposed by the conveyance of those interests. Properties also include royalty interests, production payments payable in oil or gas, and other non-operating interests in properties operated by others. Properties include agreements with foreign governments or authorities under which an enterprise participates in the operation of the related properties or otherwise serves as 'producer' of the underlying reserves. However, properties do not include other supply agreements or contracts that represent the right to purchase (as opposed to extract) oil and gas.

Mineral Lease—An agreement wherein a mineral interest owner (Lessor) conveys to another party (lessee) the rights to explore for, develop, and produce specified minerals. The lessee acquires a working interest and the Lessor retains a non-operating interest in the property, referred to as the royalty interest, each in proportions agreed upon.

Mineral Rights—The ownership of the minerals beneath the earth's surface with the right to remove them. Mineral rights may be conveyed separately from surface rights.

Mineral-Matter-Free Basis—Mineral matter in coal is the parent material in coal from which ash is derived and which comes from minerals present in the original plant materials that formed the coal, or from extraneous sources such as sediments and precipitates from mineralized water. Mineral matter in coal cannot be analytically determined and is commonly calculated using data on ash and ash-forming constituents. Coal analyses are calculated to the mineral matter free basis by adjusting formulas used in calculations in order to deduct the weight of mineral matter from the total coal.

Mini Van—Small van that first appeared with that designation in 1984. Any of the smaller vans built on an automobile-type frame. Earlier models such as the Volkswagen van are now included in this category.

Minimum Equity Test (Lease)—The IRS test which tests whether the lessor maintains a minimum amount of equity throughout the lease term.

Minimum Generation—Generally, the required minimum generation level of a utility system's thermal units. Specifically, the lowest level of operation of oil-fired and gas-fired units at which they can be currently available to meet peak load needs.

Minimum Lease Payments—The payments which the lessee is obligated to make to the lessor. Includes at least rent, bargain purchase options and guarantees.

Minimum Quality Standard—Data that is free of material misstatements (*see Material Misstatement above*). Must also meet a regulator's minimum level of accuracy of at least 95%.

Minimum Stream flow—The lowest rate of flow of water past a given point during a specified period.

Mining—Any activity directed to the extraction of ore and associated rock. Included are open pit work, quarrying, auguring, alluvial dredging, and combined operations, including surface and underground operations.

Mining Operation—One mine and/or tipple at a single physical location.

Minority Carrier—A current carrier, either an electron or a hole, that is in the minority in a specific layer of a semiconductor material; the diffusion of minority carriers under the action of the cell junction voltage is the current in a photovoltaic device.

Minority Carrier Lifetime—The average time a minority carrier exists before recombination.

Minority Interest in Income—The proportional share of the minority ownership's interest (less than 50 percent) in the earnings or losses of the consolidated subsidiary. Subsidiaries are generally fully consolidated when the parent company holds an ownership-share of between 51 percent and 100 percent. In consolidation, 100 percent of revenues, expenses, assets, etc. are included in the financial statements even though, for example, the subsidiary is only 80 percent owned by the parent company. In such cases, the consolidated balance sheet must have a caption on the right-hand side titled something like 'minority interests in consolidated affiliates,' and the income statement must have a similar line to reduce net income to the pro-rata (e.g., 80 percent) share of the consolidated subsidiary's net income.

MIR—Maximum Individual Risk

MIRR—Modified Internal Rate of Return

Miscellaneous Petroleum Products—Includes all finished products not classified elsewhere (e.g., petrolatum lube refining byproducts (aromatic extracts and tars), absorption oils, ram-jet fuel, petroleum rocket fuels, synthetic natural gas feedstocks, and specialty oils).

Miscellaneous Reserves—A supply source having not more than 50 billion cubic feet of dedicated recoverable salable reserves and that falls within the definition of Supply Source.

MISF—Multiple Investment Sinking Fund. The predominant yield calculation method in leveraged leases, using non-overlapping investment and sinking fund phases with a fixed sinking fund earnings rate.

Mitigation—Steps taken to avoid or minimize negative environmental impacts. Mitigation can include: avoiding the impact by not taking a certain action; minimizing impacts by limiting the degree or magnitude of the action; rectifying the impact by repairing or restoring the affected environment; reducing the impact by protective steps required with the action; and compensating for the impact by replacing or providing substitute resources.

Mixed Occupancy Building—A building designed and constructed for more than one type of occupancy, such as a three story building with ground floor retail and second and third floor residential apartments.

Mixed Waste—Waste containing both radioactive and hazardous constituents.

Mixing Valve—A valve operated by a thermostat that can be installed in solar water heating systems to mix cold water with water from the collector loop to maintain a safe water temperature.

MJ—Megajoule

mL—Milliliter

MLE—Maximum Likelihood Estimates

MMbbl/d—One million barrels of oil per day

MMBF—One million board feet

MMBtu—Million British Thermal Units. Thermal unit of energy equal to 1,000,000 Btus, that is, the equivalent of 1,000 cubic feet of gas having a heating content of 1,000 Btus per cubic foot, as provided by contract measurement terms.

MMcf—One million cubic feet

MMcfd—Million cubic feet per day (used with natural gas)

MMgal/d—One million gallons per day

MMst—One million short tons

MMT—Million Metric Tons

MMTh—Million therms

MOA—Memorandum of Agreement

Mobil Source Emissions Reduction Credits (MSERC)—MSERCs are ERCs approved by the governing agency that are derived from measures that reduce emissions from mobile sources (e.g., converting buses to clean fuels, scrapping cars, etc.).

Mobile Home—A housing unit built on a movable chassis and moved to the site. It may be placed on a permanent or temporary foundation and may contain one room or more. If rooms are added to the structure, it is considered a single-family housing unit. A manufactured house assembled on site is a single-family housing unit, not a mobile home.

Mobile Substation—This is a movable substation which is used when a substation is not working or additional power is needed.

Model—A floor plan and house or dwelling unit design that is repeated throughout a subdivision or within a multi-family building project. To be considered the same model, dwelling units shall be in the same subdivision or multi-family housing development and have the same energy designs and features, including the same floor area and volume, for each dwelling unit, as shown on the CF-1 R. For multi-family buildings, variations in the exterior surface areas caused by location of dwelling units within the building do not cause dwelling units to be considered a different model.

Modeling Assumptions—The conditions (such as weather conditions, thermostat settings and schedules, internal gain schedules, etc.) that are used for calculating a building's annual energy consumption as specified in the ACM Manuals (in California).

Modeling Period—The entire time duration used for determining the Baseline Model for a meter component.

Moderator—A material, such as ordinary water, heavy water, or graphite, used in a reactor to slow down high-velocity neutrons, thus increasing the likelihood of further fission.

Modified Accelerated Cost Recovery system (MACRS)—The current tax depreciation system as introduced by the Tax Reform Act of 1986, generally effective for all equipment placed in service after December 31, 1986.

Modified Degree-Day Method—A method used to estimate building heating loads by assuming that heat loss and gain is proportional to the equivalent heat-loss coefficient for the building envelope.

Modified Gross Lease—A property lease in which the landlord receives stipulated rent and is obligated to pay most, but not all, of the properties operating expenses and real estate taxes.

Modified Internal Rate of Return (MIRR)—A measure of investment performance that is similar to the internal rate of return except that negative cash flows, if any, are discounted to present value at a specified safe rate and positive cash flows are presumed to be reinvested to grow with compound interest at either the same specified safe rate or at a specified market rate until the termination of the investment. Also called "adjusted internal rate of return."

Modified Sine Wave—A waveform with at least three states (positive, off, and negative) used to simulate a sine wave. It has less harmonic content than a square wave. This type of waveform is better than a square wave, but not as suitable for some appliances as a sine wave.

Modularity—The use of complete sub-assemblies to produce a larger system. Also the use of multiple inverters connected in parallel to service different loads.

Module—The smallest self-contained, environmentally protected structure housing interconnected photovoltaic cells and providing a single dc electrical output; also called a panel.

Module De-rating Factor—A factor that lowers the power output of a solar module to account for field operating conditions e.g. dirt build-up on the module.

Modules—Photovoltaic cells or an assembly of cells into panels (modules) intended for and shipped for final consumption or to another organization for resale. When exported, incomplete modules and un-encapsulated cells are also included.

MOE—Margin of Exposure

MOI—Memorandum of Intent
• Memorandum of Information

Moist (Coal) Basis—"Moist" coal contains its natural inherent or bed moisture, but does not include water adhering to the surface. Coal analyses expressed on a moist basis are performed or adjusted so as to describe the data when the coal contains only that moisture that exists in the bed in its natural state of deposition and when the coal has not lost any moisture due to drying.

Moisture Content—The water content of a substance (a solid fuel) as measured under specified conditions being the "dry basis," which equals the weight of the wet sample minus the weight of a (bone) dry sample divided by the weight of the dry sample times 100 (to get percent); "wet basis," which is equal to the weight of the wet sample minus the weight of the dry sample divided by the weight of the wet sample times 100.

Moisture Control—The process of controlling indoor moisture levels and condensation.

Mole—The quantity of a compound or element that has a weight in grams numerically equal to its molecular weight. Also referred to as "gram molecule" or "gram molecular weight."

Molten Carbonate Fuel Cell—Type of fuel cell that consists of a molten electrolyte in which $CO_3=$ is trans-

ported from the cathode to the anode.

Monitoring—Monitoring relates to the regular measurement, assessment and recording of emissions and emission reductions by an emitting firm or an emission reduction project. For example, emitting firms may monitor the actual level of emissions reduction achieved as a result of internal abatement programs.

Monitoring Wells—Specially-constructed wells used exclusively for testing water quality.

Monocrystalline Solar Cell—A form of solar cell made from a thin slice of a single large crystal of silicon.

Monoculture—The planting, cultivation, and harvesting of a single species of crop in a specified area.

Monolithic—Fabricated as a single structure, as used to describe thin film series interconnected PV cells on a single sheet substrate.

Monopoly—The only seller with control over market sales.

Monopsony—The only buyer with control over market purchases.

Monte Carlo Simulation—A gaming studies model for risk analysis in which specific investment elements are assigned probabilities and integrated into a larger theoretical population. Repeated sampling from this larger group provides a range of possible outcomes.

Montreal Gases—Ozone depleting substances covered by the Montreal Protocol, including chlorofluorocarbons, hydrochlorofluorocarbons, carbon tetrachloride, methyl chloroform, and brominated gases.

Montreal Protocol—The Montreal Protocol on Substances that Deplete the Ozone Layer (1987). An international agreement, signed by most of the industrialized nations, to substantially reduce the use of chlorofluorocarbons (CFCs). Signed in January 1989, the original document called for a 50-percent reduction in CFC use by 1992 relative to 1986 levels. The subsequent London Agreement called for a complete elimination of CFC use by 2000. The Copenhagen Agreement, which called for a complete phase out by January 1, 1996, was implemented by the U.S. Environmental Protection Agency.

Monuments—Objects or marks used to fix or establish a boundary.

Moody's—A credit rating agency.

Moratorium—Temporary suspension.

Mortality—The volume of sound wood in trees that have died from natural causes.

MOS—Margin of Safety

Motion Sensor, Lighting—A device that automatically turns lights off soon after an area is vacated. The term Motion Sensor applies to a device that controls outdoor lighting systems. When the device is used to control indoor lighting systems, it is termed an occupant sensor. The device also may be called an occupancy sensor, or occupant sensing device.

Motor—A machine supplied with external energy that is converted into force and/or motion.

Motor Gasoline (Finished)—A complex mixture of relatively volatile hydrocarbons, with or without small quantities of additives, that has been blended to form a fuel suitable for use in spark- ignition engines. Motor gasoline, as given in ASTM Specification D439 or Federal Specification VV-G-l690B, includes a range in distillation temperatures from 122 to 158 degrees Fahrenheit at the 10-percent recovery point and from 365 to 374 degrees Fahrenheit at the 90-percent recovery point. Motor gasoline includes reformulated motor gasoline, oxygenated motor gasoline, and other finished motor gasoline. Blend-stock is excluded until blending has been completed. Reformulated Motor Gasoline. Gasoline reformulated for use in motor vehicles, the composition and properties that meet the requirements of the reformulated gasoline regulations promulgated by the U.S. EPA under Section 211K of the Clean Air Act. Oxygenated Gasoline. Gasoline formulated for use in motor vehicles that has oxygen content of 1.8 percent or higher, by weight. Includes gasohol. Other Finished Gasoline. Motor Gasoline not included in the oxygenated or reformulated gasoline categories.

Motor Gasoline Blending—Mechanical mixing of motor gasoline blending components, and oxygenates when required, to produce finished motor gasoline. Finished motor gasoline may be further mixed with other motor gasoline blending components or oxygenates, resulting in increased volumes of finished motor gasoline and/or changes in the formulation of finished motor gasoline (e.g., conventional motor gasoline mixed with MTBE to produce oxygenated motor gasoline).

Motor Gasoline Blending Components—Naphthas (e.g., straight-run gasoline, alkylate, reformate, benzene, toluene, xylene) used for blending or compounding into finished motor gasoline. These components include reformulated gasoline blendstock for oxygenate blending (RBOB) but exclude oxygenates (alcohols, ethers), butane, and pentanes plus. Note: Oxygenates are reported as individual components and are included in the total for other hydrocarbons, hydrogens, and oxygenates.

Motor Gasoline, Finished Gasohol—A blend of finished motor gasoline (leaded or unleaded) and alcohol (generally ethanol but sometimes methanol), limited to 10 percent by volume of alcohol.

Motor Gasoline, Finished Leaded—Contains more than 0.05 gram of lead per gallon or more than 0.005 gram of phosphorus per gallon. Premium and regular grades are included, depending on the octane rating. Includes leaded gasohol. Blend-stock is excluded until blending has been completed. Alcohol that is to be used in the blending of gasoline is excluded.

Motor Gasoline, Finished Unleaded—Contains not more than 0.05 gram of lead per gallon and not more than 0.005 gram of phosphorus per gallons. Premium and regular grades are included, depending on the octane rating. Includes unleaded gasohol. Blend-stock is excluded until blending has been completed. Alcohol that is to be used in the blending of gasohol is also excluded.

Motor Speed—The number of revolutions that the motor turns in a given time period (i.e. revolutions per minute, rpm).

MOU—Memorandum of Understanding

Movable Insulation—A device that reduces heat loss at night and during cloudy periods and heat gain during the day in warm weather. A movable insulator could be an insulative shade, shutter panel, or curtain.

Movistor—Metal Oxide Varistor. Used to protect electronic circuits from surge currents such as those produced by lightning.

MP—Melting Point

MPE—Maximum Probable Earthquake

MPG—Miles per gallon

MPG Shortfall—The difference between actual on-road MPG and EPA laboratory test MPG. MPG shortfall is expressed as gallons per mile ratio (GPMR).

MPI—Maximum Permitted Intake

MPR—Market Price Referent. Price paid (usually per kWh) to customers producing renewable energy by utilities or local governments.

MRC—Maximum Reservoir Contact

MRET—Mandatory Renewable Energy Target

MRF—See *Materials Recovery Facility*.

MRI—Midwest Research Institute

MRL—Maximum-Residue Limit (Pesticide Tolerance)
• Minimum Risk Level

MRP—Maturity Risk Premium

MSA—Metropolitan Statistical Area: The MSA is comprised of one of more counties and of a high degree of social and economic integration as defined by the U.S. Census Bureau.

MSDS—Material Safety Data Sheet

MSEE—Major Source Enforcement Effort

MSHA—Mine Safety and Health Administration www.msha.gov

MSHA ID Number—Seven (7)-digit code assigned to a mining operation by the Mine Safety and Health Administration.

MSL—Mean Sea Level

MSW—See *Municipal Solid Waste*.

MSWL—Municipal Solid Waste Landfill
- Municipal Solid Waste Landfill Leachate

MSWLF—Municipal Solid Waste Landfill Facility

MT—Metric Ton

MTBE (Methyl tertiary butyl ether)—Ether intended for motor gasoline blending. (See definition for *Oxygenates*.)—Now banned in the state of California.

MTD—Maximum Tolerated Dose

MTEC—Maximum Theoretical Emissions Concentration

MTR—Minimum Technology Requirement

MUD—Municipal Utility District

Mullion—A vertical framing member separating adjoining window or door sections. Slender pier that forms the division between the lights of a window, a screen, or an opening. *See Dividers*

Mullion/Transom—A popular façade construction often used with curtain wall (q.v.) facades. It consists of vertical beams (mullions) and smaller, horizontal beams (transoms).

Multicrystalline—A material that has solidified at a rate such that many small crystals (crystallites) form. The atoms within a single crystallite are symmetrically arranged with a particular orientation, whereas the crystallites themselves are differently oriented. The multitude of grain boundaries in the material (between the crystallites) reduce the cell efficiency. Multicrystalline is also referred to as polycrystalline.

Multi-Family Dwelling Unit—A dwelling unit of occupancy type R, as defined by the CBC (California Building Codes), sharing a common wall and/or ceiling/floor with at least one other dwelling unit.

Multi-Function Analysis—A storage project may at different times operate as a Generation, Transmission, Distribution or Load resource. This functionality determines the jurisdictional authority that governs its markets or terms of use; i.e., FERC/transmission, CPUC/distribution.

Multijunction Device—A high-efficiency photovoltaic device containing two or more cell junctions, each of which is optimized for a particular part of the solar spectrum.

Multilateral Agency—Commonly refers to public agencies that work internationally to provide development, environmental, or financial assistance to developing countries, such as the World Bank, or to broker international agreements and treaties, such as the United Nations.

Multilateral Consultative Process (MCP)—The Multilateral Consultative Process (MCP) was proposed in Article 13 of the Framework Convention on Climate Change. The proposed purpose of the MCP was to deal with all questions and problems related to the implementation of the framework.

Multi-Level Lighting Control—A lighting control that reduces lighting power in multiple steps while maintaining a reasonably uniform level of illuminance throughout the area controlled.

Multiple Completion Well—A well equipped to produce oil and/or gas separately from more than one reservoir. Such wells contain multiple strings of tubing or other equipment that permit production from the various completions to be measured and accounted for separately. For statistical purposes, a multiple completion well is reported as one well and classified as either an oil well or a gas well. If one of the several completions in a given well is an oil completion, the well is classified as an oil well. If all of the completions in a given well are gas completions, the well is classified as a gas well.

Multiple Cropping—A system of growing several crops on the same field in one year.

Multiple Investment Sinking Fund (MISF)—Method of income allocation used to report earnings on a leveraged lease. This method is similar to the external rate of return (ERR) method of yield calculation except that it assumes a zero earnings rate during periods of disinvestment (a sinking fund rate equal to 0).

Multiple Purpose Project—The development of hydroelectric facilities to serve more than one function. Some of the uses include hydroelectric power, irrigation, water supply, water quality control, and/or fish and wildlife enhancement.

Multiple Purpose Reservoir—Stored water and its usage governed by advanced water resource conservation practices to achieve more than one water control objective. Some of the objectives include flood control, hydroelectric power development, irrigation, recreation usage, and wilderness protection.

Multiple Zone—A supply fan (and optionally a return fan) with heating and/or cooling heat exchangers (e.g. DX coil, chilled water coil, hot water coil, furnace, electric heater) that serves more than one thermostatic zone. Zones are thermostatically controlled by features including but not limited to variable volume, reheat, recool and concurrent operation of another system.

Multiscene Dimming System—A lighting control device that has the capability of setting light levels throughout a continuous range, and that has pre-established settings within the range.

Multi-stage Controller—A charge controller that allows different charging currents as the battery approaches full state of charge.

Multi-Zone System—A building heating, ventilation, and/or air conditioning system that distributes conditioned air to individual zones or rooms.

MUNI—Municipally-Owned Utility: A utility owned and operated by a city or county.

Municipal Electric Utility—A power utility system owned and operated by a local jurisdiction. A municipal utility is a non-profit utility that is owned and operated by the community it serves. Whether or not a municipal utility is open to customer choice and competition is decided by the municipality's public officials.

Municipal Lease—A conditional sales contract disguised in the form of a lease available only to municipalities, in which the interest earnings are tax-exempt by the lessor. Most municipal leases have a "fiscal funding" clause in the contract that allows the municipality to cancel the contract if for any reason the funds to pay the lease are not available for the next fiscal year of the municipality.

Municipal Solid Waste (MSW)—Garbage. Refuse offering the potential for energy recovery; includes residential, commercial, and institutional wastes.

Municipal Utility—A provider of utility services owned and operated by a municipal government.

Municipal Waste—As defined in the Energy Security Act (P.L. 96-294; 1980) as "any organic matter, including sewage, sewage sludge, and industrial or commercial waste, and mixtures of such matter and inorganic refuse from any publicly or privately operated municipal waste collection or similar disposal system, or from similar waste flows (other than such flows which constitute agricultural wastes or residues, or wood wastes or residues from wood harvesting activities or production of forest products)."

Municipal Waste to Energy Project or Plant—A facility that produces fuel or energy from municipal solid waste.

Municipality—A village town, city, county, or other political subdivision of a State.

Municipalization—The process by which a municipal entity assumes responsibility for supplying utility service to its constituents. In supplying electricity, the municipality may generate and distribute the power or purchase wholesale power from other generators and distribute it.

Muntins—Upright piece of timber in a frame, separating panels. *See Dividers and Mullion*

Must Take Resources—Generation resources including QF generating units, nuclear units, and pre-existing power purchase contracts with minimum energy take requirements that are dispatched by the ISO before PX or Direct Access Generation.

MV—Megavolt

MVA—Megavolt-amperes
  • Market Value Added

MVAR—Megavolt-ampere reactive

MW—Megawatt: One million watts or 1,000 kW.
  • Molecular Weight
  • Mixed Waste

MWa—See *Average Megawatt.*

MWC—Municipal Waste Combustor

MWD—Metropolitan Water District

MWG—Model Working Group

MWh—Megawatt-hours

MWI—Medical Waste Incinerator

MWp—Peak megawatt

MWTA—Medical Waste Tracking Act (1989). This Act amended the Solid Waste Disposal Act to require the Administrator of the EPA to promulgate regulations on the management of infectious waste.

N

NAAQS—See *National Ambient Air Quality Standards.*

NABCEP—North American Board of Certified Energy Practitioners. A professional association developing a voluntary national certification program for solar practitioners. NABCEP is a volunteer board of renewable energy industry representatives that includes representatives of the solar industry, NABCEP certificants, renewable energy organizations, state policy makers, educational institutions, and the trades. The NABCEP offers NABCEP PV installer certification and NABCEP solar thermal installer certification for renewable energy professionals.

NAC—National Agency Check
- National Asbestos Council

Nacelle—The cover for the gear box, drive train, generator, and other components of a wind turbine.

NACI—National Agency Check and Inquiry

NACORE—National Association of Corporate Real Estate Executives www.realestateagent.com

NAECA—National Appliance Energy Conservation Act of 1987 mandates minimum energy efficiency standards for most major residential appliances.

NAEP—National Association of Environmental Professionals www.naep.org

NAFTA—North American Free Trade Agreement

NAHB—National Association of Home Builders NAHB is a trade association for housing and the building industry with 235,000 members and more than 800 state and local associations. NAHB offers a Certified Green Professional designation for building and construction professionals, which is part of the NAHB National Green Building Program and is issued by the NAHB University of Housing, the educational arm of the association.

Further, NAHB currently has its National Green Home Certification Program, which is based on the NAHB Model Green Home Building Guidelines (published in 2005). In late 2008, homes certified under the national program will also have the option of being certified based on the National Green Building Standard (ICC 700-2008), which includes provisions that define green attributes for subdivisions, multifamily dwellings, remodeling projects, additions and single-family homes.

Presently, there are three green home certification levels available in the NAHB Model Green Home Building Guidelines: Bronze, Silver and Gold. An additional Emerald level will be available in the National Green Building Standard. The green levels and certifications address seven main construction areas: site, resource efficiency, energy efficiency, water efficiency, indoor environmental quality, homeowner education, and global impact. www.nahb.org

NAICS (North American Industry Classification System)—A coding system developed jointly by the United States, Canada, and Mexico to classify businesses and industries according to the type of economic activity in which they are engaged. NA-ICS replaces the Standard Industrial Classification (SIC) codes.

NAIMA—North American Insulation Manufacturers Association www.naima.org

Name Plate—A metal tag attached to a machine or appliance that contains information such as brand name, serial number, voltage, power ratings under specified conditions, and other manufacturer supplied data.

Name Plate Generating Capacity—The rated continuous load-carrying ability expressed in megawatts (MW). Also the maximum rated output of a generator under specific conditions designated by the manufacturer.

Nanohydro—Any hydro plant that produces less than 100 watts.

Naphtha—A colorless, flammable liquid obtained from crude petroleum and used as a solvent and cleaning fluid and as a raw material for manufacture of gasoline.
- A generic term applied to a petroleum fraction with an approximate boiling range between 122 degrees Fahrenheit and 400 degrees Fahrenheit.
- Refined or partly refined light distillates with an approximate boiling point range of 27 degrees to 221 degrees Centigrade. Blended further or mixed with other materials, they make high-grade motor gasoline or jet fuel. Also, used as solvents, petrochemical feedstocks, or as raw materials for the production of town gas.

Naphtha-Type Jet Fuel—A fuel in the heavy naphtha boiling range having an average gravity of 52.8 degrees API, 20 to 90 percent distillation temperatures of 290 degrees to 470 degrees Fahrenheit, and meeting Military Specification MIL-T-5624L (Grade JP-4). It is used primarily for military turbojet and turboprop aircraft engines because it has a lower freeze point than other aviation fuels and meets engine requirements at high altitudes and speeds.

NAPS—National Allocation Plans. These set out the overall emissions cap for countries in phases I and II of the EU Emissions Trading Scheme up to 2012, and the emissions allowances that each sector and individual installation within each country receives.

NAR—National Asbestos Registry
- National Association of Realtors www.realtor.org

NARA—National Air Resources Act

NARUC—The National Association of Regulatory Utility Commissioners. An advisory council composed of governmental agencies of the fifty States, the District of Columbia, Puerto Rico and the Virgin Islands engaged in the regulation of utilities and carriers. "The chief objective is to serve the consumer interest by seeking to improve the quality and effectiveness of public regulation in America." www.naruc.org

NASD—The National Association of Securities Dealers. A mandatory association of brokers and dealers in the over-the-counter securities business. Created by the Maloney Act of 1938, an amendment to the Securities Act of 1934.

NASDAQ—An automated information network which provides brokers and dealers with price quotations on securities traded over the counter.

NASEA—Northeast Sustainable Energy Association www.nasea.org

NASEO—National Association of State Energy Officials www.naseo.org

NASR—National Association of Solvent Recyclers

NASUCA—The National Association of Utility Consumer Advocates. NASUCA includes members from 38 states and the District of Columbia. It was formed "to exchange information and take positions on issues affecting utility rates before federal agencies, Congress and the courts. www.nasuca.org

National Action Plan—Plans submitted to the Conference of the Parties (COP) by all Parties outlining the steps that they have adopted to limit their anthropogenic GHG emissions. Countries must submit these plans as a condition of participating in the UN Framework Convention on Climate Change and, subsequently, must communicate their progress to the COP regularly.

National Ambient Air Quality Standards (NAAQS)—Federal standards established by the Clean Air Act.

National Association of Regulatory Utility Commissioners (NARUC)—An affiliation of the public service commissioners to promote the uniform treatment of members of the railroad, public utilities, and public service commissions' of the 50 states, the District of Columbia, the Commonwealth of Puerto Rico, and the territory of the Virgin Islands.

National Defense Authorization Act—The federal law, enacted in 1994 and amended in 1995, that required the Secretary of Energy to prepare the Baseline Report.

National Electrical Code (NEC)—The NEC is a set of regulations that have contributed to making the electrical systems in the United States one of the safest in the world. The intent of the NEC is to ensure safe electrical systems are designed and installed. The National Fire Protection Association has sponsored the NEC since 1911. The NEC changes as technology evolves and component sophistication increases. The NEC is updated every three years. Following the NEC is required in most locations.

National Electrical Manufacturers Association (NEMA)—The U.S. trade association that develops standards for the electrical manufacturing industry. www.nema.org.

National Emissions Standards for Hazardous Pollutants (NESHAPS)—Federal standards that control pollutants considered toxic to humans.

National Environmental Policy Act (NEPA)—A federal law enacted in 1969 that requires all federal agencies to consider and analyze the environmental impacts of any proposed action. NEPA requires an environmental impact statement for major federal actions significantly affecting the quality of the environment. NEPA requires federal agencies to inform and involve the public in the agency's decision making process and to consider the environmental impacts of the agency's decision.

National Forest Management Act (NFMA)—The National Forest Management Act reorganized, expanded and otherwise amended the Forest and Rangeland Renewable Resources Planning Act of 1974, which called for the management of renewable resources on national forest lands. The National Forest Management Act requires the Secretary of Agriculture to assess forest lands, develop a management program based on multiple-use, sustained-yield principles, and implement a resource management plan for each unit of the National Forest System. It is the primary statute governing the administration of national forests.

National Gas Transportation Association (NGTA)—Formerly the National Transportation & Exchange Association. A group that promotes understanding of the national pipeline grid and is working toward standardization in the industry.

National Historic Preservation Act (NHPA)—Federal legislation requiring archaeological and cultural review of areas identified for new pipeline construction and other utility right-of-way.

National Pollution Discharge Elimination System (NPDES)—The federal water quality program. It requires a permit for the discharge of pollutants to surface waters of the United States.

National Priorities List (NPL)—The Environmental Protection Agency's list of the most serious uncontrolled or abandoned hazardous waste sites identified for possible long-term remedial action under the Comprehensive Environmental Response, Compensation, and Liability Act (CERCLA). The list is based primarily on the score a site receives from the Environmental Protection Agency Hazard Ranking System. The Environmental Protection Agency is required to update the National Priorities List at least once a year.

National Rural Electric Cooperative Association (NRECA)—A national organization dedicated to representing the interests of cooperative electric utilities and the consumers they serve. Members come from the 46 states that have an electric distribution cooperative. www.nreca.org

National Self-Determination—Self determination is the process of a nation (or a firm) deciding their own framework for emission control, measurement and monitoring methodologies, without reference to the wishes of any other nation, firm or agency.

National Transportation Safety Board (NTSB)—An independent agency reporting administratively to the Secretary of Transportation, charged with the investigation of all safety-related incidents involving transportation. These include air, rail, highway, and liquid and gas pipeline transportation. The NTSB has no power to issue regulations; however, it issues reports and recommendations.

National Uranium Resource Evaluation (NURE)—A program begun by the U.S. Atomic Energy Commission (AEC) in 1974 to make a comprehensive evaluation of U.S. uranium resources and continued through 1983 by the AEC's successor agencies, the Energy Research and Development Administration (ERDA), and the Department of Energy (DOE). The NURE program included aerial radiometric and magnetic surveys, hydrogeochemical and stream sediment surveys, geologic drilling in selected areas, geophysical logging of selected boreholes, and geologic studies to identify and evaluate geologic environments favorable for uranium.

Native Gas—Gas in place at the time that a reservoir was converted to use as an underground storage reservoir in contrast to injected gas volumes.

Native Load Customers—Wholesale and retail customers that the transmission provider constructs and operates a system to provide electric needs.

Native Vegetation—A plant whose presence and survival in a specific region is not due to human intervention. Certain experts argue that plants imported to a region by prehistoric peoples should be considered native. The term for plants that are imported and then adapt to survive without human cultivation is "naturalized."

Natural Capital—A company's environmental assets and natural resources existing in the physical environment, either owned (such as mineral, forest, or energy resources) or simply utilized in business operations (such as clean water and atmosphere). Often traditional economic measures and indicators fail to take into account the development use of natural capital, although preservation of its sustainable use is essential to a business' long-term survival and growth.

Natural Cooling—Space cooling achieved by shading, natural (unassisted, as opposed to forced) ventilation, conduction control, radiation, and evaporation; also called passive cooling.

Natural Draft—Draft that is caused by temperature differences in the air.

Natural Gas—A mixture of hydrocarbon compounds and small quantities of various non-hydrocarbons existing in the gaseous phase or in solution with crude oil in natural underground reservoirs at reservoir conditions. The principal hydrocarbons usually contained in the mixture are methane, ethane, propane, butanes, and pentanes. Typical non-hydrocarbon bases that may be present in reservoir natural gas are carbon dioxide, helium, hydrogen sulfide, and nitrogen. Under reservoir conditions, natural gas and the liquefiable portions thereof occur either in a single gaseous phase in the reservoir or in solution with crude oil and are not distinguishable at that time as separate substances. Natural Gas, based on the type of occurrence in the reservoir, is classified by two categories, as follows: Non-Associated Gas natural gas that is not in contact with significant quantities of crude oil in the reservoir. Associated/Dissolved Gas is the combined volume of natural gas that occurs in crude oil reservoirs either as free gas (associated) or as gas in solution with crude oil (dissolved). Associated gas is free natural gas, commonly known as gas cap gas, which overlies and is in contact with crude oil in the reservoir. Dissolved gas is natural gas that is in solution with crude oil in the reservoir at reservoir conditions. Statistical data pertaining to natural gas production and reserves are reported in units of 1,000,000 cubic feet (i.e., MMCF) at 14.73

pounds per square inch absolute and 60 degrees Fahrenheit for FRS purposes.

Natural Gas Field Facility—A field facility designed to process natural gas produced from more than one lease for the purpose of recovering condensate from a stream of natural gas; however, some field facilities are designed to recover propane, normal butane, pentanes plus, etc., and to control the quality of natural gas to be marketed.

Natural Gas Gross Withdrawals—Full well-stream volume of produced natural gas, excluding condensate separated at the lease.

Natural Gas Hydrates—Solid, crystalline, wax-like substances composed of water, methane, and usually a small amount of other gases, with the gases being trapped in the interstices of a water-ice lattice. They form beneath permafrost and on the ocean floor under conditions of moderately high pressure and at temperatures near the freezing point of water.

Natural Gas Liquids (NGL)—Those hydrocarbons in natural gas that are separated from the gas as liquids through the process of absorption, condensation, adsorption, or other methods in gas processing or cycling plants. Generally such liquids consist of propane and heavier hydrocarbons and are commonly referred to as lease condensate, natural gasoline, and liquefied petroleum gases. Natural gas liquids include natural gas plant liquids (primarily ethane, propane, butane, and isobutane; See *Natural Gas Plant Liquids*) and Lease Condensate (primarily pentanes produced from natural gas at lease separators and field facilities).

Natural Gas Liquids Production—The volume of natural gas liquids removed from natural gas in lease separators, field facilities, gas processing plants, or cycling plants during the report year.

Natural Gas Marketed Production—Gross withdrawals of natural gas from production reservoirs, less gas used for reservoir repressuring, non-hydrocarbon gases removed in treating and processing operations, and quantities vented and flared.

Natural Gas Plant Liquids—Those hydrocarbons in natural gas that are separated as liquids at natural gas processing plants, fractionating and cycling plants, and, in some instances, field facilities. Lease condensate is excluded. Products obtained include ethane; liquefied petroleum gases (propane, butanes, propane-butane mixtures, ethane-propane mixtures); isopentane; and other small quantities of finished products, such as motor gasoline, spe-

cial naphthas, jet fuel, kerosene, and distillate fuel oil.

Natural Gas Policy Act of 1978 (NGPA)—Signed into law on November 9, 1978, the NGPA is a framework for the regulation of most facets of the natural gas industry.

Natural Gas Processing Plant—Facilities designed to recover natural gas liquids from a stream of natural gas that may or may not have passed through lease separators and/or field separation facilities. These facilities control the quality of the natural gas to be marketed. Cycling plants are classified as gas processing plants.

Natural Gas Steam Reforming Production—A two step process where in the first step natural gas is exposed to a high-temperature steam to produce hydrogen, carbon monoxide, and carbon dioxide. The second step is to convert the carbon monoxide with steam to produce additional hydrogen and carbon dioxide.

Natural Gas Supply Association (NGSA)—A trade group representing major integrated gas producers, medium-sized companies and independents. www.ngsa.org

Natural Gas Utility Demand-side Management (DSM) Program Sponsor—A DSM (demand-side management) program sponsored by a natural gas utility that suggests ways to increase the energy efficiency of buildings, to reduce energy costs, to change the usage patterns, or to promote the use of a different energy source.

Natural Gas Vehicle—Vehicles that are powered by compressed or liquefied natural gas.

Natural Gas, Dry—The marketable portion of natural gas production, which is obtained by subtracting extraction losses, including natural gas liquids removed at natural gas processing plants, from total production.

Natural Gasoline—A term used in the gas processing industry to refer to a mixture of liquid hydrocarbons (mostly pentanes and heavier hydrocarbons) extracted from natural gas. It includes isopentane.

Natural Gasoline and Isopentane—A mixture of hydrocarbons, mostly pentanes and heavier, extracted from natural gas, that meets vapor pressure, endpoint, and other specifications for natural gasoline set by the Gas Processors Association. Includes isopentane which is a saturated branch-chain hydrocarbon, (C_5H_{12}), obtained by fractionation of natural gasoline or isomerization of normal pentane.

Natural Monopoly—A situation where one firm can

produce a given level of output at a lower total cost than can any combination of multiple firms. Natural monopolies occur in industries which exhibit decreasing average long-run costs due to size (economies of scale). According to economic theory, a public monopoly governed by regulation is justified when an industry exhibits natural monopoly characteristics.

Natural Reservoir Pressure—The energy within an oil or gas reservoir that causes the oil or gas to rise (unassisted by other forces) to the earth's surface when the reservoir is penetrated by an oil or gas well. The energy may be the result of "dissolved gas drive," "gas cap drive," or "water drive." Regardless of the type of drive, the principle is the same: the energy of the gas or water, creating a natural pressure, forces the oil or gas to the well bore.

Natural Stream flow—The rate of flow of water past a given point of an uncontrolled stream or regulated stream flow adjusted to eliminate the effects of reservoir storage or upstream diversions at a set time interval.

Natural Uranium—Uranium with the U-235 isotope present at a concentration of 0.711 percent (by weight), that is, uranium with its isotopic content exactly as it is found in nature.

Natural Ventilation—Ventilation that is created by the differences in the distribution of air pressures around a building. Air moves from areas of high pressure to areas of low pressure with gravity and wind pressure affecting the airflow. The placement and control of doors and windows alters natural ventilation patterns.

NAV—Net Asset Value

NAVAIR—Naval Air Systems Command

NAWC—National Association of Water Companies www.nawc.org

NBAR—Nonbinding Preliminary Allocation of Responsibility

NBS—National Bureau of Standards

NCA—Noise Control Act (1972). The Act establishes a national policy to promote an environment for all Americans free from noise that jeopardizes their health and welfare.

NCAQ—National Commission on Air Quality

NCAR—National Center for Atmospheric Research

NCC—National Climatic Center

NCEA—National Center for Exposure Assessment
- National Center for Environmental Assessment

NCI—National Cancer Institute www.cancer.gov

NCO—Negotiated Consent Order

NCR—Noncompliance Report
- Nonconformance Report

NCS—National Compliance Strategy

NCSL—The National Conference of State Legislatures. A national advisory council which provides services to state legislatures "by bringing together information from all states to forge workable answers to complex policy questions." www.ncsl.org.

NCTR—National Center for Toxicological Research

NCWQ—National Commission on Water Quality

NDD—Negotiated Decision Document

NDS—National Disposal Site

NDWAC—National Drinking Water Advisory Council

NEA—National Energy Act (1978). Legislative response by congress to the 1973 energy crisis.

NEAC—National Environmental Justice Advisory Council

Neat Fuel—Fuel that is free from admixture or dilution with other fuels.

NEC—US National Electrical Code which contains guidelines for all types of electrical installations which should be followed when installing a PV system.

NEDS—National Emissions Data System

NEEA—Northwest Energy Efficiency Alliance www.nwalliance.org

Negative Declaration—Also known as "Neg Dec." A California Environmental Quality Act (CEQA) document issued by the lead regulatory agency when the initial environmental study reveals no substantial evidence that the proposed project will have a significant adverse effect on the environment, or when any significant effects would be avoided or mitigated by revisions agreed to by the applicant.

Negative Feedback—A process that results in a reduction in the response of a system to an external influence. For example, increased plant productivity in response to global warming would be a negative feedback on warming, because the additional growth would act as a sink CO_2, reducing the atmospheric CO_2 concentration.

NEHA—National Environmental Health Association www.neha.org

NEI—Nuclear Energy Institute: Policy organization of the nuclear energy and technologies industry and participates in both the national and global policy-making process. NEI's objective is to ensure the formation of policies that promote the beneficial uses of nuclear energy and technologies in the U.S. and around the world. On the internet at http://www.nei.org

NEMA—US National Electrical Manufacturers Association—sets standards for some non-electronic products e.g. junction boxes.

NEMS—National Energy Modeling System

NEP—National Energy Plan
- National Estuary Program

NEPA—See *National Environmental Policy Act.* (1969)

Nephelometric Turbidity Unit—A measurement unit of the clarity of water, dependent on the amount of suspended matter.

NEPI—National Environmental Policy Institute

NEPOOL Generation Information System, or NEPOOL GIS—An accounting system for certificates. For each megawatt hour of power, the system creates an electronic certificate, which may be sold or otherwise transferred off line. Certificates for generation are created on a monthly basis and put into the system on a quarterly basis. The certificate describes: when, where, and who produced the power; the type of fuel source used; renewable portfolio standard eligibility (MA, CT, ME); the amount and type of certain pollutant emissions released; and other characteristics (vintage and green-e eligibility). Retail electric suppliers use the information to report compliance with requirements set by certain New England states, such as: minimum renewable power purchase levels; disclosure of fuel source and other characteristics of the power that they sell; and maximum levels of certain emissions.

NER—National Emissions Report

NERC—North American Electric Reliability Council www.nerc.com

NERO—National Energy Resources Organization. www.nationalenergyresources.com.

NES—National Energy Savings

NESA—National Emissions Standards Act
- National Energy Services Association. www.nesa.net.org.

NESHAP(S)—National Emission Standards for Hazardous Air Pollutants

Net (Lower) Heating Value (NHV)—The potential energy available in a fuel as received, taking into account the energy loss in evaporating and superheating the water in the fuel. Equal to the higher heating value minus 1050W where W is the weight of the water formed from the hydrogen in the fuel, and 1050 is the latent heat of vaporization of water, in Btu, at 77 degrees Fahrenheit.

Net Asset Value (NAV)—NAV is calculated by adding the value of all of the investments in the fund and dividing by the number of shares of the fund that are outstanding. NAV calculations are required for all mutual funds (or open-end funds) and closed-end funds. The price per share of a closed-end fund will trade at either a premium or a discount to the NAV of that fund, based on market demand. Closed-end funds generally trade at a discount to NAV.

Net Benefit—The energy cost savings less the cost of the energy conservation measure or service provided.

Net Billing—The arrangement by which the Bonneville Power Administration financed the cost of nuclear power plants. Utilities that owned shares in the projects, and paid a share of the costs, assigned to BPA all or part of the generating capability of the power plants. BPA, in turn, credited the wholesale power bills of those utilities to cover the costs of their shares in the projects. BPA then sold the power output of the plants by averaging their higher cost with lower cost hydropower.

Net Building Demand—The configuration uses current and voltage sensors located between the grid connection to the building and the inverter connection to the building. This measures the import or export of energy relative to the grid.

Net Capability—The maximum load-carrying ability of the equipment, exclusive of station use, under specified conditions for a given time interval, independent of the characteristics of the load. Capability is determined by design characteristics, physical conditions, adequacy of prime mover, energy supply, and operating limitations such as cooling and circulating water supply and temperature, headwater and tailwater elevations, and electrical use.

Net Cash Flow (Real Property)—Generally determined by net income plus depreciation less principal payments on long-term obligations.

Net Cell Shipments—Represents the difference between cell shipments and cell purchases.

Net Electricity Consumption—Consumption of electricity computed as generation, plus imports, minus exports, minus transmission and distribution losses.

Net Energy for Load—Net generation of main generating units that are system-owned or system-operated, plus energy receipts minus energy deliveries.

Net Energy for System—The sum of energy an electric utility needs to satisfy their service areas, including full and partial requirements consumers.

Net Energy Metering Agreement—An agreement with the local utility which allows customers to reduce their electric bill by exchanging surplus electricity generated by certain renewable energy systems. Under net metering, the electric meter runs backwards as the customer-generator feeds extra electricity back to the utility.

Net Energy Production (or Balance)—The amount of useful energy produced by a system less the amount of energy required to produce the fuel.

Net Financing Cost—Also called the cost of carry or, simply, carry, the difference between the cost of financing the purchase of an asset and the asset's cash yield. Positive carry means that the yield earned is greater than the financing cost; negative carry means that the financing cost exceeds the yield earned.

Net Generation—The amount of gross generation less the electrical energy consumed at the generating station(s) for station service or auxiliaries. In the case of co-generators, this value is intended to include internal consumption of electricity for the purposes of a production process, as well as power put on the grid or used by the facility when the system is located.

Note: Electricity required for pumping at pumped-storage plants is regarded as electricity for station service and is deducted from gross generation.

Net Head—The gross head minus all hydraulic losses except those chargeable to the turbine.

Net Heating Value—(NHV) The potential energy available in the fuel as received, taking into account the energy loss in evaporating and superheating the water in the sample. Expressed as:
$$NVH = (HHV \times (1 - MC/100))\ (LH(2)O \times MC/100)$$

Net Income—Operating income plus other income and extraordinary income less operating expenses, taxes, interest charges, other deductions, and extraordinary deductions.

Net Interstate Flow of Electricity—The difference between the sum of electricity sales and losses within a state and the total amount of electricity generated within that state. A positive number indicates that more electricity (including associated losses) came into the state than went out of the state during the year; conversely, a negative number indicates that more electricity (including associated losses) went out of the state than came into the state.

Net Investment in Place—The sum of net property, plant, and equipment (PP&E) plus investment and advances to unconsolidated affiliates.

Net Lease—A lease in which additional costs such as maintenance, insurance, and property taxes are paid separately by the lessee. Contrast with full lease.

Net Metering—The practice of using a single meter to measure consumption and generation of electricity by a small generation facility (such as a house with a wind or solar photovoltaic system). The net energy produced or consumed is purchased from or sold to the power provider, respectively.

Net Module Shipments—Represents the difference between module shipments and module purchases. When exported, incomplete modules and un-encapsulated cells are also included.

Net Operating Income (Utility)—The amount of revenue from utility operations remaining after operation and maintenance expenses, depreciation expenses, and taxes are deducted. The revenues and expenses that produce net operating income are commonly referred to as "above-the-line" items.

Net Photovoltaic Module Shipment—The difference between photovoltaic module shipments and photovoltaic module purchases.

Net Plant—In accounting, Utility Plant less Accumulated Provision for Depreciation (including Depletion) and Amortization.

Net Present Value—An approach used in capital budgeting where the present value of cash inflow is subtracted from the present value of cash outflows. NPV compares the value of a dollar today versus the value of that same dollar in the future after taking inflation and return into account.
• The value of a personal portfolio, product, or investment after depreciation and interest on debt capital are subtracted from operating income. It can also be thought of as the equivalent worth of all cash flows relative to a base point called the present. It is the total discounted value of all cash inflows and outflows from a project or investment.

Net Profits Interest—A contractual arrangement under which the beneficiary, in exchange for consideration paid, receives a stated percentage of the net profits. That type of arrangement is considered a non-operating interest, as distinguished from a working interest, because it does not involve the rights and obligations of operating a mineral property (costs of exploration, development, and operation). The net profits interest does not bear any part of net losses.

Net Receipts—The difference between total movements into and total movements out of each PAD District by pipeline, tanker, and barge.

Net Salvage—In accounting, the difference between

gross salvage and cost of removal resulting from the removal, abandonment or other disposition of retired plant. Positive net salvage results when gross salvage value exceeds removal costs. Negative net salvage results when removal costs exceed gross salvage value. Positive net salvage decreases the cost to be recovered through depreciation expense and negative net salvage increases it.

Net Summer Capacity—The maximum output, commonly expressed in megawatts (MW), that generating equipment can supply to system load, as demonstrated by a multi-hour test, at the time of summer peak demand (period of May 1 through October 31). This output reflects a reduction in capacity due to electricity use for station service or auxiliaries.

Net Winter Capacity—The maximum output, commonly expressed in megawatts (MW), that generating equipment can supply to system load, as demonstrated by a multi-hour test, at the time of peak winter demand (period of November 1 through April 30). This output reflects a reduction in capacity due to electricity use for station service or auxiliaries.

Netback Purchase—Refers to a crude oil purchase agreement wherein the price paid for the crude is determined by sales prices of the types of products that are derivable from that crude as well as other considerations (e.g., transportation and processing costs). Typically, the price is calculated based on product prices extant on or near the cargo's date of importation.

NETL—National Energy Technology Laboratory www.netl.doe.gov

Netting—Netting regulations allow an entity to use emissions reductions achieved at a permitted facility to avoid some of the pre-construction review requirements that would normally apply to a proposed major modification at that same facility. Under netting provisions, emission reductions created over a specified time at a particular source can be balanced against emissions increases expected from the modification at the same source. If the net emissions increase is below the regulatory threshold level, the modification can be exempted from certain new source review requirements.

Network—A system of transmission and distribution lines cross-connected and operated to permit multiple power supply to any principal point on it. A network is usually installed in urban areas. It makes it possible to restore power quickly to customers by switching them to another circuit.

Network Customers—Customers receiving service under the terms of the Transmission Provider's Network Integration Tariff.

Network Integration Transmission Service—A service that allows the customer to integrate, plan, dispatch, and regulate its Network Resources.

Network Load—Designated load of a transmission customer.

Neutrals—Organic compounds that have a relatively neutral pH (are neither acid nor base), complex structure and, due to their carbon bases, are easily absorbed into the environment. Naphthalene, pyrene and trichlorobenzene are examples of neutrals.

Neutron—An uncharged particle found in the nucleus of every atom except that of hydrogen.

nEV—Neighborhood EV, for short trips.

New England Power Exchange (NEPEX)—This is the operating arm of the New England Power Pool.

New England Power Pool (NEPOOL)—Formed in 1971, NEPOOL is a voluntary association of entities engaged in the electric power business in New England. NEPOOL members include investor-owned utility systems, municipal and consumer owned systems, joint marketing agencies, power marketers, load aggregators, generation owners and end users.

New Field—A field discovered during the report year.

New Field Discoveries—The volumes of proved reserves of crude oil, natural gas, and/or natural gas liquids discovered in new fields during the report year.

New Gas—Gas produced from wells drilled on production leases acquired on or after February 19, 1977.

New Issue—A stock or bond offered to the public for the first time. New issues may be initial public offerings by previously private companies or additional stock or bond issues by companies already public. New public offerings are registered with the Securities and Exchange Commission. (See Securities and Exchange Commission and Registration.)

New Renewable (Green-e program)—The Green-e Program's new renewable requirement defines new renewables as renewables that are generated from solar electric, wind, biomass and geothermal facilities which have come online since 1997, and in New England since 1998.

Green-e detailed definition: An eligible new renewable generation facility must either be:

- placed in operation (generating electricity) on

or after January 1, 1997;

- repowered on or after January 1, 1997 such that at 80% of the fair market value of the project derives from new generation equipment installed as part of the repowering;

- a separable improvement to or enhancement of an operating existing facility that was first placed in operation prior to January 1, 1997, such that the proposed incremental generation is contractually available for sale and metered separately than existing generation at the facility; or

- a separately metered landfill gas resource that was not being used to generate electricity prior to January 1, 1997.

Any enhancement of fuel source that increases generation at an existing facility, without the construction of a new or repowered, separately metered generating unit, is not eligible to participate, with the exception of new landfill gas resources identified above. An eligible "new renewable generation facility" must meet the eligibility guidelines for "renewable resources" described in the Green-e Code of Conduct. Hydroelectric facilities may not contribute toward achievement of the standard for "new" renewables at this time.

New Renewable Resource (Green-e program)—Only new renewable resources are eligible to participate in Green-e certified TRC products. The term "new" is defined to include any eligible renewable facility beginning operation after January 1, 1999, or repowered after that date. There are state-based exceptions to the date by which a generator must have begun operation.

New Reservoir—A reservoir discovered during the report year.

New Source Performance Standards (NSPS)—Federal standards for very large new sources of air pollution.

New Source Review (NSR)—NSR rules are administered by the controlling federal, state, or local entity and govern the construction of new major sources and major modifications of existing major sources. NSR rules often mandate that acquisition of emission offsets in order for the project to be built or facility modified.

Newco—The typical label for any newly organized company, particularly in the context of a leveraged buyout.

Newly Conditioned Space—Any space being convert-ed from unconditioned to directly conditioned, or indirectly conditioned space. Newly conditioned space must comply with the requirements for an addition. See Section 149 for nonresidential occupancies and Section 152 for residential occupancies (in California).

Newly Constructed Building—A building that has never been used or occupied for any purpose.

Newton—A unit of force. The amount of force it takes to accelerate one kilogram at one meter per second per second.

NFCA—Noncore Fixed Cost Account: A balancing account that matches the authorized base revenue requirement and certain other costs for noncore customers with recorded revenues intended to recover that revenue requirement and the other specified costs. The balances in the NFCA have been recoverable through rates in varying degrees over time, as determined by PUC.

NFMA—See *National Forest Management Act*.

NFRAP—No Further Remedial Action Planned

NFRC—The National Fenestration Rating Council. This is a national organization of fenestration product manufacturers, glazing manufacturers, manufacturers of related materials, utilities, state energy offices, laboratories, home builders, specifiers (architects), and public interest groups.

This organization is designated by the Commission as the Supervisory Entity, which is responsible for rating the U-factors and solar heat gain coefficients of manufactured fenestration products (i.e., windows, skylights, glazed doors) that must be used in compliance calculations.

See also Fenestration Area and Fenestration Product

NFRC 100—The National Fenestration Rating Council document entitled "NFRC 100: Procedure for Determining Fenestration Product U-factors" (November 2002).

NFRC 200—The National Fenestration Rating Council document entitled "NFRC 200: Procedure for Determining Fenestration Product Solar Heat Gain Coefficients at Normal Incidence" (November 2002).

NFRC 400—The National Fenestration Rating Council document entitled °NFRC 400: Procedure for Determining Fenestration Product Air Leakage" (January 2002).

ng—Nanograms

NG—Natural gas (mainly methane)

ng/dscm—Nanograms per dry standard cubic meter

NGA—National Governors Association www.nga.org
  - National Gas Association
  - Natural Gas Act: Federal legislation enacted in 1938 that established regulatory control over companies engaged in the interstate sale and transmission of natural gas.

NGL—Natural gas liquids (ethane, propane, butane, isobutene and natural gasoline)

NGO—Non-Governmental Organization. A private, non-profit organization that is independent of business and government, that works toward some specific social, environmental, or economic goal through research, activism, training, promotion, advocacy, lobbying, community service, etc.

NGPA—National Gas Policy Act (1978) From 1938 to 1978, the Federal government regulated only the interstate natural gas market. The Natural Gas Policy Act of 1978 (NGPA) granted the Federal Energy Regulatory Commission (FERC) authority over intrastate as well as interstate natural gas production. The NGPA established price ceilings for wellhead first sales of gas that vary with the applicable gas category and gradually increase over time. Second, it established a three-stage elimination of price ceilings for certain categories: the price ceilings for certain "old" intrastate gas were eliminated in 1979, for certain "old" interstate gas and "new" gas in 1985, and for certain other "new" gas in 1987.

NGV (Natural Gas Vehicle)—Vehicles that are powered by compressed or liquefied natural gas.

NHPA—National Historic Preservation Act

NIC—Notification of Intent to Comply

Nickel—A metal used in alloys to provide corrosion and heat resistance for products in the iron, steel and aerospace industries. Nickel is used as a catalyst in the chemical industry. It is toxic and, in some forms, is listed as a cancer-causing agent under Proposition 65 in California.

Nickel Cadmium Battery (NiCad)—A form of rechargeable battery, having higher storage densities than that of lead-acid batteries, that uses a mixture of nickel hydroxide and nickel oxide for the anode, and cadmium metal for the cathode. The electrolyte is potassium hydroxide.

NICT—National Incident Coordination Team

NIEHS—National Institute of Environmental Health Sciences

Night Flushing—The process of removing hot air from a building during the cool evening hours, to cool elements with thermal mass within the building and flush stale air.

NIM—National Impact Model

NIMBY—"Not in My Back Yard" An acronym used frequently to stop a project from proceeding forward because the community or individual does not want the project to be built or expanded near their property or place of occupancy.

NIOSH—National Institute of Occupational Safety and Health www.cdc.gov/niosh

NIPDWR—National Interim Primary Drinking Water Regulations

NIST—National Institute of Standards and Technology www.nist.gov

Nitrate—Formed when ammonia is degraded by microorganisms in soil or groundwater. This compound is usually associated with fertilizers.

Nitroaromatics—Common components of explosive maters, which will explode if activated by very high temperatures or shocks.

Nitrogen Dioxide (NO_2)—A compound of nitrogen and oxygen formed by the oxidation of nitric oxide (NO) which is produced by the combustion of solid fuels. Nitrogen Dioxide is a form of air pollution that is emitted from power plants. It is a brownish gas that can damage trees, lead to acid rain, and contribute to smog.

Nitrogen Fixation—The transformation of atmospheric nitrogen into nitrogen compounds that can be used by growing plants.

Nitrogen Oxides (NOx)—Regulated air pollutants, primarily NO and NO_2. Nitrogen oxides are precursors to the formation of smog and contribute to the formation of acid rain. Gases consisting of one molecule of nitrogen and varying numbers of oxygen atoms. Nitrogen oxides are produced in the emissions of vehicle exhausts and from power stations. In the atmosphere, nitrogen oxides can contribute to the formation of photochemical ozone (smog), which is a greenhouse gas.

Nitrous Oxide (N_2O)—A colorless gas, naturally occurring in the atmosphere. N_2O is among the six greenhouse gases to be curbed under the Kyoto Protocol. N_2O is produced by natural processes, but there are also substantial emissions from human activities such as agriculture and fossil fuel combustion. The atmospheric lifetime of N_2O is approximately 100 years, and its 100-year GWP is currently estimated to be 296 times that of CO_2. A powerful greenhouse gas emitted through soil cultivation practices, especially the use of commercial and organic fertilizers, fossil fuel combustion, nitric acid production, and biomass burning. This GHG is listed in Annex A of the Kyoto Protocol.

NL—No-load losses

NLT—Not Later Than

NMFS—National Marine Fisheries Service

NMHC—Non-methane Hydrocarbons

NMTC—New Markets Tax Credit. Federal tax credits generated from the investment in businesses located in low-income communities.

NMVOC—Non-methane Volatile Organic Chemicals

NNC—Notice of Noncompliance

NNEMS—National Network of Environmental Management Studies

NNPSPP—National Non-point Source Pollution Program

NO—Nitrogen Oxide

No Shop, No Solicitation Clauses—A no shop, no solicitation, or exclusivity, clause requires the company to negotiate exclusively with the investor, and not solicit an investment proposal from anyone else for a set period of time after the term sheet is signed. The key provision is the length of time set for the exclusivity period.

No. 1 Diesel Fuel—A light distillate fuel oil that has distillation temperatures of 550 degrees Fahrenheit at the 90-percent point and meets the specifications defined in ASTM Specification D 975. It is used in high-speed diesel engines, such as those in city buses and similar vehicles. See *No. 1 Distillate* below.

No. 1 Distillate—A light petroleum distillate that can be used as either a diesel fuel (*See No. 1 Diesel Fuel* above) or a fuel oil (See *No. 1 Fuel Oil* (below).

No. 1 Fuel Oil—A light distillate fuel oil that has distillation temperatures of 400 degrees Fahrenheit at the 10-percent recovery point and 550 degrees Fahrenheit at the 90-percent point and meets the specifications defined in ASTM Specification D 396. It is used primarily as fuel for portable outdoor stoves and portable outdoor heaters.

No. 2 Diesel Fuel—A fuel that has distillation temperatures of 500 degrees Fahrenheit at the 10-percent recovery point and 640 degrees Fahrenheit at the 90-percent recovery point and meets the specifications defined in ASTM Specification D 975. It is used in high-speed diesel engines, such as those in railroad locomotives, trucks, and automobiles.

No. 2 Distillate—A petroleum distillate that can be used as either a diesel fuel (See *No. 2 Diesel Fuel* above) or a fuel oil (See No. 2 Fuel Oil below).

No. 2 Fuel Oil (heating oil)—A distillate fuel oil that has distillation temperatures of 400 degrees Fahrenheit at the 10-percent recovery point and 640 degrees Fahrenheit at the 90-percent recovery point and meets the specifications defined in ASTM Specification D 396. It is used in atomizing type burners for domestic heating or for moderate capacity commercial/industrial burner units.

No. 2 Fuel Oil and No. 2 Diesel Sold to Consumers for All Other End Uses—Those consumers who purchase fuel oil or diesel fuel for their own use including: commercial/institutional buildings (including apartment buildings), manufacturing and non-manufacturing establishments, farms (including farm houses), motor vehicles, commercial or private boats, military, governments, electric utilities, railroads, construction, logging or any other nonresidential end-use purpose.

No. 2 Fuel Oil Sold to Private Homes for Heating—Private household customers who purchase fuel oil for the specific purpose of heating their home, water heating, cooking, etc., excluding farm houses, farming and apartment buildings.

No. 4 Fuel Oil—A distillate fuel oil made by blending distillate fuel oil and residual fuel oil stocks. It conforms with ASTM Specification D 396 or Federal Specification VV-F-815C and is used extensively in industrial plants and in commercial burner installations that are not equipped with preheating facilities. It also includes No. 4 diesel fuel used for low- and medium-speed diesel engines and conforms to ASTM Specification D 975.

No. 5 and No. 6 Fuel Oil Sold Directly to the Ultimate Consumer—Includes ships, mines, smelters, manufacturing plants, electric utilities, drilling, and railroad.

No. 5 and No. 6 Fuel Oil Sold to Refiners or Other Dealers Who Will Resale the Product—Includes all volumes of No. 5 and No. 6 fuel oil purchased by a trade or business with the intent of reselling the product to the ultimate consumers.

NO^2—Nitrogen Dioxide

NOA—Notice of Availability

NOAA—National Oceanic and Atmospheric Administration. www.noaa.gov

NOAA Division—One of the 345 weather divisions designated by the National Oceanic and Atmospheric Administration (NOAA) encompassing the 48 contiguous states. These divisions usually follow county borders to encompass counties with similar weather conditions. The NOAA division does not follow county borders when weather conditions vary considerably within a county; such is likely to happen when the county borders the ocean or contains high mountains. A state contains an average of seven NOAA divisions; a NOAA division contains an average of nine counties.

NOAEL—No Observed Adverse Effect Level

NOC—Notification of Compliance

Nocturnal Cooling—The effect of cooling by the radiation of heat from a building to the night sky.

NOHSCP—National Oil and Hazardous Substances Contingency Plan

NOI—Notice of Intent
- Not Otherwise Indicated
- Not Otherwise Indexed

Noise—Unwanted electrical signals produced by electric motors and other machines that can cause circuits and appliances to malfunction.

NOL—Notice of opportunity to lease
- Net operating loss

Noload Loss—Power and energy lost by an electric system when not operating under demand.

Nominal Capacity—The approximate energy producing capacity of a power plant, under specified conditions, usually during periods of highest load.

Nominal Dollars—A measure used to express nominal price.

Nominal Interest Rate—Interest rate stated as an annual percentage without including the effect of interest during the year.

Yields are annual. However, yields are usually compounded monthly, or whenever there is a cash flow, and yield interest is applied to each period, rather than to the whole year at once. The nominal yield is an annual expression of the yield interest rate for each period.

Nominal Price—The price paid for a product or service at the time of the transaction. Nominal prices are those that have not been adjusted to remove the effect of changes in the purchasing power of the dollar; they reflect buying power in the year in which the transaction occurred.

Nominal Voltage—A rounded voltage value used to describe batteries, modules, or systems based on their specification (e.g. a 12V, 24V or 48V battery, module, or system).

Nomination—The notification to exercise a contract or part of a contract. This is used to inform the pipeline owner about the amount of gas needed to transport or store on a certain day.

Nominee—A person designated to take the place of another.

NON—Notice of Noncompliance

Non Profit—A Non-Profit institution is an entity not conducted or maintained for the purpose of making a profit, and is registered as a 501(c)3 or 501(c)6 (trade association) corporation. No part of the net earnings of such entity accrues or may lawfully accrue to the benefit of any private shareholder or individual.

Non Recourse—A type of borrowing in which the lender does not have recourse to the borrower (lessor). Rather, the lender expects to be repaid from the rents and/or the value of the equipment.

Non-Accredited Investor—An investor not considered accredited for a Regulation D offering. (See "Accredited Investor.")

Non-Annex B Parties—Countries that are not listed in Annex B of the Kyoto Protocol.

Non-Annex I Parties—Countries that have ratified or acceded to the UNFCCC that are not listed in Annex I of the UNFCCC.

Non-Appropriation Clause—Contractual provision found in municipal leases that provides that if the governmental lessee fails to appropriate or make available funds to make the lease payments called for under the agreement for the next appropriation period, the agreement terminates at the end of the current appropriation period. Such a clause is used to prevent lease payment obligations in future years from being classified as debt. Exercise of the non-appropriation clause is not an event of default.

Nonassociated Natural Gas—Natural gas that is not in contact with significant quantities of crude oil in the reservoir. See *Natural Gas* above.

Non-Attainment Area (NAA)—Any area that does not meet the national primary or secondary ambient air quality standard established by the Environmental Protection Agency for designated pollutants, such as carbon monoxide and ozone.

Non-Attainment Pollutants—See *Criteria Pollutants* If any of the criteria pollutants exceed established health-based levels in a given air basin, they are identified as "non-attainment pollutants."

Non-basic Service—Any category of service not related to basic services (generation, transmission, distribution and transition charges).

Nonbranded Product—Any refined petroleum product that is not a branded product.

Non-bypassable Wires Charge—A charge generally placed on distribution services to recover utility costs incurred as a result of restructuring (stranded costs—usually associated with generation facilities and services) and not recoverable in other ways.

Noncoincident Demand—Sum of two or more demands on individual systems that do not occur in the same demand interval.

Noncoincidental Peak Load—The sum of two or more peak loads on individual systems that do not occur in the same time interval. Meaningful only when considering loads within a limited period of time, such as a day, week, month, a heating or cooling season, and usually for not more than 1 year.

Noncommercial Species—Tree species that do not normally develop into suitable trees for conventional forest products because of small size, poor form, or inferior quality.

Noncondensing, Controlled Extraction Turbine—A turbine that bleeds part of the main steam flow at one (single extraction) or two (double extraction) points.

Nonconformity—The use of noncompliant methods or emission factors for estimating emissions.

Nonconventional Plant (Uranium)—A facility engineered and built principally for processing of uraniferous solutions that are produced during in situ leach mining, from heap leaching, or in the manufacture of other commodities, and the recovery, by chemical treatment in the plant's circuits, of uranium from the processing solutions.

Nondedicated Vehicle—A motor vehicle capable of operating on an alternative fuel and/or on either gasoline or diesel.

Non-Depletable Energy Sources—Energy that is not obtained from depletable energy sources. [See California Code of Regulations, Title 24, Section 2-5302]

Nonfirm Energy—Energy produced by the hydropower system that is available when water conditions exceed worst historic conditions and after reservoir refill is assured.

Nonfirm Power—Power or power-producing capacity supplied or available under a commitment having limited or no assured availability.

Non-Firm Transmission Service—Point-to-point service reserved and/or scheduled on an as-available basis.

Nonfuel Components—Components that are not associated with a particular fuel. These include, but are not limited to, control spiders, burnable poison rod assemblies, control rod elements, thimble plugs, fission chambers, primary and secondary neutron sources, and BWR (boiling water reactor) channels.

Nonfuel Use (of Energy)—Use of energy as feedstock or raw material input.

Nonfungible Product—A gasoline blend or blendstock that cannot be shipped via existing petroleum product distribution systems because of incompatibility problems. Gasoline/ethanol blends, for example, are contaminated by water that is typically present in petroleum product distribution systems.

Non-Generator Resources (NGRs)—Grid resources, other than electrical generation units, such as energy storage devices and demand response.

Non-Governmental Organization (NGO)—Registered non-profit organizations and associations from business and industry, environmental groups, cities and municipalities, academics, social and activist organizations, etc. Usually carry a tax status of 501-C3 or 501-C6 from the Internal Revenue Service.

Nonhydrocarbon Gases—Typical non-hydrocarbon gases that may be present in reservoir natural gas, such as carbon dioxide, helium, hydrogen sulfide, and nitrogen.

Non-jurisdictional—Utilities, ratepayers and regulators (and impacts on those parties) other than state-regulated utilities, regulators and ratepayers in a jurisdiction considering restructuring. Examples include utilities in adjacent state and non-state regulated, publicly owned utilities within restructuring states.

Nonlinearities—Occur when changes in one variable cause a more than proportionate impact on another variable.

Non-Market Benefits—Benefits of a climate policy that can be measured in terms of avoided non-market impacts such as human-health impacts (e.g., increased incidence of tropical diseases) and damages to ecosystems (e.g., loss of biodiversity).

Nonmethane Volatile Organic Compounds (NMVOC)—Organic compounds, other than methane, that participate in atmospheric photochemical reactions.

Nonoperating Interest—Any mineral lease interest (e.g., royalty, production payment, net profits interest) that does not involve the rights and obligations of operating a mineral property.

Non-Party—A state that has not ratified the UNFCCC. Non-parties may attend talks as observers.

Non-Performance—Failure to deliver product or service under a contract or agreement.

Non-Point Source Emission—Point source emissions occur in concentrated locations, such as at a factory or generating plant. By contrast, non-point source emissions are spread more evenly across a wider area. The fertilizer or pesticide runoff from grain farms is a non-point source emission.

Nonproducing Reservoir—Reservoir in which oil and/or gas proved reserves have been identified, but which did not produce during the report year to the owned or contracted interest of the reporting company regardless of the availability and/or operation of production, gathering, or transportation facilities.

Nonrenewable Fuels—Fuels that cannot be easily made or "renewed," such as oil, natural gas, and coal.

Non-Renewable Resource—A natural resource that is unable to be regenerated or renewed fully and without loss of quality once it is used, i.e., fossil fuels or minerals.

Nonrequirements Consumer—A wholesale consumer (unlike a full or partial requirements consumer) that purchases economic or coordination power to supplement their own or another system's energy needs.

Nonresidential Building—Any building which is heated or cooled in its interior, and is of an occupancy type other than Type H, I, or J, as defined in the Uniform Building Code, 1973 edition, as adopted by the International Conference of Building Officials.

Nonresidential Manual—(in California) The manual developed by the Commission, under Section 25402.1 (e) of the Public Resources Code, to aid designers, builders and contractors in meeting the energy efficiency requirements for nonresidential, high-rise residential, and hotel/motel buildings.

Non-Road Alternative Fuel Vehicle (non-road AFV)—An alternative fuel vehicle designed for off-road operation and use for surface/air transportation, industrial, or commercial purposes. Non-road AFVs include forklifts and other industrial vehicles, rail locomotives, self-propelled electric rail cars, aircraft, airport service vehicles, construction vehicles, agricultural vehicles, and marine vessels. Recreational AFV's (golf carts, snowmobiles, pleasure watercraft, etc.) are excluded from the definition.

Nonspinning Reserve—The generating capacity not currently running but capable of being connected to the bus and load within a specified time.

Non-Substitution Clause—Found in a municipal lease agreement. A clause providing that if the lessee in a lease that has a non-appropriation clause exercises the clause to terminate the agreement, the lessee,

within a specified time period after such termination, cannot purchase or use property similar in function to the property being leased.

Nonutility Generation—Electric generation by end-users, or small power producers under the Public Utility Regulatory Policies Act, to supply electric power for industrial, commercial, and military operations, or sales to electric utilities.

Nonutility Power Producer—A corporation, person, agency, authority, or other legal entity or instrumentality that owns or operates facilities for electric generation and is not an electric utility. Non-utility power producers include qualifying Co-generators, qualifying small power producers, and other non-utility generators (including independent power producers). Non-utility power producers are without a designated franchised service area and do not file forms listed in the Code of Federal Regulations, Title 18, Part 141.

NOO—No Order Observed

NOP—Notice of Preparation (of EIR)

NOPAT—Net Operating Profit After Taxes

NOPR—A Notice of Proposed Rulemaking. A designation used by the FERC for some of its dockets.

NORA—National Oil Recyclers Association www.noranews.org

No-regrets Mitigation Options—Measures whose benefits "such as improved performance or reduced emissions of local/regional pollutants, but excluding the benefits of climate change mitigation" equal or exceed their costs. They are sometimes known as "measures worth doing anyway."

Normal Operating Cell Temperature (NOCT)—The estimated temperature of a solar PV module when it is operating under 800 W/m2 irradiance, 20°C ambient temperature and a wind speed of 1 meter per second. NOCT is used to estimate the nominal operating temperature of a module in the field.

Normal Recovery Capacity—A characteristic applied to domestic water heaters that is the amount of gallons raised 100 degrees Fahrenheit per hour (or minute) under a specified thermal efficiency.

Normalized Savings—The reductions in energy use that occurred during the performance period relative to what would have been used during the baseline period, but adjusted to a normal set of conditions (such as typical weather conditions).

North American Electric Reliability Council (NERC)—A council formed in 1968 by the electric utility industry to promote the reliability and adequacy of bulk power supply in the electric utility systems of North America. NERC consists of regional reliability councils and encompasses essentially all the power regions of the contiguous United States, Canada, and Mexico. www.nerc.com

North American Industrial Classification System (NAICS)—A new classification scheme, developed by the Office of Management and Budget to replace the Standard Industrial Classification (SIC) System, that categorizes establishments according to the types of production processes they primarily use.

Northeast Power Coordinating Council (NPCC)—One of the ten regional reliability councils that make up the North American Electric Reliability Council (NERC).

NOS—Not Otherwise Specified

Nosecone—The pointed piece farthest toward the wind on a wind generator, designed primarily for cosmetic purposes, but also protects the blade attachment points and generator from the weather.

Notary Public—An official appointed by the Secretary of State to administer oaths, to authenticate Contracts, to acknowledge Deeds, etc. A California Notary may act as such in any part of the State.

Notice of Cessation—A notice that work has ceased; a notice recorded within the time periods which is specified by statute, limiting the time for filing Mechanic's Liens on an incomplete project.

Notice of Completion—A notice recorded within the time period specified by statute, after completion of a work improvement, signaling commencement of the time period within which claims for Mechanic's Liens must be recorded.

Notice of Default—Recorded notice that a Default has occurred under the terms of a contract.

Notice of Nonresponsibility—Notice recorded by an Owner of Real Property that he will not be responsible for payment of Mechanic's Liens for work contracted by another.

Notice of Proposed Rulemaking (NOPR)—A designation used by the Federal Energy Regulatory Commission for some of its dockets.

NOV—Notice of Violation (CAA, CWA, FIFRA)

NOWC—Net Operating Working Capital

NO_X—Nitrogen oxides or oxides of nitrogen that are a chief component of air pollution produced by fossil fuel burning.

NO_X Affected Source—In the OTC NO_X Budget program, a fossil fuel fired, indirect heat exchange combustion unit[s] with a maximum rated heat input capacity of 250 MMBtu/hour or more, and all fossil fuel fired electric generating facilities rated at 15 megawatts or greater, or any other source that voluntarily opts to become a NO_X affected source.

NO_X Allocation—In the OTC NO_X Budget program, assignment of NO_X Allowances to a NO_X affected source and recorded by the NO_X budget administrator to a NO_X Allowance Tracking System account. In the RECLAIM program, NO_X sources have NO_X RTC allocations issued by the SCAQMD (South Coast Air Quality Management District located in Southern California).

NO_X Allowance—An emissions right issued by the governing state participating in the Ozone Transport Commission NO_X Budget program that gives authorization to emit one ton of NO_X during a specified year pursuant to the rules of the State's NO_X Budget program. These Allowances may sometimes be banked into later years.

NO_X Allowance Tracking System—In the OTC NO_X Budget program, the NATS is the computerized system used to track the number of OTC NO_X allowances held and used by any person.

NO_X Allowance Transfer—In the OTC NO_X Budget program, the conveyance to another NATS account of one or more NO_X Allowances from one person to another by whatever means, including, but not limited to, purchase, trade, auction or gift.

NO_X Allowance Transfer Deadline—In the OTC NO_X Budget program, the deadline by which NO_X Allowances may be submitted for recording in a NO_X affected source's compliance account for purposes of meeting NO_X Allowance requirements.

NO_X Budget—In the OTC NO_X Budget program, the total tons of NO_X emissions that may be released from NO_X affected sources.

NO_X Emissions—The sum of nitric oxides and nitrogen dioxides emitted, calculated as nitrogen dioxide.

NO_X Emissions Tracking System—In the OTC Budget program, the NETS is the computerized system used to track NO_X emissions from NO_X affected sources.

NPAA—Noise Pollution and Abatement Act (1970) Requires the FAA to establish noise standards through consultation with the EPA, and to apply them in connection with issuance of civil aircraft certificates.

NPDES—National Pollutant Discharge Elimination System

NPDWR—National Primary Drinking Water Regulation

NPDWS—National Primary Drinking Water Standards

NPL—National Priorities List

NPM—National Program Manager

NPMA—National Property Management Association www.npma.org

NPPC—National Pollution Prevention Center

NPRM—Notice of Proposed Rulemaking

NPS—Non-point Source

NPTOR—Net-Plant Turnover Ratio: Gross operating revenue divided by net-plant. Net plant is total utility plant less accumulated depreciation reserve.

NPV—*Net Present Value*: Form of calculating discounted cash flow. It encompasses the process of calculating the discount of a series of amounts of cash at future dates, and summing them.

NRC—Nuclear Regulatory Commission: An independent agency established by the Energy Reorganization Act of 1974 to regulate civilian use of nuclear materials. The NRC is headed by a five-member Commission. www.nrc.gov

NRD—National Resource Damage

NRDC—National Resources Defense Council: An environmental action organization. www.nrdc.org

NRECA—National Rural Electric Cooperative Association www.nreca.org

NREL—The National Renewable Energy Laboratory. A national lab that concentrates on studying and developing renewable energy sources. www.nrel.gov

NRSE—New & Renewable Source of Energy

NRTA—Northwest Regional Transmission Association. A sub-regional transmission group within the Western Regional Transmission Association.

NRTL—Nationally Recognized Testing Laboratory

NSDWR—National Secondary Drinking Water Regulations

NSEP—National System for Emergency Preparedness

NSHP—New Solar Homes Partnership (in California) www.newsolarhomes.ca.gov

NSPE—National Society for Professional Engineers www.nspe.org

NSPS—New Source Performance Standards

NSR—New Source Review

NSR/PSD—New Source Review and Prevention of Significant Deterioration Permitting

NSRL—No Significant Risk Level

NSWMA—National Solid Waste Management Association www.nswma.org

NTE—Not to Exceed

NTIS—National Technical Information Service

NTU—See *Nephelometric Turbidity Unit*.

N-Type Semiconductor—A semiconductor produced by doping an intrinsic semiconductor with an electron-donor impurity (e.g., phosphorous in silicon).

N-Type Silicon—Silicon doped with an element that has more electrons in its atomic structure than does silicon (e.g. phosphorus).

Nuclear Electric Power (Nuclear Power)—Electricity generated by the use of the thermal energy released from the fission of nuclear fuel in a reactor.

Nuclear Energy—Power obtained by splitting heavy atoms (fission) or joining light atoms (fusion). A nuclear energy plant uses a controlled atomic chain reaction to produce heat. The heat is used to make steam run conventional turbine generators.

Nuclear Fuel—Fissionable materials that have been enriched to such a composition that, when placed in a nuclear reactor, will support a self-sustaining fission chain reaction, producing heat in a controlled manner for process use.

Nuclear Fuel Operations—All nuclear fuel operations, excluding reactor and reactor component manufacturing or containment construction. Includes exploration and development; mining; milling; conversion; enrichment; fabrication; reprocessing; and spent fuel storage.

Nuclear Power Plant—A facility in which heat produced in a reactor by the fissioning of nuclear fuel is used to drive a steam turbine.

Nuclear Reactor—An apparatus in which a nuclear fission chain reaction can be initiated, controlled, and sustained at a specific rate. A reactor includes fuel (fissionable material), moderating material to control the rate of fission, a heavy-walled pressure vessel to house reactor components, shielding to protect personnel, a system to conduct heat away from the reactor, and instrumentation for monitoring and controlling the reactor's systems.

Nuclear Regulatory Commission (NRC)—An independent federal agency that ensures that strict standards of public health and safety, environmental quality and national security are adhered to by individuals and organizations possessing and using radioactive materials. The NRC is the agency that is mandated with licensing and regulating nuclear power plants in the United States. It was formally established in 1975 after its predecessor, the Atomic Energy Commission, was abolished.

NUG—A non-utility generator. A generation facility owned and operated by an entity who is not defined as a utility in that jurisdictional area.

Null Hypothesis—The assumption that any observed difference between two samples of a statistical population is purely accidental and not due to systematic causes.

Number of Mines—The number of mines, or mines co-located with preparation plants or tipples, located in a particular geographic area (State or region). If a mine is mining coal across two counties within a State, or across two States, then it is counted as two operations. This is done so that EIA can separate production by State and county.

Number of Mining Operations—The number of mining operations includes preparation plants with greater than 5,000 total direct labor hours. Mining operations that consist of a mine and preparation plant, or a preparation plant only, will be counted as two operations if the preparation plant processes both underground and surface coal. Excluded are silt, culm, refuse bank, slurry dam, and dredge operations except for Pennsylvania anthracite. Excludes mines producing less than 10,000 short tons of coal during the year.

NWP—Nationwide Permit Program

NWPA—Nuclear Waste Policy Act (1982). In 1982, the U.S. Congress enacted a law called the Nuclear Waste Policy Act. The Act established a comprehensive national program for the safe, permanent disposal of highly radioactive wastes. This law is based on the principle that our society is responsible for safely disposing of the nuclear wastes we create.

NWPP—Northwest Power Pool

NWPPA—Northwest Public Power Association

NWS—National Weather Service

NYMEX—A division of the New York Mercantile Exchange that trades crude oil, heating oil, gasoline, natural gas, electricity futures and options, and propane futures.

NYSE—The New York Stock Exchange. Founded in 1792, the largest organized securities market in the United States. The Exchange itself does not buy, sell, own, or set prices of stocks traded there. The prices are determined by public supply and demand. Also known as the Big Board.

O

O&M—Operations and Maintenance. Expenditures necessary to keep an energy facility operating efficiently.

O₃—Ozone

OAPEC—Acronym for Organization of Arab Petroleum Exporting Countries founded in 1968 for cooperation in economic and petroleum affairs. See *OPEC*.

OAQPS—Office of Air Quality Planning and Standards

OAR—Office of Air and Radiation

Obligation to Serve—The obligation of a utility to provide electric service to any customer who seeks that service, and is willing to pay the rates set for that service. Traditionally, utilities have assumed the obligation to serve in return for an exclusive monopoly franchise.

Obligor—One who places himself under a legal obligation.

OC—Office of the Comptroller

OCB—Oil Circuit Breaker

Occupancy Sensor—A control device that senses the presence of a person in a given space. It is commonly used to control lighting systems in buildings to reduce the amount of energy consumption.

Occupancy Type—Is one of the following (in California Building Codes):

Auditorium is the part of a public building where an audience sits in fixed seating, or a room, area, or building with fixed seats used for public meetings or gatherings not specifically for the viewing of dramatic performances.

Auto repair is the portion of a building used to repair automotive equipment and/or vehicles, exchange parts, and may include work using an open flame or welding equipment.

Civic meeting space is a city council or board of supervisors meeting chamber, courtroom, or other official meeting space accessible to the public.

Classroom, lecture, or training is a room or area where an audience or class receives instruction.

Commercial and industrial storage is a room, area, or building used for storing items.

Convention, conference, multipurpose and meeting centers are assembly rooms, areas, or buildings used for meetings, conventions and multiple purposes, including but not limited to, dramatic performances, and that has neither fixed seating nor fixed staging.

Corridor is a passageway or route into which compartments or rooms open.

Dining is a room or rooms in a restaurant or hotel/motel (other than guest rooms) where meals that are served to the customers will be consumed.

Dormitory is a building consisting of multiple sleeping quarters and having interior common areas such as dining rooms, reading rooms, exercise rooms, toilet rooms, study rooms, hallways, lobbies, corridors, and stairwells, other than high-rise residential, low-rise residential, and hotel/motel occupancies.

Electrical/mechanical room is a room in which the building's electrical switchbox or control panels, and/or HVAC controls or equipment is located.

Exercise center/gymnasium is a room or building equipped for gymnastics, exercise equipment, or indoor athletic activities.

Exhibit is a room or area that is used for exhibitions that has neither fixed seating nor fixed staging.

Financial transaction is a public establishment used for conducting financial transactions including the custody, loan, exchange, or issue of money, for the extension of credit, and for facilitating the transmission of funds.

General commercial and industrial work is a room, area, or building in which an art, craft, assembly or manufacturing operation is performed.

   High bay: Luminaires 25 feet or more above the floor.

   Low bay: Luminaires less than 25 feet above the floor.

Grocery sales is a room, area, or building that has as its primary purpose the sale of foodstuffs requiring additional preparation prior to consumption.

Kitchen/food preparation is a room or area with cooking facilities and/or an area where food is prepared.

Laundry is a place where laundering activities occur.

Library is a repository for literary materials, such as books, periodicals, newspapers, pamphlets and prints, kept for reading or reference.

Lobby, Hotel is the contiguous space in a hotel/motel between the main entrance and the front desk, including reception, waiting and seating areas.

Lobby, Main entry is the contiguous space in buildings other than hotel/motel that is directly located by the main entrance of the building through which persons must pass, including reception, waiting and seating areas.

Locker/dressing room is a room or area for changing clothing, sometimes equipped with lockers.

Lounge/recreation is a room used for leisure activities which may be associated with a restaurant or bar.

Mall is a roofed or covered common pedestrian area within a mall building that serves as access for two or more tenants.

Medical and clinical care is a room, area, or building that does not provide overnight patient care and that is used to promote the condition of being sound in body or mind through medical, dental, or psychological examination and treatment, including, but not limited to, laboratories and treatment facilities.

Museum is a space in which works of artistic, historical, or scientific value are cared for and exhibited.

Office is a room, area, or building of CBC Group B Occupancy other than restaurants.

Parking garage is a covered building or structure for the purpose of parking vehicles, which consists of at least a roof over the parking area, often with walls on one or more sides. Parking garages may have fences or rails in place of one or more walls. The structure has an entrance(s) and exit(s), and includes areas for vehicle maneuvering to reach the parking spaces.—If the roof of a parking structure is also used for parking, the section without an overhead roof is considered a parking lot instead of a parking garage.

Precision commercial or industrial work is a room, area, or building in which an art, craft, assembly or a manufacturing operation is performed involving visual tasks of small size or fine detail such as electronic assembly, fine woodworking, metal lathe operation, fine hand painting and finishing, egg processing operations, or tasks of similar visual difficulty.

Religious worship is a room, area, or building for worship. Restaurant is a room, area, or building that is a food establishment as defined in Section 27520 of the Health and Safety Code (California).

Restroom is a room or suite of rooms providing personal facilities such as toilets and washbasins.

Retail merchandise sales is a room, area, or building in which the primary activity is the sale of merchandise.

School is a building or group of buildings that is predominately classrooms and that is used by an organization that provides instruction to students.

Senior housing is housing other than Occupancy Group I that is specifically for habitation by seniors, including but not limited to independent living quarters, and assisted living quarters. Commons areas may include dining, reading, study, library or other community spaces and/or medical treatment or hospice facilities.

Stairs, active/inactive, is a series of steps providing passage from one level of a building to another.

Support area is a room or area used as a passageway, utility room, storage space, or other type of space associated with or secondary to the function of an occupancy that is listed in these regulations.

Tenant lease space is a portion of a building intended for lease for which a specific tenant is not identified at the time of permit application.

Theater, motion picture, is an assembly room, a hall, or a building with tiers of rising seats or steps for the showing of motion pictures.

Theater, performance, is an assembly room, a hall, or a building with tiers of rising seats or steps for the viewing of dramatic performances, lectures, musical events and similar live performances.

Transportation function is the ticketing area, waiting area, baggage handling areas, concourse, or other areas not covered by primary functions in Table 146-C in an airport terminal, bus or rail terminal or station, subway or transit station, or marine terminal.

Vocational room is a room used to provide training in a special skill to be pursued as a trade.

Waiting area is an area other than a hotel lobby or main entry lobby normally provided with seating and used for people waiting.

Wholesale showroom is a room where samples of merchandise are displayed.

Occupant Sensor, Lighting—A device that automatically turns lights off soon after an area is vacated. The term Occupant Sensor applies to a device that controls interior lighting systems, but can be used interchangeably with occupancy sensor, occupant sensing device, and motion sensor.

Occupational Safety and Health Act (OSHA)—A federal law, Public law 91-596, enacted in 1970, comprising federal standards for safety and health for people at work. The regulations issued under this Act can be found in Title 29, Part 1910, and Part 1926 of the Code of Federal Regulations.

Occupied Space—The space within a building or structure that is normally occupied by people, and that may be conditioned (heated, cooled and/or ventilated).

OCE—Office of Criminal Enforcement

Ocean Energy Systems—Energy conversion technologies that harness the energy in tides, waves, and thermal gradients in the oceans.

Ocean Thermal Energy Conversion (OTEC)—The process or technologies for producing energy by harnessing the temperature differences (thermal gradients) between ocean surface waters and that of ocean depths. Warm surface water is pumped through an evaporator containing a working fluid in a closed Rankine-cycle system. The vaporized fluid drives a turbine/generator. Cold water from deep below the surface is used to condense the working fluid. Open-Cycle OTEC technologies use ocean water itself as the working fluid. Closed-Cycle OTEC systems circulate a working fluid in a closed loop. A working 10 kilowatt, closed-cycle prototype was developed by the Pacific International Center for High Technology Research in Hawaii with U.S. Department of Energy funding, but was not commercialized.

Ocean Thermal Gradient (OTG)—Temperature differences between deep and surface water. Deep water is likely to be 25 to 45 degrees Fahrenheit colder. The term also refers to experimental technology that could use the temperature differences as a means to produce energy.

OCESL—Office of Criminal Enforcement and Special Litigation

OCI—Office of Criminal Investigation
• Organizational Conflicts of Interest

OCIR—Office of Community and Intergovernmental Relations

OCM—Office of Compliance Monitoring

Octane—A rating scale used to grade gasoline as to its antiknock properties. Also any of several isometric liquid paraffin hydrocarbons, C_8H_{18}. Normal octane is a colorless liquid found in petroleum boiling at 124.6 degrees Celsius.

Octane Enhancer—Any substance that is added to gasoline to increase octane.

Octane Rating—A number used to indicate gasoline's antiknock performance in motor vehicle engines. The two recognized laboratory engine test methods for determining the antiknock rating, i.e., octane rating, of gasolines are the Research method and the Motor method. To provide a single number as guidance to the consumer, the antiknock index $(R + M)/2$, which is the average of the Research and Motor octane numbers, was developed.

OD—Office of Director
- Operations Division
- Organizational Development
- Outside Diameter

ODC—Ozone-depleting Chemical

Odorant—Any material added to natural or LP gas in small concentrations to impart a distinctive odor. Odorants in common use include various mercaptans, organic sulfides, and blends of these.

ODP—Ozone-depleting Potential

ODS—Ozone-depleting Substance

OE—Office of Enforcement

OEA—Office of External Affairs

OECA—Office of Enforcement and Compliance Assurance

OECD—Refers to the Organization for Economic Co-operation and Development. It includes Australia, Austria, Belgium, Canada, the Czech Republic, Denmark, Finland, France, Germany, Greece, Hungary, Iceland, Ireland, Italy, Korea, Japan, Luxemborg, Mexico, the Netherlands, New Zealand, Norway, Poland, Portugal, Spain, Sweden, Switzerland, Turkey, the United Kingdom, and the United States. www.oecd.org

OECM—Office of Enforcement and Compliance Monitoring

OEE—Office of Energy Efficiency

OEHHA—Office of Environmental Health Hazard Assessment

OEL—Occupational Exposure Limit

OEM—Original equipment manufacturer

OEP—Office of Enforcement Policy
- Office of Environmental Policy
- Office of External Programs

OEPC—Operating Expense Per Customer: Total operating and maintenance expense divided by total number of customers.

OEPR—Office of Environmental Processes and Effects Research

OERR—Office of Emergency and Remedial Response

OES—Office of Emergency Services: Coordinates overall state agency response to major disasters in support of local government. The office is responsible for assuring the state's readiness to respond to and recover from natural, manmade, and war-caused emergencies, and for assisting local governments in their emergency preparedness, response and recovery efforts.

Off Balance Sheet Financing—Any form of financing, such as an operating lease, that, for financial re-

porting purposes, is not required to be reported on a firm's balance sheet.

Off Peak—Period of relatively low system demand. These periods often occur in daily, weekly, and seasonal patterns; these off-peak periods differ for each individual electric utility.

Off Peak Gas—Gas that is to be delivered and taken on demand when demand is not at its peak.

Offer—Price at which the owner of an emission reduction, credit, or allowance is willing to sell (a.k.a. Ask)

Offering Documents—Documents evidencing a private-placement transaction. Include some combination of a purchase agreement and/or subscription agreement, notes or stock certificates, warrants, registration-rights agreement, stockholder or investment agreement, investor questionnaire, and other documents required by the particular deal.

Off-Highway Use—Includes petroleum products sales for use in:
- Construction. Construction equipment including earthmoving equipment, cranes, stationary generators, air compressors, etc.
- Other. Sales for off-highway uses other than construction. Sales for logging are included in this category. Volumes for off-highway use by the agriculture industry are reported under "Farm Use" (which includes sales for use in tractors, irrigation pumps, other agricultural machinery, etc.)

Off-Hours Equipment Reduction—A conservation feature where there is a change in the temperature setting or reduction in the use of heating, cooling, domestic hot water heating, lighting or any other equipment either manually or automatically.

Official Development Assistance (ODA)—Official Development Assistance is funding provided governments of developed countries to developing countries to assist in various community, health and commercial projects.

Off-Peak—The period of low energy demand, as opposed to maximum or peak demand.

Off-Peak Gas—Gas that is to be delivered and taken on demand when demand is not at its peak.

Off-Peak Service—Service made available on special schedules or contracts but only for a specified part of the year during the off-peak season.

Off-Peak Service—Service made available on special schedules or contracts on a firm basis but only for a specified time during the portion of the load cycle of lease use.

Off-Road—Any non-stationary device, powered by an internal combustion engine or motor, used primarily off the highways to propel, move, or draw persons or property, and used in any of the following applications: marine vessels, construction/farm equipment, locomotives, utility and lawn and garden equipment, off-road motorcycles, and off-highway vehicles.

Offset—Offsets are a form of credit-based emissions trading. Offsets are created when a source makes voluntary, permanent emissions reductions that are in surplus to any required reductions. Existing sources that create offsets can trade them to new sources to cover growth or relocation. Regulators approve each trade. Regulators normally require a portion of the offsets to be retired to ensure an overall reduction in emissions. Offsets are an open system.

One Offset is an emissions reduction that a pollution source has achieved in excess of permitted levels and or required reductions. The excess amount is the credit and can be sold on the market. See *Emission Offset*.

Offset Rate Increase—A rate increase to pass on to customers the cost of increases that are beyond a utility's control, e.g., to cover increased Ad Valorem Taxes (property taxes collected by a county on the valuation of a utility's assets).

Offset Ratio—The amount of pollutant that must be secured relative to the on-site emission increase. Often, new sources must offset their emissions at a greater than 1:1 ratio, especially if the offsetting emission reductions are derived from an off-site source.

Offshore—That geographic area that lies seaward of the coastline. In general, the coastline is the line of ordinary low water along with that portion of the coast that is in direct contact with the open sea or the line marking the seaward limit of inland water. If a State agency uses a different basis for classifying onshore and offshore areas, the State classification should be used. (Cook Inlet in Alaska is classified as offshore.)

Offshore Reserves and Production—Unless otherwise dedicated, reserves and production that are in either state or Federal domains, located seaward of the coastline.

Offsite-Produced Energy for Heat, Power, and Electricity Generation—This measure of energy consumption, which is equivalent to purchased energy includes energy produced off-site and consumed onsite. It excludes energy produced and consumed onsite, energy used as raw material input, and electricity losses.

Off-System—Any point not on, or directly interconnected with, a transportation, storage, and/or distribution system operated by a natural gas company within a state.

Off-System (natural gas)—Natural gas that is transported to the end user by the company making final delivery of the gas to the end user. The end user purchases the gas from another company, such as a producer or marketer, not from the delivering company (typically a local distribution company or a pipeline company).

OGJ—The Oil & Gas Journal

OHEA—Office of Health and Environmental Assessment
- Office of Health Effects Assessment

Ohm—A measure of the electrical resistance of a material equal to the resistance of a circuit in which the potential difference of 1 volt produces a current of 1 ampere.

Ohm's Law—In a given electrical circuit, the amount of current in amperes is equal to the pressure in volts divided by the resistance, in ohms. The principle is named after the German scientist George Simon Ohm. Commonly stated as $E = I \times R$, or Voltage = Amperage x Resistance.

OHRV—Off Highway Recreational Vehicle

OI—Office of Investigations

OIG—Office of Inspector General

Oil—A mixture of hydrocarbons usually existing in the liquid state in natural underground pools or reservoirs. Gas is often found in association with oil.

Oil Company Use—Includes sales to drilling companies, pipelines or other related oil companies not engaged in the selling of petroleum products. Includes fuel oil that was purchased or produced and used by company facilities for the operation of drilling equipment, other field or refinery operations, and space heating at petroleum refineries, pipeline companies, and oil-drilling companies. Oil used to bunker vessels is counted under vessel bunkering. Sales to other oil companies for field use are included, but sales for use as refinery charging stocks are excluded.

Oil Reservoir—An underground pool of liquid consisting of hydrocarbons, sulfur, oxygen, and nitrogen trapped within a geological formation and protected from evaporation by the overlying mineral strata.

Oil Shale—A type of rock containing organic matter that produces large amounts of oil when heated to high temperatures.

Oil Stocks—Oil stocks include crude oil (including strategic reserves), unfinished oils, natural gas plant liquids, and refined petroleum products.

Oil Well—A well completed for the production of crude oil from at least one oil zone or reservoir.

Oil Well (Casinghead) Gas—Associated and dissolved gas produced along with crude oil from oil completions.

Oil-Gas Parity Pricing—Conversion of costs per gallon oil price to an equivalent gas price in dollars per Mcf by application of appropriate oil/gas heat (Btu) conversion factors.

Old Field—A field discovered prior to the report year.

Old Growth—Timber stands with the following characteristics: large mature and over-mature trees in the overstory, snags, dead and decaying logs on the ground and a multi-layered canopy with trees of several age classes.

Old Reservoir—A reservoir discovered prior to the report year.

Oligopoly—A few sellers who exert market control over prices.

OMB—Office of Management and Budget

OMMSQA—Office of Modeling, Monitoring Systems and Quality Assurance

OMPC—Operating Margin Per Customer: Operating revenue less operating expense divided by total number of customers.

On Peak—Periods of relatively high system demand. These periods often occur in daily, weekly, and seasonal patterns; these on-peak periods differ for each individual electric utility.

On Peak Energy—Energy supplied during periods of relatively high system demands as specified by the supplier.

One Sun—Natural solar insulation falling on an object without concentration or diffusion of the solar rays.

One-Axis Tracking—A PV System structure that is capable of rotating on a single axis in order to track the movement of the sun.

One-Time Fee—The fee assessed a nuclear utility for spent nuclear fuel (SNF) or solidified high-level radioactive waste derived from SNF, which fuel was used to generate electricity in a civilian nuclear power reactor prior to April 7, 1983, and which is assessed by applying industry-wide average dollar-per-kilogram charges to four distinct ranges of fuel burnup so that equivalent to an industry-wide

average charge of 1.0 mill per kilowatt hour.

On-Highway Use (Diesel)—Includes sales for use in motor vehicles. Volumes used by companies in the marketing and distribution of petroleum products are also included.

On-Site Generation—Generation of energy at the location where all or most of it will be used.

Onsite Transportation—The direct non-process end use that includes energy used in vehicles and transportation equipment that primarily consume energy within the boundaries of the establishment. Energy used in vehicles that are found primarily offsite, such as delivery trucks, is not measured by the MECS (an EIA survey).

On-System—Any point on or directly interconnected with a transportation, storage, or distribution system operated by a natural gas company.

On-System (Natural Gas)—Natural gas that is sold (and transported) to the end user by the company making final delivery of the gas to the end user. Companies that make final delivery of natural gas are typically local distribution companies or pipeline companies.

On-System Sales—Sales to customers where the delivery point is a point on, or directly interconnected with, a transportation, storage, and/or distribution system operated by the reporting company.

OOIP—Original Oil In Place

OPA—Oil Pollution Act (1990) The act streamlined and strengthened EPA's ability to prevent and respond to catastrophic oil spills. A trust fund financed by a tax on oil is available to clean up spills when the responsible party is incapable or unwilling to do so. The OPA requires oil storage facilities and vessels to submit to the Federal government plans detailing how they will respond to large discharges.

OPAC—Overall Performance Appraisal Certification

Opacity—The degree to which smoke or particles emitted into the air reduce the transmission of light and obscure the view of an object in the background.

OPEC (Organization of Petroleum Exporting Countries)—The acronym for the Organization of Petroleum Exporting Countries that have organized for the purpose of negotiating with oil companies on matters of oil production, prices, and future concession rights. Current members (as of the date of writing this definition) are Algeria, Indonesia, Iran, Iraq, Kuwait, Libya, Nigeria, Qatar, Saudi Arabia, the United Arab Emirates, and Venezuela. www.opec.org

Open Access—A regulatory mandate to allow others to

use a utility's transmission and distribution facilities to move bulk power from one point to another on a nondiscriminatory basis for a cost-based fee.

Open Circuit—When an electrical circuit is interrupted by breaking the path at one or more points, stopping the electrons from flowing. A light switch opens an electrical circuit when it turns off the light.

Open Circuit Voltage (V_{oc})—The maximum possible voltage across a PV array, module, or cell. The voltage across the terminals of a photovoltaic cell, module, or array with no load applied when the cell is exposed to standard insolation conditions, measured with a voltmeter.

Open Loop System—A fresh water or "direct" solar hot water system, generally for use in freeze-free climates.

Open Market Coal—Coal is sold in the open market, i.e., coal sold to companies other than the reporting company's parent company or an operating subsidiary of the parent company.

Open Refrigeration Unit—Refrigeration in cabinets (units) without covers or with flexible covers made of plastic or some other material, hung in strips or curtains (fringed material, usually plastic, that push aside like a bead curtain). Flexible covers stop the flow of warm air into the refrigerated space.

Open-Circuit Voltage—The maximum voltage produced by an illuminated solar PV cell, module, or array when no load is connected. OCV increases as the temperature of the PV material decreases.

Open-end Lease—A lease in which the lessee guarantees the amount of the future residual value to be realized by the lessor at the end of the lease. If the equipment is sold for less than the guaranteed value, the lessee must pay the amount of any deficiency to the lessor. This lease is referred to as open-end because the lessee does not know its actual cost until the equipment is sold at the end of the lease term.

Open-Loop Biomass—Solid, nonhazardous, cellulosic waste material—lignin material or agricultural livestock waste nutrients as defined in IRC Section 45(s)(3).

Open-Loop Geothermal Heat Pump System—Open-loop (also known as "direct") systems circulate water drawn from a ground or surface water source. Once the heat has been transferred into or out of the water, the water is returned to a well or surface discharge (instead of being re-circulated through the system). This option is practical where there is an adequate supply of relatively clean water, and all local codes and regulations regarding groundwater discharge are met.

Open-Loop Recycling—A recycling process in which materials from old products are made into new products in a manner that changes the inherent properties of the materials, often via a degradation in quality, such as recycling white writing paper into cardboard rather than more premium writing paper. Often used for steel, paper, and plastic. Open-loop recycling is also known as down-cycling or reprocessing.

Operable Capacity—The amount of capacity that, at the beginning of the period, is in operation; not in operation and not under active repair, but capable of being placed in operation within 30 days; or not in operation but under active repair that can be completed within 90 days. Operable capacity is the sum of the operating and idle capacity and is measured in barrels per calendar day or barrels per stream day.

Operable Generators/units—Electric generators or generating units that are available to provide power to the grid or generating units that have been providing power to the grid but are temporarily shut down. This includes units in standby status, units out of service for an indefinite period, and new units that have their construction complete and are ready to provide test generation. A nuclear unit is operable once it receives its Full Power Operating License.

Operable Nuclear Unit (Foreign)—A nuclear generating unit outside the United States that generates electricity for a grid.

Operable Nuclear Unit (U.S.)—A U.S. nuclear generating unit that has completed low-power testing and is in possession of a full-power operating license issued by the Nuclear Regulatory Commission.

Operable Refineries—Refineries that were in one of the following three categories at the beginning of a given year: in operation; not in operation and not under active repair, but capable of being placed into operation within 30 days; or not in operation, but under active repair that could be completed within 90 days.

Operable Shading Device—A device at the interior or exterior of a building or integral with a fenestration product, which is capable of being operated, either manually or automatically, to adjust the amount of solar radiation admitted to the interior of the building.

Operable Unit—A unit available to provide electric power to the grid. See definition for operating unit below.

Operable Utilization Rate—Represents the use of the atmospheric crude oil distillation units. The rate is calculated by dividing the gross input to these units by the operable refining capacity of the units.

Operated—Exercised management responsibility for the day-to-day operations of natural gas production, gathering, treating, processing, transportation, storage, and/or distribution facilities and/or a synthetic natural gas plant.

Operating Budget—A budget that lists the amount of noncapital goods and services a firm is authorized by management to expend during the operating period.

Operating Capacity—The component of operable capacity that is in operation at the beginning of the period.

Operating Costs—Recurring costs related to day-to-day operations of a facility that are paid out of current revenue.

Operating Cycle—The processes that a work input/output system undergoes and in which the initial and final states are identical.

Operating Day—A normal business day. Days when a company conducts business due to emergencies or other unexpected events are not included.

Operating Expenses—Segment expenses related both to revenue from sales to unaffiliated customers and revenue from inter-segment sales or transfers, excluding loss on disposition of property, plant, and equipment; interest expenses and financial charges; foreign currency translation effects; minority interest; and income taxes.

Operating Income—Operating revenues less operating expenses. Excludes items of other revenue and expense, such as equity in earnings of unconsolidated affiliates, dividends, interest income and expense, income taxes, extraordinary items, and cumulative effect of accounting changes.

Operating Lease—From a financial reporting perspective, a lease that has the characteristics of a usage agreement and also meets certain criteria established by the FASB. Such a lease is not required to be shown on the balance sheet of the lessee (see *Off Balance Sheet Financing*). The term is also used to refer to leases in which the lessor has taken a significant residual position in the lease pricing and, therefore, must salvage the equipment for a certain value at the end of the lease term in order to earn

its expected rate of return.

Operating Point—Defined by the current and voltage that a module or array produces when connected to a load. It is dependent on the load or the batteries connected to the output terminals.

Operating Ratio—The ratio of operating expenses to pertinent revenues.

Operating Revenues—Segment revenues both from sales to unaffiliated customers (i.e., revenue from customers outside the enterprise as reported in the company's consolidated income statement) and from intersegment sales or transfers, if any, of product and services similar to those sold to unaffiliated customers, excluding equity in earnings of unconsolidated affiliates; dividend and interest income; gain on disposition of property, plant, and equipment; and foreign currency translation effects.

Operating Subsidiary—Company that operates a coal mining operation and is owned by another company (i.e., the parent company).

Operating Unit—A unit that is in operation at the beginning of the reporting period.

Operating Utilization Rate—Represents the use of the atmospheric crude oil distillation units. The rate is calculated by dividing the gross input to these units by the operating refining capacity of the units.

Operation and Maintenance Expense (O&M)—Costs that relate to the normal operating, maintenance and administrative activities of a business.

Operational Control—The authority to introduce and implement operating, environmental, health and safety policies; or, whole ownership of the facility.

Operational Energy Storage Considerations—A description of how an energy storage project is used; i.e., on a defined basis, what application is it being employed for; what resource solution is it providing, who is deciding, etc.

Operator—The company or organization having operational control of the facility that is the subject of the report.

Operator, Gas Plant—The person responsible for the management and day-to-day operation of one or more natural gas processing plants as of December 31 of the report year. The operator is generally a working-interest owner or a company under contract to the working-interest owner(s). Plants shut down during the report year are also to be considered "operated" as of December 31.

Operator, Oil and/or Gas Well—The person responsible for the management and day-to-day operation of one or more crude oil and/or natural gas wells as of December 31 of the report year. The operator is generally a working-interest owner or a company under contract to the working-interest owner(s). Wells included are those that have proved reserves of crude oil, natural gas, and/or lease condensate in the reservoirs associated with them, whether or not they are producing. Wells abandoned during the report year are also to be considered "operated" as of December 31.

Opportunity (Energy)—An energy asset upgrade investment for consideration by a Facility manager or owner. Opportunity records can include costs, expected impacts on meters (utility savings) and physical descriptions. The central purpose of an Investment Grade Energy Audit is to identify and determine both the technical and financial feasibility of an Opportunity.

Opportunity Cost—A method to determine the cost of common equity component of return using the cost of capital of other investments of similar risk.

OPPTS—Office of Prevention, Pesticides and Toxic Substances

Optimize—A process of solving for the mathematically best (optimal) solution to a transaction.

Optimized Debt—A method of setting up the debt structure to minimize the lease rental rate to the lessee and to maximize returns to the equity participant.

Option—Contract that gives the buyer the right, but not the obligation, to buy or sell the underlying commodity at an established price on or before the agreed date.

A security granting the holder the right to purchase a specified number of a Company's securities at a designated price at some point in the future. The term is generally used in connection with employee benefit plans as Incentive Stock Options ("ISOs" or "statutory options") and Non-qualified stock options ("NSOs" or "Nonquals"). However "stand-alone options" may be issued outside of any plan. Generally non-transferable, in distinction to warrants.

Option Pool—The number of shares set aside for future issuance to employees of a private company.

Optional Delivery Commitment—A provision to allow the conditional purchase or sale of a specific quantity of material in addition to the firm quantity in the contract.

Optionality—A value derived from certain characteristics of a resource that may provide flexibility in terms of scale, function, location, time of deployment, and risk-reduction.

Options—An option is a contractual agreement that gives the holder the right to buy (call option) or sell (put option) a fixed quantity of a security or commodity (for example, a commodity or commodity futures contract), at a fixed price, within a specified period of time. May either be standardized, exchange-traded, and government regulated, or over-the-counter customized and non-regulated.

Options are contracts that give the option buyer the right but not the obligation to enter into a specific transaction purchase (a Call) or sale (Put) up to a certain date. The price (Strike Price), quantity and terms of delivery are locked in at the trade date. The expiration or exercise date (Strike Dates) is also locked in at that time, that is the date after which the option buyer's rights to enter into the transaction terminate. The option seller must live by the decision of the buyer, and is paid a premium for selling the optionality or flexibility to the buyer.

Option buyers may be either the buyers or seller of the underlying commodity. If they wish to buy the commodity they purchase a Call option i.e. that is the right but not the obligation to purchase the commodity at the specified terms. If they wish to guarantee the sale of the commodity they purchase a put option i.e. right but not the obligation to enter into a sale of the commodity at specified terms.

ORA—Office of Ratepayer Advocates

Order—A ruling issued by a utility commission granting or denying an application in whole or in part. The order explains the basis for the decision, noting any dispute with the factual assertions of the applicant. Also applied to a final regulation of a utility commission.

Organic—Derived from living organisms. A term signifying the absence of pesticides, hormones, synthetic fertilizers and other toxic materials in the cultivation of agricultural products. "Organic" is also a food labeling term that denotes the product was produced under the authority of the Organic Foods Production Act.

Organic Compounds—Chemical compounds based on carbon chains or rings and also containing hydrogen, with or without oxygen, nitrogen, and other elements.

Organic Content—The share of a substance that is of animal or plant origin.

Organic Level—The amount of organic matter prescribed to be left after logging.

Organic Waste—Waste material of animal or plant origin.

Organization for Economic Cooperation and Development (OECD)—An international organization helping governments tackle the economic, social and governance challenges of a globalized economy. Its membership comprises about 30 member countries. With active relationships with some 70 other countries, NGOs and civil society, it has a global reach. www.oecd.org

Organization of Petroleum Exporting Countries (OPEC)—Countries that have organized for the purpose of negotiating with oil companies on matters of oil production, prices, and future concession rights. Current members (as of the date of writing this definition) are Algeria, Indonesia, Iran, Iraq, Kuwait, Libya, Nigeria, Qatar, Saudi Arabia, the United Arab Emirates, and Venezuela. www.opec.org

Organochlorides—A group of organic (carbon-containing) insecticides that also contain chlorine. These chemicals tend not to break down easily in the environment. DDT, Toxaphene and Endosulfan are all organochlorides.

Orientation—The alignment of a building along a given axis to face a specific geographical direction. The alignment of a solar collector, in number of degrees east or west of true south. Position with respect to the cardinal directions, N, S, E, W.

A term used to describe the direction that the surface of a solar module faces. The two components of orientation are the tilt angle (the angle of inclination a module makes from the horizontal) and the azimuth (the compass angle that the module faces, with north equal to 0 degrees and south equal to 180 degrees).

Original Cost—The initial amount of money spent to acquire an asset. It is equal to the price paid, or present value of the liability incurred, or fair value of stock issued, plus normal incidental costs necessary to put the asset into its initial use.

Original Equipment Manufacturer (OEM)—Refers to the manufacturers of complete vehicles or heavy-duty engines, as contrasted with remanufacturers, converters, retrofitters, up-fitters, and re-powering or rebuilding contractors who are overhauling engines, adapting or converting vehicles or engines obtained from the OEMs, or exchanging or rebuilding engines in existing vehicles.

Original Equipment Manufacturer Vehicle—A vehicle produced and marketed by an original equipment manufacturer (OEM), including gasoline and diesel vehicles as well as alternative-fuel vehicles. A vehicle manufactured by an OEM but converted to an alternative-fuel vehicle before its initial delivery to an end-user (for example, through a contract between a conversion company and the OEM) is considered to be an OEM vehicle as long as that vehicle is still covered under the OEM's warranty.

Original Issue Discount—Some maturities of a new bond issue that have an offering price substantially below par; the appreciation from the original price to par over the life of the bonds is treated as tax-exempt income and is not subject to capital gains tax.

Ornamental Chandeliers—Ceiling-mounted, close-to-ceiling, or suspended decorative luminaires that use glass, crystal, ornamental metals, or other decorative material and that typically are used in hotel/motels, restaurants, or churches as a significant element in the interior architecture.

ORNL—Oak Ridge National Laboratory

Orphan Site—An orphan site is a contaminated site where the owners are either bankrupt or cannot be located. It is a large problem in the US and Canada.

ORSAT Apparatus—A device for measuring the combustion components of Boiler or furnace flue gasses.

OSB—Oversight Board

OSHA—Occupational Safety and Health Administration

OTAG—Ozone Transport Assessment Group

OTC—Ozone Transport Commission
  • Over-the-Counter. A market for securities made up of dealers who may or may not be members of a formal securities exchange. The over-the-counter market is conducted over the telephone and is a negotiated market rather than an auction market such as the NYSE.

OTC NO$_X$ Budget Program—The cap and trade program administered by the OTC MOU signing states.

Other—The "other" category is defined as representing electricity consumers not elsewhere classified. This category includes public street and highway lighting service, public authority service to public authorities, railroad and railway service, and inter-departmental services.

Other Capital Costs—Costs for items or activities not included elsewhere under capital-cost tabulations, such as for and decommissioning, dismantling, and reclamation.

Other Demand-side Management (DSM) Assistance Programs—A DSM program assistance that includes alternative-rate, fuel-switching, and any other DSM assistance programs that are offered to consumers to encourage their participation in DSM programs.

Other End Users—For motor gasoline, all direct sales to end users other than those made through company outlets. For No. 2 distillate, all direct sales to end users other than residential, commercial/institutional, industrial sales, and sales through company outlets. Included in the "other end users" category are sales to utilities and agricultural users.

Other Energy Operations—Energy operations not included in Petroleum or Coal. Other Energy includes nuclear, oil shale, tar sands, coal liquefaction and gasification, geothermal, solar, and other forms of non-conventional energy.

Other Finished—Motor gasoline not included in the oxygenated or reformulated gasoline categories.

Other Gas—Includes manufactured gas, coke-oven gas, blast-furnace gas, and refinery gas. Manufactured gas is obtained by distillation of coal, by the thermal decomposition of oil, or by the reaction of steam passing through a bed of heated coal or coke.

Other Generation—Electricity originating from these sources: biomass, fuel cells, geothermal heat, solar power, waste, wind, and wood.

Other Hydrocarbons—Materials received by a refinery and consumed as a raw material. Includes hydrogen, coal tar derivatives, gilsonite, and natural gas received by the refinery for reforming into hydrogen. Natural gas to be used as fuel is excluded.

Other Industrial Plant—Industrial users, not including coke plants, engaged in the mechanical or chemical transformation of materials or substances into new products (manufacturing); and companies engaged in the agriculture, mining, or construction industries.

Other Load Management—Demand-Side Management (DSM) program other than Direct Load Control and Interruptible Load that limits or shifts peak load from on-peak to off-peak time periods. It includes technologies that primarily shift all or part of a load from one time-of-day to another and secondarily may have an impact on energy consumption. Examples include space heating and water heating storage systems, cool storage systems, and load limiting devices in energy management systems. This category also includes programs that aggressively promote time-of-use rates and other innovative rates such as real time pricing. These rates are intended to reduce consumer bills and shift hours of operation of equipment from on-peak to off-peak periods through the application of time-differentiated rates.

Other Oils Equal to or Greater Than 401 Degrees Fahrenheit—Oils with a boiling range equal to or greater than 401 degrees Fahrenheit that are intended for use as a petrochemical feedstock.

Other Operating Costs—Costs for other items or activities not included elsewhere in operating-cost tabulations, but required to support the calculation of a cutoff grade for ore reserves estimation.

Other Oxygenates—Other aliphatic alcohols and aliphatic ethers intended for motor gasoline blending (e.g., isopropyl ether (IPE) or n-propanol).

Other Power Producers—Independent power producers that generate electricity and cogeneration plants that are not included in the other industrial, coke and commercial sectors.

Other Refiners—Refiners with a total refinery capacity in the United States and its possessions of less than 275,000 barrels per day as of January 1, 1982.

Other Service to Public Authorities—Electricity supplied to municipalities, divisions or agencies of state or Federal governments, under special contracts or agreements or service classifications applicable only to public authorities.

Other Single-Unit Truck—A motor vehicle consisting primarily of a single motorized device with more than two axles or more than four tires.

Other Supply Contracts—Any contracted gas supply other than owned reserves, producer-contracted reserves, and interstate pipeline purchases that are used for acts and services for which the company has received certificate authorization from FERC. Purchases from intrastate pipelines pursuant to Section 311(b) of the NGPA of 1978 are included with other supply contracts.

Other Trucks/Vans—Those trucks and vans that weigh more than 8,500 lbs GVW.

Other Unavailable Capability—Net capability of main generating units that are unavailable for load for reasons other than full-forced outage or scheduled maintenance. Legal restrictions or other causes make these units unavailable.

OTR—Ozone Transport Region

Outage—The period during which a generating unit, transmission line, or other facility is out of service.

Outage (Electric Utility)—An interruption of electric service that is temporary (minutes or hours) and affects a relatively small area (buildings or city blocks). See *Blackout*.

Outdoor Air—Air taken from outdoors and not previously circulated in the building.

Outdoor Lighting—Definitions include the following (in California):

Building entrance is any operable doorway in or out of a building, including overhead doors.

Building façade is the exterior surfaces of a building, not including horizontal roofing, signs, and surfaces not visible from any reasonable viewing location.

Canopy is a permanent structure consisting of a roof and supporting building elements, with the area beneath at least partially open to the elements. A canopy may be freestanding or attached to surrounding structures. A canopy roof may serve as the floor of a structure above.

Hardscape is an improvement to a site that is paved and has other structural features, including but not limited to, curbs, plazas, entries, parking lots, site roadways, driveways, walkways, sidewalks, bikeways, water features and pools, storage or service yards, loading docks, amphitheaters, outdoor sales lots, and private monuments and statuary.

Landscape lighting is lighting that is recessed into the ground or paving; mounted on the ground; mounted less than 42″ above grade; or mounted onto trees or trellises, and that is intended to be aimed only at landscape features.

Lantern is an ornamental outdoor luminaire that uses an electric lamp to replicate a pre-electric lantern, which used a flame to generate light.

Lighting zone is a geographic area designated by the California Energy Commission that determines requirements for outdoor lighting, including lighting power densities and specific control, equipment or performance requirements. Lighting zones are numbered LZ1, LZ2, LZ3, and LZ4.

Marquee lighting is a permanent lighting system consisting of one or more rows of many small lights attached to a canopy.

Ornamental lighting is post-top luminaires, lanterns, pendant luminaires, chandeliers, and marquee lighting.

Outdoor lighting is all electrical lighting for parking lots, signs, building entrances, outdoor sales areas, outdoor canopies, landscape lighting, lighting for building facades and hardscape lighting.

Outdoor sales frontage is the portion of the perimeter of an outdoor sales area immediately adjacent to a street, road, or public sidewalk.

Outdoor sales lot is an uncovered paved area used exclusively for the display of vehicles, equipment or other merchandise for sale. All internal and adjacent access drives, walkway areas, employee and customer parking areas, vehicle service or storage areas are not outdoor sales lot areas, but are considered hardscape.

Parking lot is an uncovered area for the purpose of parking vehicles. Parking lot is a type of hardscape.

Paved area is an area that is paved with concrete, asphalt, stone, brick, gravel, or other improved wearing surface, including the curb.

Pendant is a mounting method in which the luminaire is suspended from above.

Post Top Luminaire is an ornamental outdoor luminaire that is mounted directly on top of a lamppost.

Principal viewing location is anywhere along the adjacent highway, street, road or, sidewalk running parallel to an outdoor sales frontage

Public monuments are statuary, buildings, structures, and/or hardscape on public land.

Sales canopy is a canopy specifically to cover and protect an outdoor sales area.

Vehicle service station is a gasoline or diesel dispensing station.

Outdoor Reset—A temperature control strategy commonly used in water heating systems, in which the supply water temperature is automatically adjusted up or down according to the outside air temperature. On a cold day, supply water temperature will be high, but on a mild day when less heat is required the supply water temperature will be reduced. Outdoor Reset conserves energy by reducing overheating in the occupied space, by reducing heat losses at the Boilers and by reducing heat losses in the distribution piping. When implementing Outdoor Reset, system designers must remain aware of Boiler operating requirements to avoid damaging the Boilers.

Outer Continental Shelf (OCS)—The submerged lands extending from the out limit of the historic territorial sea (typically three miles) to some undefined outer limit, usually a depth of 600 feet. In the United States, this is the portion of the shelf under federal jurisdiction. See *Continental Shelf.*

Outgassing—The process by which materials expel or release gasses. They may result from finishes, binders, or adhesives used to create industrial building maters. Some of these chemicals have been shown to cause cancers. Chemically sensitive individuals can have a negative reaction from just being exposed to them.

Output—The amount of power or energy produced by a generating unit, station, or system.

Outside Air—Air taken from outdoors and not previously circulated through the HVAC system.

Outside Coil—The heat-transfer (exchanger) component of a heat pump, located outdoors, from which heat is collected in the heating mode, or expelled in the cooling mode.

Outstanding Stock—The amount of common shares of a corporation which are in the hands of investors. It is equal to the amount of issued shares less treasury stock.

Oven—An appliance that is an enclosed compartment supplied with heat and used for cooking food. Toaster ovens are not considered ovens. The range stove top or burners and the oven are considered two separate appliances, although they are often purchased as one appliance.

Oven Dry—See *Bone Dry.*

Oven Dry Ton (ODT)—An amount of wood that weighs 2,000 pounds at zero percent moisture content.

Over Collateralization—A technique to protect the investor on a securitized transaction. Basically, the lessor can set aside additional funds in a reserve or transfer more than sufficient assets to the investor vehicle. By doing so, the lessor has effectively provided a cushion that can generate cash flow in the event the lessee receivables are not sufficient to provide enough cash to satisfy investors' requirements.

Over Generation—A condition that occurs when total PX participant demand is less than or equal to the sum of regulatory must-take generation, regulatory must-run generation, and reliability must-run generation.

Overall Heat Gain—The total heat gain through all portions of the building envelope calculated as specified in Section 143 (b) 2 for determining compliance with the Overall Envelope Approach (in California).

Overall Heat Loss—(in California) The total heat loss through all portions of the building envelope calculated as specified in Section 143 (b) 1 for determining compliance with the Overall Envelope Approach.

Overburden—Any material, consolidated or unconsolidated, that overlies a coal deposit.

Overburden Ratio—Overburden ratio refers to the amount of overburden that must be removed to excavate a given quantity of coal. It is commonly expressed in cubic yards per ton of coal, but is sometimes expressed as a ratio comparing the thickness of the overburden with the thickness of the coal bed.

Overcharge—Applying current to a fully charged battery. This can damage the battery.

Overcurrent—Current that exceeds the rated current of the equipment or the ampacity of a conductor, resulting from overload, short circuit, or ground fault.

Overcurrent Device—A safety fuse or breaker designed to open a circuit when an overcurrent occurs.

Overhang—A building element that shades windows, walls, and doors from direct solar radiation and protects these elements from precipitation.

Overload—To exceed the design capacity of a device.

Overnight Capital Cost—The capital cost of a project if it could be constructed overnight. This cost does not include the interest cost of funds used during construction.

Overpacking—Process used for isolating waste by jacketing or encapsulating waste-holding containers to prevent further spread or leakage of contaminating materials. Leaking drums may be contained within oversized ones as an interim measure prior to removal and final disposal.

Overriding Royalty—A royalty interest, in addition to the basic royalty, created out of the working interest; it is, therefore, limited in its duration to the life of the lease under which it is created.

Overstory—The portion of the trees forming the upper or uppermost canopy in a forest stand.

Oversubscription—Occurs when demand for shares exceeds the supply or number of shares offered for sale. As a result, the underwriters or investment bankers must allocate the shares among investors. In private placements, this occurs when a deal is in great demand because of the company's growth prospects.

Oversubscription Privilege—In a rights issue, arrangement by which shareholders are given the right to apply for any shares that are not purchased.

Ovonic—A device that converts heat or sunlight directly to electricity, invented by Standford Ovshinsky, that has a unique glass composition that changes from an electrically non-conducting state to a semi-conducting state.

OWEC—Office of Water Regulations and Standards

Owned Reserves—Any reserve of natural gas that the reporting company owns as a result of oil and gas leases, fee-mineral ownership, royalty reservations, or lease or royalty reservations and assignments committed to services under certificate authorizations by FERC. Company-owned recoverable natural gas in underground storage is classified as owned reserves.

Owned/Rented—(As used in EIA's consumption surveys.) The relationship of a housing unit's occupants to the structure itself, not the land on which the structure is located. "Owned" means the owner or co-owner is a member of the household and the housing unit is either fully paid for or mortgaged. A household is classified "rented" even if the rent is paid by someone not living in the unit. "Rent-free" means the unit is not owned or being bought and no money is paid or contracted for rent. Such units are usually provided in exchange for services rendered or as an allowance or favor from a relative or friend not living in the unit. Unless shown separately, rent-free households are grouped with rented households.

Owner Occupied—(As used in EIA's consumption surveys.) Having the owner or the owner's business represented at the site. A building is considered owner occupied if an employee or representative of the owner (such as a building engineer or building manager) maintains office space in the building. Similarly, a chain store is considered owner occupied even though the actual owner may not be in the building but headquartered elsewhere. Other examples of the owner's business occupying a building include State-owned university buildings, elementary and secondary schools owned by a public school district, and a post office where the building is owned by the U.S. Postal Service.

Owner Trustee—The party that holds the title to the equipment under a leveraged lease.

Owners Equity—Interest of the owners in the assets of the business represented by capital contributions and retained earnings.

Ownership—(See *Owned/Rented* above.) The right to the use and enjoyment of property, or an interest therein generally to the exclusion of others.

Ownership of Building—(As used in EIA's consumption surveys.) The individual, agency, or organization that owns the building. For certain EIA consumption surveys, building ownership is grouped into the following categories: Federal, State, or local government agency; a privately owned utility company; a church, synagogue, or other religious group; or any other type of individual or group.

Oxides of Nitrogen—See NO_x.

Oxidize—To chemically transform a substance by combining it with oxygen.

Oxidizer—A group of chemicals that are very reactive, often but not always supplying oxygen to a reaction. Some oxidation reactions can release large amount of heat and gases, and, under the right conditions, cause an explosion. Others can cause rapid corrosion of metal, damage to tissue, burns and other serious effects. Examples of oxidizers include chlorine gas, nitric acid, sodium perchlorate, and ammonium nitrate.

Oxygen (O_2)—A gas which forms about 21%, by volume, of the atmosphere. It is chemically very active and is necessary for combustion. The combination of oxygen with other substances generally produces heat.

Oxygenate—A term used in the petroleum industry to denote octane components containing hydrogen, carbon and oxygen in their molecular structure. Includes ethers such as MTBE and ETBE and alcohols such as ethanol or methanol. The oxygenate is a prime ingredient in reformulated gasoline. The increased oxygen content given by oxygenates promotes more complete combustion, thereby reducing tailpipe emissions.

Oxygenated Gasoline—Finished motor gasoline, other than reformulated gasoline, having an oxygen content of 2.7 percent or higher by weight and required by the U.S. Environmental Protection Agency (EPA) to be sold in areas designated by EPA as carbon monoxide (CO) non-attainment areas. See *Non-attainment Area*. Note: Oxygenated gasoline excludes oxygenated fuels program reformulated gasoline (OPRG) and reformulated gasoline blendstock for oxygenate blending (RBOB). Data on gasohol that has at least 2.7 percent oxygen, by weight, and is intended for sale inside CO non-attainment areas are included in data on oxygenated gasoline. Other data on gasohol are included in data on conventional gasoline.

Oxygenates—Any substance which, when added to gasoline, increases the amount of oxygen in that gasoline blend. Through a series of waivers and interpretive rules, the Environment Protection Agency (EPA) has determined the allowable 'limits' for oxygenates in unleaded gasoline.

OY—Operating Year

OYG—Operating Year Guidance

Ozone—A molecule made up of three atoms of oxygen. Occurs naturally in the stratosphere and provides a protective layer shielding the Earth from harmful ultraviolet radiation. In the troposphere, it is a chemical oxidant, a greenhouse gas, and a major component of photochemical smog.

Ozone Depleting Substance (ODS)—Ozone depleting substances covered by the Montreal Protocol, including chlorofluorocarbons, hydrochlorofluorocarbons, carbon tetrachloride, methyl chloroform, and brominated gases.

Ozone Precursors—Chemical compounds, such as carbon monoxide, methane, non-methane hydrocarbons, and nitrogen oxides, which in the presence of solar radiation react with other chemical compounds to form ozone.

Ozone Transport Commission Memorandum of Understanding (OTC MOU)—The memorandum of understanding (MOU) signed by representatives of ten states and the District of Columbia as members of the Ozone Transport Commission (OTC) on September 27, 1994.

P

P & P—Preparedness & Prevention

P/N—A semiconductor (photovoltaic) device structure in which the junction is formed between a p-type layer and an n-type layer.
   • A semiconductor device structure that layers an intrinsic semiconductor between a p-type semiconductor and an n-type semiconductor; this structure is most often used with amorphous silicon devices.

P2—Pollution Prevention

PA—Permit Authority
   • Policy Analyst
   • Preliminary Assessment
   • Property Administrator

PA/SI—Preliminary Assessment/Site Inspection

PAC—Powdered Activated Carbon
   • Public Advisory Committee

PACE—Property-assessed Clean Energy-A type of financing using government bonds that are used to finance energy efficiency and energy generation projects. The source of repayment is a payment that is added to the real estate tax bill on the real property on which the capital assets are located.

Packaged Air Conditioning Units—Usually mounted on the roof or on a slab beside the building. (These are known as self-contained units, or Direct Expansion (DX). They contain air conditioning equipment as well as fans, and may or may not include heating equipment.) These are self-contained units that contain the equipment that generates cool air and the equipment that distributes the cooled air. These units commonly consume natural gas or electricity. The units are mounted on the rooftop, exposed to the elements. They typically blow cool air into the building through duct work, but other types of distribution systems may exist. The units usually serve more than one room. There are often several units on the roof of a single building. Also known as: Packaged Terminal Air Conditioners (PTAC). These packaged units are often constructed as a single unit for heating and for cooling.

Packaged Units—Units built and assembled at a factory and installed as a self-contained unit to heat or cool all or portions of a building. Packaged units are in contrast to engineer-specified units built up from individual components for use in a given building. Packaged Units can apply to heating equipment, cooling equipment, or combined heating and cooling equipment. Some types of electric packaged units are also called "Direct Expansion" or DX units.

Packing Factor—The ratio of solar collector array area to actual land area.

PADD (Petroleum Administration for Defense Districts)—The United States is divided by the U.S. Department of Energy into five PADD regions for planning purposes. The states within PADD V are Alaska, Arizona, California, Hawaii, Nevada, Oregon and Washington, which are linked closely by their oil supply network. Since very little petroleum product is export outside the district, PADD V is essentially a self-contained oil supply system with Alaska and California the main producers and California refining the majority of the crude oil consumed in the PADD.

PAGM—Permit Applicants Guidance Manual

PAI—Performance Audit Inspection
   • Pure Active Ingredient compound

Paid or Incurred—Paid or Accrued—The terms are construed according to the method of accounting upon the basis of which the taxable income is computed under subtitle A of the IRC.

Paid-in Capital—That portion of shareholder's equity which has been paid-in directly as opposed to earned profits retained in the business.

PAIR—Preliminary Assessment Information Rule

Pallet—A small wooden platform on which cargo is stored for ease of loading and unloading. Cargo shipped on pallets is referred to as palletized cargo.

Pane (Window)—The area of glass that fits in the window frame.

Panel (Solar)—A term generally applied to individual solar collectors, and typically to solar photovoltaic collectors or modules.

Panel Radiator—A mainly flat surface for transmitting radiant energy.

Panemone—A drag-type wind machine that can react to wind from any direction.

Parabolic Aluminized Reflector Lamp—A type of lamp having a lens of heavy durable glass that focuses the light. They have longer lifetimes with less lumen depreciation than standard incandescent lamps.

Parabolic Dish—A solar energy conversion device that has a bowl shaped dish covered with a highly reflective surface that tracks the sun and concentrates sunlight on a fixed absorber, thereby achieving high temperatures, for process heating or to operate a heat (Stirling) engine to produce power or electricity.

Parabolic Trough—A solar energy conversion device that uses a trough covered with a highly reflective surface to focus sunlight onto a linear absorber containing a working fluid that can be used for medium temperature space or process heat or to operate a steam turbine for power or electricity generation.

Paraffin (Oil)—A light-colored, wax-free oil obtained by pressing paraffin distillate.

Paraffin (Wax)—The wax removed from paraffin distillates by chilling and pressing. When separating from solutions, it is a colorless, more or less translucent, crystalline mass, without odor and taste, slightly greasy to touch, and consisting of a mixture of solid hydrocarbons in which the paraffin series predominates.

Paraffinic Hydrocarbons—Straight-chain hydrocarbon compounds with the general formula C_nH_{2n+2}.

Parallel—A configuration of an electrical circuit in which the voltage is the same across the terminals. The positive reference direction for each resistor current is down through the resistor with the same voltage across each resistor.

Parallel Connected—A method of connection in which positive terminals are connected together and negative terminals are connected together. Current output adds and voltage remains the same.

Parallel Connection—An electrical circuit with more than one possible path for electron flow. When wiring PV modules, this wiring configuration increases amperage (current), while voltage remains the same. Parallel wiring is positive to positive (+ to +) and negative to negative (- to -). Opposite of a series connection.

Parallel Operation—The simultaneous operation of a self-generator with power delivered or received by the electrical utility while interconnected to the grid. Parallel Operation includes only those PV systems that are interconnected with the Electrical Utility distribution system for more than 60 cycles.

Parallel Path Flow—As defined by NERC, this refers to the flow of electric power on an electric system's transmission facilities resulting from scheduled electric power transfers between two other electric systems. (Electric power flows on all interconnected parallel paths in amounts inversely proportional to each path's resistance.)

Parallel Power—Power generated by a system operating side by side with the power grid. This power gives you significant savings and enhances power reliability.

Parapet—Low wall to protect any place where there is a drop, as at the edge of a roof, balcony, terrace, etc.

Parent—A firm that directly or indirectly controls another entity.

Parent Company—An affiliated company that exercises ultimate control over a business entity, either directly or indirectly, through one or more intermediaries.

Pari Passu—At an equal rate or pace, without preference.

Parol—Oral; verbal.

Part 6—Is Title 24, Part 6 of the California Code of Regulations. See *Building Energy Efficiency Standards*

Partial Cut—A harvest method in which portions of a stand of timber are cut during a number of entries over time. Compare Pre-commercial Thinning.

Partial Load—An electrical demand that uses only part of the electrical power available. [See California Code of Regulations, Title 24, Section 2-5342(e) 2]

Partial Requirements Consumer—A wholesale consumer with generating resources insufficient to carry all its load and whose energy seller is a long-term firm power source supplemental to the consumer's own generation or energy received from others. The terms and conditions of sale are similar to those for a full requirements consumer.

Participating Preferred—A preferred stock in which the holder is entitled to the stated dividend and also to additional dividends on a specified basis upon payment of dividends to the common stockholders.

Participating Preferred Stock—Preferred stock that has the right to share on a pro-rata basis with any distributions to the common stock upon liquidation, after already receiving the preferred-liquidation preference.

Participating Source—In the Illinois ERMS program, a Participating Source is a source operating prior to May 1, 1999, located in the Chicago non-attainment area, that is required to obtain a Clean Air and Permit Program permit and has baseline emissions of at least 10 tons, as specified in the regulations, or seasonal emissions of at least 10 tons in any seasonal allotment period beginning in 1999.

Participating Supplier (Green-e program)—A retail supplier selling a Green-e certified product who has agreed to abide by the Code of Conduct and who sells products conforming to the environmental and consumer protection standards set by the Green Power Board.

Participation Agreement—Same as financing agreement. An agreement that states the obligations of all parties under a leveraged lease transaction.

Particulate—A small, discrete mass of solid or liquid matter that remains individually dispersed in gas or liquid emissions. Particulates take the form of aerosol, dust, fume, mist, smoke, or spray. Each of these forms has different properties. Particles below 10 microns (10 one-millionths of a meter, 0.0004 inch) (also known as PM10) in diameter are considered potential health risks because, when inhaled, they are taken deep into the lungs. Regulations require that an incinerator emit no more than 180 milligrams of total particulates per day standard cubic meter per minute.

Particulate Emissions—Fine liquid or solid particles discharged with exhaust gases. Usually measured as grains per cubic foot or pounds per million Btu input.

Particulate Matter (PM)—This type of air pollution includes soot, dust, dirt and aerosols. It causes apparent effects on visible and exposed surfaces, which can create or intensify breathing and heart problems and result to cancer and premature death.

Partner and Partnership—The term "partnership" includes a syndicate, group, pool, joint venture, or other unincorporated organization, through or by means of which any business, financial operation, or venture is carried on, and which is not, within the meaning of this title, a trust or estate or a corporation, and the term "partner" includes a member in such a syndicate, group, pool, joint venture or organization.

Partnership—An association of two or more persons as Co-Owners to carry on a business for profit.

Partnership Flip—A term used to describe an ownership structure often used to facilitate investment in renewable energy projects.

Passenger-miles Traveled—The total distance traveled by all passengers. It is calculated as the product of the occupancy rate in vehicles and the vehicle miles traveled.

Passivation—A chemical reaction that eliminates the detrimental effect of electrically reactive atoms on a photovoltaic cell's surface.

Passive Activity Rule—Prohibits the use of deductions and credits from passive activities, those in which the taxpayer is not involved on a regular and substantial basis, to offset income and taxes owed from non-passive activities.

Passive Solar (Building) Design—A building design that uses structural elements of a building to heat and cool a building, without the use of mechanical equipment. This requires careful consideration of the local climate and solar energy resource, building orientation, and landscape features, to name a few. The principal elements include proper building orientation, proper window sizing and placement and design of window overhangs to reduce summer heat gain and ensure winter heat gain, and proper sizing of thermal energy storage mass (for example a Trombe wall or masonry tiles). The heat is distributed primarily by natural convection and radiation, though fans can also be used to circulate room air or ensure proper ventilation.

Passive Solar Energy—Use of the sun to help meet a building's energy needs by means of architectural design (such as arrangement of windows) and materials (such as floors that store heat, or other thermal mass). It uses the natural heat transfer processes (radiation, conduction, and convection) to collect, distribute and store useable heat or coolness without the help of mechanical devices.

Passive Solar Heater—A solar water or space-heating system in which solar energy is collected, and/or moved by natural convection without using pumps or fans. Passive systems are typically integral collector/storage (ICS; or batch collectors) or thermosyphon systems. The major advantage of these systems is that they do not use controls, pumps, sensors, or other mechanical parts, so little or no maintenance is required over the lifetime of the system.

Passive Solar Heating—A solar heating system that uses no external mechanical power, such as pumps or blowers, to move the collected solar heat.

Passive Solar Home—A house built using passive solar design techniques.

Passive Solar System—A solar heating or cooling system that uses no external mechanical power to move the collected solar heat.

Passive/Natural Cooling—To allow or augment the natural movement of cooler air from exterior, shaded areas of a building through or around a building.

Pass-Through Securitization—A structure that represents the sale of lease receivables to a special purpose vehicle (SPV). Generally, the receivables are sold to a grantor trust which issues securities backed by the assets of the SPV. Distributions of funds from the SPV are made to the investors who provided the original capital.

PAT—Permit Assistance Team

Patent—A conveyance of the title to government lands by the government.

Payables to Municipality—The amounts payable by the utility department to the municipality or its other departments that are subject to current settlement.

Payback—A method of calculating how long it will take to recover the difference in cost between two different energy systems by using the energy and maintenance-cost savings from the more efficient system.

Payback Period—The amount of time required before the savings resulting from your system equal the system cost.

Payment Method for Utilities—The method by which fuel suppliers or utility companies are paid for all electricity, natural gas, fuel oil, kerosene, or liquefied petroleum gas used by a household. Households that pay the utility company directly are classified as "all paid by household." Households that pay directly for at least one but not all of their fuels used and that has at least one fuel charge included in the rent were classified as "some paid,

some included in rent." Households for which all fuels used are included in rent were classified as "all included in rent." If the household did not fall into one of these categories, it was classified as "other." Examples of households falling into the "other" category are: (1) households for which fuel bills were paid by a social service agency or a relative, and (2) households that paid for some of their fuels used but paid for other fuels through another arrangement.

Payout Ratio—The ratio of cash dividends on common stock to earnings available for common stock.

PAYT—Pay-As-You-Throw

Pay-through Securitization—A structure that allows the lessor to transfer lease receivables to a special purpose vehicle (SPV) in the form of a loan. The SPV then issues bonds to the investors and the receivables are pledged as collateral.

pb—Lead

PBA—Preliminary Benefit Analysis

PBI—Performance-Based Incentives. A rebate method for installing solar in California that is based upon the actual output of a solar system. The amount is paid monthly over a five year period.

PBOP—Post-Retirement Benefits Other Than Pensions: A term used in calculating revenue and expenditures for utilities.

PBP—Payback Period

PBR—Performance-Based Ratemaking: Under performance-based ratemaking, non-fuel costs are indexed to changes in inflation and customer growth, minus a productivity target, such that base rates to customers would be expected to decline in real terms.

PC—Pulverized Coal

PCBs (Polychloronated Biphenyls)—A group of organic compounds used in the manufacture of plastics and formerly used as a coolant in electric transformers. In the environment, PCBs are highly toxic to aquatic life. They persist in the environment for long periods of time and are biologically accumulative.

PCE—Perchlorethylene
  • Pollution Control Equipment

PCI—Per Capita Income

PCMD—Procurement and Contracts Management Division

PCS—Permit Compliance System
  • Program Coordination Staff

PDD—Project Design Document. The official application drawn up by an entity applying for project approval under the Clean Development Mechanism (CDM). PDDs must be validated by an independent third party, then approved and registered by the CDM Executive Board before a project qualifies as a CER carbon credit earner.

PE—Performance Evaluation
- Professional Engineer
- Program Element
- Private Equity

PEA—Preliminary Endangerment Assessment
- Preliminary Exposure Analysis

Peak—Periods of relatively high system demands.

Peak Clipping/Shaving—The process of implementing measures to reduce peak power demands on a system.

Peak Day Withdrawal—The maximum daily withdrawal rate (Mcf/d) experienced during the reporting period.

Peak Demand—The maximum level of use by customers of a system during a specified period of time; may apply to a class of customers only or the entire company, depending on the intended use of the data. See *Peak Load*

Peak Demand/Load—The maximum energy demand or load in a specified time period.

Peak Flow—The highest flow of water attained during a particular flood for a given stream or river.

Peak Kilowatt—One thousand peak watts.

Peak Load—The highest electrical demand within a particular period of time. Daily electric peaks on weekdays occur in late afternoon and early evening. Annual peaks occur on hot summer days.

Peak Load Month—The month of greatest plant electrical generation during the winter heating season (Oct-Mar) and summer cooling season (Apr-Sept), respectively.

Peak Load Plant—A plant usually housing old, low-efficiency steam units, gas turbines, diesels, or pumped-storage hydroelectric equipment normally used during the peak-load periods.

Peak Load Power Plant—A power generating station that is normally used to produce extra electricity during peak load times.

Peak Megawatt—One million peak watts.

Peak Power—Power generated that operates at a very low capacity factor; generally used to meet short-lived and variable high demand periods.

Peak Power Current—Current in Amperes produced by a module or array operating at the voltage on the I-V curve that will produce its maximum power.

Peak Power Point—Operating point of the IV (current-voltage) curve for a photovoltaic cell or module where the product of the current value times the voltage value is a maximum. Also called the "maximum power point."

Peak Shaving—During high demand periods, energy rates peak. Because the energy produced by a power system is typically less expensive, businesses can seamlessly switch to this more efficient source whenever it is needed.

Peak Shifting—The process of moving existing loads to off-peak periods.

Peak Sun Hours—The equivalent number of hours per day when solar irradiance averages 1 kW/m². For example, six peak sun hours means that the energy received during total daylight hours equals the energy that would have been received had the irradiance for six hours been 1 kW/m².

Peak Watt—A unit used to rate the performance of a solar photovoltaic (PV) cells, modules, or arrays; the maximum nominal output of a PV device, in Watts (Wp) under standardized test conditions, usually 1000 Watts per square meter of sunlight with other conditions, such as temperature specified.

Peak Wind Speed—The maximum instantaneous wind speed (or velocity) that occurs within a specific period of time or interval.

Peaking Capacity—Capacity of generating equipment normally reserved for operation during the hours of highest daily, weekly, or seasonal loads. Some generating equipment may be operated at certain times as peaking capacity and at other times to serve loads on an around-the-clock basis.

Peaking Hydropower—A hydropower plant that is operated at maximum allowable capacity for part of the day and is either shut down for the remainder of the time or operated at minimal capacity level.

Peaking Unit—A power generator used by a utility to produce extra electricity during peak load times.

Peat—Peat consists of partially decomposed plant debris. It is considered an early stage in the development of coal. Peat is distinguished from lignite by the presence of free cellulose and a high moisture content (exceeding 70 percent). The heat content of air-dried peat (about 50 percent moisture) is about 9 million Btu per ton. Most U.S. peat is used as a soil conditioner. The first U.S. electric power plant fueled by peat began operation in Maine in 1990.

Peer Group Benchmarking—Comparison of a company with the group of most relevant competitors based on sector-specific key figures and averages.

PEFCO—Private Export Funding Corporation

PEI—Petroleum Equipment Institute www.pei.org

PEIS—Programmatic Environmental Impact Statement

PEL—Permissible Exposure Limit
  • Personal Exposure Limit

Pellet Stove—A space heating device that burns pellets; are more efficient, clean burning, and easier to operate relative to conventional cord wood burning appliances.

Pellets—Solid fuels made from primarily wood sawdust that is compacted under high pressure to form small (about the size of rabbit feed) pellets for use in a pellet stove.

Pelton Turbine—A type of impulse hydropower turbine where water passes through nozzles and strikes cups arranged on the periphery of a runner, or wheel, which causes the runner to rotate, producing mechanical energy. The runner is fixed on a shaft, and the rotational motion of the turbine is transmitted by the shaft to a generator. Generally used for high head, low flow applications.

Pelton Wheel—A common impulse turbine runner—the wheel that receives the water, changing the pressure and flow of the water to circular motion to drive an alternator, generator, or machine. Pelton wheels (named after inventor Lester Pelton) are made with a series of cups or "buckets" cast onto a hub.

Penny Stocks—Low-priced issues, often highly speculative, selling at less than $5/share.

Penstock—Steep gradient pipe or channel for gravity acceleration of water for driving electric generators in hydroelectric plants.

Pentanes Plus—A mixture of hydrocarbons, mostly pentanes and heavier, extracted from natural gas. Includes isopentane, natural gasoline, and plant condensate.

People, Planet, Profit—The expanded set of values for companies and individuals to use in measuring organizational and societal success, specifically economic, environmental and social values. "People, planet profit" are also referred to as the components of the "triple bottom line." See *Triple Bottom Line*

PERC—Perchloroethylene

Percent Difference—The relative change in a quantity over a specified time period. It is calculated as follows: the current value has the previous value subtracted from it; this new number is divided by the absolute value of the previous value; then this new number is multiplied by 100.

Percent Utilization—The ratio of total production to productive capacity, times 100.

Perched Groundwater—Water that accumulates beneath the earth's surface but above the main water bearing zone (aquifer). Typically, perched groundwater occurs when a limited zone (or lens) of harder, less permeable soil is "perched" in otherwise porous soils. Rainwater moving downward through the soil stops at the lens, flows along it, then seeps downward toward the aquifer.

Perchlorate—A white or colorless powder that can dissolve easily in water. Many industries use perchlorate to make products such as explosives and solid rocket fuel. Though it has not been linked to cancer in humans, exposure to perchlorate could affect a person's ability to process iodine in the thyroid gland. The most likely way you can be affected by it is if you drink water contaminated with perchlorate at levels higher than what is considered safe. You are not likely to be affected through breathing air or through touching soil that has perchlorate in it.

Perchloroethylene (PCE)—A volatile organic compound used primarily as a dry-cleaning agent. Also known as "Perc" It is toxic and listed as a cancer-causing chemical under Proposition 65 in California.

Percolation—The filtering of a liquid passed through a medium with many fine spaces.

Perfluorocarbon Tracer Gas Technique (PFT)—An air infiltration measurement technique developed by the Brookhaven National Laboratory to measure changes over time (one week to five months) when determining a building's air infiltration rate. This test cannot locate exact points of infiltration, but it does reveal long-term infiltration problems.

Perfluorocarbons (PFCs)—PFCs are among the six types of greenhouse gases to be curbed under the Kyoto Protocol. PFCs are synthetic industrial gases generated as a by-product of aluminum smelting and uranium enrichment. They also are used as substitutes for CFCs in the manufacture of semiconductors. There are no natural sources of PFCs. PFCs have atmospheric lifetimes of thousands to tens of thousands of years and 100-year GWPs thousands of times that of CO_2, depending on the gas.

Perfluoromethane—A compound (CF_4) emitted as a by-product of aluminum smelting.

Performance Attributes—Performance attributes measure the quality of service and operating efficiency. Loss of load probability, expected energy curtailment, and reserve margin are all performance attributes.

Performance Based Ratemaking (PBR)—Regulated rates based on performance objectives, not actual costs.

Performance Bond—A bond supplied by one party to protect another against loss in the event of default of an existing contract. A bond to motivate a contractor to perform a contract. Other performance bonds provide for payment of a sum of money for failure of the contractor to perform under a contract.

Performance Contracting—Performance Contracting allows for the use of existing budgets to fund or help fund needed facility improvements from operational and utility savings. The Capital Improvements and all associated cost are financed over a period of time allowing for immediate implementation of improvements and a guarantee of the annual utility and operational savings. The Performance Contract offers a building owner a no risk approach to facility improvements.

Performance Indicator—(GRI Reporting Framework) Qualitative or quantitative information about results or outcomes associated with the organization that is comparable and demonstrates change over time.

Performance Period—The time period spanning from approval of the projection installation to the end of the contract, or for a specific time-frame such as 1-year within that period.

Performance Period Energy Use or Demand—The calculated energy usage (or demand) by a piece of equipment or a site after implementation of the energy project.

Performance Ratings—Solar collector thermal performance ratings based on collector efficiencies, usually expressed in Btu per hour for solar collectors under standard test or operating conditions for solar radiation intensity, inlet working fluid temperatures, and ambient temperatures.

Performance-Based Incentives (PBI)—The California Solar Initiative Program pays PBI incentives in monthly payments based on recorded kilowatt-hours of solar power produced over a five-year period. Solar projects receiving PBI incentives are paid a flat per kWh payment monthly for PV system output that is serving on-site load. The monthly PBI incentive payment is calculated by multiplying the incentive rate by the measured kWh output.

Performance-Based Regulation (PBR)—Any rate-setting mechanism which attempts to link rewards (generally profits) to desired results or targets. PBR sets rates, or components of rates, for a period of time based on external indices rather than a utility's cost-of-service. Other definitions include light-handed regulation which is less costly and less subject to debate and litigation. A form of rate regulation which provides utilities with better incentives to reduce their costs than does cost-of-service regulation.

Perimeter Heating—A term applied to warm-air heating systems that deliver heated air to rooms by means of registers or baseboards located along exterior walls.

Period of Analysis—The number of years considered in the study.

Periodicity—The length of a period in a contract. Usually annual, semiannual, quarterly, or monthly.

Perm—The measurement of water vapor through different materials measured in perm-inch (mass of water vapor moving through a unit area in unit time). Equal to 1 grain of water vapor transmitted per 1 square foot per hour per inch of mercury pressure difference.

Permanence—A key pre-requisite for the credibility of any carbon sequestration activity, particularly tree planting; that it have in place safeguards to cover the possibility that carbon removed from the atmosphere may be released in the future, for example, due to fire, disease or logging. In practice, ongoing verification of planted trees must take place where carbon offset credits have been generated for those carbon reductions.

Permanently Attached—Attached with fasteners that require additional tools to remove (as opposed to clips, hooks, latches, snaps, or ties).

Permanently Discharged Fuel—Spent nuclear fuel for which there are no plans for reinsertion in the reactor core.

Permeability—The ease with which fluid flows through a porous medium.

Permeance—A unit of measurement for the ability of a material to retard the diffusion of water vapor at 73.4 F (23 C). A perm, short for permeance, is the number of grains of water vapor that pass through a square foot of material per hour at a differential vapor pressure equal to one inch of mercury.

Permit—Permits are certificates of operation that allow the holder to operate a facility provided they do not exceed a specified rate (e.g. kilograms/tonnes per day). Permits are often designated as an upper limit. Because few systems operate at 100% of capacity at all times actual emissions are usually a fraction of the theoretical upper limit of allowed emissions. However, as new permits become harder to obtain, existing operations are motivated to increase their level of operations under their existing permits (e.g. adding a second shift thereby legally increasing the overall quantity of emissions).

Perpetuity—An annuity forever. Periodic equal payments or receipts on a continual basis.

Persian Gulf—The countries that surround the Persian Gulf are: Bahrain, Iran, Iraq, Kuwait, Qatar, Saudi Arabia, and the United Arab Emirates.

Person—An individual, a corporation, a partnership, an association, a joint-stock company, a business trust, or an unincorporated organization.

Personal Property—Movable property; all property which is not Real Property. Property consisting of Chattels as contrasted to real estate; e.g., furniture, car, equipment, machinery, or solar system.

Person-Year—One whole year, or fraction thereof, worked by an employee, including contracted manpower. Expressed as a quotient (to two decimal places) of the time units worked during a year (hours, weeks, or months) divided by the like total time units in a year. For example: 80 hours worked is 0.04 (rounded) of a person-year; 8 weeks worked is 0.15 (rounded) of a person-year; 12 months worked is 1.0 person-year. Contracted manpower includes survey crews, drilling crews, consultants, and other persons who worked under contract to support a firm's ongoing operations.

Petitioner—Any individual or firm petitioning the regulatory to be recognized as verifier or verification firm.

Petrex Method—A method for collecting vapor samples from surface soil.

Petrochemical Feedstocks—Chemical feedstocks derived from petroleum principally for the manufacture of chemicals, synthetic rubber, and a variety of plastics.

Petrochemicals—Organic and inorganic compounds and mixtures that include but are not limited to organic chemicals, cyclic intermediates, plastics and resins, synthetic fibers, elastomers, organic dyes, organic pigments, detergents, surface active agents, carbon black, and ammonia.

Petrodollars—Money paid to other countries for oil imported to the United States.

Petroleum—A broadly defined class of liquid hydrocarbon mixtures. Included are crude oil, lease condensate, unfinished oils, refined products obtained from the processing of crude oil, and natural gas plant liquids. Note: Volumes of finished petroleum products include non-hydrocarbon compounds, such as additives and detergents, after they have been blended into the products.

Petroleum Administration for Defense District (PADD)—A geographic aggregation of the 50 States and the District of Columbia into five Districts, with PADD I further split into three sub districts. The PADDs include the States listed below:

PADD I (East Coast):

- PADD IA (New England): Connecticut, Maine, Massachusetts, New Hampshire, Rhode Island, and Vermont.

- PADD IB (Central Atlantic): Delaware, District of Columbia, Maryland, New Jersey, New York, and Pennsylvania.

- PADD IC (Lower Atlantic): Florida, Georgia, North Carolina, South Carolina, Virginia, and West Virginia.

PADD II (Midwest): Illinois, Indiana, Iowa, Kansas, Kentucky, Michigan, Minnesota, Missouri, Nebraska, North Dakota, Ohio, Oklahoma, South Dakota, Tennessee, and Wisconsin.

PADD III (Gulf Coast): Alabama, Arkansas, Louisiana, Mississippi, New Mexico, and Texas.

PADD IV (Rocky Mountain): Colorado, Idaho, Montana, Utah, and Wyoming.

PADD V (West Coast): Alaska, Arizona, California, Hawaii, Nevada, Oregon, and Washington.

Petroleum Coke, Catalyst—The carbonaceous residue that is deposited on and deactivates the catalyst used in many catalytic operations (e.g., catalytic cracking). Carbon is deposited on the catalyst, thus deactivating the catalyst. The catalyst is reactivated by burning off the carbon, which is used as a fuel in the refining process. That carbon or coke is not recoverable in a concentrated form.

Petroleum Coke, Marketable—Those grades of coke produced in delayed or fluid cokers that may be recovered as relatively pure carbon. Marketable petroleum coke may be sold as is or further purified by calcining.

Petroleum Consumption—The sum of all refined petroleum products supplied. For each refined petroleum product, the amount supplied is calculated by adding production and imports, then subtracting changes in primary stocks (net withdrawals are a plus quantity and net additions are a minus quantity) and exports.

Petroleum Imports—Imports of petroleum into the 50 states and the District of Columbia from foreign countries and from Puerto Rico, the Virgin Islands, and other U.S. territories and possessions. Included are imports for the Strategic Petroleum Reserve and withdrawals from bonded warehouses for onshore consumption, offshore bunker use, and military use. Excluded are receipts of foreign petroleum into bonded warehouses and into U.S. territories and U.S. Foreign Trade Zones.

Petroleum Jelly—A semi-solid oily product produced from de-waxing lubricating oil basestocks.

Petroleum Products—Petroleum products are obtained from the processing of crude oil (including lease condensate), natural gas, and other hydrocarbon compounds. Petroleum products include unfinished oils, liquefied petroleum gases, pentanes plus, aviation gasoline, motor gasoline, naphtha-type jet fuel, kerosene-type jet fuel, kerosene, distillate fuel oil, residual fuel oil, petrochemical feedstocks, special naphthas, lubricants, waxes, petroleum coke, asphalt, road oil, still gas, and miscellaneous products.

Petroleum Refinery—An installation or facility that manufactures finished petroleum products from crude oil, unfinished oils, natural gas liquids, other hydrocarbons, and alcohol.

Petroleum Stocks, Primary—For individual products, quantities that are held at refineries, in pipelines and at bulk terminals that have a capacity of 50,000 barrels or more, or that are in transit thereto. Stocks held by product retailers and resellers, as well as tertiary stocks held at the point of consumption, are excluded. Stocks of individual products held at gas processing plants are excluded from individual product estimates but are included in other oils estimates and total.

PETS—Proposed, endangered, threatened, or sensitive species.

PF—Project Facilitator
  • Potency Factor
  • Protection Factor

PFCRA—Program Fraud Civil Remedies Act (1986) (1) establishes administrative procedures for imposing civil penalties and assessments against persons who make, submit, or present, or cause to be made, submitted, or presented, false, fictitious, or fraudulent claims or written statements to authorities or to their agents, and (2) specifies the hearing and appeal rights of persons subject to allegations of liability for such penalties and assessments.

PG&E—Pacific Gas & Electric. An investor owned utility found in California. It is the largest utility in California.

PGA—Purchased Gas Account: A rate-adjustment mechanism that allows an energy utility to recover through rates its fuel and fuel-related expenses.

PGC—Public Goods Charge: A non-bypassable surcharge imposed on all retail sales to fund public goods research, development and demonstration, and energy efficiency activities, and support low-income assistance programs.

PGP—Public Generating Pool

pH—A measure of acidity or alkalinity. A pH of 7 represents neutrality. Acid substances have lower pH. Basic substances have higher pH.

PHA—Process Hazard Analysis

Phantom Load—Any appliance that consumes power even when it is turned off. Examples of phantom loads include appliances with electronic clocks or timers, appliances with remote controls, and appliances with wall cubes (a small box that plugs into an AC outlet to power appliances).

Phase—Alternating current is carried by conductors and a ground to residential, commercial, or industrial consumers. The waveform of the phase power appears as a single continuous sine wave at the system frequency whose amplitude is the rated voltage of the power.

Phase Change—The process of changing from one physical state (solid, liquid, or gas) to another, with a necessary or coincidental input or release of energy.

Phase-Change Material—A material that can be used to store thermal energy as latent heat. Various types of materials have been and are being investigated such as inorganic salts, eutectic compounds, and paraffins, for a variety of applications, including solar energy storage (solar energy heats and melts the material during the day and at night it releases the stored heat and reverts to a solid state).

PHE—Public Health Evaluation

Phenols—Organic compounds used in plastics manufacturing, tanning, and textile, dye and resin manufacturing. They are by-products of petroleum refining. In general, they are highly toxic.

PHEV—Plug-in hybrid EV.

Phosphorous (P)—A chemical element, atomic number

15, used as a dopant in making n-semiconductor layers in photovoltaic cells.

Photobiological Hydrogen Production—A hydrogen production process that process uses algae. Under certain conditions, the pigments in certain types of algae absorb solar energy. An enzyme in the cell acts as a catalyst to split water molecules. Some of the bacteria produces hydrogen after they grow on a substrate.

Photocell—A device that produces an electric reaction to visible radiant energy (light).

Photocontrol—An electric control that detects changes in illumination then controls its electric load at predetermined illumination levels.

Photocurrent—An electric current induced by radiant energy.

Photoelectric Cell—A device for measuring light intensity that works by converting light falling on, or reach it, to electricity, and then measuring the current; used in photometers.

Photoelectrochemical Cell—A type of photovoltaic device in which the electricity induced in the cell is used immediately within the cell to produce a chemical, such as hydrogen, which can then be withdrawn for use.

Photoelectrolysis Hydrogen Production—The production of hydrogen using a photoelectrochemical cell.

Photogalvanic Processes—The production of electrical current from light.

Photon—A particle of light that acts as an individual unit of energy. Light is composed of energy particles called photons which have variable energy but constant speed.

Photosynthesis—The manufacture by plants of carbohydrates and oxygen from carbon dioxide and water in the presence of chlorophyll, with sunlight as the energy source. Carbon is sequestered and oxygen and water vapor are released in the process.

Photovoltaic—Pertaining to the direct conversion of light into electricity.

Photovoltaic (Conversion) Efficiency—The ratio of the electric power produced by a photovoltaic device to the power of the sunlight incident on the device.

Photovoltaic (PV; Solar) Array—A group of solar photovoltaic modules connected together.

Photovoltaic (Solar) Cell—Treated semiconductor material that converts solar irradiance to electricity. The smallest discrete element in a PV module that performs the conversion of light into electrical energy to produce a DC current and voltage.

Photovoltaic (Solar) Module or Panel—A solar photovoltaic product that generally consists of groups of PV cells electrically connected together to produce a specified power output under standard test conditions, mounted on a substrate, sealed with an encapsulant, and covered with a protective glazing. Maybe further mounted on an aluminum frame. A junction box, on the back or underside of the module is used to allow for connecting the module circuit conductors to external conductors.

Photovoltaic (Solar) System—A complete PV power system composed of the module (or array), and balance-of-system (BOS) components including the array supports, electrical conductors/wiring, fuses, safety disconnects, and grounds, charge controllers, inverters, battery storage, etc.

Photovoltaic and Solar Thermal Energy (as used at Electric Utilities)—Energy radiated by the sun as electromagnetic waves (electromagnetic radiation) that is converted at electric utilities into electricity by means of solar (photovoltaic) cells or concentrating (focusing) collectors.

Photovoltaic Cell (PVC)—An electronic device consisting of layers of semiconductor materials fabricated to form a junction (adjacent layers of materials with different electronic characteristics) and electrical contacts and being capable of converting incident light directly into electricity (direct current).

Photovoltaic Conversion Efficiency—The ratio of the electrical power generated by a PV device to the power of the light incident on it. This is typically in the range 5% to 20% for commercially available modules.

Photovoltaic Device—A solid-state electrical device that converts light directly into direct current electricity of voltage-current characteristics that are a function of the characteristics of the light source and the materials in and design of the device. Solar photovoltaic devices are made of various semi-conductor materials including silicon, cadmium sulfide, cadmium telluride, and gallium arsenide, and in single crystalline, multi-crystalline, or amorphous forms.

Photovoltaic Effect—The phenomenon that occurs when photons, the particles in a beam of light, knock electrons loose from the atoms they strike. When this property of light is combined with the properties of semiconductors, electrons flow in one direction across a junction, setting up a voltage. With the addition of circuitry, electrons will flow and electrical energy will be available.

Photovoltaic Generator—The total of all PV strings of

a PV power supply system, which are electrically interconnected.

Photovoltaic Module—An integrated assembly of interconnected photovoltaic cells designed to deliver a selected level of working voltage and current at its output terminals, packaged for protection against environmental degradation, and suited for incorporation in photovoltaic power systems.

Photovoltaic Panel—A term often used interchangeably with PV module (especially in single module systems).

Photovoltaic Peak Watt—Maximum "rated" output of a cell, module, or system. Typical rating conditions are 0.645 watts per square inch (1,000 watts per square meter) of sunlight, 68°F (20°C) ambient air temperature and 6.2×10^{-3} mi/s (1 m/s) wind speed. See *Peak Watt*.

Photovoltaic-Thermal (PV/T) Systems—A photovoltaic system that, in addition to converting sunlight into electricity, collects the residual heat energy and delivers both heat and electricity in usable form. Also called a total energy system.

PHSA—Public Health Service Act (1946)

Physical Units—Actual energy usage quantities. They are the actual measurements recorded by utility meters. Examples for electricity include kWh (kilowatt-hour), kW (kilowatt) and PF (power factor measurement). Examples for natural gas include m^3 (cubic meter), therm, mcf (1,000 cubic feet) or gj (gigajoule).

Physical Vapor Deposition—A method of depositing thin semiconductor photovoltaic) films. With this method, physical processes, such as thermal evaporation or bombardment of ions, are used to deposit elemental semiconductor material on a substrate.

PI—Preliminary Injunction
  • Program Information

PIC—Prior Informed Consent
  • Public Information Center

Pickle Lease—A lease by a U.S. lessor to a foreign lessee authorized under the 1984 Tax Reform Act. Accelerated depreciation is slowed down to a straight-line basis over a longer recovery period. Named after Congressman Pickle of Texas.

PIER—Public Interest Energy Research

Piezometers—Small-diameter wells used to measure groundwater levels.
  • An instrument for measuring pressure or compressibility.

PIG—Program Implementation Guide
  • A device used to clean the internal surface of a pipeline.

Pig Iron—Crude, high-carbon iron produced by reduction of iron ore in a blast furnace.

Piggyback Registration—A situation when a securities underwriter allows existing holdings of shares in a corporation to be sold in combination with an offering of new public shares.

PIK Debt Securities—(Payment in Kind) PIK Debt are bonds that may pay bondholders compensation in a form other than cash.

Piling Un-merchantable Material (PUM)—A logging contract requirement to remove and pile un-merchantable woody material of a specified size.

Pilot—A utility program offering a limited group of customers their choice of certified or licensed energy suppliers on a one year minimum trial basis.

Pilot Scale—The size of a system between the small laboratory model size (bench scale) and a full-size system.

P-I-N—A semiconductor (photovoltaic) device structure that layers an intrinsic semiconductor between a p-type semiconductor and an n-type semiconductor; this structure is most often used with amorphous silicon PV devices.

PIP—Public Involvement Program

PIPE—["Private Investment for Public Equity"] Private offering followed by a resale registration.

Pipe Loss (Frictional Head Loss)—The amount of energy or pressure lost due to friction between a flowing liquid and the inside surface of a pipe.

Pipeline—A line of pipe with pumping machinery and apparatus (including valves, compressor units, metering stations, regulator stations, etc.) for conveying a liquid or gas.

Pipeline (Natural Gas)—A continuous pipe conduit, complete with such equipment as valves, compressor stations, communications systems, and meters for transporting natural and/or supplemental gas from one point to another, usually from a point in or beyond the producing field or processing plant to another pipeline or to points of utilization. Also refers to a company operating such facilities.

Pipeline (Petroleum)—Crude oil and product pipelines used to transport crude oil and petroleum products, respectively (including interstate, intrastate, and intra-company pipelines), within the 50 states and the District of Columbia.

Pipeline Capacity—The maximum quantity of gas that can be moved through a pipeline system at any given time based on existing service conditions such as available horsepower, pipeline diameter(s), maintenance schedules, regional demand for natural gas, etc.

Pipeline Freight—Refers to freight carried through pipelines, including natural gas, crude oil, and petroleum products (excluding water). Energy is consumed by various electrical components of the pipeline, including, valves, other, appurtenances attaches to the pipe, compressor units, metering stations, regulator stations, delivery stations, holders and fabricated assemblies.

Pipeline Fuel—Gas consumed in the operation of pipelines, primarily in compressors.

Pipeline Purchases—Gas supply contracted from and volumes purchased from other natural gas companies as defined by the Natural Gas Act, as amended (52 Stat. 821), excluding independent producers, as defined in Paragraph 154.91(a), Chapter I, Title 18 of the Code of Federal Regulations.

Pipeline Quality Natural Gas—A mixture of hydrocarbon compounds existing in the gaseous phase with sufficient energy content, generally above 900 British thermal units, and a small enough share of impurities for transport through commercial gas pipelines and sale to end-users.

Pipeline, Distribution—A pipeline that conveys gas from a transmission pipeline to its ultimate consumer.

Pipeline, Gathering—A pipeline that conveys gas from a production well/field to a gas processing plant or transmission pipeline for eventual delivery to end-use consumers.

Pipeline, Transmission—A pipeline that conveys gas from a region where it is produced to a region where it is to be distributed.

Pipelines, Rate Regulated—FRS (Financial Reporting System Survey) establishes three pipeline segments: crude/liquid (raw materials); natural gas; and refined products. The pipelines included in these segments are all federally or State rate-regulated pipeline operations, which are included in the reporting company's consolidated financial statements. However, at the reporting company's option, intrastate pipeline operations may be included in the U.S. Refining/Marketing Segment if: they would comprise less than 5 percent of U.S. Refining/Marketing Segment net PP&E, revenues, and earnings in the aggregate; and if the inclusion of such pipelines in the consolidated financial statements adds less than $100 million to the net PP&E reported for the U.S. Refining/Marketing Segment.

Pipelines, Rate Regulated—FRS establishes three pipeline segments: crude/liquid (raw materials); natural gas; and refined products. The pipelines included in these segments are all Federal or State rate-regulated pipeline operations, included in the reporting company's consolidated financial statements. At the reporting company's discretion, however, intrastate pipeline operations may be included in the U.S. Refining/Marketing Segment if: (1) they comprise less than 5 percent of U.S. Refining/Marketing Segment net PP&E, revenues, and earnings in the aggregate; and (2) if the inclusion of such pipelines in the consolidated financial statements adds less than $100 million to the net PP&E reported for the U.S.

PIS—Public Information Specialist
• Placed in Service. The date an energy facility is completed and ready for its intended use.

PIT—Permit Improvement Team

Pitch Control—A method of controlling a wind turbine's speed by varying the orientation, or pitch, of the blades, and thereby altering its aerodynamics and efficiency.

Pitcheblende—Uranium oxide (U_3O_8). It is the main component of high-grade African or domestic uranium ore and also contains other oxides and sulfides, including radium, thorium, and lead components.

PIV—Pooled Investment Vehicle. A legal entity that pools various investors' capital and deploys it according to a specific investment strategy.

Placed in Service—Delivered and available for use, although the equipment may still be subject to final installation and/or assembly.
• A vehicle is placed in service if that vehicle is new to the fleet and has not previously been in service for the fleet. These vehicles can be acquired as additional vehicles (increases the size of the company fleet), or as replacement vehicles to replace vehicles that are being retired from service (does not increase the size of the company fleet).

Plaintiff—The Party who brings a Court Action.

Planetary Albedo—The fraction of incident solar radiation that is reflected by the Earth-atmosphere system and returned to space, mostly by backscatter from clouds in the atmosphere.

Planned Generator—A proposal by a company to install electric generating equipment at an existing or planned facility or site. The proposal is based on the owner having obtained either (1) all environmental and regulatory approvals, (2) a signed contract for the electric energy, or (3) financial closure for the facility.

Plant—A term commonly used either as a synonym for an industrial establishment or a generating facility or to refer to a particular process within an establishment.

Plant Acquisition Adjustments—Represents the difference between the cost to the utility of plant acquired as operating units or systems by purchase, merger, consolidation, liquidation or otherwise, and the Original Cost (defined herein) of such plant less the amount(s) credited at the time of acquisition to Accumulated Provision for Depreciation and Amortization and Contributions in Aid of Construction.

Plant Association—A grouping of plant species, or a plant community, that recurs across the landscape. Plant associations are used as indicators of environmental conditions such as temperature, moisture, light, etc.

Plant Condensate—One of the natural gas liquids, mostly pentanes and heavier hydrocarbons, recovered and separated as liquids at gas inlet separators or scrubbers in processing plants.

Plant Hours Connected to Load—The number of hours the plant is synchronized to load over a time interval usually of 1 year.

Plant Liquids—Those volumes of natural gas liquids recovered in natural gas processing plants.

Plant or Gas Processing Plant—A facility designated to achieve the recovery of natural gas liquids from the stream of natural gas, which may or may not have been processed through lease separators and field facilities, and to control the quality of the natural gas to be marketed.

Plant Products—Natural gas liquids recovered from natural gas processing plants (and in some cases from field facilities), including ethane, propane, butane, butane-propane mixtures, natural gasoline, plant condensate, and lease condensate.

Plant Use—The electric energy used in the operation of a plant. Included is the energy required for pumping at pump-storage plants.

Plant-Use Electricity—The electric energy used in the operation of a plant. This energy total is subtracted from the gross energy production of the plant.

Plates—The electrodes in a battery, usually take the form of metal plates.

Plenum—An air compartment or chamber, including uninhabited crawl space, areas above a ceiling or below a floor, including air spaces below raised floors of computer/data processing centers, or attic spaces, to which one or more ducts are connected and which forms part of either the supply-air, return air or exhaust air system, other than the occupied space being conditioned.

PLIRRA—Pollution Liability Insurance and Risk Retention Act Amends the Comprehensive Environmental Response, Compensation, and Liability Act of 1980 (CERCLA) (Superfund) to specify how such program operates within Indian lands. Exempts remedial actions on Indian lands from the cost-sharing and future maintenance requirements imposed on States and requires the President to assure the availability of off-site disposal capability. Authorizes Indian tribes to recover damages for injury to or loss of natural resources resulting from releases of hazardous substances. Treats Indian tribes substantially like States for certain information, notification, and planning provisions. Includes releases of hazardous substances from Department of Defense munitions production under CERCLA.

Plot—A forest area defined by its condition.

Plug Flow Digester—A type of anaerobic digester that has a horizontal tank in which a constant volume of material is added and forces material in the tank to move through the tank and be digested.

Plugged-Back Footage—Under certain conditions, drilling operations may be continued to a greater depth than that at which a potentially productive formation is found. If production is not established at the greater depth, the well may be completed in the shallower formation. Except in special situations, the length of the well bore from the deepest depth at which the well is completed to the maximum depth drilled is defined as "plugged-back footage." Plugged-back footage is included in total footage drilled but is not reported separately.

Plume—A body of contaminated groundwater flowing from a specific source. The movement of the groundwater is influenced by such factors as local groundwater flow patterns, the character of the aquifer in which the groundwater is contained, and the density of contaminants. A plume may also be a cloud of smoke or vapor. It defines the area where exposure would be dangerous.

Plutonium (Pu)—A heavy, fissionable, radioactive, metallic element (atomic number 94) that occurs naturally in trace amounts. It can also result as a byproduct of the fission reaction in a uranium-fuel nuclear reactor and can be recovered for future use.

PM—Particulate Matter

- Product Manager
- Program Manager
- Project Manager
- Average Profit Margin: Operating revenues less operating expense divided by operating revenue.

PM 10—Particulate that is less than 10 microns in diameter. These particulates are present in the smoke created by burning wood.

PM 2.5—Particulate that is 2.5 microns in diameter or less.

PMA—Power Marketing Authority

PMR—Proportionate Mortality Rate

P-N Junction—The semiconductor junction in a photovoltaic cell that shunts electrons into a circuit. Electrons are bumped across this junction by photons (light particles).

Pneumatic Device—A device moved or worked by air pressure.

PO—Peak Oil
- Purchase Order

POC—Point of Compliance
- Point of Contact
- Program Office Contacts

Pocket Plate—A plate for a battery in which active materials are held in a perforated metal pocket.

POE—Point of Exposure

POGO—Privately Owned/Government-Operated

POI—Point of Interception
- Point of Interconnection

Point—One percent, or one percentage point (1.00%). A point also represents 100 basis points.

Point Balancing—A process by which the interconnected operators will transfer a quantity greater or less than the confirmed nominations scheduled quantity for various contracts at a point in an attempt to make the total energy received or delivered at the point as close as possible to the scheduled quantity during a specific billing period.

Point of Delivery—Point(s) for interconnection on the Transmission Provider's System where capacity and/or energy are made available to the end user.

Point of Receipt—Point(s) of connection to the transmission system where capacity and/or energy will be made available to the transmission providers.

Point Source—Any separately identifiable stationary point from which emissions are emitted.

Point-Contact Cell—A high efficiency silicon photovoltaic concentrator cell that employs light trapping techniques and point-diffused contacts on the rear surface for current collection.

Point-to-Point Transmission Service—Reservation and/or transmission of energy from point(s) of receipt to point(s) of delivery.

Poison Pill—A right issued by a corporation as a preventative to a takeover measure. It allows right holders to purchase shares in either their company or in the combined target and bidder entity at a substantial discount, usually 50%. This discount may make the takeover prohibitively expensive.

POL—Petroleum, Oils and Lubricants

Pole Mount—A PV mount that is installed on the top or side of a pole usually set in concrete. Can be fixed or seasonally tilted.

Pole-Mile—A unit of measuring the simple length of an electric transmission/distribution line/feeder carrying electric conductors, without regard to the number of conductors carried.

Policies and Measures (GHG Emissions)—Countries must decide what policies and measures to adopt in order to achieve their emissions targets. Some possible policies and measures, which Parties could implement, are listed in the Kyoto Protocol and could offer opportunities for intergovernmental cooperation.

Polishing Treatment—The final sewage treatment process to further reduce BOD5, suspended solids, and other pollutants.

Pollutant—Substances that enter the environment or become concentrated within it, and that has or may have a detrimental biological effect, whether by natural causes or resulting from human activity.

Polluter Pays Principle—The idea that polluters should pay in proportion to their contribution to pollution and resulting damages. Note that some policies which violate the principle (abatement subsidies for example) could still be cost-effective.

POLR—Provider of Last Resort: Serves as a back-up electric service provider in the competitive retail electric market in the event a retail electric provider leaves the market for any reason.

Polychlorinated Biphenyls (PCBs)—A group of toxic chemicals used for a variety of purposes including electrical applications, carbonless copy paper, adhesives, hydraulic fluids, and caulking compounds. PCBs do not breakdown easily and are listed as cancer-causing agents under Proposition 65 in California.

Polycrystalline—A semiconductor (photovoltaic) material composed of variously oriented, small, individual crystals. See *Multi-crystalline*

Polycrystalline Cell—A wafer of silicon with a multi-grained structure. All grains have the same atomic crystal lattice, however, each grain has a unique orientation in space thereby producing a unique reflection of light.

Polycrystalline Silicon—A material used to make solar PV cells which consists of many crystals, compared to single crystal silicon.

Polyethylene—A registered trademark for plastic sheeting material that can be used as a vapor retarder. This plastic is used to make grocery bags. It is a long chain of carbon atoms with 2 hydrogen atoms attached to each carbon atom.

Polymer—Any of numerous natural and synthetic compounds of usually high molecular weight consisting of up to millions of repeated linked units, each a relatively light and simple molecule.

Polystyrene—A polymer of styrene that is a rigid, transparent thermoplastic with good physical and electrical insulating properties, used in molded products, foams, and sheet materials. (See *Foam Insulation*)

Polyvinyl Chloride (PVC)—A polymer of vinyl chloride. Tasteless. odorless, insoluble in most organic solvents. A member of the family vinyl resin, used in soft flexible films for food packaging and in molded rigid products, such as pipes, fibers, upholstery, and bristles.

POM—Particulate Organic Matter
• Polycyclic Organic Matter

Pondage—The amount of water stored behind a hydroelectric dam of relatively small storage capacity; the dam is usually used for daily or weekly control of the flow of the river.

Pool—In general, a reservoir. In certain situations, a pool may consist of more than one reservoir.

Pool Site—One or more spent fuel storage pools that has a single cask loading area. Each dry cask storage area is considered a separate site.

POOLCO—Poolco refers to a specialized, centrally dispatched spot market power pool that functions as a short-term market. It establishes the short-term market clearing price and provides a system of long-term transmission compensation contracts. It is regulated to provide open access, comparable service and cost recovery. A Poolco would make ancillary generation services, including load following, spinning reserve, backup power, and reactive power, available to all market participants on comparable terms. In addition, the Poolco provides settlement mechanisms when differences in contracted volumes exist between buyers and sellers of energy and capacity.

Pooled Funds—A funding technique used by lessors in which several forms of borrowing are pooled, or grouped, for use in funding leases and are not tied specifically to the purchase of any one piece of leased equipment.

Pooling—An informal name for a settlement process in which all participating companies first receive their costs and then earn the same rate of return.

Poor Quality Lighting Tasks—Visual tasks that require Illuminance Category E or greater, because of the choice of a writing or printing method that produces characters that are of small size or lower contrast than good quality alternatives that are regularly used in offices.

Population-Weighted Degree-Days—Heating or cooling degree-days weighted by the population of the area in which the degree-days are recorded. To compute national population-weighted degree-days, the Nation is divided into nine Census regions comprised of from three to eight states that are assigned weights based on the ratio of the population of the region to the total population of the Nation. Degree-day readings for each region are multiplied by the corresponding population weight for each region, and these products are then summed to arrive at the national population weighted degree-day figure.

POR—Program of Requirements

Pore Space—The open spaces or voids of a rock taken collectively. It is a measure of the amount of liquid or gas that may be absorbed or yielded by a particular formation.

Porous Media—A solid that contains pores; normally, it refers to interconnected pores that can transmit the flow of fluids. (The term refers to the aquifer geology when discussing sites for CAES.)

Portable Electric Heater—A heater that uses electricity and that can be picked up and moved.

Portable Fan—Box fans, oscillating fans, table or floor fans, or other fans that can be moved.

Portable Kerosene Heater—A heater that uses kerosene and that can be picked up and moved.

Portable Power—Power generation systems that can easily be transported from one site to another site.

Portfolio Management—The functions of resource planning and procurement under a traditional utility structure. Portfolio management can also be defined as the aggregation and management of a diverse portfolio of supply (and demand-reduction) resources which will act as a hedge against various risks that may affect specific resources (i.e., fuel price fluctuations and certainty of supply, common mode failures, operational reliability, changes in environmental regulations, and the risk of health, safety, and environmental damages that may occur as a result of operating some supply resources). Under a more market-driven power sector with a "powerpool" or POOLCO wholesale market structure, a portfolio manager would: aggregate and manage a diverse portfolio of spot-market purchases, contracts-for-differences, futures contracts and other market-hedging-type contracts and mechanisms.

Portfolio Standard—The requirement that an electric power provider generate or purchase a specified percentage of the power it supplies/sells from renewable energy resources, and thereby guarantee a market for electricity generated from renewable energy resources.

Portland Cement—Hydraulic cement (cement that not only hardens by reacting with water but also forms a water-resistant product) produced by pulverizing linkers consisting essentially of hydraulic calcium silicates, usually containing one or more of the forms of calcium sulfate as an inter-ground addition.

Positive Feedback—A process that results in an amplification of the response of a system to an external influence. For example, increased atmospheric water vapor in response to global warming would be a positive feedback on warming, because water vapor is a GHG.

Post and Beam Construction—A traditional building technique in which post and beam framing units are the basic load-bearing members. Post and beams may be of wood, structural steel, or concrete. In this system, there are fewer framing members, leaving more open space for in-fill. Often used in straw bale construction.

Post-Aeration—The introduction of oxygen into waste water to further reduce BOD and COD after secondary or advanced treatment.

Post-Consumer Content—Percent of materials salvaged for reuse from the waste stream of a manufacturing process rather than from consumers.

Post-Installation Conditions—The physical and operational conditions present during the time period following the installation of an energy project.

Post-Installation Report—The report that provides results of post-installation M&V activities, documents any changes in the project scope that may have occurred during projection implementation, and provides energy savings estimates for the first year of performance.

Post-Mining Emissions—Emissions of methane from coal occurring after the coal has been mined, during transport or pulverization.

Post-Money Valuation—The valuation of a company immediately after the most recent round of financing. This value is calculated by multiplying the company's total number of shares by the share price of the latest financing.

Potable Water—Water that is suitable for drinking, as defined by local health officials.

Potential Consumption—The total amount of consumption that would have occurred had the intensity of consumption remained the same over a period of time.

Potential Energy—Energy available due to position.
• Stored energy. Energy possessing the power of doing work but not actually performing such work.

Potential Pareto Improvement Criterion—(often called the PPIC criterion) The criterion that gainers from a policy change (or project) could compensate the losers from the change and still be better off. In particular note that a policy that passes this criterion does not need to include the compensation. The compensation merely has to be possible.

Potential Peak Reduction—The potential annual peak load reduction (measured in kilowatts) that can be deployed from Direct Load Control, Interruptible Load, Other Load Management, and Other DSM Program activities. (Note that Energy Efficiency and Load Building are not included in Potential Peak Reduction.) It represents the load that can be reduced either by the direct control of the utility system operator or by the consumer in response to a utility request to curtail load. It reflects the installed load reduction capability, as opposed to the Actual Peak Reduction achieved by participants, during the time of annual system peak load.

Potentially Responsible Party (PRP)—An individual, company or government body identified as potentially liable for a release of hazardous substances to the environment. By federal law, such parties may include generators, transporters, storers and disposers of hazardous waste, as well as present and past site owners and operators.

POU—Public-Owned Utility: A utility owned by its customers, who elect the organization's board of directors.

Poultry Waste—Poultry manure and litter, including

wood shavings, straw, rice hulls and other bedding material for the disposition of manure.

Pound—Pound mass (sometimes abbreviated lb(m)). A unit of mass equal to 0.454 kilograms.

Pound Force—(sometimes abbreviated lb(f)) A force which will accelerate one pound mass at a rate of 32.2 ft/second^2.

Pound of Steam—One pound of water in vapor phase; is NOT steam pressure, which is expressed as pounds per square inch (psi).

Pound Per Square Inch Absolute (psia)—A unit of pressure [hydraulic (liquid) or pneumatic (gas)] that does not include atmospheric pressure.

Pounds (District Heat)—A weight quantity of steam, also used to denote a quantity of energy in the form of steam. The amount of usable energy obtained from a pound of steam depends on its temperature and pressure at the point of consumption and on the drop in pressure after consumption.

Power—Energy that is capable or available for doing work; the time rate at which work is performed, measured in horsepower, Watts, or Btu per hour. Electric power is the product of electric current and electromotive force.

• Basic unit of electricity equal to the product of current and voltage (in DC circuits).

• The rate of doing work. Expressed as Watts (W). For example, a generator rated at 800 watts can provide that amount of power continuously. 1 Watt = 1 joule/sec.

• The rate at which energy is transferred. Electrical energy is usually measured in watts. Also used for a measurement of capacity.

Power (Electrical)—An electric measurement unit of power called a voltampere is equal to the product of 1 volt and 1 ampere. This is equivalent to 1 watt for a direct current system, and a unit of apparent power is separated into real and reactive power. Real power is the work-producing part of apparent power that measures the rate of supply of energy and is denoted as kilowatts (kW). Reactive power is the portion of apparent power that does no work and is referred to as kilovars; this type of power must be supplied to most types of magnetic equipment, such as motors, and is supplied by generator or by electrostatic equipment. Voltamperes are usually divided by 1,000 and called kilovolt amperes (kVA). Energy is denoted by the product of real power and the length of time utilized; this product is expressed as kilowatt-hours.

Power (Output) Curve—A plot of a wind energy con-

version devices power output versus wind speed.

Power (Solar) Tower—A term used to describe solar thermal, central receiver, power systems, where an array of reflectors focus sunlight onto a central receiver and absorber mounted on a tower.

Power Ascension—The period of time between a plant's initial fuel loading date and its date of first commercial operation (including the low-power testing period). Plants in the first operating cycle (the time from initial fuel loading to the first refueling), which lasts approximately 2 years, operate at an average capacity factor of about 40 percent.

Power Authorities—Quasi-governmental agencies that perform all or some of the functions of a public utility.

Power Broker—An entity authorized by FERC to engage in market-based wholesale electricity transactions.

Power Coefficient—The ratio of power produced by a wind energy conversion device to the power in a reference area of the free windstream.

Power Conditioning—The process of modifying the characteristics of electrical power (for e.g., inverting dc to ac).

Power Conditioning Equipment—Electrical equipment, or power electronics, used to convert power from a photovoltaic array into a form suitable for subsequent use. A collective term for inverter, converter, battery charge regulator, and blocking diode.

Power Content Label (Green-e program)—Much like a nutrition label, the power content label (also known as a resource disclosure label), shows an electricity service provider's generation type in a standardized format. The label may also include prices, terms of contracts with customers, air emissions and labor practices. Some states require standard disclosure labels.

Power Conversion Efficiency—The ratio of output power to input power e.g. of an inverter. Efficiency of stand-alone inverters will vary significantly with the load. Values found in manufacturers' specifications are the maximum that can be expected.

Power Density—The ratio of power output to weight or volume.

• The amount of power per unit area of a free windstream.

• The ratio of the power available from a battery to its volume (Watts per liter) or weight (Watts per kilogram).

Power Exchange (PX)—An entity providing a competitive spot market for electric power through day-

and/or hour-ahead auction of generation and demand bids.

Power Exchange Generation—Generation scheduled by the power exchange. See definition for *Power Exchange* above.

Power Exchange Load—Load that has been scheduled by the power exchange and is received through the use of transmission or distribution facilities owned by participating transmission owners.

Power Factor (PF)—The ratio of actual power being used in a circuit, expressed in watts or kilowatts, to the power that is apparently being drawn from a power source, expressed in volt-amperes or kilovolt-amperes.
- The cosine of the phase angle between the voltage and the current waveforms in an AC circuit. A measure of inverter performance.

Power Generation Mix—The proportion of electricity distributed by a power provider that is generated from available sources such as coal, natural gas, petroleum, nuclear, hydropower, wind, or geothermal.

Power Grid—A network of power lines and associated equipment used to transmit and distribute electricity over a geographic area.

Power Loss—The difference between electricity input and output as a result of an energy transfer between two points.

Power Marketer—Wholesale power entity that is registered with the Federal Energy Regulatory Commission that resells electric power to other utilities or retail consumers.

Power Marketers—Business entities engaged in buying and selling electricity. Power marketers do not usually own generating or transmission facilities. Power marketers, as opposed to brokers, take ownership of the electricity and are involved in interstate trade. These entities file with the Federal Energy Regulatory Commission (FERC) for status as a power marketer.

Power of Attorney—A written authorization to an Agent to perform specified acts on behalf of his principal. May be a "General Power" or a "Limited Power."

Power Plant—A central station generating facility that produces energy.

Power Pool—An entity established to coordinate short-term operations to maintain system stability and achieve least-cost dispatch. The dispatch provides backup supplies, short-term excess sales, reactive power support, and spinning reserve. Historically, some of these services were provided on an unpriced basis as part of the members' utility franchise obligations. Coordinating short-term operations includes the aggregation and firming of power from various generators, arranging exchanges between generators, and establishing (or enforcing) the rules of conduct for wholesale transactions. The pool may own, manage and/or operate the transmission lines ("wires") or be an independent entity that manages the transactions between entities. Often, the power pool is not meant to provide transmission access and pricing, or settlement mechanisms if differences between contracted volumes among buyers and sellers exist.

Power Production Plant—All the land and land rights, structures and improvements, boiler or reactor vessel equipment, engines and engine-driven generator, turbogenerator units, accessory electric equipment, and miscellaneous power plant equipment are grouped together for each individual facility.

Power Provider—A company or other organizational unit that sells and distributes electrical power (e.g., private or public electrical utility), either to other distribution and wholesale businesses or to end-users. Sometimes power providers also generate the power they sell.

Power Purchase Agreement (PPA)—An agreement for the sale of electricity from one party to another, where the electricity is generated and consumed on the Host Customer site. Agreements that entail the export and sale of electricity from the Host Customer site do not constitute on-site use of the generated electricity and therefore are normally not eligible for rebates by utilities.

Power Quality—A measure of the electric system's ability to deliver energy at a steady and predictable voltage level.

Power Reliability—A measure of the electric system's ability to deliver uninterrupted service.

Power Transfer Limit—The maximum power that can be transferred from one electric utility system to another without overloading any facility in either system.

Powerhouse—A structure at a hydroelectric plant site that contains the turbine and generator.

PP—Priority Pollutants
- Pollution Prevention
- Program Planning

PP&E, Additions To—The current year's expenditures on property, plant, and equipment (PP&E). The amount is predicated upon each reporting company's accounting practice. That is, accounting practices with regard to capitalization of certain items may differ across companies, and therefore this figure in FRS (Financial Reporting System) will be a function of each reporting company's policy.

PP&E, Net—The original cost of property, plant, and equipment (PP&E), less accumulated depreciation.

PPA—Power Purchase Agreement. A contractual agreement between the owner of an energy facility and a Host in which the Host agrees to purchase the energy produced by the energy facility for a specified duration and at a specified price. The term of a PPA is often between seven and twenty years.

- Pollution Prevention Act (1990). The Pollution Prevention Act focused industry, government, and public attention on reducing the amount of pollution through cost-effective changes in production, operation, and raw materials use. Opportunities for source reduction are often not realized because of existing regulations, and the industrial resources required for compliance, focus on treatment and disposal. Source reduction is fundamentally different and more desirable than waste management or pollution control.

ppb—Parts per Billion

PPC—Public Power Council

PPHH—Persons Per Household

PPIC—Pollution Prevention Information Clearinghouse

ppm—Parts per million

ppmvd—Parts per Million by Volume Dry

ppmw—Parts per Million by Weight

PPP—Purchasing Power Parity

PPP—Purchasing Power Parity

PPPA—Poison Prevention Packaging Act (1970) The act was enacted to prevent young children from accidentally ingesting hazardous substances ordinarily stored about the house. The law requires toxic, corrosive, or irritative substances to be packaged in such a way that it will be difficult for children less than 5 years to open them, yet not difficult for adults to open.

ppq—Parts per Quadrillion

PPRS—Program Planning and Review Staff

PPSP—Power Plant Siting Program

Ppt (Parts Per Trillion)—The unit commonly used to represent the degree of pollutant concentration where the concentrations are small.

ppth—Parts per Thousand

PR—Preliminary Review
- Procurement Request
- Proposed Rule(s)

PRA—Paperwork Reduction Act
- Planned Regulatory Action

PRC—Public Resources Code (California)

Pre Tax Return—Is equal to the debt component of the rate of return plus the equity component adjusted upwards to reflect the impact of Federal Income Taxes.

Precautionary Principle—In reference to the Kyoto Protocol, the idea that action to forestall large-scale, irreversible damage from climate change is warranted even though the risks of climate change are not yet fully understood.

Under the GRI Reporting Framework, the Precautionary Principle refers to the approach taken to address potential environmental impacts. In order to protect the environment, the precautionary approach is widely applied by States according to their capabilities. Where there are threats of serious or irreversible damage, lack of full scientific certainty shall not be used as a reason for postponing cost effective measures to prevent environmental degradation.

When information about potential risks is incomplete, basing decisions about the best ways to manage or reduce risks on a preference for avoiding unnecessary health risks instead of on unnecessary economic expenditures.

Pre-Commercial Thinning—Thinning for timber stand improvement purposes, generally in young, densely stocked stands. Pre-commercial thinning operations are not considered partial cuts.

Pre-Discovery Costs—All costs incurred in an extractive industry operation prior to the actual discovery of minerals in commercially recoverable quantities; normally includes prospecting, acquisition, and exploration costs and may include some development costs.

Preemptive Right—A shareholder's right to acquire an amount of shares in a future offering at current prices per share paid by new investors, whereby his/her percentage ownership remains the same as before the offering.

Preference Item—Certain tax benefits that may create additional tax liability under the alternative minimum tax.

Preferred Day-Ahead Schedule—A Scheduling Coordinator's preferred schedule for the ISO day-ahead scheduling process.

Preferred Dividend—A dividend ordinarily accruing on preferred shares payable where declared and superior in right of payment to common dividends.

Preferred Hour-Ahead Schedule—A Scheduling Coordinator's preferred schedule for the ISO hour-ahead scheduling process.

Preferred Schedule—The initial schedule produced by a Scheduling Coordinator that represents its preferred mix of generation to meet demand. The schedule includes the quantity of output (generators) and consumption (loads), details of any adjustment bids, and the location of each generator and load. The schedule also specifies the quantities and location of trades between the Scheduling Coordinator and all other Scheduling Coordinators, and is balanced with respect to generation, transmission losses, load, and trades.

Preferred Stock—A class of capital stock that may pay dividends at a specified rate and that has priority over common stock in the payment of dividends and the liquidation of assets. Many venture capital investments use preferred stock as their investment vehicle. This preferred stock is convertible into common stock at the time of an IPO.

Pregnant Solution—A solution containing dissolved extractable mineral that was leached from the ore; uranium leach solution pumped up from the underground ore zone though a production hole.

Preheater (Solar)—A solar heating system that preheats water or air that is then heated more by another heating appliance.

Preliminary Permit (Hydroelectric Power)—A single site permit granted by the FERC (Federal Energy Regulatory Commission), which gives the recipient priority over anyone else to apply for a hydroelectric license. The preliminary permit enables the recipient to prepare a license application and conduct various studies for economic feasibility and environmental impacts. The period for a preliminary permit may extend to 3 years.

Premium—A put or call buyer who is purchasing an option must pay to a put or call seller a premium for an option contract. This premium is determined by market supply and demand forces.

Premium Gasoline—Gasoline having an antiknock index (R+M/2) greater than 90. Includes both leaded premium gasoline as well as unleaded premium gasoline.

Premium on Capital Stock—The excess of the amount received by the company from the sale of an issue of the capital stock over the par or stated value of the stock. A premium also arises when a company issues a stock dividend and the market price of such stock exceeds its par or stated value. In this instance, an amount equal to the difference is transferred from retained earnings to premium on capital stock.

Pre-Money Valuation—The valuation of a company prior to a round of investment. This amount is determined by using various calculation models, such as discounted P/E ratios multiplied by periodic earnings or a multiple times a future cash flow discounted to a present cash value and a comparative analysis to comparable public and private companies.

Preparation Plant—A mining facility at which coal is crushed, screened, and mechanically cleaned.

Preproduction Costs—Costs of prospecting for, acquiring, exploring, and developing mineral reserves incurred prior to the point when production of commercially recoverable quantities of minerals commences.

Prescription—Specific written directions for forest management activities.

Present Value—The worth of future receipts or costs expressed in current value. To obtain present value, an interest rate is used to discount future receipts or costs.

- The discounted value of a payment or stream of payments to be received in the future, taking into consideration a specific interest or discount rate. Present value represents a series of future cash flows expressed in today's dollars.

Present Worth Factor—The adjustment factor that discounts a sum of future dollars back to the current year. A calculation based on interest rate and expected life of an energy retrofit Opportunity (i.e., how long the equipment will last, in years). The Present Worth Factor is an intermediate calculation used in determining other indicators of the attractiveness of an investment. Present Worth Factor = (1 - (1 + Interest Rate) - (Estimated Life))/Interest Rate.

Pressure—The "push" behind liquid or gas in a tank, reservoir, or pipe. Water pressure is directly related to "head"—the height of the top of the water over the bottom. Every 2.31 feet of vertical head gives 1 psi (pound per square inch) of water pressure.

Pressure Boundary—The primary air enclosure boundary separating conditioned air and unconditioned air. Typically, it is defined by the Air Control Layer System.

Pressure Drop—The loss in static pressure of a fluid (liquid or gas) in a system due to friction from obstructions in pipes, from valves, fittings, regulators, burners, etc, or by a breech or rupture of the system.

Pressure Reducing Valve—A valve designed to reduce a facility's water consumption by lowering supply-line pressure.

Pressure Swing Adsorption (PSA)—A gas purification process which selectively concentrates target gas molecules using porous, high surface area solid adsorbents and elevated pressure.

Pressure, Absolute (PSIA)—Pressure in excess of a perfect vacuum. Absolute pressure is obtained by algebraically adding gauge pressure to atmosphere pressure. Pressures reported in "Atmospheres" are understood to be absolute. Absolute pressure must be used in equations of state and in all gas-law calculations. Gauge pressures below atmospheric pressure are called "vacuum."

Pressurization Testing—A technique used by energy auditors, using a blower door, to locate areas of air infiltration by exaggerating the defects in the building shell. This test only measures air infiltration at the time of the test. It does not take into account changes in atmospheric pressure, weather, wind velocity, or any activities the occupants conduct that may affect air infiltration rates over a period of time.

Pressurized Tank Toilet—A toilet that uses a Facility's waterline pressure by pressurizing water held in a vessel within the tank, thereby compressing a pocket of trapped air. The water releases at a force 500 times greater than a conventional gravity toilet.

Pressurized Water Reactor (PWR)—A nuclear reactor in which heat is transferred from the core to a heat exchanger via water kept under high pressure, so that high temperatures can be maintained in the primary system without boiling the water. Steam is generated in a secondary circuit.

Presumption—That which may be assumed without proof.

Pretreatment Unit—A wastewater treatment unit that is designed to treat wastewater that does not meet the sewage discharge standards so that it meets or exceeds those standards. Pretreatment units usually require a permit from a local agency.

Prevention of Significant Deterioration (PSD)—Under the Clean Air Act, a planning and management process for air quality when a new source of air pollution is proposed in an area where ambient air quality is better than applicable standards (areas of special importance).

Preventive Maintenance—Examination of plant and equipment on a schedule basis and the replacement or repair of parts that are worn by prescribed amounts or that are in such condition that further use will involve the risk of their failure while in service. It is designed to prevent operating breakdown.

PRF—Peak Responsibility Factor

Price—The amount of money or consideration-in-kind for which a service is bought, sold, or offered for sale.

Price Cap—Situation where a price has been determined and fixed.

Price Earnings Ratio—Market price divided by the annual earnings per share of common stock. The market price used may be a spot price, or an average of closing or the high and low prices for a period; the earnings are for the corresponding period.

Price Elasticity of Demand—A measurement of the sensitivity of demand to changes in price. Technically, the ratio between the percentage change in volumes demanded and the corresponding percentage change in price.

Prima Facie—Assumed correct until overcome by further proof.

Primary Air—The air that is supplied to the combustion chamber of a furnace.

Primary Battery—A battery that cannot be re-charged.

Primary Cell—A primary cell is an electrochemical cell (battery) that cannot be recharged. The chemical process within the primary cell is only one way—discharge. When a primary cell is discharged it is discarded. Common flashlight batteries are primary cells; they are disposable batteries that should be avoided.

Primary Circuit—This is the distribution circuit (less than 69,000 volts) on the high voltage side of the transformer.

Primary Coal—All coal milled and, when necessary, washed and sorted.

Primary Energy—All energy consumed by end users, excluding electricity but including the energy consumed at electric utilities to generate electricity. (In estimating energy expenditures, there are no fuel-associated expenditures for hydroelectric power, geothermal energy, solar energy, or wind energy, and the quantifiable expenditures for process fuel and intermediate products are excluded.)

Primary Energy Consumption—Primary energy consumption is the amount of site consumption, plus losses that occur in the generation, transmission, and distribution of energy.

Primary Energy Consumption Expenditures—Expenditures for energy consumed in each of the four major end-use sectors, excluding energy in the form of electricity, plus expenditures by the electric utilities sector for energy used to generate electricity. There are no fuel-associated expenditures for associated expenditures for hydroelectric power, geothermal energy, photovoltaic and solar energy, or wind energy. Also excluded are the quantifiable consumption expenditures that are an integral part of process fuel consumption.

Primary Fuels—Fuels that can be used continuously. They can sustain the boiler sufficiently for the production of electricity.

Primary Market—The exchange of emission reductions, offsets, or allowances between buyer and seller where the seller is the originator of the supply. The exchange of greenhouse gas emission reductions is currently conducted only in the primary market (vs. the secondary market).

Primary Metropolitan Statistical Area (PMSA)—A component area of a Consolidated Metropolitan Statistical Area consisting of a large urbanized county or cluster of counties (cities and towns in New England) that demonstrate strong internal economic and social links in addition to close ties with the central core of the larger area. To qualify, an area must meet specified statistical criteria that demonstrate these links and have the support of local opinion.

Primary Recovery—The crude oil or natural gas recovered by any method that may be employed to produce them where the fluid enters the well bore by the action of natural reservoir pressure (energy or gravity).

Primary Transportation—Conveyance of large shipments of petroleum raw materials and refined products usually by pipeline, barge, or ocean-going vessel. All crude oil transportation is primary, including the small amounts moved by truck. All refined product transportation by pipeline, barge, or ocean-going vessel is primary transportation.

Prime Mover—The engine, turbine, water wheel, or similar machine that drives an electric generator; or, for reporting purposes, a device that converts energy to electricity directly (e.g., photovoltaic solar and fuel cells).

Prime Rate—The rate at which banks lend to their best customers. The all-in-cost of a bank loan to a prime credit equals the prime rate plus the cost of holding compensating balances.

Prime Supplier—A firm that produces, imports, or transports selected petroleum products across State boundaries and local marketing areas, and sells the product to local distributors, local retailers, or end users.

Principal—One who employs an Agent to act on his behalf; or the chief or foremost Party in a particular transaction; or the amount of a loan exclusive of interest; or the assets constituting a Trust Estate.

Prior Period Correction—Restatement of a production month's measurement allocation or contract quantities in subsequent months. Also called prior month's adjustments.

Priorities of Service—A predetermined schedule of service obligations or contracts which specifies where one such service or contract takes precedence over another for deliveries of energy.

Priority—That which is earlier or previous in point of time or right, such as first Deed of Trust is ahead of or has priority over a second Deed of Trust.

Private Equity—In contrast to public equity, private equity denotes a participation in a private—that is, not publicly listed—established company.

Equity securities of companies that have not "gone public" (are not listed on a public exchange). Private equities are generally illiquid and thought of as a long-term investment. As they are not listed on an exchange, any investor wishing to sell securities in private companies must find a buyer in the absence of a marketplace. In addition, there are many transfer restrictions on private securities. Investors in private securities generally receive their return through one of three ways: an initial public offering, a sale or merger, or a recapitalization.

Private Fueling Facility—A fueling facility which normally services only fleets and is not open to the general public.

Private Investment or Private Financing—Obtaining project funds by other than capital appropriation or governmental grants.

Private Letter Ruling—A ruling by the IRS requested by parties to a finance transaction that is applicable only to the assumed facts stated in the opinion.

Private Offering—Sale of unregistered, restricted securities by a company.

Private Office or Work Area—An office bounded by 72-inch or higher permanent partitions and is no more than 200 square feet. *See Occupancy Type*

Private Placement—Also known as a Reg. D offering. The sale of a security directly to a limited number of investors in a private transaction.

Private Placement Memorandum—Also known as an Offering Memorandum. A document that outlines the terms of securities to be offered in a private placement. Resembles a business plan in content and structure.

Private Securities—Private securities are securities that are not registered and do not trade on an exchange. The price per share is set through negotiation between the buyer and the seller or issuer.

Privately Owned Electric Utility—A class of ownership found in the electric power industry where the utility is regulated and authorized to achieve an allowed rate of return.

Pro Forma Invoice—An invoice provided by a supplier prior to the shipment of merchandise, informing the buyer of the kinds and quantities of goods to be sent, their value, and important specifications (weight, size, etc.).

Probable (indicated) Reserves, Coal—Reserves or resources for which tonnage and grade are computed partly from specific measurements, samples, or production data and partly from projection for a reasonable distance on the basis of geological evidence. The sites available are too widely or otherwise inappropriately spaced to permit the mineral bodies to be outlined completely or the grade established throughout.

Probable Energy Reserves—Estimated quantities of energy sources that, on the basis of geologic evidence that supports projections from Proved Reserves (See definition below), can reasonably be expected to exist and be recoverable under existing economic and operating conditions. Site information is insufficient to establish with confidence the location, quality, and grades of the energy source. Note: This term is equivalent to "Indicated Reserves" as defined in the resource/reserve classification contained in the U.S. Geological Survey Circular 831, 1980. Measured and indicated reserves, when combined, constitute Demonstrated Reserves.

Process—An activity or treatment that is not related to the space conditioning, lighting, service water heating, or ventilating of a building as it relates to human occupancy.

Process Audit (Green-e program)—A process audit is the type of audit performed annually to verify Green-e certified products meet the Green-e Standard. A process audit is different from a traditional financial audit in that the auditor only reviews those materials and processes dictated by a set of

agreed-upon procedures developed by the Center for Resource Solutions.

Process Cooling and Refrigeration—The direct process end use in which energy is used to lower the temperature of substances involved in the manufacturing process. Examples include freezing processed meats for later sale in the food industry and lowering the temperature of chemical feedstock below ambient temperature for use in reactions in the chemical industries. Not included are uses such as air-conditioning for personal comfort and cafeteria refrigeration.

Process Emissions—GHG emissions other than combustion emissions occurring as a result of intentional and unintentional reactions between substances or their transformation, including the chemical or electrolytic reduction of metal ores, the thermal decomposition of substances, and the formation of substances for use as product or feedstock.

Process Fuel—All energy consumed in the acquisition, processing, and transportation of energy. Quantifiable process fuel includes three categories: natural gas lease and plant operations, natural gas pipeline operations, and oil refinery operations.

Process Heat—Heat used in an industrial process rather than for space heating or other housekeeping purposes.

Process Heating or Cooling Demand-side Management (DSM) Program—A DSM program designed to promote increased electric energy efficiency applications in industrial process heating or cooling.

Process Heating or Cooling Waste Heat Recovery—An energy conservation system whereby some space heating or water heating is done by actively capturing byproduct heat that would otherwise be ejected into the environment. In nonresidential buildings, sources of waste heat include refrigeration/air-conditioner compressors, manufacturing or other processes, data processing centers, lighting fixtures, ventilation exhaust air, and the occupants themselves. Not to be considered is the passive use of radiant heat from lighting, workers, motors, ovens, etc., when there are no special systems for collecting and redistributing heat.

Process Load—A load resulting from a process.

Process Vent—An opening where a gas stream is continuously or periodically discharged during normal operation. Process vents include openings where gas streams are discharged to the atmosphere directly or are discharged to the atmosphere after being routed to a control device or a product recovery device.

Processed Gas—Natural gas that has gone through a processing plant.

Processing—Uranium-recovery operations whether at a mill, an in situ leach, byproduct plant, or other type of recovery operation.

Processing Beds—Vessels containing catalyst.

Processing Gain—The volumetric amount by which total output is greater than input for a given period of time. This difference is due to the processing of crude oil into products which, in total, have a lower specific gravity than the crude oil processed.

Processing Loss—The volumetric amount by which total refinery output is less than input for a given period of time. This difference is due to the processing of crude oil into products which, in total, have a higher specific gravity than the crude oil processed.

Processing of Uranium—The recovery of uranium produced by non-conventional mining methods, i.e., in situ leach mining, as a byproduct of copper or phosphate mining, or heap leaching.

Processing Plant—A surface installation designed to separate and recover natural gas liquids from a stream of produced natural gas through the processes of condensation, absorption, adsorption, refrigeration, or other methods and to control the quality of natural gas marketed and/or returned to oil or gas reservoirs for pressure maintenance, repressuring, or cycling.

Producer (Natural Gas)—A company engaged in the production and sale of natural gas from gas or oil wells with delivery generally at a point at or near the wellhead, the field, or the tailgate of a gas processing plant. For the purpose of company classification, a company primarily engaged in the exploration for, development of, and/or production of oil and/or natural gas.

Producer and Distributor Coal Stocks—Producer and distributor coal stocks consist of coal held in stock by producers/distributors at the end of a reporting period.

Producer Contracted Reserves—The volume of recoverable salable gas reserves committed to or controlled by the reporting pipeline company as the buyer in gas purchase contracts with the independent producer as seller, including warranty contracts, and which are used for acts and services for which the company has received certificate authorization from the Federal Energy Regulatory Commission.

Producer Gas—Fuel gas high in carbon monoxide (CO)

and hydrogen (H_2), produced by burning a solid fuel with insufficient air or by passing a mixture of air and steam through a burning bed of solid fuel.

Producing Property—A term often used in reference to a property, well, or mine that produces wasting natural resources. The term means a property that produces in paying quantities (that is, one for which proceeds from production exceed operating expenses).

Product (Green-e program)—A product is a retail electricity service package (also known as offering or option) sold by an electric service provider. Green-e defines a product by its resource content mix. Pricing variations that do not change a given product's resource mix, SO_x, NO_x, or greenhouse gas emissions, do not constitute different products.

Product Supplied, Crude Oil—Crude oil burned on leases and by pipelines as fuel.

Production—See production terms associated with specific energy types.

Production Capacity—The amount of product that can be produced from processing facilities.

Production Costing—A method used to determine the most economical way to operate a given system of power resources under given load conditions.

Production Costs—Costs incurred to operate and maintain wells and related equipment and facilities, including depreciation and applicable operating costs of support equipment and facilities and other costs of operating and maintaining those wells and related equipment and facilities. They become part of the cost of oil and gas produced. The following are examples of production costs (sometimes called lifting costs):

- Costs of labor to operate the wells and related equipment and facilities; repair and maintenance costs; the costs of materials, supplies, and fuels consumed and services utilized in operating the wells and related equipment and facilities; the costs of property taxes and insurance applicable to proved properties and wells and related equipment and facilities; the costs of severance taxes.

- Depreciation, depletion, and amortization (DD&A) of capitalized acquisition, exploration, and development costs are not production costs, but also become part of the cost of oil and gas produced along with production (lifting) costs identified above. Production costs include the following subcategories of costs:

— well workers and maintenance;
— operating fluid injections and improved recovery programs;
— operating gas processing plants;
— ad valorem taxes;
— production or severance taxes;
— other, including overhead.

Production Expenses—Costs incurred in the production of electric power that conform to the accounting requirements of the Operation and Maintenance Expense Accounts of the FERC Uniform System of Accounts.

Production Payments—A contractual arrangement providing a mineral interest that gives the owner a right to receive a fraction of production, or of proceeds from the sale of production, until a specified quantity of minerals (or a definite sum of money, including interest) has been received.

Production Plant Liquids—The volume of liquids removed from natural gas in natural gas processing plants or cycling plants during the year.

Production Tax Credit—An inflation-adjusted 1.5 cents per kilowatthour payment for electricity produced using qualifying renewable energy sources. (Rate changes annually)
• Provides the investor or owner of qualifying property with an annual tax credit based on the amount of electricity generated by that facility.

Production, Crude Oil—The volumes of crude oil that are extracted from oil reservoirs. These volumes are determined through measurement of the volumes delivered from lease storage tanks or at the point of custody transfer, with adjustment for (1) net differences between opening and closing lease inventories and (2) basic sediment and water. Crude oil used on the lease is considered production.

Production, Lease Condensate—The volume of lease condensate produced. Lease condensate volumes include only those volumes recovered from lease or field separation facilities.

Production, Natural Gas—The volume of natural gas withdrawn from reservoirs less (1) the volume returned to such reservoirs in cycling, repressuring of oil reservoirs, and conservation operations; less (2) shrinkage resulting from the removal of lease condensate; and less (3) non-hydrocarbon gases where they occur in sufficient quantity to render the gas unmarketable. Volumes of gas withdrawn from gas storage reservoirs and native gas, which has been transferred to the storage category, are not considered production. Flared and vented gas

is also considered production. (This differs from "Marketed Production" which excludes flared and vented gas.)

Production, Natural Gas Liquids—Production of natural gas liquids is classified as follows:

Contract Production—Natural gas liquids accruing to a company because of its ownership of liquids extraction facilities that it uses to extract liquids from gas belonging to others, thereby earning a portion of the resultant liquids.

Leasehold Production—Natural gas liquids produced, extracted, and credited to a company's interest.

Contract Reserves—Natural gas liquid reserves corresponding to the contract production defined above.

Leasehold Reserves—Natural gas liquid reserves corresponding to leasehold production defined above.

Production, Natural Gas, Dry—The volume of natural gas withdrawn from reservoirs during the report year less (1) the volume returned to such reservoirs in cycling, repressuring of oil reservoirs, and conservation operations; less (2) shrinkage resulting from the removal of lease condensate and plant liquids; and less (3) non-hydrocarbon gases where they occur in sufficient quantity to render the gas unmarketable. Volumes of gas withdrawn from gas storage reservoirs and native gas, which has been transferred to the storage category, are not considered production. This is not the same as marketed production, because the latter also excludes vented and flared gas, but contains plant liquids.

Production, Natural Gas, Wet after Lease Separation—The volume of natural gas withdrawn from reservoirs less (1) the volume returned to such reservoirs in cycling, repressuring of oil reservoirs, and conservation operations; less (2) shrinkage resulting from the removal of lease condensate; and less (3) non-hydrocarbon gases where they occur in sufficient quantity to render the gas unmarketable.

Note: Volumes of gas withdrawn from gas storage reservoirs and native gas that has been transferred to the storage category are not considered part of production. This production concept is not the same as marketed production, which excludes vented and flared gas.

Production, Oil and Gas—The lifting of the oil and gas to the surface and gathering, treating, field processing (as in the case of processing gas to extract liquid hydrocarbons), and field storage. The production function typically terminates at the outlet valve on the lease or field production storage tank. If unusual physical or operational circumstances exist, it may be more appropriate to regard the production function as terminating at the first point at which oil, gas, or gas liquids are delivered to a main pipeline, a common carrier, a refinery, or a marine terminal. Gross Company-Operated Production. Total production from all company-operated properties, including all working and non-working interests. Net Working Interest Production. Total production accruing to the reporting company's working interests less royalty oil and volumes due others.

Productive Capacity—The maximum amount of coal that a mining operation can produce or process during a period with the existing mining equipment and/or preparation plant in place, assuming that the labor and materials sufficient to utilize the plant and equipment are available, and that the market exists for the maximum production.

Productive Machine Hours—That portion of scheduled operating hours during which a machine performs its designated functions, excluding time to transport the machine and operational or mechanical delays.

Products of Combustion—The elements and compounds that result from the combustion of a fuel.

Products Supplied—Approximately represents consumption of petroleum products because it measures the disappearance of these products from primary sources, i.e., refineries, natural gas-processing plants, blending plants, pipelines, and bulk terminals. In general, product supplied of each product in any given period is computed as follows: field production, plus refinery production, plus imports, plus unaccounted-for crude oil (plus net receipts when calculated on a PAD District basis) minus stock change, minus crude oil losses, minus refinery inputs, and minus exports.

Professional Judgment—The ability to render sound decisions based on professional qualifications and relevant experience.

Profit—The income remaining after all business expenses are paid.

Profit Test (Tax)—The IRS test of whether the lessor achieves a significant profit in a lease, apart from tax benefits.

Program Administrator (PA)—Refers to a third party who performs administration of a government or other energy generation program.

Program Cost—Utility costs that reflect the total cash expenditures for the year, reported in nominal dollars, that flowed out to support DSM (demand-side management) programs. They are reported in the year they are incurred, regardless of when the actual effects occur.

Program Life—The length of time that the utility will be actively involved in promoting a demand-side management program (i.e. financing the marketing activities and the incentives of the program.)

Program Maturity—The time it takes for the full benefits of a demand-side management measure or program to be realized.

Programmable Controller—A device that controls the operation of electrical equipment (such as air conditioning units and lights) according to a preset time schedule.

Programmable Thermostat—A type of thermostat that allows the user to program into the devices' memory a pre-set schedule of times (when certain temperatures occur) to turn on HVAC equipment.

Programmed Timber Harvest—A timber harvest scheduled by a management plan to occur at a certain rate.

Project Financing—This is the most commonly used method to finance the construction of independent power facilities. Typically, the developer pledges the value of the plant and part or all of its expected revenues as collateral to secure financing from private or institutional lenders.

Project Scenario—Project Scenario is an emission reduction project's emission forecast. In some cases the project scenario may be nil but its operations may reduce emissions in the existing operations of the business. The Project Scenario is compared with the existing operations Business as Usual (or baseline) scenario to determine the emission reductions achieved by the emission reduction project.

Project Specific M&V Plan—Plan providing details on how a specific project's savings will be verified based on the general M&V options described in the project plan.

Projected Area (Wind)—The net south-facing glazing area projected on a vertical plane. Also, the solid area covered at any instant by a wind turbines blades from the perspective of the direction of the windstream (as opposed to the swept area).

Proof—A measure of ethanol content. 1 percent ethanol content equals 2 proof.

Prop 65—California's Safe Drinking Water and Toxic Enforcement Act (1986) The Proposition was intended by its authors to protect California citizens and the State's drinking water sources from chemicals known to cause cancer, birth defects or other reproductive harm, and to inform citizens about exposures to such chemicals.

Propane (C_3H_8)—A normally gaseous straight-chain hydrocarbon. It is a colorless paraffinic gas that boils at a temperature of -43.67 degrees Fahrenheit. It is extracted from natural gas or refinery gas streams. It includes all products designated in ASTM Specification D1835 and Gas Processors Association Specifications for commercial propane and HD-5 propane.

Propane Air—A mixture of propane and air resulting in a gaseous fuel suitable for pipeline distribution.

Propane, Consumer Grade—A normally gaseous paraffinic compound (C3H8), which includes all products covered by Natural Gas Policy Act Specifications for commercial and HD-5 propane and ASTM Specification D 1835. Excludes: feedstock propanes, which are propanes not classified as consumer grade propanes, including the propane portion of any natural gas liquid mixes, i.e., butane-propane mix.

Propeller (Hydro) Turbine—A turbine that has a runner with attached blades similar to a propeller used to drive a ship. As water passes over the curved propeller blades, it causes rotation of the shaft.

Proportional Interest in Investee Reserves—The proportional interest at the end of the year in the reserves of Investees that are accounted for by the equity method.

Proposed Design—The proposed building design which must comply with the standards before receiving a building permit. See *also Energy Budget and Standard Design*

Proposed Rates—New electric rate schedule proposed by an applicant to become effective at a future date.

Proposed Savings—Proposed savings are those estimated in the contract prior to project implementation determined from metering and/or calculations performed in accordance with the provisions of the approved measurement and verification plan.

Proprietary Capital—A group of balance sheet accounts which includes common capital stock, preferred capital stock, other paid-in capital installments received on capital stock, discount on capital stock, capital stock expense, appropriated retained earnings, unappropriated retained earnings, and reacquired capital stock.

Proprietary Information (Trade Secret)—Regulatory agencies will classify information as proprietary provided the owner demonstrates the following: the business has asserted a business confidentiality claim; the business has shown it has taken reasonable measures to protect the confidentiality of the information both within the company and fro outside entities; the information is not, and has not been reasonably obtainable without the business' consent; no statute specifically requires disclosure of the information; and either the business has shown that disclosure of the information is likely to cause substantial harm to its competitive position, or the information is voluntarily submitted and its disclosure would likely impair the government's ability to obtain necessary information in the future.

Propylene (C_3H_6)—An olefinic hydrocarbon recovered from refinery processes or petrochemical processes.

Prorate—To divide, distribute or assess proportionately.

Prorated Bill—The computation of a bill based upon proportionate distribution of the applicable billing schedule. A prorated bill is less than 25 days or more than 38 days.

Prospecting—The search for an area of probable mineralization; the search normally includes topographical, geological, and geophysical studies of relatively large areas undertaken in an attempt to locate specific areas warranting detailed exploration. Prospecting usually occurs prior to the acquisition of mineral rights.

Prospecting Costs—Direct and indirect costs incurred to identify areas of interest that may warrant detailed exploration. Such costs include those incurred for topographical, geological, and geophysical studies; rights of access to properties in order to conduct such studies, salaries, equipment, instruments, and supplies for geologists, including geophysical crews, and others conducting such studies; and overhead that can be identified with those activities.

Prospectus—A formal written offer to sell securities that provides an investor with the necessary information to make an informed decision. A prospectus explains a proposed or existing business enterprise and must disclose any material risks and information according to the securities laws. A prospectus must be filed with the SEC and be given to all potential investors. Companies offering securities, mutual funds, and offerings of other investment companies including Business Development Companies are required to issue prospectuses describing their history, investment philosophy or objectives, risk factors, and financial statements. Investors should carefully read them prior to investing.

Protestant—A party who files a protest to an application or other filing by another party.

Protocol—A protocol is linked to an existing convention. It is a separate, additional agreement that must be signed and ratified by the Parties to the convention. Protocols are a way of strengthening a convention by adding new, more detailed commitments. See *Kyoto Protocol*

Proton—The positively charged component of the nucleus of an atom. The positively charged hydrogen ion which remains when an electron is removed from a hydrogen atom. The proton's positive charge is equal in magnitude to that of the electron's negative charge.

Proton Exchange Membrane (PEM)—A solid polymer membrane (a thin plastic film), which serves as the electrolyte in a PEM fuel cell.

Proved (Measured) Reserves, Coal—Reserves or resources for which tonnage is computed from dimensions revealed in outcrops, trenches, workings, and drill holes and for which the grade is computed from the results of detailed sampling. The sites for inspection, sampling, and measurement are spaced so closely and the geologic character is so well defined that size, shape, and mineral content are well established. The computed tonnage and grade are judged to be accurate within limits that are stated, and no such limit is judged to be different from the computed tonnage or grade by more than 20 percent.

Proved Energy Reserves—Estimated quantities of energy sources that analysis of geologic and engineering data demonstrates with reasonable certainty are recoverable under existing economic and operating conditions. The location, quantity, and grade of the energy source are usually considered to be well established in such reserves. Note: This term is equivalent to "Measured Reserves" as defined in the resource/reserve classification contained in the U.S. Geological Survey Circular 831, 1980. Measured and indicated reserves, when combined, constitute Demonstrated Reserves.

Proved Reserves—The estimated quantities of all natural gas or crude oil that geological and engineering data demonstrate with reasonable certainty to be recoverable in future years from known reservoirs under existing economic and operating conditions.

Provider of Last Resort—A legal obligation (traditionally given to utilities) to provide service to a customer where competitors have decided they do not want that customer's business.

Proximate Analysis—A commonly used analysis for reporting fuel properties; may be on a dry (moisture free) basis, as "fired," or on an ash and moisture free basis. Fractions usually reported include: volatile matter, fixed carbon, moisture, ash, and heating value (higher heating value).

PS—Planning Staff
- Point Source
- Preparedness Staff

PSA—Pipeline Safety Act (1992)

PSA Off-gas—The impurity stream resulting from the sequential PSA pressurization/depressurization purification process. (see *Pressure Swing Adsorption*)

PSD—See *Prevention of Significant Deterioration*.

PSI—Pollutant Standards Index
- Process Safety Information

Psi—Pounds force of pressure per square inch.

Psia—Pounds force of pressure per square inch absolute (including atmospheric pressure).

Psig—Pounds force of pressure per square inch gauge (excluding atmospheric pressure).

PSM—Point Source Modeling
- Point Source Monitoring
- Process Safety Management

PSNS—Pretreatment Standards for New Sources

Psychrometer—An instrument for measuring relative humidity by means of wet and dry-bulb temperatures.

Psychrometrics—The analysis of atmospheric conditions, particularly moisture in the air.

PT—Potential Transformer

PTC—PVUSA Test Conditions
- Production Tax Credit (USA). A federal tax credit for businesses generating renewable energy from wind, biomass, geothermal, irrigation, municipal solid waste, hydropower or marine and hydrokinetic (tidal) sources. The amount of PTCs available is based on the amount of electricity generated by a facility (system) and is adjusted annually to account for inflation.
- Permit to Construct

PTCF (Financial)—Pre-tax Cash Flows

PTE—Potential to Emit

PTFE—Polytetrafluoroethylene (Teflon)

PTI—Permit to Install

PTO—Permit to Operate

PTO Letter—Permission to Operate Letter. Letter from a utility that constitutes "express written permission" for the facility to operate after the utility has inspected and tested the facility.

P-Type Semiconductor—A semiconductor in which holes carry the current; produced by doping an intrinsic semiconductor with an electron acceptor impurity (e.g., boron in silicon).

Public Adviser—An appointee of the governor who attends all meetings of the California Energy Commission and provides assistance to members of the public and intervenors in cases before the Commission.

Public Areas—Spaces generally open to the public at large, customers, congregation members, or similar spaces, where occupants need to be prevented from controlling lights for safety, security, or business reasons.

Public Authorities—Electricity supplied to municipalities, divisions, or agencies of state and Federal governments, usually under special contracts or agreements that are applicable only to public authorities.

Public Authority Service to Public Authorities—Public authority service includes electricity supplied and services rendered to municipalities or divisions or agencies of State or Federal governments under special contracts, agreements, or service classifications applicable only to public authorities.

Public Benefit Energy Program (PBE)—Refers to a program that encourages and in some cases offers incentives for energy efficiency and renewable energy generation via a pool of money collected from utility customers, ratepayers, or other means. PBE programs are generally authorized on a state level.

Public Company—A company that has securities that have been sold in a registered offering and that are traded on a stock exchange or NASDAQ. Must be a Reporting Company under SEC rules. Often used incorrectly to describe companies that are only Reporting Companies and that have not conducted a registered offering under Securities Act.

Public Entity—Includes the United States, the state and any county, city, public corporation, or public district of the state or local government, and any department entity, agency or authority of any thereof.

Public Goods Charge (PGC)—A universal charge applied to each electric utility customer's bill to support the provision of public goods. Public goods covered by California's electric PGC include public purpose energy efficiency programs, low-income services, renewables, and energy-related research and development.

Public Interest Goals—Public interest goals of electric utility regulation include: 1) inter-and intra-class and intergenerational equity); 2) the equal treatment of equals (horizontal equity); 3) balancing long- and short-term goals that have the potential to affect intergenerational balance; 4) protecting against the abuse of monopoly power; and 5) general protection of the health and welfare of the citizens of the state, nation, and world. Environmental and other types of social costs are subsumed under the equity and health and welfare responsibilities.

Public Owned Utilities (POUs)—Non-profit utility providers owned by a community and operated by municipalities, counties, states, public power districts, or other public organizations.

Public Street and Highway Lighting—Electricity supplied and services rendered for the purpose of lighting streets, highways, parks, and other public places; or for traffic or other signal system service, for municipalities or other divisions or agencies of State or Federal governments.

Public Utility—Enterprise providing essential public services, such as electric, gas, telephone, water, and sewer under legally established monopoly conditions.

Public Utility Commissions—State agencies that regulate investor-owned utilities operating in the state.

Public Utility District (PUD)—A publicly owned energy producer or distributor. PUDs operate as special government districts under the authority of elected commissions. They are not regulated by public utility commissions.

Public Utility Holding Company Act of 1935 (PUHCA)—This act prohibits acquisition of any wholesale or retail electric business through a holding company unless that business forms part of an integrated public utility system when combined with the utility's other electric business. The legislation also restricts ownership of an electric business by non-utility corporations.

Public Utility or Services Commissions (PUC or PSC)—These are state government agencies responsible for the regulation of public utilities within a state or region. A state legislature oversees the PUC by reviewing changes to power generator laws, rules and regulations and approving the PUC's budget. The commission usually has five Commissioners appointed by the Governor or legislature. PUCs typically regulate: electric, natural gas, water, sewer, telephone services, trucks, buses, and taxicabs within the commission's operating region. The PUC tries to balance the interests of consumers, environmentalists, utilities, and stockholders. The PUC makes sure a region's citizens are supplied with adequate, safe power provider service at reasonable rates.

Public Utility Regulatory Policies Act (PURPA) of 1978—One part of the National Energy Act, PURPA contains measures designed to encourage the conservation of energy, more efficient use of resources, and equitable rates. Principal among these were suggested retail rate reforms and new incentives for production of electricity by co-generators and users of renewable resources. The Commission has primary authority for implementing several key PURPA programs.

Publicly Owned Electric Utility—A class of ownership found in the electric power industry. This group includes those utilities operated by municipalities and State and Federal power agencies.

Publicly Owned Treatment Works (POTW) Permit—A permit that regulates discharges to publicly owned sewage treatment plants.

PUC—Public Utility Commission A public agency that usually is tasked with setting tariffs for utility companies.

PUHCA—The Public Utility Holding Company Act of 1935. This act prohibits acquisition of any wholesale or retail electric business through a holding company unless that business forms part of an integrated public utility system when combined with the utility's other electric business. The legislation also restricts ownership of an electric business by non-utility corporations.

Pulp Chips—Timber or residues processed into small pieces of wood of more or less uniform dimensions with minimal amounts of bark.

Pulp Wood—Roundwood, whole-tree chips, or wood residues.

Pulping Liquor (Black Liquor)—The alkaline spent liquor removed from the digesters in the process of chemically pulping wood. After evaporation, the liquor is burned as a fuel in a recovery furnace that permits the recovery of certain basic chemicals.

Pulse-Width-Modulated (PWM) Wave Inverter—A type of power inverter that produce a high quality (nearly sinusoidal) voltage, at minimum current harmonics.

PUM—See *Piling Un-merchantable Material*.

Pumped Hydroelectric Storage—Commercial method used for large-scale storage of power. During off-peak times, excess power is used to pump water to a reservoir. During peak times, the reservoir releases water to operate hydroelectric generators.

Pumped Storage Facility—A type of power generating facility that pumps water to a storage reservoir during off-peak periods, and uses the stored water (by allowing it to fall through a hydro turbine) to generate power during peak periods. The pumping energy is typically supplied by lower cost base power capacity, and the peaking power capacity is of greater value, even though there is a net loss of power in the process.

Pumped-Storage Hydroelectric Plant—A plant that usually generates electric energy during peak load periods by using water previously pumped into an elevated storage reservoir during off-peak periods when excess generating capacity is available to do so. When additional generating capacity is needed, the water can be released from the reservoir through a conduit to turbine generators located in a power plant at a lower level.

Punch List—A list of items to be completed that is prepared during final inspection of a structure or equipment installation.

Purchase—A complete bid.

Purchase Agreement Assignment—An agreement in which some or all of the lessee's rights under a purchase agreement (including the right to take title to the equipment) are assigned by the lessee to the owner trustee prior to the delivery of the property by the manufacturer. A consent of the manufacturer or supplier that confirms the availability to the owner trustee of the rights of purchaser under the contract. This assignment usually is annexed to the agreement.

Purchase Option—An option in a lease agreement that allows the lessee to purchase the leased equipment at the end of the lease term for either a fixed amount or at the future fair market value of the leased equipment.

Purchase Order Assignment—A document which transfers all rights contained in a purchase order for equipment (i.e., to purchase the equipment at a certain price, with certain terms) from the lessee to the lessor, enabling the lessor to purchase the equipment from the manufacturer and lease it to the lessee.

Purchase Price Allocation—In an asset sale, the process

of assigning agreed upon values to all major assets and liabilities of a business. The allocation separates the purchase prince into two primary categories: (1) tangible assets, and (2) intangible assets, for tax purposes. The two primary categories are then separated into secondary categories that describe each asset and specify the agreed upon values. This item is heavily negotiated throughout the Merger & Acquisition process.

Purchase-Contract Imports of Uranium—The amount of foreign-origin uranium material that enters the United States during a survey year as reported on the "Uranium Industry Annual Survey (UIAS), Form EIA-858, as purchases of uranium ore, U3O8, natural UF6, or enriched UF6. The amount of foreign-origin uranium materials that enter the country during a survey year under other types of contracts, i.e., loans and exchanges, is excluded.

Purchased—Receipts into transportation, storage, and/or distribution facilities within a state under gas purchase contracts or agreements whether or not billing or payment occurred during the report year.

Purchased Power—Power purchased or available for purchase from a source outside the system.

Purchased Power Adjustment—A clause in a rate schedule that provides for adjustments to the bill when energy from another electric system is acquired and its cost varies from a specified unit base amount.

Pure Pumped-Storage Hydroelectric Plant—A plant that produces power only from water that has previously been pumped to an upper reservoir.

Purge—To free a gas conduit of air, gas or a mixture of same.

Purge Gas—Nitrogen, carbon dioxide, liquefied petroleum gas, or natural gas used to maintain a non-explosive mixture of gases in a flare header or provide sufficient exit velocity to prevent regressive flame travel back into the flare header.

PURPA—The Public Utility Regulatory Policy Act of 1978. Among other things, this federal legislation requires utilities to buy electric power from private "qualifying facilities," at an avoided cost rate. This avoided cost rate is equivalent to what it would have otherwise cost the utility to generate or purchase that power themselves. Utilities must further provide customers who choose to self-generate a reasonably priced back-up supply of electricity. Under PURPA each electric utility is required to offer to purchase available electric energy from co-generation and small power production facilities.

Put Option—An option in a lease (e.g., for equipment purchase or lease renewal) in which the exercise of the option is at the lessor's, not the lessee's, discretion.
 • The right to sell a security at a given price (or range) within a given time period.

PV—Photovoltaic
 • Project Verification
 • Present Value

PV Array—Two or more photovoltaic panels wired in series and or parallel.

PVA—Photovoltaic array
 • Polyvinyl Alcohol

PVC—Polyvinyl Chloride

PVCs That Convert Sunlight Directly Into Energy—A method for producing energy by converting sunlight using photovoltaic cells (PVCs) that are solid-state single converter devices. Although currently not in wide usage, commercial customers have a growing interest in usage and, therefore, DOE has a growing interest in the impact of PVCs on energy consumption. Economically, PVCs are competitive with other sources of electricity.

PWSA—Ports and Waterways Safety Act

PX—The California Power Exchange Corporation, a state chartered, non-profit corporation charged with providing Day-Ahead and Hour-Ahead markets for energy and ancillary services, if it chooses to self-provide, in accordance with the PX tariff. The PX is a Scheduling Coordinator, and is independent of both the ISO and all other market participants.

PX Generation—Generation being scheduled by the PX.

PX Load—Load that has been scheduled by the PX, and which is received through the use of transmission or distribution facilities owned by participating transmission owners.

PX Participant—An entity that is authorized to buy or sell energy or ancillary services through the PX, and any agent authorized to act on behalf of such an entity.

PY—Prior Year
 • Program Year

Pyranometer—A device used to measure total incident solar radiation (direct beam, diffuse, and reflected radiation) per unit time per unit area.
 • An instrument for measuring total hemispherical solar irradiance on a flat surface, or "global" irradiance; thermopile sensors have been generally identified as pyranometers, however, silicon sensors are also referred to as pyranometers.

Pyrheliometer—A device that measures the intensity of direct beam solar radiation.

Pyrolysis—The transformation on a compound or material into one or more substances by heat alone (without oxidation). Often called destructive distillation. Pyrolysis of biomass is the thermal degradation of the material in the absence of reacting gases, and occurs prior to or simultaneously with gasification reactions in a gasifier. Pyrolysis products consist of gases, liquids, and char generally. The liquid fraction of pyrolized biomass consists of an insoluble viscous tar, and pyroligneous acids (acetic acid, methanol, acetone, esters, aldehydes, and furfural). The distribution of pyrolysis products varies depending on the feedstock composition, heating rate, temperature, and pressure.

Pyrometer—An instrument for the measurement of high temperatures.

Q

QA—Quality Assurance

QAC—Quality Assurance Coordinator

QC—Quality Control

QF—Qualifying Facility. A cogeneration facility or energy production plant that can sell its electricity and qualifies for production tax credits as defined in Section 45 of the Internal Revenue Code.

QFER—Quarterly Fuel and Energy Reports

QNCR—Quarterly Noncompliance Report

QRA—Quantitative Risk Assessment/Analysis

QUA—Qualitative Use Assessment

Quad—One quadrillion (10^{15} or 1,000,000,000,000,000) British thermal units (Btus). An amount of energy equal to 170 million barrels of oil. Total U.S. consumption of all forms of energy is (in the 1990s) about 83 quads in an average year.

Quadrillion—The quantity 1,000,000,000,000,000 (10 to the 15th power).

Qualification Test (PV)—A testing procedure for PV modules relating to electrical, mechanical, or thermal stress. Test results are subject to a list of defined requirements.

Qualifying Facility (QF)—QFs are non-utility power producers that often generate electricity using renewable and alternative resources, such as hydro, wind, solar, geothermal or biomass (solid waste). QFs must meet certain operating, efficiency, and fuel-use standards set forth by the Federal Energy Regulatory Commission (FERC). If they meet these FERC standards, utilities must buy power from them. QFs usually have long-term contracts with utilities for the purchase of this power, which is among the utility's highest-priced resources.

• Under PURPA, QFs were allowed to sell their electric output to the local utility at avoided cost rates. To become a QF, the independent power supplier had to produce electricity with a specified fuel type (cogeneration or renewables), and meet certain ownership, size, and efficiency criteria established by the Federal Energy Regulatory Commission.

Quality or Grade (of Coal)—An informal classification of coal relating to its suitability for use for a particular purpose. Refers to individual measurements such as heat value, fixed carbon, moisture, ash, sulfur, major, minor, and trace elements, coking properties, petrologic properties, and particular organic constituents. The individual quality elements may be aggregated in various ways to classify coal for such special purposes as metallurgical, gas, petrochemical, and blending usages.

Quantified Emission Limitation and Reduction (QELRC)—Also known as QELRO (Quantified Emission Limitation and Reduction Objective): The quantified commitments for GHG emissions listed in Annex B of the Kyoto Protocol. QELRCs are specified in percentages relative to 1990 emissions. Legally-binding targets and timetables under the Kyoto Protocol for the limitation or reduction of greenhouse gas emissions for developed countries. Also referred to as Quantified Emissions Limitation and Reduction Objectives (QELROs).

Quantity Wires Charge—A fee for moving electricity over the transmission and/or distribution system that is based on the quantity of electricity that is transmitted.

Quartz-Halogen Light—An incandescent lamp filled with halogen gas. Somewhat more efficient than standard incandescents.

Quasi Sine-wave—A description of the type of waveform produced by some Inverters.

Quaternary—The present period, forming the latter part of the Cenozoic Era, originating about 1 million years ago and including the Recent and Pleistocene epochs.

Quench Tower—A gas cooling and pollution control device in which heated gases are showered with water. Gases are cooled and particulates "drop out" of the gases. They can generate a waste called "quench tower dropout."

Quick Ratio—Cash, accounts receivable, government securities and cash equivalents to current liabilities. This ratio nets out all current asset items of questionable liquidity such as inventories. This is a measure of how quickly a company can pay off its short-term obligations without relying on the sale of relatively illiquid assets. A ratio of at least 1:1 is desireable.

Quiet Enjoyment Clause—A contractual provision that permits the lessee to use the leased property free from unreasonable interference from the lessor.

Quiet Title—An Action to establish Title to Real Property or to remove a cloud on Title.

Quitclaim Deed—A Deed which conveys whatever present Right, Title or Interest the grantor may have.

R

R&D—Research and Development

R2—A measure of the overall closeness of a meter Baseline Model to historical data, with a value between 0.0 and 1.0. The Baseline Model is a best-fit straight line (Linear Regression) through the given data points. A high R2, above about 0.70, indicates that the model fits the data quite well. A low value indicates that the model does not fit the data very well and therefore that the selection of Independent Variables may not be appropriate.

RA—Reasonable Alternative
  • Regional Administrator
  • Registered Architect
  • Regulatory Alternative
  • Regulatory Analysis
  • Relative Accuracy
  • Remedial Action
  • Resource Allocation
  • Return Authorization
  • Risk Analysis
  • Risk Assessment
  • Run Average

Rack Sales—Wholesale truckload sales or smaller of gasoline where title transfers at a terminal.

RACM—Reasonably Available Control Measures

RACT—Reasonably Available Control Technology

Rad—A unit of measure of absorbed radiation. Acronym for radiation absorbed dose. One rad equals 100 ergs of radiation energy per gram of absorbing material.

Radiant Barrier—A thin, reflective foil sheet that exhibits low radiant energy transmission and under certain conditions can block radiant heat transfer; installed in attics to reduce heat flow through a roof assembly into the living space.

Radiant Ceiling Panels—Ceiling panels that contain electric resistance heating elements embedded within them to provide radiant heat to a room.

Radiant Energy—Energy transferred by the exchange of electromagnetic waves from a hot or warm object to one that is cold or cooler. Direct contact with the object is not necessary for the heat transfer to occur.

Radiant Floor—A type of radiant heating system where the building floor contains channels or tubes through which hot fluids such as air or water are circulated. The whole floor is evenly heated. Thus, the room heats from the bottom up. Radiant floor heating eliminates the draft and dust problems associated with forced air heating systems.

Radiant Heating System—A heating system where heat is supplied (radiated) into a room by means of heated surfaces, such as electric resistance elements, hot water (hydronic) radiators, etc.

Radiation—The flow of energy across open space via electromagnetic waves such as light. Passage of heat from one object to another without warming the air space in between. The process of emitting energy in the form of energetic particles (such as alpha particles or gamma radiation), light or heat.

Radiative Cooling—The process of cooling by which a heat absorbing media absorbs heat from one source and radiates the heat away.

Radiative Forcing—The term radiative forcing refers to changes in the energy balance of the earth-atmosphere system in response to a change in factors such as greenhouse gases, land-use change, or solar radiation. The climate system inherently attempts to balance incoming (e.g., light) and outgoing (e.g. heat) radiation. Positive radiative forcings increase the temperature of the lower atmosphere, which in turn increases temperatures at the Earth's surface. Negative radiative forcings cool the lower atmosphere. Radiative forcing is most commonly measured in units of watts per square meter (W/m2). Also see *Greenhouse Gases*.

Radiatively Active Gases—Gases that absorb incoming solar radiation or outgoing infrared radiation, affecting the vertical temperature profile of the atmosphere. Also see *Radiative Forcing* above.

Radiator—A heating unit usually exposed to view within the room or space to be heated; it transfers heat by radiation to objects within visible range and by conduction to the surrounding air, which in turn is circulated by natural convection; usually fed by steam or hot water.
  • A room heat delivery (or exchanger) component

of a hydronic (hot water or steam) heating system; hot water or steam is delivered to it by natural convection or by a pump from a boiler.

Radiator Vent—A device that releases pressure within a radiator when the pressure inside exceeds the operating limits of the vent.

Radioactive Waste—Materials left over from making nuclear energy. Radioactive waste can destroy living organisms if it is not stored safely.

Radioactivity—The spontaneous emission of radiation from the nucleus of an atom. Radionuclides lose particles and energy through this process.

Radioisotope—A radioactive isotope.

Radiosondes—Sensors carried aboard weather balloons that have been in continuous use since 1979 for the monitoring of tropospheric temperatures.

Radium—A radioactive element with a half-life of 1,600 years that emits alpha particles as it is transformed into radon. In the past, radium was mixed with special paints to make watch faces and instrument dials glow in the dark.

Radon—A naturally occurring radioactive gas found in the United States in nearly all types of soil, rock, and water. It can migrate into most buildings. Studies have linked high concentrations of radon to lung cancer.

RAF—Risk Assessment Forum

Rafter—A construction element used for ceiling support.

RAGS—Risk Assessment Guidance

Rail (method of transportation to consumers)—Shipments of coal moved to consumers by rail (private or public/commercial). Includes coal hauled to or away from a railroad siding by truck.

Railroad and Railway Electric Service—Electricity supplied to railroads and interurban and street railways, for general railroad use, including the propulsion of cars or locomotives, where such electricity is supplied under separate and distinct rate schedules.

Railroad Locomotive—Self-propelled vehicle that runs on rails and is used for moving railroad cars.

Railroad Use—Sales to railroads for any use, including that used for heating buildings operated by railroads.

Raised Floor—A floor (partition) over a crawl space, or an unconditioned space, or ambient air.

Rammed Earth—A construction material made by compressing earth in a form; used traditionally in many areas of the world and widely throughout North Africa and the Middle East.

Ramp Rate—The rate at which you can increase load on a power plant. The ramp rate for a hydroelectric facility may be dependent on how rapidly water surface elevation on the river changes.

Ramp Up (Demand Side)—Implementing a demand-side management program over time until the program is considered fully installed.

Ramp Up (Supply Side)—Increasing load on a generating unit at a rate called the ramp rate.

Range Top—The range burners or stove top and the oven are considered two separate appliances. Counted also with range tops are stand-alone "cook tops."

Rankine Cycle—The steam-Rankine cycle employing steam turbines has been the mainstay of utility thermal electric power generation for many years. The cycle, as developed over the years uses superheat, reheat and regeneration. Modern steam Rankine systems operate at a cycle top temperature of about 1,073 degrees Celsius with efficiencies of about 40 percent.

Rankine Cycle Engine—The Rankine cycle system uses a liquid that evaporates when heated and expands to produce work, such as turning a turbine, which when connected to a generator, produces electricity. The exhaust vapor expelled from the turbine condenses and the liquid is pumped back to the boiler to repeat the cycle. The working fluid most commonly used is water, though other liquids can also be used. Rankine cycle design is used by most commercial electric power plants. The traditional steam locomotive is also a common form of the Rankine cycle engine. The Rankine engine itself can be either a piston engine or a turbine.

RAO—Remedial Action Objectives

RAP—Registration Assessment Panel
- Remedial Accomplishment Plan
- Response Action Plan

RAPS (Remote Area Power Supply)—A power generation system used to provide electricity to remote and rural homes, usually incorporating power generated from renewable sources such as solar panels and wind generators, as well as non-renewable sources such as petrol-powered generators.

RAS—Risk Assessment Study

RAT—Relative Accuracy Test

Ratchet Clause, Demand—A clause in a rate schedule which provides that maximum past or future demands are taken into account to establish billings for previous or subsequent periods.

Rate Base—The value of property upon which a utility is permitted to earn a specified rate of return as established by a regulatory authority. The rate base generally represents the value of property used by the utility in providing service and may be calculated by any one or a combination of the following accounting methods: fair value, prudent investment, reproduction cost, or original cost. Depending on which method is used, the rate base includes cash, working capital, materials and supplies, deductions for accumulated provisions for depreciation, contributions in aid of construction, customer advances for construction, accumulated deferred income taxes, and accumulated deferred investment tax credits.

Rate Case—A proceeding, usually before a regulatory commission, involving the rates to be charged for a public utility service.

Rate Class—A group of customers identified as a class and subject to a rate different from the rates of other groups.

Rate Design—The various rates per class of customer or service charged by a utility.

The term "rate design" refers to the method of classifying fixed and variable costs between demand and commodity components. Examples of different rate designs are single rate (100% commodity or volumetric), two-part rates (demand and commodity rates), three-part rates (two demand and one commodity rate) and multiple rates (zone rates).

Rate Factor—A percentage amount that, when multiplied by the original equipment cost, produces the monthly payment on a finance contract.

Rate Features—Special rate schedules or tariffs offered to customers by electric and/or natural gas utilities.

Rate of Charge—The amount of energy per unit time that is being added to the battery. Rate of charge is commonly expressed as a ratio of the battery or cell's rated capacity to charge duration in hours. Example: A C/20 rate on a 100 AH battery would be 5 amps, the capacity of the battery divided by 20.

Rate of Return—The ratio of net operating income earned by a utility is calculated as a percentage of its rate base.

Rate of Return On Rate Base—The ratio of net operating income earned by a utility, calculated as a percentage of its rate base.

Rate Schedule—A price list showing how the electric bill of a particular type of customer will be calculated by an electric utility company.

Rate Schedule (Electric)—A statement of the financial terms and conditions governing a class or classes of utility services provided to a customer. Approval of the schedule is given by the appropriate rate-making authority.

Rate Schedule, Zone—A rate schedule restricted in its availability to a particular geographic area.

Rate Structure—The design and organization of billing charges by customer class to distribute the revenue requirement among customer classes and rating period.

Rate-Basing—Refers to practice by utilities of allotting funds invested in utility Research Development Demonstration and Commercialization and other programs from ratepayers, as opposed to allocating these costs to shareholders.

Rated Battery Capacity (Ah)—Term used by battery manufacturers to indicate the maximum amount of energy that can be withdrawn from a battery at a specified discharge rate and temperature.

Rated Capacity—The manufacturer's specification for the amount of charge that may be stored in a battery, commonly expressed in amp-hours at a specific rate of discharge.

Rated Life—The length of time that a product or appliance is expected to meet a certain level of performance under nominal operating conditions; in a Luminaire, the period after which the lumen depreciation and lamp failure is at 70% of its initial value.

Rated Module Current (A)—The current output of a PV module measured under standard test conditions of 1000 W/m2 and 25°C cell temperature.
• Nominal power output of an inverter, some units cannot produce rated power continuously.

Rated Power—The power output of a device under specific or nominal operating conditions.

Rated Watt—The manufacturer's specification for power output of a generating device. In most cases, this is not the most accurate measure to look at, since it predicts output only for ideal circumstances.

Ratemaking Authority—A utility Commission's legal authority to fix, modify, approve, or disapprove rates, as determined by the powers given the commission by a State or Federal legislature.

Ratemaking Authority—A utility commission's legal authority to fix, modify, approve, or disapprove rates as determined by the powers given the commission by a State or Federal legislature.

Rate-of-Return Rates—Rates set to the average cost of electricity as an incentive for regulated utilities to operate more efficiently at lower rates where costs are minimized.

Ratepayer—This is a retail consumer of the electricity distributed by an electric utility. This includes residential, commercial and industrial users of electricity.

Rates—The authorized charges per unit or level of consumption for a specified time period for any of the classes of utility services provided to a customer.

Rates, Demand—The term "demand rate" applies to any method of charge for energy which is based upon, or is a function of, the rate of use or size of the customer's installation or maximum demand during a given period of time.

Ratification—After signing the UNFCCCC or the Kyoto Protocol, a country must ratify it, often with the approval of its parliament or other legislature. In the case of the Kyoto Protocol, a Party must deposit its instrument of ratification with the UN Secretary General in New York.

Rating (Electricity)—A manufacturer's guaranteed performance of a machine, transmission line, or other electrical apparatus, based on design features and test data. The rating will specify such limits as load, voltage, temperature, and frequency. The rating is generally printed on a nameplate attached to equipment and is commonly referred to as the nameplate rating or nameplate capacity.

Ratio Estimate—The ratio of two population aggregates (totals). For example, "average miles traveled per vehicle" is the ratio of total miles driven by all vehicles, over the total number of vehicles, within any subgroup. There are two types of ratio estimates: those computed using aggregates for vehicles and those computed using aggregates for households.

Ratoon Crop—A crop cultivated from the shoots of a perennial plant.

Raw Fuel—Coal, natural gas, wood or other fuel that is used in the form in which it is found in nature, without chemical processing.

Rayleigh Frequency Distribution—A mathematical representation of the frequency or ratio that specific wind speeds occur within a specified time interval.

RBC—Risk Based Concentration

RCO—Regional Compliance Officer
- Regional Contracting Officer

RCRA—See *Resource Conservation and Recovery Act.* (1976)

RCRIS—Resource Conservation and Recovery Information System

RD&D—Research, Development and Demonstration
- Research, Development and Deployment

RD/RA—Remedial Design/Remedial Action

RDF (Refuse Derived Fuel)—The fuel component of municipal solid waste (MSW), which is the by-product of shredding MSW to a uniform size, screening out oversized materials and isolating ferrous material in magnetic separation. The resulting RDF can be burned as a fuel source.

RDL—Reliable Detection Limit (Level)

RDT&E—Research, Development, Testing and Evaluation

RE—Renewable Energy
- Reasonable Efforts
- Removal Efficiency
- Reportable Event

REA—Rural Electrification Administration

Reactance—A phenomenon associated with AC power characterized by the existence of a time difference between voltage and current variations.

Reactive—A class of compounds which are normally unstable and readily undergo violent change, react violently with water, can produce toxic gases with water, or possess other similar properties. Reactivity is one characteristic that can make a waste hazardous.

Reactive Organic Gas—In the SCAQMD (South Coast Air Quality Management District in Southern California) New Source Review program, ROG is any gaseous chemical compound which contains the element of carbon excluding a variety of compounds.

Reactive Power—The portion of electricity that establishes and sustains the electric and magnetic fields of alternating-current equipment. Reactive power must be supplied to most types of magnetic equipment, such as motors and transformers. It also must supply the reactive losses on transmission facilities. Reactive power is provided by generators, synchronous condensers, or electrostatic equipment such as capacitors and directly influences electric system voltage. It is usually expressed in kilovars (kvar) or megavars (Mvar).
- The electrical power that oscillates between the magnetic field of an inductor and the electrical field of a capacitor. Reactive power is never converted to non-electrical power. It is calculated as the square root of the difference between the square of the kilovolt-amperes and the square of the kilowatts and is expressed as reactive volt-amperes.
- The sine of the phase angle between the current

and voltage waveforms in an AC system.

Reactor—A device in which a controlled nuclear chain reaction can be maintained, producing heat energy.

Readily Accessible—Capable of being reached quickly for operation, repair or inspection, without requiring climbing or removing obstacles, or resorting to access equipment.

Real Dollars—These are dollars that have been adjusted for inflation.

Real Price—A price that has been adjusted to remove the effect of changes in the purchasing power of the dollar. Real prices, which are expressed in constant dollars, usually reflect buying power relative to a base year.

Real Property—Land and buildings as opposed to Personal Property or Chattels.

Real Time Market—The competitive generation market controlled and coordinated by the ISO for arranging real-time imbalance energy.

Real Time Pricing—When prices are set on a short-term basis, usually one day in advance, reflecting a utility's marginal costs.

REAP—Renewable Electricity Action Plan

Reasonable Assurance—The assurance risk by the depth and breadth of analysis and substantive testing through the verification process that has been reduced to an acceptably low level.

Reasonably Assured Resources (RAR)—The uranium that occurs in known mineral deposits of such size, grade, and configuration that it could be recovered within the given production cost ranges, with currently proven mining and processing technology. Estimates of tonnage and grade are based on specific sample data and measurements of the deposits and on knowledge of deposit characteristics. RAR correspond to DOE's Reserves category.

Rebate Program—A utility company-sponsored conservation program whereby the utility company returns a portion of the purchase price cost when a more energy-efficient refrigerator, water heater, air conditioner, or other appliance is purchased.

Reburn—An advanced co-firing technique using natural gas to reduce pollution from electric power plants.

REC—Renewable Energy Credit. A tradable commodity that represents 1 megawatt hour of electricity generated by a renewable energy facility. Purchasers of RECs are often utilities subject to renewable portfolio standards that require a certain portion of electricity produced and sold to its customers be from renewable sources. Often required to meet mandatory Renewable Portfolio Standards (RPSs) in a state.

REC Banking—An administrative means by which RECs can be stored for later user or sale. For example, Texas RECs have a 3-year life. If a REC is not used in the year of its creation, it may be banked and used in either of the next 2 compliance periods (years). The issue date of the RECs coincides with the beginning of the compliance year in which the RECs were generated.

REC Trading Program—The process of awarding, trading, tracking, and submitting RECs.

Recapitalization—The reorganization of a company's capital structure. A company may seek to save on taxes by replacing preferred stock with bonds in order to gain interest deductibility. Recapitalization can be an alternative exit strategy for venture capitalists and leveraged-buyout sponsors. (See also: Exit Strategy and Leveraged Buyout.)

Recapture—Return of tax benefits to an owner of an asset if a financial contract terminates early.

Receipts—Deliveries of fuel to an electric plant
- Purchases of fuel
- All revenues received by an exporter for the reported quantity exported
- Also see *Received* (below).

Receivables From Municipality—All charges by the utility department against the municipality or its other departments that are subject to current settlement.

Received—Gas (and other fuels) physically transferred into the responding company's transportation, storage, and/or distribution facilities.

Receiver—The component of a central receiver solar thermal system where reflected solar energy is absorbed and converted to thermal energy

Receiving Party—Entity receiving the capacity and/or energy transmitted by the transmission provider to the point(s) of delivery.

Rechargeable Battery—A type of battery that uses a reversible chemical reaction to produce electricity, allowing it to be re-used many times. The chemical reaction is reversed by forcing electricity through the battery in the opposite direction to normal discharge.

Recirculated Air—Air that is returned from a heated or cooled space, reconditioned and/or cleaned, and returned to the space. This process reduces energy usage by recirculating a large portion of the air supplied to a space, which normally uses much less energy than bringing in the full airflow from outside and conditioning it to the space requirements. The mechanism used to control recirculation and mixing in an air handler is known as an Economizer.

Recirculation—Returning a fraction of the effluent outflow to the inlet to dilute incoming wastewater.

Recirculation Systems—A type of solar heating system that circulate warm water from storage through the collectors and exposed piping whenever freezing conditions occur; obviously a not very efficient system when operating in this mode.

RECLAIM—The Regional Clean Air Incentives Market established by the SCAQMD, a cap and trade rule covering sources with 4 tons per year or more of NO_X or SO_X emissions and resulting in a reduction of NO_X and SO_X emissions of 75% and 61%, respectively, from affected sources.

RECLAIM Pollutants—In the RECLAIM program, oxides of nitrogen (NO_X) and oxides of sulfur (SO_X), excluding any NO_X emissions from on-site, off-road mobile sources and any SO_X emissions from equipment burning natural gas.

RECLAIM Trading Credit (RTC)—In the RECLAIM program, the RTC is a limited authorization to emit a RECLAIM pollutant in accordance with the restrictions and requirements of the RECLAIM rules. Each RTC has a denomination of one pound of RECLAIM pollutant and a term of one year, and can be held as part of a facility's Allocation or alternatively may be evidenced by an RTC Certificate.

Reclaimed Oil—Lubricating oil that is processed to be used over again.

Reclaimed Water—Former waste water (sewage) that has been treated to remove solids and certain impurities and reintroduced into the aquifer for non-potable use, such as irrigation, dust control, and fire suppression.

Reclamation—Process of restoring surface environment to acceptable pre-existing conditions. Includes surface contouring, equipment removal, well plugging, re-vegetation, etc.

Reclamation Expenses—In the context of the coal operation statement of income, refers to all payments made by the company attributable to reclamation, including taxes.

Recognition (cash vs. accrual)—The method of calculating the effect on taxable income of cash events. Usually accrual, as opposed to cash.

Recombination—The action of a free electron falling back into a hole. Recombination processes are either radiative, where the energy of recombination results in the emission of a photon, or non-radiative, where the energy of recombination is given to a second electron which then relaxes back to its original energy by emitting phonons. Recombination can take place in the bulk of the semiconductor, at the surfaces, in the junction region, at defects, or between interfaces.

Reconciliation Period—The period of time after the quarter, half-year, or compliance year ends during which affected sources may true-up their accounts. Credit or allowances holders that are long (e.g., they have more credits or allowances in their account than are needed to cover their emissions over the just completed period) have the opportunity to sell the surplus credits or allowances during the reconciliation period. Those that are short (e.g., they have an insufficient quantity of credits or allowances in their compliance accounts than are needed to cover their emissions over the just completed period) may acquire credits or allowances to account for any shortfalls during the reconciliation period. Each program has its own reconciliation period.

Reconfirmation—The act a broker/dealer makes with an investor to confirm a transaction.

Recool—The cooling of air that has been previously heated by space conditioning equipment or systems serving the same building.

Record Drawings—Drawings that document the as installed location and performance data on all lighting and space conditioning system components, devices, appliances and equipment, including but not limited to wiring sequences, control sequences, duct and pipe distribution system layout and sizes, space conditioning system terminal device layout and air flow rates, hydronic system and flow rates, and connections for the space conditioning system. Record drawings are sometimes called "as builts."

Record of Decision (ROD)—The decision document for an environmental impact statement (EIS). Separate from the EIS itself, this document states the decision, states the reasons for the decision, identifies all alternatives and states compliance with applicable laws.

Recoverability—In reference to accessible coal resources, the condition of being physically, technologically, and economically minable. Recovery rates and recovery factors may be determined or estimated for coal resources without certain knowledge of their economic minability; therefore, the availability of recovery rates or factors does not predict recoverability.

Recoverable Coal—Coal that is, or can be, extracted from a coal bed during mining.

Recoverable Heat—That portion of thermal input to a prime mover that is not converted to mechanical power and can be reclaimed for utilization.

Recoverable Proved Reserves—The proved reserves of natural gas as of December 31 of any given year are the estimated quantities of natural gas which geological and engineering data demonstrates with reasonable certainty to be recoverable in the future from known natural oil and gas reservoirs under existing economic and operating conditions.

Recoverable Reserves—The amount of coal that can be recovered (mined) from the coal deposits at active producing mines as of the end of the year.

Recovered Energy—Energy used in a building that (1) is mechanically recovered from space conditioning, service water heating, lighting, or process equipment after the energy has performed its original function; (2) provides space conditioning, service water-heating, or lighting; and (3) would otherwise be wasted.

Recovery Boiler—A pulp mill boiler in which lignin and spent cooking liquor (black liquor) is burned to generate steam.

Recovery Efficiency—(Thermal efficiency) In a water heater, a measure of the percentage of heat from the combustion of gas that is transferred to the water as measured under specified test conditions. California Code of Regulations, Section 2-1602(e) (7).

Recovery Factor (Coal)—The percentage of total tons of coal estimated to be recoverable from a given area in relation to the total tonnage estimated to be in the demonstrated reserve base. The estimated recovery factors for the demonstrated reserve base generally are 50 percent for underground mining methods and 80 percent for surface mining methods. More precise recovery factors can be computed by determining the total coal in place and the total recoverable in any specific locale.

Recovery Percentage (Coal)—The percentage of coal that can be recovered from the coal deposits at existing mines.

Rectifier—An electrical device for converting alternating current to direct current. The chamber in a cooling device where water is separated from the working fluid (for example ammonia).

Recuperator—A heat exchanger in which heat is recovered from the products of combustion.

Recurrent Costs—Costs that are repetitive and occur when an organization produces similar goods or services on a continuing basis.

Recycled—A material that is used or reused, or reclaimed.

Recycled Feeds—Feeds that are continuously fed back for additional processing.

Recycling—The process of converting materials that are no longer useful as designed or intended into a new product.

Red Herring—The common name for a preliminary prospectus, due to the red SEC required legend on the cover. (See also: Prospectus.)

REDD—Reduced Emissions from Deforestation and Degradation. An initiative to cut greenhouse gas emissions associated with forest clearing by the inclusion of "avoided deforestation" in carbon market mechanisms. More simply, payment in return for the active preservation of existing forests.

Redeemable Preferred Stock—Redeemable preferred stock, also known as exploding preferred, at the holder's option after (typically) five years, which in turn gives the holders (potentially converting to creditors) leverage to induce the company to arrange a liquidity event. The threat of creditor status can move the founders off the dime if a liquidity event is not occurring with sufficient rapidity.

Redox Potential—A measurement of the state of oxidation of a system.

Redrill Footage—Occasionally, a hole is lost or junked and a second hole may be drilled from the surface in close proximity to the first. Footage drilled for the second hole is defined as "redrill footage." Under these circumstances, the first hole is reported as a dry hole (explanatory or developmental) and the total footage is reported as dry hole footage. The second hole is reported as an oil well, gas well, or dry hole according to the result. The redrill footage is included in the appropriate classification of total footage, but is not reported as a separate classification.

Reduced Flicker Operation—The operation of a light, in which the light has a visual flicker less than 30% for frequency and modulation.

Reduced Use-off Hours—A conservation feature consisting of manually or automatically reducing the amount of heating or cooling produced during the hours a building is not in full use.

Reduction—A verified decrease in GHG emissions caused by a project, as measured against an appropriate forward-looking estimate of baseline emissions for the project.

REEP—Reasonable Extra Efforts Program
• Review of Environmental Effects of Pollutants

REF—Rural Electrification Fund

Reference Computer Program—The reference method against which other methods are compared. For the nonresidential standards, the reference computer program is DOE 2.1 E. For the low-rise residential standards the reference computer program is CALRES (in California)

Reference Month—The calendar month and year to which the reported cost, price, and volume information relates.

Reference Year—The calendar year to which the reported sales volume information relates.

Refinance—To convert interest payable on a debt into additional principal owed.

Refined Coal—A liquid, gaseous or solid fuel produced from coal or high carbon fly ash meeting the requirements of IRC Section 45(c)(7).

Refined Petroleum Products—Refined petroleum products include but are not limited to gasolines, kerosene, distillates (including No. 2 fuel oil), liquefied petroleum gas, asphalt, lubricating oils, diesel fuels, and residual fuels.

Refiner—A firm or the part of a firm that refines products or blends and substantially changes products, or refines liquid hydrocarbons from oil and gas field gases, or recovers liquefied petroleum gases incident to petroleum refining and sells those products to resellers, retailers, reseller/retailers or ultimate consumers. "Refiner" includes any owner of products that contracts to have those products refined and then sells the refined products to resellers, retailers, or ultimate consumers.

Refiner Acquisition Cost of Crude Oil—The cost of crude oil, including transportation and other fees paid by the refiner. The composite cost is the weighted average of domestic and imported crude oil costs. Note: The refiner acquisition cost does not include the cost of crude oil purchased for the Strategic Petroleum Reserve (SPR).

Refinery—A facility that separates crude oil into varied oil products. The refinery uses progressive temperature changes to separate by vaporizing the chemical components of crude oil that have different boiling points. These are distilled into usable products such as gasoline, fuel oil, lubricants and kerosene.

Refinery Capacity Utilization—Ratio of the total amount of crude oil, unfinished oils, and natural gas plant liquids run through crude oil distillation units to the operable capacity of these units.

Refinery Fuel—Crude oil and petroleum products consumed at the refinery for all purposes.

Refinery Fuel Gas—Also known as "still gas." Gas generated at a petroleum refinery or any gas generated by a refinery process unit, which is combusted separately or in any combination with any type of gas or used as a chemical feedstock.

Refinery Gas—Non-condensate gas collected in petroleum refineries.

Refinery Input, Crude Oil—Total crude oil (domestic plus foreign) input to crude oil distillation units and other refinery processing units (cokers, etc.).

Refinery Input, Total—The raw materials and intermediate materials processed at refineries to produce finished petroleum products. They include crude oil, products of natural gas processing plants, unfinished oils, other hydrocarbons and oxygenates, motor gasoline and aviation gasoline blending components and finished petroleum products.

Refinery Losses and Gains—Processing gain and loss that takes place during the refining process itself. Excludes losses that do not take place during the refining process, e.g., spills, fire losses, and contamination during blending, transportation, or storage.

Refinery Output—The total amount of petroleum products produced at a refinery. Includes petroleum consumed by the refinery.

Refinery Production—Petroleum products produced at a refinery or blending plant. Published production of these products equals refinery production minus refinery input. Negative production will occur when the amount of a product produced during the month is less than the amount that is reprocessed (input) or reclassified to become another product during the same month. Refinery production of unfinished oils and motor and aviation gasoline blending components appear on a net basis under refinery input.

Refinery Utilization Rate—Represents the use of the atmospheric crude oil distillation units. The rate is calculated by dividing the gross input to these units by the operable refining capacity of the units.

Refinery Yield—Refinery yield (expressed as a percentage) represents the percent of finished product produced from input of crude oil and net input of unfinished oils. It is calculated by dividing the sum of crude oil and net unfinished input into the individual net production of finished products. Before calculating the yield for finished motor gasoline, the input of natural gas liquids, other hydrocarbons and oxygenates, and net input of motor gasoline blending components must be subtracted from the net production of finished aviation gasoline.

Refining/Market Segment Power—Energy. Electricity for use as energy.

Reflectance—The amount (percent) of light that is reflected by a surface relative to the amount that strikes it.

Reflectance, Solar—The ratio of the reflected solar flux to the incident solar flux.

Reflective Coatings—Materials with various qualities that are applied to glass windows before installation. These coatings reduce radiant heat transfer through the window and also reflects outside heat and a portion of the incoming solar energy, thus reducing heat gain. The most common type has a sputtered coating on the inside of a window unit. The other type is a durable "hard-coat" glass with a coating, baked into the glass surface.

Reflective Film—Transparent covering for glass that helps keep out heat from the sun.

Reflective Glass—A window glass that has been coated with a reflective film and is useful in controlling solar heat gain during the summer.

Reflective Insulation (See also *Radiant Barrier*)—An aluminum foil fabricated insulator with backings applied to provide a series of closed air spaces with highly reflective surfaces.

Reflective Window Films—A material applied to window panes that controls heat gain and loss, reduces glare, minimizes fabric fading, and provides privacy. These films are retrofitted on existing windows.

Reflectivity—The ratio of the energy carried by a wave after reflection from a surface to its energy before reflection.

Reflector Lamps—A type of incandescent lamp with an interior coating of aluminum that reflects light to the front of the bulb. They are designed to spread light over specific areas.

Reforestation—Replanting of forests on lands that have recently been harvested or otherwise cleared of trees.

Reformate—Fluid that exits the fuel reformer and supplies fuel to the fuel cell stack.

Reformation Action—A Legal Action to correct a mistake in a Deed or other document.

Reformer—A component in which fuel gas is reacted with water vapor and heat, usually in the presence of a catalyst, to produce hydrogen-rich gas for use in a fuel cell.

Reformulated Gasoline (RFG)—Finished motor gasoline formulated for use in motor vehicles, the composition and properties of which meet the requirements of the reformulated gasoline regulations promulgated by the U.S. Environmental Protection Agency under Section 211(k) of the Clean Air Act. Note: This category includes oxygenated fuels program reformulated gasoline (OPRG) but excludes reformulated gasoline blendstock for oxygenate blending (RBOB).

Refraction—The change in direction of a ray of light when it passes through one media to another with differing optical densities.

Refractory Lining—A lining, usually of ceramic, capable of resisting and maintaining high temperatures.

Refrigerant—The compound (working fluid) used in air conditioners, heat pumps, and refrigerators to transfer heat into or out of an interior space. This fluid boils at a very low temperature enabling it to evaporate and absorb heat.

Refrigerant Charge—The amount of refrigerant that is installed or "charged" into an air conditioner or heat pump. The refrigerant is the working fluid. It is compressed and becomes a liquid as it enters the condenser. The hot liquid is cooled in the condenser and flows to the evaporator where it released through the expansion valve. When the pressure is released, the refrigerant expands into a gas and cools. Air is passed over the evaporator to provide the space cooling. When an air conditioner or heat pump has too much refrigerant (overcharged) the compressor may be damaged. When an air conditioner has too little refrigerant (undercharged), the efficiency of the unit is reduced. A thermostatic *expansion* valve (TXV) can mitigate the impact of improper refrigerant charge.

Refrigerated Case—A manufactured commercial refrigerator or freezer, including but not limited to display cases, reach-in cabinets, meat cases, and frozen food and soda fountain units.

Refrigeration—The process of the absorption of heat from one location and its transfer to another for rejection or recuperation.

Refrigeration Capacity—A measure of the effective cooling capacity of a refrigerator, expressed in Btu per hour or in tons, where one (1) ton of capacity is equal to the heat required to melt 2,000 pounds of ice in 24 hours or 12,000 Btu per hour.

Refrigeration Cycle—The complete cycle of stages (evaporation and condensation) of refrigeration or of the refrigerant.

Refrigeration Ton—12,000 Btu per hour or 200 Btu per minute of heat removal. Originally, the amount of heat required to melt a ton of ice in 24 hours.

Refrigeration Unit—Lowers the temperature through a mechanical process. In a typical refrigeration unit, electricity powers a motor that runs a pump to compress the refrigerant to maintain proper pressure. (A "refrigerant" is a substance that changes between liquid and gaseous states under desirable temperature and pressure conditions.) Heat from the compressed liquid is removed and discharged from the unit and the refrigerant then evaporates when pressure is reduced. The refrigerant picks up heat as it evaporates and it returns to the compressor to repeat the cycle. A few refrigeration units use gas (either natural gas or LPG) in an absorption process that does not use a compressor. The gas is burned to heat a chemical solution in which the refrigerant has been absorbed. Heating drives off the refrigerant which is later condensed. The condensed refrigerant evaporates by a release of pressure, and it picks up heat as it evaporates. The evaporated refrigerant is then absorbed back into the chemical solution, the heat is removed from the solution and discharged as waste heat, and the process repeats itself. By definition, refrigerators, freezers, and air-conditioning equipment all contain refrigeration units.

Refundable Security Deposit—An amount paid by the lessee to the lessor as security for fulfillment of all obligations outlined in the lease agreement that is subsequently refunded to the lessee once all obligations have been satisfied. Security deposits typically are returned at the end of the lease term but, according to mutual agreement, can be refunded at any point during the lease.

Refunding—Retirement of one security issue with proceeds received from selling another. Refunding provides for retiring maturing debt by taking advantage of favorable money market conditions.

Refuse Bank—A repository for waste material generated by the coal cleaning process.

Refuse Mine—A surface mine where coal is recovered from previously mined coal. It may also be known as a silt bank, culm bank, refuse bank, slurry dam, or dredge operation.

Refuse-Derived Fuel (RDF)—A solid fuel produced by shredding municipal solid waste (MSW). Noncombustible materials such as glass and metals are generally removed prior to making RDF. The residual material is sold as-is or compressed into pellets, bricks, or logs. RDF processing facilities are typically located near a source of MSW, while the RDF combustion facility can be located elsewhere. Existing RDF facilities process between 100 and 3,000 tons per day.

REG NEG—Negotiating Rulemaking

Regeneration Harvest—A timber harvest method that removes selected trees in the existing stand to a density that allows for the establishment of a new even-aged stand below.

Regeneration with Reserves—Similar to a regeneration harvest, except that a number of green trees are left standing to meet other resource needs such as wildlife habitat. The number of trees left is usually specified as a certain number of trees per acre.

Regenerative Braking—Electricity can be generated by most electric motors. In this instance, it is energy captured and stored as drivers use their EV's brakes.

Regenerative Cooling—A type of cooling system that uses a charging and discharging cycle with a thermal or latent heat storage subsystem.

Regenerative Heating—The process of using heat that is rejected in one part of a cycle for another function or in another part of the cycle.

Regional Groups—The five regional groups meet privately to discuss issues and nominate bureau members and other officials. They are Africa, Asia, Central and Eastern Europe (CEE), Latin America and the Caribbean (GRULAC), and the Western Europe and Others Group (WEOG).

Regional Power Exchange—An entity established to coordinate short-term operations to maintain system stability and achieve least-cost dispatch. The dispatch provides back-up supplies, short-term excess sales, reactive power support, and spinning reserve. The pool may own, manager and/or operate the transmission lines or be an independent entity that manages the transactions between entities.

Regional Reliability Councils—Regional organizations charged with maintaining system reliability even during abnormal bulk power conditions such as outages and unexpectedly high loads.

Regional Transmission Group—A utility industry concept that the Federal Energy Regulatory Commission (FERC) embraced for the certification of voluntary groups that would be responsible for transmission planning and use on a regional basis.

Registered Offering—["Public Offering"] A transaction in which a Company sells specified securities to the public under a Registration Statement which has been declared effective by the SEC.

Registration—Process of registering emission reduction data with a third-party registry.

The SEC's review process of all securities intended to be sold to the public. The SEC requires that a registration statement be filed in conjunction with any public securities offering. This document includes operational and financial information about the company, the management, and the purpose of the offering. The registration statement and the prospectus are often referred to interchangeably. Technically, the SEC does not "approve" the disclosures in prospectuses.

Registration Obligation—The obligation of Company to register the shares issued to an investor in a private offering for resale to the public through a Registration Statement which the SEC has declared effective.

Registration Rights—The right to require that a company register restricted shares. Demand Registered Rights enable the shareholder to request registration at any time, while Piggy Back Registration Rights enable the shareholder to request that the company register his or her shares when the company files a registration statement (for a public offering with the SEC).

Registration Statement—The document filed by a Company with the SEC under the Securities Act in order to obtain approval to sell the securities described in the Registration Statement to the public. [S-1, S-2, S-3, S-4, SB-1, SB-2, S-8, etc.] Includes the Prospectus.

Registry—This term is used on the Trading Floor to indicate which emission reductions on offer have been registered with an independent emission reduction registry. Each registry applies its own criteria to determine what emissions merit registering. The registration process can vary widely from verification against rigorous measurement protocol to simply dating the registrant's, as yet, unverified claims.

Regression Analysis—A technique used to develop a mathematical model from a set of data that describes the correlation of measured variables.

Regular Grade Gasoline—A grade of unleaded gasoline with a lower octane rating (approximately 82) than other grades. Octane boosters are added to gasoline to control engine pre-ignition or "knocking" by slowing combustion rates.

Regulated Public Utility—A corporation engaged in the furnishing or sale of:
- Electric energy, gas, water, or sewerage disposal services, or
- Transportation on an intrastate, suburban, municipal, or interurban electric railroad, or an intrastate, municipal, or suburban trackless trolley system, or on a municipal or suburban bus system, or
- Transportation by motor vehicle—

If the rates for such furnishing or sale, as the case may be, have been established or approved by a State or political subdivision thereof, by an agency or instrumentality of the United States, by a public service or public utility commission or other similar body of the District of Columbia or of any State or political subdivision thereof, or by a foreign country or an agency or instrumentality of political subdivision thereof.

A corporation engaged as a common carrier in the furnishing or sale of transportation of gas by pipe line, if subject to the jurisdiction of the Federal Energy Regulatory Commission (FERC).

A corporation engaged as a common carrier:
- In the furnishing or sale of transportation by railroad, if subject to the jurisdiction of the Surface Transportation Board, or
- In the furnishing or sale of transportation of oil or other petroleum products (including shale oil) by pipe line, if subject to the jurisdiction of the Federal Regulatory Commission or if the rates for such furnishing or sale ore subject to the jurisdiction of a public service or public utility commission or other similar body of the District of Columbia or of any State.

A corporation engaged in the furnishing or sale of telephone or telegraph service, if the rates for such furnishing or sale meet the requirements above.

A corporation engaged in the furnishing or sale of transportation as a common carrier by air, subject to the jurisdiction of the Secretary of Transportation.

A corporation engaged in the furnishing or sale of transportation by a water carrier subject to jurisdiction under subchapter II of chapter 135 of title 49 or the IRC.

A rail carrier subject to part A of subtitle IV of the title 49, if:
- Substantially all of its railroad properties have been leased to another such railroad corporation or corporations by an agreement or agreements entered into before January 1, 1954,
- Each lease is for a term of more than 20 years, and

- At least 80 percent or more of its gross income (computed without regard to dividends and capital gains and losses) for the taxable year is derived from such leases and from sources described in subparagraphs (A) through (F), inclusive. For purposes of the preceding sentence, an agreement for lease of railroad properties entered into before January 1, 1954, shall be considered to be a lease including such term as the total number of years of such agreement may, unless sooner terminated, be renewed or continued under the terms of the agreement, and any such renewal or continuance under such agreement shall be considered part of the lease entered into before January 1, 1954.

A common parent corporation which is a common carrier by railroad subject to part A of subtitle IV of title 49 if at least 80 percent of its gross income (computed without regard to capital gains or losses) is derived directly or indirectly from sources described above. For purposes of the preceding sentence, dividends and interest, and income from leases described above, received from a regulated public utility shall be considered as derived from sources described above, if the regulated public utility is a member of an affiliated group (as defined in IRC section 1504) which includes the common parent corporation.

Regulated Stream Flow—The rate of flow past a given point during a specified period that is controlled by reservoir water release operation.

Regulation—The service provided by generating units equipped and operating with automatic generation controls that enables the units to respond to the ISO's direct digital control signals to match real-time demand and resources, consistent with established operating criteria.

Regulation A—SEC provision for simplified registration for small issues of securities. A Reg. A issue may require a shorter prospectus and carries lesser liability for directors and officers for misleading statements.

Regulation C—The regulation that outlines registration requirements for Securities Act of 1933.

Regulation D—Regulation D is the rule (Reg. D is a "regulation" comprising a series of "rules") that allow for the issuance and sale of securities.

Regulation Down—Regulation reserve provided by a resource that can decrease its actual operating level in response to a direct electronic (AGC) signal from the CAISO to maintain standard frequency in accordance with established reliability criteria.

Regulation S—The rules relating to Offers and Sales made outside the US without SEC Registration.

Regulation S-B—Reg. S-B of the Securities Act of 1933 governs the Integrated Disclosure System for Small Business Issuers.

Regulation S-K—The Standard Instructions for Filing Forms Under Securities Act of 1933, Securities Exchange Act of 1934, and Energy Policy and Conservation Act of 1975.

Regulation S-X—The regulation that governs the requirements for financial statements under the Securities Act of 1933 and the Securities Exchange Act of 1934.

Regulation Up—Regulation provided by a resource that can increase its actual operating level in response to a direct electronic (AGC) signal from the CAISO to maintain standard frequency in accordance with established reliability criteria.

Regulation XIII—The new source review and offset program administered by the SCAQMD (South Coast Air Quality Management District in Southern California).

Regulation, Procedures, and Practices—A utility commission carries out its regulatory functions through rulemaking and adjudication. Under rulemaking, the utility commission may propose a general rule of regulation change. By law, it must issue a notice of the proposed rule and a request for comments is also made; the Federal Energy Regulatory Commission publishes this in the Federal Register. The final decision must be published. A utility commission may also work on a case-by-case basis from submissions from regulated companies or others. Objections to a proposal may come from the commission or intervenors, in which case the proposal must be presented to a hearing presided over by an administrative law judge. The judge's decision may be adopted, modified, or reversed by the utility commissioners, in which case those involved can petition for a rehearing and may appeal a decision through the courts system to the U.S. Supreme Court.

Regulator—A device used to limit the current and voltage in a circuit, normally to allow the correct charging of batteries from power sources such as solar panels and wind generators.

Regulatory Compact—Under this compact, utilities are granted service territories in which they have the exclusive right to serve retail customers. In exchange for this right, utilities have an obligation to serve all consumers in that territory on demand.

Regulatory Must-Run Generation—Utilities will be allowed to generate electricity when hydro resources are spilled for fish releases, irrigation, and agricultural purposes, and to generate power that is required by federal or state laws, regulations, or jurisdictional authorities. Such requirements include hydrological flow requirements, irrigation and water supply, solid-waste generation, or other generation contracts in effect on December 20, 1995.

Regulatory Must-Take Generation—Utilities will be allowed to generate electricity from those resources—identified by the CPUC—that are not subject to competition. These resources will be scheduled with the ISO on a must-take basis. Regulatory Must-Take Generation includes QF generating units under federal law, nuclear units and pre-existing power-purchase contracts that have minimum-take provisions.

Reheat—The heating of air that has been previously cooled either by mechanical refrigeration or economizer cooling systems. The application of sensible heat to supply air that has been previously cooled below the temperature of the conditioned space by either mechanical refrigeration or the introduction of outdoor air to provide cooling. Electric, hot water or steam reheat coils are often found near the discharge end of supply duct branches, to provide heat for a zone. Thermostats located in the occupied space control these coils.

Reheating Coils—A part of some air-conditioning systems. Electric coils in air ducts used primarily to raise the temperature of circulated air after it was over-cooled to remove moisture. Some buildings have reheating coils as their sole heating source.

Reid Vapor Pressure (RVP)—A standard measurement of a liquid's vapor pressure in pounds per square inch at 100 degrees Fahrenheit. It is an indication of the propensity of the liquid to evaporate.

Reinjected—The forcing of gas under pressure into an oil reservoir in an attempt to increase recovery.

Reinjection—The feeding of unburned char and fly ash obtained from mechanical collectors into the furnace for further combustion.

Reinserted Fuel—Irradiated fuel that is discharged in one cycle and inserted in the same reactor during a subsequent refueling. In a few cases, fuel discharged from one reactor has been used to fuel a different reactor.

Reinsertion—The process of returning nuclear fuel that has been irradiated and then removed from a reactor back into a reactor for further irradiation. Re-inserted assemblies are assemblies that have been irradiated in a cycle, were not in the core in the prior cycle (cycle N), and which are in the core in the current cycle (cycle N+1).

Reinstatement—The curing of all defaults by a Borrower, i.e., the restoration of a loan to current status through payment of arrearages.

REL—Recommended Exposure Level
• Renewable Energy Law

Relamping—The replacement of a non-functional or ineffective lamp with a new, more efficient lamp.

Relative Humidity—A measure of the percent of moisture actually in the air compared with what would be in it if it were fully saturated at that temperature. When the air is fully saturated, its relative humidity is 100 percent.

Relative Solar Heat Gain—The ratio of solar heat gain through a fenestration product (corrected for external shading) to the incident solar radiation. Solar heat gain includes directly transmitted solar heat and absorbed solar radiation, which is then reradiated, conducted, or convected into the space.

Release Clause—A clause in a contract providing for release of specified portions of the property upon compliance with certain conditions.

Reliability—Electric system reliability has two components—adequacy and security. Adequacy is the ability of the electric system to supply the aggregate electrical demand and energy requirements of the customers at all times, taking into account scheduled and unscheduled outages of system facilities. Security is the ability of the electric system to withstand sudden disturbances such as electric short circuits or unanticipated loss of system facilities.

Reliability (Electric System)—A measure of the ability of the system to continue operation while some lines or generators are out of service. Reliability deals with the performance of the system under stress.

Reliability Councils—Regional reliability councils were organized after the 1965 northeast blackout to coordinate reliability practices and avoid or minimize future outages. They are voluntary organizations of transmission-owning utilities and in some cases power cooperatives, power marketers, and non-utility generators. Membership rules vary from region to region. They are coordinated through the North American Electric Reliability Council.

Reliability Must-run Generation—The ISO will allow

utilities to generate power that is needed to ensure system reliability. This includes generation:

- Required to meet the reliability criteria for interconnected systems operation.
- Needed to meet load (demand) in constrained areas.
- Needed to provide voltage or security support of the ISO or of a local area.

Reliability Must-run Unit—In return for payment, the ISO may call upon the owner of a generating unit to run the unit when required for grid reliability.

Relocatable Public School Building—A relocatable building as defined by Title 24, Part 1, Section 4-314, which is subject to Title 24, Part 1, Chapter 4, Group 1. (in California)

Relocation of Tailings—Relocation of tailings is sometimes necessary if the pile poses a threat to inhabitants or the environment, for example, through being situated too close to populated areas, on top of aquifers or other sources of water, or in unstable areas such as flood plains or faults near earthquake zones.

Remaining Life—The expected future service life of a physical asset at any given age.

Remarketing—The process of selling or leasing the leased equipment to another party upon termination of the original lease term. The lessor can remarket the equipment or contract with another party, such as the manufacturer, to remarket the equipment in exchange for a remarketing fee.

Remedial Action Plan (RAP)—A plan that outlines a specific program leading to the remediation of a contaminated site.

Remediation—Cleanup of a site to levels determined to health-protective for its intended use.

Remote Site—A site with no electrical utility grid connection.

Remote System—Energy production system that is off of the utility grid.

Removal Costs—The costs of disposing of a physical asset, whether by demolishing, dismantling, abandoning, sale, or other. Removal costs increase the amount to be recovered as depreciation expense.

Renewable Energy—Resources that constantly renew themselves or that are regarded as practically inexhaustible. These include solar, wind, geothermal, hydro and wood. Although particular geothermal formations can be depleted, the natural heat in the earth is a virtually inexhaustible reserve of potential energy. Renewable resources also include some experimental or less-developed sources such as

tidal power, sea currents and ocean thermal gradients.

Renewable Energy Certificate or Credit (REC)—The term "REC" is generally synonymous with Green Tags and Transferable Renewable Energy Credits (TRECs). A REC is not electricity. It represents the renewable or "green" aspect of electric power generated through the use of renewable fuels, such as wind, hydro, solar, and biomass that produce one MWh or kWh of electricity from a certified renewable generator. Depending on the program under which they are generated, RECs can be bought and sold separate from the power from which they are derived. REC buyers include power generators and users that are required, or elect, to provide or use a certain percentage of green power. REC sellers include power generators and traders that hold more RECs than they require.

Renewable Energy Resource—Renewable energy resources are naturally replenishable, but flow-limited. They are virtually inexhaustible in duration but limited in the amount of energy that is available per unit of time. Some (such as geothermal and biomass) may be stock-limited in that stocks are depleted by use, but on a time scale of decades, or perhaps centuries, they can probably be replenished. Renewable energy resources include: biomass, hydro, geothermal, solar and wind. In the future they could also include the use of ocean thermal, wave, and tidal action technologies. Utility renewable resource applications include bulk electricity generation, on-site electricity generation, distributed electricity generation, non-grid-connected generation, and demand-reduction (energy efficiency) technologies.

Renewable Energy Technology (RET)—A technology that exclusively relies on an energy source that is naturally regenerated over a short time and derived directly from the sun, indirectly from the sun, or from moving water or other natural movements and mechanisms of the environment. Renewable energy technologies include those that rely on energy derived directly from the sun, on wind, geothermal, hydroelectric, wave, or tidal energy, or on biomass or biomass-based waste products, including landfill gas. A renewable energy technology does not rely on energy resources derived from fossil fuels, or waste products from inorganic sources.

Renewable Obligation Certificate (ROC)—A ROC represents a unit of electricity generated from renewable energy in the UK. One ROC is issued for each megawatt hour (or 1000 kilowatt hours) of renewable electricity generated. Suppliers can purchase ROCs with the power either from a generator, or they can buy them on the market separately from the power, to meet their mandatory targets for renewable energy in the UK.

Renewable Portfolio Standard (RPS)—Requirement for electricity retailers to purchase a specific % of sales from renewable energy generators. Obligated utilities are required to ensure that the target is met, either through their own generation, power purchase from other producers, or direct sales from third parties to the utility's customers.

Renewable Power—Electricity generated from biomass, wind, solar thermal, photovoltaic, geothermal, and LIHI certified hydroelectric resources. Also called renewable resources, renewable supply, and renewable electricity.

Renewable Resources—Naturally, but flow-limited resources that can be replenished. They are virtually inexhaustible in duration but limited in the amount of energy that is available per unit of time. Some (such as geothermal and biomass) may be stock-limited in that stocks are depleted by use, but on a time scale of decades, or perhaps centuries, they can probably be replenished. Renewable energy resources include: biomass, hydro, geothermal, solar and wind. In the future, they could also include the use of ocean thermal, wave, and tidal action technologies. Utility renewable resource applications include bulk electricity generation, on-site electricity generation, distributed electricity generation, non-grid-connected generation, and demand-reduction (energy efficiency) technologies.

Renewables—Energy sources that are, within a short timeframe relative to the Earth's natural cycles, sustainable, and include non-carbon technologies such as solar energy, hydropower and wind as well as carbon-neutral technologies such as biomass.

Renewal Option—An option in the lease agreement that allows the lessee to extend the lease term for an additional period beyond the expiration of the initial lease term, in exchange for lease renewal payments.

Rent Holiday—A period of usage, usually up front, in which the lessee is not required to pay rents. Typically, the rents are capitalized into the remaining lease payments.

REP—Reasonable Efforts Program

Repair—The reconstruction or renewal of any part of an existing building for the purpose of its maintenance. NOTE: Repairs to low-rise residential buildings are not within the scope of these standards. (in California)

**Repetition/
Replication**—There are four plots in a repetition/replication, the early, mid and late seral treatment plots and a control plot. A repetition/replication is also called a "block." There should be at least three repetitions/replications in a research study to obtain statistical reliability.

REPI—Renewable Energy Production Incentive (federal)

Replacement—The substitution of a unit for another unit generally of a like or improved character.

Replacement Allowance—An allowance that provides for periodic replacement of certain components of a project that wear out more rapidly than the project itself and must be replaced during the economic life of the project.

Replacement Energy Source for Primary Heating—For the CBECS (an EIA consumption survey), the heating energy source to which the building could switch within one week without major modifications to the main heating equipment, without substantially reducing the area heated, and without substantially reducing the temperature maintained in the heated area.

Replacement Vehicle—A vehicle which is acquired in order to take the place of a vehicle which is being retired from service. These acquisitions do not increase the size of the company fleet.

Report Boundary—(GRI Reporting Network) Boundary refers to the range of entities (e.g., subsidiaries, joint ventures, sub-contractors, etc.) whose performance is represented in the report. In setting the Boundary for its report, an organization must consider the range of entities over which it exercises control (often referred to as the 'organizational boundary' and usually linked to definitions used in financial reporting) and over which it exercise influence)often called the 'operational boundary'). In assessing influence, the organization will need to consider its ability to influence entities upstream (e.g., in its supply chain) as well as entities downstream (e.g., distributors and users of its products and services). The Boundary may vary based on the specific Aspect or type of information being reported.

Report State—The State, including adjacent offshore

continental shelf areas in the Federal domain, in which a company operated natural gas gathering, transportation, storage, and/or distribution facilities or a synthetic natural gas plant covered by the individual report.

Report Week—A calendar week beginning at 12:01 a.m. on Sunday and ending at midnight on Saturday.

Report Year (Calendar)—The 12-month period, January 1 through December 31.

Report Year (Fiscal)—A 12-month period for which an organization plans the use of its funds. The fiscal year is designated by the calendar year in which it ends.

Reporting—The average number of Btu per cubic foot of gas at 60 degrees Fahrenheit and 14.73 psia delivered directly to consumers. Where billing is on a thermal basis, the heat content values used for billing purposes are to be used to determine the annual average heat content.

Reporting Company—A company that is registered with the SEC under the Exchange Act.

Reporting Principle—(GRI Reporting Network) Concepts that describe the outcomes a report should achieve and that guide decisions made throughout the reporting process, such as which Indicators to respond to, and how to respond to them.

Repossession—A situation in which a lessor reclaims and physically removes the leased equipment from the control of the lessee; usually caused by payment default.

Repowered Plant—An existing power facility that has been substantially rebuilt to extend its useful life.

Repowering—Refurbishment of a plant by replacement of the combustion technology with a new combustion technology, usually resulting in better performance and greater capacity.

Repressuring—The injection of gas into oil or gas formations to effect greater ultimate recovery.

Reprocessing—Synonymous with chemical separations.

REPS—Regional Emissions Projection System

Request for Qualifications—The document which communicates information to prospective contractors and should include, but not be limited to:

- A description of the problem;
- Expected results from the project;
- Extent and nature of anticipated contract services; and
- Criteria for evaluating statements of qualifications.

Requirements Power—The firm service needs required by designated load plus losses from the points of supply.

Reregulation—The design and implementation of regulatory practices to be applied to the remaining regulated entities after restructuring of the vertically-integrated electric utility. The remaining regulated entities would be those that continue to exhibit characteristics of a natural monopoly, where imperfections in the market prevent the realization of more competitive results, and where, in light of other policy considerations, competitive results are unsatisfactory in one or more respects. Regulation could employ the same or different regulatory practices as those used before restructuring.

RES—Renewable Energy Standard

Resale (Wholesale) Sales—Resale or wholesale sales are electricity sold (except under exchange agreements) to other electric utilities or to public authorities for resale distribution. (This includes sales to requirements and non-requirements consumers.)

Resale Registration—Registration by a Company of the investor's sale of the shares purchased by the investor in a private offering.

Rescission—The act of canceling or annulling the effect of a document.

Research and Development (R&D)—Research is the discovery of fundamental new knowledge. Development is the application of new knowledge to develop a potential new service or product. Basic power sector R&D is most commonly funded and conducted through the Department of Energy (DOE), its associated government laboratories, university laboratories, the Electric Power Research Institute (EPRI), and private sector companies.

Reseller—A firm (other than a refiner) that is engaged in a trade or business that buys refined petroleum products and then sells them to a purchaser who is not the ultimate consumer of those refined products.

Reservation—The right or an interest retained by a Grantor in a conveyance.

Reserve—That portion of the demonstrated reserve base that is estimated to be recoverable at the time of determination. The reserve is derived by applying a recovery factor to that component of the identified coal resource designated as the demonstrated reserve base.

Reserve Additions—The estimated original, recoverable, salable, and new proved reserves credited to new fields, new reservoirs, new gas purchase contracts, amendments to old gas purchase contracts, or purchase of gas reserves in-place that occurred during the year and had not been previously reported. Reserve additions refer to domestic in-the-ground natural gas reserve additions and do not refer to interstate pipeline purchase agreements; contracts with foreign suppliers; coal gas, SNG, or LNG purchase arrangements.

Reserve Capacity—The amount of generating capacity a central power system must maintain to meet peak loads.

Reserve Cost Categories of $15, $30, $50, and $100 per Pound U3O8—Classification of uranium reserves estimated by using break-even cutoff grades that are calculated based on forward-operating costs of less than $15, $30, $50, and $100 per pound U_3O_8.

Reserve Generating Capacity—The amount of power that can be produced at a given point in time by generating units that are kept available in case of special need. This capacity may be used when unusually high power demand occurs, or when other generating units are off-line for maintenance, repair or refueling.

Reserve Margin—The differences between the dependable capacity of a utility's system and the anticipated peak load for a specified period.

Reserve Margin (Operating)—The amount of unused available capability of an electric power system (at peak load for a utility system) as a percentage of total capability.

Reserve Rates—Rates used internally by the company to adjust the rates billed and subject to refund to an estimate of the final or settled rate. Accounting principles require such a reserve.

Reserve Revisions—Changes to prior year-end proved reserves estimates, either positive or negative, resulting from new information other than an increase in proved acreage (extension). Revisions include increases of proved reserves associated with the installation of improved recovery techniques or equipment. They also include correction of prior year arithmetical or clerical errors and adjustments to prior year-end production volumes to the extent that these alter reserves estimates.

Reserve, Change in—For FRS reporting, the following definitions should be used for changes in reserves. Revisions of Previous Estimates. Changes in previous estimates of proved reserves, either upward or downward, resulting from new information normally obtained from development drilling and production history or resulting from a change in economic factors. Revisions do not include changes in reserve estimates resulting from increases in proved acreage or from improved recovery tech-

niques. Improved Recovery. Changes in reserve estimates resulting from application of improved recovery techniques shall be separately shown, if significant. If not significant, such changes shall be included in revisions of previous estimates. Purchases or Sales of Minerals-in-Place. Increase or decrease in the estimated quantity of reserves resulting from the purchase or sale of mineral rights in land with known proved reserves. Extensions, Discoveries, and Other Additions. Additions to an enterprise's proved reserves that result from: (1) extension of the proved acreage of previously discovered (old) reservoirs through additional drilling in periods subsequent to discovery and (2) discovery of new fields with proved reserves or of new reservoirs of proved reserves in old fields.

Reserves (Coal)—Coal reserve estimates comprising the demonstrated coal reserve base include only proved (measured) and probable (indicated). Proved (Measured) Reserves. Reserves or resources for which tonnage is computed from dimensions revealed in outcrops, trenches, workings, and drill holes and for which the grade is computed from the results of detailed sampling. The sites for inspection, sampling, and measurement are spaced so closely and the geologic character is so well defined that size, shape, and mineral content are well established. The computed tonnage and grade are judged to be accurate within limits which are stated, and no such limit is judged to be different from the computed tonnage or grade by more than 20 percent. Probable (Indicated) Reserves. Reserves or resources for which tonnage and grade are computed partly from specific measurements, samples, or production data and partly from projection for a reasonable distance on geologic evidence. The sites available are too widely or otherwise inappropriately spaced to permit the mineral bodies to be outlined completely or the grade established throughout.

Reserves, Energy—Refers to the bank of natural resources, such as natural gas, natural gas liquids, petroleum, coal, lignite, energy available from water power, and solar and geothermal energy. Estimated Potential Natural Gas Resources. Refers to an estimate of the remaining natural gas in a specified area which are judged to be recoverable. Estimated Proved Natural Gas Reserves. An estimated quantity of natural gas which analysis of geologic and engineering data demonstrates with reasonable certainty to be recoverable in the future from known oil and gas reservoirs under anticipated economic and current operating conditions. Reservoirs that have demonstrated the ability to produce by either actual production or conclusive formation test are considered proved.

Reserves, Net—Includes all proven reserves associated with the company's net working interests. (See definition for *Working Interest*.)

Reserves, Proved (Oil and Gas)—The estimated quantities of crude oil, natural gas, and natural gas liquids which geological and engineering data demonstrate with reasonable certainty to be recoverable in future years from known reservoirs under existing economic and operating conditions. Reservoirs are considered proved if supported economically by one or more of the following: actual production; conclusive formation test; core analysis; and/or electric or other log interpretations. The area of a reservoir considered proved includes: (1) portion delineated by drilling and defined by gas-oil and/or oil-water contacts, if any; and (2) the immediately adjoining portions not yet drilled, but which can be reasonably judged as economically productive on the basis of available geological and engineering data. In the absence of information on fluid contacts, the lowest known structural occurrence of hydrocarbons controls the lower proven limited of the reservoir. Volumes of oil and gas placed in underground storage are not to be considered proven reserves, but should be classified as inventory.

Reservoir—A porous and permeable underground formation containing an individual and separate natural accumulation of producible hydrocarbons (crude oil and/or natural gas) which is confined by impermeable rock or water barriers and is characterized by a single natural pressure system.

Reservoirs are greenhouse gas storage locations within the biosphere such as oceans, soils, and forests. The reaction of these reservoirs to global climate change is difficult to predict.

Reservoir Capacity—The present total developed capacity (base and working) of the storage reservoir, excluding contemplated future development.

Reservoir Repressuring—The injection of a pressurized fluid (such as air, gas, or water) into oil and gas reservoir formations to effect greater ultimate recovery.

Resident Alien—An alien individual is treated as a resident of the United States with respect to any calendar tax year if (and only if) such individual meets

requirement any of the following:

Lawfully admitted for permanent residence. Such individual is a lawful permanent resident of the United States at any time during such calendar tax year.

Substantial presence test. Such individual meets the substantial presence test (see IRC regulations).

First year election. (see IRC regulations)

Resident Fish—Fish species that complete their entire life cycle in freshwater. Non-anadromous fish. An example is rainbow trout.

Residential Building—Means any hotel, motel, apartment house, lodging house, single and dwelling, or other residential building which is heated or mechanically cooled.

Residential Consumers—Consumers using gas or electricity for heating, air conditioning, cooking, water heating, and other residential uses in single and multi-family dwellings and apartments and mobile homes.

Residential Energy Consumption Survey (RECS)—A national multistage probability sample survey conducted by the Energy End Use Division of the Energy Information Administration. The RECS provides baseline information on how households in the United States use energy. The Residential Transportation Energy Consumption Survey (RTECS) sample is a subset of the RECS. Household demographic characteristics reported in the RTECS publication are collected during the RECS personal interview.

Residential Heating Oil Price—The price charged for home delivery of No. 2 heating oil, exclusive of any discounts such as those for prompt cash payment. Prices do not include taxes paid by the consumer.

Residential Manual—The manual developed by the Commission (in California), under Section 25402.1 of the Public Resources Code, to aid designers, builders, and contractors in meeting energy efficiency standards for low-rise residential buildings.

Residential Propane Price—The "bulk keep full" price for home delivery of consumer-grade propane intended for use in space heating, cooking, or hot water heaters in residences.

Residential Sector—An energy-consuming sector that consists of living quarters for private households. Common uses of energy associated with this sector include space heating, water heating, air conditioning, lighting, refrigeration, cooking, and running a variety of other appliances. The residential sector excludes institutional living quarters. Note: Various EIA programs differ in sectoral coverage.

Residential Type Central Air Conditioner—There are four basic parts to a residential central air-conditioning system: (1) a condensing unit, (2) a cooling coil, (3) ductwork, and (4) a control mechanism such as a thermostat. There are two basic configurations of residential central systems: (1) a "split system" where the condensing unit is located outside and the other components are inside, and (2) a packaged-terminal air-encased in one unit and is usually found in a "utility closet."

Residential Vehicles—Motorized vehicles used by U.S. households for personal transportation. Excluded are motorcycles, mopeds, large trucks, and buses. Included are automobiles, station wagons, passenger vans, cargo vans, motor homes, pickup trucks, and jeeps or similar vehicles. In order to be included (in the EIA survey), vehicles must be (1) owned by members of the household, or (2) company cars not owned by household members but regularly available to household members for their personal use and ordinarily kept at home, or (3) rented or leased for 1 month or more.

Residential/Commercial (consumer category)—Housing units, wholesale or retail businesses (except coal wholesale dealers); health institutions (hospitals, social and educational institutions (schools and universities); and Federal, state, and local governments (military installations, prisons, office buildings, etc.). Excludes shipments to Federal power projects, such as TVA, and rural electrification cooperatives, power districts, and state power projects.

Residual—The value of an asset at the end of a lease.

Residual Asset—An asset representing the right to an underlying asset retained by a lessor during a lease.

Residual Fuel Oil—A general classification for the heavier oils, known as No. 5 and No. 6 fuel oils, that remain after the distillate fuel oils and lighter hydrocarbons are distilled away in refinery operations. It conforms to ASTM Specifications D 396 and D 975 and Federal Specification VV-F-815C. No. 5, a residual fuel oil of medium viscosity, is also known as Navy Special and is defined in Military Specification MIL-F-859E, including Amendment 2 (NATO Symbol F-770). It is used in steam-powered vessels in government service and inshore power plants. No. 6 fuel oil includes Bunker C fuel oil and is used for the production of electric power, space heating, vessel bunkering, and various industrial purposes.

Residual Position—The amount of residual value built into the lease pricing. In other words, the residual value amount for which the lessor is at-risk, and must receive at lease termination in order to earn its pre-targeted rate of return.

Residual Value—The value, either actual or expected, of leased equipment at the end, or termination, of the lease.

Residual Value Guarantee—A guarantee, obtained by the lessor from another party, that the residual value will be worth a certain preset amount at the end of the lease term.

Residual Value Insurance—An insurance policy stating the guaranteed residual value on leased equipment. The insurance company pays if the residual is not realized.

Residue—Any organic matter left as residue, such as agricultural and forestry residue, including, but not limited to, conifer thinnings, dead and dying trees, commercial hardwood, noncommercial hardwoods and softwoods, chaparral, burn, mill, agricultural field, and industrial residues, and manure.

Residue Gas—Natural gas from which natural gas processing plant liquid products and, in some cases, non-hydrocarbon components have been extracted.

Residuum—Residue from crude oil after distilling off all but the heaviest components, with a boiling range greater than 1,000 degrees Fahrenheit.

Resistance (Electrical)—The ability of all conductors of electricity to resist the flow of current, turning some of it into heat. Resistance depends on the cross section of the conductor (the smaller the cross section, the greater the resistance) and its temperature (the hotter the cross section, the greater its resistance).

Resistance (Thermal)—The reciprocal of thermal conductance. See *R-Value*.

Resistance Heating—A type of heating system that provides heat from the resistance of an electrical current flowing through a conductor.

Resistive Voltage Drop—The voltage developed across a cell by the current flow through the resistance of the cell which may result from the bulk resistance of the materials in the cell and at interfaces between them.

Resistor—An electronic component used to restrict the flow of current in a circuit. Sometimes used specifically to produce heat, such as in a water heater element.

Resource Adequacy (RA)—A requirement that load-serving entities ensure they have 115 percent of the generation capacity necessary to meet expected peak load. This is also an annual proceeding at the CPUC to determine what the RA need will be for the following year, and associated policies.

Resource Conservation and Recovery Act (RCRA)—A federal law regulating solid and hazardous waste. RCRA governs the generation, storage, treatment, transport, and disposal of hazardous waste.

Resource Efficiency—The use of smaller amounts of physical resources to produce the same product or service. Resource efficiency involves a concern for the use of all physical resources and materials used in the production and use cycle, not just the energy input.

Resource Recovery—The process of converting municipal solid waste to energy and/or recovering materials for recycling.

Resource Values—A resource, natural or social, that is found in an area. Resource values may have varying levels of significance.

Respondent—A company or individual who completes and returns a report or survey form.

Responsible Party—An individual or corporate entity considered legally liable for contamination found at a property and, therefore, responsible for cleanup of the site.

Rest Voltage—The voltage of a fully charged cell or battery that is neither being charged or discharged.

Restoration Time—The time when the major portion of the interrupted load has been restored and the emergency is considered to be ended. However, some of the loads interrupted may not have been restored due to local problems.

Restricted Securities—Public securities that are not freely tradable due to SEC regulations. (See also: Securities and Exchange Commission.)

Restricted Shares—Shares acquired in a private placement are considered restricted shares and may not be sold in a public offering absent registration or after an appropriate holding period has expired. Non-affiliates must wait one year after purchasing the shares, after which time they may sell less than 1% of their outstanding shares each quarter. For affiliates, there is a two-year holding period.

Restricted-Universe Census—This is the complete enumeration of data from a specifically defined subset of entities including, for example, those that exceed a given level of sales or generator nameplate capacity.

Restriction—Limitations on the use and enjoyment of real or personal property.

Restructuring—The process of replacing a monopoly system of electric utilities with competing sellers, allowing individual retail customers to choose their electricity supplier but still receive delivery over the power lines of the local utility. It includes the reconfiguration of the vertically-integrated electric utility.

Retail—Sales covering electrical energy supplied for residential, commercial, and industrial end-use purposes. Other small classes, such as agriculture and street lighting, also are included in this category.

Retail Company—A company that is authorized to sell electricity directly to industrial, commercial and residential end-users.

Retail Competition—A system under which more than one electric provider can sell to retail customers, and retail customers are allowed to buy from more than one provider. (See also *Direct Access*)

Retail Market—A market in which electricity and other energy services are sold directly to the end-use customer.

Retail Motor Gasoline Prices—Motor gasoline prices calculated each month by the Bureau of Labor Statistics (BLS) in conjunction with the construction of the Consumer Price Index.

Retail Power—This is the sale of wholesale power to industrial, commercial, government, and residential customers.

Retail Provider—(In California) Any electric corporation as defined in Public Utilities Code Section 218, electric service provider as defined in Public Utilities Code Section 218.3, public owned electric utility as defined in Public Resources Code Section 9604, community choice aggregator as defined in Public Utilities Code Section 331.1, or the Western Area Power Administration.

Retail Transaction—The sale of electric power from a generating company or wholesale entity to the customer.

Retail Wheeling—A term for the process of transmitting electricity over transmission lines not owned by the supplier of the electricity to a retail customer of the supplier. With retail wheeling, an electricity consumer can secure their own supply of electricity from a broker or directly from the generating source. The power is then wheeled at a fixed rate, or at a regulated "non-discriminatory" rate set by a utility commission.

Retailer—A firm (other than a refiner, reseller, or reseller/retailer) that carries on the trade or business of purchasing refined petroleum products and re-selling them to ultimate consumers without substantially changing their form.

Retained Earnings (Utility)—The balance, either debit or credit, of appropriated or un-appropriated retained earnings of the utility department arising from earnings.

Retained Transaction—An investment kept for one's own portfolio; a retained transaction is not sold to another party or investor.

RETC—Renewable Energy Tax Credit. General term for Federal tax credits available for entities producing renewable energy. Includes energy credits and production tax credits (PTCs).

RETI—Renewable Energy Transmission Initiative

Retire from Service—A vehicle is retired from service if that vehicle is placed out of service and there are no future plans to return that vehicle to service.

Retired Hydropower Plant Sites—The site of a plant that formerly produced electrical or mechanical power but is now out of service. Includes plants that have been abandoned, damaged by flood or fire, inundated by new reservoirs, or dismantled.

RETO—Reasonably Expected to Occur

Retorting—The heating of oil shale to get the oil out from it.

Retrocommissioning (RCx)—Retrocommissioning, or "existing building commissioning," is a process to ensure the functionality of a building that has not previously commissioned. It is a systematic investigation of how a building's subsystems are being operated and maintained, and it is being used to identify and solve optimization and integration issues. A "building tune-up" is similar but often goes one step further, such as HVAC or water heating systems.

Retrofit—Broad term that applies to any change after the original purchase, such as adding equipment not a part of the original purchase. As applied to alternative fuel vehicles, it refers to conversion devices or kits for conventional fuel vehicles. (Same as "Aftermarket.") A replacement or modification of existing building equipment, often undertaken as part of an effort to reduce energy usage or improve energy efficiency.

Retrofitting—The application of conservation, efficiency, or renewable energy technologies to existing structures.

Return—Generally, interest on debt and the profit the company is allowed over and above the recovery of its operating expenses, depreciation and taxes.

Return Air—Air that is returned to a heating or cooling appliance from a heated or cooled space.

Return Duct—The central heating or cooling system contains a fan that gets its air supply through these ducts, which ideally should be installed in every room of the house. The air from a room will move towards the lower pressure of the return duct.

Return on Assets (ROA)—A common measure of profitability based upon the amount of assets invested. ROA is equal to the ratio of either: (1) net income to total assets, or (2) net income available to common stockholders to total assets.

Return on Common Equity—The net income less preferred stock dividends, divided by the average common stock equity.

Return on Common Stock Equity—An equity's earnings available for common stockholders calculated as a percentage of its common equity capital.

Return on Equity (ROE)—A measure of profitability related to the amount of invested equity. ROE is equal to the ratio of either: (1) net income to owners' equity or (2) net income available to common stockholders to common equity.

Return on Investment (ROI)—The interest rate at which the net present value of a project is zero. Multiple values are possible.

Revealed Preference Methods—A method of putting dollar values on something (e.g. air quality or risks) which relies on observing choices people make when in a related setting. Examples are housing studies to determine the value of cleaner air or travel studies to evaluate the value of tourism benefits.

Revenue—(Electricity)—The total amount of money received by an entity from sales of its products and/or services; gains from the sales or exchanges of assets, interest, and dividends earned on investments; and other increases in the owner's equity, except those arising from capital adjustments.

Revenue Requirement—The amount of revenues the utility or any company needs to receive in order to cover operating expenses, pay debt service, and provide a fair return to common equity investors.

Revenue-Raising Instruments—In environmental policy includes emissions taxes, which are levied against producers of pollution; and tradable emissions permits, which can be bought or sold by coal-burning electric utilities and other industries.

Revenue-Recycling—Occurs when the revenue raised by an environmental policy is used to reduce other distortionary taxes or government deficits or is rebated to households. If permits are auctioned, this gives considerable sums of money to be recycled back into the economy, either through a lump sum payment of offsetting other taxes. If the existing taxes that are correspondingly reduced were very inefficient, this allows this allows the possibility of both environmental and economic benefits from the trading system, commonly called the 'double dividend'.

Reverse Bias—Condition where the current producing capability of a PV cell is significantly less than that of other cells in its series string. This can occur when a cell is shaded, cracked, or otherwise degraded or when it is electrically poorly matched with other cells in its string.

Reverse Current Protection—Any means of preventing current flow from the battery to the solar PV array (e.g. at night) that would discharge the battery.

Reverse Thermosiphoning—When heat seeks to flow from a warm area (e.g., heated space) to a cooler area, such as a solar air collector at night without a reverse flow damper.

Reversible Turbine—A hydraulic turbine, normally installed in a pumped-storage plant, which can be used alternatively as a pump or as an engine, turbine, water wheel, or other apparatus that drives an electrical generator.

Reversing Valve—A component of a heat pump that reverses the refrigerant's direction of flow, allowing the heat pump to switch from cooling to heating or heating to cooling.

Reversion Capitalization Rate—The capitalization rate used to determine reversion value.

Revisions and Additions (gross change in reserves)—The difference (plus or minus) between the year-end reserves plus production for a given year and the year-end reserves for the previous year.

RF (Radio Frequency)—Any radiation of a frequency that may be received or radiated by radios. Common usage: RF interference (RFI); refers to the interference of radio frequency radiation with the operation of devices or appliances such as radios, televisions, computers, etc.

RFA—Regulatory Flexibility Act (1980) Requires federal agencies to consider the effects of their regulatory actions on small businesses and other small entities and to minimize any undue disproportionate burden. The chief counsel for advocacy of the U.S. Small Business Administration is charged with monitoring federal agencies' compliance with the act and with submitting an annual report to Congress.

- Remedial Facility Assessment

RFI—Request for information

RFP—Request for Proposal

- An acronym for "Request for Proposal." RFPs are part of a common bidding process in the public sector and for large projects. They normally list the project specifications and procedures, and are issued to agencies or organizations that might be qualified to participate.
- Reasonable Further Progress

RFQ—Request for Quote

- Request for Qualifications

RGGI—Regional Greenhouse Gas Initiative RGGI is a cooperative effort by Northeastern and Mid-Atlantic states to reduce carbon dioxide emissions. To address this important environmental issue, the RGGI participating states will be developing a regional strategy for controlling emissions. Central to this initiative is the implementation of a multi-state cap-and-trade program with a market-based emissions trading system. Similar initiatives are set up in the Midwest, through the Midwest Greenhouse Gas Accord, and in the West through the Western Climate Initiative.

RHA—Rivers and Harbors Act (1899) Prohibits the construction of any bridge, dam, dike or causeway over or in navigable waterways of the U.S. without Congressional approval.

RI—Reconnaissance Inspection

- Remedial Investigation

RIA—Regulatory Impact Analysis

- Regulatory Impact Assessment

Ribbon (Photovoltaic) Cells—A type of solar photovoltaic device made in a continuous process of pulling material from a molten bath of photovoltaic material, such as silicon, to form a thin sheet of material.

Ribbon Silicon—Crystalline silicon that is used in photovoltaic cells. Ribbon silicon is fabricated by a variety of solidification (crystallization) methods that withdraw thin silicon sheets from pools of relatively pure molten silicon.

Richter Scale—A scale, ranging from 1 to 10, for indicating the intensity of an earthquake.

Rider—A supplement to, an addition to, an Endorsement to a document.

Right of First Refusal—The right of first refusal gives the holder the right to meet any other offer before the proposed contract is accepted.

Right of Suvivorship—If one owner dies, that owner's interest in the property will pass by inheritance to that owner's heirs. Joint Tenants have a right of survivorship and Tenants in Common do not.

Right of Way—The right to cross over a parcel of land.

Right Side—The right side of the building as one faces the front facade from the outside (*See Front*). This designation is used to indicate the orientation of fenestration and other surfaces, especially in model homes that are constructed in multiple orientations.

Right-of-Use Asset—An asset that represents a lessee's right to use an underlying asset for the lease term.

Right-of-Way—The land and legal right to use and service the land along which a transmission line is located. Transmission line right-of-way is usually acquired in widths that vary with the kilovolt (kV) size of the line.

Rights Offering—Issuance of "rights" to current shareholders allowing them to purchase additional shares, usually at a discount to market price. Shareholders who do not exercise these rights are usually diluted by the offering. Rights are often transferable, allowing the holder to sell them on the open market to others who may wish to exercise them. Rights offerings are particularly common to closed-end funds, which cannot otherwise issue additional ordinary shares.

Rigid Insulation Board—An insulation product made of a fibrous material or plastic foams, pressed or extruded into board-like forms. It provides thermal and acoustical insulation strength with low weight, and coverage with few heat loss paths.

Rigorous—A method of calculation, especially of termination values, which solves for each value exactly without taking any shortcuts such as using pre-calculated investment or tax balances. Contrast with dominant.

RILS—Regulatory Interpretation Letters

RIM—Regulatory interpretation Memorandum

Rip Rap—Cobblestone or coarsely broken rock used for protection against erosion of embankment or gully.

Riparian Area—An area of land directly influenced by water. An ecosystem that is transitional between land and water ecosystems. Riparian areas usually have visible vegetative or physical characteristics reflecting the influence of water. Riversides, lake borders, and marshes are typical riparian areas.

Riparian Buffer—Riparian areas that are managed to protect the aquatic and riparian ecosystem. A riparian buffer protects water quality and temperature, habitat along the banks, upland habitat for aquatic and riparian species, and some or all of the floodplain.

Risk—The chance of loss on an investment due to many factors, including inflation, interest rates, default, politics, foreign exchange, call provisions, etc. In Private Equity, risks are outlined in the Risk Factors section of the Placement Memorandum.

Risk Assessment (Environmental)—A risk assessment looks at the chemicals detected at a site, the frequency and concentration of detected chemicals, the toxicity of the chemicals and how people can be exposed, and for how long. Routes of exposure to people are generally through ingestion, such as eating, contact with the skin, or inhalation. The most significant potential routes of exposure are through ingestion and contact with the skin. The health risk assessment cannot predict health effects; it only describes the increased possibility of adverse health effects, based on the best scientific information available.

Risk of Default—The risk of the partial or complete loss of receivables in general as well as of receivables and share prices losses for securities subsequent to the insolvency or danger of insolvency on the part of the debtor.

Risk Premium (Financial)—An additional required rate of return that must be paid to investors who invest in risky investments to compensate for the risk they are taking.

• A method to determine the cost of common equity component of return using the bond yield plus a risk premium based on selected stock market yields to bond yields.

Risk Rate—The annual rate of return on capital that is commensurate with the risk assumed by the investor. The rate of interest or yield necessary to attract capital.

River (method of transportation to consumers— Coal)—Shipments of coal moved to consumers via river by barge. Shipments to Great Lakes coal loading docks or Tidewater pier or coastal points are not included.

RME—Reasonable Maximum Exposure

RMI—Rocky Mountain Institute www.rmi.org

RMO—Records Management Officer

• Resource Management Office

RMPP—Risk Management Prevention Plan

• Risk Management and Prevention Program

RMS—Root mean square; defines a time averaged value of a varying sinusoidal parameter, such as AC voltage, amperage, or wattage. The square root of the average of the squares of a set of numbers.

RMU—Removal Units. Credits earned from land use,

land-use change and forestry projects (LULUCF) in industrialized countries, including such projects under the Kyoto Protocol's JI mechanism.

RO—Renewable Obligation
- Regional Office
- Reverse Osmosis

ROA—Return on Assets. A measure of the income based on the assets employed.

Road Oil—Any heavy petroleum oil, including residual asphaltic oil used as a dust pallative and surface treatment on roads and highways. It is generally produced in six grades, from 0, the most liquid, to 5, the most viscous.

ROC—Renewable Obligation Certificate
- Reactive Organic Compound
- Record of Communication
- Report of Conversation

Rock Bin—A container that holds rock used as the thermal mass to store solar energy in a solar heating system.

Rock Wool—A type of insulation made from virgin basalt, an igneous rock, and spun into loose fill or a batt. It is fire resistant and helps with soundproofing.

ROD—See *Record of Decision*.

Rodlet or GAD basket—An open garbage and debris (GAD) basket that may have contain pieces of fuel rods, disassembled fuel rods, and other fuel and non-fuel components.

ROE—*Return on Equity*: The profits distributed to common shareholders after all expenses, interest costs, and preferred stock dividends have been paid. In ratemaking, it represents the level of revenue needed that will permit equity stockholders the opportunity to earn a fair return on their investment in the utility.

ROG—Reactive Organic Gas

ROI—See *Return on Investment*.

Roll Front—A type of uranium deposition localized as a roll or interface separating an oxidized interior from a reduced exterior. The reduced side of this interface is significantly enriched in uranium.

Roll Up—To collect, especially tax payments, occurring before commencement and apply them in a lump sum on the first available date.

Rolling Blackout—A controlled and temporary interruption of electrical service. These are necessary when a utility is unable to meet heavy peak demands because of an extreme deficiency in power supply.

Rollover—A change in lease term and/or payment resulting from a change in equipment, such as in a takeout or upgrade. The rollover finances those costs associated with the change in equipment and may result in the lessor financing an amount greater than the equipment value.

ROM—Rough Order Magnitude

Roof—A building element that provides protection against the sun, wind, and precipitation.

Roof (Coal)—The rock immediately above a coal seam. The roof is commonly a shale, often carbonaceous and softer than rocks higher up in the roof strata.

Roof Insulation—Insulating materials placed underneath the roof or on the roof (building).

Roof Mount (Solar)—A PV or solar collector rack intended to be installed on a roof. For PVs, its elevation angle can be fixed or seasonally adjustable.

Roof or Ceiling Insulation—A building shell conservation feature consisting of insulation placed in the roof (below the waterproofing layer) or in the ceiling of the top floor in the building.

Roof or Ceiling Insulation, Insulation in Exterior Walls—Any material that when placed between the interior surface of the building and the exterior surface of the building, reduces the rate of heat loss to the environment or heat gain from the environment. Roof or ceiling insulation refers to insulation placed in the roof or ceiling of the top occupied floor in the building. Wall insulation refers to insulation placed between the exterior and interior walls of the building.

Roof Pond—A solar energy collection device consisting of containers of water located on a roof that absorb solar energy during the day so that the heat can be used at night or that cools a building by evaporation at night.

Roof Ventilator—A stationary or rotating vent used to ventilate attics or cathedral ceilings; usually made of galvanized steel, or polypropylene.

Roof/Ceiling Type—A type of roof/ceiling assembly that has a specific framing type and U-factor.

Room Air Conditioner—Air-conditioning units that typically fit into the window or wall and are designed to cool only one room.

Room Heater Burning Gas, Oil, and Kerosene—Any of the following heating equipment: circulating heaters, convectors, radiant gas heaters, space heaters, or other non-portable room heaters that may or may not be connected to a flue, vent, or chimney.

Room-and-Pillar Mining—The most common method of underground mining in which the mine roof is supported mainly by coal pillars left at regular intervals. Rooms are places where the coal is mined; pillars are areas of coal left between the rooms. Room-and-pillar mining is done either by conventional or continuous mining.

Root Mean Square—The square root of the average square of the instantaneous values of an ac output. For a sine wave the RMS value is 0.707 times the peak value. The equivalent value of alternating current, I, that will produce the same heating in a conductor with resistance, R, as a dc current of value I.

ROP—Rate of Progress
  • Regional Oversight Policy

ROR—Rate of Return: This figure, which is expressed as a percentage, reflects the utility's weighted cost of capital.

RORTA—Return on Total Assets: Net income divided by total assets.

ROSA—Regional Ozone Study Area

Rotary Kiln—An incinerator with a rotating combustion chamber. The rotation helps mix the wastes and promotes more complete burning. They can accepts gases, liquids, sludges, tars and solids, either separately or together, in bulk or in containers.

Rotary Rig—A machine used for drilling wells that employs a rotating tube attached to a bit for boring holes through rock.

Rotation (Forestry)—The number of years allotted to establish and grow a forest stand to maturity.

Rotor—An electric generator consists of an armature and a field structure. The armature carries the wire loop, coil, or other windings in which the voltage is induced, whereas the field structure produces the magnetic field. In small generators, the armature is usually the rotating component (rotor) surrounded by the stationary field structure (stator). In large generators in commercial electric power plants the situation is reversed. In a wind energy conversion device, the blades and rotating components.

Rough Trees—Live trees of commercial species which do not contain a sawlog because of roughness, poor form, splits, or cracks. Includes all living trees of noncommercial species.

Round Test Mesh—A sieving screen with round holes, the dimensions of which are of specific sizes to allow certain sizes of coal to pass through while retaining other sizes.

Round Trip Efficiency—The ratio of total energy that can be discharged by a storage system divided by the amount of energy needed to fully charge the system.

Roundwood—Logs, bolts, or other round sections cut from trees.
  • Wood cut specifically for use as a fuel.

ROV—Report of Violation

ROW—Right of Way
  • Rest of World

Royalty (Energy)—A contractual arrangement providing a mineral interest that gives the owner a right to a fractional share of production or proceeds there from, that does not contain rights and obligations of operating a mineral property, and that is normally free and clear of exploration, developmental and operating costs, except production taxes.

Royalty Cost—A share of the profit or product reserved by the grantor of a mining lease, such as a royalty paid to a lessee.

Royalty Interest—An interest in a mineral property provided through a royalty contract.

Royalty Interest (including overriding royalty)—These interests entitle their owner(s) to a share of the mineral production from a property or to a share of the proceeds there from. They do not contain the rights and obligations of operating the property and normally do not bear any of the costs of exploration, development, and operation of the property.

RP—Respirable Particulates
  • Responsible Party
  • Reserves/production Ratio
  • Risk Premium

RPF—Relative Potency Factor

rpm—Revolutions per Minute

RPS—Renewable Portfolio Standard. Requirement for a certain percentage of power generated by a utility to come from renewable energy sources.

RQG—Reduced Quantity Generator

RR—Retention Rate

RRP—Regional Response Plan

RRT—Regional Response Team

RSA—Rural Statistical Area: Changed to "Micropolitan Statistical Area," after the U.S. Census Bureau unveiled the term in 2003 justified by the fact that some parts of the U.S. are neither rural nor metropolitan.

RSC—Relative Source Contribution

RTC Certificates—In the RECLAIM program, certificates issued by the SCAQMD (California) and constituting evidence of RTCs held by any person and are used for information only.

RTCM—Reasonable Transportation Control Measure

RTG—A Regional Transmission Group. A voluntary organization of transmission owners, users, and other entities interested in coordinating transmission planning, expansion, operation, and use on a regional and inter-regional basis. Such groups are subject to FERC approval.

RTK—Right-to-Know

RTP—Real-time Pricing

RUL—Regular Unleaded Gasoline

Rule 144—Rule 144 provides for the sale of restricted stock and control stock. Filing with the SEC is required prior to selling restricted and control stock, and the number of shares that may be sold is limited.

Rule 144A—A safe-harbor exemption from the registration requirements of Section 5 of the 1933 Act for resales of certain restricted securities to qualified institutional buyers, which are commonly referred to as "QIBs." In particular, Rule 144A affords safe-harbor treatment for reoffers or resales to QIBs—by persons other than issuers—of securities of domestsic and foreign issuers that are not listed on a U.S. securities exchange or quoted on a U.S. automated inter-dealer quotation system. Rule 144A provides that reoffers and resales in compliance with the rule are not "distributions" and that the reseller is therefore not an "underwriter" within in the meaning of Section 2(a)(11) of the 1933 Act. If the reseller is not the issuer or a dealer, it can rely on the exemption provided by Section 4(1) of the 1933 Act. If the reseller is a dealer, it can rely on the exemption provided by Section 4(3) of the 1933 Act.

Rule 144A Exchange Offer—A transaction in which one class of securities that were issued in a private placement are exchanged for another, unusually almost identical, class of securities, in a transaction registered with the SEC on a Form S-4 Registration Statement.

Rule 501—Rule 501 of Regulation D defines Accredited Investor, among other definitions and regulations.

Rule 505—Rule 505 of Regulation D is an exemption for limited offers and sales of securities.

Rule 506—Rule 506 of Regulation D is considered a "safe harbor" for the private-offering exemption of Section 4(2) of the Securities Act of 1933. Companies using the Rule 506 exemption can raise an unlimited amount of money if they meet certain exemptions.

Rule of 78—An accelerated method of allocating periodic earnings in a financial contract based upon the sum-of-the-years method.

Rulemaking (Regulations)—The authority delegated to administrative agencies by Congress or State legislative bodies to make rules that have the force of law. Frequently, statutory laws that express broad terms of a policy are implemented more specifically by administrative rules, regulations, and practices.

Rules of Conduct—Rules set in advance to delineate acceptable activities by participants, particularly participants with significant market power.

Run Off—That portion of the precipitation that flows over the land surface and ultimately reaches streams to complete the water cycle. Melting snow is an important source of this water as well as all amounts of surface water that move to streams or rivers through any given area of a drainage basin.

Runner—The part of a hydro turbine that accepts the water and turns its energy into rotating motion.

Running and Quick-start Capability—The net capability of generating units that carry load or have quick-start capability. In general, quick-start capability refers to generating units that can be available for load within a 30-minute period.

Running Rate—The rate of return to the lessor, or cost to the lessee, in a lease based solely upon the initial equipment cost and the periodic lease payments, without any reliance on residual value, tax benefit, deposits or fees. This rate also is referred to as the street or stream rate.

Run-of-Mine Coal—Coal as it comes from the mine prior to screening or any other treatment.

Run-of-River Hydroelectric Plant—A low-head plant using the flow of a stream as it occurs and having little or no reservoir capacity for storage.

Runout—Piping that is no more than 12 feet long and that is connected to a fixture or an individual terminal unit.

Rural Electric Cooperative—A nonprofit, customer-owned electric utility that distributes power in a rural area. As of June 1990, there are three rural electric cooperatives in California:

• Anza Electric Cooperative in Anza

• Plumas-Sierra Rural Electric Cooperative in Portola

• Sunrise Valley Electrification Corporation in Alturas

Rural Electrification Administration (REA)—A lending agency of the U. S. Department of Agriculture, the REA makes self-liquidating loans to qualified borrowers to finance electric and telephone service to rural areas. The REA finances the construction and operation of generating plants, electric transmission and distribution lines, or systems for the furnishing of initial and continued adequate electric services to persons in rural areas not receiving central station service.

RV—Residual Volume

R-Value—A measure of thermal resistance of a material, equal to the reciprocal of the U-Value. The R-Value is expressed in terms of degrees Fahrenheit times hours, times square feet per Btu.

R-Value—A measure of a material's resistance to heat flow in units of Fahrenheit degrees x hours x square feet per Btu. The higher the R-value of a material, the greater it's insulating capability. The R-value of some insulating materials is 3.7 per inch for fiberglass and cellulose, 2.5 per inch for vermiculite, and more than 4 per inch for foam. All building materials have some R-value. For example, a 4-inch brick has an R-value of 0.8, and half-inch plywood has an R-value of 0.6. The below table converts the most common "R" values to inches. For other "R" values, divide the "R" value by 3 to get the number of inches.

"R"-Value	Inches
3	1
11	3.5
19	6
52	18

RVP—Reid Vapor Pressure

RWQCB—Regional Water Quality Control Board (California)

S

S&A—Sampling and Analysis
- Surveillance and analysis

SA—Special Assistant
- Sunshine Act
- Surface Area

Sacrificial Anode—A metal rod placed in a water heater tank to protect the tank from corrosion. Anodes of aluminum, magnesium, or zinc are the more frequently metals. The anode creates a galvanic cell in which magnesium or zinc will be corroded more quickly than the metal of the tank giving the tank a negative charge and preventing corrosion.

SAE Viscosity Number—A system established by the Society of Automotive Engineers for classifying crankcase oils and automotive transmission and differential lubricants according to their viscosities.

Safety Disconnect—An electronic (automatic or manual) switch that disconnects one circuit from another circuit. These are used to isolate power generation or storage equipment from conditions such as voltage spikes or surges, thus avoiding potential damage to equipment.

Safety Engineering—The planning, development, improvement, coordination and evaluation of the safety component of integrated systems of people, materials, equipment and environments to achieve optimum safety effectiveness in terms of protection of people and property.

SAI—Solar America Initiative

SAIC—Special-Agents-In-Charge

SAIDI—System Average Interruption Duration Index: Statistical data on utility service interruptions.

SAIFI—System Average Interruption Frequency Index: Statistical data on utility service interruptions.

Salable Coal—The shippable product of a coal mine or preparation plant. Depending on customer specifications, salable coal may be run-of-mine, crushed-and-screened (sized) coal, or the clean coal yield from a preparation plant.

Salable Natural Gas—Natural gas marketed under controlled quality conditions.

Sale-Leaseback—A transaction that involves the sale of equipment to a leasing company and a subsequent lease of the same equipment back to the original owner, who continues to use the equipment.

Sales Agreement—An agreement between a purchaser/buyer and seller (e.g., producer, marketer, pipeline, energy provider, design firm, engineering company, developer, LDC) which defines the terms and conditions of a purchase/sale and title transfer of energy quantities.

Sales for Resale—A type of wholesale sales covering electric energy and natural gas supplied to other utilities, cooperatives, municipalities, and Federal and state agencies for resale to ultimate consumers.

Sales to End Users—Sales made directly to the consumer of the product. Includes bulk consumers, such as agriculture, industry, and utilities, as well as residential and commercial consumers.

Sales Type—Sales categories of sales to end-users and sales for resale.

Sales Volume (Coal)—The reported output from Federal and/or Indian lands, the basis of royalties. It is approximately equivalent to production, which includes coal sold, and coal added to stockpiles.

Sales-Type Lease—A capital lease from the lessor's perspective (per FASB 13) that gives rise to manufacturer's or dealer's profit to the lessor.

Salt Dome—A domical arch (anticline) of sedimentary

rock beneath the earth's surface in which the layers bend downward in opposite directions from the crest and that has a mass of rock salt as its core.

Salt Gradient Solar Ponds—Consist of three main layers. The top layer is near ambient and has low salt content. The bottom layer is hot, typically 160°F to 212 F (71°C to 100°C), and is very salty. The important gradient zone separates these zones. The gradient zone acts as a transparent insulator, permitting the sunlight to be trapped in the hot bottom layer (from which useful heat is withdrawn). This is because the salt gradient, which increases the brine density with depth, counteracts the buoyancy effect of the warmer water below (which would otherwise rise to the surface and lose its heat to the air). An organic Rankine cycle engine is used to convert the thermal energy to electricity.

Salvage (Proceeds)—The value realized from plant removed or otherwise disposed. This value may be in the form of cash, debits to the materials and supplies accounts, trade-in allowance, or other consideration.

Salvage Logging—The harvest of dead, dying, damaged or weak trees after a forest fire to prevent the spread of disease or insects and to reduce the risk of high intensity fire.

Salvage Value—The amount received for property retired, less any expenses incurred in connection with the sale or in preparing the property for sale; or, if retained, the amount at which the material recoverable is chargeable to Materials and Supplies, or other appropriate account.

Salvage, Net—The difference between value of salvage and cost of removal resulting from the removal, abandonment, or other disposition of plant. Positive net salvage results when salvage value exceeds removal costs. Negative net salvage results when removal costs exceed salvage value. Positive net salvage decreases the cost to be recovered through depreciation expense, and negative net salvage increases it.

SAM—Sustainable Asset Management

Sample (Coal)—A representative fraction of a coal bed collected by approved methods, guarded against contamination or adulteration, and analyzed to determine the nature; chemical, mineralogic, and (or) petrographic composition; percentage or parts-per-million content of specified constituents; heat value; and possibly the reactivity of the coal or its constituents.

SAMSON—Solar and Meteorological Surface Observation Network

Sanitary Landfill—A landfill which does not take hazardous waste, often called a "garbage dump." It must be covered with dirt each day to maintain sanitary conditions.

SAP—Sampling and Analysis Plan
- Scientific Advisory Panel
- Special Access Program

SARA—Superfund Amendments and Reauthorization Act (1986) Also known as the Emergency Planning and Community Right-to-Know Act of 1986—EPCRA.

Satellite Power System (SPS)—Concept for providing large amounts of electricity for use on the Earth from one or more satellites in geosynchronous Earth orbit. A very large array of solar cells on each satellite would provide electricity, which would be converted to microwave energy and beamed to a receiving antenna on the ground. There, it would be reconverted into electricity and distributed the same as any other centrally generated power, through a grid.

Satellite Remote Sensing—The collection of data on land use, industrial activity, weather, climate, geology and other processes through Earth observations from satellites in outer space.

Satisfaction—Performance of the terms of an obligation.

Saturated Air—Air containing all the water vapor it can hold at its temperature and pressure.

Saturated Steam—Steam at the temperature that corresponds to its boiling temperature at the same pressure.

Savings Fraction—The percentage of consumption from using the old technology that can be saved by replacing it with the new, more efficient demand-side management technology. For example, if a 60-watt incandescent lamp were replaced with a 15-watt compact fluorescent lamp, the savings fraction would be 75 percent because the compact fluorescent lamp uses only 25 percent of the energy used by the incandescent lamp.

Sawlog—A log meeting minimum commercial requirements of diameter, length, and defect. The usual commercial requirements are a minimum of 8' long with an inside bark diameter of 6" for softwoods and 8" for hardwoods.

Sawtimber—Live trees of commercial species containing at least one 12' sawlog or two noncontiguous 8' logs. Softwoods must be at least 9" in diameter and hardwoods at least 11" in diameter.

Sawtooth Rents—Rents that vary throughout the term

of the lease, usually to match debt payments and tax payments in a leveraged lease so as to lessen the need for a sinking fund.

SB—Senate Bill

SBA—Small Business Act
  • Small Business Administration

SBC—Sustainable Building Coalition
  • System Benefits Charge. Program set forth in various jurisdictions to help improve its renewable energy production and infrastructure via a surcharge on customer utility bills.

SBI—Special Background Investigation

SBIC—Sustainable Buildings Industry Council SBIC is an independent, nonprofit organization of architects, engineers, product manufacturers and professional building associations promoting sustainable design in the building industry through education, outreach and advocacy. The organization provides a number of green building programs, green building guidelines and ENERGY-10 software for modeling homes and small commercial buildings to optimize energy, materials and siting. The organization's programs focus on home design, construction, performance and maintenance for a number of building types including residential, K-12 schools, small commercial, federal and large commercial. www.sbicouncil.org

SBIR—Small Business Innovation Research Program. See Small Business Innovation Development Act of 1982.

SBIS—Sustainable Building Information System

SBO—Small Business Ombudsman

SBREFA—Small Business Regulatory Enforcement Fairness Act (1996) SBREFA was signed into law on March 29, 1996, and contains five distinct sections:
  • **Subtitle A–Regulatory Compliance Simplification**: Among other things, requires the agency to publish Small Entity Compliance Guides that are written in plain language and explain the actions a small entity must take to comply with a rule or group of rules.
  • **Subtitle B–Regulatory Enforcement Reforms**: Requires agencies to support the rights of small entities in enforcement actions, specifically providing for the reduction, and in certain cases, the waiver of civil penalties for violations by small entities.
  • **Subtitle C–Equal Access to Justice**: Provides small businesses with expanded authority to go to court to be awarded attorneys' fees and costs when an agency has been found to be ex-

cessive in enforcement of federal regulations.
  • **Subtitle D–Regulatory Flexibility Act Amendments**: Provides small entities with expanded opportunities to participate in the development of certain regulations.
  • **Subtitle E–Congressional Review of Agency Rulemaking**: Agencies generally must provide Congress and the General Accounting Office with copies of all final rules and supporting analyses. Congress may decide not to allow a rule to take effect.

SBS—Sick Building Syndrome

SBTF—Sustainable Building Task Force (California)

SCA—Secured Creditor Assessment
  • Specific Collection Area

SCAG—Southern California Association of Governments

Scaled Sale—A type of timber sale contract that specifies measuring or scaling of the included timber after removal. Scaling determines the number of board feet or c-units to be paid for at contract rates.

SCAQMD—South Coast Air Quality Management District (California)

SCE—Southern California Edison Company: One of the largest electric utilities in the U.S., and the largest subsidiary of Edison International. On the internet at http://www.sce.com

SCF—Standard cubic foot.

SCFM—Standard cubic foot per minute

SCH—State Clearing House

Schedule—A statement of the pricing format of electricity and the terms and conditions governing its applications.

Scheduled Outage—The shutdown of a generating unit, transmission line, or other facility for inspection or maintenance, in accordance with an advance schedule.

Scheduling Coordinator—Scheduling coordinators (SCs) submit balanced schedules and provide settlement-ready meter data to the ISO. Scheduling coordinators also:
  1. Settle with generators and retailers, the PX and the ISO
  2. Maintain a year-round, 24-hour scheduling center
  3. Provide non-emergency operating instructions to generators and retailers
  4. Transfer schedules in and out of the PX. (The PX is a marketplace. As bids are accepted, power is being bought and sold. Once a bid is accepted, the power sold is "transferred out"

of the PX, since is it no longer available. Power that is available for sale is "transferred in" to the PX. These transfers may also take place directly between the buyer and seller, without involvement of the PX.)

5. The PX is considered a scheduling coordinator

Scheduling Coordinators—Entities certified by the Federal Energy Regulatory Commission (FERC) that act on behalf of generators, supply aggregators (wholesale marketers), retailers, and customers to schedule the distribution of electricity.

Schematic—An outline, systematic arrangement, diagram, scheme, or plan. An orderly combination of events, persons, or things according to a definite plan. A diagram showing the relative position and/or function of different components or elements of an object or system.

Schottky Barrier—A cell barrier established as the interface between a semiconductor, such as silicon, and a sheet of metal.

Scientific Equipment—Measurement, testing or metering equipment used for scientific research or investigation, including but not limited to manufactured cabinets, carts and racks.

Sconce—Wall mounted ornamental Luminaire.

Scoop Loading—An underground loading method by which coal is removed from the working face by a tractor unit equipped with a hydraulically operated bucket attached to the front; also called a front-end loader.

Scoping—A first step in the NEPA process and in the river planning process. Scoping is a means of identifying issues and concerns, their significance, and the range of alternatives.

SCR—Selective Catalytic Reduction. A method used to reduce emissions from the burning of natural gas using ammonia injection.

Screening Value—The instrument reading (ppmv) obtained when components, including but not limited to valves, pump seals, connectors, flanges, open-ended lines and other equipment components, are evaluated for leakage as described in USEPA Method 21—Determination of Volatile Organic Component Leaks.

Screenings—The undersized coal from a screening process, usually one-half inch or smaller.

Scribing—The cutting of a grid pattern of grooves in a semiconductor material, generally for the purpose of making interconnections.

Scrubber—A device to clean combustible gas or stack gas by the spraying of water.

SCS (Scientific Certification Systems)—SCS is a third-party provider of certification, auditing and testing services, and standards. Currently certified products include office furniture systems, components, and seating; building materials; carpets and rugs; hard surface flooring; paints; finishes; wood products; and cleaning products.

There are several categories of SCS product certification that can garner a project LEED credits. These certifications include:
—SCS Indoor Advantage, an indoor air emissions certification program for office furniture systems, components and seating.
—SCS Indoor Advantage Gold, an indoor air emissions certification for building materials such as adhesives and sealants, paints and coatings, textiles and wall coverings, and composite wood, as well as classroom and office furniture systems, components and seating.
—FloorScore, an indoor air emissions certification program for hard surface flooring and flooring adhesives that was developed by the Resilient Floor Covering Institute (RFCI) and is managed by SCS.
—Recycled Material Content refers to products certified by SCS that have recycled content. (A product with at least 10 percent post-consumer or 20 percent pre-consumer recycled material qualifies for LEED MR 4.1, while material certified for at least 20 percent post-consumer or 40 percent pre-consumer recycled material qualifies for LEED MR 4.2.)
—SCS Sustainable Choice refers to low-emitting carpets and rugs.
—No Formaldehyde refers to composite wood, laminate and adhesive products certified by SCS either for No Added Urea Formaldehyde, No Added Formaldehyde or Formaldehyde Free.
—Forest Stewardship Council (FSC) Chain-of-Custody means that the wood product has earned FSC Chain-of-Custody certification from SCS.

Most recently, SCS has introduced its SCS Sustainable Choice label for furniture, which ensures that products earning the label have met the sustainability requirements of the new Business and Institutional Furniture Manufacturer's Association Sustainability Standard (BIFMA SS).

SCTL—Single Circuit Transmission Line

SD—Standard Deviation

SDBE—Small and Disadvantaged Business Enterprise

SDCM—Standard Dry Cubic Meter

SDG&E—San Diego Gas & Electric. The third largest investor owned utility in California.

SDREO—San Diego Regional Energy Office (California) www.sdreo.org

SDWA—Safe Drinking Water Act (as amended in 1986)

SE (Seasonal Efficiency)—A measure of the percentage of heat from the combustion of gas and from associated electrical equipment that is transferred to the space being heated during a year under specified conditions. California Code of Regulations, Section 2-1602(d)(11).

Seaboard Method—A classification method that allocates fixed costs equally between the demand and commodity components of the rate.

Seal—An impression upon a document which lends authenticity to its execution, i.e. a Corporate Seal or Notary Seal.

Sealed Battery—A battery with a captive electrolyte and a re-sealing vent cap to which electrolyte cannot be added. Also called a valve-regulated battery.

Sealed Combustion—Combustion whereby a combustion appliance, such as a furnace, water heater or fireplace, acquires all air for combustion though a dedicated sealed passage from the outside. Combustion occurs in a sealed combustion chamber and all combustion products are vented to the outside through a separate dedicated sealed vent.

Sealed Combustion Heating System—A heating system that uses only outside air for combustion and vents combustion gases directly to the outdoors. These systems are less likely to backdraft and to negatively affect indoor air quality.

Sealed Lead Acid Battery—A form of lead-acid battery where the electrolyte is immobilized, either by being contained in an absorbent fiber separator or gel between the batteries plates.

Seam—A bed of coal lying between a roof and floor. Equivalent term to bed, commonly used by industry.

Seasonal Allotment Period—In the Illinois ERMs program, the period from May 1 through September 30 of each year.

Seasonal Curtailment—Curtailment imposed on a seasonal summer (April-October) or winter (November-March) basis because of gas or electric supply deficiency.

Seasonal Depth of Discharge—An adjustment factor used in some system sizing procedures which "allows" the battery to be gradually discharged over a 30- to 90-day period of poor solar insolation. This factor results in a slightly smaller photovoltaic ar-

ray.

Seasonal Efficiency (SE)—A measure of the percentage of heat from the combustion of gas and from associated electrical equipment that is transferred to the space being heated during a year under specified conditions.

Seasonal Emissions—In the Illinois ERMs program, the actual volatile organic material emissions at a Participating Source that occurs during a Seasonal Allotment Period.

Seasonal Energy Efficiency Ratio (SEER)—Ratio of the cooling output divided by the power consumption. It is the Btu of cooling output during its normal annual usage divided by the total electric energy input in watt hours during the same period. This is a measure of the cooling performance for rating central air conditioners and central heat pumps. The appliance standards required a minimum SEER of 10 for split-system central air conditioners and for split-system central heat pumps in 1992. (The average heat pump or central air conditioner sold in 1986 had an SEER of about 9.)

Seasonal Method—An allocation method which allocates demand and/or commodity costs to customer classes by seasonal usage.

Seasonal Performance Factor (SPF)—Ratio of useful energy output of a device to the energy input, averaged over an entire heating season.

Seasonal Pricing—A special electric rate feature under which the price per unit of energy depends on the season of the year.

Seasonal Rates—Different seasons of the year are structured into an electric rate schedule whereby an electric utility provides service to consumers at different rates. The electric rate schedule usually takes into account demand based on weather and other factors.

Seasonal Units—Housing units intended for occupancy at only certain seasons of the year. Seasonal units include units intended only for recreational use, such as beach cottages and hunting cabins. It is not likely that this type of unit will be the usual residence for a household, because it may not be fit for living quarters for more than half of the year.

Seasoned Wood—Wood, used for fuel, that has been air dried so that it contains 15 to 20 percent moisture content (wet basis).

SEC—Securities and Exchange Commission. The SEC is the United States' highest securities and exchange commission. It supervises the entire U.S. public securities market.

Second Assessment Report (SAR)—The Second Assessment Report, prepared by the Intergovernmental Panel on Climate Change, reviewed the existing scientific literature on climate change. Finalized in 1995, it is comprised of three volumes: Science; Impacts, Adaptations and Mitigation; and Economic and Social Dimensions of Climate Change.

Second Growth—A second generation of timber of merchantable age.

Second Law Efficiency—The ratio of the minimum amount of work or energy required to perform a task to the amount actually used.

Second Law of Thermodynamics—This law states that no device can completely and continuously transform all of the energy supplied to it into useful energy.

Secondary Battery—A battery that can be recharged; a rechargeable battery.

Secondary Cell—Secondary cells are batteries (electrochemical cells) that are rechargeable. The chemical reaction within the secondary cell is reversible, allowing the cell to be recharged many times.

Secondary Containment—A structure designed to capture spills or leaks, as from a container or tank. For containers and aboveground tanks, it is usually a bermed area of coated concrete. For underground tanks, it may be a second, outer, wall or a vault. Construction of such containment must meet certain requirements, and periodic inspections are required.

Secondary Energy—See *Non-Firm Energy*.

Secondary Heating Equipment—Space-heating equipment used less often than the main space-heating equipment.

Secondary Heating Fuel—Fuels used in secondary space-heating equipment.

Secondary Market—The exchange of emission reductions, offsets, or allowances between buyer and seller where the seller is not the originator of the supply. The exchange of greenhouse gas emission reductions currently involves only the primary market.

Secondary Succession—The progression of plant communities following disturbances such as fire, windthrow and timber harvesting. See *Succession*.

Secondary Use—The use of PHEV and EV batteries for stationary electric grid storage after they can no longer meet the demands of charging vehicles.

Secretariat (UNFCCC)—Staffed by international civil servants and responsible for servicing the COP and ensuring its smooth operation, the secretariat makes arrangements for meetings, compiles and prepares reports, and co-ordinates with other relevant international bodies. The Climate Change secretariat is institutionally linked to the United Nations.

Secretariat of the UN Framework Convention—The United Nations staff assigned the responsibility of conducting the affairs of the UNFCCC. In 1996 the Secretariat moved from Geneva, Switzerland, to Bonn, Germany.

Section 1031—Section of the Internal Revenue Code dealing with tax-free exchanges of like-kind property.

Secured Creditor—A creditor whose obligation is backed by the pledge of some asset. In liquidation, the secured creditor receives the cash from the sale of the pledged asset to the extent of its extension of the credit to the debtor.

Securities Act of 1933—A federal law governing the issuance of securities to the public.

Securities Act of 1934—A federal law governing the operations of stock and securities exchanges. Also governs over-the-counter trading.

Securities and Exchange Commission (SEC)—The SEC is an independent, nonpartisan, quasi-judicial regulatory agency that is responsible for administering the federal securities laws. These laws protect investors in securities markets and ensure that investors have access to all material information concerning publicly traded securities. Additionally, the SEC regulates firms that trade securities, people who provide investment advice, and investment companies.

Securitization—A proposal for issuing bonds that would be used to buy down existing power contracts or other obligations. The bonds would be repaid by designating a portion of future customer bill payments. Customer bills would be lowered, since the cost of bond payments would be less than the power contract costs that would be avoided.

Securitize—The aggregation of contracts for the purchase of the power output from various energy projects into one pool that then offers shares for sale in the investment market. This strategy diversifies project risks from what they would be if each project were financed individually, thereby reducing the cost of financing. Fannie Mae performs such a function in the home mortgage market.

Security—Collateral; property pledged to secure repayment of a debt.

Security Agreement—Document now used in place of a

Chattel Mortgage as evidence of a Lien on Personal Property. A financing Statement may be recorded to give Constructive Notice of the Security Agreement.

Security Deposit—A dollar amount held by the lessor to protect against default by the lessee. Refundable at the end of the lease. A deposit made to assure performance of any other obligation in a contract or other agreement.

Security Interest—An interest in property acquired by contract for the purpose of securing payment or performance of an obligation.

Sediment—The soil, sand and minerals at the bottom of surface waters, such as streams, lakes and rivers. Sediments capture or adsorb contaminants. The term may also refer to solids that settle out of any liquid.

Sedimentation—A process in which material carried in suspension by water flows into streams and rivers, increasing turbidity and eventually settling to the bottom.

SEDS—State Energy Data System

Seebeck Effect—The generation of an electric current, when two conductors of different metals are joined at their ends to form a circuit, with the two junctions kept at different temperatures.

Seed Money—The first round of capital for a start-up business. Seed money usually takes the structure of a loan or an investment in preferred stock or convertible bonds, although sometimes it is common stock. Seed money provides startup companies with the capital required for their initial development and growth. Angel investors and early-stage venture capital funds often provide seed money.

Seed State Financing—An initial state of a company's growth characterized by a founding management team, business-plan development, prototype development, and beta testing.

SEER (Seasonal Energy Efficiency Ratio)—The total cooling output of a central air conditioning unit in Btu's during its normal usage period for cooling divided by the total electrical energy input in watt-hours during the same period, as determined using specified federal test procedures. [See California Code of Regulations, Title 20, Section 1602(c)(11)] The higher the SEER rating, the more efficient the unit is.

SEGS—Solar Electric Generating Station

SEIA—Socioeconomic Impact Analysis
• Solar Energy Industries Association www.seia.org

SEIA (Solar Energy Industries Association)—SEIA is the national trade association for the solar industry. The organization works to expand markets, strengthen research and development, impact and guide regulations and laws impacting solar energy, and improve education and outreach for solar. It also holds the annual Solar Power Conference and Expo.

SEIS—Supplemental Environmental Impact Statement

Seismic Stability—The likelihood that soils or structures will stay in place during an earthquake.

Seismograph—A device for detecting vibrations in the earth. It is used in prospecting for probable oil or gas bearing structures. In this application, the vibrations are created by discharging explosives in shallow bore holes. The nature and velocity of the vibrations as recorded by the seismograph indicate the general nature of the section of earth through which the vibrations pass.

Selectable Load—Any device, such as lights, televisions, and power tools, which is plugged into your central power source and used only intermittently.

Selection Cutting—The periodic removal at short intervals of the oldest and largest trees in the stand, individually or in small groups.

Selective Absorber—A solar absorber surface that has high absorbtance at wavelengths corresponding to that of the solar spectrum and low emittance in the infrared range.

Selective Catalytic Reduction (SCR)—A post combustion control which taps flue gas off the boiler or other equipment and injects ammonia with nitrogen oxide gas to reduce emissions.

Selective Oxidation Step—A small amount of air is mixed with the process gas, and using a suitable catalyst, oxidizes the carbon monoxide remaining after the shift converter which would otherwise preferentially absorb on the anode catalyst. The process avoids hydrogen oxidation.

Selective Surface Coating—A material with high absorbance and low emittance properties applied to or on solar absorber surfaces.

Selenium—This metal is a nutritionally essential trace element that is toxic at higher doses. High levels of selenium have been shown to cause reproductive failure and birth defects in birds.

Self Discharge—Self discharge represents energy lost to internal chemical reactions within the cell.

Self Discharge Rate—The rate at which a battery will lose its charge when at open circuit (with no load connected).

Self-Generation—A generation facility dedicated to serving a particular retail customer, usually located on the customer's premises. The facility may either be owned directly by the retail customer or owned by a third party with a contractual arrangement to provide electricity to meet some or all of the customer's load.

Self-Generation Facility—A co-generation facility dedicated to serving a particular retail customer, usually located on the customer's premises. The facility may either be owned directly by the retail customer or owned by a third party with a contractual arrangement to provide electricity to meet some or all of the customer's load.

Self-Service Wheeling—Primarily an accounting policy comparable to net-billing or running the meter backwards. An entity owns generation that produces excess electricity at one site, that is used at another site(s) owned by the same entity. It is given billing credit for the excess electricity (displacing retail electricity costs minus wheeling charges) on the bills for its other sites.

Seller—A legally recognized entity (individual, corporation, not-for-profit organization, government, etc.) who sells energy, emission reductions, emission credits or allowances to another legally recognized entity through a sale, lease, trade, or other means of transfer. A legal entity who has contractual signatory authority and able to provide warranty of title. The seller may have legal authority to sell as agent for or on behalf of other owners.

Seller Type—Categories of major refiners and other refiners and gas plant operators.

SEM—Standard Error of the Means

Semiconductor—Any material that has a limited capacity for conducting an electric current. Certain semiconductors, including silicon, gallium arsenide, copper indium diselenide, and cadmium telluride, are uniquely suited to the photovoltaic conversion process.

Semi-Crystalline—See *Multi-Crystalline*

Senior Creditor—A creditor with a claim on income or assets prior to that of general creditors.

Senior Debt—All debt, both short-term and long-term, which is not subordinated to any other liability.

Senior Securities—Securities that have a preferential claim over common stock on a company's earnings and in the case of liquidation. Generally, preferred stock and bonds are considered senior securities.

Sensible Cooking Capacity—See *Cooling Capacity, Sensible*.

Sensible Cooling—The removal of the heat from air to reduce the temperature, without removing moisture.

Sensible Cooling Effect—The difference between the total cooling effect and the dehumidifying effect.

Sensible Cooling Load—The interior heat gain due to heat conduction, convection, and radiation from the exterior into the interior, and from occupants and appliances.

Sensible Heat—The heat absorbed or released when a substance undergoes a change in temperature.

Sensible Heat Storage—A heat storage system that uses a heat storage medium, and where the additional or removal of heat results in a change in temperature.

Sensitive Species—See *Threatened, Endangered, and Sensitive Species*.

Sensor (Solar)—Sensing device that changes its electrical resistance according to temperature. Used in the control system of a solar thermal system to measure collector and storage tank temperatures.

SEP—State Energy Program
• Superior Energy Performance

SEPA—Solar Electric Power Association www.solarelectricpower.org

Separate Metering—Measurement of electricity or natural gas consumption in a building using a separate meter for each of several tenants or establishments in the building.

Separative Work Unit (SWU)—The standard measure of enrichment services. The effort expended in separating a mass F of feed of assay x_f into a mass P of product assay x_p and waste of mass W and assay x_w is expressed in terms of the number of separative work units needed, given by the expression $SWU = WV(x_w) + PV(x_p) - FV(x_f)$, where $V(x)$ is the "value function," defined as $V(x) = (1 - 2x) 1n((1 - x)/x)$.

Separator—A piece of equipment for separating one substance from another when they are intimately mixed, such as removing oil from water, oil from gas, ash from flue gas, or tramp iron from coal.

Separator Plate—A solid piece of electrically conductive material that is inserted between cells in a stack.

Septic Tank—A tank in which the solid matter of continuously flowing sewage is disintegrated by bacteria.

Sequestration—Opportunities to remove atmospheric CO_2, either through biological processes (e.g. plants and trees), or geological processes through storage of CO_2 in underground reservoirs.
• The process of increasing the carbon content of

a carbon reservoir other than the atmosphere. Biological approaches to sequestration include direct removal of carbon dioxide from the atmosphere through land-use change, afforestation, reforestation, and practices that enhance soil carbon in agriculture. Physical approaches include separation and disposal of carbon dioxide from flue gases or from processing fossil fuels to produce hydrogen- and carbon dioxide-rich fractions and long-term storage in underground in depleted oil and gas reservoirs, coal seams, and saline aquifers.

Seral Stages—The series of relatively transitory plant communities that develop during ecological succession from bare ground to the climax stage.

Series—A configuration of an electrical circuit in which the positive lead is connected to the negative lead of another energy producing, conducting, or consuming device. The voltages of each device are additive, whereas the current is not.

Series A Preferred Stock—The first round of stock offered during the seed or early-stage round by a portfolio company to the venture investor or fund. This stock is convertible into common stock in certain cases such as an IPO or the sale of the company. Later rounds of preferred stock in a private company are called Series B, Series C, and so on.

Series Connection—A way of joining photovoltaic cells by connecting positive leads to negative leads; such a configuration increases the voltage.

Series Fan-Powered Terminal Unit—A terminal unit that combines a VAV damper in series with a downstream fan which runs at all times that the terminal unit is supplying air to the space.

Series Regulator—A type of battery charge controller or regulator in which the charging current is controlled by a switch, transistor, or field-effect transistor connected in series with the PV module or array.

Series Resistance—Parasitic resistance to current flow in a cell due to mechanisms such as resistance from the bulk of the semiconductor material, metallic contacts, and interconnections.

Series String—A device that prevents overcharging of a battery by disconnecting the charging source as the battery voltage approaches some upper limit.

Service Agreement—An agreement entered into by the transmission customer and transmission provider.

Service Area—The territory in which a utility system or distributor is authorized to provide service to consumers.

Service Drop—The lines running to a customer's house.

Usually a service drop is made up of two 120 volt lines and a neutral line, from which the customer can obtain either 120 or 240 volts of power. When these lines are insulated and twisted together, the installation is called triplex cable.

Service Lease—A lease for equipment that assigns the lessor the responsibility for maintaining the leased property.

Service Life—The length of time a piece of equipment can be expected to perform at its full capacity.

The time between the date plant is includible in plant in service, or plant leased to others, and the date of its retirement. If depreciation is accounted for on a production basis rather than on a time basis, then service life should be measured in terms of the appropriate unit of production.

Service Territory—The state, area or region served exclusively by a single electric utility.

Service Value—The difference between original cost and net salvage value of a utility plant.

Service Water Heating—Heating of water for sanitary purposes for human occupancy, other than for comfort heating.

Service Well—A well drilled, completed, or converted for the purpose of supporting production in an existing field. Wells of this class also are drilled or converted for the following specific purposes: gas injection (natural gas, propane, butane or fuel-gas); water injection; steam injection; air injection; salt water disposal; water supply for injection; observation; and injection for in-situ combustion.

Servicer—The party that performs collections activities, makes appropriate disbursements, handles terminations/buyouts and provides required reports to investors on a securitized transaction.

Servient Tenement—An Estate burdened by an Easement.

Setpoint—Scheduled operating level for each generating unit or other resource scheduled to run in the Hour-ahead Schedule.

Setback—Zoning regulations that designate the distance a building must be set back from the front, rear, and sides of the property lines.

• The height at which the upper floors of a building are recessed, or set back, from the face of the lower structure.

Setback Thermostat—See *Thermostat, Setback*

SETP—DOE Solar Energy Technologies Program

SETS—Site Enforcement Tracking System

• Superfund Enforcement Tracking System

Settlement—The process of financial settlement for

products and services purchased and sold. Each settlement involves a price and quantity. Both the ISO and PX may perform settlement functions.

Severability—A provision in a lease agreement which states that if any part or provision of a lease shall be found unenforceable, it alone shall be discarded and the remaining provisions shall be given their full force and effect.

Sewage—The waste water from domestic, commercial and industrial sources carried by sewers.

SF—Slope Factor
- Standard Form
- Superfund

SF₆—Sulfur Hexafluoride. A highly stable non-conducting chemical used for and emitted from various industrial processes and in the manufacturing of electrical circuitry. An extremely high CO_2 compound.

SFR—Sinking Fund Rate. The annual after-tax rate at which the sinking fund earns.

SGIP—Self Generation Incentive Program (for solar rebates in California)
- Small Generator Interconnection Procedure

Shade Screen—A screen affixed to the exterior of a window or other glazed opening, designed to reduce the solar radiation reaching the glazing.

Shading—The protection from heat gains due to direct solar radiation;
Shading is provided by (a) permanently attached exterior devices, glazing materials, adherent materials applied to the glazing, or an adjacent building for nonresidential bui5ldings, hotels, motels and high-rise apartments, and by (b) devices affixed to the structure for residential buildings. [See California Code of Regulations, Title 24, Section 2-5302]

Shading Coefficient (CF)—The ratio of solar heat gain through a specific glazing system to the total solar heat gain through a single layer of clear, double-strength glass.
A measure of window glazing performance that is the ratio of the total solar heat gain through a specific window to the total solar heat gain through a single sheet of double-strength glass under the same set of conditions; expressed as a number between 0 and 1.

Shaft Horsepower—A measure of the actual mechanical energy per unit time delivered to a turning shaft. 1 shaft horsepower = 1 electric horsepower = 550 ft-lb/second.

Shaft Mine—A mine that reaches the coal bed by means of a vertical shaft.

Shakes/Shingles—Flat pieces of weatherproof material laid with others in a series of overlapping rows as covering for roofs and sometimes the sides of buildings. Shakes are similar to wood shingles, but instead of having a cut and smoothly planed surface, shakes have textured grooves and a rough or "split" appearance to give a rustic feeling.

Shale Oil—A liquid similar to conventional crude oil but obtained from oil shale by conversion of organic matter (kerogen) in oil shale.

Shallow Cycle Battery—A battery with small plates that cannot withstand many deep discharges (i.e. to a low state of charge).

Shallow Pitting—Testing a potential mineral deposit by systematically sinking small shafts into the earth and analyzing the material recovered.

Shared Savings—A program in which the sole source of payment for energy conservation measures or services provided by a company is a predetermined percentage of the energy cost savings of the user resulting from the energy conservation measure or service.

Shareholder—The term "shareholder" includes a member in an association, joint-stock company, or insurance company.

Shareholder Resolution—A corporate policy recommendation proposed by a shareholder holding at least $2,000 market value or 1% of the company's voting shares presented for a vote by other shareholders at the company's annual meeting. An increasing number of shareholder resolutions request a company and/or its board of directors to carry out responsible business practices, especially regarding social, environmental and human rights issues.

SHC—Solar heating and cooling

Sheathing—A construction element used to cover the exterior of wall framing and roof trusses.

Shed—Kind of flat-roof skylight.

Shelf Life—The time for which a device can be stored and still retain its specified performance.

Shell Corporation—A corporation with no assets and no business. Typically, shell corporations are designed for the purpose of going public and later acquiring existing businesses. Also known as Specified Purpose Acquisition Companies (SPACs).

Shell Storage Capacity—The design capacity of a petroleum storage tank which is always greater than or equal to working storage capacity.

Shift Converter—A catalyst bed at a temperature favoring production of hydrogen by the water gas shift reaction (carbon monoxide plus water reacting to produce carbon dioxide and hydrogen).

Short Circuit—An electric current taking a shorter or different path than intended. A circuit in which two source leads of opposite polarity or dissimilar potential are connected directly to each other with no regulation or load in between, allowing the full energy potential of the source to flow through the circuit. A short circuit will trip the breaker or fuse, and may damage components, or even cause a fire.

Short Circuit Current (I$_{sc}$)—The current flowing freely through an external circuit that has no load or resistance; the maximum current possible.

• The current generated by an illuminated solar PV cell, module, or array when its output terminals are shorted; the maximum current possible.

Short Purchases—A single shipment of fuel or volumes of fuel purchased for delivery within 1 year. Spot purchases are often made by a user to fulfill a certain portion of energy requirements, to meet unanticipated energy needs, or to take advantage of low-fuel prices.

Short Rotation Energy Plantation—Plantings established and managed under short-rotation intensive culture practices.

Short Rotation Intensive Culture—Intensive management and harvesting at 2- to 10-year intervals of cycles of specially selected fast- growing hardwood species for the purpose of producing wood as an energy feedstock.

Short-term Debt—An obligation maturing in less than one year.

Short-term Sales—Any short-term purchase covering a time period of 2 years or less. Purchases from intrastate pipelines pursuant to Section 311(b) of the NGPA of 1978 are classified as short-term sales, regardless of the stated contract term.

Short Ton—2000 pounds. A ton, as commonly used in the U.S. and Canada.

Short-Term Debt or Borrowings—Debt securities or borrowings having a maturity of less than one year.

Short-Term Lease—A lease that, at the commencement date, has a maximum possible term under the contract, including any options to extend, of 12 months or less. Any lease that contains a purchase option is not a short-term lease.

Short-Term Purchase—A purchase contract under which all deliveries of materials are scheduled to be completed by the end of the first calendar year following the contract-signing year. Deliveries can be made during the contract year, but deliveries

are not scheduled to occur beyond the first calendar year thereafter.

Shortwall Mining—A form of underground mining that involves the use of a continuous mining machine and movable roof supports to shear coal panels 150 to 200 feet wide and more than half a mile long. Although similar to longwall mining, shortwall mining is generally more flexible because of the smaller working area. Productivity is lower than with longwall mining because the coal is hauled to the mine face by shuttle cars as opposed to conveyors.

Shrinkage—The volume of natural gas that is transformed into liquid products during processing, primarily at natural gas liquids processing plants.

Shunt Controller—A controller or regulator that re-directs, or shunts, the charging current away from the battery. Generally used for smaller systems.

Shunt Load—An electrical load used to safely use excess generated power when not needed for its primary uses. A shunt load in a residential photovoltaic system might be domestic water heating, such that when power is not needed for typical building loads, such as operating lights or running HVAC system fans and pumps, it still provides value and is used in a constructive, safe manner.

Shunt Regulator—Type of a battery charge regulator where the charging current is controlled by a switch connected in parallel with the photovoltaic (PV) generator. Shorting the PV generator prevents overcharging of the battery.

Shut In—Closed temporarily; wells and mines capable of production may be shut in for repair, cleaning, inaccessibility to a market, etc.

Shutdown Date—Month and year of shutdown for fuel discharge and refueling. The date should be the point at which the reactor became subcritical.

Shut-in Royalty—A royalty paid by a lessee as compensation for a Lessor's loss of income because the lessee has deferred production from a property that is known to be capable of producing minerals. Shut in may be caused by a lack of a ready market, by a lack of transportation facilities, or by other reasons. A shut-in royalty may or may not be recoverable out of future production.

Shutter—An interior or exterior movable panel that operates on hinges or slides into place, used to protect windows or provide privacy.

SIC—Standard Industrial Classification Code—also "SICC."

Sick Building Syndrome—A situation where the inhabitants of a building (often a newly constructed building) are affected adversely by chemicals give off by the building materials (new building or new car smell). *See Off-Gassing* Older buildings can be affected by molds and fungus growing in the mechanical system ducting or under the carpet.

Side Fins—Vertical shading elements mounted on either side of a glazed opening that blocks direct solar radiation from the lower, lateral portions of the sun's path.

Side Fins—Vertical shading elements mounted on either side of a glazed opening that can protect the glazing from lateral low angle sun penetration.

Side-of-Pole Mount—A PV mount installed on the side of a pole. May be fixed or seasonally adjustable.

Sidetrack Drilling—This is a remedial operation that results in the creation of a new section of well bore for the purpose of (1) detouring around junk, (2) redrilling lost holes, or (3) straightening key seats and crooked holes. Directional "side-track" wells do not include footage in the common bore that is reported as footage for the original well.

Siding—An exterior wall covering material made of wood, plastic (including vinyl), or metal. Siding is generally produced in the shape of boards and is applied to the outside of a building in overlapping rows.

SIEFA—Source Inventory Emission Factor Analysis

Sigma Heat—The sum of sensible heat and latent heat in a substance above a base temperature, typically 32 degrees Fahrenheit.

Sign—Definitions include the following:

Illuminated face is a side of a sign that has the message on it. For an exit sign it is the side that has the word "EXIT" on it.

Sign, cabinet is an internally illuminated sign consisting of frame and face(s), with a continuous translucent message panel, also referred to as a panel sign

Sign, channel letter is an internally illuminated sign with multiple components, each built in the shape of an individual three dimensional letter or symbol that are each independently illuminated, with a separate translucent panel over the light source for each element.

Sign, double-faced is a sign with two parallel opposing faces.

Sign, externally illuminated is any sign or a billboard that is lit by a light source that is external to the sign directed towards and shining on the face of the sign.

Sign, internally illuminated is a sign that is illuminated by a light source that is contained inside the sign where the message area is luminous, including cabinet signs and channel letter signs. Sign, traffic is a sign for traffic direction, warning, and roadway identification.

Sign, unfiltered is a sign where the viewer perceives the light source directly as the message, without any colored filter between the viewer and the light source, including neon, cold cathode, and LED signs.

Silica Gel—A desiccant, hygroscopic material that readily absorbs substantial quantities of moisture and is used to reduce the relative humidity of air or gas.

Silicon (Si)—A chemical element with atomic number 14, a dark gray semi-metal. Occurs in a wide range of silicate minerals and makes up approximately 28% of the earth's crust (by weight). Silicon has a face-centered cubic lattice structure like diamond. The most common semiconductor material used in making PV cells either traditionally in its crystalline form or more recently as an amorphous thin film.

Silt—Waste from Pennsylvania anthracite preparation plants, consisting of coarse rock fragments containing as much as 30 percent small-sized coal; sometimes defined as including very fine coal particles called silt. Its heat value ranges from 8 to 17 million Btu per short ton. Synonymous with *Culm*.

Silt, Culm, Refuse Bank, or Slurry Dam Mining—A mining operation producing coal from these sources of coal.

Silviculture—The theory and practice of forest stand establishment and management.

Simple CS (Caulk and Seal)—A technique for insulating and sealing exterior walls that reduces vapor diffusion through air leakage points by installing pre-cut blocks of rigid foam insulation over floor joists, sheet subfloor, and top plates before drywall is installed.

Simple Interest—Interest on the original principal only. Ignores impact of compounding interest. Accumulated interest is not included in subsequent calculations.

Simple Payback—The time required to recover the capital investment out of the savings of the installed energy efficiency or production equipment. It does not take into consideration any savings beyond the payback period; therefore it tends to penalize long life projects and favor projects that offer high savings over a short time. It is commonly used when funds are limited.

Sine Wave—A waveform that has is defined by an equation in which one variable is proportional to the sine of the other, as generated by an oscillator in simple harmonic motion. The sine wave is the most ideal form of electricity for running more sensitive appliances, such as radios, TVs, computers and the like.

Sine Wave Inverter—An inverter that produces grid-quality, sine wave AC electricity.

Single Crystal Cell—A wafer of silicon that has a perfect, continuous, crystal lattice (on the atomic level).

Single Crystal Silicon—An extremely pure form of crystalline silicon produced by dipping a single crystal seed into a pool of molten silicon under high vacuum conditions and slowly withdrawing a solidifying single crystal boule (rod) of silicon. The boule is sawed into thin silicon wafers and fabricated into single-crystal photovoltaic cells.

Single Glaze or Pane—One layer of glass in a window frame. It has very little insulating value (R-1) and provides only a thin barrier to the outside and can account for considerable heat loss and gain.

Single Investor Lease—A lease in which the lessor is fully at-risk for all funds (both equity and pooled funds) used to purchase the leased equipment.

Single Phase Line—This carries electrical loads capable of serving the needs of residential customers, small commercial customers, and streetlights. It carries a relatively light load as compared to heavy duty three phrase constructs.

Single Purpose Project—A hydroelectric project constructed only to generate electricity.

Single Zone—An HVAC system with a supply fan (and optionally a return fan) and heating and/or cooling heat exchangers (e.g. DX coil, chilled water coil, hot water coil, furnace, electric heater) that serves a single thermostatic zone. This system may or may not be constant volume.

Single-Circuit Line—A transmission line with one electric circuit. For three-phase supply, a single circuit requires at least three conductors, one per phase.

Single-Crystal Material—In reference to solar photovoltaic devices, a material that is composed of a single crystal or a few large crystals.

Single-Package System—A year 'round heating and air conditioning system that has all the components completely encased in one unit outside the home. Proper matching of components can mean more energy-efficient operation compared to components purchased separately.

Single-Phase—A generator with a single armature coil, which may have many turns and the alternating current output consists of a succession of cycles.

Sinkhole—A depression formed when the surface collapses into a cavern.

Sinking Fund—A reserve set aside for the future payment of taxes, or for the purpose of payment of any liability anticipated to become due at a future date.

Sinking Fund Rate—The earnings rate allocated to a sinking fund.

Sinks—Any process, activity or mechanism that results in the net removal of greenhouse gases, aerosols, or precursors of greenhouse gases from the atmosphere.

Sinter—A chemical sedimentary rock deposited by precipitation from mineral waters, especially siliceous sinter and calcareous sinter.

SIOR—Society of Industrial and Office Realtors www.sior.com

SIP—See *State Implementation Plan.*

SIR—Saving investment ratio

Site—Any location on which a facility is constructed or is proposed to be constructed.

Site Characterization—An onsite investigation at a known or suspected contaminated waste or release site to determine the extent and type(s) of contamination.

Site Energy—The Btu value of energy at the point it enters the home, sometimes referred to as "delivered" energy. The site value of energy is used for all fuels, including electricity.

Site Evaluation—An estimation of a location for its potential for solar, hydro, or wind power.

Site Mitigation Process—The regulatory and technical process by which hazardous waste sites are identified and investigated, and cleanup alternatives are developed, analyzed, decided up and applied.

Site Potential Tree—A tree that has attained the average maximum height possible given site conditions where it occurs.

Site Preparation—Various treatments applied to a harvested area to promote regeneration of the site.

Site Solar Energy—Natural daylighting, or thermal, chemical, or electrical energy derived from direct conversion of incident solar radiation at the building site.

Site-Built Fenestration—Fenestration designed to be field-glazed or field assembled units using specific factory cut or otherwise factory formed framing and glazing units that are manufactured with the intention of being assembled at the construction site and are provided with an NFRC label certificate for site-built fenestration. Examples of site-built fenestration include storefront systems, curtain walls, and atrium roof systems.

Site-specific Information DSM Program Assistance—A DSM (demand-side management) assistance program that provides guidance on energy efficiency and load management options tailored to a particular customer's facility; it often involves an on-site inspection of the customer facility to identify cost-effective DSM actions that could be taken. They include audits, engineering design calculations on information provided about the building, and technical assistance to architects and engineers who design new facilities.

SIU—Significant Industrial User

Six Sigma—A statistical term that equates to 3.4 defects per one million opportunities. Typical manufacturers operate at around three sigma or 67,000 defects per million. Six Sigma can achieve dramatic improvement in business performance through a precise understanding of customer requirements and the elimination of defects from existing processes, products and services. Key tenets of Six Sigma:
- Define
- Measure
- Analyze
- Improve
- Control

Sizing—The process of designing a solar system to meet a specified load given the solar resource and the nominal or rated energy output of the solar energy collection or conversion device.

Skidder—A self-propelled machine to transport harvested trees or logs from the stump area to the landing or work deck.

Skipped Payment Lease—A lease that contains a payment stream requiring the lessee to make payments only during certain periods of the year.

Sky Temperature—The equivalent temperature of the clouds, water vapor, and other atmospheric elements that make up the sky to which a surface can radiate heat.

Skylight—Any opening in the roof surface that is glazed with a transparent or translucent material. [See California Code of Regulations, Title 24, Section 2-5302]
- Frame containing class or translucent/transparent material, set in a roof—fixed or opening.

Skylight Area—The area of the rough opening for the skylight.

Skylight Type—A type of skylight assembly having a specific solar heat gain coefficient and 1-1-factor, whether glass mounted on a curb, glass not mounted on a curb or plastic (assumed to be mounted on a curb).

Slab—A concrete pad that sits on gravel or crushed rock, well-compacted soil either level with the ground or above the ground.

Slab on Grade—A slab floor that sits directly on top of the surrounding ground.

Slack Capacity—The amount of unused transmission capacity divided by the total firm capacity.

Slash—The un-merchantable material left on site subsequent to harvesting a timber stand, including tops, limbs, cull sections.

SLD—Special Litigation Division

SLERA—Screening Level Ecological Risk Assessment

Slinky™ Ground Loop—In this type of closed-loop, horizontal geothermal heat pump installation, the fluid-filled plastic heat exchanger pipes are coiled like a Slinky™ to allow more pipe in a shorter trench. This type of installation cuts down on installation costs and makes horizontal installation possible in areas it would not be with conventional horizontal applications.

Slope Mine—A mine that reaches the coal bed by means of an inclined opening.

Slot—A physical position in a rack in a storage pool that is intended to be occupied by an intact assembly or equivalent (that is, a canister or an assembly skeleton).

Slow Pyrolysis—Thermal conversion of biomass to fuel by slow heating to less than 450°C in the absence of oxygen.

Sludge—A dense, slushy, liquid-to-semifluid product that accumulates as an end result of an industrial or technological process designed to purify a substance. Industrial sludges are produced from the processing of energy-related raw materials, chemical products, water, mined ores, sewerage, and other natural and man-made products. Sludges can also form from natural processes, such as the run off produced by rain fall, and accumulate on the bottom of bogs, streams, lakes, and tidelands.

Slurry—A viscous liquid with a high solids content.

Slurry Dam—A repository for the silt or culm from a preparation plant.

Slurry Wall—Barriers used to contain the flow of contaminated groundwater or subsurface liquids. Slurry walls are constructed by digging a trench around a contaminated area and filling the trench with a material that tends not to allow water to pass through it. The groundwater or contaminated liquids trapped within the area surrounded by the slurry wall can be extracted and treated.

SLV—Stipulated Loss Value. The value of the asset, or the amount required to be reimbursed to the lessor, if it is lost or destroyed.

SMACNA—Sheet Metal and Air-conditioning Contractors National Association www.smacna.org

Small Business Innovation Development Act of 1982—The Small Business Innovation Research (SBIR) program is a set-aside program for domestic small-business concerns to engage in Research/Research and Development (R/R&D) that has the potential for commercialization. The SBIR program was established under the Small Business Innovation Development Act of 1982, reauthorized until September 30, 2000 by the Small Business Research and Development Enhancement Act, and reauthorized again until September 30, 2008 by the Small Business Reauthorization Act of 2000.

Small Irrigation Power—Power generated without any dam or impoundment of water.

Small Pickup Truck—A pickup truck weighing under 4,500 lbs GVW.

Small Power Producer (SPP)—Under the Public Utility Regulatory Policies Act (PURPA), a small power production facility (or small power producer) generates electricity using waste, renewable (biomass, conventional hydroelectric, wind and solar, and geothermal) energy as a primary energy source. Fossil fuels can be used, but renewable resource must provide at least 75 percent of the total energy input. (See Code of Federal Regulations, Title 18, Part 292.)

SMANCA Residential Comfort System Install Standards Manual—The Sheet Metal Contractors' National Association document entitled "Residential Comfort System Installation Standards Manual, Seventh Edition" (1998).

Smart Window—A term used to describe a technologically advanced window system that contains glazing that can change or switch its optical qualities when a low voltage electrical signal is applied to it, or in response to changes in heat or light.

SML—Security Market Lane

Smog—A mixture of smoke and fog generally used as an equivalent of air pollution, particularly associated with oxidants.

SMUD—Sacramento Municipal Utility District (California)

Snag—Any standing dead, partially dead or defective tree. A hard snag is composed primarily of sound wood. A soft snag is composed primarily of wood in advanced stages of decay and deterioration.

SNCR—Selective Non-catalytic Reduction

SNG—Synthetic Natural Gas

SNL—Sandia National Laboratory

SO—System Operator
 • Standard Offer: A series of posted offers by electric companies for the purchase of electricity from qualifying facilities.

Social Entrepreneurship—An entrepreneurial endeavor that focuses on sustainable social change, rather than merely the generation of profit.

Social Return on Investment (SROI)—A monetary measure of the social value for a community or society yielded by a specific investment.

Socially Responsible Investing (SRI)—An investment practice that gives preference to companies that value social and environmental impacts in addition to financial gain. Socially responsible investments, also known as "ethical investment," involve companies and practices that cause little or no depletion of natural assets or environmental degradation, and that do not infringe the rights of workers, women, indigenous people, children nor animals.

Socioeconomic—Relating to social or economic factors or to a combination of both social and economic factors.

Sodium Lights—A type of high intensity discharge light that has the most lumens per watt of any light source.

Sodium Silicate—A grey-white powder soluble in alkali and water, insoluble in alcohol and acid. Used to fireproof textiles, in petroleum refining and corrugated paperboard manufacture, and as an egg preservative. Also referred to as liquid gas, silicate of soda, sodium metasilicate, soluble glass, and water glass.

Sodium Tripolyphosphate—A white powder used for water softening and as a food additive and texturizer.

Soffit—A panel which covers the underside of an roof overhang, cantilever, or mansard.

Soft Coat—A low emissivity metallic coating applied to glass, which will be installed in a fenestration product through a sputter process where molecules of metals such as stainless steel or titanium are sputtered onto the surface of glass. Soft coats generally have lower emissivity than hard coats.

Soil Boring—Soil samples taken by drilling a hole in the ground.

Soil Gas Survey—Soil gas or soil vapor is air existing in void spaces in the soil between the groundwater and the ground surface. These gases may include vapor or hazardous chemicals as well as air and water vapor. A soil-gas survey involves collecting and analyzing soil-gas samples to determine the presence of chemicals and to help map the spread of contaminants within soil.

Soil Vapor Extraction—A process in which chemical vapors are extracted from the soil by applying a vacuum to wells. The vapors are then safely burned off without causing further damage to the air environment in which they are released.

Solar Access or Rights—The legal issues related to protecting or ensuring access to sunlight to operate a solar energy system, or use solar energy for heating and cooling.

Solar Air Heater—A type of solar thermal system where air is heated in a collector and either transferred directly to the interior space or to a storage medium, such as a rock bin.

Solar Altitude Angle—The angle between a line from a point on the earth's surface to the center of the solar disc, and a line extending horizontally from the point.

Solar Array—A group of solar collectors or solar modules connected together.

Solar Azimuth—The angle between the sun's apparent position in the sky and true south, as measured on a horizontal plane.

Solar Cell—A photovoltaic cell that can convert light directly into electricity. A typical solar cell uses semiconductors made from silicon. See *Photovoltaic Cell.*

Solar Collector—A device used to collect, absorb, and transfer solar energy to a working fluid. Flat plate collectors are the most common type of collectors used for solar water or pool heating systems. In the case of a photovoltaics system, the solar collector could be crystalline silicon panels or thin-film roof shingles, for example.

Solar Constant—The average amount of solar radiation that reaches the earth's upper atmosphere on a surface perpendicular to the sun's rays; equal to 1353 Watts per square meter or 492 Btu per square foot.

Solar Cooling—The use of solar thermal energy or solar electricity to power a cooling appliance. There are five basic types of solar cooling technologies: absorption cooling, which can use solar thermal energy to vaporize the refrigerant; desiccant cooling, which can use solar thermal energy to regenerate (dry) the desiccant; vapor compression cooling, which can use solar thermal energy to operate a Rankine-cycle heat engine; and evaporative coolers ("swamp" coolers), and heat-pumps and air conditioners that can be powered by solar photovoltaic systems.

Solar Declination—The apparent angle of the sun north or south of the earth's equatorial plane. The earth's rotation on its axis causes a daily change in the declination.

Solar Dish—See *Parabolic Dish.*

Solar Distillation—The process of distilling (purifying) water using solar energy. Water can be placed in an air tight solar collector with a sloped glazing material, and as it heats and evaporates, distilled water condenses on the collector glazing, and runs down where it can be collected in a tray.

Solar Energy—Electromagnetic energy transmitted from the sun (solar radiation). The amount that reaches the earth is equal to one billionth of total solar energy generated, or the equivalent of about 420 trillion kilowatt-hours.

Solar Energy Industries Association (SEIA)—A national trade association of solar energy equipment manufacturers, retailers, suppliers, installers, and consultants. www.seia.org

Solar Energy Research Institute (SERI)—Established in 1974 and funded by the federal government, the institute's general purpose is to support U.S. Department of Energy's solar energy program and foster the widespread use of all aspects of solar technology, including photovoltaics, solar heating and cooling, solar thermal power generation, wind ocean thermal conversion and biomass conversion. www.nrel.gov

Solar Film—A window glazing coating, usually tinted bronze or gray, used to reduce building cooling loads, glare, and fabric fading.

Solar Fraction—The percentage of a building's seasonal energy requirements that can be met by a solar energy device(s) or system(s).

Solar Furnace—A device that achieves very high temperatures by the use of reflectors to focus and concentrate sunlight onto a small receiver.

Solar Gain—The amount of energy that a building absorbs due to solar energy striking its exterior and conducting to the interior or passing through windows and being absorbed by materials in the building.

Solar Grade Silicon—Intermediate-grade silicon used in the manufacture of solar cells. Less expensive than electronic-grade silicon.

Solar Heat Gain—Heat added to a space due to transmitted and absorbed solar energy.

Solar Heat Gain Coefficient (SHGC)—The ratio of the solar heat gain entering the space through the fenestration area to the incident solar radiation. Solar heat gain includes directly transmitted solar heat and absorbed solar radiation, which is then reradiated, conducted, or conducted into the space.

Solar Heat Gain Factor—An estimate used in calculating cooling loads of the heat gain due to transmitted and absorbed solar energy through 1/8"-thick, clear glass at a specific latitude, time and orientation.

Solar Heating and Hot Water Systems—Solar heating or hot water systems provide two basic functions: (a) capturing the sun's radiant energy, converting it into heat energy, and storing this heat in insulated storage tank(s); and (b) delivering the stored energy as needed to either the domestic hot water or heating system. These components are called the collection and delivery subsystems.

Solar Irradiation—The amount of radiation, both direct and diffuse, that can be received at any given location.

Solar Mass—A term used for materials used to absorb and store solar energy.

Solar Module (Panel)—A solar photovoltaic device that produces a specified power output under defined test conditions, usually composed of groups of solar cells connected in series, in parallel, or in series-parallel combinations.

• A device used to convert light from the sun directly into DC electricity by using the photovoltaic effect. Usually made of multiple solar cells bonded between glass and a backing material. A typical Solar Module would be 100 Watts of power output (but module powers can range from 1 Watt to 300 Watts) and have dimensions of 2 feet by 4 feet.

Solar Noon—The time of the day, at a specific location, when the sun reaches its highest, apparent point in the sky; equal to true or due, geographic south.

Solar One—A solar thermal electric central receiver power plant ("power tower") located in Barstow, California, and completed in 1981. The Solar One had a design capacity of 10,000 peak kilowatts, and was composed of a receiver located on the top of a tower surrounded by a field of reflectors. The concentrated sunlight created steam to drive a steam turbine and electric generator located on the ground.

Solar Pond—A body of water that contains brackish (highly saline) water that forms layers of differing salinity (stratifies) that absorb and trap solar energy. Solar ponds can be used to provide heat for industrial or agricultural processes, building heating and cooling, and to generate electricity.

Solar Power—Electricity generated by conversion of sunlight, either directly through the use of photovoltaic panels, or indirectly through solar-thermal processes.

Solar Power Satellite—A solar power station investigated by NASA that entailed a satellite in geosynchronous orbit that would consist of a very large array of solar photovoltaic modules that would convert solar generated electricity to microwaves and beam them to a fixed point on the earth.

Solar Power Tower—A solar energy conversion system that uses a large field of independently adjustable mirrors (heliostats) to focus solar rays on a near single point atop a fixed tower (receiver). The concentrated energy may be used to directly heat the working fluid of a Rankine cycle engine or to heat an intermediary thermal storage medium (such as a molten salt).

Solar Radiation—A general term for the visible and near visible (ultraviolet and near-infrared) electromagnetic radiation that is emitted by the sun. It has a spectral, or wavelength, distribution that corresponds to different energy levels; short wavelength radiation has a higher energy than long-wavelength radiation. Solar radiation has a distinctive spectrum (i.e., range of wavelengths) governed by the temperature of the Sun. The spectrum of solar radiation is practically distinct from that of infrared (q.v.) or terrestrial radiation because of the difference in temperature between the Sun and the Earth-atmosphere system.

Solar Resource—The amount of solar insolation received at a site, normally measured in units of kWh/m2/day which equates to the number of peak sun hours.

Solar Simulator—An apparatus that replicates the solar spectrum, and used for testing solar energy conversion devices.

Solar Space Heater—A solar energy system designed to provide heat to individual rooms in a building.

Solar Spectrum—The total distribution of electromagnetic radiation emanating from the sun. The different regions of the solar spectrum are described by their wavelength range. The visible region extends from about 390 to 780 nanometers (a nanometer is one billionth of one meter). About 99 percent of solar radiation is contained in a wavelength region from 300 nm (ultraviolet) to 3,000 nm (near-infrared). The combined radiation in the wavelength region from 280 nm to 4,000 nm is called the broadband, or total, solar radiation.

Solar System, Active—A system that uses natural convective currents or other nonmechanical means for collecting, storing, and distributing solar energy.

Solar Thermal—The process of concentrating sunlight on a relatively small area to create the high temperatures needs to vaporize water or other fluids to drive a turbine for generation of electric power, heat water for domestic or industrial hot water, or space conditioning (heating or cooling).

Solar Thermal Collector—A device designed to receive solar radiation and convert it to thermal energy. Normally, a solar thermal collector includes a frame, glazing, and an absorber, together with appropriate insulation. The heat collected by the solar collector may be used immediately or stored for later use. Solar collectors are used for space heating; domestic hot water heating; and heating swimming pools, hot tubs, or spas.

Solar Thermal Collector, High Temperature—A collector that generally operates at temperatures above 180 degrees Fahrenheit.

Solar Thermal Collector, Low Temperature—A collector that generally operates at temperatures below 110 degrees Fahrenheit. Typically, it has no glazing or insulation and is made of plastic or rubber, although some are made of metal.

Solar Thermal Collector, Medium Temperature—A collector that generally operates at temperatures of 140 degrees F to 180 degrees Fahrenheit, but can also operate at temperatures as low as 110 degrees Fahrenheit. Typically, it has one or two glazings, a metal frame, a metal absorption panel with integral flow channels or attached tubing (liquid collector) or with integral ducting (air collector) and insulation on the sides and back of the panel.

Solar Thermal Collector, Special—An evacuated tube collector or a concentrating (focusing) collector. Special collectors operate in the temperature range from just above ambient temperature (low concentration for pool heating) to several hundred degrees Fahrenheit (high concentration for air conditioning and specialized industrial processes).

Solar Thermal Electric Systems—Solar energy conversion technologies that convert solar energy to electricity, by heating a working fluid to power a turbine that drives a generator. Examples of these systems include central receiver systems, parabolic dish, and solar trough.

Solar Thermal Panels—A system that actively concentrates thermal energy from the sun by means of solar collector panels. The panels typically consist of fat, sun-oriented boxes with transparent covers, containing water tubes of air baffles under a blackened heat absorbent panel. The energy is usually used for space heating, for water heating, and for heating swimming pools.

Solar Thermal Parabolic Dishes—A solar thermal technology that uses a modular mirror system that approximates a parabola and incorporates two-axis tracking to focus the sunlight onto receivers located at the focal point of each dish. The mirror system typically is made from a number of mirror facets, either glass or polymer mirror, or can consist of a single stretched membrane using a polymer mirror. The concentrated sunlight may be used directly by a Stirling, Rankine, or Brayton cycle heat engine at the focal point of the receiver or to heat a working fluid that is piped to a central engine. The primary applications include remote electrification, water pumping, and grid-connected generation.

Solar Thermal Power Plant—Means a thermal power plant in which 75 percent or more of the total energy output is from solar energy and the use of backup fuels, such as oil, natural gas, and coal, does not, in the aggregate, exceed 25 percent of the total energy input of the facility during any calendar year period.

Solar Thermal Systems—Solar energy systems that collect or absorb solar energy for useful purposes. Can be used to generate high temperature heat (for electricity production and/or process heat), medium temperature heat (for process and space/water heating and electricity generation), and low temperature heat (for water and space heating and cooling).

Solar Time—The period marked by successive crossing of the earth's meridian by the sun; the hour angle of the sun at a point of observance (apparent time) is corrected to true (solar) time by taking into account the variation in the earth's orbit and rate of rotation. Solar time and local standard time are usually different for any specific location.

Solar Transmittance—The amount of solar energy that passes through a glazing material, expressed as a percentage.

Solar Trough Systems (See also *Parabolic Trough*,

above)—A type of solar thermal system where sunlight is concentrated by a curved reflector onto a pipe containing a working fluid that can be used for process heat or to produce electricity. The world's largest solar thermal electric power plants use solar trough technology. They are located in California, and have a combined electricity generating capacity of 240,000 kilowatts.

Solar Two—Solar Two is a retrofit of the Solar One project (see above). It is demonstrating the technical feasibility and power potential of a solar power tower using advanced molten-salt technology to store energy. Solar Two retains several of the main components of Solar One, including the receiver tower, turbine, generator, and the 1,818 heliostats.

Solar-Cooling Fraction—The percentage of cooling needs supplied by the passive solar system.

Solar-Heating Fraction—The percentage of heating needs in the building supplied by the passive solar system.

Solarium—A glazed structure, such as greenhouse or "sunspace."

Solenoid—An electromechanical device composed of a coil of wire wound around a cylinder containing a bar or plunger, that when a current is applied to the coil, the electromotive force causes the plunger to move; a series of coils or wires used to produce a magnetic field.

Solenoid Valve—An automatic valve that is opened or closed by an electromagnet.

Solid Fuels—Any fuel that is in solid form, such as wood, peat, lignite, coal, and manufactured fuels such as pulverized coal, coke, charcoal, briquettes, pellets, etc.

Solid Oxide Fuel Cell—A type of fuel cell that typically uses a hard ceramic material instead of a liquid electrolyte.

Solidity—In reference to a wind energy conversion device, the ratio of rotor blade surface area to the frontal, swept area that the rotor passes through.

Solstice—The two times of the year when the sun is apparently farthest north and south of the earth's equator; usually occurring on or around June 21 (summer solstice in northern hemisphere, winter solstice for southern hemisphere) and December 21 (winter solstice in northern hemisphere, summer solstice for the southern hemisphere).

Solution—Mixture in which the components lose their identity and are uniformly dispersed. All solutions are composed of a solvent (water or other fluid) and the substance dissolved called the "solute." A true solution is homogeneous, as salt in water. Air is a solution of oxygen and nitrogen.

Solvent—A liquid capable of dissolving another substance to form a solution. Water is sometimes called "the universal solvent" because it dissolves so many things, although often to only a very small extent. Organic solvents are used in paints, varnishes, lacquers, industrial cleaners and printing inks, for example. The use of such solvents in coatings and cleaners has declined over the last several years, because the common ones are toxic, contribute to air pollution and may be fire hazards.

Solvent Extraction—A method of separation used to purify vegetable oils.

Sorbent—A material which extracts one or more substances present in an atmosphere or mixture of gases or liquids with which it is in contact due to an affinity for such substances.

Sound Attenuation—A reduction in the sound level.

Sour Gas—Gas having a high sulphur content.

Source—A point source or a collection of point sources of the same type on the same facility. Any process or activity which releases a greenhouse gas, an aerosol or a precursor of a greenhouse gas or other criteria pollutant into the atmosphere.

Source Emission Reduction Plan (SERP)—A contingency plan developed to reduce emissions during an air quality emergency.

Source Energy—All the energy used in delivering energy to a site, including power generation and transmission and distribution losses, to perform a specific function, such as space conditioning, lighting, or water heating. Approximately three watts (or 10.239 Btu's) of energy is consumed to deliver one watt of usable electricity.

Source Material—The term "source material" means (1) uranium, thorium, or any other material that is determined by the Atomic Energy Commission pursuant to the provisions of section 61 of the Atomic Energy Act of 1954, as amended, to be source material; or (2) ores containing one or more of the foregoing materials, in such concentration as the Commission may by regulation determine from time to time.

Source Stream—A specific fuel type, raw material or product giving rise to emissions of relevant gases at one or more emission sources as a result of its consumption or production.

South Coast Air Quality Management District (SCAQMD)—The SCAQMD is the regulatory entity that governs stationary sources emitting air pollution in the four county Los Angeles metropolitan area and administers the RECLAIM cap and trade program and the Regulation XIII new source review program.

Southeastern Electric Reliability Council (SERC)— One of the ten regional reliability councils that make up the North American Electric Reliability Council (NERC).

Southwest Power Pool (SPP)—One of the ten regional reliability councils that make up the North American Electric Reliability Council (NERC).

SOW—Scope of work

SO$_X$—Sulfur Oxides

SO$_x$ Allowance—An emission right issued by the US EPA under Title IV of the Clean Air Act Amendments of 1990 that gives authorization to emit one ton of SO$_2$ emissions on or after the vintage pursuant to the rules of the Acid Rain program. SO$_2$ Allowances may be used during their vintage year or banked and used in subsequent years.

SO$_x$ Emissions—Emissions of sulfur dioxides.

Spa—A vessel that contains heated water, in which humans can immerse themselves, is not a pool, and is not a bathtub.

Space Conditioning System—A system that provides either collectively or individually heating, ventilating, or cooling within or associated with conditioned spaces in a building. The system may operate alone or in conjunction with other systems.

Space Heater—A movable or fixed heater used to heat individual rooms.

Space Heating—The use of energy to generate heat for warmth in housing units using space-heating equipment. The equipment could be the main space-heating equipment or secondary space-heating equipment. It does not include the use of energy to operate appliances (such as lights, televisions, and refrigerators) that give off heat as a byproduct.

Spacer (Window)—Strips of material used to separate multiple panes of glass within the windows.

Spacer, Aluminum—A metal channel that is used either against the glass (sealed along the outside edge of the insulated glass unit), or separated from the glass by one or more beads of caulk, which is used to separate panes of glass in an insulated glass unit.

Spacer, Insulating—A non-metallic, relatively non-conductive material, usually of rubber compounds, that is used to separate panes of glass in an insulated glass unit.

Spacer, Other—A wood, fiberglass, or composite material that is used as a spacer between panes of glass in insulated glass units.

Spacer, Squiggle—A flexible material, usually butyl, formed around a thin corrugated aluminum strip that is used as a spacer in insulated glass units.

Spawning Gravel—Sorted, clean gravel patches of a size appropriate for the needs of resident or anadromous fish.

SPB—Simple payback

Special Collector—An evacuated tube collector or a concentrating (focusing) collector. Special collectors operate in the temperature range from just above ambient temperature (low concentration for pool heating) to several hundred degrees Fahrenheit (high concentration for air conditioning and specialized industrial processes).

Special Contract Rate Schedule—An electric rate schedule for an electric service agreement between a utility and another party in addition to, or independent of, any standard rate schedule.

Special Contracts—Any contract that provides a utility service under terms and conditions other than those listed in the utility's tariffs. For example, an electric utility may enter into an agreement with a large customer to provide electricity at a rate below the tariffed rate in order to prevent the customer from taking advantage of some other option that would result in the loss of the customer's load. This generally allows that customer to compete more effectively in their product market.

Special Naphthas—All finished products within the naphtha boiling range that are used as paint thinners, cleaners, or solvents. These products are refined to a specified flash point. Special naphthas include all commercial hexane and cleaning solvents conforming to ASTM Specification D1836 and D484, respectively. Naphthas to be blended or marketed as motor gasoline or aviation gasoline, or that are to be used as petrochemical and synthetic natural gas (SNG) feedstocks are excluded.

Special Nuclear Material—The term "special nuclear material" means (1) plutonium, uranium enriched in the isotope 233 or in the isotope 235, and any other material that the Atomic Energy Commission, pursuant to the provisions of section 51 of the Atomic Energy Act of 1954, as amended, determines to be special nuclear material, but does not include source material; or (2) any material artificially enriched by any of the foregoing, but does not include source material.

Special Purpose Property—Property that is uniquely valuable to the lessee and not valuable to anyone else except as scrap. Also referred to as limited use property.

Special Purpose Rate Schedule—An electric rate schedule limited in its application to some particular purpose or process within one, or more than one, type of industry or business.

Special Purpose Vehicle (SPV)—A separate legal entity.

Specific Gravity—The ratio of the weight of a solution to the weight of an equal volume of water at a specified temperature; used with reference to the sulfuric acid electrolyte solution in a lead acid battery as an indicator of battery state of charge. More recently called relative density.

Specific Heat—The amount of heat required to raise a unit mass of a substance through one degree, expressed as a ratio of the amount of heat required to raise an equal mass of water through the same range.

Specific Heat Capacity—The quantity of heat required to change the temperature of one unit weight of a material by one degree.

Specific Humidity—The weight of water vapor, per unit weight of dry air.

Specific Performance—A Legal Action to compel performance of an agreement.

Specific Purchases—Electricity transactions which are traceable to specific generation sources by an auditable contract trail or equivalent, such as a tradable commodity system, that provides commercial verification that the electricity source claimed has been sold once and only once to retail consumers.

Specific Volume—The volume of a unit weight of a substance at a specific temperature and pressure.

Specified Source of Power—A particular generating facility that a retail provider can confidently track to its own load due to full or partial ownership of a firm contractual relationship, such as a long-term power purchase agreement.

Specified Wholesale Sales—Wholesale electric power sales made by retail providers that can be matched to a specified source of power.

Spectral Energy Distribution—A curve illustrating the variation or spectral irradiance with wavelength.

Spectral Irradiance—The monochromatic irradiance of a surface per unit bandwidth at a particular wavelength, usually expressed in Watts per square meter-nanometer bandwidth.

Spectral Reflectance—The ratio of energy reflected from a surface in a given waveband to the energy incident in that waveband.

Spectrally Selective Coatings—A type of window glazing films used to block the infrared (heat) portion of the solar spectrum but admit a higher portion of visible light.

Spectrum—See *Solar Spectrum* above.

Specular Reflectors—Specular reflectors have mirror like characteristics (the word "specular" is derived from the Greek word meaning mirror). The most common materials used for ballasts, the devices that turn on and operate Fluorescent tubes, are aluminum and silver. Silver has the highest reflectivity; aluminum has the lowest cost. The materials and shape of the reflector are designed to reduce absorption of light within the fixture while delivering light in the desired angular pattern. Adding (or retrofitting) specular reflectors to an existing light fixture is frequently implemented as a conservation measure.

Speculative Resources (Coal)—Undiscovered coal in beds that may occur either in known types of deposits in a favorable geologic setting where no discoveries have been made, or in deposits that remain to be recognized. Exploration that confirms their existence and better defines their quantity and quality would permit their reclassification as identified resources.

Speculative Resources (Uranium)—Uranium in addition to Estimated Additional Resources (EAR) that is thought to exist, mostly on the basis of indirect evidence and geological extrapolations, in deposits discoverable with existing exploration techniques. The locations of deposits in this category can generally be specified only as being somewhere within given regions or geological trends. The existence and size of such deposits are speculative. The estimates in this category are less reliable than estimates of EAR. SR corresponds to DOE's Possible Potential Resources plus Speculative Potential Resources categories.

Spent Fuel—Irradiated fuel that is permanently discharged from a reactor. Except for possible reprocessing, this fuel must eventually be removed from its temporary storage location at the reactor site and placed in a permanent repository. Spent fuel is typically measured either in metric tons of heavy metal (i.e., only the heavy metal content of the spent fuel is considered) or in metric tons of initial heavy metal (essentially, the initial mass of the fuel before irradiation). The difference between these two quantities is the weight of the fission products.

Spent Fuel Disassembly Hardware—The skeleton of a fuel assembly after the fuel rods have been removed. Generally, SFD hardware for PWR assemblies includes guide tubes; instrument tubes, top and bottom nozzles; grid spacers; hold-down springs; and attachment components, such as nuts and locking caps. For BWR fuel assemblies, SFD hardware includes the top and bottom tie plates, compression springs for individual fuel rods, grid spacers, and water rods.

Spent Liquor—The liquid residue left after an industrial process; can be a component of waste materials used as fuel.

Spill Energy—See *Dump*.

Spillway—A passage for surplus water to flow over or around a dam.

Spin Off—A company sells a division or turns it into a subsidiary. Often companies also list spun-off subsidiaries on the exchange.

Spinning Reserve—The portion of unloaded synchronized generating capacity that is immediately responsive to system frequency and that is capable of being loaded in ten minutes. It must be capable of running for at least two hours.

Split Spectrum Photovoltaic Cell—A photovoltaic device where incident sunlight is split into different spectral regions, with an optical apparatus, that are directed to individual photovoltaic cells that are optimized for converting that spectrum to electricity.

 • A compound photovoltaic device in which sunlight is first divided into spectral regions by optical means. Each region is then directed to a different photovoltaic cell optimized for converting that portion of the spectrum into electricity. Such a device achieves significantly greater overall conversion of incident sunlight into electricity.

Split System—When applied to electric air-conditioning equipment, it means a two-part system—an indoor unit and an outdoor unit. The indoor unit is an evaporator coil mounted in the indoor circulating air system, and the outdoor unit is an air-cooled condensing unit containing an electric motor-driven compressor, a condenser fan, and a fan motor.

Split System Air Conditioner—An air conditioning system that comes in two to five pieces: one piece contains the compressor, condenser, and a fan; the others have an evaporator and a fan. The condenser, installed outside the house, connects to several evaporators, one in each room to be cooled, mounted inside the house. Each evaporator is individually controlled, allowing different rooms or zones to be cooled to varying degrees.

Split Tails—Use of one tails assay for transaction of enrichment services and a different tails assay for operation of the enrichment plant. This mode of operations typically increases the use of uranium, which is relatively inexpensive, while decreasing the use of separative work, which is expensive.

Splits—Standing bids or offers that have been divided into two or more different bids or offers, new bids and offers may be included in a final deal or re-positioned for sale.

Split-the-Savings (Electric Utility)—The basis for settling economy-energy transactions between utilities. The added costs of the supplier are subtracted from the avoided costs of the buyer, and the difference is evenly divided.

Spontaneous Combustion, or Self-heating, of Coal—A naturally occurring process caused by the oxidation of coal. It is most common in low-rank coals and is a potential problem in storing and transporting coal for extended periods. Factors involved in spontaneous combustion include the size of the coal (the smaller sizes are more susceptible), the moisture content, and the sulfur content. Heat buildup in stored coal can degrade the quality of coal, cause it to smolder, and lead to a fire.

Spot Market—A commodity market for the purchase and sale of electric energy for a short-term basis (often one day or less.)

Spot Market (Natural Gas)—A market in which natural gas is bought and sold for immediate or very near-term delivery, usually for a period of 30 days or less. The transaction does not imply a continuing arrangement between the buyer and the seller. A spot market is more likely to develop at a location with numerous pipeline interconnections, thus allowing for a large number of buyers and sellers. The Henry Hub in southern Louisiana is the best known spot market for natural gas.

Spot Market (Uranium)—Buying and selling of uranium for immediate or very near-term delivery. It typically involves transactions for delivery of up to 500,000 pounds U_3O_8 within a year of contract execution.

Spot Price—The price for a one-time open market transaction for immediate delivery of a specific quantity of product at a specific location where the commodity is purchased "on the spot" at current market rates.

Spot Pricing—The price of commodities is established by the market's supply and demand for short-term transactions. Price fluctuations reflect the continuously changing supply and demand.

Spot Purchases—A single shipment of fuel or volumes of fuel purchased for delivery within 1 year. Spot purchases are often made by a user to fulfill a certain portion of energy requirements, to meet unanticipated energy needs, or to take advantage of low-fuel prices.

SPR—Strategic Petroleum Reserve

Spray Pyrolysis—A deposition process whereby heat is used to break molecules into elemental sources that are then spray deposited on a substrate.

Spread—The difference between two values. In lease transactions, the term generally is used to describe the difference between the interest rate of the lease and the interest on the debt used to fund the lease.

Spreader Stocker—A type of furnace in which fuel is spread, automatically or mechanically, across the furnace grate.

Spreader Stoker Furnace—A furnace in which fuel is automatically or mechanically spread. Part of the fuel is burned in suspension. Large pieces fall on a grate.

Sputtering—A process used to apply photovoltaic semi-conductor material to a substrate by a physical vapor deposition process where high-energy ions are used to bombard elemental sources of semiconductor material, which eject vapors of atoms that are then deposited in thin layers on a substrate.

Square Wave—A train of rectangular voltage pulses that alternate between two fixed values for equal lengths of time.

Square Wave Inverter—A type of inverter that produces square wave output.; consists of a DC source, four switches, and the load. The switches are power semiconductors that can carry a large current and withstand a high voltage rating. The switches are turned on and off at a correct sequence, at a certain frequency. The square wave inverter is the simplest and the least expensive to purchase, but it produces the lowest quality of power.

Squirrel Cage Motors—This is another name for an induction motor. The motors consist of a rotor inside a stator. The rotor has laminated, thin flat steel discs, stacked with channels along the length. If the casting composed of bars and attached end rings were viewed without the laminations the casting would appear similar to a squirrel cage.

SRCC—Solar Rating and Certification Council

SREC—Solar Renewable Energy Credit

SRES Scenarios—A suite of emissions scenarios developed by the Intergovernmental Panel on Climate Change in its Special Report on Emissions Scenarios (SRES). These scenarios were developed to explore a range of potential future greenhouse gas emissions pathways over the 21st century and their subsequent implications for global climate change.

SRI—Sustainable and Responsible Investing (SRI) is an investment process that integrates traditional financial analysis with analysis of corporate responsibility and sustainability.

SSO—Senior sustainability officer

Stability—The property of a system or element by virtue of which its output will ultimately attain a steady state. The amount of power that can be transferred from one machine to another following a disturbance. The stability of a power system is its ability to develop restoring forces equal to or greater than the disturbing forces so as to maintain a state of equilibrium.

Stabilization—Changing active organic matter in sludge into inert, harmless material. The term also refers to physical activities such as compacting and capping at sites that limits the further spread of contamination without actual reduction of toxicity.

Stabilization Lagoon—A shallow artificial pond used for the treatment of wastewater. Treatment includes removal of solid material through sedimentation, the decomposition of organic material by bacteria, and the removal of nutrients by algae.

Stabilized Net Operating Income—Projected income less expenses that are subject to change but have been adjusted to reflect equivalent, stable property operations.

Stable Prices—Prices that do not vary greatly over short time periods.

Stack—A tall, vertical structure containing one or more flues used to discharge products of combustion to the atmosphere.

Stack (Heat) Loss—Sensible and latent heat contained in combustion gases and vapor emitted to the atmosphere.

Stack Effect—The tendency of a heated gas to rise in a vertical passage as in a chimney, small enclosure, or stairwell.

Staebler-Wronski Effect—The tendency of the sunlight to electricity conversion efficiency of amorphous silicon photovoltaic devices to degrade (drop) upon initial exposure to light.

Stagnation Temperature—A condition that can occur in a solar collector if the working fluid does not circulate when sun is shining on the collector.

Stakeholder—An individual, company or group potentially affected by the activities of a regulator, company or organizations. In sustainable business models the term includes financial shareholders as well as those affected by environmental or social factors such as suppliers, consumers, employees, the local community, and the natural environment.

Stakeholder Engagement—The ongoing process of soliciting feedback regarding a company's business practices or major decisions from financial shareholders, as well as individuals or groups effected by corporate environmental or social practices such as suppliers, consumers, employees, and the local community.

Stall—In reference to a wind turbine, a condition when the rotor stops turning.

Stand—(tree stand, timber stand) A community of trees managed as a unit. Trees or other vegetation occupying a specific area, sufficiently uniform in species composition, age arrangement, and condition as to be distinguishable from the forest or other cover on adjoining areas.

Stand Alone System—A solar system that operates without connection to a grid or another supply of electricity. A battery bank stores unused daylight production for nighttime power. Commonly used in remote regions such as mountains, ocean platforms or communication towers.

Stand Conversion—The conversion of a noncommercial stand of timber to a commercial stand.

Stand Density—The number or mass of trees occupying a site. It is usually measured in terms of stand density index or basal area per acre.

Stand-Alone Generator—A power source/generator that operates independently of or is not connected to an electric transmission and distribution network; used to meet a load(s) physically close to the generator.

Stand-Alone Inverter—An inverter that operates independent of or is not connected to an electric transmission and distribution network.

Standalone Price—The price at which a lessee would purchase a component of a contract separately.

Stand-Alone System—An energy system that operates independent of or is not connected to an electric transmission and distribution network.

Standard & Poor's—A credit rating agency.

Standard (Green-e program)—The Green-e Standard is the criteria which electricity products must meet in order to be Green-e certified.

Standard Air—Air with a weight of 0.075 pounds per cubic foot with an equivalent density of dry air at a temperature of 86 degrees Fahrenheit and standard barometric pressure of 29.92 inches of mercury.

Standard Conditions—In refrigeration, an evaporating temperature of 5 degrees Fahrenheit (F), a condensing temperature of 86 degrees F., liquid temperature before expansion of 77 degrees F., and suction temperature of 12 degrees F.
• A temperature of 20 degrees C (68 degrees F) and an absolute pressure of 760 mm (30 inches) of mercury.

Standard Contract—The agreement between the Department of Energy (DOE) and the owners or generators of spent nuclear fuel and high-level radioactive waste, under which DOE will make available nuclear waste disposal services to those owners and generators.

Standard Cubic Foot—A column of gas at standard conditions of temperature and pressure (32 degrees Fahrenheit and one atmosphere).

Standard Design—A hypothetical building that is used to calculate the custom budget for nonresidential and residential buildings. A new building or addition alone complies with the standards if the predicted source energy use of the proposed design is the same or less than the annual budget for space conditioning and water heating of the Standard Design. The Standard Design is substantially similar to the Proposed Design, except it is in exact compliance with the prescriptive requirements and the mandatory measures. (in California)

Standard Fluorescent—A light bulb made of a glass tube coated on the inside with fluorescent material, which produces light by passing electricity through mercury vapor causing the fluorescent coating to glow or fluoresce.

Standard Industrial Classification (SIC)—Replaced with North American Industry Classification System.

Standard Rate—The basic rate customers would take service under if they were not on real-time pricing.

Standard Reporting Conditions (SRC)—A fixed set of conditions (including meteorological) to which the electrical performance data of a photovoltaic module are translated from the set of actual test conditions.

Standard Test Conditions (STC)—Conditions under which a solar module is typically tested in a laboratory: (1) Irradiance intensity of 1000 W/square meter (0.645 watts per square inch), AM1.5 solar reference spectrum, and (3) a cell (module) temperature of 25 degrees C, plus or minus 2 degrees C (77 degrees F, plus or minus 3.6 degrees F). [IEC 1215]

Standards and Guidelines—Bounds or constraints within which all practices in a given area will be carried out, in achieving the goals and objectives for that area. Standards and guidelines provide environmental safeguards and also describe constraints prescribed by law.

Standby Charge—A charge for the potential use of a utility service, usually done by an agreement with another electric utility service. These services include system backup support and other running and quick-start capabilities.

Standby Current—The current used by the inverter when no load is active, corresponding to lost power.

Standby Electricity Generation—Involves use of generators during times of high demand on utilities to avoid extra "peak-demand" charges.

Standby Facility—A facility that supports a utility system and is generally running under no-load. It is available to replace or supplement a facility normally in service.

Standby Heat Loss—A term used to describe heat energy lost from a water heater tank.

Standby Loss—A measure of the losses from a water heater tank. When expressed as a percentage, standby loss is the ratio of heat loss per hour to the heat content of the stored water above room temperature. When expressed in watts, standby loss is the heat lost per hour, per square foot of tank surface area. [See California Code of Regulations, Title 20, Section 1602(f)(5)]

Standby Loss, Percent—The ratio of heat lost per hour to the heat content of the stored water above room temperature. It is one of the measures of efficiency of water heaters required for water heating energy calculations for some types of water heaters. Standby loss is expressed as a percentage.

Standby Power—For the consumer, this is the electricity that is used by your TVs, stereos, and other electronic devices that use remote controls. When you press "off" to turn off your device, minimal power (dormant mode) is still being used to maintain the internal electronics in a ready, quick-response mode. This way, your device can be turned on with your remote control and be immediately ready to operate.

Standby Service—Support service that is available as needed to supplement a customer, a utility system, or another utility if a schedule or an agreement authorizes the transaction. The service is not regularly used.

Stand-off Mounting—Technique for mounting a PV array on a sloped roof, which involves mounting the modules a short distance above the pitched roof and tilting them to the optimum angle.

STARS—Developed by the Association for the Advancement of Sustainability in Higher Education, the Sustainability Tracking, Assessment & Rating System (STARS) is a voluntary, self-reporting framework for gauging relative progress toward sustainability for colleges and universities.

Starting Delivery Date—Under the Kyoto Protocol, this date is the first specified future date when the Forward Contract emission reductions will be delivered.

Starting Surge—Power, often above an appliance's rated wattage, required to bring any appliance with a motor up to operating speed.

Starting Torque—The torque at the bottom of a speed (rpm) versus torque curve. The torque developed by the motor is a percentage of the full-load or rated torque. At this torque the speed, the rotational speed of the motor as a percentage of synchronous speed is zero. This torque is what is available to initially get the load moving and begin its acceleration.

Startup Test Phase of Nuclear Power Plant—A nuclear power plant that has been licensed by the Nuclear Regulatory Commission to operate but is still in the initial testing phase, during which the production of electricity may not be continuous. In general, when the electric utility is satisfied with the plant's performance, it formally accepts the plant from the manufacturer and places it in commercial operation status. A request is then submitted to the appropriate utility rate commission to include the power plant in the rate base calculation.

Startup/Flame Stabilization Fuel—Any fuel used to initiate or sustain combustion or used to stabilize the height of flames once combustion is underway.

Starved Electrolyte Cell—A battery containing little or no free fluid electrolyte.

State—One of the 50 States, including adjacent outer continental shelf areas, or the District of Columbia.

State Agency—All departments, boards, commissions, colleges, community and technical colleges, and universities who own and operate or who have some responsibility for the ownership and operation of state facilities, related structures, and/or appurtenances.

State Implementation Plan (SIP)—A state plan required by the Clean Air Act to bring nonattainment areas into compliance with federal ambient air quality standards.

State of Charge (SOC)—The capacity of a battery at a particular time expressed at a percentage of its rated capacity.

State Permit/License/Mine Number—Code assigned to a mining operation by the state in which the operation is located.

State Severance Taxes—Any severance, production, or similar tax, fee, or other levy imposed on the production of crude oil, natural gas, or coal by any State, local government acting under authority of State law, or by an Indian tribe recognized as eligible for services by the Secretary of the Interior.

Stated Preference Methods—A method of putting dollar values on something (e.g. air quality or risks) which relies on asking people to reveal their values through some type of survey or questionnaire.

State-Owned Facilities—Those facilities which are owned outright by the state, those facilities which are being purchased by the state, and those facilities which the state has provided full or partial construction funding or provides full or partial operations funding.

Static Baseline—A static baseline assumes that business conditions are to remain constant and apply throughout the lifetime of the project.

Static Head—The height of the water level above the point of free discharge of the water, normally measured when the pump is off.

Static Pressure—The force per unit area acting on the surface of a solid boundary parallel to the flow.
  • The force exerted per unit area by a gas or liquid, measured at right angles to the direction of flow, or the pressure when no liquid is flowing..

Static Stability—The likelihood that soils at rest will remain at rest.

Station (Electric)—A plant containing prime movers, electric generators, and auxiliary equipment for converting mechanical, chemical, and/or nuclear energy into electric energy.

Station Use—Energy that is used to operate an electric generating plant. It includes energy consumed for plant lighting, power, and auxiliary facilities, regardless of whether the energy is produced at the plant or comes from another source.

Stationary Combustion Source—A source of emissions from the production of electricity, heat, or steam, resulting from combustion of fuels in boilers, furnaces, turbines, backup generators, and other facility equipment.

Stationary Power Plant—A source of electricity that remains in one location.

Statutory Voting—A method of voting for members of the Board of Directors of a corporation. Under this method, a shareholder receives one vote for each share and may cast those votes for each of the directorships. For example: An individual owning 100 shares of stock of a corporation that is electing six directors could cast 100 votes for each of the six candidates. This method tends to favor the larger shareholders.

STC—Standard Test Conditions

Steady State Efficiency—A performance rating for space heaters; a measure of the percentage of heat from combustion of gas that is transferred to the space being heated under specified steady state conditions. [See California Code of Regulations, Title 20, Section 1602(e)(13)]

Steam—Water in vapor form; used as the working fluid in steam turbines and heating systems.

Steam (Purchased)—Steam, purchased for use by a refinery, that was not generated from within the refinery complex.

Steam Boiler—A type of furnace in which fuel is burned and the heat is used to produce steam.

Steam Coal—All non-metallurgical coal.

Steam Conversion Factors—(approximations) 1 pound of steam = 1,000 Btu = .3 kW. 10,000 lbs/hr steam = 300 boiler horsepower.

Steam Electric Plant—A power station in which steam is used to turn the turbines that generate electricity. The heat used to make the steam may come from burning fossil fuel, using a controlled nuclear reaction, concentrating the sun's energy, tapping the earth's natural heat or capturing industrial waste heat.

Steam Electric Power Plant (Conventional)—A plant in which the prime mover is a steam turbine. The steam used to drive the turbine is produced in a boiler where fossil fuels are burned.

Steam Expenses—The cost of labor, materials, fuel, and other expenses incurred in production of steam for electric generation.

Steam From Other Sources—Steam purchased, transferred from another department of the utility, or acquired from others under a joint-facility operating agreement.

Steam or Hot Water Radiators or Baseboards—A distribution system where steam or hot water circulates through cast-iron radiators or baseboards. Some other types of equipment in the building may be used to produce the steam or hot water or it may enter the building already heated as part of a district hot water system. Hot water does not include domestic hot water used for cooking and cleaning.

Steam or Hot-water System—Either of two types of a central space-heating system that supplies steam or hot water to radiators, convectors, or pipes. The more common type supplies either steam or hot water to conventional radiators, baseboard radiators, convectors, heating pipes embedded in the walls or ceilings, or heating coils or equipment that are part of a combined heating/ventilating or heating/air-conditioning system. The other type supplies radiant heat through pipes that carry hot water and are held in a concrete slab floor.

Steam Reforming—A process for separating hydrogen from a hydrocarbon fuel in the presence of steam.

Steam Transferred-Credit—The expenses of producing steam are charged to others or to other utility departments under a joint operating arrangement.

Steam Trap—A device for allowing the passage of condensate or air and condensate and preventing the passage of steam.

Steam Turbine—A device that converts high-pressure steam, produced in a boiler, into mechanical energy that can then be used to produce electricity by forcing blades in a cylinder to rotate and turn a generator shaft.

Steam, Saturated—Steam at a temperature and pressure such that any lowering of the temperature or increase in pressure will cause condensation.

Steam, Super-Heated—Water vapor heated beyond the point at which complete vaporization occurs (100% quality).

Step-Payment Contract—A contract that contains a payment stream requiring the borrower to make payments that either increase (step-up) or decrease (step-down) in amount over the term of the contract.

Stepped Dimming—A lighting control method that varies the light output of lamps in one or more predetermined discrete steps between full light output and off.

Stepped Rents—A rental structure in which the amounts step up or down (usually once or twice) during the lease term.

Stepped Switching—A lighting control method that varies the light output of a lighting system with the intent of maintaining approximately the relative uniformity of illumination by turning off alternate groups of lamps or luminaires.

Stevedore—A worker who loads and unloads ships.

Stewardship—Related to the environment, the concept of responsible caretaking based on the premise that we do not own resources, but are managers and are responsible to future generations for their condition.

STIG—Steam Injected Gas Turbine

Still Bottoms—Residues left over from the process of recovering spent solvents in a distillation unit.

Still Gas (Refinery Gas)—Any form or mixture of gases produced in refineries by distillation, cracking, reforming, and other processes. The principal constituents are methane, ethane, ethylene, normal butane, butylene, propane, propylene, etc. Still gas is used as a refinery fuel and a petrochemical feedstock. The conversion factor is 6 million Btu's per fuel oil equivalent barrel.

Stillage—The grains and liquid effluent remaining after distillation.

Stipulated Loss Value Table—A schedule included in a financing agreement, generally used for purposes of minimum insurance coverage, that sets forth the agreed-upon value of the financed equipment at various points throughout the lease term. This value establishes the liability of the borrower to the lender in the event the financed equipment is lost or becomes unusable due to casualty loss during the contract term.

Stipulated Savings—Energy savings values which are stipulated based on engineering calculations using typical equipment characteristics and operating schedules developed for particular applications, without on-site testing or metering. This may be a cost effect approach for lighting efficiency and controls projects. Also known as "deemed savings."

Stipulation—An agreement between parties to a proceeding generally relating to procedural matters or minor issues.

Stirling Engine—A heat engine of the reciprocating (piston) where the working gas and a heat source are independent. The working gas is compressed in one region of the engine and transferred to another region where it is expanded. The expanded gas is then returned to the first region for recompression. The working gas thus moves back and forth in a closed cycle.

Stock—The term "stock" includes shares in an association, joint-stock company, or insurance company.

Stock Change—The difference between stocks at the beginning of the reporting period and stocks at the end of the reporting period.

Note: A negative number indicates a decrease (i.e., a drawdown) in stocks and a positive number indicates an increase (i.e., a buildup) in stocks during the reporting period.

Stock Options—The right to purchase or sell a stock at a specified price within a stated period. Options are a popular investment medium, offering an opportunity to hedge positions in other securities, to speculate on stocks with relatively little investment, and to capitalize on changes in the market value of options contracts themselves through a variety of options strategies.

• A widely used form of employee incentive and compensation. The employee is given an option to purchase its shares at a certain price (at or below the market price at the time the option is granted) for a specified period of years.

Stocks—Supplies of fuel or other energy source(s) stored for future use. Stocks are reported as of the end of the reporting period.

Stoichiometric Condition—That condition at which the proportion of the air-to-fuel is such that all combustible products will be completely burned with no oxygen remaining in the combustion air.

Stoichiometric Ratio—The ratio of chemical substances necessary for a reaction to occur completely.

Stoichiometry—Chemical reactions, typically associated with combustion processes; the balancing of chemical reactions by providing the exact proportions of reactant compounds to ensure a complete reaction; all the reactants are used up to produce a single set of products.

Storage—Storing energy in a battery or battery stack. In water pumping, storage can be achieved by pumping water to a storage tank.

Storage Additions—Volumes of gas injected or otherwise added to underground natural gas reservoirs or liquefied natural gas storage.

Storage Agreement—Any contractual arrangement between the responding company and a storage operator under which gas was stored for, or gas storage service was provided to, the responding company by the storage operator, irrespective of any responding company ownership interest in either the storage facilities or stored gas.

Storage Capacity—The amount of energy an energy storage device or system can store.

Storage Density—The capacity of a battery, in amp-hours compared to its weight. Measured in Watt-hours per kilogram.

Storage Field Capacity (Underground Gas Storage)—The presently developed maximum capacity of a field (as collected on EIA Survey Form 191).

Storage Gas—Natural gas that is kept in storage to balance supply and demand.

Storage Hydroelectric Plant—A hydroelectric plant with reservoir storage capacity for power use.

Storage Hydropower—A hydropower facility that stores water in a reservoir during high-inflow periods to augment water during low-inflow periods. Storage projects allow the flow releases and power production to be more flexible and dependable. Many hydropower project operations use a combination of approaches.

Storage Site—Spent nuclear fuel storage pool or dry cask storage facility, usually located at the reactor site, as licensed by (or proposed to be licensed by) the Nuclear Regulatory Commission (NRC).

Storage Tank—Any container, reservoir, or tank used for the storage of organic liquids, excluding tanks which are permanently affixed to mobile vehicles such as railroad tank cars, tanker trucks or ocean vessels. The tank of a water heater.

Storage Type Water Heater—A water heater that heats and stores water at a thermostatically controlled temperature for delivery on demand. [See California Code of Regulations, Title 20, Section 1602(f)(6)]

Storage Water Heater—A water heater that releases hot water from the top of the tank when a hot water tap is opened. To replace that hot water, cold water enters the bottom of the tank to ensure a full tank.

Storage Withdrawals—Total volume of gas withdrawn from underground storage or from liquefied natural gas storage over a specified amount of time.

Storm Door—A second door installed outside or inside a prime door creating an insulating air space. Included are sliding glass doors made of double glass or of insulating glass such as thermopane and sliding glass doors with glass or Plexiglas placed on either the outside or inside of the door to create an insulating air space. Not included are doors or sliding glass doors covered by plastic sheets or doors with storm window covering on just the glass portion of the door.

Storm or Multiple Glazing—A building shell conservation feature consisting of storm windows, storm doors, or double- or triple-paned glass that are placed on the exterior of the building to reduce the rate of heat loss.

Storm Water—Water discharge generated by precipitation and runoff from land, pavements, building rooftops and other surfaces that accumulate pollutants such as oil and grease and chemicals as it travels across land.

Storm Window—A window or glazing material placed outside or inside a window creating an insulating air space. Plastic material over windows is counted as a storm window if the same plastic material can be used year after year or if the plastic is left in place year-round and is in good condition (no holes or tears). If the plastic material must be put up new each year, it is not counted as a storm window. It is counted as "plastic coverings." Glass or Plexiglas placed over windows on either the interior or exterior side is counted as storm windows.

Story Credit—A credit that looks poor from the standpoint of its financial statements but that looks more favorable in the light of the story about its prospects for the future.

Straight Line Depreciation—A method of depreciation (for financial reporting and tax purposes) in which the owner of the equipment claims an equal amount of depreciation in each year of the equipment's recovery period.

Stranded Assets—Assets that cannot be sold for some reason. Stranded costs exceed market prices.

Stranded Benefits—Public interest programs and goals which could be compromised or abandoned by a restructured electric industry. These potential "stranded benefits" might include: environmental protection, fuel diversity, energy efficiency, low-income ratepayer assistance, and other types of socially beneficial programs.

Stranded Commitment—Assets and contracts associated with shifting to competition which are above market prices and result in non-competitive conditions for the utility.

Stranded Costs—Costs incurred by a utility which may not be recoverable under market-based retail competition. Examples include un-depreciated generating facilities, deferred costs, and long-term contract costs.

Stranded Investment (Costs and Benefits)—An investment in a power plant or demand side management measures or programs, that become uneconomical due to increased competition in the electric power market. For example, an electric power plant may produce power that is more costly than what the market rate for electricity is, and the power plant owner may have to close the plant, even though the capital and financing costs of building the plant have not been recovered through prior sales of electricity from the plant. This is considered a Stranded Cost. Stranded Benefits are those power provider investments in measures or programs considered to benefit consumers by reducing energy consumption and/or providing environmental benefits that have to be curtailed due to increased competition and lower profit margins.

Strategic Conservation—Strategic conservation results from load reductions occurring in all or nearly all time periods. This strategy can be induced by price of electricity, energy-efficient equipment, or decreasing usage of equipment.

Strategic Investors—Corporate or individual investors that add value to investments they make through industry and personal ties that can assist companies in raising additional capital as well as provide assistance in the marketing and sales process.

Strategic Load Growth—A form of load building designed to increase efficiency in a power system. This load shape objective can be induced by the price of electricity and by the switching of fuel technologies (from gas to electric).

Strategic Petroleum Reserve (SPR)—The strategic petroleum reserve consists of government owned and controlled crude oil stockpiles stored at various locations in the Gulf Coast region of the country. These reserves can be drawn down in response to sever oil supply disruptions. The target is to have a reserve of 750 million barrels of oil. Use of the reserve must be authorized by the President of the United States.

Strategic Philanthropy—A corporate philanthropy or community giving program that maximizes positive impact in the community as well as for the company, including bolstered employee recruitment, retention and a stronger company brand.

Stratification—Occurs in a liquid electrolyte solution when its concentration varies from top to bottom. Can be solved by periodic controlled charging at voltages that produce gassing to mix the electrolyte solution.

Stratigraphic Test Well—A geologically directed drilling effort to obtain information pertaining to a specific geological condition that might lead toward the discovery of an accumulation of hydrocarbons. Such wells are customarily drilled without the intention of being completed for hydrocarbon production. This classification also includes tests identified as core tests and all types of expendable holes related to hydrocarbon exploration.

Stratosphere—The region of the upper atmosphere extending from the tropopause (8 to 15 kilometers altitude) to about 50 kilometers. Its thermal structure, which is determined by its radiation balance, is generally very stable with low humidity.

Stream Class—Classification of streams based on the present and foreseeable uses made of the water, and the potential effects of on-site changes on downstream uses. Four classes are defined:

- **Class I:** Perennial or intermittent streams that provide a source of water for domestic use. Class I streams are used by large numbers of anadromous fish or significant sports fish for spawning, rearing, or migration, or are major tributaries to other Class I streams.

- **Class II:** Perennial or intermittent streams that are used by fish for spawning, rearing, or migration. Class II streams may be tributaries to Class I streams or other Class II streams.

- **Class III:** All other perennial streams not meeting higher class criteria.

- **Class IV:** All other intermittent streams not meeting higher class criteria.

Stream-Flow—The rate at which water passes a given point in a stream, usually expressed in cubic feet per second.

Strike Price—Price at which the stock or commodity underlying a put or call option can be purchased (call) or sold (put) over the specified period.

String—A number of cells, modules or panels interconnected electrically in series to produce the required operating voltage.

Strip Mine—An open cut in which the overburden is removed from a coal bed prior to the removal of coal.

Strip Mining (surface)—A method used on flat terrain to recover coal by mining long strips successively; the material excavated from the strip being mined is deposited in the strip previously mined.

Strip or Stripping Ratio—The amount of overburden that must be removed to gain access to a unit amount of coal. A stripping ratio may be expressed as (1) thickness of overburden to thickness of coal, (2) volume of overburden to volume coal, (3) weight of overburden to weight of coal, or (4) cubic yards of overburden to tons of coal. A stripping ratio commonly is used to express the maximum thickness, volume, or weight of overburden that can be profitably removed to obtain a unit amount of coal.

Stripper Well—An oil or gas well that produces at relatively low rates. For oil, stripper production is usually defined as production rates of between 5 and 15 barrels of oil per day. Stripper gas production would generally be anything less than 60 thousand cubic feet per day.

Structural Glazing—A system of retaining glass or other materials to the aluminum members of a curtain wall using silicone sealant. These systems use no mechanical fasteners and, as a result, have no profiles that cast shadows on the glazing surface.

Structural Insulated Panels (SIPs)—A no-cavity solid building system of wall and roof panels "sandwiching" polystyrene insulation between an outer and inner sheathing panel (typically oriented strand board (OSB) or metal).

Structure (Forrestry)—Arrangement of components in a forest. Vertical structural layers include overstory trees, understory trees, snags, shrubs, and herbs. Horizontal structural layers include number and species of trees, spacing of trees, and number and species of shrubs and herbs. Forest structure affects ecological processes and biodiversity.

Structuring (Financial)—Pulling together the many components of a financial agreement to arrive at a single transaction. Structuring includes, but is not limited to, pricing, end-of-term options, documentation issues, indemnification clauses, funding and residual valuations.

Stud—A popular term used for a length of wood or steel used in or for wall framing.

Stumpage—(1) Standing live or dead uncut trees. (2) The value or rate paid to purchase standing trees for harvest.

Styrene—A colorless, toxic liquid with a strong aromatic aroma. Insoluble in water, soluble in alcohol and ether; polymerizes rapidly; can become explosive. Used to make polymers and copolymers, polystyrene plastics, and rubber.

Sub-Bituminous Coal—A coal whose properties range from those of lignite to those of bituminous coal and used primarily as fuel for steam-electric power generation. It may be dull, dark brown to black, soft and crumbly, at the lower end of the range, to bright, jet black, hard, and relatively strong, at the upper end. Sub-bituminous coal contains 20 to 30 percent inherent moisture by weight. The heat content of sub-bituminous coal ranges from 17 to 24 million Btu per ton on a moist, mineral-matter-free basis. The heat content of sub-bituminous coal consumed in the United States averages 17 to 18 million Btu per ton, on the as-received basis (i.e., containing both inherent moisture and mineral matter).

Subcompact Compact Passenger Car—A passenger car containing less than 109 cubic feet of interior passenger and luggage volume.

Subcooling—In the cryogenic area, e.g. LNG, subcooling is the cooling of liquid to below its saturation temperature for the pressure under consideration. In practice, subcooling has the effect of reducing boil-off in LNG storage and transportation.

Subdivision—A prescribed portion of a given State or other geographical region.

Subdrainage—A land area (basin) bounded by ridges or similar topographic features, encompassing only part of a watershed.

Sublease—A lease of less than all of the interest held by an original Lessee to a Sublessee.

Submeter—A meter that records energy usage or water usage by a specific process, a specific part of a building or a building within a larger Facility. The consumption measured at a Submeter is a portion of a parent meter.

Submetered Data—End-use consumption data obtained for individual appliances when a recording device has been attached to the appliance to measure the amount of energy consumed by the appliance.

Submetering—The practice of remetering purchased energy beyond the customer's utility meter, generally for distribution to building tenants through privately owned or rented meters.

Subordinate Occupancy—Any occupancy type, in mixed occupancy buildings, that is not the dominant occupancy.

Subordinated Note—Debt which by its terms has no right to be paid until another debt holder is paid.
• Also referred to as "junior" debt.

Subordination—A contractual arrangement in which a party with a claim to certain assets agrees to make their claim junior, or subordinate, to the claims of another party.

Subordination Agreement—An agreement under which a lien is made inferior to another lien.

Subrogate—To substitute one person in the place of another with reference to an obligation.

Subscription Agreement—The application submitted by an investor wishing to join a limited partnership. All prospective investors must be approved by the General Partner prior to admission as a partner.

Subscription Rights—A privilege to the stockholders of a corporation to purchase proportionate amounts of a new issue of securities at an established price, usually below the current market price; also, the negotiable certificate or warrant evidencing such privilege.

Subsidence—Sinking or settling of soils so that the surface is disrupted, creating a shallow hole or even an entire region if enough groundwater or other liquid below the surface is removed.

Subsidiary—An entity directly or indirectly controlled by a parent company which owns 50% or more of its voting stock.

Subsidiary Body for Implementation (SBI)—The role of the SBI is to develop recommendations to assist the Conference of Parties in assessing and reviewing the implementation of the Climate Convention (under the Kyoto Protocol).

Subsidiary Body for Scientific & Technical Advice: (SBSTA)—A permanent body established by the UNFCCC that serves as a link between expert information sources such as the IPCC and the COP.

Subsidized Energy Financing—Not defined specifically for the purposes of Section 45—it is defined under Section 48 as "financing provided under a federal, state, or local program a principal purpose of which is to provide subsidized financing for projects designed to conserve or produce energy."

Substack—A group of stacked fuel cells that make up the base unit number of cells per full system stack.

Substance vs. Form—A concept which implies that the form of a document is subordinate to the intent of the parties involved in the document.

Substation—A facility that steps up or steps down the voltage in utility power lines. Voltage is stepped up where power is sent through long-distance transmission lines. It is stepped down where the power is to enter local distribution lines.

Substitution—The economic process of trading off inputs and consumption due to changes in prices arising from a constraint on greenhouse gas emissions. How the extremely flexible U.S. economy adapts to available substitutes and/or finds new methods of production under a greenhouse gas constraint will be critical in minimizing overall costs of reducing emissions.

Substrate—The material forming the underlying layer of streams. Substrates may be bedrock, gravel, boulders, sand, or clay.
• The physical material upon which a photovoltaic cell is made. Sub-system: Any one of several components in a PV system (i.e., array, controller, batteries, inverter, load).

Subtransmission—A set of transmission lines of voltages between transmission voltages and distribution voltages. Generally, lines in the voltage range of 69 kV to 138 kV.

Succession—A series of dynamic changes by which one group of organisms succeeds another through stages leading to potential natural community or climax. An example is the development of series of plant communities called seral stages following a major disturbance such as fire or timber harvesting.

• The receiving of property by inheritance.

Suction Head—The height of pump above the surface of the water source when the pump is located above the water level.

Suction Line—The refrigerant line that leads from the evaporator to the condenser in a split system air conditioner or heat pump. This line is insulated since it carries refrigerant at a low temperature.

Sulfate Aerosols—Sulfur-based particles derived from emissions of sulfur dioxide (SO_2) from the burning of fossil fuels (particularly coal). Sulfate aerosols reflect incoming light from the sun, shading and cooling the Earth's surface (see "radiative forcing") and thus offset some of the warming historically caused by greenhouse gases.

Sulfation—The formation of lead-sulfate crystals on the plates of a lead-acid battery; large crystals of lead sulfate grow on the plate, instead of the usual tiny crystals, making the battery extremely difficult to recharge. If the crystals get large enough, shorting of the cell may occur.

Sulfer Recovery Unit—A refinery unit that removes sulfur from distillate fuel.

Sulfur—A yellowish nonmetallic element, sometimes known as "brimstone." It is present at various levels of concentration in many fossil fuels whose combustion releases sulfur compounds that are considered harmful to the environment. Some of the most commonly used fossil fuels are categorized according to their sulfur content, with lower sulfur fuels usually selling at a higher price. Note: No. 2 Distillate fuel is currently reported as having either a 0.05 percent or lower sulfur level for on-highway vehicle use or a greater than 0.05 percent sulfur level for off-highway use, home heating oil, and commercial and industrial uses. Residual fuel, regardless of use, is classified as having either no more than 1 percent sulfur or greater than 1 percent sulfur. Coal is also classified as being low- sulfur at concentrations of 1 percent or less

or high-sulfur at concentrations greater than 1 percent.

Sulfur Dioxide (SO_2)—A toxic, irritating, colorless gas soluble in water, alcohol, and ether. Used as a chemical intermediate, in paper pulping and ore refining, and as a solvent.

Sulfur Hexafluoride (SF_6)—A colorless gas soluble in alcohol and ether, and slightly less soluble in water. It is used as a dielectric in electronics. SF_6 is among the six types of greenhouse gases to be curbed under the Kyoto Protocol. SF_6 is a synthetic industrial gas largely used in heavy industry to insulate high-voltage equipment and to assist in the manufacturing of cable-cooling systems. There are no natural sources of SF_6. SF_6 has an atmospheric lifetime of 3,200 years. Its 100-year GWP is currently estimated to be 22,200 times that of CO_2.

Sulfur Oxides (SO_x)—Compounds containing sulfur and oxygen, such as sulfur dioxide (SO_2) and sulfur trioxide (SO_3).

Summer and Winter Peaking—Having the annual peak demand reached both during the summer months (May through October) and during the winter months (November through April).

Summer, Winter, Off Peak, On Peak—These terms refer to different time of day when energy costs more or less. Generally, the highest level of energy use occurs during hot, summer days when air conditioners are fully utilized. Conversely, winter nights have the least energy demand on the grid. When demand goes up, so does the price of electricity.

Sump—A pit or tank that catches liquid runoff for drainage or disposal.

Sun Path Diagram—A circular projection of the sky vault onto a flat diagram used to determine solar positions and shading effects of landscape features on a solar energy system.

Sun Tempered Building—A building that is elongated in the east-west direction, with the majority of the windows on the south side. The area of the windows is generally limited to about 7% of the total floor area. A sun-tempered design has no added thermal mass beyond what is already in the framing, wall board, and so on. Insulation levels are generally high.

Sunk Cost—Part of the capital costs actually incurred up to the date of reserves estimation minus depreciation and amortization expenses. Items such as exploration costs, land acquisition costs, and costs of financing can be included.

Sunspace—A room that faces south (in the northern

hemisphere), or a small structure attached to the south side of a house.

Super Insulated Houses—A type of house that has massive amounts of insulation, airtight construction, and controlled ventilation without sacrificing comfort, health, or aesthetics. Combined with other efficient and solar design practices, it eliminates the need for mechanical heating and cooling, and thus pays for the extra costs of the insulation.

Super Window—A popular term for highly insulating window with a heat loss so low it performs better than an insulated wall in winter, since the sunlight that it admits is greater than its heat loss over a 24 hour period.

Superadequacy—An excess in the capacity or quality of a structure or structural component that is determined by market standards.

Supercompressibility Factor—A factor used to account for the following effect: Boyle's law for gases states that the specific weight of a gas is directly proportional to the absolute pressure, the temperature remaining constant. All gases deviate from this law by varying amounts, and within the range of conditions ordinarily encountered in the natural gas industry, the actual specific weight under the higher pressure is usually greater than the theoretical. The factor used to reflect this deviation from the ideal gas law in gas measurement with an orifice meter is called the "supercompressibility factor Fpv." The factor is used to calculate actual volumes from volumes at standard temperatures and pressures from actual volumes. The factor is of increasing importance at high pressures and low temperatures.

Superconducting Magnetic Energy Storage (SMES)—SMES technology uses the superconducting characteristics of low-temperature materials to produce intense magnetic fields to store energy. SMES has been proposed as a storage option to support large-scale use of photovoltaics and wind as a means to smooth out fluctuations in power generation.

Superconductivity—The abrupt and large increase in electrical conductivity exhibited by some metals as the temperature approaches absolute zero.

Superconductor—A synthetic material that has very low or no electrical resistance. Such experimental materials are being investigated in laboratories to see if they can be created at near room temperatures. If such a superconductor can be found, electrical transmission lines with no little or no resistance may be built, thus conserving energy usually

lost in transmission. Superconductors could also have uses in computer chips, solid state devices and electrical motors or generators.

Superfund—The program operated under the legislative authority of CERCLA and SARA that funds and carries out the EPA solid waste emergency and long-term removal remedial activities. These activities include establishing the National Priorities List investigating sites for inclusion on the list, determining their priority level on the list, and conducting and/or supervising the ultimately determined cleanup and other remedial actions.

Superheated Steam—Steam at a given pressure which is above the temperature which corresponds to boiling temperature at that given pressure.

Superstrate—The covering on the sun side of a PV module, providing protection for the PV materials from impact and environmental degradation while allowing maximum transmission of the appropriate wavelengths of the solar spectrum.

Supertanker—A very large ship designed to transport more than 500,000 deadweight tonnage of oil.

Supervisory Control and Data Acquisition (SCADA)—A system of remote control and telemetry used to monitor and control the transmission and/or distribution system.

Supplemental Gas—Any gaseous substance introduced into or commingled with natural gas that increased the volume available for disposition. Such substances include, but are not limited to, propane-air, refinery gas, coke-oven gas, still gas, manufactured gas, biomass gas, or air or inerts added for Btu stabilization.

Supplemental Gaseous Fuels Supplies—Synthetic natural gas, propane-air, coke oven gas, refinery gas, biomass gas, air injected for Btu stabilization, and manufactured gas commingled and distributed with natural gas.

Supplementary—Under Kyoto, the Protocol does not allow Annex I parties to meet their emission targets entirely through use of emissions trading and the other Kyoto Mechanisms; use of the mechanisms must be supplemental to domestic actions to limit or reduce their emissions.

Supplementary Heat—A heat source, such as a space heater, used to provide more heat than that provided by a primary heating source.

Supply—The components of petroleum supply are field production, refinery production, imports, and net receipts when calculated on a PAD District basis.

Supply Bid—A bid into the PX indicating a price at

which a seller is prepared to sell energy or ancillary services.

Supply Cost—All costs of the production of electric energy as measured at the point the electric energy is transferred to the local distribution utility for delivery to a customer.

Supply Duct—The duct(s) of a forced air heating/cooling system through which heated or cooled air is supplied to rooms by the action of the fan of the central heating or cooling unit.

Supply Side—Technologies that pertain to the generation of electricity.

Supply Source—May be a single completion, a single well, a single field with one or more reservoirs, several fields under a single gas-purchase contract, miscellaneous fields, a processing plant, or a field area; provided, however, that the geographic area encompassed by a single supply source may not be larger than the state in which the reserves are reported.

Supply, Petroleum—A set of categories used to account for how crude oil and petroleum products are transferred, distributed, or placed into the supply stream. The categories include field production, refinery production, and imports. Net receipts are also included on a Petroleum Administration for Defense (PAD) District basis to account for shipments of crude oil and petroleum products across districts.

Supply-side—Activities conducted on the utility's side of the customer meter. Activities designed to supply electric power to customers, rather than meeting load though energy efficiency measures or on-site generation on the customer side of the meter.

Supply-side Economics—The use of policies such as tax cuts and business incentives to control the supply of certain goods or services.

Support Equipment and Facilities—These include, but are not limited to, seismic equipment, drilling equipment, construction and grading equipment, vehicles, repair shops, warehouses, supply points, camps, and division, district, or field offices.

Supporting Structure—The main supporting unit (usually a pole or tower) for transmission line conductors, insulators, and other auxiliary line equipment.

Surety—A Party that binds itself with another, called the principal, for the performance of an obligation.

Surface Drilling Expenses (Uranium)—These include drilling, drilling roads, site preparation, geological and other technical support, sampling, and drill-hole logging costs.

Surface Mine—A coal-producing mine that is usually within a few hundred feet of the surface. Earth above or around the coal (overburden) is removed to expose the coal bed, which is then mined with surface excavation equipment, such as draglines, power shovels, bulldozers, loaders, and augers. It may also be known as an area, contour, open-pit, strip, or auger mine.

Surface Orientation Factor—The ratio of the annual incident solar radiation on a surface for a specific tilt and orientation ($MJ/M^2/year$) divided by the annual incident solar radiation on a surface for a south-facing surface with optimal tilt ($MJ/M^2/year$).

Surface Rights—Fee ownership in surface areas of land. Also used to describe a lessee's right to use as much of the surface of the land as may be reasonably necessary for the conduct of operations under the lease.

Surface Water Loop—In this type of closed-loop geothermal heat pump installation, the fluid-filled plastic heat exchanger pipes are coiled into circles and submerged at least eight feet below the surface of a body of surface water, such as a pond or lake. The coils should only be placed in a water source that meets minimum volume, depth, and quality criteria.

Surge—An excessive amount of power drawn by an appliance when it is first switched on. An unexpected flow of excessive current, usually caused by excessive voltage, that can damage appliances and other electrical equipment.

Surge Capacity—The ability of an inverter or generator to deliver instantaneous high currents when starting motors, for example.

Surge Tank—A tank used to absorb irregularities in flow of liquids, including liquid waste materials, so that the flow out of the tank is constant.

Surplus—(Electric utility) Excess firm energy available from a utility or region for which there is no market at the established rates.

Surplus Electricity—Electricity produced by cogeneration equipment in excess of the needs of an associated factory or business.

Surplus Energy—Energy generated that is beyond the immediate needs of the producing system. This energy may be supplied by spinning reserve and sold on an interruptible basis.

Survey—A map or plat containing a statement of courses, distances and quantity of land and showing lines of possession.

Suspended Films—Low-e coated plastic films stretched between the elements of the spacers between panes of glazing; acts as a reflector to slow the loss of heat from the interior to the exterior.

Suspended Rates—New rates that have been accepted for review by a utility commission. When these rates are suspended, they do not go into effect for a designated period of time. Charges under the new rate may be refunded after the resolution of the rate proceeding.

Suspended Solids—Waste particles suspended in water. Suspended solids can harbor harmful microorganisms and toxic chemicals. Suspended solids cloud the water and make disinfection more difficult and costly.

Sustainable—An ecosystem condition in which biodiversity, renewability, and resource productivity are maintained over time. Meeting the needs of the present without compromising the ability of future generations to meet their own needs. The successful meeting of present social, economic, and environmental needs without compromising the ability of future generations to meet their own needs. Derived from the most common definition of sustainability, created in 1987 at the World Commission on Environment and Development.

Sustainable Design—A process of product, service, or organizational design that complies with the principles of social economic, and environmental sustainability.

Sustainable Development—A principle which says that a development plan must not compromise the welfare of future generations for the benefit of present generations. The principle can be applied in a number of ways, but they all relate to guaranteeing intergenerational equity (somehow defined). That is, the principle concerns fairness between generations.

Sustainable Energy—Energy produced both from renewable resources or by use of clean production technology.

Sustained Orderly Development—A condition in which a growing and stable market is identified by orders that are placed on a reliable schedule. The orders increase in magnitude as previous deliveries and engineering and field experience lead to further reductions in costs. The reliability of these orders can be projected many years into the future, on the basis of long-term contracts, to minimize market risks and investor exposure. (See also *Commercialization*.)

Sustained Yield—The maintenance in perpetuity of regular, periodic harvest of wood resources from forest land without damaging the productivity of the land.

Swamp Coolers (Evaporative Coolers)—Air-conditioning equipment that removes heat by evaporating water. Evaporative cooling techniques are most commonly found in warm, dry climates such as in the Southwest, although they are found throughout the country. They usually work by spraying cool water into the air ducts, cooling the air as the spray evaporates.

Sweet Gas—Natural gas not contaminated by corrosion inducing impurities such as hydrogen sulfide, or with a low level of impurities.

Swept Area—In reference to a wind energy conversion device, the area through which the rotor blades spin, as seen when directly facing the center of the rotor blades.

SWH—Solar water heating

Switch—A common device which breaks an electrical circuit thereby halting the flow electricity through the circuit.

Switch Mode—A form of converting one form of electricity to another by rapidly switching it on and off and feeding it through a transformer to effect a voltage change.

Switching Station—Facility equipment used to tie together two or more electric circuits through switches. The switches are selectively arranged to permit a circuit to be disconnected or to change the electric connection between the circuits.

SWRCB—State Water Resources Control Board (California)

SWRTA—The Southwest Regional Transmission Association. a sub-regional RTG within WRTA, and awaiting FERC approval.

SYD—Sum of the Years' Digits. A method of amortization or depreciation.

Synchronous Generator—An electrical generator that runs at a constant speed and draws its excitation from a power source external or independent of the load or transmission network it is supplying.

Synchronous Inverter—An electrical inverter that inverts direct current electricity to alternating current electricity, and that uses another alternating current source, such as an electric power transmission and distribution network (grid), for voltage and frequency reference to provide power in phase and at the same frequency as the external power source.

Synchronous Motor—A type of motor designed to operate precisely at the synchronous speed with no slip in the full-load speeds (rpm).

Syncrude—Synthetic crude oil made from coal of from oil shale.

Syndicate—A group of banks making a syndicated loan. A group of bond houses which act together in underwriting and distributing a new securities issue.

Syndicated—Underwriters or broker/dealers who sell a security as a group.

Syndication—A pooling arrangement or association of persons investing in real or personal property by buying an interest in an enterprise.

Synfuel—Synthetic gas or synthetic oil. Fuel that is artificially made as contrasted to that which is found in nature. Synthetic gas made from coal is considered to be more economical and easier to produce than synthetic oil. When natural gas supplies in the earth are being depleted, it is expected that synthetic gas will be able to be used widely as a substitute fuel.

Syngas—A syntheses gas produced through gasification of biomass. Syngas is similar to natural gas and can be cleaned and conditioned to form a feedstock for production of methanol.

Synthetic Coal—Coal that has been processed by a coal synfuel plant; and coal-based fuels such as briquettes, pellets, or extrusions, which are formed by binding materials and processes that recycle material.

Synthetic Lease—An equipment lease that qualifies as an off-balance sheet operating lease for financial accounting purposes but as a loan or conditional sale for tax purposes, thus enabling the lessee to retain tax benefits associated with equipment ownership.

Synthetic Natural Gas (SNG)—(Also referred to as substitute natural gas) A manufactured product, chemically similar in most respects to natural gas, resulting from the conversion or reforming of petroleum hydrocarbons that may easily be substituted for or interchanged with pipeline-quality natural gas.

System—A combination of equipment and/or controls, accessories, interconnecting means and terminal elements by which energy is transformed to perform a specific function, such as climate control, service water heating, or lighting. [See California Code of Regulations, Title 24, Section 2-5302]

System (Electric)—Physically connected generation, transmission, and distribution facilities operated as an integrated unit under one central management or operating supervision.

System (Gas)—An interconnected network of pipes, valves, meters, storage facilities, and auxiliary equipment used in the transportation, storage, and/or distribution of natural gas or commingled natural and supplemental gas.

System Autonomy—The reserve capacity of batteries in a solar photovoltaic system. Batteries in applications that require autonomy form a critical component of a solar power system.

System Availability—The proportion of time (usually expressed in hours per year) that a solar PV system will be able to meet fully the load demand.

System Benefits Charge—A non-bypassable fee on transmission interconnection; funds are allocated among public purposes, including the development and demonstration of renewable energy technologies.

System Integration (of new technologies)—The successful integration of a new technology into the electric utility system by analyzing the technology's system effects and resolving any negative impacts that might result from its broader use.

System Interconnection—A physical connection between two electric systems that permits the transfer of electric energy in either direction.

System Mix—The proportion of electricity distributed by a power provider that is generated from available sources such as coal, natural gas, petroleum, nuclear, hydropower, wind, or geothermal.

System Operating Voltage—The output voltage of a solar PV array under load, dependent on the electrical load and size of the battery stack connected to the output terminals.

System Peak Demand—The highest demand value that has occurred during a specified period for the utility system.

System Power—The mix of electricity fuel sources consumed in the state or region that are not disclosed or marketed as specific purchases or as defined by relevant state agency.

T

T&M—Time and Materials

TAC—Toxic Air Contaminant

TAD—Technical Assistance Document

Tag-Along Rights—A minority-shareholder protection affording the right to include their shares in any sale of control and at the offered price.

Tail—The part of a wind generator that makes the rotor

face into the wind. Often the tail is also involved in governing the machine, by folding down or sideways to swing the rotor out of the wind.

Tail Block—The last or lowest priced block of energy in a declining utility block rate structure.

Tailgate—The outlet of a natural gas processing plant where dry residue gas is delivered or re-delivered for sale or transportation.

Tailings—The remaining portion of a metal-bearing ore consisting of finely ground rock and process liquid after some or all of the metal, such as uranium, has been extracted.

Taillores Declaration—The first official statement made by university administrators of a commitment to sustainability in higher education. Consists of a 10-point action plan for incorporating sustainability and environmental literacy in teaching, research, operations, and outreach.

Tailrace—The pipe, flume, or channel in a hydroelectric system that carries the water from the turbine runner back to the stream or river.

Take-and-Pay—A clause that requires a minimum quantity to be physically taken and paid for, usually gas in association with oil, or wells that will be damaged by failure to produce.

Take-and-pay Contract—A take-and-pay contract is sometimes used to describe a contract in which payment is contingent upon delivery (electricity for example) and the obligation to pay is not unconditional, as in a take-or-pay contract.

Takedown Schedule—A takedown schedule means the timing and size of the capital contributions from the limited partners of a venture fund.

Take-Out Point—The metering points at which a metered entity takes delivery of energy.

Taking—Reducing the value of someone's property through government action without just compensation.

Tall Oil—The oily mixture of rosin acids, fatty acids, and other materials obtained by acid treatment of the alkaline liquors from the digesting (pulping) of pine wood.

Tame (Tertiary Amyl Methyl Ether)—Another oxygenate that can be used in reformulated gasoline. It is an ether based on reactive C5 olefins and methanol.

TAMS—Toxic Air Monitoring System

Tangible Development Costs—Costs incurred during the development stage for access, mineral-handling, and support facilities having a physical nature. In mining, such costs would include tracks, lighting equipment, ventilation equipment, other equipment installed in the mine to facilitate the extraction of minerals, and supporting facilities for housing and care of work forces. In the oil and gas industry, tangible development costs would include well equipment (such as casing, tubing, pumping equipment, and well heads), as well as field storage tanks and gathering systems.

Tangible Net Worth—Tangible net worth is a shareholder's equity adjusted for intangible assets. A conservative practice is to subtract intangible assets from stockholder's equity to more truly represent (in terms of physical assets) the amount of equity which shareholders have invested in a company. Though intangible assets may have some value, this value is often quite different from the amount appearing on financial statements.

Tank Farm—An installation used by trunk and gathering pipeline companies, crude oil producers, and terminal operators (except refineries) to store crude oil.

Tanker and Barge—Vessels that transport crude oil or petroleum products.

Tankless Water Heater—A water heater that heats water before it is directly distributed for end use as required; a demand water heater.

Tar Sands—Naturally occurring bitumen-impregnated sands that yield mixtures of liquid hydrocarbon and that require further processing other than mechanical blending before becoming finished petroleum products.

Tare Loss—Loss caused by a charge controller. One minus tare loss, expressed as a percentage, is equal to the controller efficiency.

Tare Weight—The weight of a container and packing materials without the weight of the goods it contains.

Target Market—A specific group of people or geographical area that has been identified as the primary buyers of a product or service.

Target Usage Profile—The expected energy use at a meter, reflecting the impact(s) of implemented Energy Conservation Measures. Target usage = Adjusted Baseline - Measure Impacts

Targets and Timetables—(Kyoto Protocol) Targets refer to the emission levels or emission rates set as goals for countries, sectors, companies, or facilities. When these goals are to be reached by specified years, the years at which goals are to be met are referred to as the timetables. In the Kyoto Protocol, a target is the percent reduction from the 1990 emissions baseline that the country has agreed to.

On average, developed countries agreed to reduce emissions by 5.2% below 1990 emissions during the period 2008-2012, the first commitment period.

Tariff—A document, approved by the responsible regulatory agency, listing the terms and conditions, including a schedule of prices, under which utility services will be provided.

   • Public schedules of utility rates, rules, service territory and terms that are filed for official approval with a regulatory agency.

   • A regulatory-agency-approved document listing the terms and conditions, including a price schedule under which the utility services will be provided.

   • A published volume of rate schedules and general terms and conditions under which a product or service will be supplied.

Task Lighting—Any light source designed specifically to direct light a task or work performed by a person or machine.

Task Lighting (task-oriented lighting)—Lighting designed specifically to illuminate one or more task locations, and generally confined to those locations. [See California Code of Regulations, Title 24, Section 2- 5302]

TAT—Technical Assistance Team

Tax Credits—Credits established by the federal and state government to assist the development of the alternative energy industry.

Beginning in 1976, California had a solar tax credit. From 1978 to 1985, both California and the federal government offered tax credits for alternative energy equipment. The state provided a 55 percent tax credit on solar, wind, geothermal and biomass for residential applications. However, the residential tax credits were reduced by applicable federal credits. State commercial tax credits for alternative energy systems in commercial and industrial sectors ranged from 10-15 percent. During this same time, the federal government offered a 40 percent tax credit on residential applications and a 10-15 percent credit on commercial and industrial applications. California in 1990 instituted a new 10 percent tax credit for commercial solar systems in excess of 30 watts of electricity per device. This credit expired December 31, 1993.

Tax Exempt Energy Service Agreement—Similar to the Energy Service Agreement, this fixed payment shared savings agreement can combine the rate benefits of a tax exempt financing vehicle with the marketing advantages of a shared savings agreement, and can often be popular with schools.

Tax Exempt Obligation—Bond, debt or another obligation for which the interest paid is exempt from federal income taxes under Section 103 of the U.S. Internal Revenue Code.

Tax Exempt Organization—An organization that may issue or incur tax-exempt obligations. State and local governments are tax-exempt organizations.

Tax Lease—A generic term for a lease in which the lessor takes on the risks of ownership (as determined by various IRS pronouncements) and, as the owner, is entitled to the benefits of ownership, including tax benefits.

Tax Life—The facility life permitted by the tax law for use in determining the tax depreciation deduction.

Taxable Year—The term "taxable year" means the calendar year, or the fiscal year ending during such calendar year, upon the basis of which the taxable income is computer under subtitle A or the IRC. A "Taxable Year" means, in the case of a return made for a fractional part of a year under the provisions of subtitle A or under regulations prescribed by the Secretary of the Treasury, the period for which such return is made.

Tax-Cost—A deduction (allowance) under U.S. Federal income taxation normally calculated under a formula whereby the adjusted basis of the mineral property is multiplied by a fraction, the numerator of which is the number of units of minerals sold during the tax year and the denominator of which is the estimated number of units of unextracted minerals remaining at the end of the tax year plus the number of units of minerals sold during the tax year.

Tax-Free Reorganization—Types of business combinations in which shareholders do not incur tax liabilities. There are four types—A, B, C, and D reorganizations. They differ in various ways in the amount of stock/cash that can be offered.

Taxpayer—The term "taxpayer" means any person or entity subject to any tax—local, state or federal.

TB—Trillions of barrels

TBC—To Be Considered

TBtu—Trillion Btu

TC—Target Concentration
   • Technical Center
   • Toxic Concentration
   • Toxicity Characteristic

TCE—Trichloroethylene

TCF—Measurement of one trillion cubic feet of natural gas.

TCM—Transportation Control Measure

tCO₂e, MtCO₂e—Tons of carbon dioxide equivalent, and millions of tons of carbon dioxide equivalent. This is the metric measurement unit for greenhouse emissions. The global warming impact of all greenhouse gases is measured in terms of equivalency to the impact of carbon dioxide (CO_2). For example, one million tons of emitted methane, a far more potent greenhouse gas than carbon dioxide, is measured as 23 million tons of CO_2-equivalent, or 23 $MtCO_2e$.

TCP—Transportation Control Plan

TCRI—Toxic Chemical Release Inventory

TCU—Transportation, Communications and Utilities

TD—Toxic Dose

TDM—Tariff Data Model

TDS—Total Dissolved Solids

TDV Energy—*See Time dependent Valuation (TDV) Energy*

TE—Transmission Engineering

TE&S Species—See *Threatened, Endangered and Sensitive Species*.

TEAM—Traffic Engineering and Management
- Total Exposure Assessment Model

TEC—Technical Evaluation Committee

Technological Change—How much technological change will be additionally induced by climate policies is a crucial, but not well quantified, factor in assessing the costs of long-term mitigation of greenhouse gas emissions.

Technology Transfer—The process by which energy-efficient or low emission intensive technologies developed by industrialized nations are made available to less industrialized nations. Technology transfer may occur through the sale of technology by private entities, through government programs, non-profit arrangements, or other means.

Technology-Based Emissions Standard—There is some confusion about this term. Often a TBES merely refers to a technological standard (where firms are required to adopt a given technology). This dictionary uses this definition:

- The term TBES is also used to refer to a performance standard (for example the emissions per kilometre of a car). The standard is, however, based on the performance of a specific technology. These are slightly more flexible than technological standards (which require firms to use a specific technology), and are thus more cost-effective than a technological standard. They are still unlikely to be as cost-effective as an incentive-based approach.

TEF—Toxic Equivalency Factors

TEI—Tax Equity Investor. Investor in a renewable energy facility that makes a capital contribution to a renewable energy operating partnership in exchange for a limited partnership interest (usually 99 percent at inception of the partnership) and receives the majority of the energy credits or production tax credits, tax losses, and some amount of cash flow.

Telemetering—Use of an electrical apparatus transmitting data to a distant point for indicating, recording, or integrating the values of a variable quantity.

Temperature—Degree of hotness or coldness measured on one of several arbitrary scales based on some observable phenomenon (such as the expansion).

Temperature Coefficient (of a solar photovoltaic cell)—The amount that the voltage, current, and/or power output of a solar cell changes due to a change in the cell temperature.

Temperature Compensation—Adjustment via the use of electronic circuitry to change the charge controller activation points depending on battery temperature. This is desirable if the battery temperature is expected to vary by more than 5 deg C from the ambient temperature. The temperature coefficient for lead acid batteries is typically -3 to -5 millivolts/deg C per cell.

Temperature Factors—Are used to decrease battery capacity at cold temperatures, to decrease PV module voltage at high temperatures and to increase the resistance of wire at high temperatures.

Temperature Humidity Index—An index that combines sensible temperature and air humidity to arrive at a number that closely responds to the effective temperature; used to relate temperature and humidity to levels of comfort.

Temperature Zones—Individual rooms or zones in a building where temperature is controlled separately from other rooms or zones.

Temperature, Ambient—The temperature of the air, atmosphere or other fluid that completely surrounds the apparatus, equipment or the workpiece under consideration. For devices which do not generate heat, this temperature is the same as the temperature of the medium at the point of device location when the device is not present. For devices which do generate heat, this temperature is the temperature of the medium surrounding the device when the device is present and generating heat. Allowable ambient-temperature limits are based on the assumption that the device in question is not exposed to significant radiant-energy sources such as sunlight or heated surfaces.

Temperature, Dew-Point—The temperature at which a vapor begins to condense and deposit as a liquid.

Temperature, Dry Bulb—Technically, the temperature registered by the dry bulb thermometer of a psychrometer. It is identical with the temperature of the air.

Temperature, Effective—An arbitrary index which combines into a single value the effect of temperature, humidity, and air movement on the sensation of warmth or cold felt by the human body. The numerical value is that of the temperature of still, saturated air which would induce an identical sensation.

Temperature, Wet Bulb—The temperature an air parcel would have if cooled adiabatically to saturation at constant pressure by evaporation of water from it, all latent heat being supplied by the parcel.

Temperature/Pressure Relief Valve—A component of a water heating system that opens at a designated temperature or pressure to prevent a possible tank, radiator, or delivery pipe rupture.

Tempering Valve—A valve used to mix heated water with cold in a heating system to provide a desired water temperature for end use.

Temporarily Discharged Fuel—Fuel that was irradiated in the previous fuel cycle (cycle N) and not in the following fuel cycle (cycle N+1) and that will be irradiated in a subsequent fuel cycle.

Temporary Lighting—A lighting installation where temporary connections, such as cord and plug, are used for electric power, and for which the installation does not persist beyond 60 consecutive days or more than 120 days per year.

Tenancy In Common—Ownership of property by two or more parties in equal or unequal undivided interests without right of survivorship.

Tender—An unconditional offer of payment of a debt or to fully satisfy an obligation.

Tender Offer—An offer to purchase stock directly to the shareholders. One of the more common ways hostile takeovers are implemented.

Tennessee Valley Authority (TVA)—A federal agency established in 1933 to develop the Tennessee river valley region of the southeastern U.S., and which is now the nation's largest power producer.

Tentative Minimum Tax (TMT)—A separately figured tax that eliminates many deductions and credits to arrive at the minimum tax that a taxpayer must pay. If the TMT is greater that the regular tax, then the TMT is referred to as the Alternative Minimum Tax (AMT) and must be paid in lieu of the regular tax.

TEOR—Thermally Enhanced Oil Recovery

TEP—Technical Evaluation Panel
- Toxicity Extraction Procedure
- Typical End-use Product

TEQ—Toxicity Equivalent Quotient

Terawatthour—One trillion watthours.

Term Agreement—Any written or unwritten agreement between two parties in which one party agrees to supply a commodity on a continuing basis to a second party for a price or for other considerations.

Term Sheet—A summary of the terms the investor is prepared to accept. A non-binding outline of the principal points which the Stock Purchase Agreement and related agreements will cover in detail.

Term/Tenor—Period of time (usually measured in years) during which the conditions of a contract will be carried out.

Terminal Rental Adjustment Clause (TRAC)—A lessee guaranteed residual value for vehicle leases, the inclusion of which will not, in and of itself, disqualify the tax lease status of a tax-oriented vehicle lease.

Termination Agreement—The liability of the lessee in the event of termination is set forth in a termination schedule that values the equipment at various times during the lease term. This value is designed to protect the lessor from loss of investment. If the equipment is sold at a price lower than the amount set forth in the schedule, the lessee pays the difference. In the event the resale is at a price higher than in the termination schedule, such excess amounts belong to the lessor. The termination schedule is not the same as the casualty value schedule, insured value schedule or stipulated loss value schedule.

Termination Value—The amount the lessee must pay the lessor in order to terminate the lease before the scheduled end.

Termite Shield—A construction element that inhibits termites from entering building foundations and walls.

TES—Thermal Energy Storage

Test Well Contribution—A payment made to the owner of an adjacent or nearby tract who has drilled an exploratory well on that tract in exchange for information obtained from the drilling effort.

Tetrachloroethylene (TCE)—Volatile organic compound that is commonly used as an industrial degreasing solvent. TCE affects the central nervous system and is listed as a cancer-causing chemical under Proposition 65 in California.

TFS—Total Fuel Hydrocarbons

TGC—Tradable Green Certificate. European equivalent to RECs (Renewable Energy Certificate).

TGO—Total Gross Output

THC—Total hydrocarbons

The Carbon Trust—The Carbon Trust certifies and therefore improves confidence in claims by corporations to carbon mitigation. www.carbontrust-standard.com

The Climate Group—The Climate Group works at an international level with governments and business leaders to advance the smart policies, technologies and finance needed to cut global greenhouse gas emissions, and unlock a 'clean industrial revolution'. www.theclimategroup.com

The Climate Registry—The Climate Registry through its support of both voluntary and mandatory reporting programs and providing comprehensive, accurate data to reduce greenhouse gas emissions. www.theclimateregistry.org

The International Integrated Reporting Committee—The International Integrated Reporting Committee promotes integrated reporting to help business make more sustainable decisions, and enable investors and shareholders to understand organizations actual performance. www.theiirc.org

The National Physical Laboratory—The National Physical Laboratory is a UK world-leading center of excellence in developing and applying the most accurate standards, science and technology available. www.npl.co.uk

The OECD—The OECD advocates policies for innovation and the transition to a low-carbon economy to improve economic and social well-being for people round the world while meeting the challenge of limiting greenhouse gas emissions. www.oecd.org

The United Nations Conference on Trade and Development (UNCTAD)—UNCTAD is home to the Intergovernmental Panel of Experts on International Standards on Accounting and Reporting (ISAR) which supports a standardized reporting framework. www.unctad.org

The World Economic Forum—The World Economic Forum engages business, political, academic and other influential thought leaders to shape global, regional and industry agendas. www.weforum.org

Therm—One hundred thousand (100,000) British thermal units (1 therm = 100,000 Btu).

Thermal—A term used to identify a type of electric generating station, capacity, capability, or output in which the source of energy for the prime mover is heat.

Thermal (Energy) Storage—A technology that lowers the amount of electricity needed for comfort conditioning during utility peak load periods. A buildings thermal energy storage system might, for example, use off-peak power to make ice or to chill water at night, later using the ice or chilled water in a power saving process for cooling during the day. See *Thermal Mass*.

Thermal Balance Point—The point or outdoor temperature where the heating capacity of a heat pump matches the heating requirements of a building.

Thermal Break (Thermal Barrier)—An element of low heat conductivity placed in such a way as to reduce or prevent the flow of heat. Some metal-framed windows are designed with thermal breaks to improve their overall thermal performance.

Thermal Break Window Frame—Metal fenestration frames that are not solid metal from the inside to the outside, but are separated in the middle by a material, usually urethane, with a lower conductivity.

Thermal Capacitance—The ability of a material to absorb and store heat for use later.

Thermal Chimney—Heat rising in a tall structure provides a constant suction that may be used to vent the house, bring warm air from collectors, or pull cool air from rock storage. This creates what is called a "stack effect."

Thermal Conductivity—The quantity of heat that will flow through a unit area of the material per hour when the temperature difference through the material is one degree.

Thermal Control Layer—The component (or components) installed in an assembly to control the transfer of thermal energy (heat). Typically, these are comprised of insulation products, radiant barriers or trapped gaps filled with air or other gases. One quantitative measure of a Thermal Control Layer's resistance to heat flow is the R-Value. R-Values are limited in that they deal with conduction, which is only one of three modes of heat flow (the other two being convection and radiation), and in that their range of applicability is typically limited to materials and not assemblies.

Thermal Cover—Vegetative condition, generally with greater than 70% canopy closure and 40 feet in height, that can significantly ameliorate weather effects such as wind, heat, cold, and snow. Used by wildlife in winter.

Thermal Cracking—A refining process in which heat and pressure are used to break down, rearrange, or combine hydrocarbon molecules. Thermal-cracking includes gas oil, visbreaking, fluid coking, delayed coking, and other thermal cracking processes (e.g., flexicoking).

Thermal Efficiency—A measure of the efficiency of converting a fuel to energy and useful work; useful work and energy output divided by higher heating value of input fuel times 100 (for percent).

Thermal Electric—Electric energy derived from heat energy, usually by heating a working fluid, which drives a turbogenerator.

Thermal Energy—The energy developed through the use of heat energy.

Thermal Energy Storage—The storage of heat energy during power provider off-peak times at night, for use during the next day without incurring daytime peak electric rates. A building's thermal energy storage system might, for example, use off-peak power to make ice or to chill water at night, later using the ice or chilled water in a power-saving process for cooling during the day. Often found in central plant systems in large buildings or complexes.

Thermal Envelope Houses—An architectural design (also known as the double envelope house), sometimes called a "house-within-a-house," that employs a double envelope with a continuous airspace of at least 6 to 12 inches on the north wall, south wall, roof, and floor, achieved by building inner and outer walls, a crawl space or sub-basement below the floor, and a shallow attic space below the weather roof. The east and west walls are single, conventional walls. A buffer zone of solar-heated, circulating air warms the inner envelope of the house. The south-facing airspace may double as a sunspace or greenhouse.

Thermal Expansion—Expansion of a substance as a result of the addition of heat. In the context of climate change, thermal expansion of the world's oceans in response to global warming is considered the predominant driver of current and future sea-level rise.

Thermal Limit—The maximum amount of power a transmission line can carry without suffering heat-related deterioration of line equipment, particularly conductors.

Thermal Mass—A material used to store heat, thereby slowing the temperature variation within a space. Typical thermal mass materials include concrete, brick, masonry, tile and mortar, water, and rock or other materials with high heat capacity.

Thermal Power Plant—Any stationary or floating electrical generating facility using any source of thermal energy, with a generating capacity of 50 megawatts or more, and any facilities appurtenant thereto. Exploratory, development, and production wells, resource transmission lines, and other related facilities used in connection with a geothermal exploratory project or a geothermal field development project are not appurtenant facilities for the purposes of this division. Thermal power plant does not include any wind, hydroelectric, or solar photovoltaic electrical generating facility.

Thermal Resistance (R-Value)—This designates the resistance of a material to heat conduction. The greater the R-value the larger the number. The resistance of a material or building component to the massage of heat in (hr. x ft.^{2} x °F/Btu.

Thermal Resource—A facility that produces electricity by using a heat engine to power an electric generator. The heat may be supplied by the combustion of coal, oil, natural gas, biomass, or other fuels, including nuclear fission, solar, or geothermal resources.

Thermal Storage—Storage of heat or heat sinks (coldness) for later heating or cooling. Examples are the storage of solar energy for night heating; the storage of summer heat for winter use; the storage of winter ice for space cooling in the summer; and the storage of electrically-generated heat or coolness when electricity is less expensive, to be released in order to avoid using electricity when the rates are higher. There are four basic types of thermal storage systems: ice storage; water storage; storage in rock, soil or other types of solid thermal mass; and storage in other materials, such as glycol (antifreeze).

Thermal Storage Walls (Masonry or Water)—A thermal storage wall is a south-facing wall that is glazed on the outside. Solar heat strikes the glazing and is absorbed into the wall, which conducts the heat into the room over time. The walls are at least 8 in thick. Generally, the thicker the wall, the less the indoor temperature fluctuates.

Thermally Enhanced Oil Recovery (TEOR)—Injection of steam to increase the amount of petroleum that may be recovered from a well.

Thermalphotovoltaic Device (TPV)—A device in which solar energy is concentrated on to a radiator which reaches a high temperature and emits the energy in a different part of the spectrum, better matched to the bandgap of the matched solar cell. This approach should enable high cell efficiencies to be obtained.

Thermo photovoltaic Cell—A device where sunlight concentrated onto a absorber heats it to a high temperature, and the thermal radiation emitted by the absorber is used as the energy source for a photovoltaic cell that is designed to maximize conversion efficiency at the wavelength of the thermal radiation.

Thermo Siphon (Thermosyphon)—The natural, convective movement of air or water due to differences in temperature. In solar passive design a thermosyphon collector can be constructed and attached to a house to deliver heat to the home by the continuous pattern of the convective loop (or thermosyphon).

Thermochemical Conversion Process—Chemical reactions employing heat to produce fuels.

Thermocouple—A device consisting of two dissimilar conductors with their ends connected together. When the two junctions are at different temperatures, a small voltage is generated.

Thermodynamic Cycle—An idealized process in which a working fluid (water, air, ammonia, etc) successively changes its state (from a liquid to a gas and back to a liquid) for the purpose of producing useful work or energy, or transferring energy.

Thermodynamics—A study of the transformation of energy into other manifested forms and of their practical applications. The three laws of thermodynamics are:

• Law of Conservation of Energy—energy may be transformed in an isolated system, but its total is constant

• Heat cannot be changed directly into work at constant temperature by a cyclic process

• Heat capacity and entropy of every crystalline solid becomes zero at absolute zero (0 degrees Kelvin)

Thermoelectric Conversion—The conversion of heat into electricity by the use of thermocouples.

Thermoelectric Power Plant—A term used to identify a type of electric generating station, capacity, capability, or output in which the source of energy for the prime mover is heat.

Thermography—A building energy auditing technique for locating areas of low insulation in a building envelope by means of a thermographic scanner.

Thermohaline Circulation (THC)—A three-dimensional pattern of ocean circulation driven by wind, heat and salinity that is an important component of the ocean-atmosphere climate system. In the Atlan-tic, winds transport warm tropical surface water northward where it cools, becomes more dense, and sinks into the deep ocean, at which point it reverses direction and migrates back to the tropics, where it eventually warms and returns to the surface. This cycle or "conveyor belt" is a major mechanism for the global transport of heat, and that has an important influence on the climate. Global warming is projected to increase sea-surface temperatures, which may slow the THC by reducing the sinking of cold water in the North Atlantic. In addition, ocean salinity also influences water density, and thus decreases in sea-surface salinity from the melting of ice caps and glaciers may also slow the THC.

Thermopile—A large number of thermocouples connected in series.

Thermosiphon System—This passive solar hot water system consists of and relies on warm water rising, a phenomenon known as natural convection, to circulate water through the collectors and to the tank. In this type of installation, the tank must be above the collector. As water in the collector heats, it becomes lighter and rises naturally into the tank above. Meanwhile, cooler water in the tank flows down pipes to the bottom of the collector, causing circulation throughout the system. The storage tank is attached to the top of the collector so that thermosiphoning can occur.

Thermostat—A device used to control temperatures; used to control the operation of heating and cooling devices by turning the device on or off when a specified temperature is reached.

Thermostat, Setback—A device, containing a clock mechanism, which can automatically change the inside temperature maintained by the HVAC system according to a preset schedule. The heating or cooling requirements can be reduced when a building is unoccupied or when occupants are asleep. [See California Code of Regulations, Title 24, Section 2- 5352(h)]

Thermostatic Expansion Valve (TXV)—A refrigerant metering valve, installed in an air conditioner or heat pump, which controls the flow of liquid refrigerant entering the evaporator in response to the superheat of the gas leaving it.

Thick Cells—Conventional solar cells in most types of PV modules, such as crystalline silicon cells, which are typically from 200-400 micrometers thick. In contrast, thin-film cells are several microns thick.

Thick-crystalline Materials—Semiconductor material, typically measuring from 200-400 micrometers thick, that is cut from boules, ingots or ribbons.

Thin Film PV Module—A solar PV module constructed with sequential layers of thin film semiconductor materials usually only micrometer thick. Currently, thin film technologies account for around 12% of all solar modules sold around the world. This share is expected to increase, since thin film technologies represent a potential route to lower costs.

Thin film—A layer of semiconductor material, such as copper indium diselenide, cadmium telluride, gallium arsenide, or amorphous silicon, a few microns or less in thickness, used to make photovoltaic cells.

Third Assessment Report (TAR)—The most recent Assessment Report prepared by the Intergovernmental Panel on Climate Change, which reviewed the existing scientific literature on climate change, including new information acquired since the completion of the Second Assessment report (SAR). Finalized in 2001, it is comprised of three volumes: Science; Impacts and Adaptation; and Mitigation.

Third-party DSM Program Sponsor—An energy service company (ESCO) which promotes a program sponsored by a manufacturer or distributor of energy products such as lighting or refrigeration whose goal is to encourage consumers to improve energy efficiency, reduce energy costs, change the time of usage, or promote the use of a different energy source.

Third-party Transactions—Third-party transactions are arms-length transactions between nonaffiliated firms. Producing country-to-company transactions are not considered to be third-party transactions.

Thorium—An element that is a byproduct of the decay of uranium.

Threatened, Endangered and Species—Formal classifications of species. The U.S. Fish and Wildlife Service, makes the following designations:

- **Sensitive**: Species for which population viability is a concern. Sensitive species are not federally designated under the Endangered Species Act.
- **Candidate**: Species under consideration for listing as endangered or threatened but for which conclusive data on biological vulnerability are not currently available to support listing. Also known as Category 2 species.
- **Endangered**: A species in danger of becoming extinct throughout all or a significant portion of its range.
- **Threatened**: A species likely to become endangered in the foreseeable future.

- **Proposed**: Those species named in formal documents published in the Federal Register under the direction of the Endangered Species Act and 50 CFR 402.2 but which have not yet been listed as endangered or threatened.

Three-phase Current—Alternating current in which three separate pulses are present, identical in frequency and voltage, but separated 120 degrees in phase.

Three-phase Line—This is capable of carrying heavy loads of electricity, usually to larger commercial customers.

Three-phase Power—Power generated and transmitted from generator to load on three conductors.

Threshold Concept—The threshold concept refers to the idea that the marginal damage costs are essentially zero at low levels of emissions, and then rise dramatically once emissions exceed a certain level (called the threshold). For many pollutants (including a number of conventional pollutants for which the concept has been applied when developing policy) the concept is incorrect. That is to say that damages gradually rise over a wide range of emissions.

Threshold Limit Value (TLV)—Public health exposure level set by the National Institute for Occupational Safety and Heal for worker safety. It is the level above which a worker should not be exposed for the course of an eight-hour day, due to possible adverse health effects.

Through-put Contract—Parties to a through-put contract commit to ship certain minimum quantities of oil, refined products or gas at a fixed rate through a pipeline. Certain quantities have to be shipped in each period, such as a month or a year, to provide the cash flow to meet operating expenses and debt service of the pipeline company. In the event the product is not shipped and the pipeline company has insufficient cash to meet its expenses and debt service, the parties to the through-put agreement are unconditionally obligated to contribute additional funds in proportion to ownership. The through-put contract is to a take-or-pay contract and serves as an indirect guarantee for project financing of the pipeline. Through-put agreements are also used in connection with processing plants where some product is put through the plant.

Throw Distance—The distance between the luminaire and the center of the plane lit by the luminaire on a display.

Tidal Power—The power available from the rise and fall

of ocean tides. A tidal power plant works on the principal of a dam or barrage that captures water in a basin at the peak of a tidal flow, then directs the water through a hydroelectric turbine as the tide ebbs.

Tidewater Piers and Coastal Ports (method of transportation to consumers)—Shipments moved to tidewater piers and coastal ports for further shipments to consumers via coastal water or ocean.

TIE—Times Interest Earned

Tie Line—A transmission line connecting two or more power systems.

Tilt Angle (of a Solar Collector or Module)—The angle at which a solar collector or module is set to face the sun relative to a horizontal position. The tilt angle can be set or adjusted to maximize seasonal or annual energy collection.

Tilt-Up Tower—A nonclimbable wind generator tower that tilts up and down to allow installation and servicing of the turbine on the ground. Normally these employ a gin pole—a horizontal lever arm that helps raise and lower the tower.

Timber Stand Improvement—Intermediate pruning, weeding, and thinning of a stand of timber prior to its reaching mature rotation age to improve growing conditions and control stand composition.

Timberland—Forest land capable of producing 20 cubic feet of wood per acre per year.

Time Clocks or Timed Switches—Time clocks are automatic controls, which turn lights off and on at predetermined times.

Time Dependent Valuation (TDV) Energy—The time varying energy caused to be used at by the building to provide space conditioning and water heating and for specified buildings lighting, accounting for the energy used at the building site and consumed in producing and in delivering energy to a site, including, but not limited to, power generation, transmission and distribution losses.

Time Value of Money—The basic principle that money can earn interest; therefore, something that is worth $1 today will be worth more in the future if invested. This is also referred to as future value.

Time-of-Day Lock-Out or Limit—A special electric rate feature under which electricity usage is prohibited or restricted to a reduced level at fixed times of the day in return for a reduction in the price per kilowatt hour.

Time-of-Day Pricing—A special electric rate feature under which the price per kilowatt hour depends on the time of day.

Time-of-Day Rate—The rate charged by an electric utility for service to various classes of customers. The rate reflects the different costs of providing the service at different times of the day.

Time-of-Use (TOU) Rates—The pricing of electricity based on the estimated cost of electricity during a particular time block. Time-of-use rates are usually divided into three or four time blocks per twenty-four hour period (on-peak, mid-peak, off-peak and sometimes super off-peak) and by seasons of the year (summer and winter). Real-time pricing differs from TOU rates in that it is based on actual (as opposed to forecasted) prices which may fluctuate many times a day and are weather-sensitive, rather than varying with a fixed schedule.

Time-of-Use Meter—A measuring device that records the times during which a customer uses various amounts of electricity. This type of meter is used for customers who pay time-of-use rates.

Timer—A device that can be set to automatically turn appliances or other devices off and on at set times.

Timer (Water Heater)—This device can automatically turn the heater off at night and on in the morning.

Times Fixed Charges and Preferred Dividends Earned—The ratio of (a) income before interest charges to (b) the sum of interest charges and dividends on preferred stock. Used as a measure of preferred dividend coverage or safety.

Times Fixed Charges Earned Before Income Taxes—The ratio of (a) income before interest charges, adjusted to exclude income taxes, to (b) interest charges (principally interest on long-term debt). Used as a measure of the interest coverage or safety.

Timing Differences—Differences between the periods in which transactions affect taxable income and the periods in which they enter into the determination of pretax accounting income. Timing differences originate in one period and reverse or "turn around" in one or more subsequent periods. Some timing differences reduce income taxes that would otherwise be payable currently; others increase income taxes that would otherwise be payable currently.

TIN—The identifying number assigned to a person under IRC section 6109.

Tin Oxide—A wide band-gap semiconductor similar to indium oxide; used in heterojunction solar cells or to make a transparent conductive film, called NESA glass when deposited on glass.

Tinted or Reflective Glass or Shading Films—Types of glass or a shading film applied to glass that, when installed on the exterior of a building, reduces the rates of solar penetration into the building. Includes Low E Glass.

Tip Speed Ratio—In reference to a wind energy conversion devices blades, the difference between the rotational speed of the tip of the blade and the actual velocity of the wind.

Tipping Fee—Price charged to deliver municipal solid waste to a landfill, waste-to-energy facility, or recycling facility.

Tipple—A central facility used in loading coal for transportation by rail or truck.

TIS—Technical Information Staff
  • Tolerance Index System
  • Traffic Information System

Title—Evidence of a person's right or the extent of his interest in property.

Title 20—California Code of Regulations relating to appliance efficiency. It is also known as the Appliance Energy Efficiency Standards. Title 20 sets minimum efficiency requirements for appliances in the state of California.

Title 24—All of the building standards and associated administrative regulations published in Title 24 of the California Code of Regulations. The Building Energy Efficiency Standards are contained in Part 6. Part 1 contains the administrative regulations for the building standards. Massive changes go into effect in 2014.

Title III—SARA Community Right-to-Know Act *see SARA*

Title IV of the Clean Air Act Amendments of 1990—Sets goals for the electric utility industry to reduce annual sulfur dioxide (SO_2) emissions by 10 million tons and annual nitrogen oxides (NO_x) emissions by 2.0 million tons from 1990 levels by the year 2000. Beginning in the year 2000, total utility SO_2 emissions are then limited to 8.9 million tons and total industrial SO_2 emissions are expected to be 5.6 million tons. Title IV's control of SO_2 emissions instituted two important innovations in U.S. environmental policy. First, it introduced the SO_2 emissions trading program where firms are given permits to release a specified number of tons of SO_2. The government issues only a limited number of permits consistent with the desired level of emissions. The owners of the permits may keep them and release the pollutants, or reduce their emissions and sell the permits. The fact that the permits have value as an item to be sold or traded gives the owner an incentive to reduce their emissions. Second, it established an average annual cap on aggregate emissions by electric utilities. This cap was set at about one-half of the amount emitted in 1980. The emissions cap represents a guarantee that emissions will not increase with economic growth. Title IV used a more traditional approach in setting NO_x emission rate limitations for coal-fired electric utility units, although emission rate-averaging among commonly-owned and operated utilities provides them with some flexibility in compliance. Hence, there is no cap on NO_x emissions, but Title IV is expected to result in a 27 percent reduction of NO_x from 1990 emission levels.

TL—Total losses
  • Transmission line or lines

TLEV (Transitional Low Emission Vehicle)—A vehicle certified by the California Air Resources Board to have emissions from zero to 50,000 miles no higher than 0.125 grams/mile (g/mi) of non-methane organic gases, 3.4 g/mi of carbon monoxide, and 0.4 g/mi of nitrogen oxides. Emissions from 50,000 to 100,000 miles may be slightly higher.

T-Line—Transmission Line

TLV—Threshold Limit Value

TMDL—Total Maximum Daily Limit
  • Total Maximum Daily Load

TNRCC—Texas Natural Resource Conservation Commission www.tceq.state.tx.us

TOC—Trace Organic Analysis
  • Total Organic Carbon
  • Total Organic Compound
  • Total Owning Cost

TOD—The Oil Drum

TOG—Technical Operating Guidance
  • Total Organic Gases

Tolling Arrangement—Contract arrangement under which a raw material or intermediate product stream from one company is delivered to the production facility of another company in exchange for the equivalent volume of finished products and payment of a processing fee.

Toluene ($C_6H_5CH_3$)—Colorless liquid of the aromatic group of petroleum hydrocarbons, made by the catalytic reforming of petroleum naphthas containing methyl cyclohexane. A high-octane gasoline-blending agent, solvent, and chemical intermediate, and a base for TNT (explosive).

Tombstone—An announcement placed in a financial newspaper or journal which announces a financing or performance of some financial service.

Ton (of Air Conditioning)—A unit of air cooling capacity; 12,000 Btu per hour. The approximate amount of energy it takes to melt a ton of ice.

Ton Mile—The product of the distance that freight is hauled, measured in miles, and the weight of the cargo being hauled, measured in tons. Thus, moving one ton for one mile generates one ton mile.

Ton(s) of Air Conditioning—Units used to characterize the cooling capacity of air conditioning equipment. One ton equals 12,000 Btu/hour.

Topping and Back Pressure Turbines—Turbines which operate at exhaust pressure considerably higher than atmospheric (non-condensing turbines). These turbines are often multistage types with relatively high efficiency.

Topping Cycle—A means to increase the thermal efficiency of a steam electric generating system by increasing temperatures and interposing a device, such as a gas turbine, between the heat source and the conventional steam-turbine generator to convert some of the additional heat energy into electricity.

Topping-Cycle Plants—Energy systems which produce electricity first and heat as a by-product.

Torque (Motor)—The turning or twisting force generated by an electrical motor in order for it to operate.

Total AC Load Demand—The sum of the AC loads; its value is important to select the correct Inverter in the case of solar PV.

Total Debt to Tangible Net Worth Ratio—Total debt (including current debt, long-term senior and subordinated debt) ratio/net worth less intangible assets. The measure of the relative amount invested in a company by creditors and owners. A high ratio in comparison to industry norms indicates a greater dependence on outside financing and an unwillingness of the owners to risk their own capital. At some point, the risk of operations will switch to the creditors. Leases should be capitalized in accordance with FAS 13 in computing this ratio.

Total Discoveries—The sum of extensions, new reservoir discoveries in old fields, and new field discoveries, that occurred during the report year.

Total DSM Cost—Total utility and nonutility costs.

Total Gas in Storage—The sum of base gas and working gas.

Total Harmonic Distortion—The measure of closeness in shape between a waveform and its fundamental component.

Total Heat—The sum of the sensible and latent heat in a substance or fluid above a base point, usually 32 degrees Fahrenheit.

Total Incentives—The incentive a utility offers is expressed as a percentage of the technology cost. The utility can assume any level between 0 and 100 percent. A value greater than 100 percent is possible if the utility decides to pay for all the equipment and give a rebate as an additional incentive.

Total Incident Radiation—The total radiation incident on a specific surface area over a time interval.

Total Internal Reflection—The trapping of light within the PV cell by internal reflection of incident light at angles greater than the critical angles for the interfaces, so that the light cannot escape the cell and is therefore eventually absorbed by the semiconductor.

Total Liquid Hydrocarbon Reserves—The sum of crude oil and natural gas liquids reserves volumes.

Total Maximum Daily Load (TDML)—The total daily maximum load on a water shed is defined as the total effluent from all sources running into a given stream. In the context of 1999 US EPA rules, states are responsible for regulating the TDML into the water shed. This term has a special significance because it represents a key step forward in water quality protection from the approach that focused almost exclusively on regulating point sources to one that took account of all sources of pollution.

Total Nonutility Costs—Cash expenditures incurred through participation in a DSM program that are not reimbursed by the utility.

Total Operated Basis—The total reserves or production associated with the wells operated by an individual operator. This is also commonly known as the "gross operated" or "8/8ths" basis.

Total Organic Carbon (TOC)—A measure of the total organic carbon molecules present in a sample.

Total Resource Cost (TRC) Test—A ratio used to assess the cost effectiveness of a demand-side management program. Although this economic desirability test provides information about the relative merits of different DSM programs, several important issues are not addressed in this analysis. First, this cost-effectiveness test does not indicate the level of program participation that will be achieved. Second, the most cost-effective mix of DSM technologies is not determined by this test because this methodology only evaluates one specific measure at a time. Finally, these tests are static; they do not include a feedback mechanism to account for changes in demand due to the DSM program. The TRC Test measures the ratio of total benefits to the costs incurred by both the utility and the participant. The TRC test is applicable to conservation, load management, and fuel substitution technologies. For fuel substitution technologies, the test compares the impact from the fuel not selected to the impact of the fuel that is chosen as a result of implementing the technologies. The TRC Test includes benefits occurring to both participants and nonparticipants. Benefits include avoided supply costs (i.e. transmission, distribution, generation, and capacity costs). Costs include those incurred by both the utility and program participant.

Total Suspended Particulates (TSP)—The quantity of solid particles in a gas or exhaust stream. Any finely divided material (solid or liquid) that is airborne with a diameter smaller than a few hundred micrometers.

Total Utility Costs—Total direct and indirect utility costs.

TOU—Time of Use: A rate for electrical energy that varies with time, generally tracking the cost of generation at the time. The pricing of electricity based on the estimated cost of electricity during a particular time block. TOU Rates are usually divided into three or four time blocks per 24-hour period (on-peak, mid-peak, off-peak and sometimes super off-peak) and by seasons of the year (winter and summer). Real-time pricing differs from TOU Rates in that it is based on actual (as opposed to forecast) prices, which may fluctuate many times a day and are weather-sensitive.

Tower—A steel structure found along transmission lines which is used to support conductors.

Toxic Substances—A chemical or mixture of chemicals that presents a high risk of injury to human health or to the environment.

Toxic Substances Control Act (TSCA)—A federal law of 1976 to regulate chemical substances or mixtures that may present an unreasonable risk of injury to health or the environment.

Toxicity—Ability to harm human health or environment, such as injury, death or cancer. One of the criteria that is used to determine whether a waste is a hazardous waste (the "Toxicity Characteristic").

TP—Technical Product
  • Total Particulates

TPD—Tons per Day

TPH—Tons per Hour

TPY—Tons per Year

TQM—Total Quality Management

TRAC—Terminal Rental Adjustment Clause. A lessee residual guarantee in a vehicle lease.

Trace Gas—A term used to refer to gases found in the Earth's atmosphere other than nitrogen, oxygen, argon and water vapor. When this terminology is used, carbon dioxide, methane, and nitrous oxide are classified as trace gases. Although trace gases taken together make up less than one percent of the atmosphere, carbon dioxide, methane and nitrous oxide are important in the climate system. Water vapor also plays an important role in the climate system; its concentrations in the lower atmosphere vary considerably from essentially zero in cold dry air masses to perhaps 4 percent by volume in humid tropical air masses.

Tracker—A mounting rack for a PV array that automatically tilts to follow the daily path of the sun through the sky. A "tracking array" will produce more energy through the course of the day than a "fixed array" (non-tracking), particularly during the long days of summer. Some trackers are single-axis while others are dual-axis.

Tracking Solar Array—A PV array that is moved to follow the path of the sun in order to maintain the maximum incident solar radiation on its surface. The two most common methods are firstly single-axis tracking in which the array tracks the sun from east to west, and secondly, two-axis tracking in which the array points directly at the sun all the time. Two-axis tracking arrays capture the maximum possible daily energy. Typically, a single axis tracker will give 15% to 25% more power per day, and dual axis tracking will add a further 5%.

Tradable Renewable Certificates (TRC)—(Green-e Program)—A generic term for a bundle of attributes except the actual electrical energy associated with the generation of electricity at a renewable energy facility. Depending upon the facility, the TRC will embody various attributes with varying quantitative values (see 'Attribute,' above). Values—such as avoided emissions—are quantified according to some baseline metric, engineering estimate, or a value deemed by private or government bodies. A renewable 'tag' and a REC are the equivalent of a TRC.

• For purposes of participating in Green-e, eligible renewable generation may be characterized as having a commodity electricity attribute and a renewable attribute, sometimes referred to as a "greenness" attribute. Hence, for each MWh of eligible renewable energy generated, a corresponding MWh of renewable attribute is generated. This renewable attribute may be separately documented form the commodity electricity through the use of a certificate or a "ticket," which the generator may then sell or broker separately from the commodity energy. In general, one TRC is equivalent to the renewable attributes from one MWh of eligible renewable generation.

Trade—A transaction where a buyer and seller exchange a recognized commodity.

Trade Acceptance—A draft drawn by the seller of goods on the buyer and accepted by the buyer for payment at a specified future date.

Tradeable Emission Permits—A permit is an authorization allowing an emitter to emit a specified number of tons of emission, once those tons have been emitted, the permit expires. The total number of permits in any tradable market equals the desired level of emissions sought by the regulating authorities. Tradable permits allow emitters to determine the most economic manner to cover their emissions—by buying permits to cover emissions, taking actions to reduce emissions and selling excess permits, or a combination of those activities.

Trading Day—The 24-hour period beginning at midnight and ending at the following midnight.

Trading Zone—In the RECLAIM program, one of two areas delineated in Rule 2005—New Source Review for RECLAIM, Map 1.

Trailing Edge—The part of a wind energy conversion device blade, or airfoil, that is the last to contact the wind.

Transfer (Electric utility)—To move electric energy from one utility system to another over transmission lines.

Transfer Capability—The overall capacity of interregional or international power lines, together with the associated electrical system facilities, to transfer power and energy from one electrical system to another.

Transfer Price—The monetary value assigned to products, services, or rights conveyed or exchanged between related parties, including those occurring between units of a consolidated entity.

Transferable Renewable Energy Credits (TRECs)—See *Renewable Energy Certificate or Credit*

Transformer—A device, which through electromagnetic induction but without the use of moving parts, transforms alternating or intermittent electric energy in one circuit into energy of similar type in another circuit, commonly with altered values of voltage and current.

Transistor—A semi-conductor device used to switch or otherwise control the flow of electricity.

Transition Costs (Charge)—See *Embedded Costs Exceeding Market Prices.*

Transmaterialization—The process of substituting a service for a product in order to meet customer needs while reducing the use of materials and natural resources.

Transmission—The movement of electricity or natural gas from suppliers to distributors or large customers. These are typically comprised of large, high volume interstate pipelines and high voltage multi-state electric lines.

Transmission and Distribution (T&D) Loss—The losses that result from inherent resistance in electrical conductors and transformation inefficiencies in distribution transformers in a transmission and distribution network.

Transmission Charge—Part of the basic service charges on every customer's bill for transporting electricity from the source of supply to the electric distribution company. Public Utility Commissions regulate retail transmission prices and services. The charge will vary with source of supply.

Transmission Circuit—A conductor used to transport electricity from generating stations to load.

Transmission Company, Gas—A company which obtains at least 90% of its gas operating revenues from sales for resale and/or transportation of gas for others and/or main line sales to industrial customers and classifies at least 90% of its mains (other than service pipe) as field and gathering, storage, and/or transmission.

Transmission Line—A set of conductors, insulators, supporting structures, and associated equipment used to move large quantities of power at high voltage, usually over long distances between a generating or receiving point and major substations or delivery points.

Transmission Network—A system of transmission or distribution lines so cross-connected and operated as to permit multiple power supply to any principal point.

Transmission Owner—An entity that owns transmission facilities or has firm contractual right to use transmission facilities.

Transmission System (Electric)—An interconnected group of electric transmission lines and associated equipment for moving or transferring electric energy in bulk between points of supply and points at which it is transformed for delivery over the distribution system lines to consumers or is delivered to other electric systems.

Transmission Type (Engine)—The transmission is the part of a vehicle that transmits motive force from the engine to the wheels, usually by means of gears for different speeds using either a hydraulic "torque-converter" (automatic) or clutch assembly (manual). On front-wheel drive cars, the transmission is often called a "transaxle." Fuel efficiency is usually higher with manual rather than automatic transmissions, although modern, computer-controlled automatic transmissions can be efficient.

Transmission Upgrade Deferral—The avoided cost of deferred infrastructure on the high-voltage transmission grid.

Transmission-Dependent Utility—A utility that relies on its neighboring utilities to transmit to it the power it buys from its suppliers. A utility without its own generation sources, dependent on another utility's transmission system to get its purchased power supplies.

Transmittance—The time rate of heat flow per unit area under steady conditions from the air (or other fluid) on the warm side of a barrier to the air (or fluid) on the cool side, per unit temperature difference between the two sides.

Transmitting Utility (TRANSCO)—A regulated entity which owns and may construct and maintain wires used to transmit wholesale power. It may or may not handle the power dispatch and coordination functions. It is regulated to provide non-discriminatory connections, comparable service, and cost recovery. According to the Energy Policy Act of 1992, it includes any electric utility, qualifying cogeneration facility, qualifying small power production facility, or Federal power marketing agency which owns or operates electric power transmission facilities which are used for the sale of electric energy at wholesale.

Transom—Horizontal bar dividing a window or opening into two or more segments.

Transparency—A measure of increased accountability and decreased corruption in which a business reports on its ethics and performance results through accessible publication of the business' practices and behavior. There is a strong movement to increase the transparency of business processes via independently-verified corporate responsibility reporting.

Transparent Price—The most recent price contract available to any buyer or seller in the market.

Transport—Movement of natural, synthetic, and/or supplemental gas between points beyond the immediate vicinity of the field or plant from which produced except (1) for movements through well or field lines to a central point for delivery to a pipeline or processing plant within the same state or (2) movements from a citygate point of receipt to consumers through distribution mains.

Transportation Agreement—Any contractual agreement for the transportation of natural and/or supplemental gas between points for a fee.

Transportation Sector—An energy-consuming sector that consists of all vehicles whose primary purpose is transporting people and/or goods from one physical location to another. Included are automobiles; trucks; buses; motorcycles; trains, subways, and other rail vehicles; aircraft; and ships, barges, and other waterborne vehicles. Vehicles whose primary purpose is not transportation (e.g., construction cranes and bulldozers, farming vehicles, and warehouse tractors and forklifts) are classified in the sector of their primary use.

Transported Gas—Natural gas physically delivered to a building by a local utility, but not purchased from that utility. A separate transaction is made to purchase the volume of gas, and the utility is paid for the use of its pipeline to deliver the gas. Also called "Direct-Purchase Gas," "Spot Market Gas," "Spot Gas," "Gas for the Account of Others," and "Self-Help Gas."

Transporter—The party or parties, other than buyer or seller, owning the facilities by which gas or LNG is physically transferred between buyer and seller.

Transshipment—A method of ocean transportation whereby ships off-load their oil cargo to a deep-water terminal, floating storage facility, temporary storage, or to one or more smaller tankers from which or in which the oil is then transported to a market destination.

Trash Rack—A large strainer at the input to a hydro system. Used to remove debris from the water before it enters the pipe.

Traveling Grate—A type of furnace in which assembled links of grates are joined together in a perpetual belt arrangement. Fuel is fed in at one end and ash is discharged at the other.

TRC—Technical Review Committee
• Tradable Renewable Credits

TRC Provider—Any company that sells TRCs to residential, non-residential/commercial, or wholesale customers.

TRD—Technical Review Document

TRE—Toxicity Reduction Evaluation

Treasury Stock—Stock issued by a company but later reacquired. It may be held in the company's treasury indefinitely, reissued to the public, or retired. Treasury stock receives no dividends and does not carry voting power while held by the company.

Treating Plant—A plant designed primarily to remove undesirable impurities from natural gas to render the gas marketable.

Tree Measurement Sale—A type of timber sale contract in which the buyer and seller agree upon the volume at the time of the sale.

Trellis—An architectural feature used to shade exterior walls; usually made of a lattice of metal or wood; often covered by vines to provide additional summertime shading.

TRI—Toxic Release Inventory

Trial Burn—A test of incinerators or boilers and industrial furnaces in which emissions are monitored for the presence of specific substances, such as organic compounds, particulates, criteria pollutants, metals and hydrogen chloride.

Trickle (Solar) Collector—A type of solar thermal collector in which a heat transfer fluid drips out of header pipe at the top of the collector, runs down the collector absorber and into a tray at the bottom where it drains to a storage tank.

Trickle Charge—A small charging current designed to keep a battery fully charged.

Trillion Btu—Equivalent to 1,000,000,000,000 or 10 to the 12th power Btu.

TRIP—Toxic Release Inventory Program

Triple Bottom Line—An expansion of the tradition company reporting framework of net financial gains or losses to take into account environmental and social performance. See *People, Planet, Profit*

Triple Pane (Window)—This represents three layers of glazing in a window with an airspace between the middle glass and the exterior and interior panes.

Triple top Line—A phrase describing a company's improved top-line financial performance over the long term due to sustainable business practices, including less capital investment and increased revenues.

TRIS—Toxic Chemical Release Inventory System

TRO—Temporary Restraining Order

Trombe Wall—A wall with high thermal mass used to store solar energy passively in a solar home. The wall absorbs solar energy and transfers it to the space behind the wall by means of radiation and by convection currents moving through spaces under, in front of, and on top of the wall.

Troposphere—The inner layer of the atmosphere below about 15 kilometers, within which there is normally a steady decrease of temperature with increasing altitude. Nearly all clouds form and weather conditions manifest themselves within this region. Its thermal structure is caused primarily by the heating of the earth's surface by solar radiation, followed by heat transfer through turbulent mixing and convection.

Trough—High-temperature (180°+) concentrator with one axis-tracking.

TRPH—Total Recoverable Petroleum Hydrocarbons

TRS—Total Reduced Sulfur

TRSC—Total Reduced Sulfur Compounds

True Lease—Another term for a tax lease in which, for IRS purposes, the lessor qualifies for the tax benefits of ownership and the lessee is allowed to claim the entire amount of the lease rental as a tax deduction. Must meet FAS 13 guidelines to qualify.

True Power—The actual power rating that is developed by a motor before losses occur.

True South—The direction, at any point on the earth that is geographically in the northern hemisphere, facing toward the South Pole of the earth. Essentially a line extending from the point on the horizon to the highest point that the sun reaches on any day (solar noon) in the sky.

Trunk Line—A main pipeline.

Truss—Combination of timbers or other material to form a frame, placed at intervals, carrying the purlins. As well as a frame, of timber or metal, the term means a project from the face of a wall, or a large console.

Trust—A fiduciary relationship in which a Trustee holds Title to property for the benefit of others.

Trust Indenture—Agreement between the Company, the debt holders, and the trustee for the debt holders. Required for registered offerings of debt securities.

Trustee—The person to whom property is conveyed in Trust.

Trustor—The person who conveys property in Trust.

TRV—Toxicity Reference Value

TS—Total Solids
• Toxic Substance

TS&N—Transmission Safety and Nuisance

TSA—Transportation System Administration

TSCA—Toxic Substances Control Act (1976) Enacted by Congress to give EPA the ability to track the 75,000 industrial chemicals currently produced or imported into the United States. EPA repeatedly screens these chemicals and can require reporting or testing of those that may pose an environmental or human-health hazard. EPA can ban the manufacture and import of those chemicals that pose an unreasonable risk.

TSD—Technical Support Division
- Technical Support Document
- Treatment, Storage and Disposal

TSDR—Treatment, Storage, Disposal and Recycling Facility

TSE—Transmission System Engineering

TSIN—Transmission Services Information Network

TSP—See *Total Suspended Particulates.*

TSS—Technical Services Staff
- Terminal Security System
- Total Suspended (non-filterable) Solids
- Technical Support Staff

T-Statistic—The significance or amount of influence of a particular Independent Variable within a meter Baseline Model. The higher the value, the stronger the indication that the Independent Variable really does have an impact on the utility usage. Normally, the energy analyst looks for a T-Statistic greater than 2.0 as an indication of relevance. If the value is less than 2.0, he or she would conclude that the Independent Variable has no effect on the meter and so would not select that variable to be included in the Baseline Model.

TTLC—Total Threshold Limit Concentration

TTN—Technology Transfer Network

TTO—Total Toxic Organics

Tube (Fluorescent Light)—A fluorescent lamp that has a tubular shape.

Tube-in-Plate-Absorber—A type of solar thermal collector where the heat transfer fluid flows through tubes formed in the absorber plate.

Tube-Type Collector—A type of solar thermal collector that has tubes (pipes) that the heat transfer fluid flows through that are connected to a flat absorber plate.

TUHC—Total Unburned Hydrocarbons

Tungsten Halogen Lamp—A type of incandescent lamp that contains a halogen gas in the bulb, which reduces the filament evaporation rate increasing the lamp life. The high operating temperature and need for special fixtures limits their use to commercial applications and for use in projector lamps and spotlights.

Tunneling—Quantum mechanical concept whereby an electron is found on the opposite side of an insulating barrier without having passed through or around the barrier.

TUR—Toxic Use Reduction

Turbidity—The relative clarity of water, which may be affected by material in suspension in the water.

Turbine—A machine for generating rotary mechanical power from the energy of a stream of fluid (such as water, steam, or hot gas). Turbines convert the kinetic energy of fluids to mechanical energy through the principles of impulse and reaction, or a mixture of the two.

Turbine Generator—A device that uses steam, heated gases, water flow or wind to cause spinning motion that activates electromagnetic forces and generates electricity.

Turgo—In hydroelectric systems, a type of impact hydro runner optimized for lower heads and higher volumes than a Pelton runner.

Turn Down Ratio—The lowest load at which a boiler will operate efficiently as compared to the boiler's maximum design load.

Turn of Logs—A group of logs yarded at the same time by the same machine.

Turnaround Time—The time it takes from beginning to end of a project.

Turnkey System—A system which is built, engineered, and installed to the point of readiness for operation by the owner.

TVOC—Total Volatile Organic Compounds

TWA—Time Weighted Average

Two-Axis Tracking—A solar array tracking system capable of rotating independently about two axes (e.g., vertical and horizontal).

Two-Tank Solar System—A solar thermal system that has one tank for storing solar heated water to preheat the water in a conventional water heater.

TY—Test Year

Type of Drive (Vehicle)—Refers to which wheels the engine power is delivered to, the so-called "drive wheels." Rear-wheel drive has drive wheels on the rear of the vehicle. Front-wheel drive, a newer technology, has drive wheels on the front of the vehicle. Four-wheel drive uses all four wheels as drive wheels and is found mostly on Jeep-like vehicles and trucks, though it is becoming increasingly more common on station wagons and vans.

Typical Usage Profile—Expected energy usage at a meter for a typical year, calculated from the Baseline Model. It does not reflect the impact of current weather, Baseline Adjustments or any implemented Energy Conservation Measures.

TZ—Treatment Zone

U

U.S. Refiner Acquisition Cost of Imported Crude oil— The average price paid by U.S. refiners for imported, that is, non-U.S., crude oil booked into their refineries in accordance with accounting procedures generally accepted and consistently and historically applied by the refiners concerned. The refiner acquisition cost of imported crude oil includes transportation and other fees paid by the refiner.

UA—A measure of the amount of heat that would be transferred through a given surface or enclosure (such as a building envelope) with a one degree Fahrenheit temperature difference between the two sides. The UA is calculated by multiplying the U-Value by the area of the surface (or surfaces).

UAQI—Uniform Air Quality Index

UBC—Uniform Building code

UCC—Ultra Clean Coal
- Uniform Commercial Code

UCC 1—A UCC document filed by a lender informing the public that the filing party legally owns the equipment. The filed document remains in effect for a five year period upon which the document filing will need to be renewed.

UCC 2—A UCC document that amends an existing UCC 1 document.

UCC 3—A UCC document that releases the lien position held by the lender or other secured party.

UCC Financing Statement—A document, under the UCC (Uniform Commercial Code), filed with the county and/or the Secretary of State to provide public notice of a security interest in personal property.

UCL—Upper Control Limit
- Upper Confidence Limit

UCR—Upper Confidence Range

UCS—Union of Concerned Scientists

UDC—Utility Distribution Company. An entity that owns a distribution system for the delivery of energy to and from the ISO-controlled grid, and that provides regulated, retail service to eligible end-use customers who are not yet eligible for direct access, or who choose not to arrange services through another retailer.
- Utility Displacement Credits

UDF—Utility Displacement Factor

UEC—Average Unity of Energy Consumption for each end use

UEDS—Utility External Disconnect Switch

UEG—Utility Electric Generation: In the natural gas industry, a gas-fired electric utility power plant.

UEL—Upper Explosive Limit

UESC—Utility energy services contract

UF—Uncertainty Factor

U-Factor—The quantity of heat transmitted per hour through one square foot of a building section (wall, roof, window, etc.) for each degree Fahrenheit of temperature difference between the air on the warm side and the air on the cold side of the building section.

U-Factor—The overall coefficient of thermal transmittance of a construction assembly, in Btu/(hr. x ft^2 x °F), including air film resistance at both surfaces.

UFC—Uniform Fire Code

UFL—Upper Flammability Limit

UL—Underwriters Laboratories www.ul.com

UL 1598—The Underwriters Laboratories document entitled "Standard for Luminaires," 2000.

UL 181—The Underwriters Laboratories document entitled "Standard for Factory Made Air Ducts and Air Connectors," 1996.

UL 181A—The Underwriters Laboratories document entitled "Standard for Closure Systems for Use With Rigid Air Ducts and Air Connectors," 1994.

UL 181B—The Underwriters Laboratories document entitled "Standard for Closure Systems for Use With Flexible Air Ducts and Air Connectors," 1995.

UL 723—The Underwriters Laboratories document entitled "Standard for Test for Surface Burning Characteristics of Building Materials," 1996.

UL 727—The Underwriters Laboratories document entitled "Standard for Oil-Fired Central Furnaces," 1994.

UL 731—The Underwriters Laboratories document entitled "Standard for Oil-Fired Unit Heaters," 1995.

ULEV (Ultra-Low Emission Vehicle)—A vehicle certified by the California Air Resources Board to have emissions from zero to 50,000 miles no higher than 0.040 grams/mile (g/mi) of non-methane organic gases, 1.7 g/mi of carbon monoxide, and 0.2 g/mi of nitrogen oxides. Emissions from 50,000 to 100,000 miles may be slightly higher.

ULI—Urban Land Institute www.uli.org

ULPA—Uniform Limited Partnership Act, see also the RULPA, Revised Uniform Limited Partnership Act U.L.P.A. § 101 et seq. (1976), as amended in 1985 (R.U.L.P.A.).

ULSD—Ultra-low Sulfur Diesel

Ultimate Analysis—A procedure for determining the primary elements in a substance (carbon, hydrogen, oxygen, nitrogen, sulfur, and ash).

Ultimate Customer—A customer that purchases electricity for its own use and not for resale.

Ultrahigh Voltage Transmission—Transporting electricity over bulk-power lines at voltages greater than 800 kilovolts.

Ultraviolet—Electromagnetic radiation in the wavelength range of 4 to 400 nanometers.
• Zone of invisible radiations beyond the violet end of the spectrum of visible radiations. Since UV wavelengths are shorter than the visible, their photons have more energy, enough to initiate some chemical reactions and to degrade most plastics.

Umbrella Group—Negotiating group within the UNFCCC process comprising the United States, Canada, Japan, Australia, New Zealand, Norway, Iceland, Russia, and Ukraine.

UMG-Si—Upgraded Metallurgical-grade Silicon

UMRA—Unfunded Mandates Reform Act (1995) The Act requires Congress and federal agencies to consider the costs and benefits to state, local and tribal governments and to the private sector before imposing federal requirements that necessitate spending by these governments or the private sector.

UN Framework Convention on Climate Change: (UNFCCC)—A treaty signed at the 1992 Earth Summit in Rio de Janeiro that calls for the "stabilization of greenhouse gas concentrations in the atmosphere at a level that would prevent dangerous anthropogenic interference with the climate system." The treaty includes a non-binding call for developed countries to return their emissions to 1990 levels by the year 2000. The treaty took effect in March 1994 upon ratification by more than 50 countries. The United States was the first industrialized nation to ratify the Convention.

Unaccounted For (Crude Oil)—Represents the arithmetic difference between the calculated supply and the calculated disposition of crude oil. The calculated supply is the sum of crude oil production plus imports minus changes in crude oil stocks. The calculated disposition of crude oil is the sum of crude oil input to refineries, crude oil exports, crude oil burned as fuel, and crude oil losses.

Unaccounted For (Natural Gas)—Represents differences between the sum of the components of natural gas supply and the sum of components of natural gas disposition. These differences may be due to quantities lost or to the effects of data reporting problems. Reporting problems include differences due to the net result of conversions of flow data metered at varying temperatures and pressure

bases and converted to a standard temperature and pressure base; the effect of variations in company accounting and billing practices; differences between billing cycle and calendar-period time frames; and imbalances resulting from the merger of data reporting systems that vary in scope, format, definitions, and type of respondents.

Unbundled—The separation of a service into its distinct elements, e.g., the provisions of natural gas to residential or industrial customers is now split into the transportation of natural gas via pipeline to the utility and the delivery of the gas from the utility to the user.

Unbundling—Disaggregating electric or gas utility service into its basic components and offering each component separately for sale with separate rates for each component. For example, generation, transmission and distribution could be unbundled and offered as discrete services.

Uncertainty—Uncertainty is a prominent feature of the benefits and costs of climate change. Decision makers need to compare risk of premature or unnecessary actions with risk of failing to take actions that subsequently prove to be warranted. This is complicated by potential irreversibilities in climate impacts and long-term investments.

Uncompleted Wells, Equipment, and Facilities Costs—The costs incurred to (1) drill and equip wells that are not yet completed, and (2) acquire or construct equipment and facilities that are not yet completed and installed.

Unconditioned Space—A space that is neither directly nor indirectly conditioned space, which can be isolated from conditioned space by partitions and/or closeable doors. [See California Code of Regulations, Title 24, Section 2-5302]
• Enclosed space within a building that is not directly conditioned or indirectly conditioned.

Unconsolidated Entity—A firm directly or indirectly controlled by a parent but not consolidated with the parent for purposes of financial statements prepared in accordance with generally accepted accounting principles. An unconsolidated entity includes any firm consolidated with the unconsolidated entity for purposes of financial statements prepared in accordance with generally accepted accounting principles historically and consistently applied. An individual shall be deemed to control a firm that is directly or indirectly controlled by him or by his father, mother, spouse, children, or grandchildren.

Unconventional Fuels Tax Credit—An incentive tax credit applying to a variety of more costly energy production including, for natural gas, coalbed methane, tight sands, and Devonian shale production.

Unconventional Gas—Natural gas that can not be economically produced using current technology.

Underground Feeder (UF)—May be used for photovoltaic array wiring if sunlight resistant coating is specified; can be used for interconnecting balance-of-system components but not recommended for use within battery enclosures.

Underground Gas Storage—The use of sub-surface facilities for storing gas that has been transferred from its original location. The facilities are usually hollowed-out salt domes, geological reservoirs (depleted oil or gas fields) or water-bearing sands topped by an impermeable cap rock (aquifer).

Underground Gas Storage Reservoir Capacity—Interstate company reservoir capacities are those certificated by the Federal Energy Regulatory Commission. Independent producer and intrastate company reservoir capacities are reported as developed capacity.

Underground Home—A house built into the ground or slope of a hill, or which has most or all exterior surfaces covered with earth.

Underground Mine—A mine where coal is produced by tunneling into the earth to the coal bed, which is then mined with underground mining equipment such as cutting machines and continuous, longwall, and shortwall mining machines. Underground mines are classified according to the type of opening used to reach the coal, i.e., drift (level tunnel), slope (inclined tunnel), or shaft (vertical tunnel).

Underground Service Entrance (USE)—May be used within battery enclosures and for interconnecting balance-of-systems.

Underground Storage—The storage of natural gas in underground reservoirs at a different location from which it was produced.

Underground Storage Injections—Gas from extraneous sources put into underground storage reservoirs.

Underground Storage Withdrawals—Gas removed from underground storage reservoirs.

Underlying Asset—An asset that is the subject of lease for which a right to use that asset has been conveyed to a lessee. The underlying asset could be a physically distinct portion of a single asset.

Understory—The trees and other woody species growing under a relatively continuous cover of branches and foliage formed by the overstory trees.

Underwriter—In jargon, the "underwriters" are the investment banks selling the securities in an underwritten registered offering. But beware, under the Securities Act, the class of persons who are considered "underwriters" is far more expansive and problematic.

Underwritten Offering—Registered offering that is sold through a consortium of investment banks assembled by one or more lead investment banks.

Undeveloped Property—Refers to a mineral property on which development wells or mines have not been drilled or completed to a point that would permit the production of commercial quantities of mineral reserves.

Undifferentiated/Unspecified Reserves and Production—Reserves and production that are not separable by FERC production areas or by states. Undifferentiated and unspecified reserves consist only of company-owned gas in underground storage.

Undiscovered Recoverable Reserves (Crude Oil and Natural Gas)—Those economic resources of crude oil and natural gas, yet undiscovered, that are estimated to exist in favorable geologic settings.

Undiscovered Resources (Coal)—Unspecified bodies of coal surmised to exist on the basis of broad geologic knowledge and theory. Undiscovered resources include beds of bituminous coal and anthracite 14 inches or more thick and beds of sub-bituminous coal and lignite 30 inches or more thick that are presumed to occur in unmapped and unexplored areas to depths of 6,000 feet. The speculative and hypothetical resource categories comprise undiscovered resources.

Unearned Income—The portion of income from a lease or loan that must be earned over the life of the contract in accordance with GAAP.

UNEP—United Nations Environmental Program

Uneven Rent Test (Tax)—The IRS test of the smoothness of the rents in a stepped or sawtooth lease.

UNFCCC—United Nations Framework Convention on Climate Change Also referred to informally as the UN climate change convention. It is the international agreement for action on climate change and was drawn up in 1992. A framework was agreed for action aimed at stabilizing atmospheric concentrations of greenhouse gases. The UNFCCC entered into force on March 1994 and currently has 192 signatory parties. The UNFCCC in turn agreed the Kyoto Protocol in 1997 to implement emission reductions in industrialized countries up to 2012 and is currently seeking the negotiation of a new treaty to extend commitments beyond 2012.

Unfilled Requirements—Requirements not covered by usage of inventory or supply contracts in existence as of January 1 of the survey year.

Unfinished Oils—All oils requiring further processing, except those requiring only mechanical blending. Unfinished oils are produced by partial refining of crude oil and include naphthas and lighter oils, kerosene and light gas oils, heavy gas oils, and Residuum.

Un-Fractionated Streams—Mixtures of un-segregated natural gas liquid components, excluding those in plant condensate. This product is extracted from natural gas.

Unglazed Solar Collector—A solar thermal collector that has an absorber that does not have a glazed covering. Solar swimming pool heater systems usually use unglazed collectors because they circulate relatively large volumes of water through the collector and capture nearly 80 percent of the solar energy available.

Unguaranteed Residual Value—The portion of residual value in a lease transaction for which the lessor is at-risk. The lessor takes on the risk that the equipment may or may not be worth this expected value at the end of the lease term.

Uniform Building Code (UBC)—A building code published by the International Conference of Building Officials, adopted and amended by Oregon Department of Commerce. The UBC covers the fire, life and structural safety aspects of all buildings and related structures.

Uniform Commercial Code (UCC)—A set of standard rules, adopted by 49 states, that governs commercial transactions.

Uniform Mechanical Code (UMC)—A code sponsored by the International Association of Plumbing and Mechanical Officials and the International Conference of Building Officials, adopted and amended by the Oregon Department of Commerce. The UMC contains requirements for the installation and maintenance of heating, ventilating, cooling, and refrigeration systems.

Uniform System of Accounts—Prescribed financial rules and regulations established by the Federal Energy Regulatory Commission for utilities subject to its jurisdiction under the authority granted by the Federal Power Act.

- A list of a company's account numbers and corresponding account titles, together with specific instructions for the use of individual accounts and general instructions as to the basis of accounting. For utilities, Uniform Systems of Accounts have been issued by both the National Association of Regulatory Utility Commissioners (NARUC) and the Federal Energy Regulatory Commission (FERC). These accounts differ between various utilities, i.e., gas versus electric.

Unilateral CDM Projects—Unilateral CDM projects which do not have a project investor from abroad.

Uninterruptible Power Supply (UPS)—A power supply capable of providing continuous uninterruptible service; normally containing batteries to provide energy storage.

Unit (Fuel or Pulp)—A bulk measure of hog fuel or pulp chips containing 200 cubic feet. A unit contains varying amounts of solid material depending on the amount of compaction. It is customary to weigh material, correct for moisture, and calculate the number of bone dry tons.

Unit (Timber)—A timber harvest area with defined boundaries. The area is managed as a unit for harvest and subsequent management activities.

Unit Energy Consumption (UEC)—The annual amount of energy that is used by the electrical device or appliance.

Unit Interior Mass Capacity (UIMC)—The amount of effective heat capacity per unit of thermal mass, taking into account the type of mass material, thickness, specific heat, density and surface area. *See also Thermal Mass*

Unit Offering—Private or public offering of securities in groups of more than one security. Most often a share of stock and warrant to purchase some number of shares of stock, but could be two shares of stock, a note and a share of stock, etc. Also used in some cases to refer to the sale of LP and LLC interests, since those interests are composed of more than one right.

Unit Price—Total revenue derived from the sale of product during the reference month divided by the total volume sold; also known as the weighted average price. Total revenue should exclude all taxes but include transportation costs that were paid as part of the purchase price.

Unit Value, Consumption—Total price per specified unit, including all taxes, at the point of consumption.

Unit Value, Wellhead—The wellhead sales price, including charges for natural gas plant liquids subsequently removed from the gas; gathering and compression charges; and state production, severance, and/or similar charges.

Unitary Air Conditioner—An air conditioner consisting of one or more assemblies that move, clean, cool, and dehumidify air.

United Nations Framework Convention on Climate Change (UNFCCC)—The centerpiece of global efforts to combat global warming. It was adopted in June 1992 at the Rio Earth Summit, and entered into force on March 21, 1998. The Convention's primary objective is the "stabilization of greenhouse gas concentrations in the atmosphere at a level that would prevent dangerous anthropogenic (man-made) interference with the climate system. Such a level should be achieved within a time-frame sufficient to allow ecosystems to adapt naturally to climate change, to ensure that food production is not threatened, and to enable economic development to proceed in a sustainable manner."

United Nations Global Compact—An international initiative that seeks to bring businesses together voluntarily in order to promote socially and environmentally responsible practices. Signatories pledge to uphold the Compact's 10 Principles:

- Businesses should support and respect the protection of internationally proclaimed human rights;
- Make sure that they are not complicit in human rights abuses;
- Businesses should uphold the freedom of association and the effective recognition of the right to collective bargaining;
- The elimination of all forms of forced and compulsory labor;
- The effective abolition of child labor;
- The elimination of discrimination in respect of employment and occupation;
- Businesses should support a precautionary approach to environmental challenges;
- Undertake initiatives to promote greater environmental responsibility;
- Encourage the development and diffusion of environmentally friendly technologies; and
- Businesses should work against corruption in all its forms, including extortion and bribery.

United States—The 50 States and the District of Columbia. Note: The United States has varying degrees of jurisdiction over a number of territories and other political entities outside the 50 States and the District of Columbia, including Puerto Rico, the U.S. Virgin Islands, Guam, American Samoa, Johnston Atoll, Midway Islands, Wake Island, and the Northern Mariana Islands. EIA data programs may include data from some or all of these areas in U.S. totals. For these programs, data products will contain notes explaining the extent of geographic coverage included under the term "United States."

United States Business Council on Sustainable Development (USBCSD)—A non-profit organization promoting sustainable development by establishing networks and partnerships between American companies and government entities. The USBCSD provides a voice for industry and is the U.S. branch of the World Business Council of Sustainable Development.

Unitization—A term used in connection with Continuing Property Record Unit. Unitization is the process of assigning work order costs to applicable property record units.

Universal Service—Electric service sufficient for basic needs (an evolving bundle of basic services) available to virtually all members of the population regardless of income.

Unlawful Detainer—A legal action to recover possession of property.

Unleaded Gasoline—Contains not more than 0.05 gram of lead per gallon and not more than 0.005 gram of phosphorus per gallon. Premium regular and intermediate grades are included, depending on the octane rating.

Un-Merchantable Wood—Material which is unsuitable for conversion to wood products due to poor size, form, or quality.

Unprocessed Gas—Natural gas that has not gone through a processing plant.

Unscheduled Outage Service—Power received by a system from another system to replace power from a generating unit forced out of service.

Unspecified Source of Power—Electricity generation that cannot be matched to a particular generating facility. Unspecified sources of power include power purchases from entities that own fleets of generating facilities such as independent power producers, electric retail providers, and federal power agencies. Unspecified sources of power also refer to power purchased from power marketers, brokers and markets.

Unsuitable Areas—Areas not appropriate for timber harvest due to fragile or shallow soils, scenic values, special wildlife habitat areas, and riparian or wetland values, among other possible reasons.

Unvented Heater—A combustion heating appliance that vents the combustion by-products directly into the heated space. The latest models have oxygen-sensors that shut off the unit when the oxygen level in the room falls below a safe level.

UORA—Used Oil Recycling Act (1989)

Upgrade (Electric Utility)—Replacement or addition of electrical equipment resulting in increased generation or transmission capability.

Upgradient—The direction from which water flows in an aquifer. In particular, areas that are higher than contaminated areas and, therefore, are not prone to contamination by the movement of polluted groundwater.

Uprate (Electric Utility)—An increase in the rating or stated measure of generation or transfer capability.

UPS—Uninterruptible Power System. A system that always provides continuous power regardless of what is happening on the grid. Typically in the form of a battery package. Frequently used with computer systems to keep essentially data and software safe from power outages.

Upstream—Refers to the point (or close to it) where fossil fuels enter the economy. In the U.S., it means at the input to oil refineries, at coal processing plants and where natural gas enters pipelines. Conversely, downstream refers to any point in the economy, and in particular, at the level of energy consumers rather than suppliers. It is commonly interpreted to be industrial boilers, electric utilities and other major energy users, but also applies, in theory, to all consumers of gasoline, coal, electricity etc.

Upwind—In relation to a wind turbine, toward the wind. An upwind turbine has its blades on the upwind side of the tower.

UR—Uptake Rate

Uranium (U)—A heavy, naturally radioactive, metallic element (atomic number 92). Its two principally occurring isotopes are uranium-235 and uranium-238. Uranium-235 is indispensable to the nuclear industry because it is the only isotope existing in nature, to any appreciable extent, that is fissionable by thermal neutrons. Uranium-238 is also important because it absorbs neutrons to produce a radioactive isotope that subsequently decays to the isotope plutonium-239, which also is fissionable by thermal neutrons.

Uranium Concentrate—A yellow or brown powder obtained by the milling of uranium ore, processing of in situ leach mining solutions, or as a byproduct of phosphoric acid production.

Uranium Deposit—A discrete concentration of uranium mineralization that is of possible economic interest.

Uranium Endowment—The uranium that is estimated to occur in rock with a grade of at least 0.01 percent U_3O_8. The estimate of the uranium endowment is made before consideration of economic availability of any associated uranium resources.

Uranium Enrichment—The process of increasing the percentage of pure uranium above the levels found in naturally occurring uranium ore, so that it may be used as fuel.

Uranium Hexafluoride (UF6)—A white solid obtained by chemical treatment of U_3O_8 and which forms a vapor at temperatures above 56 degrees Centigrade. UF_6 is the form of uranium required for the enrichment process.

Uranium Importation—The actual physical movement of uranium from a location outside the United States to a location inside the United States.

Uranium Mill—A plant where uranium is separated from ore taken from mines.

Uranium Mill Tailings—The sand-like materials left over from the separation of uranium from its ore. More than 99 percent of the ore becomes tailings.

Uranium Mill Tailings Radiation Control Act (UMTRA) of 1978—The act that directed the Department of Energy to provide for stabilization and control of the uranium mill tailings from inactive sites in a safe and environmentally sound manner to minimize radiation health hazards to the public. It authorized the Department to undertake remedial actions at 24 designated inactive uranium-processing sites and at an estimated 5,048 vicinity properties.

Uranium Ore—Rock containing uranium mineralization in concentrations that can be mined economically, typically one to four pounds of U_3O_8 per ton or 0.05 percent to 0.2 percent U_3O_8.

Uranium Oxide—Uranium concentrate or yellowcake. Abbreviated as U_3O_8.

Uranium Property—A specific piece of land with uranium reserves that is held for the ultimate purpose of economically recovering the uranium. The land can be developed for production or undeveloped.

Uranium Reserves—Estimated quantities of uranium in known mineral deposits of such size, grade, and configuration that the uranium could be recovered at or below a specified production cost with currently proven mining and processing technology and under current law and regulations. Reserves are based on direct radiometric and chemical measurements of drill holes and other types of sampling of the deposits. Mineral grades and thickness, spatial relationships, depths below the surface, mining and reclamation methods, distances to milling facilities, and amenability of ores to processing are considered in the evaluation. The amount of uranium in ore that could be exploited within the chosen forward-cost levels are estimated in accordance with conventional engineering practices.

Uranium Resource Categories (International)—Three categories of uranium resources defined by the international community to reflect differing levels of confidence in the existence of the resources. Reasonably assured resources (RAR), estimated additional resources (EAR), and speculative resources (SR) are described below:

- Reasonably assured resources (RAR): Uranium that occurs in known mineral deposits of such size, grade, and configuration that it could be recovered within the given production cost ranges, with currently proven mining and processing technology. Estimates of tonnage and grade are based on specific sample data and measurements of the deposits and on knowledge of deposit characteristics. Note: RAR corresponds to DOE's uranium reserves category.

- Estimated additional resources (EAR): Uranium in addition to RAR that is expected to occur, mostly on the basis of geological evidence, in extensions of well-explored deposits, in little-explored deposits, and in undiscovered deposits believed to exist along well-defined geological trends with known deposits. This uranium can subsequently be recovered within the given cost ranges. Estimates of tonnage and grade are based on available sampling data and on knowledge of the deposit characteristics, as determined in the best-known parts of the deposit or in similar deposits. Note: EAR corresponds to DOE's probable potential resources category.

- Speculative resources (SR): Uranium in addition to EAR that is thought to exist, mostly on the basis of indirect evidence and geological extrapolations, in deposits discoverable with existing exploration techniques. The location of deposits in this category can generally be specified only as being somewhere within given regions or geological trends. The estimates in this category are less reliable than estimates of RAR

and EAR. Note: SR corresponds to the combination of DOE's possible potential resources and speculative potential resources categories.

Urban Growth Boundary—A land use boundary surrounding a city within which urban land uses are allowed.

Urban Heat Island (UHI)—Refers to the tendency for urban areas to have warmer air temperatures than the surrounding rural landscape, due to the low albedo of streets, sidewalks, parking lots, and buildings. These surfaces absorb solar radiation during the day and release it at night, resulting in higher night temperatures.

URESC—Utility renewable electricity service contract

Urethanes—A family of plastics (polyurethanes) used for varnish coatings, foamed insulations, highly durable paints, and rubber goods.

URF—Inhalation Unit Risk Factor

URR—Ultimately Recoverable Resources

USACE—United States Army Corp of Engineers

Usage Agreement—Contracts held by enrichment customers that allow feed material to be stored at the enrichment plant site in advance of need.

USAO—United States Attorney's Office

USBM—United States Bureau of Mines

USBS—United States Bureau of Standards

USC—United States Code

USDA—United States Department of Agriculture

USDOE—United States Department of Energy

USDOI—United States Department of the Interior

USDOT—United States Department of Transportation

USDW—Underground Source(s) of Drinking Water

Use-Case—A document that describes a problem being solved by a particular storage system in a particular location with a clear operating regime, governance scheme, and identified stream of benefits, etc.

Used and Useful—A concept used by regulators to determine whether an asset should be included in the utility's rate base. This concept requires that an asset currently provide or be capable of providing a needed service to customers.

Useful Heat—Heat stored above room temperature (in a solar heating system).

Useful Life—A period of time during which an asset has economic value and is usable. The useful life of an asset sometimes is called the economic life of the asset.

Useful Thermal Output—The thermal energy made available in a combined-heat-and-power system for use in any industrial or commercial process, heating or cooling application, or delivered to other end users, i.e., total thermal energy made available for processes and applications other than electrical generation.

USEPA—United States Environmental Protection Agency

USFS—United States Department of Agriculture Forest Service.

USFWS—United States Fish and Wildlife Service.

USGBC (US Green Building Council)—Composed of more than 15,000 organizations from across the building industry, USGBC is a nonprofit organization committed to expanding sustainable building practices. The organization developed the LEED rating system for developing high-performance, sustainable buildings and also provides educational programs on green design, construction, and operations for professionals throughout the building industry, USGBC hosts Greenbuild, an international conference and expo focused on green building. www.usgbc.org

USGS—United States Geological Survey

USNRC—United States Nuclear Regulatory Commission

USPHS—United States Public Health Service

UST—Underground Storage Tank

Usuary—The charge of a greater rate of interest for the loan of money than is permitted by law.

Usury Law—Laws regulating the charging of interest rates. Most usury laws protect consumers from unauthorized interest rates.

Utility—A regulated entity which exhibits the characteristics of a natural monopoly (also referred to as a power provider). For the purposes of electric industry restructuring, "utility" refers to the regulated, vertically integrated electric company. "Transmission utility" refers to the regulated owner/operator of the transmission system only. "Distribution utility" refers to the regulated owner/operator of the distribution system which serves retail customers.

Utility Demand-Side Management Costs—The costs incurred by the utility to achieve the capacity and energy savings from the Demand-Side Management (DSM) Program. Costs incurred by consumers or third parties are to be excluded. The costs are to be reported in nominal dollars in the year in which they are incurred, regardless of when the savings occur. The utility costs are all the annual expenses (labor, administrative, equipment, incentives, marketing, monitoring and evaluation, and other) incurred by the utility for operation of the DSM Program, regardless of whether the costs are expensed or capitalized. Lump-sum capital costs (typically accrued over several years prior to start up) are not to be reported. Program costs associated with strategic load growth activities are also to be excluded.

Utility Distribution Companies (UDC)—The entities that will continue to provide regulated services for the distribution of electricity to customers and serve customers who do not choose direct access. Regardless of where a consumer chooses to purchase power, the customer's current utility, also known as the utility distribution company, will deliver the power to the consumer.

Utility Financing—Grants, loans, and resource acquisition payments provided by utilities for energy conservation.

Utility Gases—Natural gas, manufactured gas, synthetic gas, liquefied petroleum gas-air mixture, or mixtures of any of these gases.

Utility Generation—Generation by electric systems engaged in selling electric energy to the public.

Utility Green Pricing—A utility offers its customers a choice of power products, usually at differing prices, offering varying degrees of renewable energy content. The utility guarantees to generate or purchase enough renewable energy to meet the needs of all green power customers.

Utility Plant—Includes Plant: In service, Purchased or Sold, In Process of Reclassification, Leased to Others, Held for Future Use, Completed Construction Not Classified, Construction Work in Progress, Plant Acquisition Adjustments and Other Utility Plant. The Uniform System of Accounts prescribes for the deduction of Accumulated Provision for Depreciation and Amortization.

Utility Plant in Service—That portion of a utility's plant which is devoted to the operations of the company. Excludes plant: purchased or sold, in process of reclassification, leased to others, held for future use, under construction, and acquisition adjustments and adjustment accounts, and without deduction of Accumulated Provision for Depreciation and Amortization.

Utility, Gas—A company that is primarily a distributor of natural gas to ultimate customers in a given geographic area.

Utility-Earned Incentive—Costs paid to a utility for achieving consumer participation in DSM programs.

Utility-Interactive Inverter—An inverter that can operate only when connected to the utility grid supply and an output voltage frequency fully synchronized with the utility power.

Utility-Sponsored Conservation Program—Any program sponsored by an electric and/or natural gas utility to review equipment and construction features in buildings and advise on ways to increase the energy efficiency of buildings. Also included are utility-sponsored programs to encourage the use of more energy-efficient equipment. Included are programs to improve the energy efficiency in the lighting system or building equipment or the thermal efficiency of the building shell.

Utilization Factor—The ratio of the maximum demand of a system or part of a system to the rated capacity of the system or part of the system.

UTS—Universal Treatment Standards

U-Value or U-Factor—A measure of how well heat is transferred by the entire window—the frame, sash and glass—either into or out of the building. U-value is the opposite of R-value. The lower the U-factor number, the better the window will keep heat inside a home on a cold day. The U-Value measures the rate at which heat flows or conducts through a building assembly (wall, floor, ceiling, etc.). The smaller the U-Value, the more energy-efficient an assembly and the slower the heat transfer. Window performance labels include U-Values (calling them U-factors) to help make comparisons of energy efficiency across window products.

V

V$_{ac}$—Volts alternating current

Vacuum—A pressure less than atmospheric pressure, measured either from the base of zero pressure or from the base of atmospheric pressure.

Vacuum Distillation—Distillation under reduced pressure (less the atmospheric), which lowers the boiling temperature of the liquid being distilled. This technique with its relatively low temperatures prevents cracking or decomposition of the charge stock.

Vacuum Evaporation—The deposition of thin films of semiconductor material by the evaporation of elemental sources in a vacuum.

Vacuum Zero—The energy of an electron at rest in empty space; used as a reference level in energy band diagrams.

Valence Band—The highest energy band in a semiconductor that can be filled with electrons.

Valence Level Energy/Valence State—Energy content of an electron in orbit about an atomic nucleus. Also called bound state.

Validation—The stage in carbon offset project development where an independent third-party audits a project's design to ensure it meets the rules of a prescribed standard, such as the CDM. Checks include whether emissions reductions and other benefits are real and permanent compared to a business-as-usual baseline.
 • The process of independent evaluation of a project activity by an accredited independent entity.

Valley Filing—Valley filling is a form of load management that increases or builds, off-peak loads. This load shape objective is desirable if a utility has surplus capacity in the off-peak hours. If this strategy is combined with time-or-use rates, the average rate for electricity can be lowered.

Valley Segment—That portion of a stream network with similar morphologies and governing geomorphic processes identified by valley bottom and sideslope geomorphic characteristics.

Value (of shipments)—The value received for the complete systems at the company's net billing price, freight-on-board factory, including charges for cooperative advertising and warranties. This does not include excise taxes, freight or transportation charges, or installation charges.

Value Added by Manufacture—A measure of manufacturing activity that is derived by subtracting the cost of materials (which covers materials, supplies, containers, fuel, purchased electricity, and contract work) from the value of shipments. This difference is then adjusted by the net change in finished goods and work-in-progress between the beginning- and end-of-year inventories.

Value Engineering—Substituting building types, systems, materials or finishes that reduce costs without compromising objectives and needs. The process exposes potentially hidden building costs that may not have been anticipated for the building's operations.

Value of Service—A monetary measure of the usefulness or necessity of utility service to a customer group.
 • The concept that the value of a utility service to a consumer cannot be greater than the cost of an equally satisfactory substitute service or the consumer will switch to the substitute.

Value of Service Pricing—A method of apportioning costs among utility customers so that users who place a greater value on the service are charged higher rates than the more price sensitive customers.

Value-added Tax (VAT)—A tax assessed on the increased value of goods as they pass from the raw material stage through the production process to final consumption. The tax on processors or merchants is levied on the amount by which they increase the value of items they purchase.

Value-at-Risk (VaR)—This measures a portfolio's largest likely loss over a given period of time with a given probability. The time period is called the holding period and the probability is known as the confidence interval.

Vanadium—A toxic metal that is both mined and is a by-product of petroleum refining. Compounds of vanadium are used in the steel industry, as a catalyst in the chemical industry, in photography and in insecticides.

Vapor—The gaseous state of a substance as distinguished from permanent gases. A gaseous fluid may be classified as either a vapor or a gas. If it is near the region of condensation, it is called a vapor. If it is well above the region of condensation, it is called a gas. Vapors in general do not follow the ideal gas law, and engineers prefer to use tables and charts based on experimental data when working with vapors. Gases, however, may obey the ideal gas laws over a wide range of temperature and pressure.

Vapor Barrier—A material with a permeance of one perm or less which provides resistance to the transmission of water vapor. [See California Code of Regulations, Title 24, Section 2-5302] A material that retards the movement of water vapor through a building element (walls, ceilings) and prevents insulation and structural wood from becoming damp and metals from corroding. It is often applied to insulation batts or separately in the form of treated papers, plastic sheets and metallic foils.

Vapor Control Layer—The component (or components) installed in an assembly to control the movement of water by vapor diffusion.

Vapor Displacement—The release of vapors that had previously occupied space above liquid fuels stored in tanks. These releases occur when tanks are emptied and filled.

Vapor Retarder—A material that retards the movement of water vapor through a building element (walls, ceilings) and prevents insulation and structural wood from becoming damp and metals from cor-

roding. Often applied to insulation batts or separately in the form of treated papers, plastic sheets, and metallic foils.

Vapor-Dominated Geothermal System—A conceptual model of a hydrothermal system where steam pervades the rock and is the pressure-controlling fluid phase.

Variable Air Volume (VAV) System on the Heating and Cooling System—A means of varying the amount of conditioned air to a space. A variable air volume system maintains the air flow at a constant temperature, but supplies varying quantities of conditioned air in different parts of the building according to the heating and cooling needs.

Variable Air Volume System—A VAV System is a mechanical HVAC system capable of serving multiple zones. It maintains control over the temperature in a zone by controlling the amount of heated or cooled air supplied to the zone.

Variable Cost—Costs, such as fuel costs, that depend upon the amount of electric energy supplied.

Variable Interest Rate—Interest rate charged under a financing transaction that is subject to upward and downward adjustment during the contract term.

Variable Lease Payments—Payments made by a lessee to a lessor for the right to use an underlying asset that vary because of changes in facts or circumstances occurring after the commencement date, other than the passage of time.

Variable Price—Prices that vary frequently. Prices that are not stable.

Variable Speed Drive—A means of changing the speed of a motor in a step-less manner. In the case of an AC motor, this is accomplished by varying the frequency. A Variable Speed Drive is also called an ASD (adjustable speed drive) or VSD (variable frequency drive).

Variable Term Lease—A lease that ends when all payments are made. The term is flexible to accommodate skips and increased rents.

Variable-Energy Resources (VER)—The electrical output of some renewable energy technologies (esp. wind and solar) may vary over time or exhibit intermittency.

Variable-Speed Wind Turbines—Turbines in which the rotor speed increases and decreases with changing wind speed, producing electricity with a variable frequency.

Variance—The amount by which actual utility usage differs from target utility usage. Variance is the difference between target energy usage and actual energy usage. A positive variance indicates energy usage is below expectations. A negative variance would mean that energy usage is more than expected.

Varistor—A non-ohmic or voltage-dependent variable resistor. Normally used as over-voltage limiters to protect sensitive equipment from power spikes or lightning strikes by shunting the energy to ground.

VAT—Value Added Tax. Tax levied on goods at each stage of manufacturing and transfer that takes into account the value added to products.

Vault—An enclosed room or pit having an access opening in the top, side wall, or both. May be in a building, a separate above-ground structure, or underground.

VAV System (Variable Air Volume System)—A mechanical HVAC system capable of serving multiple zones that controls the temperature maintained in a zone by controlling the amount of heated or cooled air supplied to the zone. An HVAC system that controls temperature within a space by varying the volumetric flow of heated or cooled supply air to the space. Normally, the supply air from the air handler is cooled and passes through a thermostatically controlled VAV box (volume control box) before entering the space. The VAV box will have a minimum position setting to ensure adequate ventilation air at all times. It may be equipped with a reheat coil for those times when heat from people, lighting and equipment in the space will not be adequate.

VC—Venture Capital

VCU—Voluntary Carbon Unit. The name of carbon offset credits specifically verified to the Voluntary Carbon Standard, one of the leading independent standards established to demonstrate integrity in project-based emission reductions in the unregulated voluntary carbon market.

V_{dc}—Volts DC

VE—Visual Emissions

Vector-Borne Disease—Disease that results from an infection transmitted to humans and other animals by blood-feeding anthropods, such as mosquitoes, ticks, and fleas. Examples of vector-borne diseases include Dengue fever, viral encephalitis, Lyme disease, and malaria.

Vehicle Fuel Consumption—Vehicle fuel consumption is computed as the vehicle miles traveled divided by the fuel efficiency reported in miles per gallon (MPG). Vehicle fuel consumption is derived from the actual vehicle mileage collected and the assigned MPGs obtained from EPA certification files adjusted for on-road driving. The quantity of fuel used by vehicles.

Vehicle Fuel Expenditures—The cost, including taxes, of the gasoline, gasohol, or diesel fuel added to the vehicle's tank. Expenditures do not include the cost of oil or other items that may have been purchased at the same time as the vehicle fuel.

Vehicle Identification Number (VIN)—A set of codes, usually alphanumeric characters, assigned to a vehicle at the factory and inscribed on the vehicle. When decoded, the VIN provides vehicle characteristics. The VIN is used to help match vehicles to the EPA certification file for calculating MPGs.

Vehicle Importer—An original vehicle manufacturer (of foreign or domestic ownership) that imports vehicles as finished products into the United States.

Vehicle Miles Traveled (VMT)—The number of miles traveled nationally by vehicles for a period of 1 year. VMT is either calculated using two odometer readings or, for vehicles with less than two odometer readings, imputed using a regression estimate.

Vehicle-to-Grid (a.k.a. V2G)—The use of batteries that power plug-in electric vehicles (PEVs) as storage media capable of providing electrical services to the grid.

Vendee—Purchaser of property, goods, or services.

Vendor—A seller of property, goods, or services.

Vendor Financing—Financing provided by an equipment supplier, equipment manufacturer, company or contractor.

Vent—A component of a heating or ventilation appliance used to conduct fresh air into, or waste air or combustion gases out of, an appliance or interior space.

Vent Damper—A device mounted in the vent connector that closes the vent when the heating unit is not firing. This traps heat inside the heating system and house rather than letting it draft up and out the vent system.

Vent Pipe—A tube in which combustion gases from a combustion appliance are vented out of the appliance to the outdoors.

Vented—Gas released into the air on the production site or at processing plants.

Vented Cell—A battery with a vent to expel gases liberated during charging.

Vented Heater—A type of combustion heating appliance in which the combustion gases are vented to the outside, either with a fan (forced) or by natural convection.

Vented/Flared—Gas that is disposed of by releasing (venting) or burning (flaring).

Ventilation—Process of supplying or removing air by mechanical or natural means to or from some type of enclosed space. The air may be treated or conditioned for comfort.

Ventilation Air—That portion of supply air that is drawn from outside, plus any recirculated air that has been treated to maintain a desired air quality.

Ventilation System (Coal)—A method for reducing methane concentrations in coal mines to non-explosive levels by blowing air across the mine face and using large exhaust fans to remove methane while mining operations proceed.

Venture Capital—Investment capital for generally small and medium-sized companies with high growth potential that invest in innovative technologies and services such as environmental protection, renewable energy and computer systems. Venture capital financings are frequently conducted by a lending institution and accompanied by management consulting services.

Venture Capital Financing—An investment in a startup business that is perceived to have excellent growth prospects but does not have access to capital markets. Type of financing sought by early-stage companies seeking to grow rapidly.

Venture Philanthropy—A charitable giving model that bridges venture capital strategies with philanthropic giving, creating strategic relationships among individuals and nonprofit organizations.

VEO—Visible Emission Observation

Verification—Verification is often undertaken during a due diligence process in a buy/sell transaction. It provides independent assurance that actual or expected emission reductions have been/will be achieved from an emission reduction project during a specified period. The level of assurance provided will depend on the procedures undertaken by the independent verifier, the scope of which is usually agreed by the transacting parties and may include: assurance as to compliance with Kyoto/national regime requirements (however this will only be possible when such requirements are clearly defined), adequacy of measuring and monitoring systems for emission reduction credits, reviewing the operations of the underlying emission reductions project etc.

Verification (Green-e program)—Green-e Verification consists of the Green-e Process Audit and the Green-e Compliance Review. The Process Audit requires retail and wholesale electricity service providers to conduct an annual third-party verification of their power purchases and sales. The Compliance Review is a semiannual review of a company's marketing materials to ensure that the company is abiding by the Green-e Code of Conduct, governing the use of the Green-e logo and customer disclosure requirements.

Verification Activities—Activities undertaken during the third-party verification that include reviewing a reporting facility's reported emissions, verifying their accuracy according to the standards specified by a regulator and submitting a verification opinion to a regulator.

Verification Opinion—The final opinion rendered by a verification firm attesting whether or not a facility's reported emissions are free of material misstatement and that all verification process checklist items have been completed by the verification firm.

Verification Phase—The step of the California Energy Contingency Plan to determine the existence and scope of an energy shortage and report to Energy Commission executives, the Governor and the Legislature where required under the plan.

Verification Team—More than one verifier, including all subcontractors, acting for a verification firm to provide verification services for a client.

Verified Emission Reductions (VERs)—Emission reductions that are not covered by a regulatory scheme but are verified by an independent third party.

Verified Emissions Report—An emissions data report that has been reviewed and approved by a third-party verifier and accepted by a regulator.

Verified Savings—Verified savings are those reported in the annual report of the energy project. They are based on verification activities conducted during the performance period, and are the savings calculated for that year of the energy project.

VERs—Verified Emission Reductions. The general name given to carbon offset credits in the voluntary carbon market. These are tradable credits for greenhouse emission reductions generated to meet voluntary demand for carbon credits by organizations and individuals wanting to offset their own emissions.

Vertical Glazing—*See Window*

Vertical Ground Loop—In this type of closed-loop geothermal heat pump installation, the fluid-filled plastic heat exchanger pipes are laid out in a plane perpendicular to the ground surface. For a vertical system, holes (approximately four inches in diameter) are drilled about 20 feet apart and 100 to 400 feet deep. Into these holes go two pipes that are connected at the bottom with a U-bend to form a loop. The vertical loops are connected with horizontal pipe (i.e., manifold), placed in trenches, and connected to the heat pump in the building. Large commercial buildings and schools often use vertical systems because the land area required for horizontal ground loops would be prohibitive. Vertical loops are also used where the soil is too shallow for trenching, or for existing buildings, as they minimize the disturbance to landscaping.

Vertical Integration—An arrangement whereby the same company owns all the different aspects of making, selling, and delivering a product or service. In the electric industry, it refers to the historically common arrangement whereby a utility would own its own generating plants, transmission system, and distribution lines to provide all aspects of electric service.

Vertical Multijunction (VMJ) Cell—A compound cell made of different semiconductor materials in layers, one above the other. Sunlight entering the top passes through successive cell barriers, each of which converts a separate portion of the spectrum into electricity, thus achieving greater total conversion efficiency of the incident light. Also called a multiple junction cell. See multijunction device and split-spectrum cell.

Vertical-Axis Wind Turbine (VAWT)—A type of wind turbine in which the axis of rotation is perpendicular to the wind stream and the ground.

VES—Vertical Electric Sounding

Vessel—A ship used to transport crude oil, petroleum products, or natural gas products. Vessel categories are as follows: Ultra Large Crude Carrier (ULCC), Very Large Crude Carrier (VLCC), Other Tanker, and Specialty Ship (LPG/LNG).

Vessel Bunkering—Includes sales for the fueling of commercial or private boats, such as pleasure craft, fishing boats, tugboats, and ocean-going vessels, including vessels operated by oil companies. Excluded are volumes sold to the U.S. Armed Forces.

Vest—To give immediate, fixed right in property, with either present or future enjoyment of possession; also denotes the manner in which Title is held.

VHAP—Volatile Hazardous Air Pollutant

VHT—Vehicle Hours of Travel

Video Logging—A method for close-up inspection of the interior of a well or pipe by means of a color camera that can view the well casing and screen at 90 degrees to the well's axis.

VIN (Vehicle Identification Number)—A set of about 17 codes, combining letters and numbers, assigned to a vehicle at the factory and inscribed on a small metal label attached to the dashboard and visible through the windshield. The VIN is a unique identifier for the vehicle and therefore is often found on insurance cards, vehicle registrations, vehicle titles, safety or emission certificates, insurance policies, and bills of sale. The coded information in the VIN describes characteristics of the vehicle such as engine size and weight.

Vintage—The first year that credits may be used for compliance. Depending on the program, credits can be used during their vintage, or banked for re-issuance in a later year.

Vinyl Window Frame—A fenestration frame constructed with a polyvinyl chloride (PVC) which has a lower conductivity than metal and a similar conductivity to wood.

Virgin Coal—Coal that has not been accessed by mining.

Visbreaking—A thermal cracking process in which heavy atmospheric or vacuum-still bottoms are cracked at moderate temperatures to increase production of distillate products and reduce viscosity of the distillation residues.

Viscosity—A measure of the ease with which a liquid can be poured or stirred. The higher the viscosity, the less easily a liquid pours.

Visible Light Transmittance (VLT)—The ratio (expressed as a decimal) of visible light that is transmitted through a glazing material to the light that strikes the material.

Visible Radiation—The visible portion of the electromagnetic spectrum with wavelengths from 0.4 to 0.76 microns.

VISTTA—Visibility Impairment from Sulfur Transformation and Transport in the Atmosphere

VKT—Vehicle Kilometers Traveled

V_{mp}—The voltage at which a PV device is operating at maximum power.

VMT—Vehicle Miles Traveled
- Vertical Miles Traveled

V_{oc}—Open-circuit voltage Volt (V): The unit of electromotive force that will force a current of one ampere through a resistance of one ohm. Voltage at maximum power.

VOC—See *Volatile Organic Compounds*.

Void Space—The space in a tank between the top of a

tank and the liquid level. If the tank is used to store combustible liquids that easily evaporate, this space can fill with vapors which may reach explosive levels.

VOL—Volatile Organic Liquid

Vol—Volume

Volatile—Describes substances that readily evaporate at normal temperatures and pressures.

Volatile Matter (Coal)—Those products, exclusive of moisture, given off by a material as gas or vapor. Volatile matter is determined by heating the coal to 950 degrees Centigrade under carefully controlled conditions and measuring the weight loss, excluding weight of moisture driven off at 105 degrees Centigrade.

Volatile Organic Compounds (VOCs)—Organic compounds that participate in atmospheric photochemical (sun-driven) reactions.

Volatile Solids—A solid material that is readily decomposable at relatively low temperatures.

Volatiles—Substances that are readily vaporized.

Volatility—The tendency of a liquid to pass into the vapor state at a given temperature. Vapor pressure.

Volatilization Rate—The rate at which a chemical changes from a liquid to gas. It is also known as "air flux."

Volt (V)—The volt is the International System of Units (SI) measure of electric potential or electromotive force. A potential of one volt appears across a resistance of one ohm when a current of one ampere flows through that resistance. Reduced to SI base units, $1 V = 1$ kg times m^2 times s^{-3} times A^{-1} (kilogram meter squared per second cubed per ampere).

Volt Ampere—The volt-amperes of an electric circuit are the mathematical products of the volts and emperes of the client.

Voltage—The difference in electrical potential between any two conductors or between a conductor and ground. It is a measure of the electric energy per electron that electrons can acquire and/or give up as they move between the two conductors.

Voltage at Maximum Power (V$_{mp}$)—The voltage at which maximum power is available from a module. [UL 1703]

Voltage Drop—The voltage lost along a length of wire or conductor due to the resistance of that conductor. This also applies to resistors. The voltage drop is calculated by using Ohm's Law.

Voltage of a Circuit (Electric Utility)—The electric pressure of a circuit, measured in volts. Usually a nominal rating, based on the maximum normal effective difference of potential between any two conductors of the circuit.

Voltage Peak Power Point (Vpp)—The voltage at which a photovoltaic module or array generates at the highest power (watts). A "12 volt nominal" PV module will typically have a peak power voltage of around 17 volts. A PV array-direct solar pump should reach this voltage in full sun conditions. In a higher voltage array, it will be a multiple of this voltage.

Voltage Protection—A sensing circuit on an Inverter that will disconnect the unit from the battery if input voltage limits are exceeded.

Voltage Reduction—Any intentional reduction of system voltage by 3 percent or greater for reasons of maintaining the continuity of service of the bulk electric power supply system.

Voltage Regulator—A device that controls the operating voltage of a photovoltaic array.

Voltage Support—Services provided by generating units or other equipment such as shunt capacitors, static VAR compensators, or synchronous condensers that are required to maintain established grid voltage criteria. This service is required under normal or system emergency conditions.

Voltage, Nominal—A way of naming a range of voltage to a standard. Example: A "12 volt nominal" system may operate in the range of 10 to 20 Volts. We call it "12 volts" for simplicity.

Voltmeter—An electrical or electronic device used to measure voltage.

Volumetric Wires Charge—A type of charge for using the transmission and/or distribution system that is based on the volume of electricity that is transmitted.

Voluntary Commitment—Actions taken by an entity that reduce emissions outside of regulatory requirements. During the Kyoto Protocol negotiations, a draft article on voluntary commitments would have permitted developing countries to take on voluntary, legally binding emission reduction targets, but was dropped from the final Protocol text.

Voluntary Measures—Measures to reduce GHG or other emissions that are adopted in the absence of government mandates.

Voting Rights—The common stockholders' right to vote their stock in the affairs of the company. Preferred stock usually has the right to vote when preferred dividends are in default for a specified amount of time. The right to vote may be delegated by the stockholder to another person.

VP—Vapor Pressure

VSD—Virtually Safe Dose

VSI—Visual Site Inspection

VSS—Volatile Suspended Solids

VTSR—Verified Time of Sample Receipt

W

WAA—Warren-Alquist Act

WACC—Weighted Average Cost of Capital

Wafer—A thin sheet of crystalline semiconductor material either made by mechanically sawing it from a single-crystal boule or multicrystalline ingot or block, or made directly by casting. The wafer is "raw material" for the solar cell.

Waiver—A relinquishment or abandonment of a right.

Walk-in Refrigeration Units—Refrigeration/freezer units within a building that are large enough to walk into. They may be portable or permanent, such as a meat storage locker in a butcher store. Walk-in units may or may not have a door, plastic strips, or other flexible covers.

Wall—A vertical structural element that holds up a roof, encloses part or all of a room, or stands by itself to hold back soil.

Wall Insulation—Insulating materials within or on the walls between heated areas of the building and unheated areas or the outside. The walls may separate air-conditioned areas from areas not air-conditioned.

Wall Orientation—The geographical direction that the primary or largest exterior wall of a building faces.

Wall Type—A type of wall assembly that has a specific heat capacity, framing type, and U-factor.

WAP—Waste Analysis Plan

WAPA—Western Area Power Administration: Markets and delivers reliable, cost-based hydroelectric power and related services within a 15-state region of the central and western U.S. One of four power marketing administrations within the U.S. Department of Energy whose role is to market and transmit electricity from multi-use water projects. On the internet at http://www.wapa.gov

Warrant—A type of security that entitles the holder to buy a proportionate amount of common stock or preferred stock at a specified price for a period of years. Warrants are usually issued together with a loan, a bond, or preferred stock and act as sweeteners, to enhance the marketability of the accompanying securities. They are also known as stock-pur-

chase warrants and subscription warrants.

Warranty—A promise, either written or implied, that the material and workmanship of a product are without defect or will meet a specified level of performance over a specified period of time. A seller's guarantee to purchaser that product is what it is represented to be and, if it is not, that it will be repaired or replaced. Within the context of vehicles, refers to an engine manufacturer's guarantee that the engine will meet "certified" engine standards at 50,000 miles or the engine will be replaced. Retrofits may generally void an engine warranty.

Warranty Contracts—Gas purchase agreements for the sale of natural gas by a producer to a pipeline company wherein the producer warrants it will have available sufficient gas supplies to meet its commitments over the life of the contract. Generally, the producer does not dedicate gas reserves underlying any specific acreage, lease, or fields to the agreement. Substitution of various sources of gas supply may be permitted according to the terms of the contract. Warranty contracts, by their terms, may vary from the above.

Waste Coal—Usable coal material that is a byproduct of previous processing operations or is recaptured from what would otherwise be refuse. Examples include anthracite culm, bituminous gob, fine coal, lignite waste, coal recovered from a refuse bank or slurry dam, and coal recovered by dredging.

Waste Energy—Municipal solid waste, landfill gas, methane, digester gas, liquid acetonitrile waste, tall oil, waste alcohol, medical waste, paper pellets, sludge waste, solid byproducts, tires, agricultural byproducts, closed loop biomass, fish oil, and straw used as fuel.

Waste Feed—The flow of wastes into an incinerator, boiler or industrial furnace. The waste feed can vary from continuous to intermittent (batch) flows.

Waste Heat Boiler—A boiler that receives all or a substantial portion of its energy input from the combustible exhaust gases from a separate fuel-burning process.

Waste Heat Recovery—Any conservation system whereby some space heating or water heating is done by actively capturing byproduct heat that would otherwise be ejected into the environment. In commercial buildings, sources of water- heat recovery include refrigeration/air-conditioner compressors, manufacturing or other processes, data processing centers, lighting fixtures, ventilation exhaust air, and the occupants themselves. Not to be considered is the passive use of radiant heat from lighting, workers, motors, ovens, etc., when there are no special systems for collecting and redistributing heat.

Waste Materials—Otherwise discarded combustible materials that, when burned, produce energy for such purposes as space heating and electric power generation. The size of the waste may be reduced by shredders, grinders, or hammermills. Noncombustible materials, if any, may be removed. The waste may be dried and then burned, either alone or in combination with fossil fuels.

Waste Oils and Tar—Petroleum-based materials that are worthless for any purpose other than fuel use.

Waste Streams—Unused solid or liquid by- products of a process.

Waste-to-Energy—A recovery process in which waste is incinerated or otherwise turned into steam or electricity, and used to generate heat, light or power through the process of combustion.

Waste-to-Profit—The process of using one company's waste or by-product as the input or raw material for another company, thereby increasing business profits and decreasing waste. Also referred to as byproduct synergy.

Wastewater Separator—Equipment used to separate oils and water from locations downstream of process drains.

Wastewater, Domestic and Commercial—Wastewater (sewage) produced by domestic and commercial establishments. May contain oil, emulsified oil, or other organic compounds which are not recycled or otherwise used in a facility.

Wastewater, Industrial—Wastewater produced by industrial processes.

Water Bed Heater—An appliance that uses an electric resistance coil to maintain the temperature of the water in a water bed at a comfortable level.

Water Conditions—The status of the water supply and associated water in pondage and reservoirs at hydroelectric plants.

Water Control Layer—A sheet, spray or trowel-applied membrane or material layer that controls the passage of liquid water even after long or continuous exposure to moisture. There is currently no standard measurement or rating system for water control effectiveness.

Water Economizer—A system by which the supply air of a cooling system is cooled directly or indirectly by evaporation of water; or other appropriate fluid, in order to reduce or eliminate the need for mechanical cooling.

Water Heated in Furnace—Some furnaces provide hot water as well as heat the home. The water is heated by a coil that is part of the furnace. There is no separate hot water tank.

Water Heater—An automatically controlled, thermally insulated vessel designed for heating water and storing heated water at temperatures less than 180 degrees Fahrenheit.

• An appliance for supplying hot water for purposes other than space heating or pool heating. [See California Code of Regulations, Title 20, Section 1602(f)(8)]

Water Heater Blanket—Insulated wrap attached to a water heater which supplements the insulation contained in the water heater.

Water Heater Efficiency Measures—Energy Factor (EF)—A measure of the overall efficiency of a water heater based on its recovery efficiency, standby loss and energy input as set out in the standardized Department of Energy test procedures.

Water Heating DSM Programs—These are demand-side management (DSM) programs designed to promote increased efficiency in water heating, including water heater insulation wraps.

Water Heating Equipment—Automatically controlled, thermal insulated equipment designed for heating and storing heated water at temperatures less than 180 degrees Fahrenheit for other than space heating purposes.

Water Jacket—A heat exchanger element enclosed in a boiler. Water is circulated with a pump through the jacket where it picks up heat from the combustion chamber after which the heated water circulates to heat distribution devices. A water jacket is also an enclosed water-filled chamber in a tankless coiled water heater. When a faucet is turned on water flows into the water heater heat exchanger. The water in the chamber is heated and transfers heat to the cooler water in the heat exchanger and is sent through the hot water outlet to the appropriate faucet.

Water Pollution Abatement Equipment—Equipment used to reduce or eliminate waterborne pollutants, including chlorine, phosphates, acids, bases, hydrocarbons, sewage, and other pollutants. Examples of water pollution abatement structures and equipment include those used to treat thermal pollution; cooling, boiler, and cooling tower blowdown water; coal pile runoff; and fly ash waste water. Water pollution abatement excludes expenditures for treatment of water prior to use at the plant.

Water Pumping—Photovoltaic modules/cells used for

pumping water for agricultural, land reclamation, commercial, and other similar applications where water pumping is the main use.

Water Reservoir—A large inland body of water collected and stored above ground in a natural or artificial formation.

Water Source Heat Pump—A type of (geothermal) heat pump that uses well (ground) or surface water as a heat source. Water has a more stable seasonal temperature than air thus making for a more efficient heat source.

Water Table—In a shallow aquifer, a water table is the depth at which free water is first encountered in a monitoring well.

Water Turbine—A turbine that uses water pressure to rotate its blades; the primary types are the Pelton wheel, for high heads (pressure); the Francis turbine, for low to medium heads; and the Kaplan for a wide range of heads. Primarily used to power an electric generator.

Water Vapor—Water in a vaporous form, especially when below boiling temperature and diffused (e.g., in the atmosphere). Water vapor is the primary gas responsible for the greenhouse effect. It is believed that increases in temperature caused by anthropogenic emissions of greenhouse gases will increase the amount of water vapor in the atmosphere, resulting in additional warming (see *"positive feedback"*).

Water Vapor (H_20)—Most abundant GHG. Anthropogenic activities are not significantly increasing its concentration, but warming leads to a positive water vapor feedback. The concentration of water vapor regulates the temperature of the planet, in part, because of its relationship with the atmosphere and the water cycle.

Water Well—A well drilled to (1) obtain a water supply to support drilling or plant operations, or (2) obtain a water supply to be used in connection with an improved recovery program.

Water Wheel—A wheel that is designed to use the weight and/or force of moving water to turn it, primarily to operate machinery or grind grain.

Water Year (Utility)—Measured water flow during a 12-month period starting October 1st and continuing to September 30th of the following year.

Water-Cooled Vibrating Grate—A boiler grate made up of a tuyere grate surface mounted on a grid of water tubes interconnected with the boiler circulation system for positive cooling. The structure is supported by flexing plates allowing the grid and

grate to move in a vibrating action. Ashes are automatically discharged.

Waterless Urinal—Urinal with no water line. Most designs use a specialized material that allows fluid to drain one-way into the sewer system.

Waterless Urinal—Uses no water and has no moving parts. Urine flows into a trap, where it passes through a lighter-weight liquid and down the drain. Waterless urinals are promoted as saving water, requiring less maintenance and improving hygiene.

Watershed—The land area drained by a given river; synonymous with drainage basin (also catchment).

Waterway—A river, channel, canal, or other navigable body of water used for travel or transport.

WATSCO—The Western Association for Transmission System Coordination.

Watt (W)—A unit of measure of electric power at a point in time, as capacity or demand. One watt of power maintained over time is equal to one joule per second. Some Christmas tree lights use one watt. The Watt is named after Scottish inventor James Watt and is capitalized when shortened to w and used with other abbreviations, as in kWh.

Watt Peak—Is the Direct Current Watts output of a Solar Module as measured under an Industry standardized Light Test before the Solar Module leaves the Manufacturers facility.

• The Watt Power output of a Solar module is the number of Watts Output when it is illuminated under standard conditions of 1000 Watts/meter2 intensity, 25°C ambient temperature and a spectrum that relates to sunlight that has passed through the atmosphere (AM or Air Mass 1.5).

Watt-Hour (Wh)—The electrical energy unit of measure equal to one watt of power supplied to, or taken from, an electric circuit steadily for one hour.

Wattmeter—A device for measuring power consumption.

Watts per Square Foot—A shorthand measure of the energy use of a building, often applied to indoor lighting. Energy codes often limit the watts per square foot based on building type and function.

Wave Power—The concept of capturing and converting the energy available in the motion of ocean waves to energy.

Waveform—The shape of a wave or pattern representing a vibration. The shape characterizing an AC current or voltage output.

Waveform—The shape of the phase power at a certain frequency and amplitude.

Wavelength—The distance between similar points on successive waves.

Wax—A solid or semi-solid material derived from petroleum distillates or residues by such treatments as chilling, precipitating with a solvent, or de-oiling. It is a light-colored, more-or-less translucent crystalline mass, slightly greasy to the touch, consisting of a mixture of solid hydrocarbons in which the paraffin series predominates. Includes all marketable wax, whether crude scale or fully refined. The three grades included are microcrystalline, crystalline-fully refined, and crystalline-other. The conversion factor is 280 pounds per 42 U.S. gallons per barrel.

WB—West Bulb

WBCSD—World Business Council for Sustainable Development www.wbcsd.ch

WBE—Women's Business Enterprise

WDMA—Window & Door Manufacturers' Association www.wdma.com

Weather—Describes the short-term (i.e., hourly and daily) state of the atmosphere. Weather is not the same as climate.

Weatherization—Caulking and weather stripping to reduce air infiltration and exfiltration into/out of a building.

Weatherstripping or Caulking—Any of several kinds of crack-filling material around any windows or doors to the outside used to reduce the passage of air and moisture around moveable parts of a door or window. Weather stripping is available in strips or rolls of metal, vinyl, or foam rubber and can be applied on the inside or outside of a building.

WECC—Western Electricity Coordinating Council

Weight, Specific—Weight per unit volume of a substance.

Weighted Average Antidilution—The investor's conversion price is reduced, and thus the number of common shares received on conversion increased, in the case of a down round; it takes into account both: (a) the reduced price and, (b) how many shares (or rights) are issued in the dilutive financing.

Weighted Average Capital Costs (WACC)—An average representing the expected return on all of a company's securities. Each source of capital is weighted according to its prominence in the company's capital structure.

Weighted Debt Rate—The single debt rate equivalent to multiple rates actually used in an amortization.

Weir—A dam in a waterway over which water flows and that serves to raise the water level or to direct or regulate flow.

Well—A hole drilled in the earth for the purpose of (1) finding or producing crude oil or natural gas; or (2) providing services related to the production of crude oil or natural gas. Wells are classified as (1) oil wells; (2) gas wells; (3) dry holes; (4) stratigraphic test wells; or (5) service wells.

Well Water for Cooling—A means of cooling that uses water from a well drilled specifically for that purpose. The subterranean temperature of the water stays at a relatively constant temperature. Where water is abundant, it provides a means of getting 55-degree Fahrenheit water with no mechanical cooling. Used usually for heat rejection in a water source heat pump.

Wellhead—The point at which the crude (and/or natural gas) exits the ground. Following historical precedent, the volume and price for crude oil production are labeled as "wellhead," even though the cost and volume are now generally measured at the lease boundary. In the context of domestic crude price data, the term "wellhead" is the generic term used to reference the production site or lease property.

Wellhead Price—The value at the mouth of the well. In general, the wellhead price is considered to be the sales price obtainable from a third party in an arm's length transaction. Posted prices, requested prices, or prices as defined by lease agreements, contracts, or tax regulations should be used where applicable.

WEPX—Western Energy Power Exchange

WER—Water Effects Ratio

WESP—Wet Electrostatic Precipitator

Western Electricity Coordinating Council (WECC)—A regional forum for promoting regional electric service reliability in Western Canada and the Western United States.

WET—Waste Extraction Test
- Whole Effluent Toxicity Test

Wet Bottom Boiler—Slag tanks are installed usually at the furnace throat to contain and remove molten ash.

Wet Lease—A lease in which the lessor provides bundled services, such as the payment of property taxes, insurance, maintenance costs, fuel or provisions, and may even provide persons to operate the leased equipment. This type of lease typically is referred to in aircraft leasing and marine chargers.

Wet Natural Gas—A mixture of hydrocarbon compounds and small quantities of various non-hydrocarbons existing in the gaseous phase or in solution with crude oil in porous rock formations at reservoir conditions. The principal hydrocarbons normally contained in the mixture are methane, ethane, propane, butane, and pentane. Typical non-hydrocarbon gases that may be present in reservoir natural gas are water vapor, carbon dioxide, hydrogen sulfide, nitrogen and trace amounts of helium. Under reservoir conditions, natural gas and its associated liquefiable portions occur either in a single gaseous phase in the reservoir or in solution with crude oil and are not distinguishable at the time as separate substances. Note: The Securities and Exchange Commission and the Financial Accounting Standards Board refer to this product as Natural Gas.

Wet Shelf Life—The period over which a charged battery, filled with electrolyte, can remain unused before its performance falls below a specified.

Wet-Bulb Temperature—Wet-Bulb Temperature is a measure of combined heat and humidity. At the same temperature, air with less relative humidity has a lower Wet-Bulb Temperature.

Wet-Bulb Temperature—The temperature at which water, by evaporating into air, can bring the air to saturation at the same temperature. Wet-bulb temperature is measured by a wet-bulb psychrometer.

Wetlands—Lands where saturation with water is the primary factor determining soil development and the kinds of plant and animal communities living on or under the surface.

WETT—Whole Effluent Toxicity Testing

Weymouth Formula—A formula for calculating gas flow in large diameter pipelines.

WG—Work Group

WGA—Western Governors' Association www.westgov.org

WGBC—World Green Building Council www.worldgbc.org

Wharfage—A charge assessed by a pier or dock owner for handling incoming or outgoing cargo.

WHB—Waste Heat Boiler

Wheeling—The transmission of electricity by an entity that does not own or directly use the power it is transmitting. Wholesale wheeling is used to indicate bulk transactions in the wholesale market, whereas retail wheeling allows power producers direct access to retail customers. This term is often used colloquially as meaning transmission.

Wheeling Charge—An amount charged by one electrical system to transmit the energy of, and for, another system or systems.

Wheeling Service—The movement of electricity from one system to another over transmission facilities of interconnecting systems. Wheeling service contracts can be established between two or more systems.

White Spirit—A highly refined distillate with a boiling point range of about 150 degrees to 200 degrees Centigrade. It is used as a paint solvent and for dry-cleaning purposes.

Whole-House Cooling Fan—A mechanical/electrical device used to pull air out of an interior space; usually located in the highest location of a building, in the ceiling, and venting to the attic or directly to the outside.

Wholesale Bulk Power—Very large electric sales for resale from generation sources to wholesale market participants and electricity marketers and brokers.

Wholesale Competition—A system whereby a distributor of power would have the option to buy its power from a variety of power producers, and the power producers would be able to compete to sell their power to a variety of distribution companies.

Wholesale Electric Power Market—The purchase and sale of electricity from generators to resellers (retailers), along with the ancillary services needed to maintain reliability and power quality at the transmission level.

Wholesale Power—Any energy or capacity that is sold by one utility for resale by another utility.

Wholesale Power Market—The purchase and sale of electricity from generators to resellers (who sell to retail customers), along with the ancillary services needed to maintain reliability and power quality at the transmission level.

Wholesale Price—The rack sales price charged for No. 2 heating oil; that is, the price charged customers who purchase No. 2 heating oil free-on-board at a supplier's terminal and provide their own transportation for the product.

Wholesale Sales—Energy supplied to other electric utilities, cooperatives, municipals, and Federal and state electric agencies for resale to ultimate consumers.

Wholesale Supplier—A company that sells TRCs to a TRC Provider. This company may be the renewable generator or may be a third-party broker. May also be called a wholesaler or wholesale marketer.

Wholesale Transition—The sale of electric power from an entity that generates electricity to a utility or other electric distribution system through a utility's transmission lines.

Wholesale Transmission Services—The transmission of electric energy sold, or to be sold, in the wholesale electric power market.

Wholesale Wheeling—The wheeling of electric power in amounts and at prices that generally have been negotiated in long-term contracts between the power provider and a distributor or very large power customer.

Whole-tree Harvesting—A harvesting method in which the whole tree (above the stump) is removed.

WHWT—Water and Hazardous Waste Team

WICF—Western Interconnection Forum

WIEB—Western Interstate Energy Board www.westgov.org/wieb

Williams Act of 1968—An amendment of the Securities and Exchange Act of 1934 that regulates tender offers and other takeover-related actions such as larger share purchases.

Wind Chill Factor—The equivalent temperature resulting from the combined effect of wind and temperature. For example: At 10 degrees Fahrenheit above 0 with a 20-mile per hour wind, the effect is the same as 24 degrees Fahrenheit below 0 without wind.

Wind Energy—Energy available from the movement of the wind across a landscape caused by the heating of the atmosphere, earth, and oceans by the sun.

Wind Energy Conversion System (WECS) or Device—An apparatus for converting the energy available in the wind to mechanical energy that can be used to power machinery (grain mills, water pumps) and to operate an electrical generator.

Wind Generator—A WECS designed to produce electricity.

Wind Power Plant—A group of wind turbines interconnected to a common utility system through a system of transformers, distribution lines, and (usually) one substation. Operation, control, and maintenance functions are often centralized through a network of computerized monitoring systems, supplemented by visual inspection. This is a term commonly used in the United States. In Europe, it is called a generating station.

Wind Resource Assessment—The process of characterizing the wind resource, and its energy potential, for a specific site or geographical area.

Wind Rose—A diagram that indicates the average percentage of time that the wind blows from different directions, on a monthly or annual basis.

Wind Speed—The rate of flow of the wind undisturbed by obstacles.

Wind Speed Duration Curve—A graph that indicates the distribution of wind speeds as a function of the cumulative number of hours that the wind speed exceeds a given wind speed in a year.

Wind Speed Frequency Curve—A curve that indicates the number of hours per year that specific wind speeds occur.

Wind Speed Profile—A profile of how the wind speed changes with height above the surface of the ground or water.

Wind Turbine—A term used for a wind energy conversion device that produces electricity; typically having one, two, or three blades.

Wind Turbine Rated Capacity—The amount of power a wind turbine can produce at its rated wind speed, e.g., 100 kW at 20 mph. The rated wind speed generally corresponds to the point at which the conversion efficiency is near its maximum. Because of the variability of the wind, the amount of energy a wind turbine actually produces is a function of the capacity factor (e.g., a wind turbine produces 20% to 35% of its rated capacity over a year).

Wind Velocity—The wind speed and direction in an undisturbed flow.

Windmill—A WECS that is used to grind grain, and that typically has a high-solidity rotor; commonly used to refer to all types of WECS.

Window—A generic term for a glazed opening that allows daylight to enter into a building and can be opened for ventilation.

• A wide band gap material chosen for its transparency to light. Generally used as the top layer of a photovoltaic device, the window allows almost all of the light to reach the semiconductor layers beneath.

Window Area—The area of the surface of a window, plus the area of the frame, sash, and mullions.

Window Type—A window assembly having a specific solar heat gain coefficient relative solar heat gain and U-factor.

Window Wall Ratio—The ratio of the window area to the gross exterior wall area.

Windpower Curve—A graph representing the relationship between the power available from the wind and the wind speed. The power from the wind increases proportionally with the cube of the wind speed.

Windpower Profile—The change in the power available in the wind due to changes in the wind speed or velocity profile; the windpower profile is proportional to the cube of the wind speed profile.

Wingwall—A building structural element that is built onto a building's exterior along the inner edges of all the windows, and extending from the ground to the eaves. Wingwalls help ventilate rooms that have only one exterior wall which leads to poor cross ventilation. Wingwalls cause fluctuations in the natural wind direction to create moderate pressure differences across the windows. They are only effective on the windward side of the building.

Wintergreen Lease—A lease that requires the lessee to give notice to the lessor in order to renew for another term. Otherwise, the lease terminates on the already established termination date.

Wire (Electrical)—A generic term for an electrical conductor.

Wires Charge—A broad term referring to fees levied on power suppliers or their customers for the use of the transmission or distribution wires.

Withholding Agent—The term "withholding agent" means any person required to deduct and withhold any tax under the provisions of taxing authority or regulator.

WL—Warning Letter
- Working Level (radon measurement)

WMD—Waste Management Division
- Water Management Division

WMMA—Waste Materials Management Act (1989)

Wobbe Index—A number which indicates interchangeability of fuel gases and is obtained by dividing the heating value of a gas by the square root of its specific gravity.

Wood and Waste (as used at Electric Utilities)—Wood energy, garbage, bagasse (sugarcane residue), sewerage gas, and other industrial, agricultural, and urban refuse used to generate electricity for distribution.

Wood Conversion to Btu—Converting cords of wood into a Btu equivalent is an imprecise procedure. The number of cords each household reports having burned is inexact, even with the more precise drawings provided, because the estimate requires the respondent to add up the use of wood over a 12-month period during which wood may have been added to the supply as well as removed. Besides errors of memory inherent in this task, the estimates are subject to problems in definition and perception of what a cord is. The nominal cord as delivered to a suburban residential buyer may differ from the dimensions of the standard cord. This difference is possible because wood is most often cut in lengths that are longer than what makes a third of a cord (16 inches) and shorter than what makes a half cord (24 inches).

In other cases, wood is bought or cut in unusual units

(for example, pickup-truck load, or trunk load). Finally, volume estimates are difficult to make when the wood is left in a pile instead of being stacked. Other factors that make it difficult to estimate the Btu value of the wood burned is that the amount of empty space between the stacked logs may vary from 12 to 40 percent of the volume. Moisture content may vary from 20 percent in dried wood to 50 percent in green wood. (Moisture reduces the useful Btu output because energy is used in driving off the moisture). Finally, some tree species contain twice the Btu content of species with the lowest Btu value. Generally, hard woods have greater Btu value than soft woods. Wood is converted to Btu at the rate of 20 million Btu per cord, which is a rough average that takes all these factors into account.

Wood Energy—Wood and wood products used as fuel, including round wood (cord wood), limb wood, wood chips, bark, sawdust, forest residues, charcoal, pulp waste, and spent pulping liquor.

Wood Pellets—Sawdust compressed into uniform diameter pellets to be burned in a heating stove.

Wood Stove—A wood-burning appliance for space and/or water heating and/or cooking.

Work Function—The energy difference between the Fermi level and vacuum zero. The minimum amount of energy it takes to remove an electron from a substance into the vacuum.

Working (top storage) Gas—The volume of gas in the reservoir that is in addition to the cushion or base gas. It may or may not be completely withdrawn during any particular withdrawal season. Conditions permitting, the total working capacity could be used more than once during any season.

Working Capital—The amount of cash or other liquid assets that a company must have on hand to meet the current costs of operations until such a time as it is reimbursed by its customers. Sometimes it is used in the narrow sense to mean the difference between current and accrued assets and current and accrued liabilities.

Working Fluid—A fluid used to absorb and transfer heat energy.

Working Interest—An interest in a mineral property that entitles the owner of that interest to all of share of the mineral production from the property, usually subject to a royalty.

A working interest permits the owner to explore, develop, and operate the property. The working-interest owner bears the costs of exploration, development, and operation of the property and, in return, is entitled to a share of the mineral production from the property or to a share of the proceeds there from. It may be assigned to another party in whole or in part, or it may be divided into other special property interests.

- Gross working interest. The reporting company's working interest plus the proportionate share of any basic royalty interest or overriding royalty interest related to the working interest.
- Net working interest. The reporting company's working interest is not including any basic royalty or overriding royalty interests.

Working Storage Capacity—The difference in volume between the maximum safe fill capacity and the quantity below which pump suction is ineffective (bottoms).

Workout—A negotiated agreement between the debtor and its creditors outside the bankruptcy process.

World Business Council on Sustainable Development (WBCSD)—An association of 170 international companies that provides business leadership with support to operate, innovate, and grow through sustainable development initiatives that incorporate the "three pillars of economic growth"—environmental protection, social development and economic growth. www.wbcsd.org

World Resources Institute (WRI)—WRI works with governments, companies, and civil society to build solutions to urgent environmental challenges. www.wri.org

Wound Rotor Motors—A type of motor that has a rotor with electrical windings connected through slip rings to the external power circuit. An external resistance controller in the rotor circuit allows the performance of the motor to be tailored to the needs of the system and to be changed with relative ease to accommodate system changes or to vary the speed of the motor.

WPAG—Western Public Agency Group

WPI—Wholesale Price Index

WQ—Water Quality

WQA—Water Quality Act (1987)

WQMP—Water Quality Management Plan

WQS—Water Quality Standard

Wrap Lease—A lease in which the lessor sells the equipment to an investor for equity and a note payable over the lease term. This method effectively transfers tax benefits to an investor.

WRD—Water Resources Division

WRDA—Water Resources Development Act (1986 as amended)

WREZ—Western Renewable Energy Zones

Writ—A process of the Court under which property may be seized and sold.

Write-off—The act of changing the value of an asset to an expense or a loss. A write-off is used to reduce or eliminate the value of an asset and reduce profits.

Write-up/Write-down—An upward or downward adjustment of the value of an asset for accounting and reporting purposes. These adjustments are estimates and tend to be subjective, although they are usually based on events affecting the investee company or its securities beneficially or detrimentally.

WRL—Work Release Letter

WRTA—The Western Regional Transmission Association, an RTG.

WSCC—Western Systems Coordinating Council www.wecc.biz

WSF—Water Soluble Fraction

WSM—Watershed Model

WSPP—The Western Systems Power Pool. A FERC approved industry institution that provides a forum for short-term trades in electric energy, capacity, exchanges and transmission services. The pool consists of approximately 50 members and serves 22 states, a Canadian province and 60 million people. The WSSP is headquartered in Phoenix, Arizona.

WSSCC—The Western System Coordinating Council. A voluntary industry association created to enhance reliability among western utilities.

WSTP—Wastewater Sewage Treatment Plan

WTI—West Texas Intermediate (crude oil)

WTO—World Trade Organization

WTP—Water Treatment Plant

WTPS—Water, Toxics and Pesticides Staff

WWF—World Wide Fund for Nature

WWT—Wastewater Treatment

WWTP—Wastewater Treatment Plant

X

Xenon—A heavy gas used in specialized electric lamps.

Xeriscaping—Landscaping designed to save water.

X-ray—A type of electromagnetic radiation having low energy levels.

Xylene ($C_6H_4(CH_3)_2$)—Colorless liquid of the aromatic group of hydrocarbons made by the catalytic reforming of certain naphthenic petroleum fractions. Used as high-octane motor and aviation gasoline blending agents, solvents, chemical intermediates. Isomers are metaxylene, orthoxylene, paraxylene.

Xyloid Coal—Brown coal or lignite mostly derived from wood.

Y

Yarder—A machine used in yarding timber.

Yarding—The initial movement of logs from the point of felling to a central loading area or landing.

Yarding Unmerchantable Material—(YUM) A logging contract requirement to remove and pile unmerchantable woody material of a specified size. Usually required in timber sale contracts on publicly-owned land.

Yaw—The rotation of a horizontal axis wind turbine around its tower or vertical axis.

Yellowcake—A natural uranium concentrate that takes its name from its color and texture. Yellowcake typically contains 70 to 90 percent U_3O_8 (uranium oxide) by weight. It is used as feedstock for uranium fuel enrichment and fuel pellet fabrication.

Yield—The rate of return to the lender, investor, developer or lessor.

Yield to Maturity—In finance, the total rate of return that would be realized on an investment such as a bond if purchased at the current market price, held as an investment, and redeemed for the principal amount at maturity.

YTC—Yield to Call

YTD—Year to Date

YTF—Yet To Find

YTM—Yield to Maturity

YUM—See *Yarding Unmerchantable Material*

Yurt—An octagonal shaped shelter that originated in Mongolia, and traditionally made from leather or canvas for easy transportation.

Z

ZD—Zero Drift

Zenith Angle—The angle between directly overhead and a line through the sun. The elevation angle of the sun above the horizon is 90° minus the zenith angle.

Zero Emission Vehicle (ZEV)—A vehicle that does not produce any air pollutants such as carbon monoxide, oxides of nitrogen, unburned hydrocarbons or particulate.

Zero Gas—Gas at atmospheric pressure.

Zero Waste—A production system aiming to eliminate the volume and toxicity of waste and materials by conserving or recovering all resources.

ZEV—Zero emissions vehicle.

Zinc—A metal used for auto parts, for galvanizing, and in production of brasses and dry cell batteries. It is nutritionally essential but toxic at higher levels.

ZOI—Zone of Incorporation

Zonal Control—The practice of dividing a residence into separately controlled HVAC zones. This may be done by installing multiple HVAC systems that condition a specific part of the building, or by installing one HVAC system with a specially designed distribution system that permits zonal control. The Energy Commission (in California) has approved an alternative calculation method for analyzing the energy impact of zonally controlled space heating and cooling systems. To qualify for compliance credit for zonal control, specific eligibility criteria specified in the Residential ACM Manual must be met.

Zone—A geographical area. A geological zone, however, means an interval of strata of the geologic column that has distinguishing characteristics from surrounding strata. Also, a space or group of spaces within a building with heating and/or cooling requirements sufficiently similar so that comfort conditions can be maintained by a single controlling device.

Zone of Origination—In the RECLAIM program, the Trading Zone or Regulation XIII zone in which the RTC is originally assigned by the District.

Zoning—The combining of rooms in a structure according to similar heating and cooling patterns. Zoning requires using more than one thermostat to control heating, cooling, and ventilation equipment.

Zoning Ordinance—The set of laws and regulations controlling the use of land and construction of improvements in a given area or zone.

ZRL—Zero Risk Level

Z-statistic—The number calculated in a z-test, whose significance is evaluated by reference to a z-table.

ZT—Zulu Time

Z-test—A test of any of a number of hypotheses in inferential statistics that has validity if sample size is sufficiently large and the underlying data are normally distributed.

Z-value—The standard normal deviate. Represents the number of standard deviation units that the random variable, or observation, in a data set is above or below the mean. Tables for the area under a normal curve are constructed in terms of "z" values for a given z—the tables provide the basis for determining the probability of an outcome between two values or the probability of an outcome being greater or less than a specific value.